General Physics

The late *Oswald H. Blackwood* was Professor of Physics at the University of Pittsburgh. *William C. Kelly* is Director of the Office of Scientific Personnel in the National Research Council. He was formerly Associate Professor of Physics at the University of Pittsburgh. *Raymond M. Bell* is Professor of Physics at Washington and Jefferson College.

General Physics

Fourth Edition

OSWALD H. BLACKWOOD
WILLIAM C. KELLY
RAYMOND M. BELL

JOHN WILEY & SONS, INC.   New York • London • Sydney • Toronto

This book was set in Helvetica Light by Progressive Typographers, Inc. and printed and bound by Vail Ballou Press, Inc. The text and cover were designed by Al Wright; the editor was Alice Hopkins, Stanley G. Redfern supervised production.

Cover: Lick Observatory Photograph

Library of Congress Cataloging in Publication Data:

Blackwood, Oswald Hance, 1888-1953.
  General physics.

  Includes bibliographies.
  1. Physics. I. Kelly, William Clark, 1922-
joint author. II. Bell, Raymond Martin, 1907-   joint
author. III. Title.

QC21.2.B52 1972   530   72-6799
ISBN 0-471-07923-5

Printed in the United States of America

10 9 8 7 6 5 4 3 2 1

Preface

Fourth editions require explanations. We offer a fourth edition of *General Physics* because teachers and students of physics continue to find this book helpful. It has been revised to increase its usefulness.

This is a textbook for introductory courses in colleges. We have tried to encourage the student from the beginning by providing lower hurdles during the first few weeks of the course. References to everyday experiences are used wherever possible to introduce the student to physical principles. The increase in mathematical difficulty of the topics covered is gradual. Knowledge of calculus is not required, but alternative derivations using calculus are occasionally provided for illustrative purposes. We have attempted to keep physical ideas in the forefront at all times.

The present edition reflects changing requirements for physics textbooks. Teachers and students no longer expect a textbook to be the only book used in the course, much less the only learning aid. Supplementary readings, films, computers, and laboratory and demonstration apparatus all play a greater role than formerly in the teaching of college physics. The textbook need no longer be all-inclusive and is seldom dominant.

Rather it is the principal integrator and road map for the course.

The objectives of the course are also changing. Most physics teachers believe that an introductory course in physics should give the student three things: a knowledge of the basic principles of physics, particularly the great unifying concepts; an understanding of the ways in which physics develops; and a sense of the endless frontiers at the edges of what we now know. Physics students increasingly seek a fourth objective: relevance to their lives and to the society in which they live. This book contains, we believe, discussions that will be found helpful in reaching all these goals.

Readers of earlier editions will find the present approach generally familiar, but some of the material is quite new. The most notable change has been the addition of a chapter devoted to the special theory of relativity. Once a byword for the arcane and difficult, relativity can be successfully taught in the introductory course and, in our opinion, deserves such inclusion. Chapter 10 has been tested in the classroom and found teachable. Its addition brings to nine the number of chapters devoted to those conceptual and experimental pillars of twentieth-century physics—relativity and quantum physics.

We have examined each chapter in the book for its contribution and have made some omissions and consolidations. The section devoted to classical mechanics, for example, has been reduced from thirteen chapters to ten—without, we believe, sacrifice of continuity or unity. We have been guided in making these changes by the results of a questionnaire survey of teachers.

We have returned to the electric-current convention in the discussion of electrical circuits. The mks system of units is used almost exclusively. As in earlier editions, we state units in the worked examples, including those in electricity and magnetism.

Questions, problems, and references to outside reading have been revised. The questions are arranged in order of increasing difficulty and some of the last in each group may require outside reading. There are three levels of difficulty of problems, marked A, B, and C. In addition, superfluous data are given for some problems. Others require estimations of order of magnitude: the student must ask whether his answer makes sense from the practical viewpoint. We have been helped by the problem ideas published by H. R. Crane in *The Physics Teacher* from time to time and have included some of these interesting problems as a way of enlivening what is sometimes a dull part of the physics course.

Teachers who have used earlier editions of this book have encouraged us to think of it as a useful teaching tool. We offer this edition to them and to other teachers and students with the hope that it will continue to provide a satisfying introduction to the powerful principles and excitement of physics.

WILLIAM C. KELLY
Washington, D.C.

RAYMOND M. BELL
Washington, Pennsylvania

Acknowledgments

Many friends have helped us. Those whose comments led to improvements in earlier editions have already been named; our indebtedness to them continues. In addition, David Current of Central Michigan University and M. G. Cheney of the University of Texas at Arlington reviewed the entire third-edition text and made many helpful suggestions for its revision. Russell K. Hobbie, of the University of Minnesota read the manuscript of the fourth edition and provided perceptive and meticulous comments. William D. Foland of Washington and Jefferson College read several drafts of Chapter 10 and made constructive comments. Others who read Chapter 10 in draft form were Professor Cheney, Sister Lorraine Lawrence of San Francisco College for Women, and Arthur J. M. Johnson of Montana State University. Students of introductory physics at Washington and Jefferson College used the preliminary form of Chapter 10 and helped us improve it. Colleagues who have made helpful suggestions from time to time and have influenced the direction of the revision include O. P. Puri of Clark College, Clement L. Henshaw of Colgate University, Leonard M. Diana of the University of Texas at Arlington, Newell S. Gingrich of the University of Missouri, Henry H. Rogers of Emory University, Donald C. Martin of Marshall University, and John H. Clarke of the University of Sydney. Finally, much assistance was provided by the physics teachers who responded to our questionnaire and provided detailed information about desirable changes in the third edition. To all, our warmest thanks.

Our families have also helped. Especially we wish to thank Gertrude Blackwood Kelly, who served as the authors' production manager, typist, artist, and critic for this edition as she has for earlier ones.

W. C. K.
R. M. B.

Contents

Mechanics

CHAPTER ONE

An Introduction

As you begin your study of physics, you want to know what lies ahead of you. What is physics? How did it develop and how is it likely to change in the future? What do physicists do? To what kinds of problems can physics be successfully applied? These are large questions, and here and in the succeeding chapters of this book we can make only a modest beginning in answering them. In fact, our goal is to give you a background in the fundamentals of physics, so that your reading in later life will help you, as an educated and self-educating person, to answer these questions more completely for yourself. In this chapter, we want to reach a few preliminary understandings with you about what physics is and what it is not.

Physics is often defined as the study of matter and energy. Although "matter" and "energy" have everyday meanings, these meanings do not do justice to the precise significance of the words as we shall use them in our later discussion. For the moment, let us say that the study of matter reveals the properties of the fundamental particles and their aggregations from which we believe the universe is constructed. Energy, on the other hand, is a useful concept in studying what combinations of these particles Nature permits and how the combinations—molecules, tennis balls, stellar galaxies—"behave." You must not, however, expect us to devote the first half of this book to "Matter" and the second half to "Energy." The concepts are inseparably intermixed. You are more likely to find discussions of matter—atomic structure and the nature of solids, for example—leaning heavily upon the concept of energy.

Physics courses are often subdivided for convenience into six areas, each of which is given over to a group of related phenomena. Thus you will study in this book:

1. Mechanics (the study of forces and motion)
2. Molecular physics and heat
3. Wave motion and sound
4. Electricity and magnetism
5. Light and other electromagnetic radiations
6. Atomic, solid state, and nuclear physics

These areas, however, are not sharply separated from one another. Physics is one subject, not several. The principles studied in mechanics, for example, underlie all of the other areas of physics. Wave motion must be studied in order to understand sound, light and other electromagnetic radiations, and atomic and nuclear physics. Unifying principles run through all of physics and tie it together. You will learn how the principle of the conservation of energy, which you first encounter in mechanics, applies to molecular physics and heat, to wave motion and sound, to electricity and magnetism, to electromagnetic radiation, and to atomic and nuclear physics. A fortunate thing it is, too, for the student and the physicist that physics rests upon basic principles which make understandable and cohesive many otherwise unrelated facts. We shall emphasize these principles.

Physics is not a static subject. It has grown from the work of men through twenty-five centuries of recorded history. Its "frontiers" are changing today as contemporary physicists obtain new knowledge and fit it into the structure of knowledge that mankind has inherited. Although in this book we shall be concerned principally with the physics of today, including its frontiers, we shall discuss from time to time the historical development of important ideas in physics and the role of physicists in human history.

The history of physics is an interesting subject in its own right and one that you can study with increasing pleasure and profit as your knowledge of the fundamentals of physics increases. The reading references at the ends of the chapters will help you in this. You will be impressed by the long development that important ideas in physics have undergone and by the number of men who have contributed to this slow, painstaking development. Isaac Newton, the great British physicist of the seventeenth century, congratulated on his scientific achievements, stated the situation well: "If I have seen farther than the others, it is because I have stood on the shoulders of giants."

The acceleration of the growth of physics in the last three hundred years is also remarkable. Newton (1642–1727) and his eminent predecessor Galileo Galilei (1564–1642) stand at the beginning of an upsurge of interest and accomplishment in physics that has continued at an ever-increasing pace to our day and that shows no sign of flagging. Figure 1-1 shows schematically how the discoveries in electricity and magnetism, for example, have come at a rapidly increasing rate. Our purpose in any historical discussion will not be to give you a capsule history of physics. Instead, we want to convince you that physics is a human enterprise which has laboriously produced a structure of knowledge of the world, starting from the simplest ideas and observations and proceeding to measurements of great precision and to theories of great universality.

Physics is a worthwhile study for two reasons: it organizes our knowledge of the world about us into a more orderly and satisfying form, and it can be successfully applied to solving problems of practical importance to mankind. The physicist is more interested in the first of these. In his research, he is essentially trying to *understand* nature, not to invent a new product or an industrial process. A theoretical physicist is satisfied, for example, if he can develop a theory or a physical model which ties together a greater number of facts than previous theories did and which even predicts new facts hitherto unsuspected. The experimental physicist finds his reward when his laboratory measurements conclusively verify or disprove predictions of theory or even lead to a breakthrough into a new realm of nature which has never before been explored. We shall illustrate these activities of the physicist later in this book. For the moment, we want merely to emphasize that the goal of the physicist is to understand the physical world.

The physicist makes his discoveries in many ways. He may come upon new knowledge by checking the predictions of a theory, by following an inspired hunch, by developing a new measuring technique or refining an old one, by reasoning by analogy, and even

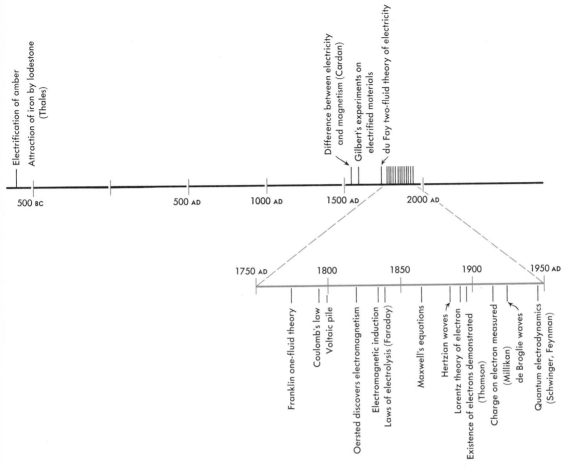

Fig. 1-1. *The acceleration of discovery in physics. The dates of some of the basic discoveries in electricity.*

by stumbling upon a discovery through sheer chance. Thus there is no single "Scientific Method"; there are scientific methods which physicists and other scientists use as they devote their minds and experimental skills, "no holds barred," to obtaining a better understanding of nature.

We shall have many occasions to discuss important applications of physics. Modern technology depends heavily upon physics, and technological progress follows advances in physics and the other basic sciences (Fig. 1-2). Greater understanding of electricity and magnetism by physicists led, as we shall see,

to the electrical power industry and the communications industry: electric lighting and heating, the telephone, radar, radio and television. In medicine and dentistry, some of the most useful diagnostic and therapeutic tools are applications of physics: X-rays, electrocardiographs, cardiac pacemakers, radio-isotopes, even stethoscopes. Work by physicists in the theory of shock waves (such as the "sonic boom" that you may have heard) led to advances in the design of jet planes and of launching vehicles used in space exploration.

Solving environmental problems requires

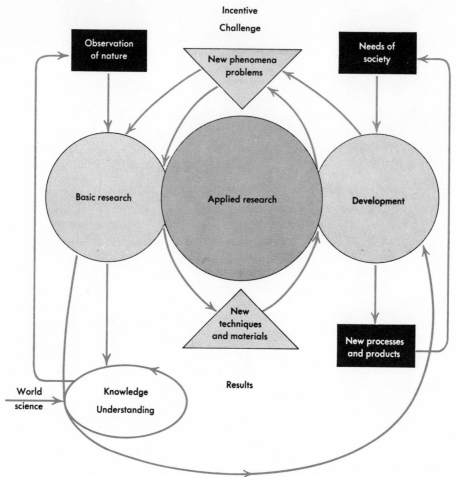

Fig. 1-2. *The relationship among basic research, applied research, and technological applications of a basic science such as physics. Each feeds the others and is fed by them. (Courtesy Naval Research Advisory Committee).*

us to understand the underlying scientific phenomena and to be able to make precise measurements. Is technological development throughout the world affecting the mean temperature of the earth? To answer this question, precise measurements based on physical principles must be made on both the CO_2 content and the concentration of small particles in the atmosphere. Will the operation of supersonic planes at high altitudes destroy the ozone layer that protects us from excessive ultraviolet radiation at the earth's surface? Again, careful studies involving precise measurements are needed.

Physics is useful, but remember that the applications of physics are only a part of the story. Like the physicist, you will be mostly concerned with understanding nature. However, we shall point out applications of physics as we go because they add to the interest which we hope you will find in your study of physics.

Measurement

Since physics deals almost exclusively with measurable quantities, it is very important to know precisely what measurement means. When we measure any quantity, we determine the ratio of that quantity to a chosen unit. For example, to measure the width of a room in yards we find the number of times the length of a yardstick, moved end to end across the room, is contained in the width of the room. If the yardstick can be fitted into the width about three and a half times, we say that the width is approximately 3.5 yards. Note that the *measurement* is 3.5 yards and the *unit of measurement*—represented physically and rather inaccurately by the wooden yardstick with "Jones Hardware Store" stamped on it—is the *yard.* But how was the yard itself determined?

In primitive societies, there was little commerce outside the village, and the units of measurement were crude and simple. The yard was often defined as one-half of the distance from finger tip to finger tip of the king's outstretched arms, and the pound-weight as the weight of 7000 "grains of barley chosen from the middle of the ear." Simple, inexact units served well enough for trade among friends and neighbors, but trouble arose when commerce developed between the cities and towns of a country. To meet the needs of merchants, units of measure were legalized throughout an entire nation. No attempts were made to interrelate them, hence the result was a hodgepodge. Our American system of measures, inherited from Great Britain, has as units of volume the cubic foot, gill, pint, quart, gallon, peck, and bushel. We find it hard to remember that 1728 cubic inches equal 1 cubic foot or that 5280 feet equal 1 mile. Few people are certain which is larger, the troy pound or the pound avoirdupois.

The difficulty increases when we cross a national boundary. When you travel in Canada, you may suppose that gasoline is much more expensive there than it is in this country. You may not know that the "imperial" gallon of Canada is about 20 per cent larger than the gallon which is legal in the United States.

Advancing science and technology have emphasized precision of measurement (Fig. 1-3). Physicists measure quantities as small as the "diameter" of the tiny atomic nucleus and as large as the distance across our galaxy, the Milky Way. Your car contains hundreds of parts machined to an accuracy of one ten-thousandth of an inch. Transistors and vacuum tubes used in your television set, gyroscopes used to control the flight of rockets and planes, nuclear reactors used in atomic-energy generating stations—all must be precisely made if they are to perform properly. Precision manufacturing requires precise measurements and these depend upon precisely defined units of measurement. Nowadays we even measure non-physical phenomena. For example, the Gallup polls estimate public opinion as to controversial questions. Increasingly, people appreciate the validity of the British physicist Lord Kelvin's statement, "When you cannot measure, your knowledge is very meager and unsatisfactory."

The Metric System

During the French Revolution, in the latter part of the eighteenth century, scientists introduced a new *metric* system of weights and measures. It is so convenient that it is used everywhere for scientific work. It has been adopted commercially by all except the English-speaking nations. The metric system is convenient because its units for measuring a quantity are related decimally. To change centimeters to meters is as easy as converting cents into dollars. Just shift the decimal point. For instance, 1.23 meters = 123 centimeters; 3.456 kilograms = 3456 grams. The metric system is an international language of measurement. It is understood by educated people everywhere. We shall stress

(a)   (b)

Fig. 1-3. (a) *Nuclear magnetic resonance of metals and alloys is studied with the aid of a tuned radio transmitter and receiver. The sample is in the probe between the poles of an electromagnet. (Courtesy of National Bureau of Standards.) (b) An 85-ft radio telescope used for detecting radio signals from ionized hydrogen regions of our galaxy and from other galaxies. (Courtesy National Radio Astronomy Observatory.)*

the metric system in this book, but shall use British units in the first few chapters when we discuss examples from daily life.

Metric and British Units of Length—the Meter and the Yard

Originally the *meter,* the metric standard of length, was defined as the ten-millionth part of the distance from the equator to the North Pole, on a line through Paris, and a platinum-iridium bar was constructed whose length, at the temperature of melting ice, was as near as possible to this. The "meter stick" was later found to be a little too short, and in 1872 the platinum bar itself was adopted as the primary standard. This bar is kept at the International Bureau of Weights and Measures at Sevres (pronounced Say´vr) near Paris. Copies of the standard meter are kept at Washington, London, and other capitals.

The distance between two marks on the bar has been measured in wavelengths of light, and the meter is now defined in these terms. The meter equals 1,650,763.73 wavelengths of a certain very sharp spectral orange-red line in the spectrum of krypton-86. Light waves form a more imperishable and precise standard than any material standard.

The yard is legally defined in the United States as 0.9144 meter. 39.37 inches approximately equal 1 meter, which is about 10 per cent longer than a yard. Also, 2.54 centimeters equal 1 inch, and 30.5 centimeters equal 1 foot (Fig. 1-4).

Table 1-1 lists metric units of length and their approximate British equivalents.

In studying atomic and nuclear physics, other units of length are convenient: the Angstrom unit ($=10^{-8}$ cm), the Bohr radius ($=5.29 \times 10^{-9}$ cm), and the fermi ($=10^{-13}$ cm). In astronomy, the astronomical unit (AU), which is equal to the mean earth-sun distance

Table 1-1. Some Metric Units of Length and Their Approximate British Equivalents[a]

1 kilometer (km) = 1000 meters ≅ 5/8 mile	
= 0.62 mile	
1 meter (m) ≅ 1.1 yard ≅ 3.3 feet	
1 centimeter (cm) = 1/100 meter = 1/2.54 inch	
= 1/30.5 foot	
1 millimeter (mm) = 1/1000 meter = 10^{-3} meter	
1 micron (μ) = 1/1,000,000 meter = $1/10^6$ meter	
= 10^{-6} meter	

[a] The symbol ≅ means "approximately equals."

of 1.4960×10^{13} cm, and the light-year, which is equal to the distance traveled by light in free space in one year (9.46×10^{17} cm), are often used.

What is Mass?

Another fundamental quantity is *mass.* Mass is a measure of the tendency of a body to resist a change of velocity. We can compare the masses of two bodies by exerting equal forces on them and seeing which is easier to "get going." Suppose that two pasteboard boxes lie on the pavement, one empty, the other containing a brick. If a boy kicks first one and then the other, he quickly determines which box has the greater mass and is harder to "speed up." Place two cars on a level table, and compress a spring between them, tying the two by a cord (Fig. 1-5). Burn the cord, releasing the cars, and the spring will exert equal forces on each of them. Then, if the masses are equal, the cars will travel equal distances in equal times. If the masses are not equal, the car of smaller mass will acquire the greater speed. Later we shall discuss quantitatively the relation between mass and change of velocity.

Units of Mass—the Gram and the Pound

The metric unit of mass, the *kilogram,* is the mass of a certain platinum-iridium cylinder which is kept with the standard meter. One *gram* is the one-thousandth part of a kilogram and is nearly equal to the mass of a cubic centimeter of water at 39.2° Fahrenheit or 4° Celsius. A new five-cent piece has a mass of about 5 grams; the mass of a dime is one-half as great. The pound is legally defined in the United States as a mass of 453.59237 grams.

The atomic mass unit (amu) is defined as $\frac{1}{12}$ of the mass of an atom of carbon-12. One amu equals approximately 1.66×10^{-24} gm.

Table 1-2. Some Metric Units of Mass and Volume and Their Approximate Equivalents in British Units[a]

1 metric ton (1000 kilograms) = 1.10 short tons	
1 gram (gm) = 1/454 pound	
= 1/28.4 avoirdupois ounce	
1 kilogram (kg) (1000 grams) = 2.20 pounds	
1 liter (l)	
(1000 cubic centimeters) = 1.06 liquid quarts	

[a] (See Appendix B for other equivalents.)

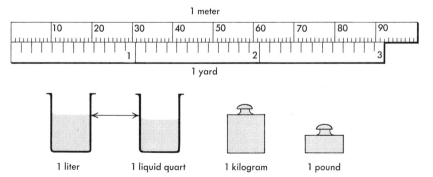

Fig. 1-4. *Metric and British units of length, volume, and mass.*

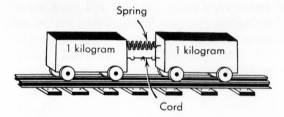

Fig. 1-5. Comparing masses.

Units of Time—the Mean Solar Day and the Second

For many ages, the motions of the heavenly bodies have served to measure time. The apparent motion of the sun gives us our day, that of the moon measures off the months, and the positions of the stars tell us of the passing of the year. About 7000 years ago Egyptian scholars noticed that in the spring the star Sirius was barely visible on the eastern horizon at sunrise, and thus they marked the beginning of their year. Since the length of the day, the time from one noon to the next, varies during the year, we use the *mean solar day,* the average time from noon to noon, as measured with a clock and a sundial. We divide this day into 24 equal hours, and each of these into 3600 seconds. The *mean solar second,* the unit of time in both the metric and the British systems, is $1/(24 \times 60 \times 60) = 1/86{,}400$ of a mean solar day. Recently a more precise definition of a second was agreed upon: the second is $1/31{,}556{,}925.9747$ of the tropical year 1900. The concept of time itself, which seems so familiar to us in everyday life, turns out to be far from simple when examined by the physicist—as we shall see in Chapter 10.

Radio station WWV (2.5, 5, 10, 15, 20, and 25 megahertz) in Fort Collins, Colorado broadcasts standard time signals for use in laboratories: pure tones (500 hertz and 600 hertz) and signals separated by precise time intervals. The signals are stabilized to one part in 10^{10}.

Derived Units

The set of units derived from the *centimeter, gram,* and *second* is known as the cgs system. In the mks system of units, the *meter* is the fundamental unit of length, the *kilogram* is the unit of mass, and the *second* is the unit of time.

The units of length, mass, and time are called the *fundamental* units in the cgs and mks systems because all other units of mechanics can be derived from them. For example, as units of area we use the *square centimeter* (cgs) and the *square meter* (mks), and as units of speed the *centimeter per second* (cgs) and the *meter per second* (mks). Some derived units are given special names. For example, the *barn,* a unit of area or cross section used in nuclear physics, equals 10^{-28} square meter.

What are Significant Figures?

Every measurement involves error. Suppose that you measure the length of a cube, using a ruler that is divided into tenths of a centimeter, and that your measured value is 2.13 cm. Other students get the values: 2.16 cm, 2.15 cm, 2.13 cm, 2.14 cm, 2.12 cm. Notice that everyone agrees as to the first two digits, but the third one is known only approximately. It is a *doubtful figure.* The digits that are exactly known and the *first* doubtful figure are called *significant figures.* We record only significant figures in taking data in the laboratory. The best value you could report for the length of the cube is the *average* or *mean value,* obtained by adding the individual values and dividing by the number of measurements: $\frac{1}{6}(2.13 \text{ cm} + 2.16 \text{ cm} + 2.15 \text{ cm} + 2.13 \text{ cm} + 2.14 \text{ cm} + 2.12 \text{ cm}) = 2.14 \text{ cm}$. The *per cent deviation* of *your* measurement—2.13 cm—from the mean value is the difference between *your* measurement and the mean, divided by the mean value and multiplied by 100%: 0.01 cm/2.14 cm × 100% = 0.47%.

When we use data in calculations, we must not introduce more digits into the result than the original data permit. The following rules of computation will help you to determine the number of significant figures you may reasonably carry in a result. The rules also will help you to save time in making calculations.

1. In adding or subtracting, retain in the sum or difference only one doubtful figure.

Example. A ball 4.234 cm in diameter rests on a desk that is 85.1 cm high. What is the vertical distance from the floor to the top of the ball?

$$
\begin{array}{r}
4.234 \text{ cm} \\
\underline{85.1 \quad \text{ cm}} \\
89.334 \text{ cm} = 89.3 \text{ cm}
\end{array}
$$

2. In multiplication or division, carry in the product or quotient one more *digit than the least number of digits* in any *factor.*

Example. A metal strip is 1.11 cm long and 0.33 cm wide. What is its area? By computation, the area is 0.3663 cm². The width has two significant figures. Keep three in the product. Record it as 0.366 cm².

What is the reason for rule 2? Suppose that we measure the length and width of a metal strip by means of a centimeter scale. In each case we estimate tenths of a millimeter. One person might judge the length to be 1.13 cm, another 1.12 cm, and another 1.11 cm. The third figure or digit is doubtful or uncertain. The fractional uncertainty is $0.01/1.12 = 1/112$ or about 1%. Similarly, the uncertainty in the width may turn out to be 0.01 cm, and the fractional uncertainty in the width is $0.01/0.33 = 1/33$ or about 3%. But because of the uncertainty of the length and width, there is an uncertainty in the area. According to the theory of errors, the percentage uncertainty of the area is approximately the sum of the percentage errors in the length and width or is about $1\% + 3\% = 4\%$. The uncertainty in the area is thus $4\% \times 0.366$ cm² or 0.0147 cm². The second figure of the area is doubtful. We should record the area as 0.37 cm² or, to be liberal, as 0.366 cm². This has one more digit than the factor (0.33) having the smallest number.

Summary

Physics is the study of *matter* and *energy.* For convenience, physics is divided into mechanics, molecular physics and heat, wave motion and sound, electricity and magnetism, light and other electromagnetic radiation, and atomic, solid state, and nuclear physics. However, broad principles run through all of physics and tie it together. Physics is growing: it is not only a body of knowledge, but a contemporary activity whose frontiers are expanding. The physicist seeks to *understand* nature; the engineer, to *apply* new knowledge of the physical world to solving the problems of mankind. Both activities are beneficial to humanity.

Fundamental in physics is the process of measurement. To measure a quantity means to find the ratio of that quantity to its chosen unit. Today the metric *meter-kilogram-second* system of units is used increasingly in science. The British Engineering *foot-pound-second* system is used in everyday life in English-speaking nations. The meter is approximately 3.3 feet; the kilogram is 2.2 pounds. Important in all measurements is the proper use and understanding of significant figures. Significant figures include the exactly known digits in a number and the first doubtful digit.

Questions

1. Define the basic units of length and mass in the metric system.

2. Define the mean solar day and the mean solar second.

3. Which is more acceptable as a basis for a standard of time: the rotation of the earth or atomic frequencies? Why?

4. Give several reasons for the importance of "significant figures" in scientific work.

5. What is your height in meters? Your mass in kilograms?

6. Which is longer: the 100-meter dash or the 100-yard dash?

7. What is the length of a football field in meters?

8. Given a pointer just one meter long, to how many significant figures could you measure the length of (a) a board that was between one and two meters long, (b) a lecture room that was between eleven and twelve meters long?

9. If you were given a meter stick and a protractor, how could you measure the width of a stream without crossing it?

10. Name several of the events reported in the news in the past month that have been related to the field of physics.

11. How does a traffic speed limit of 80 km/hr compare with one of 65 mi/hr?

12. Could the meter, kilogram, and second be used as units of measurement on another planet?

13. When are approximations or rounding off of numbers acceptable in everyday life? When are quick estimates often desirable?

14. How many significant figures has each of the following: $5000.00; 0.0019 inch; 23.008 meters; 1200 cm; 0.01 mile; Aug. 21, 1889; 1400.0089 barns; 1.5 elephants?

Problems

A

1. Estimate (to two significant figures) the number of grams in one ounce.

2. Show that 1.00 km approximately equals 0.621 mi.

3. A light year is 5.88×10^{12} miles. How many kilometers is this?

4. The distance from Paris to Brussels is 290 kilometers. How many miles is this?

5. In short-distance running, the 440-yard dash is used. How many meters is this?

6. Mount Rainier, in Washington, is 14,410 ft high. Express this height in meters.

B

7. A sprinter runs 100 yd in 10.1 sec. At the same level of performance how long would it take him to run 100 m?

8. What constant speed in mi/sec would an astronaut have to acquire to travel 2 astronomical units in 60 days?

9. If mountain A is 6558 m high and mountain B is 2.08×10^4 ft high, which is higher?

10. A car burns gasoline at the rate of 18 mi/gal. Express this rate of fuel consumption (a) in kilometers per gallon, (b) in kilometers per liter.

11. Find (a) the volume, in liters, of an aquarium that is 70 cm long, 30 cm wide, and 20 cm deep; (b) the area, in square feet, of a board that is 50 in. long and 15 in. wide.

12. The human brain is estimated to have about 10^{10} neurons. Each neuron enters into several thousand branching connections with other neurons. Estimate the total number of synaptic connections in the brain.

13. According to present estimates, the Milky Way, our galaxy, contains about 10^{11} stars. About 1% of these might possess a planet capable of supporting life, that is, having the proper size, temperature, and rate of reception of radiant energy. There are about 10^8 galaxies within range of present telescopes. Estimate the number of planets suitable for life in the observable universe, using the above assumptions.

C

Keep the proper number of significant figures in all of your calculations. Test your ability to do so in solving these problems.

14. A boy who weighs 62.5 lb drinks a glassful of milk weighing 0.43 lb. How much does he then weigh?

15. A runner sprinted a distance measured as 100.3 yd in an official time of 13.4 sec. Calculate his average velocity (distance per unit time).

16. Find the sum of each of the following pairs of quantities:

(a) 1.23 cm   (b) 1.2345 gm   (c) 1.23 sec
     35.1 cm   35   gm   35.11 sec

17. A man wishes to pour concrete to form a slab 6.0 in. thick in the form of a rectangle 6.2 ft wide and 10.5 ft long. Calculate the number of cubic yards of concrete needed.

18. Find the mean and the percentage deviation from the mean:

(a) 0.99 sec   (b) 0.95 cm   (c) 0.90 gm
    1.00 sec   1.00 cm   1.00 gm
    1.01 sec   1.05 cm   1.10 gm

19. Make acceptable scientific statements about the timekeeping qualities of clocks A, B, C, and D, whose readings at noon (as determined by C) on successive days are:

    A: 11:56   11:57   11:58   11:59
    B: 12:04   12:08   12:12   12:16
    C: 12:00   12:00   12:00   12:00
    D: 11:52   11:47   11:42   11:37

20. Calculate the velocity (distance/time) for each of these course records:

(a) Jet plane	2451 mi	2 hr 58.71 sec
(Los Angeles to New York)		
(b) Running (mile)	1.00 mi	3 min 51.1 sec
(c) Running (15 mile)	15.00 mi	1 hr 12 min 48.2 sec
(d) Horse race (mile track)	1.00 mi	1 min 53.6 sec

For Further Reading

Bondi, Hermann, *The Universe at Large,* Science Study Series, Doubleday and Company, Inc., Garden City, New York, 1960.

Dirac, P. A. M., "The Physicist's Picture of Nature," *Scientific American,* May, 1963.

Feather, Norman, *Mass, Length, and Time,* Chapters 1–3, Edinburgh University Press, Edinburgh, 1959.

Thomson, Sir George, *The Forseeable Future,* Viking Press, New York, 1960.

Weiskopf, Victor F., *Knowledge and Wonder,* Science Study Series, Doubleday and Company, Inc., Garden City, New York, 1962.

CHAPTER TWO

Pressures in Fluids at Rest

Force is a term that you will meet again and again in your study of physics. You will study gravitational forces, electric forces, and nuclear forces, to name several. How to calculate the magnitudes of forces, how to combine them, and how to describe their effect upon the motion of a body are topics to which we shall give considerable discussion. In this chapter we introduce some relatively simple forces — forces due to the weight of fluids at rest. Fluids include liquids and gases.

Why do we start with this topic? Fluids at rest make a good starting place because they illustrate the simplifying process that goes on when a physicist attempts to understand nature. As we discuss fluids, you will notice that we start with a very simple situation that can easily be described quantitatively. Then we add complexities one by one and try to understand how each modifies the original situation. Before we take a look at fluids, however, let us define some terms.

What is a Force?

The man on the street often uses the words *force* and *pressure* interchangeably, but they differ in meaning. We think of a force as a push or a pull. The scientist defines it as *that which deforms a body or changes its velocity.* For example, by exerting a force you can stretch a rubber band or you can speed up a football.

The most familiar kind of force is *weight,* or the pull of the earth. Lift a pound of sugar in a shopping bag, and you will experience the down pull of the earth's attraction. Put another pound of sugar into the bag, and the down pull will increase. Thus the earth pull on a body increases with its mass. At a given locality, weight and mass are directly proportional to each other; they are different physical quantities, however.

The weights of different bodies can be found by using a *spring balance.* In its simplest form, it is a spiral spring, fitted with a pointer and scale to indicate elongations (Fig. 2-1). Suspend a 1-pound body from the spring; the pointer will move to a certain position on the scale until the *elastic force* with which the spring resists a change in its length equals the weight of the body. Add a second body of equal mass, and the spring will be stretched twice as far. In this way the scale can be marked off or *calibrated* to indicate the weight of a body suspended from it. If a spring balance is inserted into a suitable case (Fig. 2-2*a*), it can conveniently be used

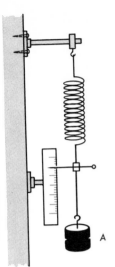

Fig. 2-1. *A spring balance to measure forces.*

to measure the weight of parcels that are to be mailed. It can also be used in a horizontal position to measure other forces besides the weights of bodies — for example, the pull necessary to drag a sled along a sidewalk.

Some common units of force are the kilogram-weight (kgwt) and the gram-weight (gwt), the newton and the dyne, and the pound-weight. The standard kilogram-weight is a force equal to the weight of the standard kilogram. (Since weight varies slightly with position on the earth's surface, as you will learn in Chapter 11, we specify that the weight shall be determined at sea level and 45° north latitude.) One kilogram-weight equals 1000 gram-weights.

The newton equals about 1/9.8 kilogram-weight, or about 102 gramweights, and the dyne equals approximately 1/980 gram-weight. Thus the newton is a little more than a tenth of a kilogram-weight and the dyne a little more than a milligram-weight. We shall discuss the newton and the dyne more fully in Chapter 5. Some spring balances used in the laboratory are calibrated in newtons (Fig. 2-2b).

The standard pound-weight (lb) is a force equal to the pull of the earth upon a standard

pound under the conditions specified above. It equals approximately 454 gram-weights.

What is Pressure?

In physics, pressure always means *force per unit area.* In Fig. 2-3a, a 2-pound block on a table top is supported by a surface 1 in.² in area, so that the average pressure at the bottom surface of the block is 2 pound-weights/in.² In Fig. 2-3b, the area of the supporting surface is 2 in.², and the average pressure is 1 pound-weight/in.² In each position the block exerts a force of 2 pound-weights against the table top.

Density

Weight density is a term that will be useful to us in this and following chapters. The weight density of a body means its weight per unit volume. For example, the weight density of water is approximately 1 gwt/cm³, 1000

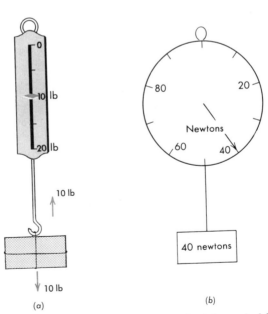

Fig. 2-2. *Spring balances.* (a) *A household spring balance.* (b) *A spring balance used in the laboratory.*

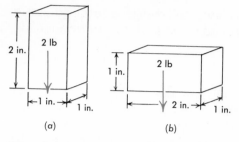

Fig. 2-3. (a) *The weight of the 2-lb block causes a pressure of 2 lb/in.*^{2} (b) *The pressure is one-half as great as in* (a).

kgwt/m^3, or 62.4 pound-weights/ft^3. (The term *mass density* is also used. It means mass per unit volume.)

The *specific gravity* of a substance is often spoken of instead of its density. By this expression we mean the *ratio* of the density of the substance to that of water. For example, the density of cork is 0.25 gwt/cm^3, which is one-fourth the density of water. Therefore, the specific gravity of cork is 0.25. Notice that in the cgs metric system the density and the specific gravity of a substance are numerically the same. For this reason computations of pressure in liquids are more convenient in the metric system than in the British system.

Equilibrium

If a body is at rest or moves with constant velocity, we say that it is *in equilibrium.* The net force on the body must be zero. For example, a 10-pound parcel held by a spring balance (Fig. 2-2a) experiences an up force of 10 lb due to the spring balance and a down force of 10 lb due to gravity. The up force and the down force cancel each other, not only if the body is at rest, but also if it moves at constant velocity.

If several forces act along a vertical line on a body in equilibrium, the sum of all of the upward forces must equal the sum of all of the downward forces.

The Pressure at a Point in a Liquid

Let us now use the terms force, pressure, and density *plus* the idea of equilibrium, *plus* a little common sense, to answer some questions about pressures, not underneath blocks, but in liquids. You will believe that you understand something about liquids if you can predict what the pressure is at *any* depth in a liquid confined in a container.

If you consider where liquids are ordinarily found, you will immediately see that we must simplify the situation if we are to begin to understand it. In the ocean, one encounters liquid pressures but the situation is seldom a simple one: tides keep the water in motion, waves cross its surface, and fresh water and salt water of different densities meet at the estuaries of rivers. Many complexities must be considered if one is to describe the pressures in these liquids: blood flowing in a blood vessel, blood plasma flowing through a looping tube during a transfusion in a hospital, gasoline in the half-full tank of a moving car, ginger ale in a conical paper cup with large bubbles of carbon dioxide rising to the surface.

We shall make a very modest beginning. We shall calculate the liquid pressure (1) at the bottom (2) of water (3) at rest (4) in a cylindrical container. The cylindrical glass vessel (Fig. 2-4*a*) is 4.0 cm^2 in cross section. It is filled with water to a depth of 10 cm. What pressure does the weight of the liquid exert at the bottom of the vessel?

The bottom must support the weight of the liquid since the vertical sides lend no support. The cross-sectional area of the column is 4.0 cm^2, and its volume is 40 cm^3. Since each cubic centimeter of water weighs 1 gwt, the down force on the bottom of the vessel is 40 gwt. The pressure at the bottom is 40 gwt/4.0 cm^2 or 10 gwt/cm^2.

Figure 2-4*b* represents a box 1 ft square and 2 ft high which is filled with water. The 2 ft^3 of water weighs 124.8 pound-weights,

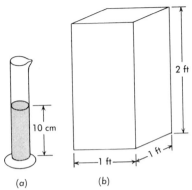

(a)   (b)

Fig. 2-4. (a) *The liquid causes a pressure of 10 gwt/cm². (b) The pressure at the bottom of the liquid is 2 × 62.4 lb/ft².*

so that the pressure at the bottom is 124.8 pound-weights/ft².

Notice also that *the pressure in gram-weights per square centimeter at any point in a body of water is numerically equal to the depth.*

A column of water 100 cm high and 1 cm² in cross section weighs 100 gwt, and its weight causes a pressure of 100 gwt/cm². A column of water 1 ft high produces a pressure of 62.4 pound-weights/ft², or 0.433 pound-weight/in.²

Pressures are often expressed in terms of the heights of the columns of water or mercury that would produce them. For example, a pressure of 100 gwt/cm² may be written 100 cm-water.

We shall now take another step in calculating pressures in liquids and derive a general formula. Let us determine the pressure, not at the bottom of a container filled with water, but at *any* point in *any* homogeneous liquid. For the moment, we shall assume that the container is a cylindrical one.

The pressure at any point in a liquid increases with the depth below the surface. In order to estimate the pressure at a point in a liquid, imagine that *a small, horizontal diaphragm is placed with its center at that point, and determine the force per unit area*

exerted at the diaphragm. We can make the area of the horizontal diaphragm as small as we please—as small as that of the head of a pin if we wish. The pressure at this depth is called the pressure at the point at which the small diaphragm is placed.

In Fig. 2-5, let the depth of the point below the surface be h and the area of the small diaphragm be A. The volume of liquid vertically above the diaphragm is Ah, and the weight of the liquid above the diaphragm is AhD_w, D_w being its weight density. The downward force exerted on the diaphragm is AhD_w. Thus the pressure at depth h is given by

$$\text{Pressure} = \frac{\text{Force}}{\text{Area}} = \frac{AhD_w}{A} = hD_w$$

Since the point at which we place the diaphragm could be any point without changing the argument, we can say that in general the liquid pressure p at a depth h below the sur-

Table 2-1. Densities of Liquids, Solids, and Gases

A. Liquids	gwt/cm³ or gm/cm³	lb/ft³
Gasoline (15°C)	0.66–0.69	41.0–43.0
Mercury (0°C)	13.596	850
Milk	1.028–1.034	64.0–64.5
Sulfuric acid	1.84	116.8
Water (20°C)	0.998	62.4
Water (4.00°C)	1.000 (by definition)	62.6
B. Solids at 20°C		
Aluminum	2.7	169
Copper	8.30–8.95	518–558
Gold, cast	19.3	1200
Ice (0°C)	0.92	58
Iron, wrought	7.8–7.9	487–493
Lead	11.3	708
Oak	0.6–0.9	37–56
Platinum	21.37	1334
Uranium	18.7	1167
Wood, balsa	0.12	7.5
Zinc, cast	7.04–7.16	440–447
C. Gases at standard conditions (0°C and 76 cm-of-mercury)		
Air	1.293×10^{-3}	0.0807
Carbon dioxide	1.977×10^{-3}	0.1234
Chlorine	3.214×10^{-3}	0.2006
Hydrogen	0.0899×10^{-3}	0.00561
Methane	0.7168×10^{-3}	0.04475
Nitrogen	1.2505×10^{-3}	0.7807
Oxygen	1.4290×10^{-3}	0.0892

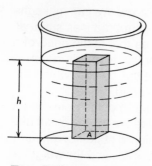

Fig. 2-5. *The pressure at a depth* h *is equal numerically to the weight of a column of liquid of unit area and depth* h.

face of a liquid of weight-density D_w is $p = hD_w$.

Example.* (*a*) What is the liquid pressure at a point 20 cm below the surface of mercury in a cylindrical beaker? (*b*) What is the downward force at that depth on a horizontal diaphragm 4.0 cm² in area?

(*a*) $\quad p = hD_w$

$\qquad p = 20 \text{ cm} \times 13.6 \text{ gwt/cm}^{\cancel{3}\, 2}$

$\qquad = 272 \text{ gwt/cm}^2$

(*b*) $\quad F = pA$

$\qquad F = 272 \text{ gwt/cm}^2 \times 4.0 \text{ cm}^2$

$\qquad = 1.09 \times 10^3 \text{ gwt}$

Pressure is Independent of Direction

So far, we have talked only about pressures that produce forces in the *downward* direc-

---

* In solving illustrative examples, we include the units of all physical quantities. Often we make use of the fact that *all conversion factors are 1*—as you will see in the following example.

Suppose that we wish to change 60 mi/hr to feet per second. One mile = 5280 ft; therefore 5280 ft/1 mi = 1. Also, 1 hr/3600 sec = 1. Now the magnitude of a quantity that is multiplied by unity is not changed. Therefore

$$60 \text{ mi/hr} = \frac{60 \text{ mi}}{1 \text{ hr}} \times \frac{5280 \text{ ft}}{1 \text{ mi}} \times \frac{1 \text{ hr}}{3600 \text{ sec}}$$

$$= \frac{88 \text{ ft}}{\text{sec}} = 88 \text{ ft/sec}$$

Stating units throughout all computations is helpful in that wrong units in the answer may indicate errors in basic assumptions. Then, too, by stating units, you clarify your ideas. By writing gwt/cm², you are reminded that pressure is force per unit area.

tion, on the bottom of a tank or on a horizontal diaphragm, for example. Do liquids exert forces in other directions?

A block of wood resting on a table exerts a downward force only. Sand contained in a box, however, exerts forces not only on the bottom of the box but also on the sides. If a hole is made in one side of the box, some of the sand escapes. Similarly, a liquid in a vessel exerts forces on the sides of the container as well as on the bottom. *The liquid is not rigid, and it transmits forces equally in all directions.* The push against unit area of the wall surface is perpendicular to the wall and equal to that exerted against a horizontal surface of unit area at the same depth.

Consider a small sphere which is at rest, submerged in a vessel of water (Fig. 2-6). The water pushes against every part of its surface. If the forces pushing it toward the left did not balance the opposite forces pushing it toward the right, the sphere would not remain in equilibrium. The down push against the upper half of the ball is a little less than the up push against the lower half. However, the difference between these forces decreases as the ball becomes smaller. If the ball were exceedingly small—a mere point—the up force and the down force would be equal. The forces at a point in a liquid at rest act in many directions, and they are equal. In other words,

Fig. 2-6. *Forces acting on a submerged ball. The pressure at a point in a liquid is independent of direction.*

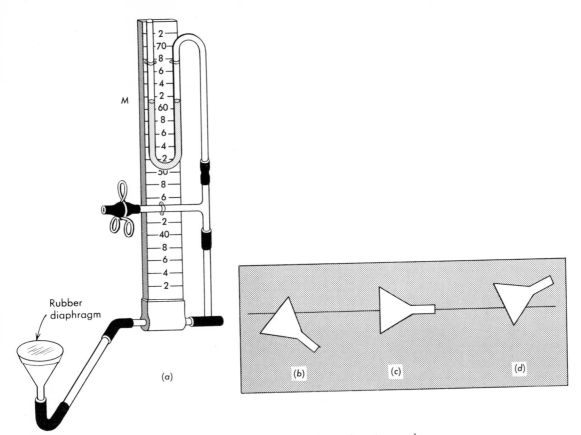

M

Rubber
diaphragm

(a)

(b)   (c)   (d)

*Fig. 2-7. The pressure at the diaphragm does not depend on its angle
of tilt.*

in a liquid at rest, the pressure at any point
has no definite direction.

We can study the forces caused by the
pressure of a liquid by means of the apparatus shown in Fig. 2-7*a*. A rubber diaphragm
closes the opening in a glass funnel which is
connected by a flexible hose and a glass
tube to a U-shaped glass tube containing
water or oil. If the rubber diaphragm is
pushed inward, the confined air is compressed and the liquid rises in the U-tube: the
device serves as a pressure indicator. If this
funnel is submerged in water (Fig. 2-7*b*), the
diaphragm is pushed inward because of the
pressure, and the liquid in the tube rises more
and more as the pressure increases with
depth. Tip the funnel sidewise as in Fig. 2-7*c*.

The elevation at **M** remains the same,
showing that the force on the diaphragm is
just as great as before. The reading does not
change when the funnel is tilted at other
angles; hence, once again we see that the
force exerted against the diaphragm is *independent of direction* (at the same depth).

Pressure is Independent of the Shape of the Container

We are now ready to drop the restriction that
the container of the liquid must have vertical
walls. To see that the pressure formula can be
applied whatever the shape of the container,
consider the three containers *A*, *B*, and *C*
(Fig. 2-8). Each is closed at the bottom with a

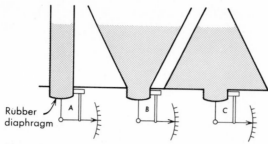

Fig. 2-8. *The pressure at the rubber diaphragm is the same in each vessel.*

rubber diaphragm attached to one end of a lever. When water is poured into the containers, the force caused by the weight of the water pushes each diaphragm downward and moves its lever. The position of each pointer measures the force against the rubber and hence the pressure at the bottom of the container. The containers are filled to the same depth, and the pressures are experimentally found to be equal. In *B*, for example, the diaphragm supports only the weight of the water that is vertically above it. The remainder of the liquid is upheld by the sloping walls of the container. In *C*, the walls push down on the liquid, increasing the force on the diaphragm above what it would have been had the walls been vertical. *The pressure in a liquid does not depend upon the shape of the container. The pressure is given by $p = hD_w$ whatever the shape.* This independence of pressure of the shape of a container is called the *hydrostatic paradox.*

However, you must be careful to measure *h* vertically to a free surface of the liquid — in hydrostatic problems, the free surface is usually that exposed to the atmosphere. Figure 2-9 represents two containers interconnected by a tube and filled with water. The pressure at *A* in the open container is numerically equal to the weight of a volume of water of unit area and height H_1. The pressure at *B* in the closed vessel is *not* equal to the weight of a unit column of water of height H_2, but is equal to the pressure at *A*. If an open vertical tube were inserted in the top of the closed

box, the water would rise in it to the same level as that in the open container.

In determining the pressure due to the weight of a column of liquid, measure the vertical distance to the free surface of the liquid.

Transmission of Pressures, Pascal's Principle

Liquids transmit changes of pressure. Suppose that the container represented in Fig. 2-10 is filled with water to a level 10 cm above the top of the box. The pressure at *C*, caused by the liquid, is 10 gwt/cm². All parts of the top of the box are at the same level; hence the water pressure there is everywhere the same, and a force of 10 gwt acts upward on each square centimeter of the box top. Pour a little water into the tube, so as to raise the level 1 cm. The water pressure at the level of *C* is now 11 gwt/cm², and a force of 11 gwt acts upward on each square centimeter of the top of the box. Therefore the increase of pressure is transmitted equally to various parts of the liquid at the level of *C*. You can show that the same increase in pressure is caused at every point in the liquid. This is one proof of a general principle stated by the French scientist, philosopher, and mystic, Blaise Pascal, over three centuries ago:

Any change of pressure in an enclosed fluid at rest is transmitted undiminished to all parts of the fluid.

When the pressure at every point of a sur-

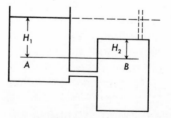

Fig. 2-9. *The pressures at* A *and* B *are equal, for each point is at the same depth below the free surface of the water.*

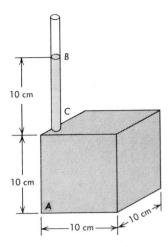

B

10 cm

C

10 cm

A

←——10 cm——→ ←10 cm→

Fig. 2-10. Pascal's principle. By raising the water level at B 1 cm, we increase the pressure 1 gwt/cm² in all parts of the fluid. Any change of pressure in a fluid at rest is transmitted undiminished to all parts of that fluid.

face is constant, we can find the total force exerted against that surface by taking the product of the pressure by the area. In Fig. 2-10, the pressure in the liquid at the top of the box is 10 gwt/cm², and the liquid exerts a total up force on the top of the box of 1000 gwt. In order to compute the total out force against the left vertical side, we must find the *average* pressure. A more complete analysis shows that the average pressure is the average of the pressure at the top and that at the bottom of the box. The force is the area times this average pressure. At the top of the box, the pressure is 10 gwt/cm². At the bottom, it is 20 gwt/cm², and the average pressure is 15 gwt/cm². Thus the water exerts against this side a total out force of 1500 gwt.

Applications of Pascal's Principle

Pascal's principle is applied in technological devices which are used to exert great forces or to control forces readily. Assume that in Fig. 2-11 a force of 10 pound-weights acts on the small piston which has an area of 1 in.² This force causes a pressure of 10 pound-

weights/in.² In accord with Pascal's principle, the increase of pressure is transmitted to all parts of the liquid so that a force of 10 pound-weights acts on each square inch of the larger piston. Its area is 100 in.²; hence the total upward force acting on it is 1000 pound-weights.

The applied force F_1 acts on a piston of area A_1, and the "load" force F_2 on a piston of area A_2. Since the pressures at the two pistons are equal,

$$p_1 = p_2$$

$$\frac{F_1}{A_1} = \frac{F_2}{A_2}$$

Automobile brakes and power steering mechanisms are operated by pistons in liquid-filled cylinders. When the driver pushes the brake pedal (Fig. 2-12a) to the "master" piston, he increases the pressure of the confined liquid. The increase of pressure is transmitted to four other cylinders, whose pistons act against the brake shoes. One advantage of hydraulic brakes is that the forces on all the brake bands are equal. A disadvantage is that, if a communicating tube breaks or if the liquid leaks out, all four brakes fail. To provide protection against this possibility, dual brake systems are now mandatory.

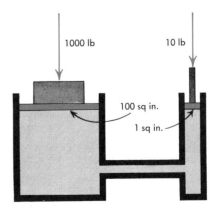

1000 lb   10 lb

100 sq in.

1 sq in.

Fig. 2-11. A hydraulic press. A small force acting on the small piston causes the large piston to exert a large force.

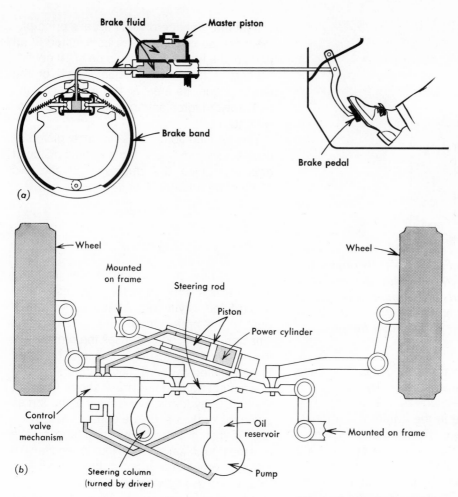

Brake fluid — Master piston — Brake band — Brake pedal

(a)

Wheel — Mounted on frame — Steering rod — Piston — Power cylinder — Wheel — Control valve mechanism — Steering column (turned by driver) — Oil reservoir — Pump — Mounted on frame

(b)

Fig. 2-12. *Pascal's principle applied in automobiles.* (a) Hydraulic brakes, *in which increase of pressure in the master cylinder is transmitted to the piston and cylinder in each brake, expanding the brake shoe against the brake lining.* (b) Power steering, *in which the rotation of the steering wheel operates a valve mechanism so that the oil under pressure is pumped into the power cylinder, turning the wheels.*

In power steering (Fig. 2-12b), rotation of the steering wheel by the driver turns the steering column and operates a valve mechanism so that oil is pumped under pressure into the right or left side of the power cylinder. The oil pressure produces a force on the cylinder and the steering rod is pushed to the right or left, turning the wheels.

Buoyancy and Archimedes' Principle

We shall now use what we have learned about pressures in liquids at rest to understand why some bodies float in liquids and why submerged bodies seem to weigh less than they do in air. A bather, walking into deeper water, notices that, as more of his body is submerged and more water is displaced, his apparent weight decreases; finally he floats with only his face above the surface. This up push of a liquid we call *buoyancy*. An empty ship floats high in the water. As it is loaded, it sinks deeper. The buoyancy of the floating ship balances the weight of the ship and its contents. If it did not, the ship would sink to the bottom. Buoyancy increases with the amount of liquid displaced.

What is the relationship between buoyancy and the amount of liquid displaced? Consider first the following experiment.

Suspend a cylinder *B* and a bucket *A* of equal volume from the left arm of a balance and balance the load by means of a weight *C* (Fig. 2-13). Submerge the cylinder in a vessel of water, and the left arm will tip upward because of the buoyancy of the liquid. Fill the bucket completely with water, and equilibrium will be restored. The cylinder and bucket have equal volumes, and the weight of the water in *A* equals the buoyancy; hence the buoyancy equals the weight of the liquid that would fill the space occupied by the cylinder. Similar experiments can be performed with other liquids.

The law of buoyancy, called *Archimedes' principle,* is as follows:

A body wholly or partly submerged in a fluid is buoyed up by a force equal to the weight of the displaced fluid.*

Now let us show that Archimedes' principle can be explained with the help of ideas developed earlier in this chapter. As usual, we shall simplify the problem by assuming that the submerged body has a geometrically simple shape. Imagine then that a rectangular

* Archimedes' principle applies to gases as well as to liquids. Note also that the volume of the displaced fluid may be greater than the volume of the fluid actually present. For example, one can float a small test tube containing 200 gm of lead shot in a slightly larger test tube containing 150 gm of water. Nevertheless, the buoyant force is still equal to the weight of the water which would be needed to fill the space occupied by the submerged part of the floating body.

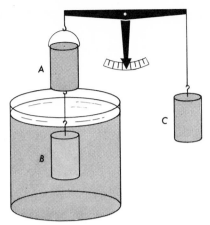

Fig. 2-13. Archimedes' principle. The weight of the water in the bucket balances the up push on the cylinder.

block of cross-sectional area *A* (Fig. 2-14) is completely submerged in a liquid of weight density D_w, and let the vertical distance of the bottom of the block from the free surface of the liquid be *h*. Let the distance of the top of the block from the free surface be *h'*. Then the pressure at the bottom of the block is hD_w, and the upward force exerted by the liquid on the block is $hD_w \times A$. By similar reasoning,

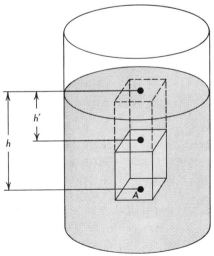

Fig. 2-14. Illustration of Archimedes' principle.

the down force exerted on the top of the block is $h'D_w \times A$. The difference of these forces, the amount by which the force pushing the block up is greater than the force pushing it down, is what we mean by buoyancy and will be denoted by F.

$$F = (h - h')D_w \times A$$

But $(h - h') \times A$ is the volume V of the block; hence

$$F = VD_w$$

VD_w is the weight of the displaced liquid; hence the buoyancy equals the weight of the displaced liquid.

If the buoyancy is greater than the weight of the submerged body, the body will rise to the surface and float unless some third force is applied to keep it submerged. The submerged part of the floating body displaces such a volume of liquid that the **weight** of the displaced liquid equals the weight of the entire floating body.

Example.* A wooden block whose volume is 100 cm³ and whose weight is 80 gwt is placed in water. (*a*) What is the volume under water when the block floats? (*b*) What third force is required to submerge the block completely?

(*a*) When the block floats, the buoyant force (up) equals the weight (down) of 80 gwt. Hence, the buoyant force is 80 gwt and, by Archimedes' principle, the weight of the displaced water is 80 gwt. Since water has a density of 1 gwt/cm³, the volume of the displaced water is 80 cm³. Thus, the volume of the block under water is 80 cm³ (and the volume *out* of water is 20 cm³). (*b*) To submerge the block completely requires that the total down force be equal to the buoyancy when the block is submerged. When the block is submerged, the volume of the displaced water is 100 cm³ and this much water has a weight of 100 gwt. Hence the buoyancy is 100 gwt. Since the weight of the block is 80 gwt,

---

* Several important ideas are involved in the solution. Go over it until you understand thoroughly each step.

an additional 20-gwt down force is needed to hold the block under water.

Archimedes

Many remarkable discoveries were made by the Greek scientist Archimedes (287–212 B.C.), not the least of which was the principle we have just discussed. Archimedes was one of the first to apply quantitative mathematical reasoning to the solution of problems in physics. His treatises "On Floating Bodies" and "On Levers" marked advances in quantitative reasoning about mechanics that were not surpassed, as we shall see, until the time of Galileo in the sixteenth century. Archimedes was also a mathematician of the first rank. His achievements in plane and solid geometry and number theory arouse the admiration of modern mathematicians. It is noteworthy that one of the first great physicists was also one of the great mathematicians; this connection between physics and mathematics has persisted to our day.

Archimedes spent most of his life in the Greek city of Syracuse in Sicily. According to one story about Archimedes, Hiero, the ruler of Syracuse, had given gold to an artisan to make a crown. Hiero suspected that the gold had been alloyed with silver so that the artisan could keep some of the gold for himself. The ruler asked his friend Archimedes to test the purity of the metal. While Archimedes was puzzling over the problem, he went to the public baths one day and noticed that, as he stepped into a tub filled to the brim with water, some of the liquid spilled over. It occurred to him that he could test Hiero's crown by submerging it in a vessel filled with water and noting the volume that was displaced. He could then submerge pure gold, equal in weight to the crown, and note the overflow. If the crown displaced more water than the gold, its volume must be greater because of the addition of silver. The philosopher was so eager to try the experiment, the story goes, that he rushed through the street shouting "Eureka!" ("I have found it!").

Fig. 2-15. *The push of the atmosphere forces the rubber inward.*

Often the volume of a body is measured by the method just described, but usually we find the apparent loss of weight on submersion and then compute the volume by means of Archimedes' principle. After weighing the body in air, one can compute the density of the body.

The Pressure of the Atmosphere

We have seen how one can calculate the pressure in a liquid at rest—beneath the surface of a quiet lake, for example, or in a gasoline storage tank. Can one similarly calculate the pressure due to the weight of the earth's atmosphere?

It is easy to verify that the atmosphere does have weight and that it exerts a pressure. If we weigh a liter flask, first when evacuated and then after it is opened to the atmosphere, the increase in weight will be more than 1 gwt. So air does have weight. The air in a medium-sized classroom weighs several hundred pounds; that in a large lecture room, more than a ton. Moreover, the atmosphere exerts a pressure: Stretch a rubber membrane over the upper opening of a glass jar as shown in Fig. 2-15. The membrane is flat because the air pressure is equal on both sides. Pump air out of the enclosure, and the diaphragm will be pushed inward because of the difference in pressure. If a varnish can is evacuated, the force of the atmosphere pushing against the sides will crush it.

The huge forces can be exerted by atmospheric pressures was demonstrated 300 years ago by Otto von Guericke, who invented the air pump. He designed two hemispherical iron vessels, which could be put together forming an airtight enclosure 22 in. in diameter (Fig. 2-16). When the interior was evacuated by means of an air pump, great forces were required to separate the hemispheres. Von Guericke arranged a public demonstration at which sixteen horses, eight on each side, pulled against each other in a tug of war, and failed to separate the hemispheres. (Eight horses and a tree would have been just as effective, but not so showy.) You

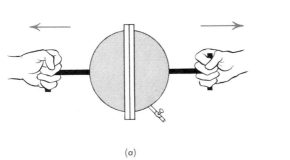
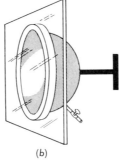

(a)   (b)

Fig. 2-16. *Magdeburg hemispheres.* (a) *The push of the atmosphere holds them together.* (b) *The force against the flat plate is equal to the excess atmospheric pressure times the area.*

can easily calculate the force required to part the hemispheres. Suppose that one of them were replaced by a flat metal plate. The area of the circular opening was about 380 in.² If the atmospheric pressure were 15 pound-weights/in.², the horizontal force against the plate would be about 5700 pound-weights. The surface area of a hemisphere is greater than that of the flat plate, but the forces do not act in the same direction.

Suppose that we have measured the pressure of the atmosphere by finding the ratio of the force required to separate two small Magdeburg hemispheres to their cross-sectional area; the result, let us say, is 1000 gwt/cm². How deep is the ocean of air above us that produces such a pressure? Shall we use the relation $p = hD_w$, inserting the density of air

at sea level? If we do, the result of our calculation—the "depth of the atmosphere"—is 26,500 ft. Something is clearly wrong: jet airliners quite commonly cruise at altitudes of more than 30,000 ft! In fact, rocket soundings and earth-satellite measurements (Fig. 2-17) show that the atmosphere has measurable pressures at much greater heights.

Our failure to obtain a reasonable value for the "depth" of the atmosphere illustrates the danger of applying uncritically to one physical situation results obtained in a different physical situation. The relation $p = hD_w$ was derived for a *liquid at rest.* Liquids are highly incompressible: to decrease the volume of a given mass of water by 1% requires an increase in pressure of around 3700 lb/in.² Gases are highly compressible, their den-

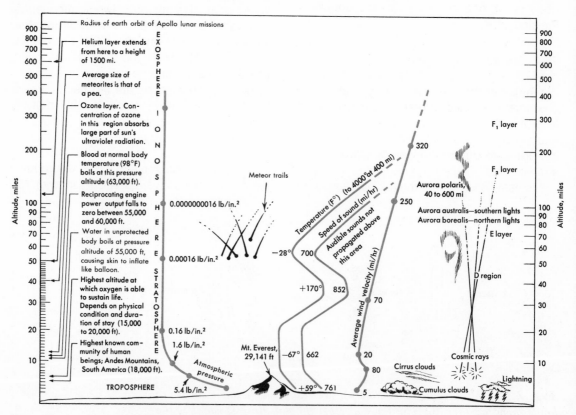

Fig. 2-17. Characteristics of the earth's atmosphere.

sities varying directly with pressure at constant temperature. The ocean of air is a fluid whose lower layers are compressed by the overlying layers. The density of the atmosphere diminishes from a maximum value at the surface of the earth to a minimum value which is really the density of matter in interplanetary space. Furthermore, changes in weather demonstrate that the atmosphere is not at rest. The lower layers of the atmosphere are in motion; cold masses of air from the poles continually collide with warm air masses from the tropics. Because of these complications, calculations dealing with atmospheric pressure require sophisticated methods of solution and high-speed computing techniques.

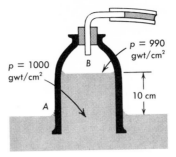

Fig. 2-18. *The pressure at* A *is equal to the pressure at* B *plus the pressure due to the liquid column 10 cm high.*

Measuring Atmospheric Pressure

Although atmospheric pressure cannot be readily calculated, it can be measured by means of a *barometer* in which atmospheric pressure supports a liquid in an evacuated tube — a more calculable situation.

Pump air out of a bell jar, the bottom of which dips into a vessel of water (Fig. 2-18). Then the water will rise higher in the enclosure as the pumping progresses. Consider the downward push of the air at the water surface outside. A force of about 1000 gwt acts on each square centimeter of the surface, and the pressure must be the same inside the jar. When the air pressure inside the jar is diminished, the water is forced up into it by the excess push of the atmosphere. If the water level in the vessel is 10 cm above that outside, the air pressure inside is 10 gwt/cm² less than atmospheric pressure, and the pressure of the air inside is 990 gwt/cm².

The ancients knew that water rises in an evacuated tube. They explained this fact by saying that nature abhors a vacuum. Over three centuries ago, the Duke of Tuscany, in Italy, dug a deep well and found that no available pump could lift the water higher than 34 ft. Torricelli, a friend and follower of the great

scientist Galileo, asked Galileo why this could be. The old man jestingly replied that probably nature's abhorrence of a vacuum did not extend beyond 34 ft! Torricelli wondered whether the "abhorrence of a vacuum" depended upon the density of the liquid used. He thought that mercury, being 13.6 times as dense as water, might be pumped to a smaller height, that is, 34/13.6 ft, or about 30 in. Instead of using a pump to exhaust the air from a pipe, he filled a long glass tube with mercury, closed the open end with his finger, inverted the tube, and inserted the end under the surface of a pool of mercury (Fig. 2-19a). When he removed his finger, liquid escaped from the tube until the upper surface was about 30 in. above the level in the dish, just as he had predicted (Fig. 2-19b). Later experiments showed that the vertical height of the column was independent of the width of the tube (Fig. 2-19c) and its angle of tilt (Fig. 2-19d).

When Pascal heard about Torricelli's invention, he suggested that, if atmospheric pressure is caused by the weight of the ocean of air, the barometric pressure should decrease with elevation, just as the water pressure in a lake decreases as a diver rises nearer the surface. Accordingly, he requested that his brother-in-law carry a barometer to the top of a mountain (Puy de Dôme) in southern France. This man "was ravished with astonishment and delight" to observe that when

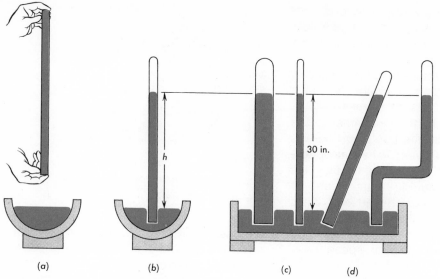

Fig. 2-19. (a) and (b): *Making a simple mercury barometer. The vertical height of the mercury in a barometer is independent of the width of the tube* (c) *and its angle of tilt* (d).

he ascended 1000 ft, the barometric column fell nearly an inch. This result confirmed the hypothesis that the weight of the air causes atmospheric pressure.

Modern mercury barometers (Fig. 2-20), used to measure atmospheric pressure in weather stations and laboratories, are quite similar to the first barometer devised by Torricelli. Atmospheric pressure can be stated in any pressure units: pound-weights per square inch or gram-weights per square centimeter, for example; it is also given in centimeters-of-mercury (cm-Hg) or atmospheres (1 standard atmosphere = 76.0 cm-Hg $\cong$ 14.7 lb/in.^{2} $\cong$ 1,013,000 dynes/cm^2). One *bar* is defined as 10^6 dynes/cm^2.

Example. Atmospheric pressure on a certain day was reported by a weather station as 75 cm-Hg. What was the pressure in gram-weights per square centimeter?

A pressure of 75 cm-Hg means a pressure equal to the liquid pressure at the base of a vertical column of mercury 75 cm high. Hence

$$p = hD_w = 75 \text{ cm} \times 13.6 \text{ gwt/cm}^3$$
$$= 1020 \text{ gwt/cm}^2$$

Summary

Force is that which deforms a body or changes its velocity. *Weight,* the gravitational force that the earth exerts on a body, is a familiar example. Weights can be measured on a spring balance. Force units include the kilogram-weight, the newton, and the pound-weight. *Pressure* is force per unit area and is measured in such units as gram-weights per square centimeter. *Weight density* is weight per unit volume. A body at rest or moving with constant velocity is in *equilibrium.* The net force on a body in equilibrium is zero.

The pressure in a fluid depends upon the depth of the fluid and its density. Liquid pressure is independent of direction and of the shape of the container. The pressure at a depth h below the surface of a liquid of weight density D_w equals the weight of a column of the liquid of height h and of unit cross-sectional area. $p = hD_w$. Any change of pressure in an enclosed fluid at rest is transmitted undiminished to all parts of the fluid (Pascal's principle).

Archimedes' principle states that a body wholly or partly submerged in a fluid is buoyed up by a force equal to the weight of the displaced fluid. Archimedes' principle is useful in determining the densities of solids and liquids.

The pressure of the atmosphere cannot be calculated as readily as can the pressure in liquids at rest. Atmospheric pressure can be measured: standard atmospheric pressure is equivalent to the pressure exerted by 34 ft of water or 76 cm of mercury.

The calculation of pressures in fluids at rest illustrates the simplifying process that often leads to understanding. One starts with the simplest possible situation and adds complexities one by one, trying to understand how each modifies the original situation.

Questions*

1. Define force, weight, and pressure. State units in which each may be expressed.

2. Define weight density and state the units in which it may be expressed. What is the difference between weight density and specific gravity?

3. State Archimedes' principle. Would you expect it to apply to both liquids and gases? How can you prove it experimentally? Give an

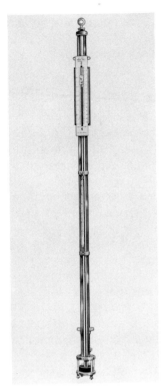

Fig. 2-20. A mercury barometer showing reservoir and column.

example of how Archimedes' principle applies to gases.

4. Compare the weights of 10-ft ladders made of wood, aluminum, and magnesium.

5. Does the pressure at the face of a dam depend on the length of the reservoir measured perpendicularly to the dam?

6. Estimate the weight of the air in your classroom.

7. Estimate how many pounds of gold you could pack into the trunk of a standard passenger car. What would be its value at $50 per ounce? Could the car move it?

8. Estimate the pressure in tons/ft² at the tip of an ice pick in ordinary use.

9. Why are our bodies not crushed by the enormous forces exerted by the atmosphere? Estimate the atmospheric force that is exerted on your body.

* Occasionally, as in question 6 or 7, you will be asked to estimate a numerical answer. For example, if you estimate the volume of a room to one significant figure and know the density of air to one significant figure, you should be able to approximate the weight of air in the room to the same precision. Is it 0.1, 1, or 10 kgwt? For some purposes, this degree of precision—a "horseback estimate"—is satisfactory. In other questions where numbers are given, try to determine the answer by rough calculation to determine the "order of magnitude"—that is, whether the result is 10^2, 10^3, or 10^{-6}. Thus, if a silver coin has a mass of 2.5 gm and a silver atom has a mass of about 2×10^{-22} gm, there are about 10^{22} silver atoms in the coin ("order-of-magnitude," "horseback," or "right-ball-park" result).

10. If two objects of equal weight and unequal volumes are balanced in air, what will happen when they are compared in a vacuum?

11. Is it possible, using a quart of water or less, to burst a large metal drum filled with water? Explain.

12. Compare the force required to hold a wooden block barely under the surface of a pool with that required to hold it at a depth of 6 ft.

13. How could you determine whether a spoon was made of solid silver or of some alloy?

14. Design a method for determining the lift of a small helium-filled balloon by attaching a long chain that drags on the floor.

15. Why does water not escape freely when a jug is suddenly inverted?

16. How could you prove that pressure is (a) independent of the shape of the container and (b) independent of direction?

17. If you were sending a balloon to the stratosphere, how should it be inflated at takeoff?

18. Can the volume of the displaced fluid be greater than the actual volume of the fluid? Give an example.

19. What is a siphon? Explain why water flows through a siphon. Over what maximum elevation can water be siphoned? If the barometric pressure increases, does the rate of flow through a siphon change? Why? Can gasoline be siphoned over a higher wall than water? Why?

20. A can full of water is suspended from a spring balance. How will the reading of the balance change (a) if a block of cork is placed in the water, (b) if a piece of lead is placed in it? Why?

21. How does a submarine operate? State clearly what happens to its displacement when it (a) submerges, (b) runs submerged, (c) surfaces.

22. To what maximum depth below the surface of salt water can a man submerge himself and still be able to breathe through a hollow reed that extends above the surface? Describe a reasonable test—one not requiring you to get wet—for obtaining data needed to solve the problem.

23. Develop a method for measuring the average density of a living man under two conditions: lungs fully inflated and lungs fully deflated. Both measurements are to be accurate to within one per cent.

24. Normal blood pressure, taken on a woman's arm, is 110 mm-Hg. Estimate the blood pressure in her leg when she is sitting.

Problems

A

1. The hatch door of a submarine has an area of 0.80 m². What force does the water exert on the door when it is at a depth of 200 m? (Specific gravity of sea water is 1.03.)

2. The water pressure in the faucet in the basement of an office building is 80 lb/in.² What is the pressure in a faucet on the fourth floor at an elevation 50 ft above the other one?

3. A vertical valve in the wall of a dam 200 m below the water surface is 0.25 m² in cross section. Find the force on it caused by the water.

4. A cubical glass vessel 30 cm high is filled one-quarter with mercury and three-quarters with water. Find the pressure at the bottom.

5. What downward force due to the atmosphere acts on the top of a square table, 36 in. on each side, if the atmospheric pressure is 34 ft-of-water?

6. In cracking a nut, a force of 80 lb is exerted on the biting edge of a tooth. If the surface of contact is a square 0.10 in. on each side, compute the pressure at the surface.

7. When a mercury barometer is vertical, the length of the liquid column is 76 cm. What would the length of the column be if the tube were tilted to an angle of 60° with the vertical?

8. The excess or gauge pressure in each tire of a 2400-lb car is 30 lb/in.² What is the

surface area of each tire touching the ground? State clearly any assumptions you make about the weight distribution of the car.

9. Compute the weight of the air in a room 45 ft × 30 ft × 12 ft. (Weight density of air: 2.2 lb/yd³.)

B

10. A boy standing in a swimming pool, submerged up to his neck, finds that when he exhales, his apparent weight increases 8 lb. What volume of air was exhaled?

11. When an 80-kgwt man stepped into a canoe, it sank 3.0 cm further into the water. Find the cross-sectional area of the canoe at the water surface.

12. A vertical tube 40 cm long is attached at its lower end to the top of a box 30 cm × 30 cm × 30 cm resting on a table top, and the box and the tube are completely filled with water. What force does the liquid exert (a) on the bottom of the box, (b) on one side?

13. A uniform wooden cylinder 20.0 cm long and 4.0 cm² in cross section barely floats in gasoline of weight density 0.80 gwt/cm³. What fraction of the cylinder's volume will be above the surface when it floats in salt water of weight density 1.10 gwt/cm³?

14. A swimming pool is 20 ft wide and 40 ft long. Its depth varies uniformly from 4.0 ft at one end to 10.0 ft at the other. Find the total force on the bottom due to the weight of the water.

15. A city reservoir is located on a hill so that the bottom of the reservoir is at an elevation (above sea level) of 2000 ft. What is the water pressure in a water tap at an elevation of 1100 ft if the depth of the water in the reservoir is 80 ft?

16. A piston of a hydraulic device for lifting automobiles is 5.0 in. in diameter. The device is driven by water from the city system. What is the necessary water pressure to raise a car if the total load lifted weighs 3142 lb?

17. A nail 0.015 in.² in cross-sectional area is pushed through a hole in the rubber stopper of a bottle filled with water. (a) If a force of 4.0 lb is exerted on the nail, what is the increase in pressure in the liquid? (b) If the bottom of the bottle has an area of 4.0 in.², what is the force exerted against it?

18. A body weighs 100 gwt in air and 80 gwt submerged in water. (a) What is its density? What will it weigh (b) in alcohol (density 0.80 gwt/cm³) and (c) in a liquid of density 1.1 gwt/cm³?

19. If the density of the earth's atmosphere were constant at 1.4 × 10⁻³ gwt/cm³ at all altitudes, to what height would the atmosphere reach?

20. A pair of Magdeburg hemispheres is 0.10 m in diameter. The barometric pressure is 76 cm-Hg. What force is required to separate them (a) if the interior is completely evacuated, (b) if the pressure inside is 29 cm-Hg?

21. A siphon like that depicted in Fig. 2-21 is used to empty a tank filled with oil of weight density 0.90 gwt/cm³. When the valve is closed, what is the pressure on each side of it? The lengths of the two vertical columns are 40 cm and 90 cm, respectively. The barometer reading is 73.5 cm-Hg.

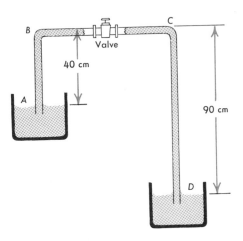

Fig. 2-21. A siphon. The pressure at C is less than that at B. When the valve is open, water is forced from B to C.

22. The average density of the air in a building is 0.00120 gwt/cm³, and the barometer reads 75.0 cm-Hg at street level. (*a*) How much will it read at the roof, 200 m above the ground? (*b*) What will be the increase in the force exerted on the drum of a man's ear, area = 0.30 cm², when he descends from the roof to the street level?

23. What is the buoyancy due to the atmosphere on an 80.0-kg man if the mean density of his body is 1.00 gwt/cm³ and that of the air is 0.00130 gwt/cm³.

C

24. A hollow stopper made of glass of specific gravity 2.5 weighs 25.0 gwt in air and 7.0 gwt submerged in water. What is the volume of the internal cavity?

25. An iron ball weighing 150 gwt and having a weight density of 7.80 gwt/cm³ floats in mercury of weight density 13.6 gwt/cm³. How many cubic centimeters of the ball are above the surface?

26. An iron sinker (density 7.8 gwt/cm³) is to be used to make a 2500-gwt oak board (density 0.60 gwt/cm³) barely float in water. What must be the weight of the sinker if it is (*a*) placed on top of the board, (*b*) tied to the bottom?

27. How heavy a stone can a boy lift under water if he can lift a 20-kgwt stone in air? The specific gravity of the stone is 6.0.

28. A crown made of silver and gold weighs 1600 gwt in air and 1510 gwt submerged in water. What is (*a*) the volume of silver, (*b*) its weight, (*c*) the percentage by weight of silver in the crown? (Density of gold; 19.32 gwt/cm³; that of silver: 10.53 gwt/cm³.)

29. The specific gravity of ice is 0.92 and that of sea water, 1.03. What is the total volume of an iceberg that floats with 1000 m³ exposed?

30. A block of cork weighs 2.00 gwt, and its weight density is 0.25 gwt/cm³. Pieces of aluminum (weight density: 2.70 gwt/cm³) are attached to it so that the cork and metal are barely submerged in water. How many grams of the metal are required?

31. A ferryboat with approximately vertical sides and horizontal cross section of 1000 m² at the water line sank 6.0 cm as a number of automobiles were driven on board. If each car weighed 1500 kgwt, how many were taken aboard?

32. A hollow copper sphere with a volume of 1500 cm³ is balanced on an equal-arm scale by weights whose combined volume is 30.0 cm³. What is the error in the measured weight of the sphere due to buoyancy? (Density of air is 1.293×10^{-3} gwt/cm³.)

33. A cone-shaped funnel weighing 200 gwt, open at the apex, is placed mouth downward on a flat glass plate and is filled with water. The diameter of the cone is 10.0 cm, and its height is 15.0 cm. What minimum force is required to hold the cone in place?

34. A block of wood (specific gravity 0.90) floats in a vessel of water. If sufficient kerosene (specific gravity 0.80) is added to submerge the block, what fraction of the block will be below the kerosene-water interface?

35. A long cylindrical stick of wood 3.0 cm in diameter weighs 0.50 kgwt. A boy pushes it straight down into water until its lower end is 1.0 m below the surface. What force must the boy exert to hold the stick in this position?

For Further Reading

Angrist, Stanley W., "Fluid Control Devices," *Scientific American,* December, 1964, pp. 80–88.

Fenn, W. O., "The Mechanism of Breathing," *Scientific American,* January, 1960, p. 138.

Knedler, John W., Jr., ed., *Masterworks of Science* (Archimedes, "On Floating Bodies"), Doubleday and Company, Garden City, 1947.

Landsberg, Helmut E., "The Origin of the Atmosphere," *Scientific American,* August, 1953, pp. 82–86.

CHAPTER THREE

Resultants and Components of Forces

In the preceding chapter you learned how fluids exert forces on bodies. It is easy to see how such forces are combined because, in simple problems, the forces act in the same line. In this chapter, you will see how two or more forces, acting in different directions on a body, may be replaced by a single force. Also, you will see how a single force may be replaced by two or more forces. Since more than one force usually acts upon a body, it is essential to know how to combine these forces in order to determine how the body will move. Finally, you will study the effects of one kind of force—friction.

Vectors and Scalars

Combining several forces correctly into a single force or replacing a force by several equivalent forces requires that we recognize the *vector* nature of forces. Vector (Latin *vehere* = to carry) quantities are those that have both a magnitude or size and a direction. Suppose a boy pulls on a rope attached to a log. If we want to describe the effect that the rope has on the log, we must give both the magnitude of the pull and its direction: 20 lb (magnitude) east (direction). A force of 20 lb east is different from one of 20 lb west in that the log will tend to move east in the first instance and west in the second; both are different from a 10-lb force north. Vector quantities then must be defined in both magnitude and direction. Arrows (vectors) are convenient for representing vector quantities; the length of the arrow is made proportional to the magnitude of the quantity and the arrow pointed in the direction or line of action of the vector quantity. Thus, a force of 20 lb east will be represented by $\xrightarrow{20\ lb}$. We shall work with many different vector quantities in physics: force, displacement, velocity, and acceleration, to name a few.

Other quantities, that require only that we state their magnitudes completely to specify them, are called *scalar* (Latin *scala* = a ladder) quantities. As their name suggests, they can be specified on a scale of sizes and do not require a directional specification. Mass, for example, is a scalar quantity. The mass of a certain body is completely specified by a magnitude of 20 kg; it makes no sense to say 20 kg east. Temperature, volume, and energy are some of the other scalar quantities we shall deal with.

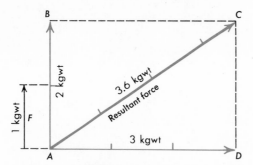

Fig. 3-1. *The parallelogram law. The diagonal* AC *represents the resultant of the forces* AB *and* AD.

Adding Forces

Scalar quantities can be easily added. If you pour 1000 cm³ of water into 1000 cm³ of water, the resulting volume of the water is 2000 cm³. Vector quantities obey a more interesting algebra: two forces which are each 1000 gwt in magnitude may have a sum of zero, 1000 gwt, 2000 gwt, or some other result, depending upon the directions of the forces. What are the rules for adding forces?

When forces act in the same direction, we add them arithmetically. Suppose that 10 sophomores pull north on a rope and that each man exerts a force of 40 kilogram-weights. Then a horse pulling north with a force of 400 kgwt would have the same effect as the 10 men. A single force which is equivalent to others we call their *resultant.*

The resultant of several forces is the single force that is equivalent to them.

In a tug of war, let the sophomores pull east with a force of +400 kgwt while the freshmen pull west with a force of −360 kgwt. Then the resultant force — in this example the algebraic sum of the two — is +40 kgwt directed east.

When two forces are not parallel, their resultant is found by the *parallelogram rule.* First, we represent the forces by vectors of proper directions and lengths. For example, the arrow or vector *F* (Fig. 3-1) of any desired length represents 1 kgwt; then *AB*, two times as long, represents a force of 2 kgwt directed north, and *AD*, three times as long, is a force of 3 kgwt directed east. (Notice that the two arrows are placed tail to tail.) Complete the parallelogram *ABCD*. The arrow *AC*, which is the diagonal of the parallelogram, represents the resultant of *AB* and *AD*. It is 3.6 times as long as *F*; hence, by this graphical method, the result is 3.6 kgwt.

The resultant of two forces is represented by the diagonal of a parallelogram of which the two adjacent sides represent the forces both in magnitude and direction. You can easily demonstrate the truth of this statement by a simple experiment in the laboratory, using pulley, string, and standard weights.

The resultant of two forces depends upon the angle between them. In Fig. 3-2*a* the two forces are oppositely directed, and their resultant equals the difference between them. In *b* and *c*, the resultant increases as the angle becomes smaller, and in *d* it equals the *sum* of the two. Thus, the resultant of two forces may be greater or less in magnitude than the larger of them.

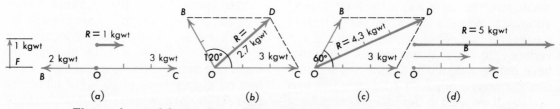

Fig. 3-2. *The resultant of the two forces* OB *and* OC *depends upon the angle between them.*

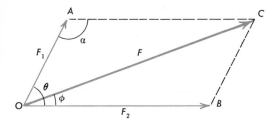

Fig. 3-3. *Resultant of two forces.*

The resultant F of two forces F_1 and F_2 that make an angle θ (Greek *theta*) with each other (Fig. 3-3) can be computed by an analytical method by applying the law of cosines.*

$$F^2 = F_1^2 + F_2^2 - 2F_1F_2 \cos \alpha$$

Since $\alpha + \theta = 180°$, $\cos \alpha = -\cos \theta$ and

$$F^2 = F_1^2 + F_2^2 + 2F_1F_2 \cos \theta$$

(If F_1 and F_2 are perpendicular to each other, $\cos \theta = 0$, and $F^2 = F_1^2 + F_2^2$.) Next, the *direction* of F can be given with reference to F_1 by finding the angle ϕ (Greek *phi*). Notice that BC equals F_1. Hence, applying the law of cosines to OBC, we obtain

$$F_1^2 = F^2 + F_2^2 - 2FF_2 \cos \phi$$

in which F, F_1, and F_2 are known and from which ϕ can be found.

To find the resultant of three or more known forces acting at a point, successively apply the parallelogram law, combining the resultant of two forces with a third force, and so on. (Another method for finding the resultant of several forces — the method of components — will be discussed later in this chapter.)

Example. Combine the forces 3.0 kgwt north, 4.0 kgwt east, and 5.0 kgwt southeast, represented in Fig. 3-4 by *AB, AD,* and *AE*. First find *AC*, the resultant of *AB* and *AD*, by completing a parallelogram with *AB* and *AD*, as sides, using some convenient scale. Then

* See Appendix C.

find *AR*, the resultant of *AB, AD,* and *AE*, by completing a parallelogram with *AC* and *AE* as sides. *AR* is about 7 kgwt in the drawing. By computation $AR \cong 7.5$ kgwt.

Equilibrium

In a tug of war, when opposing teams pull equally, the forces balance each other, their resultant is zero, and no motion is produced (Fig. 3-5). When a body is at rest or moves with *constant* velocity, it is said to be in *equilibrium.*

When a body is in equilibrium, the resultant of all the forces acting on it must be zero.†

The force that will balance several forces so as to produce equilibrium is called their *equilibrant* or *antiresultant.* The equilibrant of several forces is equal and *opposite* to their resultant. Hence, knowing the equilibrant of several unknown forces acting in known directions, you can use the parallelogram rule to find their magnitudes.

For example, a load weighing 50 kgwt is suspended as shown in Fig. 3-6a. Let us calculate the tensions in the ropes.

1. Since point O is at rest, it is in equilibrium. What forces act at O? They are the tension in rope OC (unknown in size, but directed along OC), the tension in OB (unknown in size, but directed along OB), and the tension in OA (50 kgwt, directed vertically downward).
2. Draw a vector diagram (Fig. 3-6b) representing the forces at O: 50 kgwt (drawn to some convenient scale) vertically downward, a force F_1 of unknown magnitude acting along OC, and a force F_2 of unknown magnitude acting along OB.
3. The 50-kgwt force acting downward is the *equilibrant* of forces F_1 and F_2. Hence their *resultant* is a 50-kgwt force OD acting ver-

† In Chapter 9 we will see that another condition must also be fulfilled to ensure equilibrium.

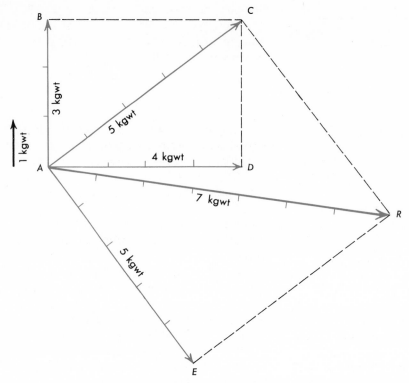

Fig. 3-4. *Resultant of three forces,* **AB,** **AD,** *and* **AE.** *Find the resultant* **AC** *of* **AB** *and* **AD.** *Then find the resultant of* **AC** *and* **AE.**

tically upward. Draw *OD* to scale and complete the parallelogram with *OD* as the diagonal. The sides are F_1 and F_2.

4. You can verify that F_1 and F_2 are 30 kgwt and 40 kgwt respectively by measuring their lengths with a ruler. To calculate their magnitudes, use the law of sines:*

* See Appendix C.

$$\frac{F_1}{\sin 37.5°} = \frac{F_2}{\sin 52.5°} = \frac{50 \text{ kgwt}}{\sin 90°}$$

$$F_1 = 30 \text{ kgwt}$$

$$F_2 = 40 \text{ kgwt}$$

[Note that *two* diagrams are necessary: one (Fig. 3-6*a*) to show the physical arrangement, and another (Fig. 3-6*b*) to show the relationships among the vectors.]

Fig. 3-5. *Equilibrium. When forces are in equilibrium, no change of motion is produced.*

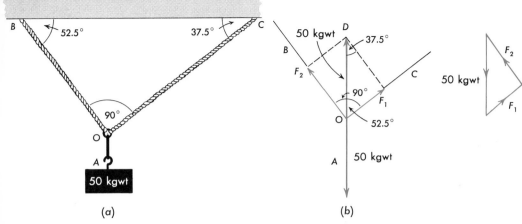

(a)

(b)

Fig. 3-6. (a) *Point* O *is in equilibrium.* (b) *Vector diagrams of forces acting on* O.

Often it is convenient to draw two sides only of the parallelogram and its diagonal, so as to form a triangle. In Fig. 3-7*a* a sign weighing 30 kgwt is supported by a wire attached to one end of a wooden rod. We wish to find the pull of the wire and the push exerted by the rod at *F*.

In Fig. 3-7*b*, let *CA* represent the known force, 30 kgwt. Through *A* draw a horizontal line (parallel to the push of the rod). Through *C* draw a second line parallel to the pull of the wire. The arrows *AB* and *BC* represent the unknown forces, 40 kgwt and 50 kgwt. You can find these values by measurement or by using the fact that the triangles *ABC* and *EFG* are similar and their corresponding sides are therefore proportional.

The triangle (or polygon) method is particu-

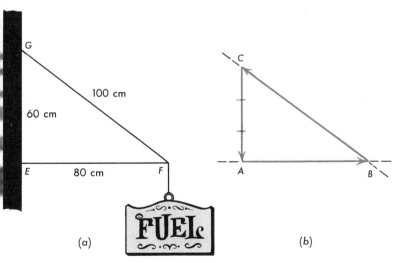

(a)

(b)

Fig. 3-7. *The triangle method.* (a) *Three forces act at* F. (b) *Force triangle.*

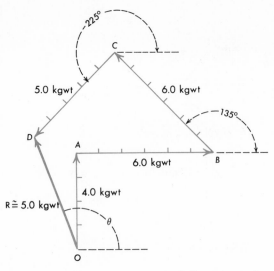

Fig. 3-8. *The polygon method. Represent the known forces by the arrows OA, AB, BC, CD, head to tail. The arrow OD represents their resultant force, approximately 5.0 kgwt.*

larly convenient when we wish to find the resultant of several forces.

Example. Four boys push on a beach ball in a game of water polo. *A* pushes north with a force of 4.0 kgwt, *B* east with a force of 6.0 kgwt, *C* northwest with a force of 6.0 kgwt, and *D* southwest with a force of 5.0 kgwt. What is the resultant force *OD*?

Draw the appropriate arrows, head to tail (Fig. 3-8). *OD*, drawn from the tail of vector *OA* to the head of vector *CD*, closes the polygon and represents the required resultant force, approximately 5.0 kgwt.

The Components of a Force

So far, you have learned to combine two or more forces into a single force called their resultant or vector sum. Sometimes it is desirable in physics to do the reverse: to start with a force and to find two or more other forces in specified directions whose resultant is that force. For example, a tractor pulls on a

horizontal chain with a force *OA* (Fig. 3-9) of 100 kgwt northeast. Can we find two other forces, one in a direction 30° north of east and the other 45° north of west, which together are equivalent to *OA*? If the two new forces are to be equivalent to *OA*, *OA* must be their resultant or vector sum. Hence, *OA* is the diagonal of a parallelogram of which the two desired forces *OB* and *OC* are sides. By calculation or by making a scale drawing, we find that the magnitude of *OC* is 103 kgwt and that of *OB* is 26 kgwt. If *OA* were replaced by forces *OB* and *OC*, the effect on *O* would be unchanged. *OB* and *OC* are *components* of *OA* in the directions specified. Of course, other components could be found in other directions; it is often convenient, as we shall see, to choose components that are perpendicular to each other.

The components of a force in any specified directions are the forces in those directions which, acting together, are equivalent to that force.

When a sled is pulled over snow by means of a rope (Fig. 3-10), the pull tends to drag the sled forward and also to lift it from the surface. Suppose the pull of 10 kgwt is exerted at an angle of 60° with the horizontal. By the parallelogram law, the applied force of 10 kgwt *AC*, is found to be equivalent to the vertical force *AB* of 8.7 kgwt, and the horizontal force

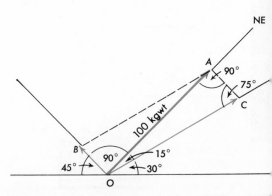

Fig. 3-9. *Components of a force. OB and OC are components of OA.*

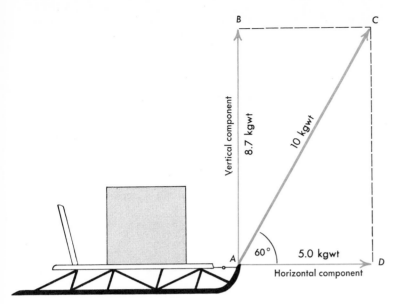

Fig. 3-10. Components. AD *and* AB *are equivalent to* AC. *They are its* horizontal and vertical components.

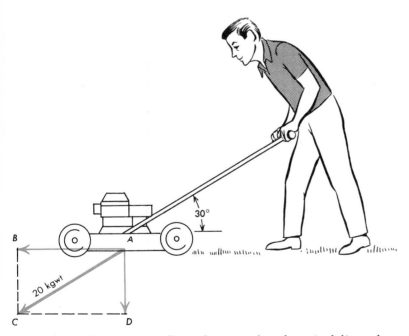

Fig. 3-11. Components. Draw horizontal and vertical lines through A. Construct the parallelogram ABCD. AB *and* AD *are the desired* components.

AD, of 5.0 kgwt. Thus the effect on the sled is the same as if one rope pulled it horizontally with a force of 5.0 kgwt and another rope exerted a vertically upward force of 8.7 kgwt. The two equivalent forces are the horizontal and vertical components of the force *AC*.

Example 1. A man exerts a force *AC* of 20 kgwt along the handle of a lawn mower (Fig. 3-11) which makes an angle of 30° with the horizontal. What are the vertical and horizontal components of *AC*?

Draw horizontal and vertical lines through the tail of the arrow *AC*. Construct the parallelogram *ABCD*. The force *AC* is equivalent to *AB* and *AD*, acting together. By measurement we find the components to be 17.3 kgwt and 10.0 kgwt, respectively.

Or, by trigonometry,

$$AB = AC \cos 30° = 20 \text{ kgwt} \times 0.866 = 17.3 \text{ kgwt}$$

$$AD = AC \sin 30° = 20 \text{ kgwt} \times 0.50 = 10.0 \text{ kgwt}$$

Example 2. What force parallel to the roadway is required to keep an automobile weighing 2000 kgwt from starting down a uniform incline which is 100 m long and 20 m high?

In Fig. 3-12, *OG* represents the weight of the car. Lines are drawn through *O* parallel to the slope and perpendicular to it. The component forces *OH* and *OE* are equivalent to *OG*. The triangles *OEG* and *ABC* are similar; hence their corresponding sides are proportional.

Therefore:

$$\frac{F_P}{2000 \text{ kgwt}} = \frac{20 \text{ m}}{100 \text{ m}}$$

F_P, the downhill component, represented by *OH*, is 400 kgwt. Or, by another method,

$$F_P = w \sin \angle CAB$$
$$= 2000 \text{ kgwt} \sin \angle CAB = 400 \text{ kgwt}$$

Hence an *uphill* force of 400 kgwt is required to hold the car on the slope. Similarly, F_N, the normal component, represented by *OE*, is the force exerted perpendicular to the hill by the car and is given by

$$F_N = w \cos \angle CAB$$
$$= 2000 \text{ kgwt} \cos \angle CAB = 1960 \text{ kgwt}$$

In Fig. 3-13, *OA* represents the force that the wind exerts on the wings of an airplane flying horizontally at constant velocity. *OC*, the horizontal component of this force, the "aerodynamic drag," is opposed by the for-

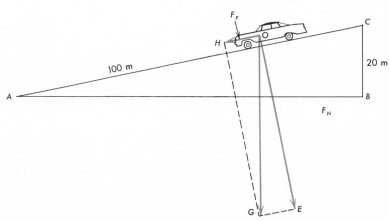

Fig. 3-12. *Components. Draw* OH *parallel to the slope and* OE *normal to it. Complete the parallelogram.* OH *is the downhill component of* OG.

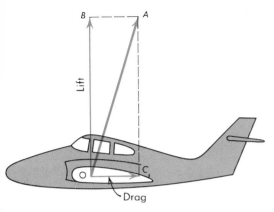

Fig. 3-13. *"Lift" and "drag." Components of the force exerted on the wings of an airplane.*

ward thrust of the jet engine. Although all the vectors are not shown in Fig. 3-13, the thrust of the jet engine equals the aerodynamic drag plus the drag of air friction on the plane. *OB*, the "lift," is equal to the weight of the plane. The airplane is in equilibrium under the action of these forces.

How Can a Boat Sail into the Wind?

Figure 3-14 represents a sailboat propelled forward by a wind blowing from the north. The wind exerts a force *AD* perpendicular to the sail. This force can be resolved into two components, *AB* and *AC*. One component, *AB*, parallel to the direction in which the boat is pointed, pushes it forward. The boat speeds up until the friction of water on its hull equals *AB*. The other component, *AC*, forces the boat sidewise, to the leeward. This motion is opposed by the large frictional force of the water against the keel or centerboard of the boat.

Resultant by Components

The resultant of several forces can be accurately computed by finding the sum of their components. Choose a convenient set of **X** and **Y** axes. Resolve each force into its **X** and **Y** components along these axes, and find the algebraic sums of all the **X** and **Y** components. The sum of all the **X** components is the **X** component of the resultant, and the sum of all the **Y** components is its **Y** component. Consider once again the problem of the four boys pushing on a beach ball, which was solved by the vector polygon method as shown in Fig. 3-8 (p. 38). This same problem is solved in Fig. 3-15 by the method of components explained above.

Friction

The forces we have discussed so far have been gravitational ones, due to the pull of the earth on objects, and elastic ones with which bodies resist a change in their shape. Now we shall consider a third kind of force—friction—which is present whenever one surface

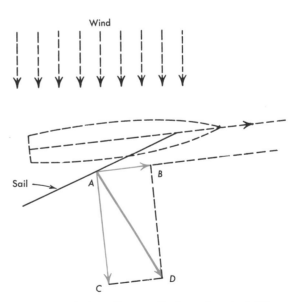

Fig. 3-14. *A sailboat sails into the wind. The component AB of AD makes the boat move forward.*

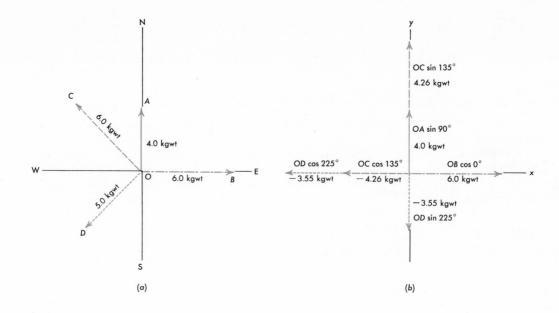

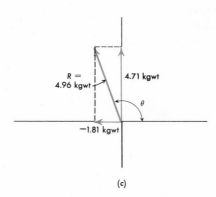

(c)

Fig. 3-15. Finding a resultant by the method of components.

Force	X Component	Y Component
OA	$OA \cos 90° = 0$	$OA \sin 90° = 4.0 \text{ kgwt} \times 1 = 4.0 \text{ kgwt}$
OB	$OB \cos 0° = 6.0 \text{ kgwt}$	$OB \sin 0° = 6.0 \text{ kgwt} \times 0 = 0$
OC	$OC \cos 135° = 6.0 \text{ kgwt} \times -0.71$ $= -4.26 \text{ kgwt}$	$OC \sin 135° = 6.0 \text{ kgwt} \times 0.71$ $= 4.26 \text{ kgwt}$
OD	$OD \cos 225° = 5.0 \text{ kgwt} \times -0.71$ $= -3.55 \text{ kgwt}$	$OD \sin 225° = 5.0 \text{ kgwt} \times -0.71$ $= -3.55 \text{ kgwt}$
	Total X component $= -1.81$ kgwt	Total Y component $= 4.71$ kgwt

$$R = \sqrt{(-1.81 \text{ kgwt})^2 + (4.71 \text{ kgwt})^2} = 4.96 \text{ kgwt}$$

$$\tan \theta = -\frac{4.71 \text{ kgwt}}{1.81 \text{ kgwt}} = -2.60$$

$$\theta = 111°$$

42   **Ch. 3   *Resultants and Components of Forces***

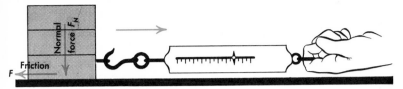

Fig. 3-16. *The friction force is proportional to the normal force pushing the blocks against the surface.*

moves over another and which always acts to oppose the motion.

A boy pushes a wooden box over a floor. Friction between the bottom of the box and the floor must be overcome by his forward push. If the boy stops pushing, friction quickly brings the box to rest. A driver steps on the brake pedal as he approaches a stop light. Friction between the brake shoe and the brake lining slows the wheels; friction between the tires and the road brings the car to a stop. A bullet speeds down the barrel of a rifle. As it moves faster and faster, friction with the barrel causes the bullet's temperature to rise until the bullet is traveling on an extremely thin film of molten copper from its jacket. In a world of moving things, frictional forces clearly must be reckoned with.

Friction has three harmful effects. First, it increases the effort necessary to operate machines; second, it causes heat, which may do damage; and third, it wears away the rubbing surfaces. Often we wish to diminish friction. The automobile manufacturer provides rollers and balls for some of the axles of his car and uses different kinds of lubricants in different places. Sometimes friction is useful. When the streets are icy, sand or cinders are scattered over the surface and chains are attached to automobile tires in order to increase the friction. Without friction, belts could not drive machines, and brakes could not be used. In a frictionless world it would be practically impossible to start an automobile on a level street. Nails could not be used to hold boards in place. We could not walk.

Three kinds of friction must be considered: sliding friction, rolling friction, and fluid friction. We will discuss the first two in this chapter and the third in Chapter 8.

Friction of Solid Surfaces

Let us investigate sliding friction by considering a simple experiment. Suppose that you measure the horizontal force required to drag a wooden block over a level table top. The force that you exert in the forward direction is equal and opposite to the force of friction. Repeat the experiment, adding equal blocks, one at a time (Fig. 3-16), thus increasing the *normal force,* which is defined as the force perpendicular to the two surfaces and pressing them together. Suppose that the results of the experiment are as follows:

F_N	Total load or normal force (gwt)	200	400	600	800	1000
F	Friction force (gwt)	51	104	150	204	249
F/F_N	Friction force/ Total load	0.255	0.260	0.250	0.255	0.249

Your experiment shows that *the ratio of the friction force to the normal force that pushes one surface against the other* is approximately constant. This ratio we call the *coefficient of kinetic, or sliding, friction, μ* (Greek *mu*).

Coefficient of kinetic friction μ

$$= \frac{\text{Friction force}}{\text{Normal force}} = \frac{F}{F_N}$$

Next, place one wooden block on top of another, and drag the two over the table top at a uniform speed such that the block moves about 20 cm in 5 sec (speed 4 cm/sec) (Fig. 3-17*a*). The necessary force and the coeffi-

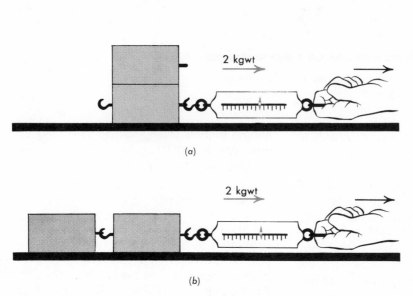

(a)

(b)

Fig. 3-17. *The friction in* a *is equal to that in* b, *although the areas of contact differ.*

cient of kinetic friction are about the same as that required when the two blocks are connected "in tandem" as in Fig. 3-17*b*. In this experiment, the total weight is the same, and the frictional force is the same, but the area of contact differs.

Measure the coefficient of kinetic friction at different speeds: first find the frictional force needed to start a block from rest, and second the frictional force when the block is moving at a uniform speed such that it travels about 20 cm in 5 sec (speed 4 cm/sec). You will find that the starting frictional force is much greater than the frictional force when the blocks are moving. However, friction at a speed of 4 cm/sec is only slightly less than that at 2 cm/sec. We define the coefficient of static, or starting, friction μ_S as

$$\mu_S = \frac{\text{Friction force just at starting}}{\text{Normal force}}$$

Finally, if you were to repeat these experiments with steel blocks sliding on a dry steel

plate (Table 3-1), you would find that the coefficients of kinetic and static friction for steel on steel would be somewhat smaller than for wood on wood. However, the frictional forces would be approximately independent of the "area of contact" and, except at starting, of the speed.

Table 3-1. Coefficients of Kinetic Friction

Oak on oak, grains parallel	0.4–0.5
Steel on steel, 20 mi/hr	0.24
Steel on steel, 60 mi/hr	0.10
Steel on steel, 700 mi/hr	0.02
Iron on automobile brake lining	0.2–0.4
Automobile tire on wet road surface at 20 mi/hr	
On clean concrete	0.7–0.9
On wood, steel, muddy concrete	0.2–0.3
On ice	0.02–0.03

Experiments like these led the Italian artist-scientist Leonardo da Vinci (1452–1519) and after him the French physicist Charles Coulomb (1736–1806) to state that the force of sliding friction:

1. Is directly proportional to the normal force pressing the two surfaces together.
2. Depends upon the nature of the surfaces in contact: their smoothness, kind of material, and degree of lubrication.
3. Is independent of the "area of contact."
4. Is independent of the speed when motion has begun.

These four laws of sliding friction led the early investigators to think of it as due to the intermeshing of surface roughness such as would occur if one piece of sandpaper were drawn over another. More recent work on friction has shown that the first two results are valid for many kinds of surfaces but that the third and fourth results and the notion of intermeshing of surface roughness are over-simplified. Before considering the modern description of sliding friction, let us note that the coefficient of kinetic friction (Table 3-1) can be computed for two surfaces if the frictional force and the corresponding normal force are known. Knowledge of the coefficient of kinetic friction can be useful in solving certain problems.

Example 1. A boy and a sled together weigh 40 kgwt. A horizontal force of 10 kgwt is required to drag the loaded sled over a level pavement. What is the coefficient of kinetic friction?

The vertical force pushing the sled against the ground is 40 kgwt and the force of friction is 10 kgwt. Hence, the coefficient of kinetic friction is

$$\mu = \frac{F}{F_N} = \frac{10 \text{ kgwt}}{40 \text{ kgwt}} = 0.25$$

Example 2. In Fig. 3-18, one boy lifts the 800-gwt block by means of the spring scale while another boy pushes the block against the wall with a force of 400 gwt. What upward pull is required (a) to raise the block and (b) to lower it at constant speed? The coefficient of kinetic friction is 0.20.

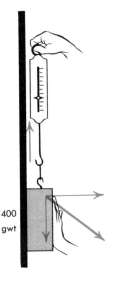

400 gwt

Fig. 3-18. *The force to pull the block upward is equal to friction plus the weight of the block.*

(a) The up pull is equal to the weight of the block, 800 gwt, plus the force of friction.

$$F_1 = 800 \text{ gwt} + 400 \text{ gwt} \times 0.20 = 880 \text{ gwt}$$

(b) The up pull is equal to the weight of the block minus the force of friction.

$$F_2 = 800 \text{ gwt} - 400 \text{ gwt} \times 0.20 = 720 \text{ gwt}$$

Example 3. In Fig. 3-19, what force *F* is required to pull the 1000-gwt block uphill at constant speed? The coefficient of kinetic friction is 0.20.

From similar triangles, the component of the weight of the block parallel to the slope, F_P, is 600 gwt, and the component normal to the slope, F_N, is 800 gwt.

The friction force μF_N is
$$0.20 \times 800 \text{ gwt} = 160 \text{ gwt}.$$

The downhill component F_P of the weight
$$\text{of the block} = 600 \text{ gwt}.$$

The total required force
$$F_P + \mu F_N = 160 \text{ gwt} + 600 \text{ gwt} = 760 \text{ gwt}.$$

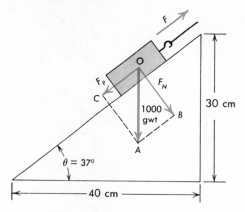

Fig. 3-19. *Block on an inclined plane.*

Sliding Friction and Surface Attraction

If sliding friction of two clean dry surfaces were due to the roughness of their surfaces, one would expect that smoothing the surfaces would reduce the amount of sliding friction. The small "hills and valleys" of the two surfaces would not be so deep and, according to this viewpoint, a smaller force would be sufficient to keep one surface moving over the other. Investigators were surprised to find that carefully polished and cleaned flat surfaces showed *more* friction rather than less. They attributed this to the attraction which the molecules of the one surface had for those of the other. Molecular attraction becomes larger as the distance between the molecules becomes less. Smoothing surfaces and making them flat, polishing them to remove fine scratches, and cleaning them to remove grease allow the molecules of the one surface to approach and grasp the molecules of the other, virtually welding the surfaces. Moving one surface over the other requires that a force be exerted to break the tiny molecular attractions which acting together constitute the force of sliding friction. So effective is the molecular attraction that tiny bits of the one smooth surface are pulled out when the other surface is moved over it.*

On less carefully polished surfaces, sur-

face attraction is still effective where the surfaces approach each other (Fig. 3-20). These small areas together, and not the geometrical area of the base of the sliding object, are the true "area of contact." As the normal force increases, the surfaces are pushed closer together, the small areas of contact grow, and the frictional force increases. Thus, the force of sliding friction is *independent* of the geometrical area of the surfaces in contact, but is *dependent* upon the small areas of real contact over which molecular attraction is effective.

Lubrication is effective in reducing sliding friction, according to this theory, because a thin film of the lubricant separates the high spots of the surfaces and reduces the areas of real contact and thus reduces the molecular attractions. Not only oils and greases act as lubricants; dust, water vapor, and even oxide coatings on metals can also reduce sliding friction. However, if the lubricant layer is thick enough to separate the surfaces at every point, sliding friction gives way to *fluid friction* which we will discuss later. Ordinary water on a wet roadway can act as a lubricant. The tire does not grip the road, but instead floats on a thin film of water. This condition, called "hydroplaning," can cause dangerous skids even at moderate speeds if the driver is not careful. Increase of temperature generally reduces sliding friction: the brakes on your car are less effective after they have been

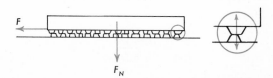

Fig. 3-20. *Two unpolished surfaces in contact.*

---

* Radioactive tracer studies in which a tiny amount of radioactive material is added to the one surface show that after each sliding contact a little of the radioactivity has been transferred to the other surface.

applied for a while and have become heated. However, very clean surfaces will weld on contact at high temperatures.

How does sliding friction depend upon velocity? We have already seen from our simple experiments that starting friction is larger than sliding friction once the two surfaces are in motion. Before starting, the areas of real contact have had a chance to grow by gradual flattening of the high spots of the surfaces; the molecular attractions are large and starting friction is high. Once in motion, the surfaces are not in contact long enough for the areas of real contact to grow, and sliding friction decreases. However, over a considerable range of speeds—the range in which most student laboratory experiments take place—the coefficient of kinetic friction is fairly constant. At high speeds—the speed of a rotary saw blade cutting metal or of a bullet traveling down the barrel of a gun—the coefficient of kinetic friction again decreases because melting takes place and the surfaces are separated by a thin lubricating layer of molten metal. Table 3-2 summarizes the results of many measurements of the coefficient of kinetic friction of steel on steel.

Table 3-2. Coefficients of Friction for Steel on Steel*

Condition	Coefficient of Static Friction	Velocity, in./sec	Coefficient of Kinetic Friction
Degassed at high temperature at high vacuum	Surfaces weld on contact	0	
Grease-free in vacuum	0.78	0	
Grease-free in air	0.39	0	
Clean and coated with oleic acid	0.11	0	
Unlubricated		0.001	0.48
"		0.01	0.39
"		0.1	0.31
"		1	0.23
"		10	0.19
"		100	0.18
Lubricated with stearic acid			0.029

* Data taken from *American Institute of Physics Handbook.*

When an automobile travels slowly, there is plenty of time for the tire surfaces to be attracted by the road surface since the area of the tire touching the road is momentarily at rest relative to the road. The friction is relatively great. At high speeds there is less opportunity for molecular attraction, and the friction is smaller. If the brakes are set strongly and the car skids, the tires slide over the road and sliding friction is much reduced. The tires do not have time to grip the road, and the driver loses control of steering. In order to get the wheels turning again, he should release the brakes for an instant and then apply them again, but *not quite strongly enough to lock the wheels.* When a car goes up an icy hill and the wheels begin to spin, the driver should release the gas pedal for an instant so that the wheels may slow down and the tire surfaces may be gripped by the road surface.

Rolling Friction

When an automobile travels along a highway, rolling friction arises because the tires and the pavement are deformed where they make contact with each other. As the wheel rolls along, the various parts of the tire are successively deformed, and effort must be expended to overcome friction within the rubber. The greater the air pressure and the smaller the deformation, the less will be the rolling friction. On a hard roadway, one saves gasoline by keeping the tires of his car well inflated; but on a soft road, the pressure should be less to prevent sinking into the mud. When a locomotive moves along a railway track, the distortion of wheel and rail is much less than that produced in an automobile tire on a highway. The rolling friction is smaller, but it can never be zero. Rolling friction is less than sliding friction; therefore roller bearings and ball bearings are used on automobile axles and the like. When an axle turns within a bearing of the kind illustrated in Fig. 3-21 sliding friction occurs where the two surfaces rub against each other. When a ball bearing is used, the sliding friction is replaced by rolling friction (Fig. 3-22).

Fig. 3-21. *An axle sliding on a bearing (space exaggerated).*

Rolling friction F depends upon the radius of the wheel r, the load on the axle of the wheel F_N, and the coefficient of rolling friction k which is characteristic of the materials and their condition:

$$F = k \frac{F_N}{r}$$

For steel on steel, $k = 5.1 \times 10^{-5}$ m; for hardwood on hardwood, $k = 5.1 \times 10^{-4}$ m. Note that this coefficient has units.

Summary

Forces are *vector* (directed) quantities and can be represented by arrows or vectors, parallel to the forces, whose lengths are proportional to the magnitudes of the forces. *Scalar* quantities have only magnitudes.

The resultant of several forces acting at a point is the single force that is equivalent to them. The resultant of two or more such forces may be found by vector addition. *Graphical* methods of vector addition rely upon the parallelogram rule or the polygon rule. *Analytical* methods rely upon the theorem of Pythagoras, the law of cosines, or the law of sines.

The components of a force in specified directions are forces in those directions which, acting together, are equivalent to the given force.

The equilibrant of one or more forces is that force which, combined with them, will produce equilibrium by reducing the net force to zero. The equilibrant of several forces is equal and opposite to their resultant.

Solid-on-solid friction is the resisting force opposing the motion of one body which slides or rolls over the surface of another.

The frictional force of two solid surfaces which are not cut or distorted (*a*) is greater when the motion starts than when the motion is established, (*b*) decreases with velocity, (*c*) is independent of the geometrical area of contact, but depends upon the area of real contact, (*d*) is directly proportional to the normal force pushing one surface against the other.

The *coefficient of kinetic friction* of two surfaces is the ratio of the friction force parallel to the surfaces to the normal force pushing one surface against the other when the surfaces move relative to each other. The *coefficient of static friction* is the ratio of the friction force parallel to the surfaces at starting to the normal force pushing the two surfaces together.

Rolling friction is caused by the deformation produced where the wheel or cylinder pushes against the surface on which it rolls. Rolling friction is less than sliding friction.

Questions

1. Define vector and scalar.
2. Define equilibrant, resultant, and component.
3. What is the physical significance of a closed polygon of vectors?
4. Make a sketch to show how three vectors can have a zero resultant.

Fig. 3-22. *A ball bearing. Sliding friction is replaced by rolling friction.*

5. When telephone wires are loaded with sleet, are they more likely to break when they are taut or when they sag? Why?

6. Is the tension greater in a wire used to hang a picture when a short wire is used or when a long wire is used? Explain.

7. It is impossible to hang an object, however small its weight, from the center of a wire and have the wire remain horizontal. Explain why with drawings.

8. Show by a drawing how a boat can sail into the wind. What determines the speed with which it moves upwind?

9. Estimate which gives a greater resultant: 1.5 kgwt north and 1.5 kgwt east, or 1.0 kgwt north and 1.0 kgwt northeast.

10. Define coefficient of kinetic friction and give a numerical example of how it can be determined.

11. Distinguish between static friction and kinetic friction. Why is the coefficient of static friction greater than that of kinetic friction for the same materials?

12. State instances in which friction is desirable and others where it should be avoided.

13. When the wheels of a car begin to skid, why should the brakes be released and then applied again?

14. Suppose that a meter stick rests horizontally on the forefingers of your hands. One finger is then moved slowly toward the other. Why must the two fingers meet at the midpoint of the meter stick?

Problems

(*Assume constant speed for moving bodies.*)

A

1. Using a graphical method, find the resultant of the following forces: 10.0 kgwt northeast, 10.0 kgwt southeast.

2. The ends of a rope 2.5 m long are tied to a horizontal beam at points 1.5 m apart. A 20-kgwt weight is suspended from the mid-point of the rope. Find the tension or pull in the rope.

3. What is the magnitude of the force that will balance the following three forces, producing equilibrium: 10.0 kgwt northeast, 6.0 kgwt east, 3.0 kgwt south? (Use a graphical method.) How does this force compare with the resultant of the three forces?

4. A 50-kgwt weight is suspended by two wires each making an angle of 30° with the horizontal. Find the tension or pull in each wire.

5. A ball weighing 1.0 kgwt pushes against a brick lying on a 30° inclined plane. Find the force exerted on the brick by the ball.

6. A football player strengthens his leg muscles by pushing a loaded sled over a level field. The total weight is 150 kgwt, and the coefficient of friction is 0.50. What horizontal force must he exert?

7. A four-wheel-drive truck weighs 5000 kgwt. What maximum pull can it exert on a trailer on a level street (*a*) in good weather when the coefficient of friction between tires and street is 0.80, (*b*) in icy weather when the coefficient is 0.05?

B

8. A picture weighing 4.0 kgwt is hung from a nail. Each half of the cord makes an angle θ with the horizontal. Make a graph plotting the pull (tension) in the cord versus θ for values of θ from 0° to 90°. What values does the tension approach as θ approaches 0° and 90°?

9. A boy weighing 45 kgwt sits in a swing whose ropes would be broken by a force of 75 kgwt. What maximum force can he exert on a horizontal rope without breaking the ropes of the swing? (The swing ropes need not remain vertical.)

10. The propeller of a 1000-kgwt airplane exerts a horizontal force of 100 kgwt. Find the equilibrant force exerted by the wind against the wings of the plane flying at constant speed. (Draw a force parallelogram.)

11. A 100-kgwt weight is attached at the

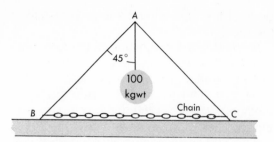

Fig. 3-23. *What force does the chain BC exert at B?*

upper end of a pole which makes an angle of 60° with the horizontal. A horizontal rope is attached to this end of the pole. Find the pull of the rope, neglecting the weight of the pole.

12. A block of wood weighing 10 kgwt rests on an inclined plane 60 cm high and 100 cm long. Find (*a*) the force pushing the block against the plane, (*b*) the force tending to make it slide downhill.

13. The boom of a crane is 5.0 m long and makes an angle of 60° with the vertical. A load of 200 kgwt is attached at the upper end of the boom. A horizontal cable supports this end. Calculate the tension in the cable, neglecting the weight of the boom.

14. The handle of a shuffleboard stick makes an angle of 30° with the vertical. What is the vertical component of a force of 2.50 kgwt directed along the handle?

15. A telephone pole is braced by a cable, one end of which is attached to the pole at a point 10 m above the level ground surface. The other end is anchored to the ground at a point 6 m from the base of the pole. If the pull of the cable is 60 kgwt, what is the horizontal component of the force in the cable?

16. A board is tilted at an angle of 25° with the horizontal. A block weighing 10.0 kgwt slides down the board at constant speed. What force parallel to the board will pull the block uphill at constant speed?

17. A normal force of 1.0 kgwt pushes a 2.0-kgwt block against a wall. The coefficient of kinetic friction between block and wall is 0.30. What vertical force does a cord exert on

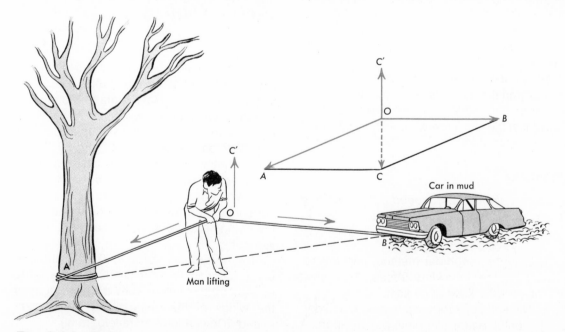

Fig. 3-24. *Pulling a car out of a ditch.*

the block when the block moves (*a*) upward, (*b*) downward, at constant speed?

18. The angle of repose θ for a block resting on an inclined plane is the angle that the plane makes with the horizontal when the block is just at the point of slipping downhill. Prove that the tangent of θ equals the coefficient of static friction.

C

19. What is the angle between the two ropes of a hammock when the tension in each is (*a*) equal to the weight of a girl sitting in the middle of the hammock, (*b*) twice her weight?

20. John and Harry sit in swings, facing each other, and pull on opposite ends of a horizontal rope, thus causing John's swing ropes to be displaced 45° from the vertical and Harry's 30°. If John weighs 40 kgwt, how much does Harry weigh?

21. A lawn mower weighing 10 kgwt rests on a hill making an angle of 30° with the horizontal. The handle of the lawn mower makes an angle of 45° with the surface of the ground. (*a*) What uphill push exerted along the handle will produce equilibrium? (*b*) What force is then exerted normal to the surface? Neglect friction.

22. A sanding block weighing 10.0 kgwt rests on a level floor. It is pushed by means of a handle making an angle of 30° with the horizontal. If the force applied parallel to the handle is 10.0 kgwt, determine (*a*) the total vertical force pushing the block against the floor, (*b*) the horizontal component of the applied force, and (*c*) the frictional force if the coefficient of static friction is 0.40. (*d*) Will the block start to move?

23. A V-shaped frame stands vertically upright and supports a load of 100 kgwt (Fig. 3-23). Calculate the stretching force on the chain. (Assume zero friction.)

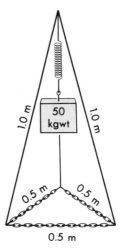

Fig. 3-25. Tripod frame supporting a load.

24. A man, wishing to drag a 1000-kgwt automobile out of a ditch, attaches a 10-m rope to the car and to a tree (Fig. 3-24). By exerting an up force of 20 kgwt, he raises the midpoint of the rope 0.50 m. What tension does he produce in the rope?

25. A tripod frame made of light metal rods, each 1.0 m long, hinged at their upper ends, supports a 50-kgwt load as shown in Fig. 3-25. The lower ends of the rods rest on a smooth surface and are held by chains, each 0.5 m long. Calculate the tension in each chain.

For Further Reading

Palmer, Fredric, "Friction," *Scientific American,* February, 1951, pp. 55–58.

Rabinowicz, Ernest, "Stick and Slip," *Scientific American,* May, 1956, p. 109.

Rabinowicz, Ernest, "Wear," *Scientific American,* February, 1962, pp. 127–136.

CHAPTER FOUR

Rectilinear Motion

We have been considering the action of forces on bodies that are at rest or that move with constant velocity. Presently we shall consider how forces accelerate bodies. First, in this chapter, we shall study *kinematics,* the laws of accelerated motion. These laws are important to the physicist as he studies the behavior of the physical world. He needs to determine not only the positions of bodies but also their motions. In this book, first you will study the *linear* motions of visible bodies such as baseballs, automobiles, and earth satellites, then the *rotary* motions of flywheels and gyroscopes. Afterwards you will consider the behavior of particles too small to be seen, molecules and atoms. Following that, you will deal with electrons, neutrons, protons, and other parts of atoms.

What is our goal in the study of motion? It is to *predict* where a body will be and how fast it will be moving at any chosen time in the future. We start from information about where a body is at present, how fast it is moving, and how its motion is changing; we construct a timetable for future motion, using the principles of physics to guide us. The "classical" principles of physics that we shall use in this part of our discussion are reliable guides in

predicting the motion of "earth-scale" objects — planets, satellites, automobiles, and baseballs. The classical principles fail, however, to describe accurately the motions of very small atomic particles or, on the other hand, motions on an astronomical scale at very large speeds approaching that of light. Other principles will be used as guides in describing motion in the domains of the atom and of deep space, as we shall see.

Many of the motions that we shall consider are very involved. As usual, we shall first consider the simplest of all, that of a body whose particles travel in parallel, straight lines. This is called rectilinear (straight-line) motion.

Velocity and Acceleration

When you drive a car, you pay attention to three measuring instruments. The speedometer tells *how fast* you are going, the odometer (roadmeter) tells *how far* you have gone, and your watch tells you how much *time* has elapsed since you started.

Suppose, to take a much simplified example, that you drive your car 60 mi along a straight road in 2 hrs, keeping the speedometer reading constant. Then, at *every instant,* your velocity is 30 mi/hr. Suppose, to be more realistic, that the speedometer reading varies. Then your *average* velocity is 30 mi/hr. *Average velocity* (v_{av}) *is distance traveled per unit time* [(Δs)* *divided by* (Δt)].

$$v_{av} = \frac{\Delta s}{\Delta t} = \frac{s - s_0}{t - t_0}$$

where s_0 is the distance of the body from a reference point at time t_0 and s is its distance at a later time t. Often in solving problems, we take $t_0 = 0$; that is, we "start the clock" at t_0. If the velocity does not change, then the velocity at any instant (the instantaneous velocity that is registered on the speedometer) is the same as the average velocity.

We often use velocity and speed interchangeably, but technically the two differ in meaning. Speed means rate of motion. It is a *scalar* (non-directed) quantity. Velocity means rate of motion *in a specified direction.* When your car goes around a curve and its speedometer reading is constant, its speed is constant, but its velocity varies in direction. The speed is the *magnitude* of the velocity at any instant.

When you push down the accelerator of your car, the speedometer needle gradually moves up the scale; the car is *accelerated.* When you release the accelerator and apply the brakes, the car is retarded or *decelerated.* Let Δv be the change in velocity in a time Δt, and a the acceleration. Then

$$a = \frac{\Delta v}{\Delta t} = \frac{v - v_0}{t - t_0}$$

where v_0 is the velocity at time t_0 and v the velocity at a later time t.

Acceleration is Change of Velocity per Unit Time

In this chapter, we shall assume that all accelerations are *constant.* Later, we shall

---

* The symbol Δ (Greek *delta*) means "a change in" and Δs is a physical quantity representing $s - s_0$, where s_0 is the initial distance of the body from a given reference point and s is its distance at some later time.

consider motions in which the acceleration changes. Accelerations can be positive or negative quantities. A positive acceleration indicates that the velocity of the body is increasing; a negative acceleration, that the velocity is decreasing.

An advertisement for an automobile claims that the car can accelerate from 0 mi/hr to 60 mi/hr in 12 sec. This is a change of velocity of 88 ft/sec in 12 sec or 7.33 ft/sec per second. Hence the acceleration is 7.33 (ft/sec)/sec. In stating accelerations, the time must be mentioned *twice,* once in specifying the *change of velocity* and again in stating the *time* in which that change occurred.

Example. The velocity of a car increases from 5.0 m/sec to 20 m/sec in 10.0 sec. What is its acceleration?

$$a = \frac{\Delta v}{\Delta t} = \frac{v - v_0}{t - t_0}$$

$$a = \frac{20 \text{ m/sec} - 5.0 \text{ m/sec}}{10.0 \text{ sec}}$$

$$= 1.50(\text{m/sec})/\text{sec}$$

This is often written 1.50 m/sec².

Distances Traveled with Constant Acceleration

Suppose that a car starts from rest and accelerates along a level highway, the driver doing his best to keep the acceleration constant (Fig. 4-1). Let the data gathered by a stopwatch, speedometer, and odometer be as represented in the first, second, and fourth rows of Table 4-1. The average velocities for each second are computed for each successive second and are tabulated in the third row. What relations can be obtained from these numbers?

In each successive second the velocity of the car increased 3.2 ft/sec; therefore the acceleration was 3.2 (ft/sec)/sec during the entire trip.

In two seconds the car traveled four times

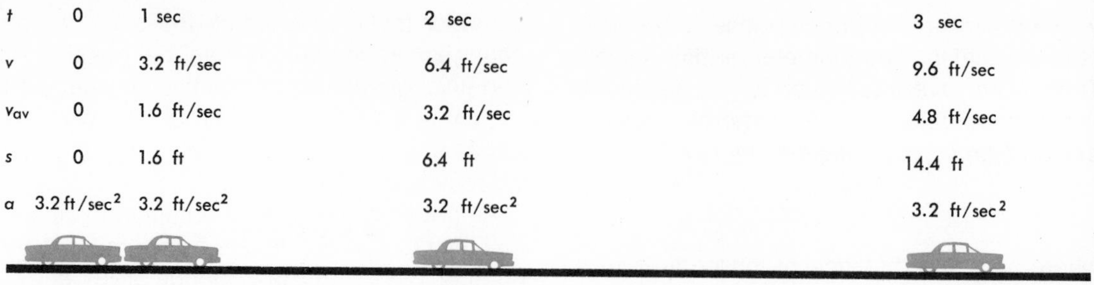

t	0	1 sec		2 sec		3 sec
v	0	3.2 ft/sec		6.4 ft/sec		9.6 ft/sec
v_{av}	0	1.6 ft/sec		3.2 ft/sec		4.8 ft/sec
s	0	1.6 ft		6.4 ft		14.4 ft
a	3.2 ft/sec²	3.2 ft/sec²		3.2 ft/sec²		3.2 ft/sec²

Fig. 4-1. *Uniformly accelerated car. The distance traveled from rest varies as the square of the time.*

Table 4-1.

Time, sec	Velocity, ft/sec	Average Velocity (from start), ft/sec	Total Distance Traveled, ft
0	0	0	0
1	3.2	1.6	1.6
2	6.4	3.2	6.4
3	9.6	4.8	14.4
4	12.8	6.4	25.6
5	16.0	8.0	40.0

as far from rest as in one second, and in three seconds it traveled nine times as far from rest. In other words: *with constant acceleration from rest, the distance traveled from the starting point varies as the square of the time.*

Finally let us ascertain how far the car advanced *during* the third and fourth seconds after starting. To do this, first we find the average velocity during these two seconds, which was $\frac{1}{2}$(6.4 ft/sec + 12.8 ft/sec) or 9.6 ft/sec. Then the distance traveled s is found to be

$$s = v_{av}t = 9.6 \text{ ft/sec} \times 2.0 \text{ sec} = 19.2 \text{ ft}$$

This result agrees with the measurements of total distance traveled: the total distance traveled from rest by the end of the fourth second (25.6 ft) minus the total distance traveled by the end of the second second yields the distance traveled *during* the third and fourth

seconds (19.2 ft). To make this result more general, let the initial velocity of a body, at zero time, be v_0; its velocity after a time t will be $v = v_0 + at$. Then its average velocity during the time is

$$v_{av} = \tfrac{1}{2}(v_0 + v_0 + at) = v_0 + \tfrac{1}{2}at$$

The distance traveled during the time t is

$$s = v_{av} \times t = (v_0 + \tfrac{1}{2}at)t = v_0t + \tfrac{1}{2}at^2 \quad (1)$$

Notice that v_0t is the distance that the body would have traveled if it had not accelerated; $\frac{1}{2}at^2$ is the distance it would have traveled if its initial velocity had been zero.

Example 1. A car, starting from rest, accelerates 4 m/sec². How far will it travel in 3.0 sec?

$$s = v_0t + \tfrac{1}{2}at^2 = 0 \text{ m/sec} \times 3 \text{ sec}$$
$$+ \tfrac{1}{2}(4 \text{ m/sec}^2)(3.0 \text{ sec})^2 = 18 \text{ m}$$

Example 2. The initial velocity of a car is 44 m/sec. After 4 sec, its velocity is 28 m/sec. (*a*) What is its acceleration? (*b*) How far did it travel during the four seconds?

(*a*) $\quad a = \dfrac{v - v_0}{t} = \dfrac{28 \text{ m/sec} - 44 \text{ m/sec}}{4 \text{ sec}}$

$$= -4.0 \text{ m/sec}^2$$

(*b*) $\quad s = v_0t + \tfrac{1}{2}at^2$
$$= 44 \text{ m/sec} \times 4 \text{ sec} -$$
$$\tfrac{1}{2}(4.0 \text{ m/sec}^2) \times 16 \text{ sec}^2$$
$$= 176 \text{ m} - 32 \text{ m} = 144 \text{ m}$$

When the initial velocity, the final velocity, and the acceleration are given, and the distance, but not the time, is involved, another relation is useful.

Since

$$s = v_{av}t = \tfrac{1}{2}(v + v_0)t$$

and

$$a = \frac{v - v_0}{t}$$

or

$$t = \frac{v - v_0}{a}$$

therefore,

$$s = \tfrac{1}{2}(v + v_0)\frac{v - v_0}{a}$$

or

$$2as = v^2 - v_0^2$$

and

$$v^2 = v_0^2 + 2as \qquad (2)$$

Hence knowing any three of the quantities v, v_0, a, and s, one can obtain the fourth quantity from this last equation.*

Example 3. One foggy night an automobile traveled at a velocity of 66 ft/sec (45 mi/hr).

The driver saw a car stalled on the road ahead and set the brakes, causing an acceleration of -22 ft/sec². How far did the car travel after the brakes were set before coming to rest?

$$v^2 = v_0^2 + 2as$$
$$0 = 4356 \text{ ft}^2/\text{sec}^2 - 2 \times 22 \text{ ft/sec}^2 \times s$$
$$s = 99 \text{ ft}$$

This is the stopping distance *after the brakes were set.* Suppose that the driver required $\tfrac{3}{4}$ sec to set them. The car would travel about 50 ft during this time; hence the total stopping distance would be about 149 ft.

A car traveling in a highly populated area at 20 mi/hr, having a fairly alert driver, has an imaginary "nose" about 40 ft long (Fig. 4-2). On the open highway at 60 mi/hr, the nose lengthens to 240 ft, for that is the approximate stopping distance.

Be careful in using kinematic equations such as (1) and (2). Remember that they were obtained under the following simplifying assumptions: (1) the motion is rectilinear and

---

A calculus proof. If you have studied calculus, you will recognize that these proofs can also be carried out quite directly in the notation of the calculus. The acceleration a is constant and is given by

$$a = \frac{dv}{dt}$$

To obtain equation (1), write this as

$$dv = a\, dt$$

and integrate twice. Integrating the first time we have

$$\int dv = a \int dt$$

$$v = at + k_1$$

where k_1 is a constant. Inserting the initial condition that $v = v_0$ when $t = 0$, we find that

$$v_0 = k_1$$

and

$$v = at + v_0$$

Now

$$v = \frac{ds}{dt}$$

and

$$ds = v\, dt = (at + v_0)\, dt$$

Integrating again,

$$\int ds = \int at\, dt + \int v_0\, dt$$

and

$$s = \tfrac{1}{2}at^2 + v_0 t + k_2$$

where k_2 is a constant. Applying the initial condition that $s = 0$ at $t = 0$, we find that $k_2 = 0$. Therefore,

$$s = v_0 t + \tfrac{1}{2}at^2$$

To obtain equation (2) we first write

$$dv = a\, dt$$

and then multiply both sides of the equation by v:

$$v\, dv = va\, dt$$

But $v = ds/dt$ and therefore

$$v\, dv = a\frac{ds}{dt}\, dt = a\, ds$$

Integrating both sides, we have

$$\tfrac{1}{2}v^2 = as + k_3$$

where k_3 is a constant. Applying the condition that $v = v_0$ when $s = 0$, we finally obtain

$$v^2 = v_0^2 + 2\, as$$

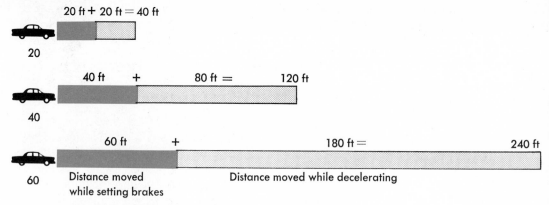

Fig. 4-2. Stopping distances for a car having excellent brakes.

(2) the acceleration is constant. In other motions that we shall consider later, one or both of these assumptions may not be realistic in describing the physical system that we will be interested in, and we shall have to derive other equations.

Graphs of Motion

Motion can be represented graphically by plotting position, velocity, or acceleration against time. The resulting graphs often help us to understand the motion more clearly, supplementing what we gain from equations such as (1) and (2) above.

Let us consider the rectilinear motion of a car whose position (distance from the point at which the clock was started) and velocity change with time as shown in Table 4-2.

Table 4-2.

Time, sec	Velocity, m/sec	Distance Traveled from Initial Position, m
0	10	0
1	10	10
2	10	20
3	14	32
4	18	48
5	22	68
6	26	92

Casual inspection of the data reveals that (1) the car was in motion at the beginning of the test interval, (2) during the first two seconds thereafter the velocity was constant, and (3) at the end of the second second the car began to accelerate at a constant acceleration (the velocity increased uniformly with time). In fact, we can easily calculate the acceleration:

$$a = \frac{18 \text{ m/sec} - 14 \text{ m/sec}}{1 \text{ sec}} = 4 \text{ m/sec}^2$$

The distance traveled and the instantaneous velocity (speedometer reading), as recorded in Table 4-2 are plotted against time in Fig. 4-3. A graph of calculated acceleration *versus* time is also given so that all three quantities—position, velocity, and acceleration—will be in front of us. Qualitatively the graphs look as we would expect them to. The acceleration graph in (*c*) is a horizontal line at zero (constant zero acceleration) during the first two seconds, rising rapidly to a second horizontal line at 4 m/sec² during the last four seconds. The velocity graph in (*b*) is horizontal at 10 m/sec during the first two seconds (corresponding to an initial velocity of 10 m/sec and zero acceleration) and then rises uniformly with an increase of 4 m/sec each second thereafter (corresponding to a constant acceleration of 4 m/sec²).

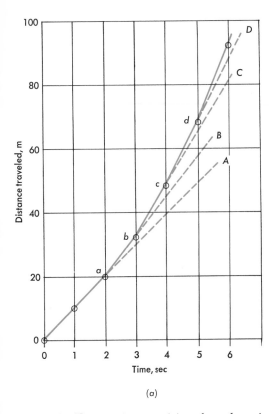

(a)

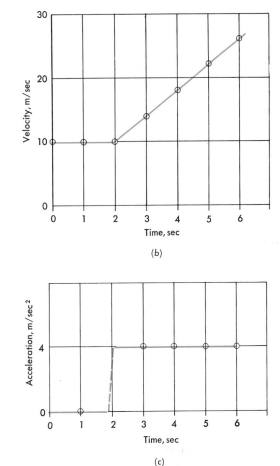

(b)

(c)

Fig. 4-3. Kinematic quantities plotted against time for the car whose motion is described in table 4-2: (a) *distance traveled* (*solid line*), (b) *velocity,* (c) *acceleration.*

The distance-time graph in (a) is the most interesting. During the first two seconds, it rises uniformly, the distance increasing 10 meters each second (corresponding to a constant velocity of 10 m/sec). But what can we say about the curved portion of the graph beyond point a? Is there a relation between the distance-time graph and the velocity-time graph during the last four seconds as there is in the first two seconds?

To answer these questions we must look more closely at the meaning of velocity. Earlier we defined average velocity as

$$ v_{av} = \frac{\Delta s}{\Delta t} $$

where Δs is the distance traveled in a time Δt. If the velocity does not change, the instantaneous velocity (the speedometer reading) is the same as the average velocity and can be found by dividing the distance traveled by the time. If the velocity is *not* constant, the instantaneous velocity and the average velocity are not the same. To find the instantaneous velocity in such cases, we find the *slope** of the distance-time graph at the point at which we want the instantaneous velocity. The slope, as can be shown rigorously by the use of cal-

* See Appendix D.

culus,* corresponds to the limiting value of the ratio of distance traveled to time as the time interval becomes smaller and approaches zero. For small time intervals, changes in the velocity will not be large enough to affect the average appreciably, and the instantaneous velocity and the average velocity for that small interval become equal. Hence

v = slope of the distance-time curve
= limiting value, as

$$\Delta t \text{ approaches zero, of } \frac{\Delta s}{\Delta t}$$

where Δv is the change in velocity in a (small) time interval Δt.*

Applying this to point b of the distance-time curve in Fig. 4-3, corresponding to $t = 3$ sec, we draw the tangent line b-B and measure its slope to find the velocity: $v_b = (46 \text{ m} - 32 \text{ m})/1$ sec = 14 m/sec. Note that this is the velocity recorded in Table 4-2 at 3 sec. Similarly, we can measure the slopes at c and d and show that the velocities are, as expected, 18 m/sec and 22 m/sec at 4 sec and 5 sec, respectively. To complete the story, note that the slope of the straight-line portion of the distance-time graph is 10 m/sec, the velocity during the first two seconds. Thus, *the slope at a point of the distance-time graph is the velocity at that instant.*

What about the relation between velocity v and acceleration a? We can show that

a = slope of the velocity-time curve
= limiting value, as

$$\Delta t \text{ approaches zero, of } \frac{\Delta v}{\Delta t}$$

where Δv is the change in velocity in a (small) time interval Δt.

---

* In the notation of the calculus:

$$v = \lim_{\Delta t \to 0} \frac{\Delta s}{\Delta t} = \frac{ds}{dt}$$

and

$$a = \lim_{\Delta t \to 0} \frac{\Delta v}{\Delta t} = \frac{dv}{dt}$$

In Fig. 4-3 the slope of the velocity-time curve during the first two seconds is zero and during the last two seconds is 4 m/sec², yielding the expected accelerations. Thus, *the slope at a point of the velocity-time graph is the acceleration at that instant.*

Falling Bodies

One of the most interesting kinds of uniformly accelerated motion is that of a body falling freely near the surface of the earth. When the Italian scientist Galileo was a young professor at the University of Pisa in the last decade of the sixteenth century, he discovered the laws of accelerated motion. Galileo released a ball at the top of an inclined plane and measured the distance that it traveled in a chosen time interval, for instance, one second. Then he observed that, in *twice* the time, the ball traveled *four times* as far from rest. He showed that, in general, the distance traveled from rest was proportional to the *square* of the time. Galileo found that the law held at each angle of inclination when the inclined plane was tilted more and more, and he reasoned that it should hold for freely falling bodies where the angle of inclination was effectively 90°. He found also that light wooden balls accelerated just as much as heavy metal balls for the same slope of the inclined plane, hence he asserted that *in a vacuum all freely falling bodies, regardless of their weights, accelerate equally.*

This law has been verified by precise experiments. However, the acceleration due to gravity, g, varies with elevation and position on the earth's surface. At sea level and 45° north latitude, the value is 9.807 m/sec². We shall use the approximate value $g \cong 9.8$ m/sec² = 980 cm/sec² $\cong$ 32 ft/sec². The equations of uniformly accelerated motion, derived earlier in this chapter, apply to freely falling bodies near the surface of the earth. The acceleration to be used in solving problems dealing with freely falling bodies is that due to gravity.

Example. A baseball is thrown vertically upward with an initial velocity of 19.6 m/sec. (*a*) What vertical height will it reach? (*b*) What time will elapse before it strikes the ground?

Let us first solve this problem by the following simple process. The ball *decelerates* when it rises just as it *accelerates* when it falls. The acceleration due to gravity has the effect that a *rising* body *loses* 9.8 m/sec of velocity each second that it rises just as a *falling* body *gains* 9.8 m/sec each second that it falls. Now the ball has 19.6 m/sec of velocity to lose before it reaches its maximum height because there its instantaneous velocity is zero. It will take it 2 sec to do this. Returning from that maximum height, the ball will fall for an equal time because the distances are the same. Thus the ball takes 2 sec plus 2 sec, or 4 sec for the entire trip.

The distance *up* traveled by the ball is equal to the distance *down* in returning. In falling from rest for 2 sec, a body travels $s = \frac{1}{2}(9.8 \text{ m/sec}^2)(2 \text{ sec})^2 = 19.6 \text{ m}$. Thus, the ball must have risen 19.6 m.

Now, let us solve the problem more formally.

(*a*) $\quad v^2 = v_0{}^2 + 2\,as$
$\quad\quad 0 = (19.6 \text{ m/sec})^2 + 2(-9.8 \text{ m/sec}^2)h$
$\quad\quad h = 19.6 \text{ m}$

(*b*) $\quad s = v_0 t + \frac{1}{2}\,at^2$

We want to know what *t* is when *s* = 0, that is, when the ball returns to the ground.

$0 = 19.6 \text{ m/sec } t + \frac{1}{2}(-9.8 \text{ m/sec}^2)t^2$

Factoring,

$0 = (19.6 \text{ m/sec } - \frac{1}{2}\,9.8 \text{ m/sec}^2\ t)t$

whence, $\quad t = 0$
$\quad\quad\quad\quad t = 4 \text{ sec}$

How do you interpret the first of these solutions, *t* = 0?

We find it hard to comprehend the difficulties of scientists in olden times. Galileo had no clock, so he measured time by the amount of water that escaped from a tube at the bottom of a vessel of water. For instance, if $\frac{1}{3}$ pint of water escaped while the ball rolled 1 ft down the inclined plane, then about $\frac{2}{3}$ pint would flow out while the ball traveled 4 ft.

Public opinion of Galileo's time was an even greater obstacle to research. People did not believe in experimenting. They preferred to rely on ancient authority. Nearly two thousand years before Galileo, the Greek philosopher Aristotle had observed that, because of air friction, feathers fall more slowly than stones. Since Aristotle could not conceive of a space devoid of matter—a vacuum—he stated that moving bodies would always encounter resistance and that heavy bodies would always fall faster than light ones. Galileo boldly challenged this view. According to indirect evidence in his writings, Galileo climbed to the topmost gallery of the Leaning Tower of Pisa and dropped a large stone and a small one. Both struck the ground at almost the same instant.

When the effects of air friction are minimized—as they would be in a tube at high vacuum (low pressure)—all bodies experience the same acceleration due to gravity at a given location. That *g* is independent of the mass of the body was crucial not only in helping to decide between the views of Aristotle and of Galileo, but in establishing the modern theory of relativity, as we shall see in a later chapter. It is impossible in a brief space to do justice to the different theories of Aristotle and Galileo. The world picture of Aristotle was not a naive one as some writers suggest. The subtle discourses in which Galileo broke the hold of Aristotelianism on the thought of his contemporaries and helped to establish a quantitative physics based on simple models are well worth your reading as masterpieces of persuasiveness and classics of science. Selections from the original works and from some of the best contemporary criticism of them are listed among the references at the end of this chapter.

Summary

The average velocity of a body equals the distance it travels in a specified direction per unit time. Velocity is a vector quantity. Speed is the magnitude of the velocity at a given instant.

The acceleration of a body is its change of velocity per unit time. Acceleration may be expressed in feet-per-second per second or in centimeters-per-second per second.

The general equations for the rectilinear motion of a particle which has constant acceleration are:

$$v_{av} = \frac{\Delta s}{\Delta t} = \frac{s - s_0}{t - t_0}$$

$$a = \frac{\Delta v}{\Delta t} = \frac{v - v_0}{t - t_0}$$

$$v = v_0 + at$$

$$s = v_0 t + \tfrac{1}{2} at^2$$

$$v^2 = v_0^2 + 2as$$

The (instantaneous) velocity v of a body is the limiting value of $\Delta s/\Delta t$ as Δt approaches zero; the velocity at a given time equals the slope of the distance-time curve at that time.

The (instantaneous) acceleration at a given time is similarly the limiting value of $\Delta v/\Delta t$ as Δt approaches zero and is equal to the slope of the velocity-time curve at that time.

The acceleration due to gravity is constant for all bodies in vacuum at a given location. At $45°$ north latitude and sea level, $g \cong 32$ ft/sec^2 $\cong 9.8$ m/sec^2.

Questions

1. Define acceleration and state units in which it may be expressed.

2. What does a speedometer measure? Can you calculate acceleration from it alone?

3. A leaf and an acorn start to fall from the same height at the same time. Why does the acorn strike the ground before the leaf?

4. Why is the word *constant* stressed so much in acceleration problems? If the acceleration were not constant, what would happen?

5. Is the acceleration due to gravity, g, a constant? Discuss.

6. Is it possible to move north with an acceleration directed to the south?

7. If ballast is dropped from a balloon, will the acceleration of the falling ballast be greater during the first second or the tenth second? Why?

8. Substitute units on both sides of this equation and prove that they are identical: $v^2 = v_0^2 + 2as$.

9. Using Table 4-1, estimate how far the car would travel during the sixth second if its motion continued as before. (*Suggestion:* Find the average velocity during that second.)

10. Fill in the blank spaces in the following table for a freely falling body. Acceleration = 9.8 m/sec^2.

Time of Fall (from Rest)	Velocity	Average Velocity	Total Distance
0 sec	_____	_____	_____
$\frac{1}{4}$ sec	_____	_____	_____
$\frac{1}{2}$ sec	_____	_____	_____
1 sec	_____	_____	_____
2 sec	_____	_____	_____
3 sec	_____	_____	_____

11. Express $2\,g$ in (km/hr)/sec.

12. Estimate the average speed for the whole trip for a car that travels 5 mi at 20 mi/hr and 5 mi at 40 mi/hr.

Problems

(*Assume that air friction is negligible in the following problems.*)

A

1. When the brakes were set to give maximum retardation, the speed of a car traveling on a straight line changed from 66 km/hr to 6

km/hr in 1.00 sec. Find (*a*) its deceleration, (*b*) the distance moved while decelerating.

2. A ball is dropped from the roof of a skyscraper. At the end of 0 sec, 3 sec, and 4 sec, what are (*a*) the velocity, (*b*) the acceleration, (*c*) the total distance traveled from rest?

3. When a player struck a golf ball, its velocity was changed from 0 m/sec to 30 m/sec in 0.020 sec. What was its average acceleration?

4. An air pilot, jumping from a plane with a parachute, lands at a velocity of 4.9 m/sec. From what elevation must he jump (without a parachute) to attain this velocity?

5. A jet airplane goes from rest to a speed of 100 m/sec in a distance of 150 m. What is its average acceleration in m/sec² and in *g*'s where 1 *g* = 9.8 m/sec²?

6. A ball is thrown vertically downward from the top of a cliff with an initial velocity of 10 m/sec. If it strikes the ground in 4.0 sec, what is the height of the cliff?

7. A baseball, thrown vertically upward, returns after 5.0 sec. What was its initial velocity?

8. An oil well is 4800 m deep. What time would be required for a tool to fall freely to the bottom of the well? Assume constant *g*.

B

9. A baseball is knocked vertically upward with a velocity of 29.4 m/sec. After 5 sec, (*a*) how far is it above the ground, (*b*) what is its velocity?

10. If, at the same instant, a ball is dropped from a height of 19.6 m and another ball is thrown vertically upward with a velocity of 19.6 m/sec, where and when will they reach the same elevation?

11. At a signal, an athlete running with a velocity of 3.00 m/sec begins to slow down, comes to rest, reverses his motion, and is running 3.00 m/sec when he reaches the point where he heard the signal. The total elapsed time is 10.0 sec. What is his deceleration, assumed to be constant?

12. An automobile traveling 90 km/hr ran into a concrete wall and stopped in 0.10 sec. (*a*) What was its deceleration? (*b*) From what elevation would the car have to fall from rest to acquire the given speed?

13. A rifle barrel is 60 cm long. The velocity of the bullet fired from it is 700 m/sec. (*a*) What is the average velocity of the bullet while it is being accelerated in the barrel? (*b*) What time is required for the bullet to travel the length of the barrel? (*c*) What is its average acceleration?

14. A golf ball is thrown vertically downward from a tall building with an initial velocity of 25 m/sec. Find (*a*) the average velocity, (*b*) the distance traveled during the first 2.5 sec.

15. If you walk at 5 ft/sec for 0.5 min and run 10 ft/sec for 0.5 min, what is your average velocity for the entire interval?

16. If you walk 50 ft at 5 ft/sec and run 50 ft at 10 ft/sec, what is your average velocity for the entire interval?

C

17. An automobile driver traveling 90 km/hr saw a fallen tree on the road 100 m ahead. To set the brakes required 0.75 sec. Afterward, the deceleration was 8 m/sec². (*a*) What was his total stopping time? (*b*) How far did he travel before the brakes were applied? (*c*) What was his stopping distance?

18. A baseball *B* is thrown vertically upward 1.0 sec after baseball *A* is thrown upward. After what time interval and at what elevation will *B* pass *A* if the initial velocity of each is 20 m/sec? Assume that each ball weighs 150 gm.

19. A boy standing on a bridge 30 m above the water throws a ball vertically upward with an initial velocity of 20 m/sec. (*a*) How far will the ball rise? (*b*) How soon will it reach the water? (*c*) With what velocity will it strike the water?

20. When the traffic lights turn green at an intersection, a motorcycle *A* starts from rest

with a constant acceleration of 3.0 m/sec². At that instant, an automobile *B* traveling 50 km/hr passes *A*. How soon will *A* overtake *B*?

21. A speeding car traveling 90 km/hr passes a motorcycle patrolman at rest. He immediately starts in pursuit. If he accelerates 3.0 m/sec² for 12 sec and thereafter travels at constant velocity, how far will he travel before overtaking the car?

22. A ball is dropped from rest at the same instant that another one is thrown vertically downward with a velocity of 400 cm/sec. After how many seconds will they be separated by 12 m?

23. A bus passes a traffic signal at a constant speed of 20 km/hr. At the same instant an automobile starts from rest at the traffic signal. (*a*) What time will be required for the car to pass the bus, if the constant acceleration of the car is 2.0 m/sec²? (*b*) What will be the velocity of the automobile when it passes the bus?

24. A car starts from rest and experiences a constant acceleration of 1.0 m/sec² for 5.0 sec. The velocity then remains constant for 10 sec. Next the brakes cause a constant deceleration of 2.0 m/sec² until the car stops. Make a graph, plotting time along the *X* axis and the velocity of the car along the *Y* axis. Show that the area under this curve numerically equals the distance traveled and check the result by computing this distance.

25. An aviator (wearing a pressurized suit) jumps from a balloon at an altitude of 3.11×10^4 m and falls to 5.34×10^3 m, where he opens his parachute. Neglecting air resistance, calculate (*a*) the time for his free fall and (*b*) his velocity just before opening the parachute. (*c*) If the measured time for his "free" fall was 4 min, 38 sec, did air resistance greatly affect his motion?

26. If a rocket ship accelerates at 9.8 m/sec² for ten days, what will be (*a*) its velocity at the end of this period and (*b*) its distance from the earth?

27. To escape from the earth, a rocket must attain a velocity of 1.03×10^4 m/sec by the

time its charge of propellant is burned out. It has been calculated that the altitude of the rocket at the end of burning will be 1075 km. Calculate (*a*) the average acceleration of the rocket, (*b*) the time for the charge to burn out. (Keep only three significant figures at each step.)

28. A boy in an elevator descending at a constant speed of 10 m/sec jumps to a height of 50 cm above the floor. How far does the elevator descend before he reaches the floor again?

29. A baseball falling freely from rest from the roof of a building requires 0.10 sec to pass from the top to the bottom of a window 2.0 m high. How high is the roof above the top of the window?

30. Car *A* is initially 200 m behind car *B*, and both are traveling at a constant speed of 30 m/sec. As *A* pulls into the left-hand lane and accelerates uniformly, *B* accelerates uniformly and reaches a velocity of 35 m/sec in the 20 sec required for *A* to catch up with *B*. Calculate (*a*) the acceleration of *A*, (*b*) the final velocity of *A*, and (*c*) the distance traveled by *A* before reaching *B*.

31. (*a*) Make a graph of the distance traveled versus time for the data given in Table 4-1. From the graph, determine the instantaneous velocities at the times given in the table and compare these with the velocities in the table. (*b*) Make a graph of the velocities versus time in the same table. From the graph, determine the distances and compare them with the distances in the table.

32. Car *A* initially has a velocity of 90 km/hr and is 30 m behind car *B* when *A* sees *B* through the fog. *B* is moving at a constant velocity of 50 km/hr. What must be the minimum deceleration of *A* to avoid a rear-end collision?

For Further Reading

Boyer, Carl, "Aristotle's Physics," *Scientific American,* May, 1950, pp. 48–51.

Cohen, I. B., "Galileo," *Scientific American,* August, 1949, pp. 40–47.

Holton, Gerald, *Introduction to Concepts and Theories in Physical Science,* Chapter 2, Addison-Wesley Publishing Company, Reading, Mass., 1952.

Knedler, John W., Jr., ed., *Masterworks of Science* (Galileo, "Dialogues Concerning Two New Sciences"), Doubleday and Company, Garden City, 1947.

Kolers, Paul A., "The Illusion of Motion," *Scientific American,* October, 1964, pp. 98–106.

CHAPTER FIVE

Force and Acceleration

When Galileo studied the laws of falling bodies, he observed that balls of different materials and different weights, rolling down the same inclined plane, had equal accelerations, but that their accelerations increased as the slope of the inclined plane increased. In this chapter, we shall begin our study of *dynamics* and see how the acceleration of a body depends both upon the force acting on it and upon its mass. We shall assume that the bodies are *mass points*—objects whose linear dimensions can be considered small. Later we shall take up the dynamics of extended bodies.

When a Body Moves at Constant Velocity

We must first consider what happens when the acceleration is *zero* and the body moves at constant velocity. Aristotle reflected upon this problem and decided that any moving body was in an unbalanced condition. Rest was the "normal" condition; motion, even at constant velocity, was "abnormal" and could continue only under the action of a force. A ball continued moving along a level surface, according to Aristotle, because the air opened before it and closed behind, creating a force that propelled the ball forward at constant velocity.

We shall come to a different conclusion, but let us admit that there are experimental difficulties in our way. It will be difficult for us to find a physical system in which there are no forces that speed up or slow down the moving body. Friction, for one thing, creates problems. However, let us consider a "thought experiment"* on a stone sliding on the smooth surface of a frozen pond. Friction, although small, creates a backward force that decelerates the stone. Ashes sprinkled on the ice would increase the friction and would cause the stone to stop in a shorter time. Suppose that the ice were treated in some way so as to decrease the friction. Then the stone would move farther before stopping. Finally, if

---

* "Thought experiments" are often helpful to physicists as the ultimate simplification of an experimental situation in constructing a theory. In a thought experiment, the physical conditions of interest stand out sharply, and the "thought experimenter" can consider them without having to worry about irrelevant conditions that must be controlled by the experimenter in a real situation. Here we imagine that the sheet of ice is infinitely large, perfectly flat, etc. Of course, the physicist soon returns to the real world to compare his theory with the results of real experiments.

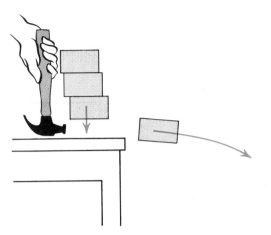

Fig. 5-1. Inertia. When the bottom block is knocked out, the other blocks oppose being accelerated, remain behind, and drop almost vertically.

the opposing force were reduced to zero, would not the stone continue to travel with constant velocity? Sir Isaac Newton took the bold step of saying that it would. His first law of motion states:

Every body persists in its state of rest or of uniform motion in a straight line unless it is acted upon by an external, unbalanced force.

Consider carefully what this statement means. Either rest or motion *at constant velocity* is the result of balanced forces. An unbalanced force is *not* required to maintain motion at constant velocity, as Aristotle thought. Matter once in motion at constant velocity retains that motion—forever—until an external unbalanced force acts upon the body to speed it up or slow it down.

An automobile rushes along a highway at a speed of 60 mi/hr. Suddenly a tire bursts and the car crashes into a concrete wall. The passengers keep moving and smash against the windshield. Again, place four smooth wooden blocks in a vertical stack at the edge of a smooth table top (Fig. 5-1). With a sharp horizontal blow of a hammer, knock the

bottom block out of the stack. The other blocks drop vertically and are displaced horizontally only slightly. With the exception of a slight amount of friction, no horizontal force acted on the top three blocks and they moved only slightly from their former horizontal position. The use of air pucks, metal disks that are supported on a thin layer of compressed gas, illustrates the motion of bodies at almost constant velocity (Fig. 5-2). When one of the air pucks is set into motion on a long horizontal glass plate, friction is extremely small, and the puck maintains a remarkably constant velocity. Air tracks and air tables (Fig. 5-3) are laboratory devices in which the moving objects glide with little friction on a layer of compressed air. All of these examples illustrate the tendency of bodies to remain at rest or in motion at constant velocity unless acted upon by some external, unbalanced force. We sometimes call this tendency *inertia.*

Can we prove Newton's first law? Obviously we cannot. To prove it we would have to set up a test range somewhere out in space where there would be no frictional forces and no gravitational forces. We would start the

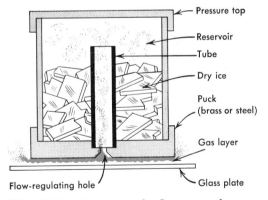

Fig. 5-2. An air puck. Compressed gas, escaping at the small hole, supports the puck on a practically frictionless layer. When set into motion on a horizontal sheet of flat glass, the puck glides along at almost constant velocity until it hits an obstacle. (Courtesy Apparatus Drawings Project).

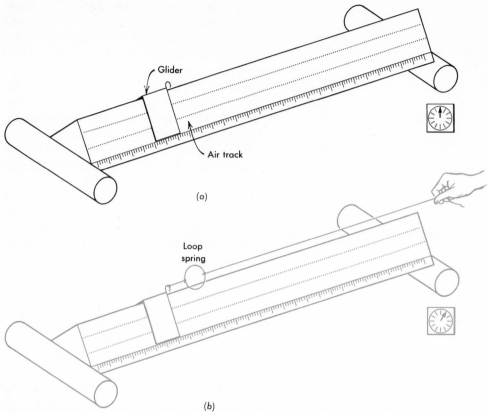

Glider

Air track

(a)

Loop
spring

(b)

Fig. 5-3. An air track.

body moving in a straight line and hope to measure its velocity thereafter without exerting a force on it! Finding a suitable frame of reference would also be a problem. The trouble with the earth as a platform on which to test Newton's law is that the earth rotates. As we shall see in the next chapter, an object moving on a circle at constant speed has an acceleration. We would have to seek a nonrotating platform—perhaps a spaceship far out in space—whose linear acceleration is zero in order to test the first law of motion.

Since direct experimental proof is not available, we must consider Newton's first law—and his second and third laws, which we will consider next—as useful postulates, much like the postulates you used in your study of plane geometry. Newton's laws have led to verifiable and useful conclusions in

science and engineering. These successful predictions strengthen our confidence in their correctness within certain physical situations. We shall soon have to recognize, however, that Newton's laws do not apply to the very small nor to the very fast.

Force, Mass, and Acceleration

The unbalanced force acting on a body is proportional to the acceleration which the force produces. Support a 1-kg block of wood by a cord so as to balance the earth-pull on it, and the block will remain at rest. Release it, and the unbalanced force of the earth's attraction will accelerate it approximately 9.8 m/sec^2 (1 g). Mount the block on small wheels and pull it along a table top, exerting a force equal to the weight of the block. Then, if there is little friction, the acceleration will again be about 1 *g*. An unbalanced force equal to twice the weight of the block will cause an acceleration of 19.6 m/sec^2 or 2 *g*.

Place a glider on a carefully leveled (horizontal) air track (Fig. 5-3*a*) and give it a small initial velocity. Measure the distances the glider travels in successive equal time intervals and compute the velocity for each of these intervals. The velocity is constant because friction and other decelerating forces are negligibly small. Now attach one end of a horizontal cord to the glider and the other end to a carefully calibrated spring whose extension is a measure of the constant unbalanced force exerted on the glider (Fig. 5-3*b*). Exert an unbalanced force on the glider and measure the distance *s* it travels from rest in time *t*. Compute the acceleration from $s = \frac{1}{2}at^2$. Now double the unbalanced force acting on the glider. Starting from rest as before, the glider will travel twice as far in the same time *t*; hence the acceleration is doubled. Experiments of this type suggest that *the acceleration of a body is proportional to the unbalanced external force acting on it.*

Next, let us consider how the unbalanced force is related to the mass when the acceleration is constant. In Chapter 1 we saw that a more massive body is more difficult to accelerate than a less massive one. For example, you must exert a greater force to give a baseball a certain acceleration than to give a tennis ball the same acceleration. You can kick an empty pasteboard box more easily than you could if a brick were hidden inside it. A heavily loaded truck is harder to accelerate or decelerate than a motorcycle. Experiments demonstrate that *the unbalanced force required to produce a certain acceleration in a body is proportional to its mass.*

You can test this law experimentally with the air track. First measure the acceleration of a single glider when a constant unbalanced force is exerted on it by the spring. Then attach a second, equal glider to the first one, doubling the mass of the system,

and exert the *same* unbalanced force on it. The acceleration becomes *half* as great. However, if the unbalanced force is also doubled, the acceleration is the *same* as in the first experiment; doubling the mass and also doubling the accelerating (unbalanced) force yields the same acceleration.

Experiments such as we have described demonstrate that the unbalanced force acting on a body is proportional both to the acceleration produced and to the mass of the body. If both the mass of the system and the acceleration are doubled, the accelerating force must be quadrupled. This result is in accord with Newton's second law of motion, one form of which is as follows:

The unbalanced force acting on a body is proportional to the product of the mass of the body and its acceleration.

An unbalanced force *F* acting on a body of mass *m* produces an acceleration *a* such that

$$F = kma$$

The value of the constant *k* depends upon the units used to measure *m* and *F*. These units can be so chosen that *k* is unity, and

$$F = ma$$

Units in Dynamics

Physicists often use the *newton* as a unit of force. *It is defined as that force which acting on a body of one kilogram mass will accelerate it one meter-per-second per second.* Inserting these quantities into the equation *F = ma,* we have

$$1 \text{ newton} = 1 \text{ kg-m/sec}^2$$

The newton is called an mks (meter-kilogram-second) *absolute* unit because it is defined in terms of the meter, kilogram, and second, and because it is an invariant, having the same value everywhere.

The weight *w* of a body of mass *m* is given by Newton's second law as

$$w = mg$$

in which g is the acceleration due to gravity.

The earth-pull of 1 kgwt on a body of 1-kg mass will accelerate it 9.8 m/sec². Therefore its weight is

$$1 \text{ kgwt} = 1 \text{ kg} \times 9.8 \text{ m/sec}^2 = 9.8 \text{ kg-m/sec}^2$$
$$= 9.8 \text{ newtons}$$

Thus 1 kgwt = 9.8 newtons.

Example 1. The velocity of a 1600-kg automobile changed from 5.00 m/sec to 17.0 m/sec in 6.00 sec. What unbalanced force accelerated it?

$$F = ma$$

$$= 1600 \text{ kg} \times \frac{17.0 \text{ m/sec} - 5.00 \text{ m/sec}}{6.00 \text{ sec}}$$

$$= 3200 \text{ kg-m/sec}^2 = 3200 \text{ newtons}$$

The dyne (Greek *dunamis,* force) is an absolute cgs (centimeter-gram-second) unit of force. The dyne is defined as that force which, acting on a body of mass 1 gm, will accelerate it 1 cm/sec².

$$1 \text{ dyne} = 1 \text{ gm-cm/sec}^2$$
$$= 1/1000 \text{ kg} \times 1/100 \text{ m/sec}^2$$
$$= 1/100{,}000 \text{ newton} = 10^{-5} \text{ newton}$$

$$1 \text{ gwt} = 1 \text{ gm} \times 980 \text{ cm/sec}^2$$
$$= 980 \text{ gm-cm/sec}^2 = 980 \text{ dynes}$$

Thus 1 dyne = 10^{-5} newton = 1/980,000 kgwt = 1/980 gwt.

Example 2. A 5-cent piece having a mass of 5.00 gm, sliding on a level table top, decelerated 98 cm/sec². What was the decelerating force of friction?

$$F = ma$$
$$= 5.00 \text{ gm} \times (-98 \text{ cm/sec}^2)$$
$$= -490 \text{ gm-cm/sec}^2$$
$$= -490 \text{ dynes}$$
$$= -490 \text{ dynes} \times 1 \text{ gwt}/980 \text{ dynes}$$
$$= -0.500 \text{ gwt}$$

Other systems of units are also possible.*

* See Appendix H for a description of consistent British units that can be used with Newton's second law.

We shall work mainly with the systems described above (Table 5-1).

Table 5-1. Systems of force units

System	Mass	Acceleration	Force
mks	1 kg	1 m/sec²	1 newton, 10^5 dynes
cgs	1 gm	1 cm/sec²	1 dyne $\cong \frac{1}{980}$ gwt

Applications of Newton's Second Law

Newton's second law provides a connection between force and acceleration that permits us to describe the motion of a body, knowing the resultant force, or to calculate the resultant force, knowing the acceleration. We shall encounter examples of the use of the second law throughout this book, but let us now consider applications to some simple physical systems. Our method consists of (1) identifying the body whose motion we want to describe, (2) finding the external forces that act *on* the body (by drawing a "free-body" diagram of the more complex systems), (3) calculating the resultant force by any of the methods of vector addition, (4) calculating the acceleration from $F = ma$ where F is the *resultant* force and where *consistent* units must be used, and (5) computing the position and velocity of the body at any desired instant by using the kinematic equations.

Example 1. What acceleration will the blocks in Fig. 5-4 experience (*a*) if the friction is zero,

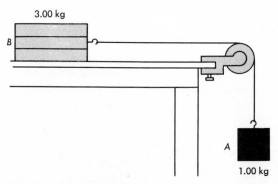

Fig. 5-4. Acceleration and unbalanced force.

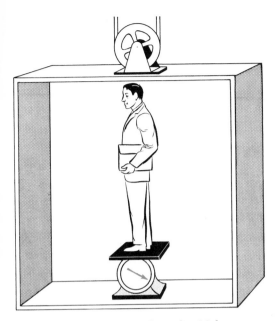

Fig. 5-5. *How much does the 98-kg man apparently weigh?*

and (*b*) if the coefficient of kinetic friction between *B* and the table is 0.20?

The bodies being accelerated are *A* and *B*. Let us consider them as a single body in this problem. (*a*) The resultant force acting on the system is the weight of *A*, 1.00 kgwt or 9.80 newtons. The total mass is 4.00 kg.

$$F = ma$$
$$9.80 \text{ newtons} = 4.00 \text{ kg} \times a$$
$$9.80 \text{ kg-m/sec}^2 = 4.00 \text{ kg} \times a$$
$$a = 2.45 \text{ m/sec}^2$$

(*b*) The force of friction is 3.00 kgwt × 0.20 = 0.60 kgwt, acting toward the left on *B*. The resultant force acting on the system is then 1.00 kgwt − 0.60 kgwt = 0.40 kgwt = 0.40 × 9.80 newtons.

$$0.40 \times 9.80 \text{ kg-m/sec}^2 = 4.00 \text{ kg} \times a$$
$$a = 0.98 \text{ m/sec}^2$$

Example 2. What up force does the spring scale exert on the 98-kg man (Fig. 5-5) (*a*) when the elevator rises or descends at con-

stant velocity, (*b*) when it accelerates upward 2.0 m/sec², (*c*) when it accelerates downward 2.0 m/sec²?

(*a*) The acceleration is zero; the resultant force acting on the man must be zero; hence the spring scale registers the weight of the man, 98 kgwt, because the up push of the scale on the man must just equal the down pull of the earth on him.

(*b*) Let the up force of the spring be *f*. The unbalanced force $F = f - w$ must accelerate him upward 2.0 m/sec². His mass is 98 kg.

$$F = ma$$
$$F = 98 \text{ kg} \cdot 2.0 \text{ m/sec}^2$$
$$= 196 \text{ kg-m/sec}^2$$
$$= 196 \text{ newtons} = 20 \text{ kgwt}$$

$$f = F + w = 20 \text{ kgwt} + 98 \text{ kgwt} = 118 \text{ kgwt}$$

(*c*) $f - 98 \text{ kgwt} = 98 \text{ kg}(-2.0 \text{ m/sec}^2)$
$$= -196 \text{ newtons}$$
$$= -20 \text{ kgwt}$$
$$f = 78 \text{ kgwt}$$

Example 3. Find (*a*) the acceleration of *A* (Fig. 5-6), neglecting friction and the mass of the pulley and the string, and (*b*) the tension

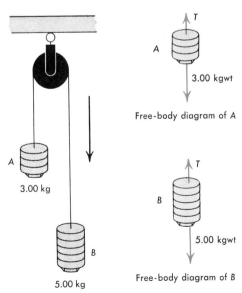

Free-body diagram of A

Free-body diagram of B

Fig. 5-6. *Atwood's machine.*

in the string when *A* and *B* are accelerating. This apparatus is called Atwood's machine.

As in Example 1, we could "lump" *A* and *B* into a single system of mass 8.00 kg acted upon by a resultant force of 2.00 kgwt or 19.6 newtons. However, this method is not readily applicable to more complex physical systems. Let us therefore use "free-body" diagrams in solving this problem. A free-body diagram shows the body in isolation from the rest of the physical system and under the action of the forces that affect only it. The free-body diagram of *A* (Fig. 5-6) shows two forces acting on *A*: the weight −3.00 kgwt or −29.4 newtons (the minus sign indicates the force acts in the negative *y*-direction) and +*T*, the unknown tension in the cord in the upward direction. Similarly, *B* is acted upon by +*T* up and −49.0 newtons down.

Let us now insert the resultant forces in absolute units into equations for Newton's second law, doing this separately for *A* and *B*:

$$F = ma$$
$$T - 29.4 \text{ newtons} = 3.00 \text{ kg } a_A \quad \text{(Body } A\text{)}$$
$$T - 49.0 \text{ newtons} = 5.00 \text{ kg } a_B \quad \text{(Body } B\text{)}$$

where a_A and a_B are the accelerations of *A* and *B*, respectively. If *B* goes down when the bodies are released, *A* must go up with an acceleration equal and *opposite in sign* to that of *B* if the string does not stretch. Hence:

$$a_A = -a_B$$

Let us substitute $-a_A$ for a_B in the equation for body *B* above and subtract that equation from the one for body *A*. We then have

$$19.6 \text{ newtons} = 8.00 \text{ kg } a_A$$
$$19.6 \text{ kg-m/sec}^2 = 8.00 \text{ kg } a_A$$
$$a_A = 2.45 \text{ m/sec}^2$$

(*b*) To obtain the tension in the cord, we insert the value of a_A into the equation for body *A*:

$$T - 29.4 \text{ newtons} = (3.00 \text{ kg})(2.45 \text{ m/sec}^2)$$
$$T = 29.4 \text{ newtons} + 7.35 \text{ newtons}$$

$$T = 36.75 \text{ newtons} = 3.75 \text{ kgwt}$$

Note that when the system is accelerating, the tension in the cord is everywhere 3.75 kgwt. What does the tension become in each side of the cord when *B* reaches the floor and stops? Why?

Mass and Weight

In the preceding sections we have used the equation

$$w = mg$$

where *w* is the weight of a body of mass *m* at a place where the value of the acceleration due to gravity is *g*. You will recall that in Chapter 4 it was pointed out that *g* varies from position to position on the earth. The weight *w* of the body varies proportionately with *g* so that *w/g*, which equals the mass *m*, is a constant.

Mass, as we have seen, is defined as the ratio of resultant force to acceleration:

$$m = \frac{F}{a}$$

Mass and weight are **different** physical quantities. The mass of a body is constant (except, as we shall see, at velocities approaching the velocity of light) and does not depend upon the position of the body. Weight, however, is the gravitational force that the earth exerts on the body; according to the law of universal gravitation (Chapter 11), both the weight of a body and the acceleration due to gravity vary with the distance of the body from the center of the earth. Mass is a scalar quantity; weight, a vector. We can measure mass with inertial cars and track (Chapter 1) or, more conveniently and precisely, with the equal-arm balance (Fig. 5-7). We measure weight with a spring balance, although at a standard location the properly calibrated spring balance could be used to determine mass. (On the moon the mass determination by the spring balance would differ greatly from that by an equal-arm bal-

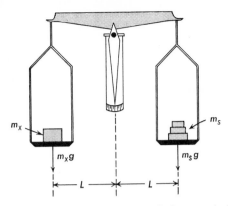

Fig. 5-7. The equal-arm balance. At balance, the product of the weight of the body of unknown mass by L equals the product of weight of the bodies of standard mass by L. Since the arms of the balance are of equal length and since g does not vary appreciably from one pan to the other, $m_x = m_s$.

ance!) Finally, the units of weight and mass are different and should be stated properly in solving problems: weight is expressed in force units such as kilogram-weights, newtons, gram-weights, and dynes; mass is expressed in mass units such as grams and kilograms.

Action and Reaction are Equal; They Act on Different Bodies

A variety of everyday experiences suggests that whenever one body exerts a force on another, the second body exerts an opposite force on the first one. *Forces always appear in pairs.* The force that the first body exerts on the second we call the *action,* and that exerted by the second body on the first is the *reaction.* When you push your forefinger against your thumb, the action force exerted by the finger on the thumb is accompanied by an opposite, reaction force exerted by the thumb on the finger. When you sit on a bench, the downward, action force exerted by your body *on the bench* is accompanied by an upward, reaction force exerted by the bench *on your body.* The action force and the reaction force *always* act on different bodies.

Although daily experience does not enable us to quantify and compare the action and reaction forces, indirect evidence obtained in laboratory experiments (Chapter 7) tells us that they are *equal* in magnitude. If you exert an action force of 80 kgwt downward on a bench, the bench exerts an upward, reaction force of 80 kgwt on you.

Action and reaction are also equal when bodies are accelerated. Hit a golf ball with a club, and the forward, action force accelerating the ball will be equal and opposite to the reaction force decelerating the club. Slap your hand against a wall. At every instant during the impact two forces will be exerted: one, your action against the wall; the other, its equal reaction against your hand. When a bullet is fired from a rifle, the forward force accelerating the bullet will be equal to the backward force that makes the gun "kick" or recoil (Fig. 5-8). When a diver jumps from a

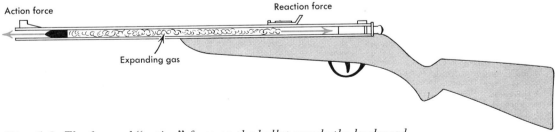

Fig. 5-8. The forward "action" force on the bullet equals the backward "reaction" force on the gas.

Fig. 5-9. *The action force on the truck equals the reaction force on the car.*

canoe, two forces are exerted. One pushes him away; the other makes the canoe move backward. When a 5-ton truck collides with an automobile, the action force and the reaction force are equal (Fig. 5-9). In Fig. 5-10, the action force accelerates the gases expelled by the rocket. The opposite and equal reaction drives the rocket upward. Action and reaction forces are *always* equal and opposite and act on different bodies.

The fact that forces are paired is expressed by Newton's third law of motion:

When one body A "acts on" or exerts a force on another body B, the second body "reacts" or exerts on the first body a force equal in magnitude but opposite in direction.

$$F_{A \text{ on } B} = -F_{B \text{ on } A}$$

That forces come in action-reaction pairs is also important in understanding the equilibrium of bodies at rest. In Fig. 5-11, a 20-pound basket rests on a picnic table. Two pairs of action-reaction forces must be considered in showing that the basket is in equilibrium: (1) the earth exerts a force (solid arrow) of 20 lb downward *on the basket* and the basket exerts an upward force (solid arrow) of 20 lb *on the earth* at its center, (2) the table pushes up with a force (dashed arrow) of 20 lb *on the basket* and the basket exerts a downward force of 20 lb *on the table.* The basket is in equilibrium under the action of the upward force exerted by the table on the basket and the downward force exerted on the earth on the basket, but these two forces are *not* action and reaction.

We shall consider Newton's third law further in Chapter 7 when we discuss momentum.

Fig. 5-10. *A rocket launching. The upward "action" thrust on the rocket equals the downward "reaction" thrust on the jet. (Courtesy General Dynamics.)*

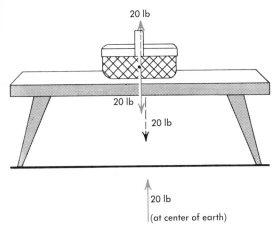

20 lb

20 lb

20 lb

20 lb
(at center of earth)

Fig. 5-11. Action-reaction pairs. Two pairs of action-reaction forces are shown, one pair by solid arrows, one by dashed arrows. The picnic basket is in equilibrium under the action of two forces: the up push of the table and the down pull of the weight, which are not an action-reaction pair.

Summary

Newton's laws of motion are:

1. Every body persists in its state of rest or of uniform motion in a straight line unless it is acted upon by some external, unbalanced force.

2. The external, unbalanced force required to accelerate a body is proportional to the product of the mass of the body and the acceleration that is produced.

$$F = ma.$$

3. When one body exerts a force on another, the second body exerts on the first body a force equal in magnitude but opposite in direction; action and reaction are equal and opposite. They act on different bodies.

The newton is that force which gives a body of 1-kg mass an acceleration of 1 m/sec².

One standard kilogram-weight gives a body of 1-kg mass an acceleration of 9.8 m/sec² (1 g); 9.8 newtons = 1 kgwt.

The dyne is that force that gives a body of 1-gram mass an acceleration of 1 cm/sec². 980 dynes = 1 gwt.

The weight w of a body is the pull exerted on it by the earth and varies from one position to another. Its mass, $w/g,$ is constant.

Questions

1. State Newton's laws of motion and give illustrations of each.

2. Distinguish between mass and weight.

3. Estimate your mass in kilograms and your weight in newtons.

4. How can you compare the masses of two boxes without weighing them?

5. How is the mass of a body measured?

6. Can Newton's first law be proved? Discuss.

7. Where does a 100-pound mass weigh 100 lb?

8. What is the best way to remove a magazine from the bottom of a pile of magazines without handling those above?

9. The pull of gravity on a 2-pound ball is twice as great as that on a 1-pound ball. Why are they accelerated equally when falling freely?

10. Why cannot a man in a canoe make it move forward by pushing against the bow?

11. A 100-kg man stands on a spring scale in an elevator. What would be the reading of the scale (*a*) if the supporting cables should break and the elevator cage should fall freely; (*b*) if the elevator were accelerated upward 9.8 m/sec²?

12. Two bodies of equal mass rest on the pans of a trip scale. Will the scale remain balanced if it is accelerated upward in an elevator?

13. Why is it difficult to play "catch" —throw a ball back and forth—on smooth ice?

14. If you fell freely while holding a 5-kg walkie-talkie, how much would it appear to you to weigh?

15. Sketch a spring balance with a circular

scale that will read both newtons and kilo-gram-weights. Mark the scale divisions appropriately.

Problems

A

1. A boy and sled weigh 30 kgwt and are at rest on a level sidewalk. If a horizontal force of 20 kgwt acts on them, what acceleration results (a) when there is no friction, and (b) when the coefficient of kinetic friction between sled and ground is 0.050?

2. A 1600-kg automobile can accelerate from 1.5 m/sec to 9.0 m/sec in 8.0 sec. What accelerating force acts on it?

3. How great an acceleration will a resultant force of 15 kgwt give to a 50-kg package on a frictionless surface?

4. What is the acceleration produced when a resultant force of 8.0 newtons is applied to a mass of 10 kg?

5. A 70-kg man stands close to the rear wall of a bus. If the bus suddenly accelerates 0.98 m/sec^2, what force does the wall exert against him?

6. A rocket has a gross weight at launching of 10,000 kgwt and a thrust of 32,000 kgwt. Calculate the initial acceleration of the rocket when launched vertically upward.

7. A pushes northward on a 10-kg water-polo ball, B pushes eastward, and C south-ward. If each person exerts a force of 5.0 kgwt, what is the acceleration of the ball? Assume that friction is negligible.

8. A force of 0.10 kgwt acts on a body for 5.0 sec; the body travels horizontally 9.8 m starting from rest. What is its mass?

9. If an 80-kg passenger in a car decel-erates from 15 m/sec to 5.0 m/sec in 10 sec, with what force will he push on the seat belt?

10. (a) What horizontal force is required to pull a 500-gm block at uniform velocity along a horizontal table top if the coefficient of kinetic friction is 0.10? (b) What would be the required force to accelerate the block 98 cm/sec^2 across the same table?

B

11. Two 10.0-kg bodies A and B are con-nected by a cord passing over a frictionless pulley. If 0.50 kg is shifted from B to A, (a) what acceleration results, (b) what is the ten-sion in the cord?

12. A 4.0-kg ball and a 6.0-kg ball are at-tached to the ends of a cord passing over a frictionless pulley. (a) What is the accelera-tion of the system? (b) What force does the cord exert on the 6.0-kg ball?

13. A 30-gm rifle bullet acquires a velocity of 400 m/sec in traversing a rifle barrel 60 cm long. Find (a) the acceleration, and (b) the accelerating force in dynes, newtons, and kilogram-weights.

14. A 1600-kg automobile is towed on a level street by means of a rope that would break if the pull were greater than 100 kgwt. If the friction force is 20 kgwt, how great an acceleration can be given to the car without breaking the rope?

15. Starting from rest, an aircraft weighing 9000 kgwt travels 1200 m on the runway before reaching its takeoff speed of 100 m/sec. (a) Calculate the magnitude of the acceleration, assuming it is constant. (b) If the jet engine provides 7000 kgwt of thrust, cal-culate the total resistance force (a combina-tion of wheel friction and aerodynamic drag).

16. A string that will be broken by a force of 30 kgwt is tied to a block weighing 16 kgwt, resting on a level surface. What maximum acceleration can the string cause (a) if the coefficient of kinetic friction is 0, (b) if it is 0.20?

17. A 1.00-kg ball falls through the air with a vertical acceleration of 7.8 m/sec^2. What frictional force acts on it?

18. A man weighing 90 kgwt stands in an elevator. What force does the floor exert on him (a) when the elevator rises at constant speed, (b) when it accelerates upward 4.9 m/sec^2, (c) when it falls freely?

19. A hollow sphere with a volume of 300 cm^3 and a mass of 98 gm is taken 60 cm

below the surface of a lake and released. What is the acceleration of the sphere at the instant of release? Neglect friction.

20. A player kicks a 1.0-kg football and gives it a velocity of 12 m/sec. The time during which the force acts is 0.10 sec. Find (a) the average acceleration, (b) the average force exerted on the ball, (c) the average force exerted on the player's foot.

21. A 5-cent piece, whose mass is 5.00 gm, slides along a table top. The initial velocity is 100 cm/sec. The coin travels 60 cm before stopping, being uniformly decelerated. What are (a) the deceleration, (b) the frictional force, (c) the coefficient of kinetic friction?

22. A 1600-kg automobile slows down from a velocity of 10 m/sec to one of 5.0 m/sec in 4.0 sec. What is the force of friction (a) if the road is level, (b) if it rises 1.0 m per 100 m?

23. A 200-gm ball, traveling 30 m/sec, strikes a catcher's mitt and is stopped in 0.050 sec. (a) Find the deceleration. (b) What force does the ball exert on the mitt?

24. An aviator and a parachute together have a mass of 150 kg. As the aviator falls through the air, at a certain instant his downward acceleration is 2.45 m/sec². What frictional force acts on the system?

25. How far would a stone (specific gravity 2.5), starting from rest, sink in sea water (specific gravity 1.03) in two seconds? Neglect the retardation due to fluid friction.

26. A body slides down a long board that makes an angle of 60° with the horizontal. The coefficient of kinetic friction is 0.20. If the body starts from rest, how much time is required for it to attain a velocity of 15 m/sec?

C

27. In Fig. 5-4, what is the downward acceleration of A if the coefficient of friction between table top and block pile B is (a) 0, (b) 0.15, (c) 0.30?

28. Two blocks, A and B, each weighing 200 gwt, rest on a horizontal surface. The coefficient of kinetic friction between blocks

and surface is 0.20. If A is attached to B by means of a cord and B is attached to a third, equal, block C by a suspension cord passing over a pulley, find (a) the acceleration and (b) the tension in each cord.

29. A normal force of 80 kgwt pushes a 30-kg block against a vertical wall. What vertical force must act on the block to accelerate it upward 2.45 m/sec² (a) if the coefficient of kinetic friction between wall and block is 0, (b) if it is 0.20?

30. A rope would be broken by a force of 30 kgwt. What is the smallest acceleration with which a boy weighing 40 kgwt can slide down the rope without breaking it?

31. An 80-kg man stands in a bus, which slows down from 10.0 m/sec to 2.0 m/sec in 1.00 sec. At what angle with the vertical must he lean in order not to fall?

32. A force of 2.0 newtons, in addition to gravity, acts on a body of mass 0.098 kg. What is the acceleration of the body if the force is directed (a) vertically upward, (b) vertically downward, and (c) horizontally? (d) In (c) what will its velocity be 3.3 sec after the body begins to move?

33. An automobile is going 15 m/sec on a level pavement. If the brakes are applied suddenly, locking the wheels, in what distance

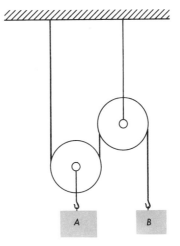

Fig. 5-12.

will the car come to rest? The coefficient of kinetic friction between wheels and pavement is 0.625.

34. What are some of the conditions necessary to land a 2000-kg plane on the deck of a carrier 150 m long without having the deceleration exceed $4g$?

35. Find the tension in the cord and the acceleration of each block (Fig. 5-12) when B is released (a) if A is 6.0 kgwt and B is 3.0 kgwt and (b) if A is 6.0 kgwt and B is 5.0 kgwt. Assume that friction is negligible.

36. Two filled boxes, A made of cardboard and B of wood, are placed touching each other at the top of a 30° inclined plane. Each box has a mass of 5.0 kg, and A is above B on the plane. The coefficient of friction between A and the plane is 0.10 and that between B and the plane is 0.20. When the boxes are released from rest, (a) what force does A exert on B, and (b) what are their accelerations?

37. Two blocks having masses of 200 gm and 300 gm are attached to a rope passing over a light, frictionless pulley fastened to the ceiling of an elevator. The elevator is accelerated upward 80 cm/sec². How much rope will move through the pulley during the first second after the blocks are released?

For Further Reading

Andrade, E. N. daC., *Sir Isaac Newton, His Life and Work,* Science Study Series, Doubleday and Company, Garden City, 1964.

Knedler, John W., Jr., ed., *Masterworks of Science* (excerpts from Newton's *Principia*), Doubleday and Company, Garden City, 1947.

Rogers, Terence A., "Physiological Effects of Acceleration," *Scientific American,* February, 1962, pp. 60–70.

Sciama, Dennis, "Inertia," *Scientific American,* February, 1957, p. 99.

CHAPTER SIX

Curvilinear Motion

In the last two chapters you studied the accelerations of bodies traveling along straight lines. Now you will consider bodies that move in curved paths. Familiar examples are the motions of projectiles such as baseballs and footballs, and of automobiles on curves. Before we can understand such motions, however, we must learn how to combine velocities.

Velocity a Vector Quantity

Velocities, like forces, are directed or vector quantities. Suppose that a man could row a boat at a speed of 3 mi/hr in still water, and that a river current would carry him southward with a velocity of 4 mi/hr if he did not row. If he rows the boat on the river, his velocity relative to the river bank is the resultant of the velocity of the river and the velocity relative to the water, produced by the rowing. In other words, the resultant velocity is the velocity relative to an observer standing on the river bank and equals the vector sum of the velocity of the boat relative to the river and the velocity of the river relative to the river bank. If the boat is pointed downstream (south), the resultant velocity will be the *sum* of the velocities of the

two, that is, 7 mi/hr, directed south. If the boat is pointed upstream (north), the resultant will be the *difference* between the two, 1 mi/hr, directed south. Lastly, if the boat is headed eastward, the resultant velocity, 5 mi/hr, may be found by the parallelogram method (Fig. 6-1). An observer standing on the bank would see the boat moving at a velocity of 5 mi/hr in the direction indicated. What would the rower see? He would see the observer moving relative to him with a velocity of 5 mi/hr in the direction opposite to that shown. In a dense fog, the rower could not see the river bank and, unless he knew the velocity of the river, could only conclude from his compass and speed indicator that he was heading east at 3 mi/hr.

An airplane pilot in flying his plane must make allowance for the wind. Suppose that a small plane has an air velocity (velocity of the *plane* relative to the *air*) of 60 mi/hr in still air and that the wind velocity (velocity of the *air* relative to the *ground*) is 30 mi/hr from the south. If the plane is headed east, it will travel along *OC* (Fig. 6-2a) with a resultant velocity (velocity of the *plane* relative to the *ground*) of 67 mi/hr. As a second example, suppose that the wind blows from the south and that

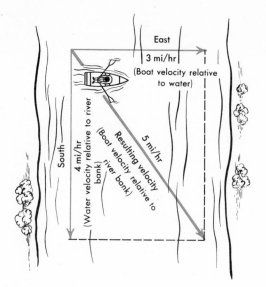

Fig. 6-1. The resultant velocity of the boat (relative to the river bank) is found by the parallelogram method.

the pilot wishes to travel east. Then he must fly his plane in the direction *OA* (Fig. 6-2*b*).

The Motion of a Projectile

In the example given above, the two velocities that were combined were constant. Let us now consider the combination of veloc-

ities, one of which is changing. A projectile furnishes a good example.

A stone thrown horizontally falls vertically just as fast as if it were dropped vertically from rest. You can verify this fact by throwing a stone horizontally from a window and dropping a second one at the same instant. The two stones will strike the level street simultaneously. Suppose that one stone is dropped from an elevation of 19.6 m, and that a second stone is thrown horizontally from the same elevation with a velocity of 5 m/sec (Fig. 6-3). Then each stone will strike the ground after 2 seconds, but the second stone will hit at *B*, 10 m from the wall. If the initial velocity of the second stone is 10 m/sec, it will strike at *C*, 20 m from the building. What initial velocity must the stone have to strike at *D*, 40 m from the wall? Such experiments show that the two motions—uniformly accelerated motion in the vertical direction and motion at constant velocity in the horizontal direction—are *independent*. Neither stone accelerates horizontally, for—if air friction is negligible—no horizontal force acts on it. Each accelerates downward equally; hence they fall equal distances in equal times.

Figure 6-4 represents an airplane, used in delivering materials to an Antarctic expedition, traveling horizontally and at constant

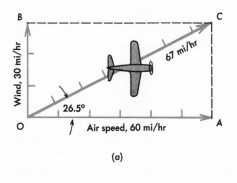

(a)

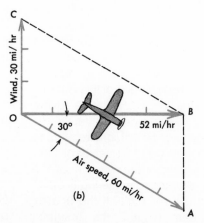

(b)

Fig. 6-2. Resultant velocity of airplane: (a) 67 mi/hr (b) 52 mi/hr.

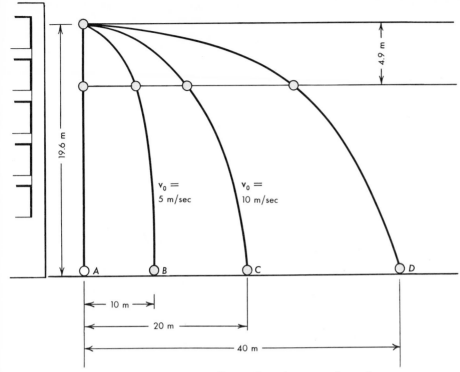

Fig. 6-3. A stone thrown horizontally strikes the ground at the same time as one dropped from rest, but at distances from the building that depend on the initial horizontal velocity v_0.

velocity. Three bales of supplies have been released at different times. Were it not for air friction, the three would be directly below the bay of the plane. Actually, they are retarded and lag behind.

Example. A bale of tent material is released from an airplane traveling horizontally at an elevation of 490 m, at a velocity of 200 m/sec. (*a*) How soon will the bale strike the earth, and (*b*) how far will it advance horizontally while falling? (Neglect air friction.)

(*a*) $s = \frac{1}{2} at^2$

$490 \text{ m} = \frac{1}{2} \times 9.8 \text{ m/sec}^2 \times t^2$

$t = 10 \text{ sec}$

(*b*) Sidewise displacement: $s = v_{av}t = v_o t =$ 200 m/sec $\times$ 10 sec = 2000 m

The velocity of a projectile at various positions on its path can be determined by the parallelogram method. Suppose that a ball is thrown horizontally with a velocity of 10 m/sec. At *B* (Fig. 6-5), the ball has been falling for 1 sec. The vertical component *BE* of its velocity is 9.8 m/sec. The horizontal component of its velocity, *BF*, is the same as the initial velocity (if we neglect friction), namely 10 m/sec. The resultant of these two vectors, *BG*, is about 14 m/sec. You can apply the same method to find the velocity when the ball reaches *C* after falling for 2 sec.

The Range of a Projectile

The path of a projectile, a punted football, for example, can be determined by a method closely resembling that just used for a stone

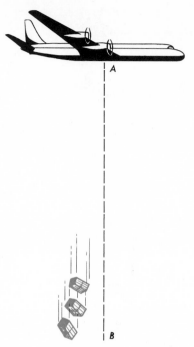

Fig. 6-4. *Supplies falling from an airplane move forward as they drop.*

Let us now try to find a more general solution. Suppose that a projectile is fired with an initial velocity v_0 at an angle θ with the horizontal (Fig. 6-8). The vertical component of the initial velocity is $v_0 \sin \theta$. The magnitude of the velocity in the vertical direction decreases 9.8 m/sec each second during the first half of the flight and then increases 9.8 m/sec each second during the second half. Thus the projectile will lose all its upward velocity and reach its highest elevation at *B* in a time $t = (v_0 \sin \theta)/g$. It will strike the level ground at *D* after a time of flight $T = 2t$, so that

$$T = \frac{2v_0 \sin \theta}{g}$$

The horizontal component of the projectile's velocity, $v_0 \cos \theta$, remains constant throughout its flight. (Why?) Thus the horizontal range *R* represented by *OD* is given by

$$R = \frac{2v_0 \sin \theta}{g} \times v_0 \cos \theta = \frac{v_0^2}{g} (2 \sin \theta \cos \theta)$$

By a theorem of trigonometry, $2 \sin \theta \cos \theta = \sin 2\theta$. Hence

$$R = \frac{v_0^2}{g} \sin 2\theta$$

which is projected horizontally. Suppose that the football traveled initially along the direction *AE* (Fig. 6-6) with a velocity of 21 m/sec. If the earth did not attract the ball, it would travel with constant velocity, according to Newton's first law, along the line *AE*. Actually, the attraction of the earth pulls it down 4.9 m during the first second, 19.6 m during the first two seconds, etc. If air friction is neglected and the projectile travels near the surface of the earth so that the acceleration due to gravity is constant, the path is a parabola as represented in the figure. The horizontal distances traveled in successive seconds are equal. The *range* of the football, the distance traveled horizontally during its flight, is 45 m. The path can be illustrated by a stream of water from a garden hose (Fig. 6-7). Notice that the water jet deviates from the symmetrical parabolic path on which it would travel if the air friction were zero.

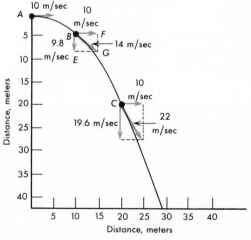

Fig. 6-5. *Determining the velocity of a projectile during its flight.*

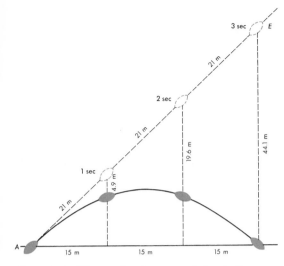

Fig. 6-6. *The path of a punted football.*

Fig. 6-7. *The nearly parabolic path of a water jet.*

Since the greatest value the sine of an angle can have is 1 when the angle is 90°, we notice that *the range is greatest when $2\theta = 90°$ and $\theta = 45°$*.

Example. A projectile is fired with a velocity $v_0 = 28.3$ m/sec at an angle $\theta = 45°$ with the horizontal (Fig. 6-8). (*a*) What maximum elevation does it reach? (*b*) What is its total time of flight? (*c*) What is its range?

The initial components of velocity in the vertical and horizontal directions respectively are

$$v_{0y} = v_0 \sin \theta = (28.3 \text{ m/sec}) \sin 45°$$
$$= (28.3 \text{ m/sec})\, 0.707 = 20 \text{ m/sec}$$
$$v_{0x} = v_0 \cos \theta = 20 \text{ m/sec}$$

The velocity in the horizontal direction (if friction is negligible) does not change, and $v_x = v_{0x} = 20$ m/sec.

(*a*) To find the maximum elevation reached, h, we first compute the time for the projectile to lose all its upward velocity.

$$t = \frac{v_{0y}}{g} = \frac{20 \text{ m/sec}}{9.8 \text{ m/sec}^2} = 2.04 \text{ sec}$$

Since it takes the projectile just as long to return to the ground from height h as a ball

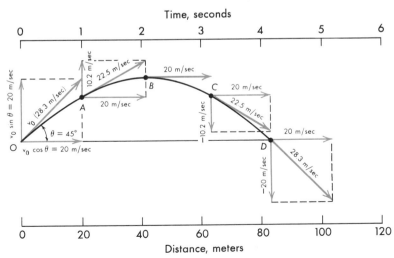

Fig. 6-8. *The range of a projectile.*

dropped from that elevation, we have

$$h = \tfrac{1}{2}gt^2$$
$$= \tfrac{1}{2}(9.8 \text{ m/sec}^2)(2.04 \text{ sec})^2 = 20.4 \text{ m}$$

(*b*) The time of flight *T* is twice the time required to reach the maximum elevation.

$$T = 2t = 4.08 \text{ sec}$$

(*c*) The range is the horizontal velocity times the time of flight.

$$R = Tv_0 \cos \theta$$
$$= 4.08 \text{ sec} \times 20 \text{ m/sec} = 81.6 \text{ m}$$

Centripetal Forces

When your car goes around a curve, you tend to move forward in a straight line, in accord with Newton's first law. If no sidewise force acted on you, you would fall out of the car. In fact, you may hold onto the seat or press up against the side of the car. The seat stretches a little or the side wall bulges slightly, setting up an elastic force which acts on you and forces you sideways toward the center of the circle on which you move (Fig. 6-9). When a body moves thus on a circle, with a velocity whose magnitude is constant but whose direction is changing, the sideways force acting toward the center of the circle is called a *centripetal* force (Latin *centrum,* the center; *petere,* to seek).

Remember that a body moving on a curved path is accelerated even if the magnitude of its velocity is constant. This acceleration is due to the change in *direction* of the motion. A car which rounds a curve at a constant speed of 35 mi/hr (Fig. 6-9) has a constantly

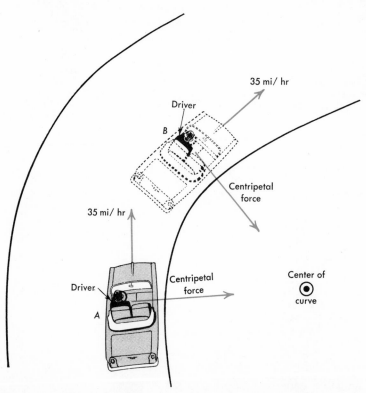

Fig. 6-9. *The side of the car pushes the driver around the curve. The magnitude of the velocity is constant, but the direction is changing.*

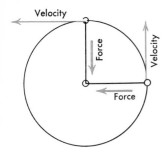

Fig. 6-10. The centripetal force accelerates the stone toward the center of the circle. The speed of the stone is constant, but its velocity varies.

changing velocity, since at one instant the car is heading north, a little later northeast, and so on. According to Newton's second law, a resultant force must act on the car to accelerate it. The centripetal force which produces this acceleration on an unbanked road is due to friction between the tires and the road. If the car hits a patch of ice on the highway, the frictional force may not be large enough to produce the necessary acceleration and the car will skid off the road. According to Newton's first law, the skidding car will move forward in a direction tangent to the curve, *not* outward at right angles to it. (Why?)

When you whirl a stone on a circle by means of a cord, the cord constantly pulls on the stone, exerting a centripetal force (Fig. 6-10). An athlete whirls a 16-pound ball in a circle. The centripetal force is the resultant of the weight of the ball and the tension in the wire (Fig. 6-11). When he releases the wire, the ball flies off tangentially. Some washing machines have spinners to remove water from the clothing. The wet clothes are placed in a vertical cylinder with perforated walls. As the cylinder rotates, the water, clinging to the cloth by adhesion, is whirled around on a circular path. When the speed increases, the pull of the adhesive force toward the center is not great enough to cause circular motion, and the water flies off on a tangent.

In a *cream separator,* used by dairymen, the rotation of whole milk in a cylindrical bowl at several hundred revolutions per minute causes the heavier constituent, the skim milk, to accumulate at the sides while the less dense cream gathers near the center. The bowl revolves so fast that the forces separating the cream from the milk are much greater than the weight of the liquid, and the separation is accomplished in a few seconds. The cream is tapped off by a pipe leading to the center of the bowl and the skim milk by a pipe leading to the edge of the bowl.

High-speed *ultracentrifuges* (Fig. 6-12) are used in physiological laboratories and biophysical laboratories to separate molecules of different molecular weights in blood and other biological materials. Some ultracentrifuges can exert centripetal forces that are a million times the weight of the objects they accelerate.

The Centripetal Force Equation

When a body of mass m moves with a velocity v on a curve of radius r, its acceleration is

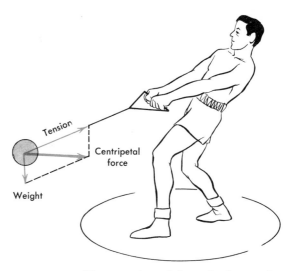

Fig. 6-11. The centripetal force is the resultant of the tension in the wire and the weight of the ball.

Fig. 6-12. An ultracentrifuge. Materials to be studied are placed in the rotor (lower center of picture) and spun in vacuum at constant speeds up to 70,000 rev/min. The ultracentrifuge can determine the molecular weights of virus particles. (Courtesy of Specialized Instruments Corporation.)

directed toward the center and equals v^2/r. Hence by Newton's second law, the centripetal force F is given by

$$F = ma = \frac{mv^2}{r}$$

To understand why this is so, consider a particle traveling with constant speed along a circular path (Fig. 6-13*a*). Its velocity at *G* is represented by the arrow or vector *v*. The velocity of the particle when it reaches *H* is represented by another arrow v_2, equal in length to *v* but differing from it in direction. Let us now displace the vectors *v* and v_2 parallel to their original directions (a permissible operation on vectors) and form a parallelogram *IJLK* of which *v* is one side, v_2 the diagonal, and *v'* another side (Fig. 6-13*b*). By the parallelogram law, v_2 is the resultant of *v* and *v'*. That is, *v* combined with *v'* produces v_2. Hence *v'* is the *vector difference* between v_2 and *v*. It represents the *change of velocity* experienced by the particle while traveling from *G* to *H*.

In Fig. 6-13*a*, the distance traveled along the arc from *G* to *H* is equal to the speed *v* of the particle times the interval *t*. Notice that the arc distance is slightly greater than the chord distance $\overline{GH}$. If the time interval and the angle θ were half as great, the two distances would be more nearly equal. Assume that *t* and θ become extremely small, approaching zero. Then, in the limit as *t* and θ approach

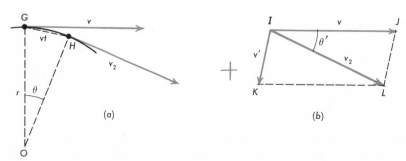

Fig. 6-13. Derivation of $F = mv^2/r$.

zero, the arc distance and the chord distance are exactly equal. That is,

$$\overline{GH} = vt$$

The triangles OGH and IJL are similar; hence their corresponding sides are proportional.

$$\frac{JL}{IJ} = \frac{GH}{OG}$$

or

$$\frac{v'}{v} = \frac{vt}{r}$$

Transposing,

$$\frac{v'}{t} = \frac{v^2}{r}$$

But v'/t is the *change of velocity per unit time,* that is, the average acceleration a. Hence

$$a = \frac{v^2}{r} \qquad (1)$$

The time interval t may be as small as we wish, so that the arc distance GH becomes more and more nearly equal to the straight-line distance GH. Therefore, equation (1) gives not merely the *average* acceleration but also the *exact* acceleration at any instant. Note further that the *direction* of a is the same as the direction of v' which is toward the *center of the circle*. The centripetal force and the acceleration that it produces—according to Newton's second law—must have the same direction.

Therefore, if the mass of the particle is m, the centripetal force F, urging it toward the center, is given by

$$F = ma = \frac{mv^2}{r} \qquad (2)$$

Example 1. An 80-kg man rides in a car which makes a sudden turn. He moves along a curve of radius 40 m with a speed of 20 m/sec. What centripetal force acts on him?

$$F = \frac{mv^2}{r} = \frac{80 \text{ kg}(20 \text{ m/sec})^2}{40 \text{ m}}$$

$$= 800 \text{ kg-m/sec}^2 = 800 \text{ newtons}$$

Example 2. An airplane travels with a speed of 200 m/sec. What is the radius of the curve along which the airplane moves if the centripetal force acting on an 80-kg pilot is 100 kgwt?

$$F = 100 \text{ kgwt} = 980 \text{ newtons} = 980 \text{ kg-m/sec}^2$$

$$980 \text{ kg-m/sec}^2 = \frac{80 \text{ kg } (200 \text{ m/sec})^2}{r}$$

$$r = 3260 \text{ m}$$

If the number of revolutions per second at which a rotor is turning is n, then the velocity v of a particle at a distance r from the center of rotation is

$$v = \frac{s}{t} = (2\pi r)\, n$$

and the centripetal force acting on a particle of mass m at this distance r from the center is

$$F = m\frac{v^2}{r} = m\frac{(2\pi rn)^2}{r} = 4\pi^2 mrn^2$$

Further Examples of Centripetal Forces

A bicycle rider, going around a curve, leans inward to provide the centripetal force required to push her sideways toward the center of the circle on which she moves (Fig. 6-14a). The forces acting on the rider are represented in Fig. 6-14b, in which AB represents her weight, AC the push of the bicycle, and AD the resultant of these two forces. This resultant is the horizontal, centripetal force which pushes the rider toward the center of the curve. The sharper the turn and the greater the speed, the greater must be the angle of tilt of the bicycle. When the cyclist rounds a curve, the wheels are likely to skid. To prevent this, the roadway should be "banked" as shown in the figure. Railway tracks are usually tilted at curves, the outer

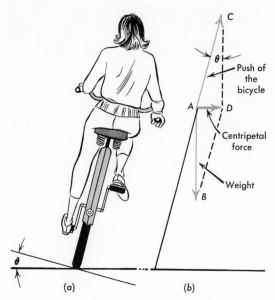

overcome gravity is directed upward, the centripetal force sidewise, and the total required force may be found by the parallelogram method. The most interesting position is at *C*, where gravity supplies part of the centripetal force and the push of the seat on the pilot is the **difference** between his weight and the required centripetal force. $R + w = mv^2/r$. At a certain speed on a loop of a given radius, he may be momentarily in the "weightless state" in which this difference is zero.

In diving, the pilot sometimes directs his airplane nearly vertically downward and attains a velocity of perhaps 500 mi/hr or more. When he pulls out of the dive, the centripetal force that the plane exerts on him may exceed eight times his weight. The walls of the arteries are unable to force blood to his brain and he may lose consciousness. To minimize the danger, he leans forward or lies horizontal, so that his head may be as low as possible. Pilots of supersonic planes often wear anti-*g* suits, rubber suits containing air at high pressure, that help to prevent the blood from draining from the brain.

Centripetal forces act on artificial satellites in their orbits. Gravitation supplies the centripetal force that keeps a satellite in its orbit. We must discuss the law of universal gravitation (Chapter 11) before we can take up the

Fig. 6-14. (a) *The bicyclist leans sidewise to provide the necessary centripetal force to accelerate herself toward the center of the circle on which she moves.* (b) *Two forces act on the rider. Their resultant,* **AD**, *accelerates her sideways.*

rail being higher than the inner one; automobile highways are banked. Unfortunately, the best angle of banking depends upon the speed. A highway curve which is safe for cars traveling 40 mi/hr may not be banked steeply enough for those traveling 60 mi/hr. In Fig. 6-14*b*, the angle of banking $\theta = ACD$ and

$$\tan \theta = \frac{\text{Centripetal force}}{\text{Weight}} = \frac{mv^2/r}{mg}$$

Centripetal forces are more important in airplanes than in automobiles. Pilots must be careful not to change direction suddenly, lest damage be done to the airplanes themselves or to the passengers. When an aviator "loops the loop," the force *R* exerted on him by the seat must overcome gravity and also supply the needed centripetal force. At *A*, Fig. 6-15, the required upward external force on him is the sum of the two. $R - w = mv^2/r$ or $R = mv^2/r + w$. At *B*, the force required to

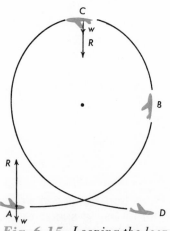

Fig. 6-15. Looping the loop.

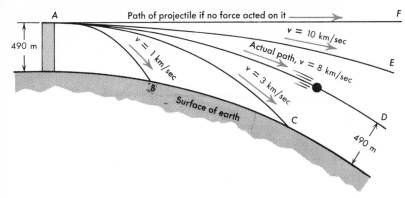

Fig. 6-16. In the absence of air friction, a ball thrown horizontally may travel off into space, descend to earth, or move in a circular path. It is always accelerated.

motion of satellites in any detail, but let us here note that the basic ideas of the present chapter apply to satellites as well as to automobiles. Suppose, for example, that a ball is dropped from the top of a tower 490 m high. If air friction is negligible, the ball reaches the ground after 10 sec. Next, suppose that it is projected horizontally, striking the earth at *B* (Fig. 6-16). If in a third trial the initial horizontal velocity is greater, the ball will strike at some more distant point *C*. If the ball were shot with a speed of 10 km/sec, it would travel along *AE* and escape from the earth. Finally, if the horizontal velocity were about 8 km/sec, the ball, accelerated as before, would be pulled down toward the center of the earth by its weight just enough so that the path would be circular and *parallel to the surface of the earth.* In other words, at *D*, the ball, though continually accelerated, would always move parallel to the earth's surface. If air friction were negligible, the ball would continue to move in a circle as the moon does, and the attraction of the earth would make it fall just enough to keep it at a constant distance from the curved surface. The speed of the ball would not increase, because, at each instant, the force would be directed at right angles to the line of motion. However, air friction on the ball at this velocity and at this height would

not be negligible. It would in fact be so large that the ball would be heated and would burn rapidly. To launch an earth satellite successfully, we must take it well outside the dense atmosphere of the earth before giving it a high velocity — as we shall see in a later chapter.

Centripetal and Centrifugal Forces

The expression "centrifugal force" (Latin *centrum,* center; *fugere,* to flee) is often used, and it is important that we understand the meaning of this term. Remember that when an automobile goes around a curve, an *inward* force pushes *the passenger* toward the center of the circle, and this is the centripetal force. The opposite *outward* force which he exerts *on the side of the car* is the centrifugal force. When you whirl a stone in a circle by means of a string, the string exerts a sidewise, centripetal force on the stone. The stone, having inertia, opposes being accelerated sidewise. It exerts on the string a centrifugal force. The centripetal force and centrifugal force form an action-reaction pair of forces. In accord with Newton's third law, *the centripetal force and the centrifugal force act on different bodies. They alone cannot produce equilibrium.*

Summary

Velocities are vector quantities. They may be combined and resolved like forces.

A ball thrown horizontally and a second ball dropped at the same instant will descend through equal vertical distances in equal times.

A body of mass m traveling on a circle with constant speed is accelerated, for the direction of its velocity v continually changes. The centripetal force F accelerating such a body moving on a circle of radius r is given by $F = mv^2/r$ and acts toward the center of the circle.

The *centripetal* force acting *on* a body and pushing it toward the center of the circle is equal and opposite to the *centrifugal* force exerted *by* the body.

Questions

1. Define centripetal force and give examples of bodies that experience centripetal forces. State the difference between centripetal force and centrifugal force.

2. Upon what does the most desirable angle of banking of a roadway depend? Why are modern roads not banked at this angle?

3. With what initial velocity would you throw a ball vertically upward so that it would just graze the ceiling of a room?

4. When an automobile went around a curve, the force exerted by a rider broke one of the windows. Was this force centripetal or centrifugal?

5. When you whirl a stone on the end of a string, the string pulls on your hand. When you release the string, why does the stone fly off tangentially, rather than radially?

6. How can you throw a stone to hit a pole as you pass it in a car under the following conditions: (a) you throw the stone just as you pass the pole; (b) you throw the stone before you reach the pole. Discuss the relationship of the direction and magnitude of the initial velocity of the stone to other variables.

7. If a bucket of water is swung in a vertical circle of radius r, what is the minimum speed v so that the water will not spill?

8. A train moves with a constant velocity of 30 m/sec. A picture, hanging against the front wall of a car, 2.45 m above the floor, falls from its support. Where will it strike the floor? What is its path with respect to the ground?

9. Why does the moon not "fall" to earth? Why does a satellite not fall?

10. If a bullet is fired horizontally, how far does it fall in the first second? Describe the motion of a bullet fired horizontally at 8 km/sec from the top of Mount Everest if air friction were negligible.

11. How does the earth's revolution affect the apparent weight of a body at the equator? At the poles? Which presses down harder on the track, an eastbound train or a westbound train traveling at the same speed over the track?

12. Suppose that you threw a stone directly at a bird perched on a telephone wire and that the bird began to fall freely at the instant the stone left your hand. Would the stone hit the bird? Discuss.

13. How should one lean to maintain his balance (a) in a bus that is stopping, (b) in a bus that is going around a curve? Discuss the forces acting.

14. Make sketches showing the paths of a golf ball with an initial velocity of 30 m/sec if it is directed (a) horizontally from the top of a cliff, (b) at an angle of 30° above the horizontal over level ground.

15. Sketch the direction of the velocity and of the acceleration (a) at the instant of release of a ball thrown upward vertically, (b) at the instant of release of a ball thrown upward at a 45° angle, (c) of a ball whirled on a string with uniform circular motion.

16. A boat goes upstream from point A to point B in time t. In going downstream from B to A, it takes $\frac{1}{2} t$. Compare the velocity of the

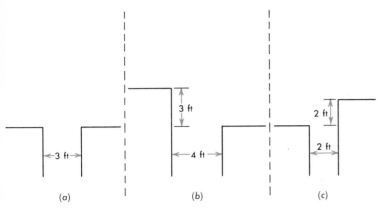

Fig. 6-17. A stunt actor jumps from one building to another.

boat in still water with the velocity of the stream, assuming that the boat maintains the same velocity with reference to the water.

17. A stunt actor in a film has to jump from the top of one building to another. Which of the three jumps—from left to right—in Fig. 6-17 would be the easiest? Discuss.

Problems

A

1. A ferry boat travels northward with a velocity of 2.0 m/sec. A girl runs on the deck with a velocity of 2.0 m/sec. What is her resultant velocity if she runs (*a*) northward, (*b*) southward, (*c*) eastward, (*d*) southwest?

2. An airplane heads north with an air speed (velocity in still air) of 240 km/hr. What is the resultant velocity of the plane in a 120-km/hr wind directed (*a*) from the west, (*b*) from the southwest?

3. A boat can travel at a rate of 1.00 m/sec in still water. (*a*) At what angle must it approach the shore in order to go straight across a river that flows southward with a velocity of 0.50 m/sec? (*b*) With what velocity will it cross the river?

4. A pennant attached to the radio antenna of a car indicates a wind blowing from a direction 30° east of south. The car is travel-

ing due south with a velocity of 20 m/sec. What is the actual wind velocity if it is directed from the east?

5. A rifle is aimed horizontally at the bullseye of a target 300 m away. The velocity of the bullet when it leaves the muzzle of the gun is 300 m/sec. How far below the bullseye will the bullet strike? Neglect air friction and the earth's rotation.

6. A baseball thrown horizontally from a window at an elevation of 44.1 m strikes the level street at a point 30 m from the building. With what velocity was it thrown?

7. A bicyclist goes around a 1.0-km circular banked track with a velocity of 15 m/sec. What should be his inclination to the vertical? Make the conservative assumption that friction is zero.

B

8. An airplane, traveling horizontally with a speed of 170 mi/hr (250 ft/sec) at an elevation of 5000 ft, drops a bale of hay. (*a*) How soon will the hay strike the earth? (*b*) How far will the airplane travel while the hay is falling?

9. A 1.0-kg ball is whirled in a horizontal circle, the radius of which is 0.50 m. With what maximum velocity can it move if the breaking strength of the cord is 4.0 kgwt?

10. An airplane dives toward the earth and

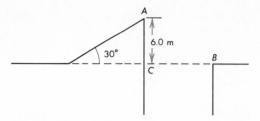

Fig. 6-18.

levels off to a horizontal direction. The radius of the curve on which it moves at a velocity of 300 m/sec is 600 m. What centripetal force acts on the pilot if his weight is 80 kgwt?

11. An automatic washing machine has a load of clothing in a cylindrical drum of 50 cm diameter. During the "spin-dry" operation, the periphery of the cylinder travels at 1000 cm/sec. Find the acceleration of a piece of cloth on the periphery of the cylinder.

12. At what maximum speed can a 1000-kg car travel safely around a curve of radius 50 m if the coefficient of kinetic friction between tires and road surface is 0.50? The road is level.

13. A boy weighing 50 kgwt sits on a horizontal rotating platform at a distance of 1.0 m from the axis of rotation. If the coefficient of static friction is 0.10, (a) what is the greatest sidewise force available to prevent his sliding? (b) At what maximum velocity can he revolve without sliding?

14. A spring gun shoots a 100-gm ball with an initial velocity of 1960 cm/sec at an angle of 30° with the horizon. Find (a) the time of flight, (b) the maximum elevation, (c) the range.

C

15. A 100-gm ball travels northward with a velocity of 400 cm/sec. An eastward force of 2000 dynes acts on it for 15 sec. What is the final velocity of the ball?

16. A baseball is thrown upward at an angle of 30° with the horizon, the initial veloc-

ity being 19.6 m/sec. (a) Find the direction and magnitude of the velocity after 1.00 sec. (b) How long will it take the ball to reach its greatest elevation? (c) How high will it rise, and (d) how far will it travel horizontally before returning to the surface of the earth?

17. If the earth revolved fast enough, the apparent weight of a 68-kg body at the equator would be zero. (a) What would the velocity of the body have to be for this to occur? (Assume $r = 6.4 \times 10^6$ m.) (b) If the earth rotated at this velocity, how long would a day be?

18. A drum majorette whirls her baton at constant speed in a vertical circle of radius 0.60 m. The ball on the end of the baton weighs 0.50 kgwt and its velocity is 3.0 m/sec. Find the force exerted on the ball by the stick (a) when the ball is at its greatest elevation, (b) when the ball is at its lowest elevation.

19. A roadway is 10 m wide. A 1500-kg automobile goes around a curve of radius 300 m with a velocity of 20 m/sec. (a) What centripetal force acts on the car? (b) Approximately how much should the outer edge of the road be elevated above the inner edge if banking alone is to provide the needed centripetal force?

20. Two balls of masses 0.040 kg and 0.080 kg, tied to the ends of a string 1.0 m long, are tossed into the air. They revolve around their center of gravity, revolving $2/\pi$ times per second. What is the tension in the cord?

21. A stone is thrown horizontally from the top of an inclined plane 30 m long and 15 m high with an initial velocity of 9.8 m/sec. How far down the incline will it land?

22. Plot the path of a golf ball if its initial velocity is (a) 30 m/sec directed horizontally from the top of a cliff, (b) 30 m/sec directed at an angle of 30° above the horizontal over level ground.

23. A stunt driver wants to ride a motorcycle across the gap *AB* (Fig. 6-18). If he

leaves *A* at 40 m/sec, how wide a horizontal gap *CB* can he cross?

24. In a botanical experiment, plants were grown on a large rotating platform that turned at a rate of 6 rev/min. At what angle with the vertical would you expect the plants to grow if their distance from the center of rotation was (*a*) 10 ft, (*b*) 20 ft? Discuss the results by reference to the effect of gravity on the growth processes of plants.

For Further Reading

Kirkpatrick, Paul, "Bad Physics in Athletic Measurements," *Scientific American,* April, 1937.

CHAPTER SEVEN

The Conservation of Energy and Momentum

This chapter introduces two of the most important and fundamental principles of physics—the law of conservation of energy and the law of conservation of momentum. We shall see how topics we considered earlier—forces, kinematics, Newton's laws of motion—lead to these powerful generalizations. First we must understand the relationship between *energy* and *work.*

What is Work?

The word work is related to the same Greek word as irksome. It is used in many senses and often means some activity for which pay is received. A golf caddy "works" when he stands idly while the player tries to hit the ball. A watchman "works" when he sits by a railroad crossing. A student "works" his problems in mathematics. In physics, as in all of the exact sciences, it is important that each term have but one meaning, and so we shall define work carefully.

The following are examples of work as the physicist uses the term. When a laborer lifts a concrete block up onto a truck, he does work against gravity. When a tractor drags a plow across a field or a locomotive pulls a train along a level track, work is done against friction. In each of these examples a force acts on a *moving* body. The work done depends upon two factors: the *force* exerted and the *distance* that the body *advances* in the direction of the force.

We define the work done in moving a body as *the product of the force exerted on the body and the distance that the body moves in the direction of the force.*

Work = Force × Distance moved in the same direction as the force

$$W = F \cdot s$$

In the mks system, the *joule* is defined as the work done when a force of one *newton* acts through a distance of one *meter.* Thus a joule equals one newton times one meter or one *newton-meter.* Since one kilogram-weight equals 9.8 newtons, one kilogram-weight-meter of work equals 9.8 newton-meters or 9.8 joules. Another metric unit much used in scientific investigations is the *erg* or *dyne-centimeter.* This unit is about equal to the work done when a mosquito crawls one centimeter up a vertical wall. Ten million ergs equal one joule. The *foot-pound* (ft-lb) equals the work done when a force of one *pound-weight* acts through a distance of one *foot.* Although the units of work just mentioned are the more common ones, note that work can be properly expressed in *any* force-times-distance units.

When the force and the motion are not parallel, we use a more general definition of work: work is the product of the distance moved and the component of the force parallel to the motion (or, vice versa, the force and the component of the displacement parallel to the force). If the force makes an angle θ with the displacement (Fig. 7-1), the work done is

$$W = (F \cos \theta)s$$

Work is a scalar quantity. Suppose that a boy pushing a lawn roller over level ground exerts a force of 10 lb parallel to the handle, which is at an angle of 30° with the horizontal. Then the horizontal component of the force, *F* cos θ, is 8.66 lb. In order to push the roller 10 ft, the boy must do 86.6 ft-lb of work.

Example. A laborer carries a 20-kg sack of cement up a slanting ladder 5 m long to the roof of a porch 3 m above the street. How much work does he do in lifting the cement?

The force which he exerts is 20 kgwt directed vertically upward, and the vertical component of the distance moved is 3 m. He must do 60 kgwt-m of work to overcome the earth's attraction. Neither the length of the ladder nor its angle of inclination makes any difference in the final result, namely, the raising of a 20-kg sack of cement 3 m.

A workman holds a 20-kg block 1 m above the ground, neither raising nor lowering it from that height. According to our definition of work, he does *no* work: he exerts a force, but the distance moved is zero, and hence the work done is zero. Why then does the workman get tired if he holds the block for five minutes? The answer lies in the structure of his muscles. In order to exert an upward force,

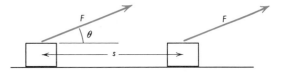

Fig. 7-1. *The work done is* (F *cos* θ)s.

the muscle fibers act in sets. First one set of fibers contracts and takes up the load. After a while, this set relaxes and another contracts and so on. During the contractions, *internal* work is done, and the body becomes heated, even though the man is doing no work on the block.

Power

The term *power,* like work, is popularly used in many ways. Often it means strength or force. In physics, the word is restricted to mean *the time rate of doing work.*

If given sufficient time, a workman can transfer five tons of brick from the street to the roof of a skyscraper. A hoisting engine can do this same amount of work more quickly; therefore its power is greater.

$$\text{Power} = \frac{\text{Work}}{\text{Time}}$$

$$P = \frac{W}{t}$$

Also, since $W = F \cdot s$,

$$P = F \cdot \frac{s}{t} = F \cdot v$$

where *v* is the velocity.

Before James Watt improved the steam engine in the days of King George III, horses were used to pump water in draining coal mines in Great Britain. When Watt tried to sell his engines, mine owners asked how many horses an engine would replace. To answer this question, Watt harnessed strong work horses to a load and found that each of them could work several hours at the average rate of about 550 foot-pounds per second, which became the standard horsepower, the British unit of power.

$$1 \text{ standard horsepower}$$
$$= 550 \text{ ft-pounds per second}$$

In the mks system:

$$1 \text{ watt} = 1 \text{ joule/sec}$$
$$= 1 \text{ newton-meter/sec}$$

Other convenient relations are

$$1 \text{ watt} = 10,200 \text{ gwt-cm/sec}$$
$$1 \text{ kilowatt} = 1000 \text{ watts}$$
$$1 \text{ horsepower} = 0.746 \text{ kilowatt}$$

Example. A man weighing 80 kgwt runs up a stairway, raising himself vertically 5 m in 10 sec. What is his power?

The work done is

$$80 \text{ kgwt} \times 5 \text{ m} = 400 \text{ kgwt-m}$$

The power is

$$400 \text{ kgwt-m/10 sec} = 40 \text{ kgwt-m/sec}$$
$$= 390 \text{ joules/sec}$$
$$= 390 \text{ watts}$$

Energy

The word *work* is one of the oldest in the English language, but the word *energy* is comparatively young. It has been used for about a century. The two are closely related, for the energy of a body or system is defined as *its capacity for doing work. Energy* is derived from the Greek expressions *en* = in, and *ergon* = work. The relationship between work and energy is like that between money and a bank account. You can increase your bank account by depositing money. You can increase the energy of a body by doing work on it. The change of energy is equal to the work done and may be expressed in work units (Table 7-1). Energy, like work, is a scalar quantity.

Table 7-1. Units of Work and Energy

1 joule = 1 newton-m = 10^7 ergs
1 kw-hr = 3,600,000 joules = 3.6×10^6 joules
1 erg = 1 dyne-cm
1 ft-lb = 1.34 joules

A pendulum clock is wound, storing up gravitational energy, and the descending weight does work in turning its wheels. Gasoline supplies chemical energy to propel an automobile. The waters of Niagara yield energy to light the buildings of Buffalo. Radiant energy from the sun makes it possible for us to live.

There are many kinds of energy. In this chapter we shall deal mostly with kinetic energy and with gravitational potential energy.

Kinetic Energy and Work

The energy of a body due to its motion is called *kinetic* (motional) energy. For example, a massive wrecking ball, swung by a crane, can do work in knocking down the wall of a building. The kinetic energy of a speeding automobile enables it to coast up a grade after the power is shut off, or to break off a fence post in a collision. The energy of a moving hammer drives a spike into a plank (Fig. 7-2). *Kinetic energy is energy due to motion.*

Since a moving body does work in stopping, an *equal* amount of work must previously have been done on it to give it kinetic energy. What is the relationship between the kinetic energy of a body and its velocity and mass? When the velocity of an automobile is doubled, the time required for a given force to stop it is doubled. The average velocity during the deceleration is also doubled. Thus the distance traveled is quadrupled. Since the stopping distance is quadrupled, the work done by the given force in stopping the car is four times as great, and it has four times as much kinetic energy. Hence, we might expect

Fig. 7-2. *Kinetic energy is energy of motion.*

the kinetic energy to depend upon the *square* of the velocity, and this turns out to be true. Let us now determine the relationship more exactly.

An automobile pulling a trailer of mass *m* gives it an acceleration *a* from rest (Fig. 7-3). By Newton's second law of motion, the average unbalanced force *F* exerted on the trailer is given by

$$F = ma$$

Let this average force be exerted during a time *t*. The distance *s* through which the force acts is given by

$$s = \tfrac{1}{2}at^2$$

The work *W* done in accelerating the trailer equals the force times the distance through which the force acts. That is,

$$W = F \cdot s = (ma)(\tfrac{1}{2}at^2) = \tfrac{1}{2}m(at)^2$$

But *at* equals the final velocity *v*; hence the work done in accelerating the trailer is

$$F \cdot s = \tfrac{1}{2}mv^2$$

Since the work done in accelerating the

trailer equals* its final kinetic energy E_K,

$$E_K = \tfrac{1}{2}mv^2$$

* *Calculus proof.* We write Newton's second law as

$$F = ma = m\,\frac{dv}{dt}$$

and multiply each side of the equation by *v*, obtaining

$$F \cdot v = mv\,\frac{dv}{dt}$$

or

$$F \cdot v\,dt = mv\,dv$$

Now

$$v\,dt = ds$$

where *ds* is the distance traveled in time *dt*. Therefore,

$$F \cdot ds = mv\,dv$$

Let us now integrate both sides of this equation:

$$\int F \cdot ds = \int mv\,dv$$

$$\int F \cdot ds = \tfrac{1}{2}mv^2 + k$$

But $\int F \cdot ds$ is the work done *W* and is zero when *v* is zero. Therefore *k* equals zero, and we have

$$W = \tfrac{1}{2}mv^2$$

or

$$E_K = \tfrac{1}{2}mv^2$$

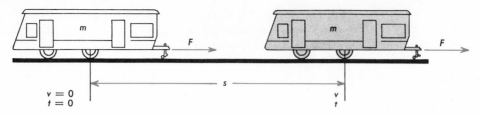

Fig. 7-3. *Work is done in giving kinetic energy to the trailer.*

Example 1. A rifle bullet having a mass of 0.010 kg has a velocity of 1.00 km/sec. What is its kinetic energy?

$E_K = \frac{1}{2}mv^2$
$\qquad = \frac{1}{2}(0.010 \text{ kg}) \times (10^3 \text{ m/sec})^2$
$\qquad = 5 \times 10^3 \text{ (kg-m/sec}^2) \text{ m} = 5000 \text{ newton-m}$
$\qquad = 5000 \text{ joules}$

Example 2. A 140-gm baseball has a velocity of 20 m/sec. How much work does the ball do in stopping?

$$E_K = \frac{1}{2}\, 0.14 \text{ kg} \times (20 \text{ m/sec})^2$$
$$= 28 \text{ newton-m} = 28 \text{ joules}$$

Therefore, the work done in stopping is also 28 joules.

Example 3. An archer pulled back a bowstring 0.80 m and discharged an arrow having a mass of 0.030 kg. The velocity of the arrow when released was 40 m/sec. What average force did the bowstring exert on the arrow?

$$E_K = \frac{1}{2}\, 0.030 \text{ kg} \times (40 \text{ m/sec})^2$$
$$= 24 \text{ newton-m}$$
$$F \cdot s = 24 \text{ newton-m}$$
$$F(0.80 \text{ m}) = 24 \text{ newton-m}$$
$$F = 30 \text{ newtons} = 3.1 \text{ kgwt}$$

Potential Energy and Work

A body can have energy that is *not* due to its motion. Examples are *gravitational potential* energy, such as that of the water in a tank on the roof of a building, and *elastic potential* energy, such as the energy of an archer's bow.

The gravitational potential energy of a body is due to the position of the body and equals the work done *on* the body against gravity in moving the body to that position.

Figure 7-4 represents two bodies, A and B, each of mass *m* and weight *w*, attached to a cord passing over a pulley. Assume that the friction is zero and the cord and pulley have zero mass. Then the body A in descending to level L_1 through a distance *h* would do work in lifting B through an equal distance. The work W done in lifting B would be

$$W = F \cdot s = wh = mgh$$

Since the body A in its initial position could do this amount of work, its gravitational potential energy, relative to level L_1, was initially

$$E_P = wh = mgh$$

and it gave this much energy to B.

The initial potential energy of body A due to the gravitational attraction of the earth depends upon the level from which the vertical height *h* is measured. If A is allowed to descend a greater distance to level L_2, more work is done on B in raising it to a greater height. Hence the potential energy of A, relative to level L_2, is greater than that relative to L_1. In stating the gravitational potential energy of a body, we must specify the level with respect to which the potential energy is measured.

Example. A 1.00-kg book is held 0.50 m above a table top that is 1.00 m above the floor. What is the potential energy of the book, relative to (*a*) the table top and (*b*) the floor?

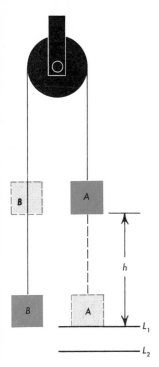

Fig. 7-4. A does work in raising B.

(a) $E_P = mgh$
$= 1.00$ kg$(9.80$ m/sec$^2) 0.50$ m
$= 4.90$ kg-m^2/sec^2
$= 4.90$ joules, relative to
the table top

(b) $E_P = 1.00$ kg$(9.80$ m/sec$^2)1.50$ m
$= 14.7$ joules, relative to the floor

Energy Transformations

We constantly witness transformations of energy. When the electric motor of your car starts the engine, the chemical energy of the storage battery is transformed into the energy of the electric current and this into the mechanical energy that turns the engine shaft. After the car starts, the energy of the gasoline is first turned into heat; part of this is then converted into mechanical energy to push the car along a level highway or to lift it to the top of the hill. In all such changes, energy is trans-

formed from one form into another. In this chapter we shall consider mainly changes from potential energy into kinetic, or vice versa.

Suspend a 2.00-kg ball by a long cord, forming a pendulum (Fig. 7-5). When the ball is at the highest point *A* of its path, 0.50 m above its lowest position *B*, its potential energy relative to its lowest position is 1.00 kgwt-m or 9.8 joules. As the ball swings downward, the potential energy diminishes and the kinetic energy increases. At *B* the potential energy is zero and the kinetic energy is 9.8 joules. When the ball reaches *C*, 0.25 m above the level of *B*, one-half of its energy is potential. At *D*, its total energy is again potential.

If the friction were zero, the ball would continue to vibrate indefinitely. Actually, the work to overcome air friction requires energy, and the vibration gradually dies down. The work done against air friction causes the thermal or heat energy of the air molecules and the molecules of the ball to increase. Thus when the ball comes to rest, its initial energy has been converted into thermal energy.

The Conservation of Energy

Throw a baseball vertically upward (Fig. 7-6). At first all its energy—20 joules—is kinetic.

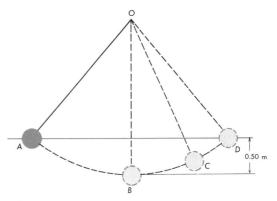

Fig. 7-5. Vibrating pendulum. Its energy is continually transformed from kinetic energy to potential energy and vice versa.

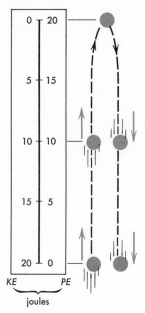

Fig. 7-6. *Conservation of energy. Neglecting loss due to air friction, the energy of the ball remains constant throughout its flight.*

As the ball rises, losing kinetic energy, its potential energy increases. At the highest point of its path, the ball is at rest for an instant, and all its energy — 20 joules — is potential. Then it speeds downward and strikes the earth. If friction is small, measurements of its velocity show that the ball returns to the point from which it was thrown with the same speed and the same energy that it had initially. Moreover, the total energy of 20 joules, kinetic and potential taken together, would be the same at every point on the path.

The kinetic energy which seems to disappear as the ball rises is actually "saved up" or conserved in the increasing potential energy. During the rise, the ball does work against gravity and the energy is stored in the form of potential energy. When the ball reaches its maximum elevation and begins to fall, gravity does work on the ball and the kinetic energy of the falling ball increases at the expense of its potential energy. The ball and the earth,

which act upon each other by a gravitational force, are said to constitute a **conservative system,** and gravity is said to be a conservative force. The total mechanical energy — kinetic plus potential — of a conservative system is constant.

Instead of throwing the ball to its maximum elevation, carry it up a flight of stairs to that elevation. The ball is in equilibrium as you carry it upward, your up force equaling the down force due to gravity. You do work on the ball, increasing its potential energy to 20 joules at the top. When you drop the ball from its maximum elevation, the potential energy is converted into kinetic energy during the fall so that the kinetic energy just before the ball hits the ground is 20 joules. Thus the energy you give the ball is conserved.

Let us consider analytically what happens when the ball falls from any height h_1 at which its velocity is v_1 to any other height h_2 at which its velocity is v_2. The work W done by gravity on the ball during this time is equal to the increase in its kinetic energy:

$$W = mg(h_1 - h_2) = \tfrac{1}{2}mv_2^2 - \tfrac{1}{2}mv_1^2$$

$$mgh_1 - mgh_2 = \tfrac{1}{2}mv_2^2 - \tfrac{1}{2}mv_1^2$$

Or

$$mgh_1 + \tfrac{1}{2}mv_1^2 = mgh_2 + \tfrac{1}{2}mv_2^2$$

Since mgh_1 is the potential energy at one arbitrarily selected position and mgh_2 is the potential energy at a second position we have

$$(E_P)_1 + (E_K)_1 = (E_P)_2 + (E_K)_2 = E$$

where E is the total mechanical energy of the ball. (Strictly speaking, E is the total energy of the *system* consisting of the ball and the earth.) This important result is the law of conservation of mechanical energy: *In a conservative system the sum of the potential energy and the kinetic energy is constant.*

Although we derived the law of conservation of mechanical energy for a falling body, it evidently applies to a rising body as well. Moreover, although we shall not give detailed proofs, we can apply this conservation law to a projectile or any other body moving in two or three dimensions as well as to bodies that rise or fall vertically in one-dimensional motion—as long as the forces are conservative.

Example. A 1000-kg car coasts from rest down a grade which descends 2.5 m in each 100 m. If friction were zero, what would be the velocity of the car after it had gone 100 m?

If friction is negligible, the system is a conservative one, and we can apply the law of conservation of mechanical energy. At the start, $v_1 = 0$; h_1 is 2.5 m if h_2 is taken as zero.

$$\tfrac{1}{2}mv_1{}^2 + mgh_1 = \tfrac{1}{2}mv_2{}^2 + mgh_2$$
$$1000 \text{ kg}(9.8 \text{ m/sec}^2)2.5 \text{ m} = \tfrac{1}{2}(1000 \text{ kg})v_2{}^2$$
$$v_2{}^2 = 49 \text{ m}^2/\text{sec}^2$$
$$v_2 = 7.0 \text{ m/sec}$$

Now let us consider what happens when the forces are *not* conservative as in frictional processes or in most collisions between bodies. For example, when the ball (Fig. 7-6) strikes the ground, it stops, and apparently all the kinetic energy is destroyed. In fact, the impact causes the molecules of the ball and of the ground to vibrate or oscillate more violently. The *orderly* motion of the ball is changed into *disorderly* motion of molecules. Its energy is changed into heat energy. However, careful measurements show that, in such situations, the total energy after the impact is the same as before the impact. Thus, mechanical energy alone is not conserved, but total energy is. Similarly, when a block slides with friction down an inclined plane, work is done against friction and a part of the mechanical energy is converted into heat energy in the block and the plane. The initial mechanical energy of the block at the top of the plane equals not the mechanical energy at the bottom, but rather the mechanical energy at the bottom plus the heat energy produced.

It is the firm belief of scientists, justified by thousands of experiments, that *energy can be transformed but it is never created or destroyed.* This is the principle of the *conservation of energy.* It exemplifies the idea that in physics, at least, we cannot get something for nothing.

The principle of the conservation of energy has been challenged by reputable physicists many times, but is firmly established because no one has ever shown experimentally that energy can be created or destroyed.* Nevertheless, thousands of inventors have wasted time and money in attempts to devise "perpetual-motion" machines that will deliver more energy than is supplied to them. The United States Patent Office protects itself from annoyance by requiring a working model of the perpetual-motion machine before the patent application is considered.

Momentum and Force

A moving body possesses not only kinetic energy, but also *momentum.* The linear momentum $\mathcal{M}$ of a body is defined as the product of its mass and its velocity.

$$\mathcal{M} = mv$$

A 4000-kg truck traveling east at 30 km/hr has momentum equal to that of a 2000-kg truck traveling east at 60 km/hr—120,000 kg-km/hr. Like velocity, linear momentum is a vector quantity and obeys the rules for vector addition. We shall use the term "momentum" for "linear momentum"; in a later chapter, we shall discuss another physical quantity called angular momentum.

What is the relationship between force and momentum? Suppose that an external resultant force F acts during a time Δt upon a body of mass m and causes an acceleration $a = (v - v_0)/\Delta t$. By Newton's second law:

$$F = ma = \frac{m(v - v_0)}{\Delta t} = \frac{(mv - mv_0)}{\Delta t} = \frac{\Delta \mathcal{M}}{\Delta t}$$

* An object's mass and its energy, however, are related to each other, as we shall see when we discuss the special theory of relativity. Mass can be converted into energy, and vice versa, but the total *mass-energy* in such events remains constant, preserving the law of conservation of energy in this modified form.

or

$$F = \frac{\Delta \mathcal{M}}{\Delta t}$$

From this equation it is evident that resultant force equals rate of change of momentum.* Newton's laws of motion may therefore be restated as follows:

1. *The momentum of a body remains constant if no external unbalanced force acts on it.* (If $F = 0$, $\Delta \mathcal{M} = 0$, and the momentum is constant.)
2. *The rate of change of momentum of a body is proportional to the unbalanced, external force acting on it.* ($F = \Delta \mathcal{M}/\Delta t$.)
3. *Whenever one system gains momentum another system loses an equal amount.* (We shall examine this statement more closely later in this chapter.)

The thrust exerted on a rocket by the escaping combustion gases in its exhaust can be computed from the rate of change of momentum. If the constant speed with which the gas escapes is *v*, both the force exerted by the rocket on the exhaust and, by Newton's third law, the force exerted by the exhaust on the rocket, are given by

$$F = \frac{\Delta \mathcal{M}}{\Delta t} = \frac{\Delta (mv)}{\Delta t} = \left(\frac{\Delta m}{\Delta t}\right) v$$

where $\Delta m/\Delta t$ is the mass per unit time expelled from the rocket in the exhaust.

Example. A small rocket burns 6000 kg of fuel in 60 sec. The exhaust velocity is 2000 m/sec. What is the average thrust exerted on the rocket?

---

* *Calculus proof.* We write Newton's second law as

$$F = m \frac{dv}{dt}$$

Making the assumption that *m* is independent of *t* (we shall challenge this assumption in Chapter 10), we have

$$F = \frac{d(mv)}{dt} = \frac{d \mathcal{M}}{dt}$$

$$F = \left(\frac{\Delta m}{\Delta t}\right) v$$

$$= \frac{6000 \text{ kg}}{60 \text{ sec}} \, 2000 \text{ m/sec}$$

$$= 2.0 \times 10^5 \text{ newtons}$$

Immediately after ignition in the launching of a rocket, the thrust builds up until it is greater than the weight of the rocket. At launching, the thrust of a 6.2-million-pound Saturn V rocket is 7.5×10^6 lb. The rocket accelerates upward, under the action of the resultant force, moving relatively slowly through the lower denser layers of the atmosphere and then more rapidly in the upper atmosphere where air friction is less. As the fuel burns and the exhaust gases are expelled, the mass of the rocket decreases so that the thrust is relatively more effective in accelerating the rocket. The greater the ratio of the initial mass of the rocket to its final mass when the fuel has been burned—the mass ratio—and the greater the exhaust velocity, the greater the final velocity of the rocket. Rocket designers seek high mass ratios by designing the rocket to carry a minimum of "excess mass" (shell of the rocket, fuel tanks, guidance equipment, and so on) and a maximum of fuel. They attempt to increase the exhaust velocity by improving the fuel and the design of the combustion chamber and the nozzle through which the exhaust leaves the rocket.

Momentum and Impulse

When a resultant force acts through a distance, we have seen that the product of the force and the distance is called work and equals the change in *kinetic energy* of the body. Let us now consider the product of the resultant force and the *time* during which it acts on the body. This product is called the *impulse.* Suppose that a resultant force *F* acts on a body of mass *m* for a time Δt. Then the impulse is

$$F(\Delta t) = \left(\frac{\Delta \mathscr{M}}{\Delta t}\right)(\Delta t)$$
$$= \Delta \mathscr{M} = mv - mv_0$$

Thus the impulse of a force is equal to the change in *momentum* of the body.*

Example 1. A golfer strikes a 0.20-kg golf ball and increases its velocity from 0 m/sec to 70 m/sec. (*a*) What was the impulse of the blow? (*b*) What change of momentum did it produce? (*c*) What average force did the club exert on the ball if the impulse lasted 0.0012 sec?

(*a*) Impulse $= F \Delta t = \Delta(mv)$
$= 0.20$ kg $\times 70$ m/sec
$= 14$ kg-m/sec
$= 14(\text{kg-m/sec}^2)$ sec
$= 14$ newton-sec

(*b*) $\quad F \Delta t = \Delta \mathscr{M}$
$\Delta \mathscr{M} = 14$ kg-m/sec, the change in momentum

(*c*) $\quad F = \dfrac{14 \text{ newton-sec}}{0.0012 \text{ sec}}$

$= 1.17 \times 10^4$ newtons

Example 2. A 2000-kg car traveling 30 m/sec ran into a concrete wall and stopped in 0.050 sec. What impulse did the wall exert on the car?

$F \Delta t = mv - mv_0$
$= 2000$ kg$(0$ m/sec $- 30$ m/sec$)$
$= -6.0 \times 10^4$ newton-sec

* *Calculus proof.* Starting with Newton's second law expressed in terms of momentum, we have

$$F = \frac{d\mathscr{M}}{dt}$$

or $\qquad\qquad Fdt = d\mathscr{M}$

Integrating both sides,

$$\int F\, dt = \mathscr{M} - \mathscr{M}_0$$

where $\int F\, dt$ is the impulse and $\mathscr{M} - \mathscr{M}_0$ is the change in momentum.

Conservation of Momentum

We have already discussed the law of the conservation of energy. Another equally important law asserts that momentum also is conserved. This means that whenever one body gains momentum, some other body must lose an equal amount. *Momentum is never created or destroyed.* Unless an external resultant force acts upon a system, its momentum is constant.

Suppose that a rapidly moving truck runs into an automobile that is at rest. The forward (action) force acting on the automobile increases its velocity and its momentum. The backward (reaction) force acting on the truck decreases its velocity and momentum. The increase in momentum of the car equals the decrease of momentum of the truck, so that no momentum is lost in the collision. When a golf player strikes a ball or a football player kicks a football, momentum is neither created nor destroyed (Fig. 7-7).

The law that momentum is indestructible is in accord with Newton's third law of motion, which states that action and reaction are equal. To see why this is so, consider a ball of mass m_1, initial velocity u_1, and initial momentum $m_1 u_1$ which strikes another ball of mass m_2, initial velocity u_2, and initial momentum $m_2 u_2$. Let the average action force on m_1 be F_1 and the average opposite reaction force on m_2 be F_2. Each force acts during a time Δt. After the collision, let the velocity of m_1 be v_1 and that of m_2 be v_2.

From Newton's third law,

$$F_1 = -F_2$$

$$\frac{m_1 v_1 - m_1 u_1}{\Delta t} = -\frac{m_2 v_2 - m_2 u_2}{\Delta t}$$

$$m_1 u_1 + m_2 u_2 = m_1 v_1 + m_2 v_2$$

Thus the total momentum before the collision equals the total momentum after the collision. In applying this law, remember that, unlike energy, momentum is a vector quantity.

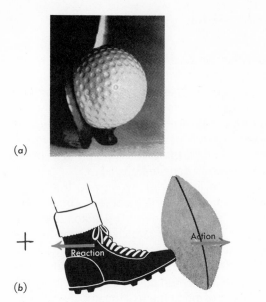

(a)

(b)

Fig. 7-7. (a) *The momentum lost by the golf club is gained by the ball. (Courtesy of Professor H. L. Edgerton.)* (b) *The momentum lost by the foot equals that gained by the ball.*

As an example, suppose that two 1.0-kg balls are attached to strings like pendulums. Let ball *A*, moving 2 m/sec, strike ball *B*, which is at rest. Suppose that both balls are perfectly elastic; no energy is converted into heat energy. Then *A* will lose all its energy and momentum. Just after the impact, *B*'s velocity will be 2 m/sec. The ball will rise to *B'* (Fig. 7-8a). Suppose that both balls are released from the same elevation at the same instant. Let *A* have a positive velocity of 2 m/sec and a momentum +2.0 kg-m/sec when it strikes *B*. The momentum of *B* will be −2.0 kg-m/sec. Just before the impact the sum of the momentums, plus and minus, is zero. After the impact, the two momentums will be reversed. *A* will move to the left with negative momentum and *B* will move to the right. Again their sum will be zero.

As a second example, suppose that a piece of wax is attached to one ball *B* at the point of impact. Let ball *A*, moving 2 m/sec, strike ball

B which is at rest. The two cling together after the impact. In Fig. 7-8b, the two balls will move forward together with a velocity of 1.0 m/sec. That is

Momentum before impact
$$= \text{Momentum after impact}$$
$$(1.0 \text{ kg})(2.0 \text{ m/sec}) = (2.0 \text{ kg})v_2$$
$$v_2 = 1.0 \text{ m/sec}$$

If the two balls are released from the same elevation at the same instant, they will stop when they hit each other and stick together. *A* has positive momentum before impact and *B* negative momentum. Again momentum is conserved. (What happens to the energy?)

The Conservation Laws and Collisions

The laws of conservation of energy and momentum are broadly applicable in physics. We shall use them again and again. When one considers the many changes that a physical system can undergo — even the simpler systems studied in an introductory course in physics — he is all the more impressed with the universality of the conservation laws. In all of these changes, however diverse, the energy of the system and its momentum remain constant if no external forces act upon it. The conservation laws add greatly to our understanding of physical systems and to our ability to predict changes in them.

Collision processes, in which bodies interact with each other and exchange energy and momentum, provide good examples of the laws of conservation of energy and momentum. Let us first consider an elastic collision, one in which not only energy is conserved, but mechanical energy is conserved, that is, no heat is produced in the collision. The collision of two highly elastic steel balls is a good approximation. Suppose that ball *A* of mass m_1 and initial velocity u_1 collides elastically with ball *B* of mass m_2 initially at rest. What are their final velocities v_1 and v_2? We apply the conservation laws separately:

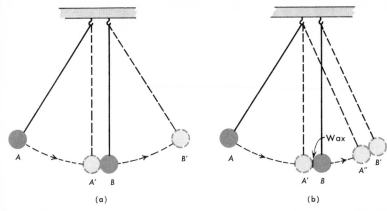

Fig. 7-8. (a) *Ball* A *gives all its momentum and energy to* B. (b) *No momentum is lost, but part of the kinetic energy is changed into heat.*

the initial and final total kinetic energies must be equal and the initial and final total momentums must be equal.
Conservation of energy:

$$\tfrac{1}{2}m_1u_1{}^2 = \tfrac{1}{2}m_1v_1{}^2 + \tfrac{1}{2}m_2v_2{}^2 \quad (1)$$

Initial total   Final total kinetic
kinetic energy   energy

Conservation of momentum:

$$\overrightarrow{m_1u_1} = \overrightarrow{m_1v_1} + \overrightarrow{m_2v_2} \quad (2)$$

Initial   Final total
total momentum   momentum

The arrows over the momentums are to remind us that momentums are vector quantities and that the addition in the second equation is a vector one. If the collision is a "head-on" collision, as we shall assume that it is, the initial and final momentums all lie in the same line and we can drop the vector signs. To solve these equations for the final velocities v_1 and v_2, treating the velocity u_1 and the masses as known quantities, we transpose the first term on the right in equation (2), square both sides of equation (2) and substitute the value of $v_2{}^2$ thus obtained into equation (1). After simplifying, we solve the quadratic equation for v_1. (See Appendix D). The result is

$$v_1 = u_1$$

$$v_1 = \left(\frac{m_1 - m_2}{m_1 + m_2}\right)u_1 \quad (3)$$

The first solution $v_1 = u_1$ corresponds to a "miss"—ball *A* goes on its way without change of velocity. The second solution is the interesting one and corresponds to a head-on hit. If we substitute this value for v_1 into equation (2), we find that v_2, the final velocity of ball *B* is

$$v_2 = \frac{2m_1}{m_1 + m_2}u_1 \quad (4)$$

Suppose that the masses of *A* and *B* are the same. Then equations (3) and (4) predict that $v_1 = 0$ and $v_2 = u_1$. In other words, ball *A* gives all of its velocity to ball *B* and stops in its tracks. Similar procedures would be followed by nuclear physicists in nuclear scattering experiments; for example, protons, the nuclei of hydrogen, are fired at high energy at other protons at rest in a target so that the proton-proton interaction can be studied.

In the example just given, the collision was elastic—kinetic energy was conserved. Let us conclude this chapter by looking at an *inelastic* collision, one in which some of the initial kinetic energy is changed into heat. The ballistic pendulum, used to measure

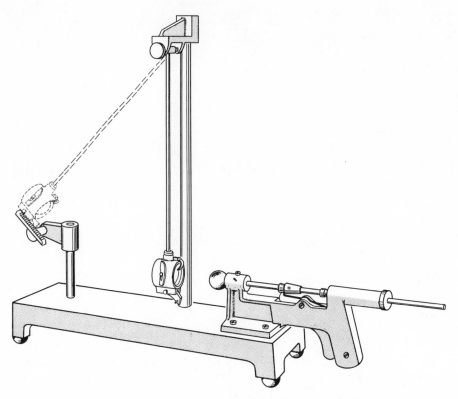

Fig. 7-9. Cenco-Blackwood ballistic pendulum. The ball, discharged by the spring gun, is caught in the bob of the pendulum, which consists of a heavy cup mounted on a light rod. In this impact, do (1) the momentum and (2) the kinetic energy of the system due to forward motion change? The ball and bob swing out and are caught by a ratchet. Thus the distance through which the center of gravity is raised can be measured. (Courtesy Central Scientific Company.)

the velocity of a projectile, provides an example of an inelastic collision. The spring gun in Fig. 7-9 discharges a ball which is trapped in the bob of the pendulum. Let the mass of the ball be m and its velocity v. Just after the impact, let the common velocity of the ball and bob be v_2. The mass of the pendulum bob is M.

No momentum is lost; therefore,

$$mv = (M + m)v_2$$

There are two unknowns, v and v_2; hence we need another equation.

From conservation of energy, the kinetic energy of the ball and bob just *after* the impact must equal the potential energy of the two when they have swung to their highest position. Suppose that their center of gravity rises through a vertical distance h.

$$\tfrac{1}{2}(m + M)v_2^2 = (m + M)gh$$

From these two equations, v may be found. In this impact, the kinetic energy of the bob and ball just after the impact does *not* equal the kinetic energy of the ball before the impact. Some kinetic energy is changed into heat energy in the collision.

Summary

Work is the product of the force exerted on a body and the distance that the body moves in the direction of the force. If the force F makes an angle θ with the displacement s, the work W is $W = Fs \cos \theta$. Commonly used units of work are the joule or newton-meter, the erg or dyne-centimeter, and the foot-pound. One joule = 10,000,000 ergs = 1 newton-meter $\cong \frac{3}{4}$ foot-pound.

Power is work done per unit time. One horsepower = 550 foot-pounds per second. One watt = 1 joule per second.

The energy of a body or system is its capacity for doing work.

The kinetic energy of a body of mass m and velocity v is given by:

$$E_K = \tfrac{1}{2}mv^2$$

The principle of the conservation of energy asserts that energy is neither created nor destroyed.

The momentum $\mathcal{M}$ of a body of mass m and velocity v is given by $\mathcal{M} = mv$. Momentum, unlike kinetic energy, is a vector quantity.

The impulse of a blow is the product of the force and the time that it acts. An impulse may be expressed in dyne-seconds or pound-seconds. Impulse is equal to the change of momentum produced.

Momentum, like energy, is conserved. The momentum of an isolated system is constant. If one body within the system gains momentum, another body must lose momentum.

Questions

1. Define work and state several units in which it is measured.

2. Compare a watt-second and a newton-meter.

3. Which is larger, one horsepower or one kilowatt?

4. Two men pick up a heavy log and carry it along. (a) Are they doing work if the height of the log above the ground does not change?

(b) Why do they get tired?

5. Why does a road wind around a hill instead of going straight up the slope?

6. Prove that pressure times volume has the dimensions of work or energy.

7. How could you measure your power without using specialized laboratory equipment?

8. Define energy and state several units in which it is measured.

9. When a ball is thrown inside a moving bus, does its kinetic energy depend on the velocity of the bus? Explain.

10. When does a baseball pitcher do "work"?

11. A certain ball appears to be made of steel but does not bounce when it is dropped on a heavy steel plate. State some hypothesis to explain this, using good physical reasoning.

12. If a hollow cylinder is half filled with sand and placed on a 30° inclined plane, it has little tendency to roll. Why?

13. If the pendulum (Fig. 7-10) is released, does it hit the wall when it swings back?

14. Define momentum and state some units in which it can be expressed.

15. Distinguish between work and energy and between potential energy and kinetic energy. Give several examples of each.

16. Discuss the law of conservation of momentum and relate it to Newton's third law of motion.

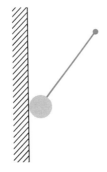

Fig. 7-10.

Fig. 7-11.

17. What is conserved in an elastic impact? In an inelastic one?

18. If you try to walk on a plank that is resting on smooth cylindrical rollers, what will happen? Which laws of conservation are involved?

19. If you were sitting in the middle of a sheet of smooth ice, how could you get off?

20. A block weighing 100 lb hangs at an elevation of 6.0 ft above the top of a table that is 2.0 ft high. What is the potential energy of the block before and after the table is removed? Is the potential energy absolute or relative?

21. A 1-kg clay ball, moving north at a velocity of 10 m/sec, strikes a ball of equal mass moving south with equal velocity. What was the momentum of each before the impact? What was the total momentum?

22. A pendulum bob *A* made of steel swings down through an angle of 30° and strikes head-on a bob *B* of equal mass. (*a*) If *B* is initially at rest, what will happen to *A* and *B* after elastic impact? (*b*) If *B* is coated with putty, what will happen?

23. A bird is in a cage that is supported by a spring balance. Is the reading of the balance when the bird is flying greater than, less than, or the same as, that when the bird rests on the floor of the cage?

24. Two bodies, whose masses are 1.0 kg and 5.0 kg, have equal kinetic energies. Which has the larger momentum?

25. If marbles *A* and *B* (Fig. 7-11) hit a row of seven other marbles at rest on a level, frictionless track, *C* and *D* move off with approximately the same velocity *v* as *A* and *B* had. Why does not marble *D* fly off with 2*v*?

26. Can a horse run up a 45° inclined plane at 60 mph (88 ft/sec) all day long? In walking uphill, can a horse raise himself more than 1 ft vertically every second and keep it up hour after hour?

Problems

A

1. An 80-kgwt man carries a 20-kgwt bag of cement up a ladder to an elevation 5.0 m above the floor. What is (*a*) the total work, (*b*) the useful work?

2. An elevator raised 8 passengers, each weighing 75.0 kgwt, to the top story of a building 300 m high. (*a*) How much useful work was done? (*b*) If the journey required 100 sec, what was the output power of the motor?

3. The output power of an electric motor is 50 kw. At what constant speed can it raise an elevator weighing 1000 kgwt?

4. A 150-gm baseball is thrown vertically upward with an initial velocity of 20 m/sec. (*a*) How much kinetic energy has it? (*b*) From energy considerations, find the elevation to which it will rise.

5. A 1000-kg car traveling 10 m/sec can be stopped by good brakes in 7.0 m. (*a*) What is the corresponding stopping distance at 20 m/sec? (*b*) What is the decelerating force acting on the car?

6. What is the momentum of a 16-kg block if its velocity is 2.0 m/sec?

7. A 2000-gm stone, sliding on an icy pavement with an initial velocity of 0.98 m/sec, is stopped in 4.0 sec. What retarding force acts on it?

8. A 10.0-kg ball is suspended from a cord that is 5.0 m long. What is the potential energy of the ball relative to its lowest position when the cord is at an angle of (*a*) 30°, (*b*) 60°, with the horizontal?

B

9. How much work is done in pumping water from a lake to fill a cubical tank 5.0 m on each edge? The water is pumped in at the bottom of the tank, which is at the level of the

lake. (*Hint:* Through what average distance above the water surface of the lake is the water lifted?)

10. The engine of a small airplane delivers energy to the propeller at a rate of 30 hp. What force is exerted by the propeller if the velocity of the plane is 60 mi/hr (88 ft/sec)?

11. Calculate the power of an 80-kgwt man who, in 10 sec, runs up stairs that have a vertical height of 5.0 m. Express his power in watts.

12. A hammer head, with a mass of 1.20 kg, strikes a spike at a velocity of 80 cm/sec and drives it vertically downward 1.00 cm into a block of wood. What average force does the hammer exert on the spike?

13. A 1600-kg car has a velocity of 30 m/sec. (*a*) What is its kinetic energy? (*b*) From how high a cliff must it fall from rest in order to acquire this kinetic energy?

14. A 30-gm bullet, traveling 500 m/sec, penetrates 10 cm into a massive wooden block. What average force does (*a*) the bullet exert on the block, (*b*) the block exert on the bullet?

15. When a bullet of mass 20 gm goes through a massive board, its velocity decreases from 1000 m/sec to 600 m/sec. (*a*) How much kinetic energy does the bullet lose? (*b*) What force does the board exert on it if the board is 2.5 cm thick?

16. A 10-gm bullet strikes a 990-gm wooden block at rest and is embedded in it. After the impact the velocity of the block is 400 cm/sec. (*a*) What was the velocity of the bullet before the impact? What was the kinetic energy (*b*) of the bullet before the impact, (*c*) of the block and bullet after the impact?

17. In an elastic collision, a 2.0-kg ball moving east at 1.0 m/sec makes a head-on collision with a 1.0-kg ball (*a*) at rest, (*b*) moving west at 2.0 m/sec. Calculate the velocities after collision in each case.

18. How much work is required to upset a cubical block of ice 0.50 m on each side, weighing 115 kgwt, and resting on a level floor? (*Hint:* How much must the center of the block be raised?)

19. Make a graph plotting momentum versus kinetic energy for a 1.0-kg body that falls vertically 122.5 m.

20. A 0.040-kg arrow was discharged from a bow with a velocity of 10 m/sec. (*a*) How much momentum did it have? (*b*) If the accelerating force acted on the arrow for 0.010 sec, what was the average force?

21. A 0.016-kg bullet, traveling northward 600 m/sec, struck a bear weighing 400 kgwt, traveling southward at 3.0 m/sec. Find the momentum of each before the collision. Remember that the momentum of the bear was negative. Did he "stop in his tracks"?

22. A 0.016-kg bullet is discharged from a rifle with a muzzle velocity of 600 m/sec. The gun has a mass of 8.0 kg. (*a*) With what velocity does it recoil? (*b*) What force will stop it in 0.25 sec?

23. An 8.0-kg model pile driver falls 1.0 m in 0.45 sec. (*a*) What is its kinetic energy when it strikes a nail and drives it into a board? (*b*) If the nail penetrates 10 cm, what is the average force exerted?

24. A 1.0-kg block slides down an inclined plane 2.0 m long and 1.5 m high. With what speed will it reach the bottom (*a*) if the friction is negligible, (*b*) if half the energy is expended in frictional losses?

C

25. Freighter *A*, heading east at a velocity of 5.0 m/sec, collides with an equally massive freighter *B*, heading north at a velocity of 10.0 m/sec. If the two wrecks lock together, in what direction and with what velocity will they drift after the collision?

26. A youngster weighing 30 kgwt starts from rest and slides down a banister 3.0 m high and 5.0 m long. His velocity at the bottom is 3.0 m/sec. Find (*a*) the energy loss due to friction and (*b*) the force of friction.

27. A 30.0-gm rifle bullet having an initial horizontal velocity of 1.0 km/sec penetrates a wooden block to a depth of 15 cm. (*a*) What is

the average opposing force? (b) If the barrel of the rifle that fires the bullet is 50 cm long, what is the average accelerating force acting on the bullet in the barrel?

28. A 20.0-gm bullet moving horizontally with a velocity of 800 m/sec embeds itself in a 3980-gm wooden ball suspended by a long cord. (a) What is the velocity of the ball immediately after it is struck? (b) To what elevation does the ball rise?

29. A signal gun of mass 300 kg and mounted on wheels shoots an 8.0-kg projectile at an angle of 30° with the horizontal with a muzzle velocity of 1000 m/sec. With what initial horizontal velocity does the gun recoil?

30. In order to launch a glider, a stationary winch rapidly draws in a light, strong cable attached to the glider. Initially, the glider is stationary on the ground. The mass of the glider is 200 kg. When the glider reaches a speed of 30 m/sec and an altitude of 10 m, the pilot detaches his glider from the cable. During the launching operation, the winch takes in 100 m of cable. (a) Find the potential energy and the kinetic energy of the glider at the end of the launching operation. (b) Find the average tension in the cable. (Neglect air resistance and friction between glider and ground.)

31. Two elastic spheres of equal masses, A and B, rolling on a frictionless horizontal surface, collide at 90°. What will be the path of B, if the path of A after the collision is at 135° from its original direction and at 135° from the original direction of B?

32. A bullet of mass m is fired by a gun of mass M. Find the recoil velocity of the gun in terms of the energy E released in the explosion.

33. A bouncing "Superball," when dropped from a height of 100 cm, returns to a height of 92 cm. (a) Determine the efficiency of the process by calculating what percentage of the kinetic energy is not converted into thermal energy in one bounce. (b) How many bounces will the ball make before coming to rest?

34. Two bodies, A and B, meet head-on in a perfectly elastic collision. A has a mass of 2.0 kg and an initial velocity of 20/3 cm/sec. B has a mass of 1.0 kg and an initial velocity of −10/3 cm/sec. What will their velocities be after impact?

35. A bullet of mass m and speed v is fired horizontally into a block of mass M. If the block rests on a rough horizontal table (coefficient of kinetic friction μ), derive an expression for the distance the block will slide.

36. A 2.00-gm bullet traveling horizontally 1.00 km/sec strikes and embeds itself in a 1000-gm bird perched on a branch 19.6 m above the ground. At what horizontal distance from its initial position on the branch does the bird strike the ground?

37. If, in ancient Egypt, it took 2000 slaves working 5 years to build a pyramid 200 m² at the base, how long would the same number of slaves have to work to build a pyramid having the same shape, but 400 m² at the base? Assume that both pyramids are solid, not hollow.

38. From what height (in stories) in a burning building can a man jump into a net, which is being held by five firemen, if the net is not to hit the pavement when the impact occurs? Estimate the answer, making a reasonable set of assumptions as to the quantities you will need. State the assumptions, as well as the answer. A fireman's net consists of a circular piece of canvas or other material, supported around the edge by a rigid metal ring. The firemen stand around the ring, holding it at about waist height.

For Further Reading

Feinberg, G. and Goldhaber, M., "Conservation Laws," *Scientific American,* October, 1963, pp. 36–45.

Lewis, H. W., "Ballistocardiography," *Scientific American,* February, 1958, p. 89.

Taylor, Lloyd W., *Physics,* Chapter 17, Dover Publications, New York, 1959.

CHAPTER EIGHT

Fluids in Motion

In Chapter 2, we studied fluids at rest. Our method was to consider the simplest physical situation first and then to add complications one by one. We concluded that the liquid pressure in a liquid at rest is given by $p = h \cdot D_w$ and is independent of direction or the shape of the container. Now we shall consider a fluid in motion and investigate the relationship between its pressure and velocity. Fluid friction, too, will be important in this discussion, and we shall consider how it differs from solid-on-solid friction.

Fluid mechanics, the branch of physics that we are about to discuss, has several subdivisions corresponding to the kinds of physical problems one considers. One can investigate the properties of an ideal frictionless fluid or, on the other hand, one can study a viscous fluid in which friction is present. If the density of a fluid changes with pressure, the fluid is said to be compressible. Air and other gases have this property, and *aerodynamics* in its most general form deals with compressible gases. Most liquids are highly incompressible on the other hand—their densities change relatively little with pressure; *hydrodynamics* deals mainly with incompressible fluids.

The principles of fluid mechanics are applied in many branches of science and engineering. To mention a few, we find applications in the design of ship hulls, planes, and rockets; in the study of the circulation of blood and other body liquids in medicine; in the production of oil and its transportation through long pipelines; in meteorology; in the study of flames and detonations; and in efforts to obtain commercial power from nuclear fusion by thermonuclear reactions. Some specific examples will be given later.

Streamlines and Tubes of Flow

Consider water flowing from a large reservoir through a horizontal pipe that has a constriction at one point (Fig. 8-1). We shall simplify our problem by assuming that the water is incompressible and that fluid friction is negligible. We shall also assume that the water is a continuous medium, that is, that it is not necessary for our present purposes to treat the water as an assembly of water molecules. To describe the flow, we want to specify the pressure and the velocity at each point in the liquid. Initially, just after the valve has been opened and the water started through the

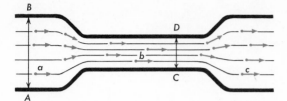

Fig. 8-1. Streamlines and a tube of flow. The streamlines indicate by their direction at each point the direction of the velocity of the flow at that point. A bundle of streamlines, such as those passing through the cross section AB, *constitutes a tube of flow.*

pipe, the velocity at a point such as *a* will fluctuate and will not be constant in time. After the flow has "settled down," however, a velocity-measuring device introduced at *a* will record a constant velocity. We shall consider only *stationary* flow of this kind in this part of our discussion. Notice that we are not saying that the velocity of a little portion of water flowing along the tube is the same at *b* and *c* as it is at *a*; only that the velocity at a given point does not change with time in stationary flow.

The path of a little portion of water that goes from *a* to *b* to *c* is called a *streamline.* Other streamlines are shown above the one passing through *abc.* A bundle of such streamlines is called a *tube of flow.* One such tube of flow contains all the streamlines that pass through the area shown in cross section as *AB.* This tube of flow narrows on entering the constriction so that it passes through the smaller area whose cross section is *CD. Notice that all of the water that flows through AB eventually goes through CD: no more and no less.*

In Fig. 8-1, let the area of the cross section of the pipe at *AB* be A_1 and the velocity of the water at *AB* be v_1. Let the cross-sectional area at *CD* be A_2 and the velocity v_2. In a time Δt, the volume of water passing *AB* is $A_1 v_1 \Delta t$. At *CD* the volume is $A_2 v_2 \Delta t$. The two volumes must be equal. (Why?)

$$A_1 v_1 \, \Delta t = A_2 v_2 \, \Delta t$$

$$A_1 v_1 = A_2 v_2$$

This is the equation of continuity of the (incompressible) flow. It says that the velocity of the liquid at different points in a tube is inversely proportional to the cross-sectional area of the tube of flow. The physical meaning of the equation is that matter is conserved. The number of cubic meters of water flowing through *AB* per unit time ($A_1 v_1$) equals the number of cubic meters flowing through *CD* per unit time ($A_2 v_2$). For example, when a river flows through a wide, level plain, the water travels slowly, but in going through a narrow gorge, its speed increases.

Pressure and Velocity

Let us insert vertical pressure-measuring tubes (Fig. 8-2) into the pipe that we have been considering. The height of rise of the water in the tubes will indicate the pressure at points *a*, *b*, and *c*. If the water rises 10 cm in the tube at *b*, the liquid pressure there is 10 cm-water or 10 gwt/cm². We notice in the drawing—and a simple experiment readily confirms this—that the pressure at *c* is less than at *b* or *d*. Why would we expect the pressure to decrease as the velocity increases?

To answer this question, suppose that a tiny cylinder of liquid is carried forward in the flow (Fig. 8-2). It will be accelerated when it enters the narrow part of the tube because the water velocity is greater there. In order to accelerate the cylinder, the force on its rear end must be greater than the opposing force acting on the front end; hence the pressure at *b* must exceed that at *c*. The cylinder is accelerated because it moves from a region of higher pressure to one of lower pressure. As the cylinder moves from *c* to *d*, it slows down, or is decelerated. Therefore the retarding force pushing against its forward end must be greater than the force tending to accelerate it. Thus the pressure at *d* is greater than that at *c*. The cylinder decelerates because it moves from a region of lower pressure to one of

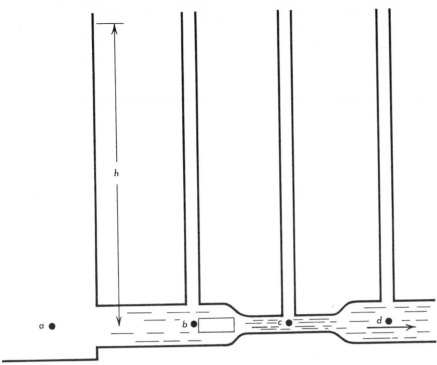

Fig. 8-2. The velocity of the liquid is greater at c *than at* b. *The pressure is greater at* b *than at* c. *At* a *the liquid is at rest.*

higher pressure. This rule is valid when frictional effects are negligibly small.

Whenever the velocity of a horizontally moving stream of fluid increases because of a constriction, the pressure must decrease. Higher velocity is associated with lower pressure, and vice versa.

Gases as well as liquids show this effect. In order to demonstrate decrease of pressure due to increase in velocity, hold one edge of a sheet of paper in front of your lips and blow past it. The pressure in the air stream above the paper will be reduced. The excess pressure below will force the paper upward.

A second example is the ordinary spray gun (Fig. 8-3). When the piston is pushed inward, air is forced past the end of the vertical tube. The velocity of the air in the jet is so great that the reduction in pressure is large. Atmospheric pressure forces liquid upward in

the tube. The liquid enters the jet and is carried forward as spray. The carburetor of an automobile operates in the same manner.

Let us now consider the quantitative relationship between velocity and pressure in a liquid. Recall that Fig. 8-2 represents a liquid flowing in stationary flow in a horizontal tube at zero elevation. Assume that friction is zero.

Let the pressures at b and c be p_1 and p_2. Let the corresponding velocities be v_1 and v_2.

Suppose that a small cylinder of liquid of cross section A and length l moves to the right along its tube of flow through a distance l. The force pushing to the right on the cylinder is p_1A and the force pushing to the left on it is p_2A. Hence, the resultant force accelerating the cylinder is $(p_1 - p_2)A$. The distance through which the force acts is l. The mass m of the cylinder of liquid of mass density D_m is the volume times the mass density or AlD_m.

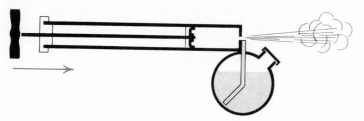

Fig. 8-3. A spray gun. The pressure in the air jet is less than the atmospheric pressure.

According to the work–kinetic energy relation (Chapter 7), the work done in accelerating the cylinder equals the increase in its kinetic energy. Therefore, neglecting frictional losses,

$$\tfrac{1}{2}mv_2{}^2 - \tfrac{1}{2}mv_1{}^2 = \tfrac{1}{2}AlD_m(v_2{}^2 - v_1{}^2)$$

$$= (p_1 - p_2)Al$$

$$p_1 - p_2 = \tfrac{1}{2}D_mv_2{}^2 - \tfrac{1}{2}D_mv_1{}^2 \qquad (1)$$

This equation was first derived by the Swiss scientist Daniel Bernoulli (1700–1782). It may be used only when the fluid flows *horizontally.* Suppose, on the contrary, that *a* is at an elevation h_1 and *b* is at an elevation h_2. Then, adding terms to account for the change of gravitational potential energy per unit volume, we have

$$p_1 - p_2 + D_mgh_1 - D_mgh_2 = \tfrac{1}{2}D_mv_2{}^2 - \tfrac{1}{2}D_mv_1{}^2$$

This is a more general form of Bernoulli's equation for incompressible fluids. To apply it to a gas, one must replace the terms containing the constant density with ones that tell how the density depends upon the pressure.

Example. The depth of the water at rest in the reservoir (Fig. 8-2) at *a* is 4.90 m and the velocity in the tube at *b* is 4.90 m/sec. (a) What is the pressure at *b*? (b) If the cross-sectional area at *c* is three-fourths that at *b*, what is the velocity at *c*? (Assume zero friction.)

(a) Take a streamline that includes *a* and *b*. The velocity at *a* is zero. Hence the pressure p_a at *a* is hD_w or hD_mg.

$$p_a - p_b = \tfrac{1}{2}Dv_b{}^2 - \tfrac{1}{2}Dv_a{}^2$$

$$1 \times 10^3 \text{ kg/m}^3 \times 9.8 \text{ m/sec}^2 \times 4.90 \text{ m} - p_b$$
$$= \tfrac{1}{2} \times 1 \times 10^3 \text{ kg/m}^3 \times (4.90 \text{ m/sec})^2$$
$$- \tfrac{1}{2} \times (1 \times 10^3 \text{ kg/m}^3)(0 \text{ m/sec})^2$$

$$p_b = \{4.90 \times 9.80 - \tfrac{1}{2} \times (4.90)^2\}$$
$$(10^3 \text{ kg-m/sec}^2)/\text{m}^2 = 3.60 \times 10^4 \text{ newtons/m}^2$$

(b) $$v_bA_b = v_cA_c$$

$$4.90 \text{ m/sec } A_b = v_c(\tfrac{3}{4}A_b)$$

$$v_c = 6.53 \text{ m/sec}$$

The Velocity of a Liquid Jet

The velocity of the liquid in a jet from an orifice can be found by applying Bernoulli's equation or by directly applying the law of conservation of energy.

Allow water to escape from an opening *B* in a tank (Fig. 8-4). The pressure at *B just inside the tank* is D_mgh and its velocity is zero. Just outside the tank the liquid pressure is zero and the velocity is the escape velocity of the jet *v*. By Bernoulli's equation:

$$D_mgh - 0 = \tfrac{1}{2}D_mv^2 - 0$$

$$v^2 = 2gh$$

To calculate the velocity by the law of conservation of energy, we must first find the potential energy of water at *B* just inside the tank. The potential energy equals the work done in forcing the water into the tank against the pressure of the overlying layers of water already in the tank. Suppose that a piston of

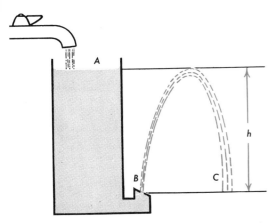

Fig. 8-4. Each unit volume of water escaping at B has kinetic energy $\frac{1}{2}D_m v^2$ equal to the pressure $D_m gh$ at the same level in the reservoir. Thus the jet rises to the level of the free surface of the liquid.

area **A** moves through a short distance Δl and forces water into the tank at a point where the liquid pressure is $p = D_w h = D_m gh$. The force exerted by the piston $F = pA$ acts through a distance Δl and does work equal to $pA(\Delta l)$ or $p(\Delta V)$ where ΔV is the volume of the water. Hence the work done per unit volume is equal to the pressure $p = D_m gh$; this is also the potential energy per unit volume. Now according to the law of conservation of energy, the potential energy per unit volume just inside the tank equals the kinetic energy per unit volume just outside the orifice if friction is negligible.

$$D_m gh = \tfrac{1}{2}D_m v^2$$

$$v^2 = 2gh$$

By either method, the velocity of the water at the orifice is equal to that which a particle would acquire in falling through a vertical distance *h*. *If friction is negligible, the velocity with which a liquid issues from an orifice equals that which it would have after falling a vertical distance equal to the depth of the liquid.* This is called *Torricelli's law* in honor of Galileo's pupil, who first stated it.

Example. With what speed will water issue from the openings *B* and *C* (Fig. 8-5), the distances below the upper water surface being 1.22 m and 4.90 m, respectively? Use Torricelli's law.

A freely falling body would require 0.5 sec to fall a vertical distance of 1.22 m and 1 sec to fall 4.90 m. The velocities acquired would be 4.90 m/sec and 9.8 m/sec, respectively. Hence these are the speeds of the water in the two jets.

The Flow of Viscous Fluids

When a liquid flows through a pipe or an airplane moves through the air, frictional forces oppose the motion; we say that the fluids are viscous. Viscosity in liquids is due to attractions among molecules within the liquid and between molecules of the liquid and those of solids in contact with it. Viscosity in gases, where the molecules are much further apart than in liquids, is due to collisions between the fast flowing molecules and those flowing at lower velocities; in the collisions,

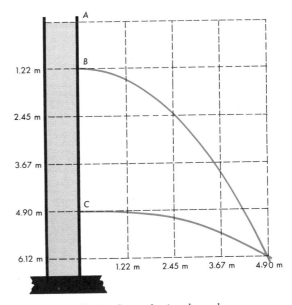

Fig. 8-5. With what velocity does the water escape (a) at B and (b) at C?

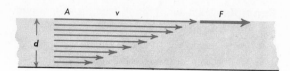

Fig. 8-6. *Viscous drag. The force required to pull the sheet of paper of area* A *along the surface of the liquid with velocity* v *is given by* F = ηAv/d, *in which* η *is the coefficient of viscosity of the liquid.*

the fast molecules give up momentum to the slow ones and are retarded in their flow. The effect of viscosity is to make the fluid sluggish in flow and resistant to the motion of objects through it. Fluid friction increases as the velocity increases and depends upon the shapes of the objects in contact with the fluid and upon the fluid itself. Usually the engineer wants to reduce fluid friction as much as possible by "streamlining" so that less power is needed at a given velocity to drive ships through the water or planes through the air. Occasionally, fluid friction is deliberately introduced into a device to suppress unwanted vibrations, "damping" such vibrations by converting their mechanical energy into heat energy of the damping fluid.

Imagine that you drag a large raft-like sheet of paper slowly across the surface of a shallow pool of water (Fig. 8-6). The layer of water in contact with the raft moves with it at the same velocity. The layer of water at the bottom of the pool is at rest. In between, the layers of water slip past each other (like the cards in a deck when the top card is pushed horizontally), each layer traveling a little more slowly than the one above it. This is called *laminar* (*lamina* = a layer) flow. Water flowing slowly through a smooth pipe of constant cross-section area also undergoes laminar flow, but the "layers" are then concentric shells which slip past each other.

If you pull the raft rapidly over the surface of the water, the orderly motion of laminar flow ceases and the water becomes *turbulent.*

Eddies form in the wake of the raft, and the frictional drag on the raft increases greatly.

The frictional force on a body in laminar flow can be calculated, although the mathematics involved is often beyond the scope of this book. Frictional drags in turbulent flow, however, are difficult or impossible to calculate, and it is usually necessary to resort to wind-tunnel or towing-tank measurements of the drag on models of the object being studied—a newly designed jet plane, for example, or the hull of a new ship.

In laminar flow at low velocities, fluid friction is directly proportional to the velocity. At higher velocities the friction is proportional to the *square* of the velocity. This change from the first power of the velocity to the square, for any given fluid and apparatus, always occurs at the same critical velocity. This fact can be demonstrated by the experiment devised by Osborne Reynolds. He fed a thin stream of colored liquid into a tube through which a stream of the uncolored liquid was flowing. When the liquid flowed slowly and laminarly, the colored stream remained unbroken (Fig. 8-7a). At a certain critical velocity, the colored stream was broken up because of turbulence

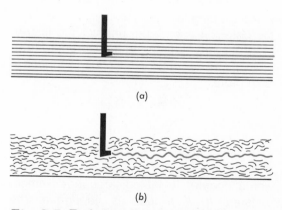

(a)

(b)

Fig. 8-7. *Turbulent flow.* (a) *Below the critical velocity, the liquid particles move on parallel lines in laminar flow and the colored liquid travels on a straight line.* (b) *Above the critical velocity, turbulence breaks up the stream.*

(Fig. 8-7*b*). Also, at this critical velocity, the friction began to vary as the square of the velocity.

The *Reynolds number R,* a dimensionless quantity, for a cylindrical pipe of diameter *d* in which a fluid of density D_m and coefficient of viscosity η (Greek *eta*) is flowing at a velocity *v* is defined as

$$R = \frac{D_m v d}{\eta}$$

When *R* is less than approximately 2000, the flow in cylindrical pipes is found experimentally to be laminar. For *R* greater than 2000, the flow is turbulent. This expression for *R* must be modified if the geometry is not that of flow through a cylindrical pipe.

Example. Water at 20°C flows at a velocity of 0.10 m/sec through a cylindrical tube whose diameter is 0.005 m. Is the flow laminar or turbulent? The coefficient of viscosity of water is 0.0010 newt-sec/m².

$$R = \frac{1.0 \times 10^3 \text{ kg/m}^3 \times 0.10 \text{ m/sec} \times 0.005 \text{ m}}{0.0010 \text{ newt-sec/m}^2}$$

$$= 500$$

Since *R* is less than 2000, the flow is laminar.

Laminar Flow

Suppose that the "raft" discussed above (Fig. 8-6) has an area *A* and that it moves over a pool of water of depth *d* with a constant velocity *v*. If the flow is laminar, the force of fluid friction *F* is found experimentally to depend upon the area *A*, the quantity *v/d* (called the *velocity gradient*), and the coefficient of viscosity η of the fluid:

$$F = \eta A \frac{v}{d}$$

Notice in Table 8-1 that the coefficient of viscosity depends upon the temperature. The expression "slow as molasses in January" reminds us that cold syrup flows sluggishly

Table 8-1. Coefficients of Viscosity in dyne-sec/cm² or poises

(1 dyne-sec/cm² = 0.1 newt-sec/m²)

A. Liquids		B. Gases	
Carbon disulfide (20° C)	0.00367	Air (20° C)	0.000178
Ethyl alcohol (20° C)	0.01192	Air (40° C)	0.000189
Glycerine (20° C)	8.33	Hydrogen (20° C)	0.000093
Castor oil (20° C)	10.280		
Water (0° C)	0.0179		
Water (20° C)	0.0100		
Blood (37° C)	~.030		

Automobile oils	At 0° C	At 20° C
SAE 40	17–200	3.5–17
30	11–60	2.3–6.1
10	3.2–13	0.8–2
10W	2–8	0.6–1.3

from a container. If heated, it flows more freely. The viscosities of all liquids decrease with increase of temperature. The viscosities of gases, unlike those of liquids, *increase* with increase of temperature. The reason for this difference is that, as pointed out above, the mechanism of viscosity is different in liquids and in gases. Increase in temperature tends to weaken the effect of the molecular attraction in liquids. In gases, increased temperature and greater molecular activity cause disorderly mixing of molecules so that the viscosity increases with increase of temperature.

When two sliding surfaces are greased or oiled, a thin film of lubricant forms between the two surfaces, forcing them apart so that rubbing of the solid surfaces is prevented and fluid or viscous friction occurs inside the liquid film. The frictional forces are nearly independent of the load; they vary with the speed, with the area of contact of the two surfaces, and with the viscosity of the lubricant. Since the friction increases with the viscosity, it is advisable, in choosing a lubricant for an axle or bearing, to use the least viscous oil that is able to keep the two rubbing surfaces separated. The lubrication of automobiles is complicated by the fact that they must operate in hot weather and cold weather. Oils for winter lubrication are less viscous than those used in summer. When an engine is first started in zero weather, the oil does not form

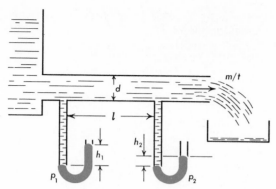

Fig. 8-8. The discharge of liquid from a cylindrical tube. Pressure-measuring tubes containing mercury indicate the pressures p_1 and p_2 at points separated by a distance l. For example, the difference of levels in the two arms of the pressure tube on the left indicates that the liquid pressure there is h_1D_w, where D_w is the weight density of mercury.

films on the bearings. The wise motorist therefore allows a cold engine to "idle" at low speed until the oil has been warmed a little.

A second example of laminar flow is the discharge of a liquid from a pipe. Suppose that liquid flows through a tube of diameter d (Fig. 8-8). Because of fluid friction, the liquid pressure at points separated by a distance l is not constant. Let the pressures be p_1 and p_2, where p_1 is greater than p_2. If the Reynolds number is less than 2000, the flow will be laminar and the mass of liquid m discharged in a time t can be calculated. The result of applying some mathematical procedures more advanced than those used in this book is (Poiseuille's law)

$$\frac{m}{t} = \frac{\pi D_m d^4}{128\eta} \frac{p_1 - p_2}{l}$$

If you are sure the flow is laminar, you can use this equation to calculate the coefficient of viscosity η of a liquid by measuring the rate of discharge through a tube of known diameter and the decrease in pressure per unit length $(p_1 - p_2)/l$. Conversely, if the coefficient of

viscosity is known, the equation can be used as the principle of a liquid flowmeter for determining the discharge rate.

Terminal Velocities

Let us consider the force of fluid friction acting on a sphere of radius a falling through a viscous liquid or gas so that the flow of fluid past the sphere is laminar. As the sphere falls it accelerates until the back drag of fluid friction plus the upward force of buoyancy equals the weight of the sphere. At that velocity, v_t, called the **terminal velocity**, the sphere is in equilibrium and its velocity will remain constant thereafter. Stokes' law states that the force of fluid friction F is given by

$$F = 6\pi\eta a v_t$$

The weight of the sphere is $\frac{4}{3}\pi a^3 Dg$ where D is the mass density of the sphere. We must also apply Archimedes' principle to calculate the buoyant force; the weight of the displaced fluid, the buoyancy, is $\frac{4}{3}\pi a^3 D_f g$, where D_f is the mass density of the fluid. Hence the sum of the forces acting on the falling sphere—upward forces are positive—is

$$\underset{\substack{\text{(Viscous} \\ \text{drag)}}}{6\pi\eta a v_t} + \underset{\text{(Buoyancy)}}{\tfrac{4}{3}\pi a^3 D_f g} - \underset{\text{(Weight)}}{\tfrac{4}{3}\pi a^3 Dg} = 0$$

or,

$$v_t = \frac{2}{9}\frac{a^2 g(D - D_f)}{\eta}$$

Hence, by measuring the terminal velocity of a sphere of known density while it is falling in a continuous fluid whose density and coefficient of viscosity are known, one can deduce its radius. We shall encounter Stokes' law again when we discuss the measurement of the electric charge on a single electron.

In parachute descents the parachutist reaches the ground with approximately the same terminal velocity whether he jumps from a elevation of 1 mile or of 5 miles. In bailing out of a plane at high altitudes, men purposely fall great distances before pulling the release cords of their parachutes in order to escape quickly from the extremely low temperatures and low air densities of the upper atmosphere. In such a jump, the parachutist falls faster and faster, encountering ever-increasing friction, until the opposing force equals his weight, after which his velocity becomes constant at about 130 mi/hr. When the parachutist pulls the release cord, his parachute opens, and it is gripped by the air. The increased back pull of friction causes the speed to diminish to about 12 mi/hr, so that the up pull of air friction again equals the weight of the jumper and his parachute. At this lower velocity—although it is uncomfortably large for a landing on rough terrain—the aviator has a reasonable chance of reaching the ground safely. Note, however, that there is considerable turbulence associated with the motion of a parachute through the air; Stokes' law does not apply.

Lift Due to Fluid Flow

When a fluid flows past a stationary object or an object moves through a fluid, the flow is altered near the object. Air streaming up and over the upper surface of an airplane wing in flight (Fig. 8-9a) accelerates. By Bernoulli's law, the air pressure is slightly lower there than at the lower surface of the wing. This small pressure difference multiplied by the large area of the wing gives the lift that supports the plane in flight. Near the wing, the air velocity gradually drops to zero relative to the wing because the viscous air "attaches" itself to the wing. This *boundary layer,* a layer only a few millimeters thick, sheathes the wing in normal flight; its behavior largely determines not only the magnitude of the lift, but also the resistance which the air flow offers to the plane. In Fig. 8-9b, the angle of attack, the angle between the wing and the air stream, has been increased. The lift increases some-

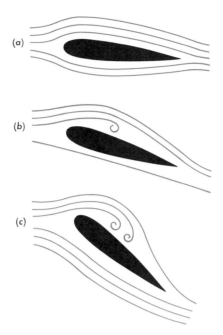

Fig. 8-9. (a) *An airplane wing at a small angle of attack allows the air to slip smoothly past it in laminar flow.* (b) *At a greater angle, turbulence begins to develop in the boundary layer near the trailing edge of the wing.* (c) *At a large angle of attack, the boundary layer separates from the wing, drag increases and lift decreases, and the plane stalls.*

what, but the flow is no longer as smooth as before; the air above the wing "stagnates," some turbulence sets in near the trailing edge, and the boundary layer starts to separate from the upper surface of the wing. In Fig. 8-9c, the angle of attack is such that the air flow above most of the wing is turbulent, the lift falls off sharply, and the plane stalls.

Aerodynamic engineers seek to control the boundary layer more carefully so that planes will have increased lift at a given speed and landing speeds can be reduced. One method is to blow air at high velocity over the upper wing, keeping the air in the boundary layer moving and delaying the onset of stagnation and turbulence.

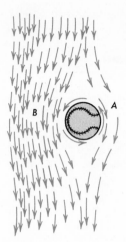

Fig. 8-10. The pressure at B *is less than at* A; *hence the ball curves to your left.*

The principles that describe the lift on an airplane wing apply to other objects as well. The hydrofoil is a boat whose hull is raised from the water by the lift produced by an underwater "wing," thus lessening the friction drag and increasing the maximum speed at which the boat can operate. Aerodynamic forces act on bridges. The collapse of the Tacoma Narrows Bridge in 1940 was due to the linking of aerodynamic forces exerted by a high wind with the vibrational motion of this suspension bridge. Finally, we can use these principles to explain why a baseball curves. Suppose that a pitcher throws a baseball and makes it spin as shown in Fig. 8-10. The air rushes past the ball as indicated by the arrows. The spinning baseball produces a small "whirlpool" of air, which opposes the wind at one side of the ball and reinforces it at the other side. In consequence, the air velocity is greater at *B* than at *A*. The pressure at *B* is lowered because of the high velocity, and the baseball curves to your left. If the direction of spin were reversed, the pressure at *A* would be diminished and the ball would curve to the right.

Drag Due to Fluid Flow

An ideal fluid—one with no viscosity—would flow through a tube without loss of energy and would not exert a drag on an object. *Superfluids,* such as liquid helium at temperatures near absolute zero, have negligibly small viscosities. Most fluids have appreciable viscosities, however. We have seen how the viscosity is responsible for the formation of boundary layers. Tubes conducting viscous fluids, like airplane wings, have boundary layers; the velocity of the flow decreases to zero at the wall of the tube.

Viscosity results in a conversion of energy into heat in the fluid. In a horizontal pipe, there is a progressive drop of pressure along the pipe in the direction of flow of a viscous fluid. A pressure difference over a section of pipe means that the fluid does work, at the expense of the energy of the system, against the frictional drag; this work is converted into heat. In order to study frictional effects in a moving liquid, connect a reservoir filled with water to a horizontal pipe stuffed with cotton and fitted with vertical tubes (Fig. 8-11). When the valve is closed, the water rises to the same elevation in each vertical tube. When the valve is opened slightly so as to permit a small rate of flow, the water level falls in each tube because of a decrease of pressure along the horizontal pipe so that the levels lie along line *b*. Open the valve further so as to double the rate of flow. Then the decrease of pressure will be doubled (line *c*). The pressure drop is proportional to the rate of flow of the water.

Frictional losses of energy are important when fluids are transported long distances. Petroleum is forced through pipelines from Texas to eastern states. Powerful pumps at the starting point raise the pressure of the liquid to several atmospheres. If there were no flow, the pressure would be the same at all points along the pipe (assuming that it were horizontal). When oil is flowing the pressure

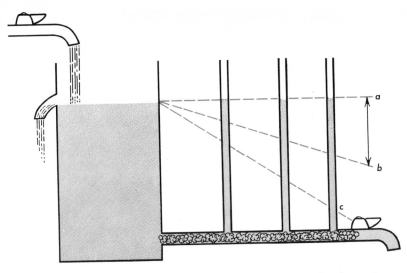

Fig. 8-11. The fall of pressure due to friction in a tube depends upon the velocity of the liquid.

decreases with distance because of the friction. Pumps are installed at intervals of 50 to 100 mi along the route in order to increase the pressure and maintain the motion. The same practice applies to the transfer of natural gas. Most familiar of all is the pumping of water to supply the residences of a city.

Frictional drag on objects moving through fluids necessitates the expenditure of power and limits the maximum velocity attainable. At low velocities, as we have seen, the frictional drag is proportional to the velocity. As the velocity increases and turbulence becomes more prominent, the fluid friction becomes proportional to the square of the velocity. At the velocities of bullets and rockets, the drag may vary as the cube or higher power of the velocity. A tramp steamer, traveling at 6 mi/hr, can traverse the Atlantic using about one-tenth as much fuel as it would if it traveled at 18 mi/hr. High speed in sea vessels is expensive because powerful engines must be used and because a great deal of fuel is required for a journey. An ordinary automobile engine is powerful enough to

drive a small airplane at 60 mi/hr. To drive a jet plane at 1000 mi/hr requires approximately 400 pounds of fuel per minute.

Drag on a moving object can be reduced by careful attention to the shape of the object and to the smoothness of its surface so that turbulence in the boundary layer will be kept to a minimum. A body shaped like a teardrop encounters little friction. The fluid flows smoothly up and over the surfaces. The boundary layer remains close to the body, and there is little turbulence, hence relatively little drag. A body of this shape experiences as little as one-fourteenth as much friction as a disk of equal diameter (Fig. 8-12). Fish are streamlined. This probably came about through the slow processes of evolution. The fish that were not streamlined either failed to capture other fish and starved or were themselves caught. Thus only the better streamlined ones survived. Early airplanes had numerous guy wires and braces which greatly increased the friction; these have now been eliminated, and the airplane approximates more closely the ideal teardrop form. The

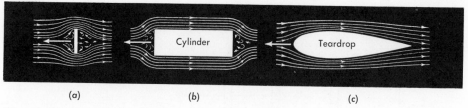

| (a) | (b) | (c) |

Fig. 8-12. *The frictional force acting on the streamlined body in* c *is about $\frac{1}{14}$ that on the disk in* a.

wheels and supports of the landing gear formerly caused about one-half of the friction drag. In most planes, the landing gear is now drawn up inside the body after the plane leaves the ground. The friction drag on an airplane wing ordinarily varies as the square of its velocity through the air. When the air speed reaches the speed of sound (750 mi/hr at 0°C), the drag suddenly increases. This increase occurs because the compressed air ahead of the plane cannot readily escape, and it piles up in a wave somewhat like snow pushed ahead by a snow plow. The viscous drag at supersonic speeds is relatively small compared to this wave drag. Bullets have pointed noses, and supersonic airplanes have slender, pointed fuselages and thin wings. Their front edges are sharply pointed so that they cut through the air (Fig. 8-13).

Summary

When the velocity of a horizontal fluid stream increases, the pressure in it decreases.

The pressure p_1 and velocity v_1 at one point in a liquid stream and p_2 and v_2 at another point at the same elevation are interrelated by

$$p_1 - p_2 = \tfrac{1}{2}D_m v_2{}^2 - \tfrac{1}{2}D_m v_1{}^2$$

The velocity of the liquid in a jet issuing from an orifice in an open tank equals that which a freely falling body would acquire in falling through a vertical distance equal to the depth of the liquid at the orifice.

When a body falls through a fluid, a constant "terminal" velocity is reached when the friction equals the weight of the body minus the buoyant force.

Fluid friction increases with the viscosity of the fluid and with the velocity. It depends upon the size and shape of the body that moves with respect to the fluid. It *decreases* with increase of temperature in liquids, but *increases* with increase of temperature in gases.

Questions

1. Discuss the following: "If the speed were sufficiently small, a steamship could be drawn across a lake by means of a thread."

2. Discuss the limiting or terminal velocity of a parachutist. Upon what two factors does the velocity principally depend?

3. State two laws of fluid pressure that apply to fluids at rest, but not to those that are moving.

4. Can two streamlines intersect?

5. When a baseball is thrown southward, rotating clockwise as viewed from above, does it curve to the pitcher's right or to his left? As viewed from the east, how must it rotate so as to curve downward more sharply?

6. Why are motor cars not designed to have a "teardrop" shape?

7. Compare viscosity in liquids with that in gases. Discuss the reasons for the differences.

8. Give units for each factor in Stokes' law and show that the units are consistent.

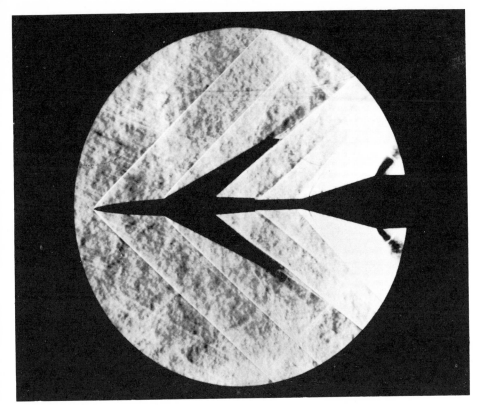

Fig. 8-13. Photograph of air streaming past a swept-wing model plane at a speed 1.5 times that of sound in a supersonic wind tunnel (NASA photo).

9. Why does the temperature of the water in a bathroom shower change when someone opens another faucet in the same house?

10. Estimate the diameter of a clay particle that takes one day to sink from the ceiling to the floor through still air in an ordinary room.

11. In Fig. 8-5, calculate two other heights for which emerging jets will intersect.

12. A reservoir of mercury and one of water have orifices each 1 m below the upper surface. Compare the velocities with which the liquids will flow from them.

13. Why can a ping-pong ball be balanced on a vertical jet of air?

14. Apply the Bernoulli effect to the motion of individuals in a football crowd that leaves a field through a narrow gate.

15. What is the significance of the Reynolds number? What are its units?

16. Interpret Bernoulli's equation from the standpoint of (*a*) work and energy and (*b*) force and acceleration.

17. Why are the pressures in the horizontal tube equal when the valve is closed (Fig. 8-11)?

18. Does a raindrop have an ideal "tear-drop" shape as it falls through air?

Problems

A

1. What (*a*) force and (*b*) power are required to drag a sheet of paper 80 cm × 100 cm along the surface of a pool of water 50

cm deep at a velocity of 50 cm/sec? (Assume no turbulence.) The temperature of the water is 20°C.

2. A flow of 3.0×10^4 cm^3 of water per minute is maintained through a pipe whose bore is 2.0 cm^2 in area. What is the velocity of the water?

3. Water flows through a pipe 2.00 cm^2 in cross-sectional area with a velocity of 2.0 m/sec. (a) How many cubic centimeters are delivered per minute? (b) If the first pipe is followed by one that is 1.00 cm^2 in cross section, what is the speed of the water in the second pipe?

4. What work is done in pumping 2.0 m^3 of water against a pressure of 1000 dynes/cm^2?

5. If water flows through a pipe (cross-sectional area, 1.0 cm^2) at 2.0 cm/sec, how many cubic centimeters flow per minute?

6. If in landing with a parachute you hit the ground at 10 km/hr, from what vertical height would you have had to jump freely to attain this speed?

B

7. An open tank 10 m high, resting on a level platform, is filled with water. Water issues from an opening 5.0 m below the water surface. At what horizontal distance from the tank does the water jet strike the platform?

8. (a) How much work is required to force 1.00 cm^3 of water into a water reservoir at a point 1.20 m below the surface? (b) If a jet of water escapes from this opening, find the kinetic energy of 1.00 cm^3 of the water at the jet.

9. The water in a reservoir is 4.90 m deep. The velocity of the water in a horizontal pipe connected to the bottom of the reservoir is 2.0 m/sec. What is the pressure in the pipe? (Neglect friction.)

10. Ethyl alcohol at 20°C flows at a velocity of 0.020 m/sec through a cylindrical tube of 0.20 cm radius. Is the flow turbulent or laminar? Specific gravity of ethyl alcohol at 20°C $= 0.79$.

11. Water flows from a section of a horizontal pipe where the cross-sectional area is 400 cm^2 and the velocity is 5.0 m/sec into a section whose cross-sectional area is 1200 cm^2. Calculate the change in pressure of the water. (Neglect friction.)

C

12. In a horizontal oil pipeline of constant cross-sectional area, the pressure decrease between two points 250 m apart is 2.50×10^4 newtons/m^2. Calculate the energy loss in joules per cubic meter per meter.

13. At point A in a tube carrying water, the elevation is 300 cm, the cross section of the bore is 1.00 cm^2, the pressure of the water is 75 cm-Hg, and its velocity is 80 cm/sec. What is the pressure, in cm-Hg, at point B, where the elevation is 200 cm, the cross section is 0.50 cm^2, and the velocity is 160 cm/sec? (Neglect friction.)

14. In Fig. 8-5, compute the horizontal distance from the tank to the water jet B at a point 5.0 m below the level of the water surface.

15. (a) What is the Reynolds number when water at 20°C flows at 10 cm/sec through a pipe of 1.0 cm diameter? (b) At what speed would the flow become turbulent?

16. A stream of water in a horizontal tube 1.00 cm^2 in cross section has a speed of 80 cm/sec. The pressure at a point A is 10.0×10^5 dynes/cm^2. Find (a) the kinetic energy per cubic centimeter at A and (b) the pressure at point B where the cross-sectional area is 0.50 cm^2. (Disregard frictional losses.)

17. Calculate the terminal velocities in air at 20°C of (a) a spherical raindrop of radius 0.020 cm and (b) a spherical lead shot of the same radius. Density of air at that temperature is 1.29×10^{-3} gm/cm^3.

18. In Fig. 8-5, (a) compute the horizontal distance from the tank to where jets emerging from 2.45 m and 3.67 m hit the ground at the 6.12-m level. (b) At what height will an emerging jet travel the greatest horizontal distance?

For Further Reading

Blanchard, Duncan C., *From Raindrops to Volcanoes,* Doubleday and Company, Garden City, 1967.

Cornish, J. J. "The Boundary Layer," *Scientific American,* August, 1954, p. 72.

Hess, Felix, "The Aerodynamics of Boomer-angs," *Scientific American,* November, 1968, p. 124.

Shapiro, Ascher H., *Shape and Flow: The Fluid Dynamics of Drag,* Doubleday and Company, Garden City, 1961.

Wiggers, C. J., "The Heart," *Scientific American,* May, 1957, p. 74.

CHAPTER NINE

Rotational Motion

In earlier chapters we considered transla-tional motion on straight or curved paths when the moving bodies do not rotate. Now we shall study the rotational motion of rigid bodies.

Forces May Cause Rotation

Thus far, we have considered bodies acted upon by concurrent forces, forces that act at the same point or whose lines of action pass through the same point. If the resultant of such forces is not zero, the body will accelerate, but it will not be made to rotate. If the forces acting upon the body are not concurrent, the body may be set into rotation. For example, if two 1-kgwt forces are applied at *C*, the center of a light, rigid meter stick, as in Fig. 9-1*a*, the meter stick is in equilibrium. However, if one 1-kgwt force is applied at *C* and one at *A* (Fig. 9-1*b*), the meter stick begins to rotate and is not in equilibrium, even though the resultant force is zero.

A body whose dimensions can be consid-ered small we have referred to as a *point mass.* Forces acting on a point mass are es-sentially concurrent forces. If the dimensions of a body are appreciably large and if the

separations of the parts of the extended body are unchanging, the body is called a *rigid body.* Forces acting on a rigid body, as in the example of the meter stick, may or may not be concurrent. In this chapter we shall consider mainly non-concurrent forces acting on rigid bodies.

A lever shows in a simple way how forces can produce rotation. It is a rigid bar or rod pivoted at a point called the *fulcrum* (Fig. 9-2). Levers have been used since prehistoric times, and the cave men, like children of today, learned that in order to lift a heavy load, the force should be applied to a lever at a large distance from the fulcrum. Archi-medes first stated the rule in a quantitative way. He showed that, when two bodies, sus-pended from a lever, balance each other and produce equilibrium, the weights of the bodies are inversely proportional to their dis-tances from the pivot or fulcrum. We can more conveniently state this relationship by saying that the weight of one body times its distance from the pivot or fulcrum is equal to the weight of the other body times its distance. In Fig. 9-3, the 2-kgwt body is suspended from the lever at a point 25 cm from its fulcrum, and the 1-kgwt body at a point 50 cm from the

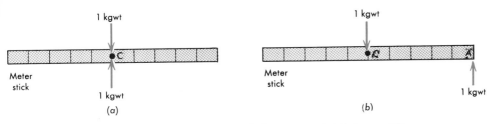

1 kgwt

Meter
stick

1 kgwt

(a)

1 kgwt

Meter
stick

•C

A

1 kgwt

(b)

Fig. 9-1. (a) *The resultant force is zero, and the meter stick is in equilibrium.* (b) *The resultant force is zero, but the meter stick begins to rotate.*

fulcrum. The products of the forces by their distances from this point are equal.

In dealing with forces that produce rotation, it is convenient to speak of the *lever arm* of the force, which means *the perpendicular distance from the fulcrum or pivot to the line along which the force acts.* In Fig. 9-3, the lever arm of the force acting at *B* is 25 cm, and that of the one acting at *A* is 50 cm.

Torques

The effectiveness of a force in producing rotation about a given pivot depends upon two factors: one is the magnitude of the force, the other its lever arm. The product of *lever arm times force* is called the *torque* τ (Greek *tau*) or sometimes the *moment* of the force about the specified pivot or axis of rotation.

$$\text{Torque} = \text{Lever arm} \times \text{Force}$$
$$\tau = l \times F$$

Torques that tend to cause counterclockwise rotation about a pivot are usually considered to be positive; those tending to produce clockwise rotation, negative. The pivot or axis of rotation must always be specified in stating a torque.

In Fig. 9-3, the torques about *C* are:

Clockwise: 25 cm × 2.0 kgwt
$$= -50 \text{ cm-kgwt}$$
Counterclockwise: 50 cm × 1.0 kgwt
$$= +50 \text{ cm-kgwt}$$

Note that the algebraic sum of the torques about *C* is zero. The sum of the torques about *any* pivot is zero so long as the body is in equilibrium.

Measuring the Lever Arm

In the levers which we have considered, the forces were directed at right angles to the

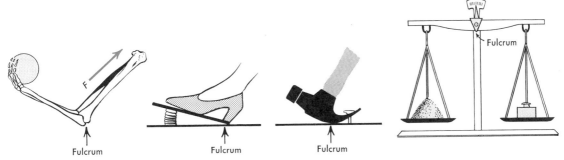

Fig. 9-2. Familiar examples of the lever.

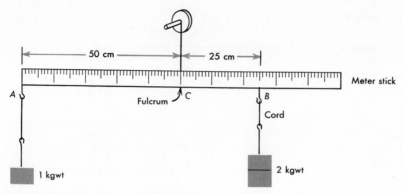

Fig. 9-3. A lever in equilibrium.

lever. As an illustration of a different situation, examine Fig. 9-4 carefully. Remember that the lever arm of a force is the *shortest distance, that is, the perpendicular distance, from the fulcrum or axis to the line on which the force acts.*

When the force is not perpendicular to the lever, we may resolve it into two components, one parallel to the lever, the other at right angles to it. Figure 9-5 represents a door, of width l, in three different positions. When the door is in position OA, the force F is perpendicular to it; hence the torque acting on the door about the hinge is $l \times F$. When the door is in the position OA', we resolve the force F into two components, $F \cos 30°$ parallel to the

door and $F \sin 30°$ at right angles to it. The torque acting on the door is $lF \sin 30°$. When the door is at OA'', the line of the force passes through the hinge; hence the torque is zero.

In general, if the angle between the line of action of the force and the line connecting the fulcrum and the point at which the force acts is θ, the torque τ exerted is given by

$$\tau = lF \sin \theta$$

The Two Conditions of Equilibrium

In Chapter 3 you learned that, when a body is in equilibrium, the resultant of all the external forces acting upon it is zero. Now we find that the resultant of all the torques must also be zero. The two conditions for complete equilibrium of a body are:

1. The resultant of all the external forces acting on the body must be zero ($\Sigma F = 0$).*
2. The sum of all the positive and negative torques acting on the body about any chosen axis must be zero ($\Sigma \tau = 0$).

These two conditions of equilibrium enable us to solve many problems.

Example. Two men, A and B, carry a 30-kgwt sack of cement by means of a pole 3.0 m long

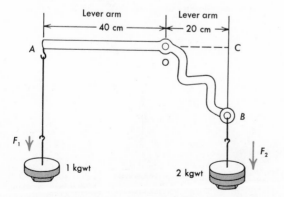

Fig. 9-4. The lever arm of the force F_1 about O is 40 cm. That of F_2 is 20 cm.

* For Σ (Greek *sigma*) read "the sum of." Thus ΣF means the sum of the forces, that is, $\Sigma F = F_1 + F_2 + F_3 + \cdots$.

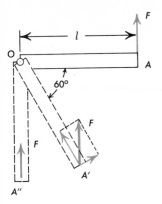

Fig. 9-5. The torque exerted by F *about* O *depends upon the lever arm.*

(Fig. 9-6). At what distance from *A* must the sack be suspended so that he may carry two-thirds of the load?

By the first condition of equilibrium, the resultant force on the pole must be zero. *A* exerts a force of $\frac{2}{3}$(30 kgwt) or 20 kgwt and *B* a force of 10 kgwt. Calling up-forces positive and down-forces negative, we have

$$\Sigma F = 20 \text{ kgwt} + 10 \text{ kgwt} - 30 \text{ kgwt} = 0$$

By the second condition, the resultant torque about *C* (or any other point) must be zero.

Counterclockwise torque about *C*
$$= (3 \text{ m} - x) \times 10 \text{ kgwt}$$
Clockwise torque about $C = -x \times 20$ kgwt
$$\Sigma\tau = (3 \text{ m} - x) \times 10 \text{ kgwt} - x \times 20 \text{ kgwt} = 0$$
$$x = 1.0 \text{ m}$$

To check our answer, let us repeat the calculation, but write the sum of the torques about *A* rather than *C*. Since the force at *A* has a zero lever arm about *A*, we have

$$\Sigma\tau = 0 \times 20 \text{ kgwt} - x \times 30 \text{ kgwt}$$
$$+ 3 \text{ m} \times 10 \text{ kgwt} = 0$$
$$x = 1.0 \text{ m, as before}$$

The axis about which torques are computed in applying the conditions of equilibrium can be chosen at *any* point. Some points—for ex-

ample, those at which forces are known to act—are more convenient than others, but any point will do.

The Center of Gravity of a Body

We can neglect the weight of the lever in solving some problems, but in others it must be taken into account. The difficulty is that there are trillions on trillions of molecules in the lever, each of which is attracted by the earth. The way out of this difficulty is to find the point where the *resultant* of all these tiny forces acts; this point we call the *center of gravity* of the body.

The center of gravity of a body is the point where the resultant of the weights of all its particles acts. A single up force equal to the weight of the body applied at its center of gravity will produce equilibrium; hence we may consider in solving certain problems that the entire weight of a body acts at its center of gravity.

In order to locate the center of gravity of a meter stick, support it on a pencil and shift it to a position where it balances horizontally. Then the down pull of the weight is equal to the up push of the pencil, and the two forces act in the same straight line. Hence the center

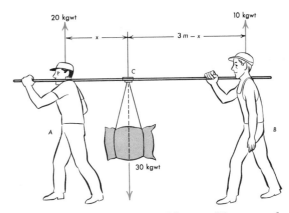

Fig. 9-6. Torques in equilibrium. The sum of the torques, positive and negative, about C *is zero.*

Mechanics 127

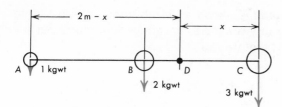

Fig. 9-7. *Center of gravity.*

of gravity of the meter stick is vertically above the pencil.

The centers of gravity of bodies may be determined by computation by applying the conditions of equilibrium.

Example. Figure 9-7 represents three spheres weighing 1.00 kgwt, 2.00 kgwt, and 3.00 kgwt, respectively. They are equally spaced along a rod 2.00 m long, of negligible weight. How far is the center of gravity D from C?

Let the distance be x.

The sum of the positive and negative torques about D must be zero.

$$\Sigma_T = (2.00 \text{ m} - x) \times 1.00 \text{ kgwt}$$
$$+ (1.00 \text{ m} - x) \times 2.00 \text{ kgwt}$$
$$- x \times 3.00 \text{ kgwt} = 0$$
$$x = 0.667 \text{ m}$$

The center of gravity of an irregular body such as a door may be found experimentally as follows: First suspend the door from a cord at A (Fig. 9-8). Hang from this same point of suspension a load on the end of a cord to serve as a plumb line. After the door and the plumb bob come to rest, the center of gravity will be somewhere on the vertical line AB. Trace this line with a pencil, and then suspend the body at some other point, C. The center of gravity is on the line CD. Therefore, it must be at G, where the two lines intersect.

A ladder problem. A uniform ladder 3 m long rests with one end on a level floor and the other against a smooth vertical wall (Fig. 9-9). The distance AB is 1.8 m and the ladder

weighs 20 kgwt. (*a*) What horizontal force, F_2, acting at B, will prevent the ladder from slipping? (*b*) What is the out push F_4 of the wall at C?

(*a*) The up push F_1 of the floor at B must be equal and opposite to the weight $F_3 = 20$ kgwt of the ladder. (First condition of equilibrium.)

Assume that the axis is at C. Any point may be chosen as the fulcrum if the body is in equilibrium, but C is particularly convenient since one of the unknown forces acts there. This unknown force will not appear in the sum of the torques about C. The sum is:

1.8 m × 20 kgwt − 0.9 m × 20 kgwt
$$+ 2.4 \text{ m} \times F_2 = 0$$
$$F_2 = -7.5 \text{ kgwt}$$
$$F_2 + F_4 = 0 \qquad \text{(Why?)}$$
$$F_4 = 7.5 \text{ kgwt}$$

Stability of Bodies

The concept of center of gravity and the conditions of equilibrium enable us to predict whether bodies will be stable or unstable if

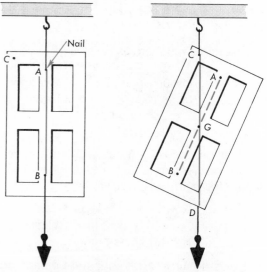

Fig. 9-8. *Locating the center of gravity of a door.*

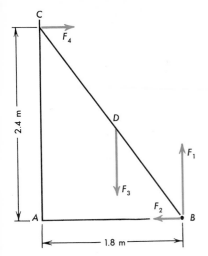

Fig. 9-9. The ladder problem. To find the force F_4, assume that the fulcrum is at C and apply the two conditions of equilibrium.

storing torque and the rotor will return to its original position. In Fig. 9-10c the rotor has been counterbalanced and is in *neutral* equilibrium: whatever the position of the rotor, torques are balanced.

The stability or instability of a boat or a ship can be explained in a similar fashion.

In Fig. 9-11a the ship, upright on the water, is acted upon (1) by a buoyant force upward at its center of buoyancy *CB*, located at the center of gravity of the displaced water, and (2) by the weight of the ship acting downward at the ship's center of gravity. If properly ballasted as shown in Figs. 9-11a and b, the ship is stable because if rocked by a wave to one side in a clockwise sense, a *counterclockwise* restoring torque, set up by the buoyant force acting at the new center of buoyancy, will bring the ship back to its upright position. However, if the ship is "topheavy" due to improper loading as in Fig. 9-11c, tipping it clockwise will set up a clockwise torque that will overturn it.

displaced slightly from their original positions. In Fig. 9-10a, a rotor consisting of a ball mounted on a light rod that is free to rotate about an axis *O* is balanced vertically above *O*. The sum of the torques and the sum of the forces acting on the rotor are both zero and it is in equilibrium — *unstable* equilibrium, however, because the slightest displacement will set up an unbalanced torque that will cause the rotor to swing down from its original position. In Fig. 9-10b the rotor is in *stable* equilibrium: a slight displacement will set up a re-

Translation and Rotation

A rigid body can undergo pure *translation* in which its particles move parallel to each other (Fig. 9-12a). In rotational motion about an axis (Fig. 9-12b), the particles move on concentric circles centered, in this example, on an axis that passes through one of the par-

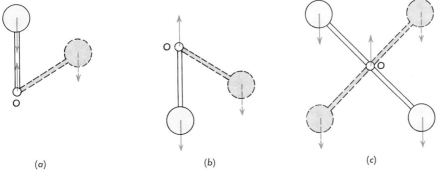

(a)   (b)   (c)

Fig. 9-10. Unstable, stable, and neutral equilibrium.

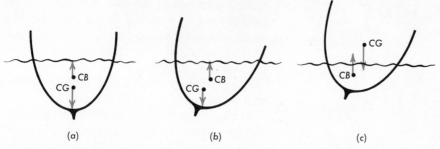

Fig. 9-11. A ship. CB *is the center of buoyancy;* CG, *the center of gravity.*

ticles. Or, the body can undergo both translation and rotation about its center of gravity (Fig. 9-12*c*). If you think about it a little, you will see that there are more general motions possible, involving rotation about a point. For example, the L-shaped body in Fig. 9-12 could undergo a tumbling motion about the outside corner of the L, not limiting its motion to the plane of the paper. Rotations about a point require more advanced mathematics for their description. We shall limit ourselves to the motions described in Fig. 9-12.

We have already discussed the quantities we need to describe the translational motion of rigid bodies. To describe rotational motion, we must define some additional quantities. Many of these closely resemble those of translational motion. For comparison, we shall state, beside some equations for rotational motion, the corresponding ones for translational motion. The proofs are quite similar to those for translational equations.

Angular Displacement, Velocity, and Acceleration

Suppose that a wheel rotates through an angle such that a point on the rim of the wheel moves through an arc distance *AB* (Fig. 9-13). Then the angle θ through which the wheel is displaced may be expressed in degrees, in revolutions, or in *radians*. If *AB* is equal to the radius *OA*, then θ is one radian. *One radian is an angle whose arc distance is equal to its radius.* In general, for an angle whose arc distance is *s* and radius *r*,

$$\theta \text{ (radians)} = \frac{s}{r}$$

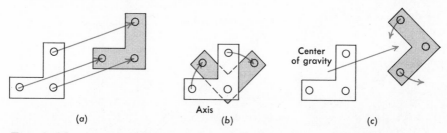

Fig. 9-12. Rotation and translation of a rigid body. (a) *The body undergoes pure translation; its particles—only three are shown—move on parallel lines.* (b) *In pure rotation, the particles move on concentric circles centered at the axis.* (c) *Translation plus rotation about the center of gravity.*

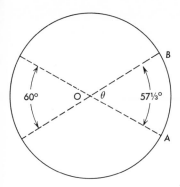

Fig. 9-13. *One radian = 360°/2π ≅ 57.3°.*

The circumference of a circle is 2π times its radius; therefore, 2π radians equal 360° and

$$1 \text{ radian} = \frac{360°}{2\pi} \cong 57.3°$$

The *angular velocity* of a body is the rate of its angular displacement. Thus, if a grindstone makes 3 revolutions per second, its average angular velocity ω (Greek *omega*) is 6π radians per second. If a body rotates through an angle $\Delta\theta$ in a time Δt, its average angular velocity is

$$\omega_{av} = \frac{\Delta\theta}{\Delta t} \qquad \left(\text{Translational } v_{av} = \frac{\Delta s}{\Delta t} \right)$$

The instantaneous angular velocity ω is the limit of the ratio $\Delta\theta/\Delta t$ as Δt approaches zero. (Translation $v = $ limit of $\Delta s/\Delta t$ as Δt approaches zero.)

If the angular velocity of a body changes from ω_0 to ω in a time Δt, its average angular acceleration α_{av} (greek *alpha*) is

$$\alpha_{av} = \frac{\omega - \omega_0}{\Delta t} = \frac{\Delta\omega}{\Delta t} \qquad \left(\text{Translational } a = \frac{\Delta v}{\Delta t} \right) \tag{1}$$

The angular acceleration is the limit of the ratio $\dfrac{\Delta\omega}{\Delta t}$ as Δt approaches zero. (Translational $a = $ limit of $\Delta v/\Delta t$ as Δt approaches zero.) If a body with an initial angular velocity ω_0 has a constant angular acceleration α, it

will turn through an angle θ in a time t such that

$$\theta = \omega_0 t + \tfrac{1}{2}\alpha t^2 \quad \text{(Translational } s = v_0 t + \tfrac{1}{2}at^2)$$

The angular velocity ω of a body and the linear velocity v of a point on that body are related by

$$\omega = \frac{v}{r} \tag{2}$$

in which r is the radius of the circle on which the point moves. Also, the angular acceleration α and the linear acceleration a are interrelated by

$$\alpha = \frac{a}{r} \tag{3}$$

To prove the relations (2) and (3), consider a wheel of radius r rotating with angular velocity ω. Assume that the wheel turns through an angle $\Delta\theta$ in a time Δt, while a point on the rim moves a distance Δs with velocity v. Then

$$\omega = \frac{\Delta\theta}{\Delta t} = \frac{\Delta s/r}{\Delta t} = \frac{v}{r}$$

From equations (1) and (2),

$$\alpha = \frac{\Delta\omega}{\Delta t} = \frac{\Delta v}{r\,\Delta t} = \frac{a}{r}$$

Torque, Moment of Inertia, and Angular Acceleration

In studying translational motion, you learned that the velocity of a body remains constant unless it is acted upon by an unbalanced, external force, and that the acceleration of the body is proportional to the force. Angular accelerations are caused by unbalanced *torques*. Exert a torque on the handle of a grindstone, and you will accelerate it. Release the handle, and the angular velocity will gradually decrease because of the torque due to the friction of the bearings.

An unbalanced torque τ acting on a body produces an angular acceleration α given by

$$\tau = I\alpha \quad \text{(translational } F = ma)$$

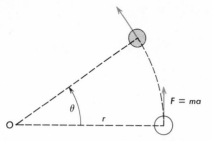

Fig. 9-14. The resultant force F *causes a linear acceleration* a $=$ F/m. *The torque* rF *causes an angular acceleration* $\alpha = \tau/mr^2$.

in which I is called the **moment of inertia** (or rotational inertia) about the axis of spin. What does the moment of inertia depend upon?

We recall that a rigid body is composed of many particles. We shall first consider the application of Newton's second law to a single rotating particle. Let a particle of mass m move on a circle of radius r. Let a resultant force F act on the particle in a direction at right angles to the radius (Fig. 9-14). This force will cause an acceleration a such that

$$F = ma$$

Multiply both sides of the equation by r.

$$r \times F = mra \qquad (5)$$

The angular acceleration α (Greek *alpha*) is given by

$$\alpha = \frac{a}{r} \qquad \text{(See equation 3)}$$

Thus, from equations (3) and (5),

$$r \times F = mr^2\alpha$$

But $r \times F$ is the torque τ, causing the angular acceleration α. We define mr^2 as the moment of inertia I of the particle about the axis through the point O so that

$$\tau = I\alpha$$

Suppose that two particles of masses m_1 and m_2 are mounted on a massless rod which can rotate about an axis through the center of

gravity of the system, perpendicular to the line between the particles (Fig. 9-15). Then the total moment of inertia of the system about the axis is given by

$$I = m_1r_1^2 + m_2r_2^2$$

An equation of this form but containing many terms can be set up for any extended body, r_1, r_2, r_3, etc. being the distances of the respective masses from the axis of rotation. Thus the moment of inertia of a body is the *sum* of the moments of inertia of the particles composing it. To take a simple example, a thin hoop of mass m and radius r can be thought of as composed of many particles of mass Δm — each at the same distance r from the center of the hoop. The moment of inertia of the hoop about an axis through the center is the sum of the moments of inertia of the particles. Thus

$$I = \Sigma(\Delta m)r^2 = r^2\Sigma(\Delta m) = mr^2$$

The moment of inertia of a thin-walled hollow cylinder about its central axis is similarly equal to mr^2.

The moment of inertia of a disk (or a solid cylinder) of radius r about an axis through the center is $\frac{1}{2}mr^2$. The moment of inertia of a uniform sphere, rotating like the earth on an axis through its center, is $\frac{2}{5}mr^2$. The value for a thin rod of length l rotating on a perpendicular axis through its midpoint is given by $I = \frac{1}{12}ml^2$. If the axis passes through one end of the rod, $I = \frac{1}{3}ml^2$ (Fig. 9-16). The units of I are mass units times distance units squared: kilogram-meter^2, for example.

Thus the moment of inertia of a body measures its resistance to angular acceleration. It differs from inertia in linear motion in one

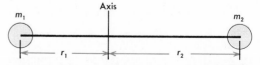

Fig. 9-15. The rotational inertia of the system about the axis shown is $I = m_1r_1^2 + m_2r_2^2$.

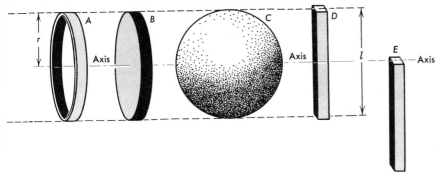

Fig. 9-16. *Moments of inertia:* (A) mr^2, (B) $\frac{1}{2}mr^2$, (C) $\frac{2}{5}mr^2$, (D) $\frac{1}{12}ml^2$, (E) $\frac{1}{3}ml^2$.

respect. The inertia of a body—its resistance to translational acceleration—depends solely upon its mass. The moment of inertia of a body depends not only upon its mass but also upon the manner in which the mass is distributed about the axis of rotation. The flywheel of a gasoline engine has a large moment of inertia so that the engine shaft may keep turning during the intervals between explosions. To this end, the wheel should be very massive, and most of the metal should be located in the rim, as far from the axis of rotation as is conveniently possible.

In Fig. 9-17 two cylinders are mounted at the ends of a rod that can revolve on a vertical axis. The load *A* exerts a force *F* on the string which is wound on the wheel *B* of radius *r*. The force causes a torque $\tau = rF$ that accelerates the system. If we release the weight and count the number of revolutions that the system makes in a time *t*, we can compute the acceleration α from $\theta = \frac{1}{2}\alpha t^2$. If we repeat the experiment after moving the cylinders closer to the axle, the system will have a greater acceleration. Thus, moving the cylinders closer to the axis of rotation decreases the

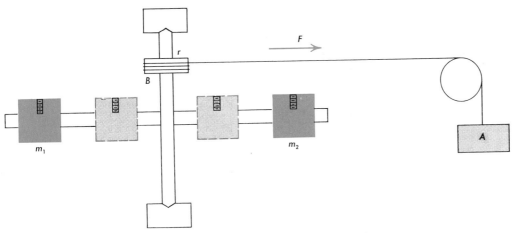

Fig. 9-17. *Angular acceleration apparatus. The unbalanced torque* $\tau = rF$ *causes an acceleration such that* $\tau = I\alpha$.

moment of inertia about that axis. Is the force *F* equal to the weight of *A*? (Be careful!)

Angular Momentum

The momentum relationships for rotational motion resemble those for translational motion. By definition, the angular momentum *L* of a rotating body is the product of its moment of inertia *I* and its angular velocity ω.

$$L = I\omega \qquad \text{(Translational } \mathscr{M} = mv\text{)}$$

Angular momentum is sometimes called moment of momentum. Suppose that the particle in Fig. 9-14 at some instant has a velocity *v* and hence a linear momentum *mv*. The moment of its linear momentum is defined as $(mv)r$. But $v = \omega r$ where ω is the angular velocity of the particle. Hence $(mv)r = (mr^2)\omega = I\omega$.

Just as the resultant force equals the rate of change of linear momentum, the resultant torque equals the rate of change of angular momentum.

$$\tau = \frac{\Delta L}{\Delta t} \qquad \left(\text{Translational } F = \frac{\Delta \mathscr{M}}{\Delta t}\right)$$

The total angular momentum remains constant if no external unbalanced torque acts upon the system. This is the law of conservation of angular momentum.

A good illustration of constant angular momentum is the earth itself. For millions of years it has continued to spin with practically constant angular velocity and momentum. (However, the earth is not a perfect sphere, and the sun's attraction exerts a torque that makes the earth's axis wobble or precess with a period of about 25,000 years.)

When the moment of inertia of a spinning body changes, due to a redistribution of its mass, its rate of spin must change so that its angular momentum is conserved. Attach a ball to one end of a cord which passes through a hollow tube or pipe, and whirl the ball in a circle. Pull the cord suddenly through the tube so as to decrease the radius of the circle on which the ball travels. Since the line of action of the force passes through the center of rotation, the force exerts no torque on the ball. Drawing the ball in closer to the axis diminishes its rotational inertia. The angular velocity must increase to keep the angular momentum constant. The ball will revolve faster as you pull it in. Let a student stand on a stool which can rotate with little friction; set him into rotation. Then he can control his rate of spin by raising or lowering his arms. When he raises them to a horizontal position, his moment of inertia increases and his spin velocity decreases. Lowering them again increases the spin velocity. A high diver jumps from a springboard and acquires a certain angular momentum. As he moves through the air, he doubles up or straightens out so as to change his moment of inertia. In this way, he controls his angular velocity so as to strike the water head first, or feet first, at will.

If you hold a cat feet up and release him at an elevation of a yard or more above the floor, he will land on his feet. How does he accomplish this? Figure 9-18 shows the cat falling. In order to turn himself in midair, kitty quickly performs the following operations: First, he folds up his front legs so that the front part of his body, *A*, has a small moment of inertia. Then he twists *A* forward through a large angle and *B* backward through a smaller angle. Next, he stretches out his front legs and folds his hind legs. Now he can turn section *B* forward while *A* turns backward a little. By repeating these operations, he quickly rotates his body enough to land on his feet. A month-old kitten can turn himself, without any practice, in a half-second. A cat does not learn this skill. He inherits his knowledge of physics!

Rotational Kinetic Energy

The rotational kinetic energy of a body of moment of inertia *I* is equal to the work done by a resultant torque in accelerating the rotating body to its final velocity and is given by

$$(E_K)_R = \tfrac{1}{2}I\omega^2 \qquad \text{(Translational } E_K = \tfrac{1}{2}mv^2)$$

To show that this is correct, consider a constant unbalanced torque τ acting on the body while it accelerates from rest to a final velocity ω while turning through an angle θ. The work done is

$$W = \tau\theta \qquad \text{(Translational } W = Fs)$$

since

$$\Delta W = F\,\Delta s = F(r\,\Delta\theta) = (Fr)(\Delta\theta) = \tau\,\Delta\theta$$

$$\tau = I\alpha$$

and hence

$$\tau\theta = (I\alpha)\theta$$

The angle θ is related to the angular acceleration and the angular velocity by

$$\omega^2 = 2\alpha\theta \qquad \text{(Translational } v^2 = 2as)$$

Hence

$$W = \tau\theta = (I\alpha)\,\frac{\omega^2}{2\alpha} = \tfrac{1}{2}I\omega^2 = (E_K)_R$$

Example. The radius of a 100-kg flywheel is 1.00 m. It revolves at a rate of 3.2 rev/sec. Assuming that all the metal is located in its very thin rim, compute (a) the rotational inertia of the wheel, (b) its angular velocity, and (c) its kinetic energy.

(a) $I = mr^2 = 100$ kg $\times 1.0$ m$^2 = 100$ kg-m^2

(b) $\omega = 3.2 \times 2\pi$ radians/sec
     $= 20.1$ rad/sec

(c) $(E_K)_R = \tfrac{1}{2} \times 100$ kg-m$^2 \times 404$ rad^2/sec^2
     $= 2.0 \times 10^4$ kg-m^2/sec^2
     $= 2.0 \times 10^4$ joules

Translation and Rotation

Often, both translational and rotational motion are present in a physics problem. We can solve such problems by considering the

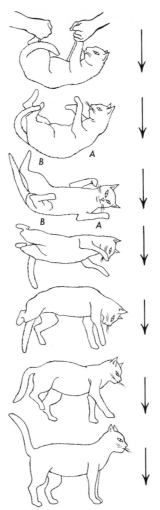

Fig. 9-18. How a cat lands on its feet.

translational motion of the center of mass of the rigid body combined with rotational motion about the center of mass. In the translational motion, the body acts as if it were a particle: the mass of the "particle" is the total mass of the body, and the resultant force acting on the "particle" is the resultant of all of the external forces acting on the body regardless of where these forces are applied. After determining the resultant force F, one applies $F = ma.$ In the rotational motion, we calculate the resultant of all external torques

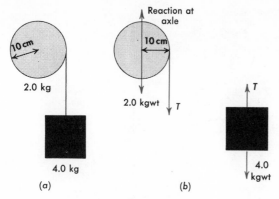

Fig. 9-19. (a) *A block attached by a cord to a massive disk.* (b) *Free-body diagrams.*

about an axis through the center of mass and calculate the moment of inertia about the same axis. The governing equation is $\tau = I\alpha$. In the final steps of the solution, one uses equations that express the spatial relationships between translational and rotational motion such as $s = r\theta$, $v = r\omega$, and $a = r\alpha$.

Conservation of total mechanical energy can also be applied. Note, however, that the kinetic energy now includes *both* the kinetic energy of translation ($\frac{1}{2}mv^2$) and the kinetic energy of rotation ($\frac{1}{2}I\omega^2$).

Example 1. A 4.0-kg block (Fig. 9-19) is attached to a light, flexible, stretchless cord which is wound around the rim of a disk of mass 2.0 kg and radius 10 cm. What is the linear acceleration of the block when it is released and what is the angular acceleration of the disk? Assume that friction on the axle of the pulley is negligible.

We apply $F = ma$ to the translational motion of the block, and $\tau = I\alpha$ to the rotational motion of the disk. Free-body diagrams, showing the forces on the disk alone and the block alone are given in Fig. 9-19b. Let the tension in the cord be T. Then for the block,

$$F = ma$$

$$(4.0)(9.8) \text{ newtons} - T = (4.0 \text{ kg})a \qquad (1)$$

and for the disk

$$\tau = I\alpha$$

$$T(0.10 \text{ m}) = (\tfrac{1}{2}2.0 \text{ kg } 0.01 \text{ m}^2)\alpha \qquad (2)$$

where we have calculated the resultant torque and the moment of inertia ($I = \frac{1}{2}mr^2$) about the center of mass of the disk.

We notice that we have two equations, (1) and (2), with three unknown quantities: T, a, and α. We need another relation. It is

$$a = r\alpha = (0.10\text{m})\alpha \qquad (3)$$

because the cord is "stretchless." Substituting equation (3) into equation (2) and solving equations (1) and (2) for a, one obtains

$a = 7.84 \text{ m/sec}^2$, the acceleration of the block.

Substituting this value for a in equation (3)

$$\alpha = 78.4 \frac{1}{\text{sec}^2}$$

$$= 78.4 \text{ rad/sec}^2,$$

the angular acceleration of the disk.

Example 2. A metal block and a cylinder of equal mass "run a race" starting from rest at the top of an inclined plane. The block slides and the cylinder rolls. (*a*) Assuming that no mechanical energy is dissipated in heat, which will win? (*b*) What are their final translational velocities?

(*a*) Note that the block and cylinder have equal potential energies at the top of the hill, and therefore they must have equal total kinetic energies at the bottom. Part of the cylinder's energy at the bottom is rotational, and hence part only is translational. Therefore, the cylinder has smaller final translational velocity than the block. The cylinder requires more time for the journey, and it loses the race.

(*b*) The velocities of the block and the cylinder may be found by applying the law of the conservation of energy. Assume that the height of the inclined plane is h and that the mass of the block is m. The potential energy of the block at the top of the plane, mgh, is equal to its kinetic energy $\frac{1}{2}mv^2$ at the bottom.

Therefore,

$$v_B = \sqrt{2gh}$$

Let the mass of the cylinder be m and its radius r so that its moment of inertia, I, is $\frac{1}{2}mr^2$. Let its linear velocity at the bottom of the plane be v and its angular velocity $\omega = v/r$.

As before,

$$E_P = E_K$$
$$mgh = \frac{1}{2}mv^2 + \frac{1}{2}I\omega^2$$
$$= \frac{1}{2}mv^2 + \frac{1}{2}(\frac{1}{2}mr^2)(v^2/r^2)$$
$$= \frac{3}{4}mv^2$$
$$v = \sqrt{\frac{4}{3}gh}$$

The gyroscope

A gyroscope is a massive wheel (Fig. 9-20) mounted on gimbals so that it can revolve around the axes *AB*, *CD*, and the vertical axis *Y*. Suppose that the end *B* of the axle points north and that the wheel spins clockwise about the spin axis *AB*. Steadily turn the system so that the end *B* points east. Then the upper half of the wheel which was moving eastward initially will continue to move eastward, the lower half westward, and the wheel will tilt on the axis *CD* so that the end *B* dips down. Thus, an attempt to rotate the system about the *Y* axis causes it to tilt on the axis *CD*. This effect is called *gyroscopic precession*.

Suspend a small body at the end *A* of the axle of the gyroscope. If the wheel is at rest, the loaded end will move downward; if spinning, the gyroscope will revolve about the vertical axis *Y*, and the end *A* will swing around in a horizontal circle, clockwise as viewed from above. Thus, an attempt to turn the system on the axis *CD* causes it to revolve on the vertical axis *Y*.

If *no* resultant torque acts on the gyroscope, its angular momentum will remain unchanged, and it will continue spinning with its axis of spin fixed in space.

The precession of a gyroscope may be

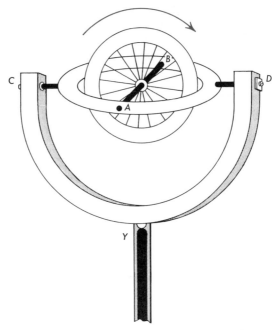

Fig. 9-20. A gyroscope. It can rotate independently on the three axes AB, CD, *and* Y. *When the wheel spins on axis* AB *and you try to tilt it on axis* CD, *it rotates on* Y.

explained by taking into account the fact that angular momentum and torque are vector quantities. The gyroscope wheel (Fig. 9-21) rotates with an angular velocity ω and an angular momentum $I\omega$. This angular momentum is represented by the vector *OA* pointing east; the direction of the angular momentum is chosen to be that in which a right-handed screw would advance if turned in the same way that the wheel is rotating. Suppose, now, that you exert a torque on the disk, tending to tilt its upper edge to the right. The direction of the torque is that in which a right-handed screw would advance if the torque acted upon it. Here the torque is directed *into* the paper or to the north in the figure. The vector *OB*, representing the *change* of angular momentum $\Delta(I\omega)$ produced by the torque in a small interval of time, points in the same direction as the

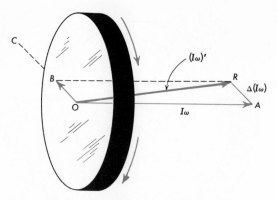

Fig. 9-21. Precession explained by vectors. The vector OA *represents the initial angular momentum of the spinning disk.* OB (*or* AR) *is the change of angular momentum produced by a torque along* OB. OR, *the final angular momentum, is turned into the paper. Thus a torque along* OB *makes the axle revolve in a horizontal plane.*

torque vector and is directed toward the north. The resultant of these two vectors, *OR*, is directed east by northeast, and is the final angular momentum $(I\omega)'$. Thus your torque along *OB* causes the axis of the gyroscope to swing around in a *horizontal* plane, that is, to rotate about a vertical axis.

Gyroscopic compasses are always employed in submarines and often on other vessels. They are especially useful in aeronautics. The gyroscope wheel is kept spinning by an air jet. A loaded ring attached to the gyroscope suspension sets up a torque that tends to make the gyroscope point its axle along a north-south line, regardless of the turning of the airplane. Gyroscopes are also used as horizon indicators, and in the automatic piloting of airplanes. Suppose that the airplane flies northward and that the gyroscope spins clockwise on an axis parallel to the motion. If the airplane deviates to the right, the forward end of the axle of the gyroscope tilts downward, closing an electric circuit which operates levers to bring the airplane back to its course.

In *inertial systems of guidance,* used increasingly in planes, submarines, and rockets, three gyroscopes with their axes of spin mutually perpendicular establish a fixed coordinate system that enables the automatic navigator to determine position with respect to the earth. The inertial navigation device determines location by finding the direction of a plumb line in the coordinate system set up by the spinning gyroscope. If the plumb line changes direction by 90°, the moving object has traveled a quarter of the way around the earth. Corrections for the rotation of the earth and the acceleration of the plumb bob during the motion are made automatically.

Similarities Between Translational and Rotational Units

Table 9-1 shows some similarities between the units of translational motion and those of rotational motion.

Table 9-1. Corresponding Units

Translational or Linear Motion		Rotational or Angular Motion	
	Units		*Units*
Distance s	m	Angle θ	rad
Velocity v	m/sec	Angular velocity $\omega = \theta/t$	rad/sec
Acceleration a	m/sec²	Angular acceleration α	rad/sec²
Mass m	kg	Moment of inertia I	kg-m²
Force $F = ma$	newton	Torque $\tau = I\alpha$	m-newton
Momentum mv	kg-m/sec	Angular momentum $I\omega$	kg-m²/sec
Work $F \times s$	joule	Work $\tau \times \theta$	joule
$E_k = \frac{1}{2}mv^2$	joule	$(E_k)_R = \frac{1}{2}I\omega^2$	joule

Summary

The tendency of a force to produce rotation about an axis is the *torque* of the force about that axis.

The *lever arm* of a force is the perpendicular distance from the axis to the line of action of the force. Torque = Lever arm × Force.

The *two conditions of equilibrium* of a body are:

1. The resultant of the forces acting on it must be zero. $\Sigma F = 0$.
2. The sum of all the positive and negative

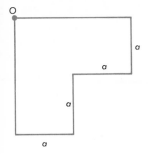

Fig. 9-22.

torques acting on it about any axis must be zero. $\Sigma_\tau = 0$.

The *center of gravity* of a body is the point through which the resultant of the weights of all its particles acts. In *unstable* equilibrium, a slight rotation of the body sets up an unbalanced torque. In *stable* equilibrium, a slight rotation sets up a restoring torque. In *neutral* equilibrium, the torques are in balance for all angular positions of the body.

The laws of rotational motion resemble those of translational motion. If no unbalanced external torque acts on a spinning body, its angular velocity ω and its angular momentum $I\omega$ remain constant.

The moment of inertia I of a ring of mass m and radius r, rotating like a flywheel on an axis through its center, is given by $I = mr^2$; that of a uniform circular disk by $I = \frac{1}{2}mr^2$. Torque $= I \times \alpha$. $(E_K)_R = \frac{1}{2}I\omega^2$.

A gyroscope acted upon by a torque precesses about an axis at right angles to the axis about which the torque acts. If no resultant torque acts on a gyroscope, the direction of its axis of spin remains fixed in space.

A radian is an angle such that the arc distance is equal to the radius: $360° = 2\pi$ radians.

Questions

1. Define torque, lever arm, and fulcrum.
2. Define the center of gravity of a body and show how it may be located experimentally.
3. Define stable equilibrium, neutral equilibrium, and unstable equilibrium, and give an example of each.
4. State the two conditions of equilibrium. When are torques considered positive and when negative? What are some common units of torque?
5. Give an illustration of a body that is not in equilibrium although the sum of the forces acting on it is zero.
6. Can the sum of the torques acting on a body be zero if the resultant force is not zero?
7. In closing a heavy door, what should be the angle between the force and the door to produce maximum torque?
8. Specify the coordinates (with reference to O) of a balance point for this thin plate (Fig. 9-22).
9. One end of a meter stick rests on a cork floating in water; the other end is attached to a string in which there is a tension directed upward so that the meter stick is at an angle of 30° with the water surface. What is the angle between the string and a line perpendicular to the water surface?
10. If the bar shown in Fig. 9-23 is in equilibrium, what must be the value of the angle θ? All the forces are in the plane of the paper.
11. State for rotating bodies three laws of motion that are analogous to Newton's three laws for linear motion.
12. State the rotational analogues to the following: $F = ma$, $W = Fs$, $E_K = \frac{1}{2}mv^2$, $P = Fv$, and $\mathcal{M} = mv$.

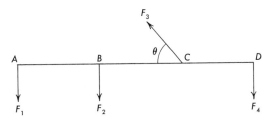

Fig. 9-23.

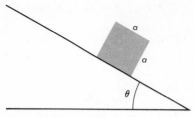

Fig. 9-24.

13. Define a radian. How many radians are equivalent to 360°?

14. Estimate the velocity of rotation, in radians per day, (*a*) of the earth on its axis, (*b*) of the earth in its orbit, (*c*) of the moon in its orbit, and (*d*) of the moon on its axis.

15. Give an example of a body that has rotational kinetic energy. Upon what does the rotational kinetic energy depend?

16. What conservation principle does a diver use in doing the somersault? Discuss.

17. Explain how an acrobat can increase his angular velocity when he doubles up his body in midair.

18. Why does a disk rolling down a hill have a smaller linear acceleration than if it were sliding without friction?

19. If a ring, a solid disk, and a sphere of the same mass and radius roll down an inclined plane, which one will get to the bottom first? Why?

20. A small airplane travels westward. Its single propeller rotates clockwise as viewed by the pilot. If the airplane turns toward the south, does it tend to nose down or to tilt upward? Why?

21. Estimate your linear speed relative to the earth's axis (*a*) at the equator, (*b*) at 45° latitude, (*c*) at a pole.

22. What condition is required for the block to slide, rather than tip over, on the rough inclined plane shown in Fig. 9-24?

23. What is precession? Does the earth precess?

24. A ball tied to a string revolves in a circle. Does the kinetic energy of the ball increase (*a*) if the string winds onto a rod so as to decrease the radius of the circle, (*b*) if someone pulls the string through the hole in a tube so as to decrease the radius? Explain.

25. If two metal spheres of equal mass and radius, one of which is hollow, roll down an inclined plane, which will reach the bottom first?

26. A solid disk and a ring of the same diameter and mass roll down from *A* to *B*, then up toward *C* (Fig. 9-25). (*a*) Which reaches *B* first? (*b*) Which reaches *C* first? Discuss.

27. As a rotating sphere is heated, its radius expands. What change takes place in its angular velocity?

28. Solid spheres *A* and *B* roll down inclined planes as shown in Fig. 9-26. Under what conditions will they reach point *O* at the same time?

Problems

A

1. In a human jaw the distance from the pivots to the front teeth is 7.5 cm, and the muscles are attached at points 2.5 cm from the pivots. What force must the muscles exert to cause a biting force of 40 kgwt?

2. A meter stick is pivoted at the 50-cm mark. If a 100-gwt load is attached at the 100-cm mark, what load must be attached at 25 cm to balance the system?

3. An airplane propeller revolving at a rate of 3.0 rev/sec has an acceleration of 5.0 rev/sec² for 20 sec. What is its final rotational velocity (*a*) in rev/sec, (*b*) in rad/sec?

4. An outboard motor is started by means of a rope wound on a drum 25 cm in diameter. The tension on the rope is 15 kgwt. (*a*) What torque is exerted on the drum? (*b*) What is the

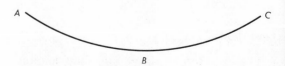

Fig. 9-25.

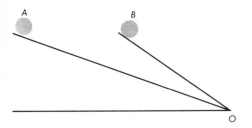

Fig. 9-26.

speed of the rope when the drum turns 5.0 rev/sec if the rope does not slip on the drum?

5. (a) What is the average angular velocity of the minute hand of a clock? (b) If the hand is 8.0 cm long, what is the average linear speed of the tip?

6. The mass of a grindstone is 50 kg and its radius is 0.50 m. (a) What is its moment of inertia about its axis? (b) When it rotates 1 rev/sec, what is its angular velocity?

7. A rope wound onto the rim of the grindstone described in problem 6 exerts a pull of 20 newtons. Find (a) the torque, (b) the angular acceleration, and (c) the angle of rotation during the first second after starting.

8. An automobile wheel has a mass of 30 kg. Assuming that it is equivalent to a ring of equal mass and 0.30 m in radius, find its moment of inertia.

9. What is the kinetic energy of a 5.0-kg cylinder rolling on a level surface at 4.0 m/sec?

B

10. What is the total linear acceleration of a point on the circumference of a wheel 1.0 m in diameter, which has an angular acceleration of −3.0 radians/sec² and an angular velocity of 1.0 radian/sec?

11. A man exerts a downward force of 20 kgwt on one of the pedals of a bicycle. The pedal is 25 cm from the axle about which it revolves. Find the torque about this axle when the pedal crank makes the following angles with the vertical: (a) 90° (b) 60°, (c) 30°, (d) 0°.

12. A 50-kgwt uniform plank 9.0 m long

rests with one end projecting over the edge of a wharf. The length of the projecting portion is 3.0 m. How far from the edge of the wharf can a 40-kgwt boy safely venture on the plank?

13. A uniform square barn door, 2.0 m on each edge, weighing 40 kgwt, is supported by two hinges at the extremities of one side. Using the conditions of equilibrium, find the force exerted by the upper hinge.

14. The center of gravity of an irregular log is 2.0 m from one end. The log is 5.0 m long and weighs 40 kgwt. If two men support it at its ends, how much force must each exert?

15. A 75-kg man and a 25-kg boy carry a 30-kg uniform pole 3.0 m long. If the boy supports one end, at what distance from this end must the man grasp the pole in order to carry two-thirds of the load?

16. A hoop weighs 0.25 kgwt and its radius is 0.50 m. It rolls along a pavement, making 1.00 rev/sec. Find (a) its moment of inertia, (b) its angular velocity, (c) its rotational kinetic energy.

17. The flywheel and other moving parts of an automobile engine have a moment of inertia equal to that of a ring whose weight is 16 kgwt and radius 0.10 m. (a) What is the moment of inertia of the system? (b) What is its kinetic energy when it makes 1800 rev/min? (c) What average power is required to bring it to this velocity in 5.0 sec?

18. A pull of 2.04 kgwt is exerted on a cord that is wound on the rim of a solid disk. The radius of the disk is 20 cm and its mass is 36 kg. Find (a) the torque acting on the disk, (b) the angular acceleration, (c) the angular displacement from rest in 5.0 sec.

19. A 1600-kg automobile accelerates from 1.0 m/sec to 8.0 m/sec in 8.00 sec. The tires have an outer radius of 40 cm. What is their angular acceleration?

20. A lawn roller weighs 50 kgwt and its radius is 0.40 m. (a) Assuming that all its mass is located in the rim, what is its moment of inertia? When it rotates once per second, what are (b) its angular momentum and (c) its rotational kinetic energy?

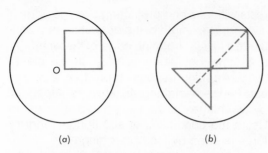

Fig. 9-27.(a) *Thin square block on a thin disk.*
(b) *Thin square block and thin triangular block on a thin disk.*

21. Two boys, each with a mass of 30 kg, are seated at the opposite ends of a plank 2.0 m long. The plank is pivoted on a vertical axis, and the system rotates in a horizontal plane about the midpoint of the plank, the angular velocity being 5.0 rev/min. What will the angular velocity be if both boys move to positions 0.50 m from the midpoint without touching the ground?

22. A 5-cent piece of mass 5.00 gm and radius 1.00 cm rolls on a level tabletop with a velocity of 3.00 cm/sec. (*a*) What is its angular velocity? (*b*) What is its total kinetic energy?

C

23. Loads of 2, 4, 6, 8, and 10 kgwt are attached at 1.0-m intervals along a uniform rod 4.0 m long, weighing 5 kgwt. At what distance from the 2-kgwt load must the rod be supported in order to produce equilibrium?

24. The down force exerted by the front wheels of an automobile is 600 kgwt; that exerted by the rear wheels is 500 kgwt. The distance from the front axle to the rear axle is 250 cm.(*a*) What is the distance of the center of gravity of the car from the front axle? (*b*) What is the weight of the car?

25. An 80-kgwt man stands on tiptoes. The distance from his toes to the point where his leg bones act is 22 cm. That from the toes to the point where the "tendon of Achilles" pulls on the end of his ankle bone is 27 cm. What

down forces do his leg bones exert on his ankles?

26. A uniform ladder 5.0 m long, weighing 20 kgwt, rests against a smooth, vertical wall, making an angle of 30° with it. A man puts his foot against the bottom of the ladder to hold it steady and seizes a rung 1.0 m from the lower end of the ladder. What horizontal force must he exert in order to tilt the ladder?

27. For both parts of Fig. 9-27, find the distance of the center of mass from a perpendicular axis through the center of the disk whose radius is 10 cm. Assume the blocks and the disks are made of the same material.

28. A square piece of sheet iron is 50 cm on each side. A square 10 cm on each side is cut from one corner. What is the distance of the center of gravity of the remainder from the opposite corner?

29. A cubical box, 1.0 m on each side, weighing 50 kgwt, rests on a level floor. (*a*) What horizontal force exerted against the upper edge of the box will cause it to tilt? (*b*) What is the *least* force that will cause the box to tilt, and how should the force be applied?

30. A uniform ladder 6.0 m long, weighing 21 kgwt, leans against a smooth vertical wall. The distance from the foot of the ladder to the wall is 3.0 m. What force does the wall exert on the ladder when a man weighing 80 kgwt stands (*a*) at the middle of the ladder, (*b*) at a point 1.0 m from its lower end?

31. What horizontal force applied at the hub is required to raise a 20-kgwt wheel of radius 50 cm over an obstacle 10 cm high?

32. An iron ring that weighs 20 kgwt is supported in a horizontal plane by three posts 120° apart. (*a*) Where should a boy apply a downward force to tip the ring with least effort? (*b*) How large should this minimum force be?

33. Ten seconds after an electric fan is turned on, the angular velocity of the blades is 480 rev/min. Find their average angular acceleration.

34. A flywheel 3.0 m in diameter is rotating

40 rev/min when thrown out of gear. If the diameter of the axle is 10 cm and the coefficient of friction is 0.05, find the time required for it to come to rest. (Assume the mass of the flywheel is concentrated in its rim.)

35. A solid cylinder of mass 0.24 kg and radius 0.020 m starts from rest and rolls down a 30° inclined plane 1.96 m long. What is (a) its final linear velocity, (b) its linear acceleration? (c) What would its final linear velocity be if it slid without friction down the plane?

36. An iron hoop of radius 0.50 m and mass 2.0 kg starts from rest and rolls downhill, acquiring a velocity of 4.9 m/sec at the bottom of the hill. Find (a) its final translational kinetic energy, (b) its final rotational kinetic energy, (c) the height of the hill. (d) What would its final velocity have been if it had slid downhill without friction?

37. A disk of mass 50 kg and radius 1.0 m is pivoted at its center. A rope wrapped around the disk supports a 10-kg block. Calculate the velocity of the block 2.0 sec after it is released.

38. A figure skater spins with arms outstretched at an angular velocity of 1.0 rad/sec. Her moment of inertia is 2.4 kg-m². When she draws in her arms, her moment of inertia becomes 0.15 kg-m². Calculate (a) her angular momentum and (b) her final angular velocity. Assume friction is negligible.

39. The axle of a yo-yo is 0.30 cm in radius, and the radius of each of its disks is 4.0 cm. The mass of the yo-yo is 200 gm. (a) What is its acceleration as it unwinds from the string, the other end of which is at rest? (b) What is the tension in the string? (c) What is the velocity of the center of gravity 0.50 sec after release?

CHAPTER TEN

Relativity

The preceding chapters have dealt with the physics of familiar objects under everyday conditions—falling bodies, accelerating cars, colliding billiard balls, and rotating wheels. Although our analysis of their motions required us to use some carefully defined concepts—acceleration, force, mass, momentum, and energy, for example—we were operating in normal everyday situations. Our concepts of space and time in particular were so ordinary that we had hardly any reason to examine them and certainly no reason to challenge them. The elegant system of mechanics that described this world was the one developed by Newton and his successors, building on the work of Galileo and others, and is usually called *Newtonian* or *classical mechanics.*

For over a century following Newton, this system of mechanics had an almost unbroken string of successes in describing the parts of the physical universe that could be studied then. Classical mechanics was able to describe many laboratory phenomena successfully and provided the basis for major engineering developments. In the next chapter we shall examine one of the greatest triumphs of Newtonian mechanics: its description of planetary motion.

Not everything could be explained by Newtonian mechanics, however. Near the beginning of the nineteenth century, and increasingly as the century wore on, areas of the natural world emerged that required "non-Newtonian" descriptions. Chemistry was one of these—the laws of combination of atoms into chemical compounds could not be derived from Newtonian mechanics. Electricity and magnetism were others.

These exceptions were difficult to reconcile with belief in a "Newtonian" world. What was even more difficult was that the familiar concepts of space and time—the keystones of Newtonian mechanics—came into question. This chapter will discuss how this came about and will introduce some of the most fundamental ideas of modern physics—those of relativity. These concepts, contained in the *special theory of relativity,* challenge older and more familiar ideas and require us to clarify our thinking about space and time.

Space and Time in Classical Mechanics

Newton distinguished between two kinds of space, relative space and absolute space.

in his great treatise on mechanics, *Mathematical Principles of Natural Philosophy,** Relative space* was described by what we would call a local *frame of reference* or *coordinate frame.* For example, to locate yourself in central New York City, you could specify your location on the grid of intersecting streets and avenues: the southwest corner of Fifth Avenue and Fifty-seventh Street, for example. A location on the earth is usually given by specifying the latitude and longitude of that point: City Hall in New York City is at latitude 40°42′43″ N and longitude 74°0′29″W.

More generally, in physics we represent the positions of bodies on a frame of reference chosen for convenience, as we did in Chapter 6 in discussing the motion of projectiles. In this chapter we shall use *x, y, z* (Cartesian) coordinates to specify locations with reference to three mutually perpendicular axes (Fig. 10-1). Notice that an operation by an observer or experimenter is implied in specifying coordinates: one *measures* the perpendicular distance from P_1 to the *YZ* plane to determine the coordinate x_1, and so on.

Finally, note that there can be many local frames of reference, each related to the others. Thus, you could locate the corner of Fifth Avenue and Fifty-seventh Street in New York City on the earth grid by specifying its latitude and longitude or, alternatively, give the location of City Hall on a street map of New York City. You could also figure out the distance from City Hall on the earth grid to the corner of Fifth Avenue and Fifty-seventh Street on the street map by some straightforward calculations if you knew how the two frames of reference were oriented with respect to each other.

By *absolute space* Newton meant that which was described by a primary, unmoving frame of reference. In his definition:

---

* Published in London in 1686. By "natural philosophy" the seventeenth-century writers meant physics.

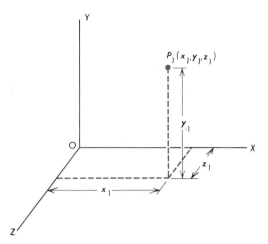

Fig. 10-1. *The location of* P_1 *with reference to axes* X, Y, *and* Z *is given by the three coordinates* x_1, y_1, z_1, *measured as shown.*

"Absolute space, in its own nature, without relation to anything external, remains always similar and immovable."

He was seeking a coordinate frame that would be independent of the moving earth and to which all relative frames could be referred.

An imprecise example is given by an "address game" that children sometimes play: "I live at 72 North Elm Street, Central City, State of Ohio, the United States of America, the Earth, the Solar System, Our Galaxy, the Universe." The difficulty with this scheme, in the Newtonian sense of absolute space, is that the earth, solar system, and galaxy are all measurably moving with respect to other bodies, and *no way has been found to specify their positions in the "Universe,"* which is assumedly the primary frame of reference.

Newton was not sure that a primary, fixed frame of reference *could* be found:

"For it may be that there is no body really at rest, to which the places and motions of others may be referred."

But he went on to say that it might be possible to find some body absolutely at rest "in the remote regions of the fixed stars, or perhaps far beyond them." No such unmoving frame of reference has been found, as we shall see when we return to this question of an absolute frame of reference.

Like space, *time* was considered in the Newtonian world in both a relative and absolute sense. Newton's description is often quoted:

"Absolute, true, and mathematical time, of itself, and from its own nature flows equably without relation to anything external, and by another name is called duration: relative, apparent, and common time, is some sensible and external (whether accurate or unequable) measure of duration by the means of motion, which is commonly used instead of true time; such as an hour, a day, a month, a year."

By this Newton meant that we measure time duration, as a practical, everyday matter, in a variety of ways. All are based on some repetitive motion; contemporary examples include a wristwatch with its vibrating balance wheel, a grandfather clock with its swinging pendulum, the daily rotation of the earth and its annual revolution around the sun, the pouring of sand in an hourglass, or the vibrations of cesium atoms in an atomic clock. Some of these are extremely precise, most of them are not, and all measure relative time. They are at best, according to the Newtonian view, imperfect means of representing absolute time, which, as Newton said and as many people still believe, flows imperturbably on, unaffected by the fact that someone forgot to wind his wristwatch or that the earth's rotation is slowing down and increasing the length of the day.

In principle, however, some super clockmaker could invent a "perfect clock" whose rate would never change. If this were then mounted in a church tower and made visible

to all citizens, it would keep absolute, public time. A boy passing the church could determine from his readings of the clock how long it took him to cross the street. A passenger in a bus passing nearby could also determine how long it took the boy to cross the street. In this classical view of things, the passenger would expect to obtain the same result from the bus as he would if he were standing on the ground near the boy.

Except for certain practical difficulties in making reliable clocks and taking readings from them, time measurements were absolute in classical mechanics — in agreement with everyday observation. All observers could expect to obtain the same result for a given time interval regardless of where they were or how fast they were moving. We shall soon find that modern physics takes a rather different view of things when it deals with new domains of experience.

Moving Systems in Classical Mechanics

Descriptions of the motions of bodies, according to classical mechanics, depend on the frame of reference of the observer. In this respect they are unlike measurements of time intervals. We encountered examples of this in Chapter 6 when we discussed the motion of projectiles.

Suppose, for example, a passenger dropped a book from a height of 1.225 m above the floor of a bus that had a constant velocity of 10 m/sec. How would the passenger on the bus and other observers outside the bus describe the motion of the book? The passenger (Fig. 10-2*a*) would see* the book *fall vertically* and strike the floor of the bus at his feet in about 0.5 sec. An observer

---

* You may wonder how this could be done. Let us equip the bus with large glass windows and provide our observers with stroboscopic cameras that allow multiple exposures of the image of the falling book at 0.10-sec intervals. They can all thus obtain objective data, and the external observers can resist the psychological tendency to put themselves into the frame of reference attached to the bus.

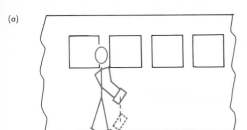

(a)

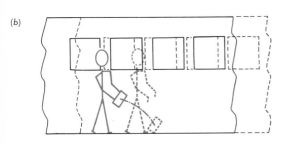

(b)

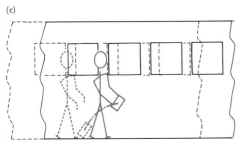

(c)

Fig. 10-2. Frames of reference. (a) To a passenger on the moving bus (and to an observer looking in from a car moving at equal velocity beside the bus), the book falls vertically. (b) To an observer at rest on the ground, the book follows a parabolic path, moving forward as it falls. (c) To an observer looking in from a car moving past the bus at twice its velocity, the book follows a parabolic path, moving backward as it falls.

standing on the street and looking in as the bus went by (Fig. 10-2*b*) would see an entirely different trajectory. He would agree with the passenger that the book struck the floor after a half-second, but he would testify that its path was a *parabola*. An observer riding in a car that traveled beside the bus at a velocity equal to that of the bus would agree entirely

with the passenger: the book fell straight down and struck the floor in a half-second. Finally, an observer looking into the bus from a car that traveled past the bus at a velocity *twice* that of the bus (Fig. 10-2*c*) would say that the book followed a parabolic path *curving backward* and struck in a half-second. Thus all observers would agree that the time interval was the same — a half-second — and that the book landed at the passenger's feet, but two of them would disagree strongly with the other two as to the trajectory of the book through space. Who is right?

Each observer is right, of course, if he has faithfully and accurately observed the phenomenon. No principle of physics exists to assert that the description offered by the passenger is any better or any nearer the "truth" than that offered by the other observers. But if classical mechanics is to be a respectable science it has to be able to reconcile one description with another so that *all* descriptions will be consistent with the basic laws of physics. It does this by recognizing that the four observers are in different frames of reference and that recognition of their relative motions will clear up the apparent differences in their descriptions.

For example, the four observers might meet at some later time and exchange information about their observations. All would agree that the book moved vertically in accord with the laws of uniformly accelerated motion and that nothing was inconsistent about their descriptions of the *vertical motion*. The outside observer standing on the street would then realize that, *relative to his frame of reference* (a frame attached to the ground), the frame of reference of the passenger (attached to the bus) was moving forward 10 m/sec. Thus in the frame of reference of the ground, the book had a constant horizontal velocity of 10 m/sec and its resulting motion in that frame, for reasons that we discussed in Chapter 6, was a parabola. So the passenger and the observer standing outside the bus would finally

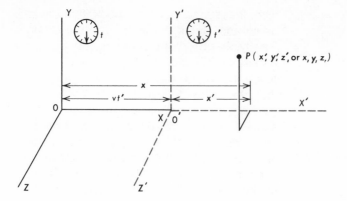

Fig. 10-3. Moving frames of reference. The point P has the coordinates x', y', z', in the X', Y', Z' frame and x, y, z, in the X, Y, Z frame. The velocity of the origin O' relative to O is v.

agree that their observations were consistent. Each would believe that if he had been at the other's observation post he would have observed the trajectory that the other observed. Similarly, the passenger and the observer in a car moving as fast as the bus would recognize that the velocity of the frame of reference of the passenger (attached to the bus) *relative to the frame of reference of the outside observer* (attached to the moving car) was zero. Thus to the outside observer the falling book had a horizontal velocity of zero and its trajectory, as viewed by him, was a vertical line. Finally, the outside observer in the car moving past at 20 m/sec (relative to the ground) would see the frame of reference of the passenger *moving backward* at a constant velocity of 10 m/sec and would observe a parabolic trajectory curving backward.

Transformation Equations

These relations can be formalized in several equations that contain the statement of Galilean* relativity. Consider an $x'y'z'$ frame of

---

* Named for Galileo, whose *Dialogue on the Great World Systems* provided the first scientific clarification of relative motion. Relative motion was an important part of Galileo's argument that the earth could both rotate and travel around the sun without having objects on the surface of the earth "fly off" into space.

reference (Fig. 10-3) which moves with a *constant velocity v* relative to an x, y, z frame. Let us specify that at time zero the origins O' and O and the Y and Y' and Z and Z' axes all coincide and that motion takes place along the X axis. Then for any point P

$$x = x' + vt' \tag{1}$$

$$y = y' \tag{2}$$

$$z = z' \tag{3}$$

$$t = t' \tag{4}$$

where the last equation expresses the belief of classical mechanics that the time coordinate of an event, as measured in the X, Y, Z system is the same as that of the same event measured in the X', Y', Z' system.

The four equations just given are called *transformation equations.* By means of them we can convert the coordinates (both space and time) of an event, as observed by an observer in the one system, into those of the same event as observed by an observer in the other system.

Example. Transform the coordinates of the event represented by the book's hitting the floor of the bus, as described in our earlier example, from the frame of reference of the passenger to that of the observer standing outside on the street. Let zero time be the instant when the book starts to fall and let us further assume that the passenger passes the outside observer at that instant. Then for the passenger the space and time coordinates of the book on impact are $x' = 0$, $y' = 0$, $z' = 0$, and $t' = 0.5$ sec. Since $v = 10$ m/sec, we can use the transformation equations to obtain the corresponding coordinates in the outside observer's frame of reference:

$$x = x' + vt'$$
$$= 0 + (10 \text{ m/sec})0.5 \text{ sec} = 5 \text{ m}$$
$$y = y' = 0$$
$$z = z' = 0$$
$$t = t' = 0.5 \text{ sec}$$

This tells us that the passenger and the outside observer assign the same y, z, and t coordinates to the event and can bring their differing x-x' coordinates into agreement by introducing the velocity of the one frame of reference relative to the other. A final important point is that if they analyze their data, they will *both* conclude that the book falls in accord with the equation $y = \frac{1}{2}gt^2$ and the acceleration g is the same as measured in *either* system. *Thus this law of physics is the same to both observers.*

The Principle of Relativity in Mechanics

Let us now generalize what we learned in the last two sections. We call a frame of reference that is moving with constant velocity an *inertial frame of reference*. As we saw in Chapter 5, Newton's three laws of motion hold true in such a frame.* Suppose that two inertial frames of reference have a velocity v with respect to each other and that an observer in each performs a number of experiments in

* The bus moving at "constant velocity" and the table in your classroom are only approximations to an inertial frame of reference because of the centripetal acceleration of the earth's motion. They are fairly good approximations, however, because the centripetal acceleration due to the earth's rotation is only about 0.3% of g at the equator.

mechanics (Fig. 10-4): for example, he measures the acceleration due to gravity, studies the motion of gliders on a level frictionless air track, observes the relationship between the length of a pendulum and the time for one complete swing, and investigates the velocities of two colliding spheres. From these experiments, each derives certain laws of physics: the constant acceleration due to gravity, Newton's first law of motion, the law of the pendulum (Chapter 18), and the law of conservation of linear momentum, respectively. When the two observers compare their results, they find that they have derived the same laws. Furthermore, if observer A looks over at the experiments going on in B's frame of reference, records data, and derives the resulting laws—*being careful to transform coordinates properly*—he again finds the same laws. B has the same experience when he observes A's experiments. In other words, these laws of physics are insensitive to the velocity of the frame of reference in which they are observed.

None of these mechanics experiments, therefore, can be used as a "velocity indicator" to tell how fast the frame is moving through "absolute" space. In a blacked-out jet airplane moving at a constant velocity of 500 miles per hour in smooth flight, the traveler has no sensations and can make no experiments or observations that would tell him the situation is any different from that when the plane is parked on the ground. To be sure, when the plane takes off or lands or encounters turbulent air, the passenger knows that he is not at rest on the ground, but in the absence of such accelerations he is unable to tell whether he is moving or at rest.

These considerations lead us to the principle of relativity in mechanics:

The laws of mechanics are independent of the inertial (uniformly moving) frame of reference in which they are derived. Any such frame of reference is as good as any other for observing the laws of mechanics.

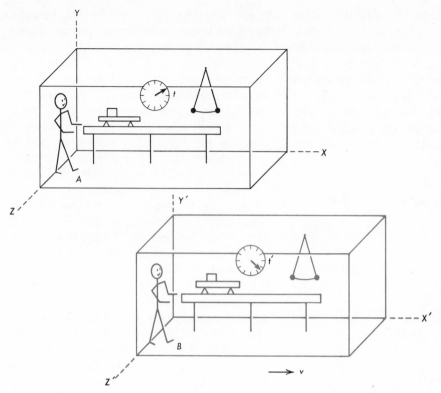

*Fig. 10-4. Each of two experimenters in two physics laboratories, one of which moves with respect to the other with a uniform velocity **v**, observes the laws of mechanics in his own and in the other's laboratory. The laws are the same.*

This principle was recognized by Newton, and classical mechanics is consistent with it. However, in trying to extend it to domains of physics other than mechanics—such as electricity and magnetism—difficulties arose for later physicists, difficulties so serious that they brought about a radical change in our ideas about space and time. Let us now see how this change came about.

The Speed of Light

Light travels through a vacuum with a speed—the magnitude of its velocity—of about 3.00×10^5 km/sec or 186,000 mi/sec. The speed of light in vacuum is *independent of the inertial frame of reference in which it is measured.* In later chapters, we shall examine the nature of light in considerable detail and, among other things, discuss how the speed of light is measured. Here we shall concentrate on the strange fact that the speed of light* is the same for all observers—a fact that is of crucial importance for the theory of special relativity.

To understand why this behavior of light is so strange, consider a more familiar situation. Suppose a gun fires a bullet (Fig. 10-5) with a muzzle velocity whose magnitude is 1000 m/sec from a jet plane which is in level flight

---

* In this chapter, when we refer to the speed of light we shall mean the magnitude of its velocity in a vacuum. In material media, such as glass and water, the speed of light differs from substance to substance.

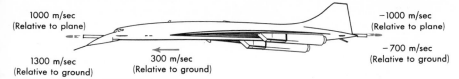

1000 m/sec
(Relative to plane)

1300 m/sec
(Relative to ground)

300 m/sec
(Relative to ground)

−1000 m/sec
(Relative to plane)

−700 m/sec
(Relative to ground)

Fig. 10-5. *Combination of velocities for material objects. The resultant velocity of the bullet (its velocity relative to the ground) is the vector sum of its muzzle velocity and the velocity of the plane relative to the ground.*

at a velocity of 300 m/sec. If the bullet is fired in the forward direction, its velocity to an observer in the frame of reference of the plane is 1000 m/sec, but its velocity to an observer in the frame of reference of the ground is 1300 m/sec. Similarly, if the bullet is fired in the backward direction, its velocity in the frame of reference of the plane is still 1000 m/sec, but in that of the ground is 700 m/sec. Material objects such as bullets, multistage rockets, and people walking down the aisles of moving planes obey the mechanical law of vector addition of velocities.

This is *not* so for light. If an observer measures the speed of a flash of light from a laboratory source (such as an intense lamp) (Fig. 10-6a), he finds that the speed is that given above, approximately 3.00×10^5 km/sec. If he measures the speed of light emitted in the forward direction by very rapidly moving atoms in a beam of particles (Fig. 10-6b),

the speed turns out to be the *same* (within the error of measurement)—3.00×10^5 km/sec—and *not* the ordinary vector sum of the velocity of light and the velocity of the source of light. Thus if we regard light as a stream of energy packets—*photons*, as we shall call them in a later chapter—we see that their behavior is different from that of material particles in that their measured speed in vacuum is always the same.

The Michelson-Morley Experiment

As revealed by some phenomena, light apparently consists of energy-carrying packets; in other phenomena, it seems to be a wave disturbance. Is the observed constancy of the speed of light consistent with the usual behavior of waves?

In Chapter 19 we shall consider waves in more detail, but here we need only note that

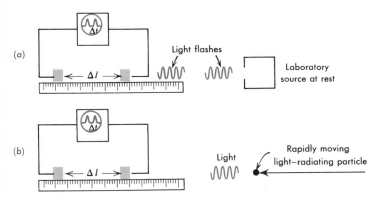

(a)

Light flashes

Laboratory
source at rest

(b)

Light

Rapidly moving
light–radiating particle

Fig. 10-6. *The speed of light is the same whether the source is at rest or is moving rapidly with respect to the observer.*

waves require a medium for their propagation and that the speed of waves of a specified kind is characteristic of that medium. For example, large waves travel over the surface of the ocean with a speed of about 1 m/sec.

The speed of a wave is independent of the speed of the source, but the wave *does* require a material medium. Also, if the medium carrying the waves is moving, the apparent speed of the waves, relative to the observer, will vary. If light is a wave it must have a medium through which to travel. What is the medium for light waves? Does the speed of light waves depend upon the motion of the observer through this medium?

Nineteenth-century physicists thought light traveled through a *luminiferous ether* that pervaded all space and served as the medium for the propagation of light, just as the surface of the ocean was the medium in which ocean waves traveled. Physicists were hopeful that the ether might also prove to be the primary, unmoving frame of reference to which Newton referred as "absolute space." If there were an ether, it was reasoned, the earth moved through it with its orbital speed of 30 km/sec, and the ether streaming past the earth must set up an ether wind. When light traveled with the wind, it might be expected to have a somewhat greater speed *relative to a stationary ether* than when it traveled against the wind. When light traveled *across* the ether wind its speed would be somewhere between these extremes.

For example, suppose that we wish to detect a wind in the stratosphere, far above the clouds, by a somewhat unconventional method. Two equally swift jet planes *A* and *B* run a race, starting from Chicago. *A* travels 300 miles east and back. *B* goes an equal distance north and back. If there is no wind, *A* and *B* will return at the same time. If the wind blows east or west, *A* will lose the race, returning after *B* does. An east wind retards *A* on the outward trip and assists it on the return. You might think that the two effects would annul each other. Actually, the loss due to the

wind when going east is greater than the gain going west.

Beginning in 1881, A. A. Michelson, joined later by E. W. Morley, conducted a series of extremely precise experiments with his new optical interferometer* invented to measure any detectable change in the speed of light with change of direction relative to the supposed ether wind. They were unable to detect any such difference even though their apparatus was quite capable of measuring it if it had existed. Michelson concluded: "The result of the hypothesis of a stationary ether is thus shown to be incorrect."

The experiments of Michelson and Morley showed that the speed of light regarded as a wave was independent of direction of propagation relative to that of the earth's motion. They also showed that the evidence for an ether and for a primary frame of reference was insufficient to support either notion, and both of these ideas gradually disappeared from physics.

The Special Theory of Relativity

The close of the nineteenth century and the opening of the twentieth were years of great change in physics. Startling experimental discoveries—such as the discovery of X-rays and of radioactivity—revealed whole new domains that needed to be studied. Equally important were the new theories that were first proposed in those years, such as the quantum theory.

One of the most far-reaching developments was the announcement of the *special theory of relativity* by Albert Einstein in 1905. The word "special" denotes the limitation of the theory to *non-accelerating* frames of reference as contrasted with the *general theory of*

---

* The interferometer compared the time for light to travel with and against the supposed "ether wind" with that for it to travel an equal distance across the wind. The principles of optical physics underlying the interferometer are discussed in Chapter 37. Here we can consider it as a special type of clock.

relativity, which deals with *accelerating* systems and the effects of gravitational fields.

Situations, such as those we described earlier in this chapter, in which one frame of reference moves with constant velocity with respect to another and in which an observer in one system observes events in the other, are what the special theory of relativity (STR) is all about. What we shall discuss now differs from Galilean relativity in that we must assume that one frame of reference may move with respect to the other with a speed *comparable to that of light.* At these high relative velocities, the concepts of space and time become inextricably connected with each other. When measurements refer to events in a frame moving with respect to the measurer, space and time are no longer separable as they were in classical mechanics. As a result, STR predicts many strange happenings—well-regulated clocks run slow, length intervals contract, the mass of an object changes—predictions that are verified experimentally. At the same time, the theory is in agreement with classical mechanics at speeds that are small compared to that of light. We would require this, because Galilean relativity is a well-verified description of nature at those speeds.

We have sketched a part of the background for STR, starting with classical or Galilean relativity and including the important and unexpected development that the speed of light is the same for all observers. We have not included a number of other reasons for a new theory, reasons which arose in areas of physics that we have not yet considered. Enough has been said, perhaps, to point to the need for the radical new approach that Einstein took in the special theory.

STR is based on just two postulates or special assumptions:

1. The laws of physics are the same for observers in all inertial frames of reference.
2. The speed of light in vacuum (denoted by c and having the value of about 3.00×10^8 m/sec) is the same for all observers.

The first postulate means that observers A and B, traveling in different inertial frames of reference, would not invent different kinds of physical laws: their observations would yield the same basic laws after the appropriate transformations of coordinates had been carried out. The first postulate could be interpreted as a special form of the principle of the uniformity of nature: as far as we know, nature obeys one set of laws, not many different sets. Another way of looking at the first postulate is that it rules out the possibility of detecting the motion of a frame of reference from measurements made entirely within that frame. If the laws of physics were different in one frame than in others, that could be taken as a means of "detecting velocity" from within that frame.

The second postulate is supported by the actual measurements of the speed of light, but there is some evidence that Einstein had come to this conclusion independently of the Michelson-Morley results. It had seemed necessary for a consistent treatment of spacetime to adopt this postulate and, like all bold and innovative thinkers, Einstein accepted a strange requirement and went on to develop his theory, one with far-reaching consequences.

Time dilation

Let us apply these ideas first to observations of time intervals by observers in two inertial frames of reference S and S' moving relative to each other with a speed v comparable to that of light.* Consider the light-flash clock shown in Fig. 10-7. Light flashes in this idealized clock, once emitted, are reflected back

* This situation exists in a laboratory when an observer makes measurements on a beam of high-energy particles whose speed may be 99% that of light. As yet, comparable speeds have not been reached by vehicles carrying living observers. Astronauts reach speeds of 24,200 miles per hour (0.004% of c) as they leave earth orbit on the way to the moon. Perhaps space travelers of the future will approach the speed of light in interstellar space, but until then we shall have to think of our observers in STR situations as traveling in imaginary space ships.

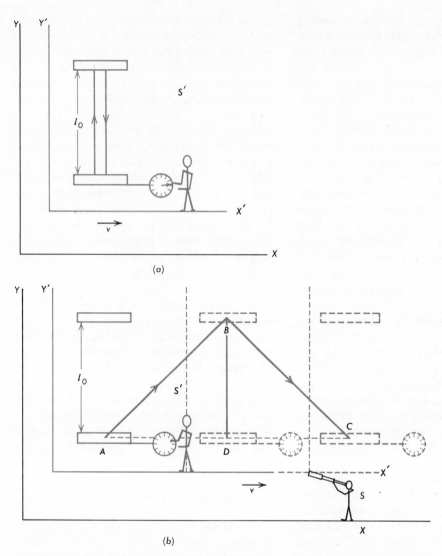

Fig. 10-7. *Time dilation.* (a) *A light-flash clock in the frame of reference* S′ *which moves relative to* S *with a velocity* **v**. (b) *The path of light flashes as viewed by the observer in* S.

and forth between parallel mirrors separated a distance l_0 (measured in the frame S′). Every time a light flash strikes the bottom mirror, after one complete up-and-down trip at speed c, the counter registers the completion of a time interval $\Delta t'$. Thus the clock is similar to a pendulum clock which records complete to-and-fro swings of the pendulum.

The light-flash clock is easier to analyze, however, so we shall use it as our timer instead of a conventional timer. Since the laws of physics must be independent of the apparatus used, according to the first postulate of STR, we are allowed to make this substitution.

As the frame S′ passes him, an observer in S, who knows how the clock operates, watches what is happening in S'. Whereas observer S' sees the light flash go straight up and down in a time $\Delta t'$ (Fig. 10-7a), observer S sees the same light flash follow the path ABC (Fig. 10-7b) in some time Δt. According to classical mechanics Δt would equal $\Delta t'$ because the speed of light in S', as observed from S, would be greater in the same proportion that the distance ABC was greater than $2l_0$. *But the second postulate of STR tells us that the speed of light as observed in S is just c,* the same as observed in S'. Therefore, since the distance ABC is greater than $2l_0$ and the speed of light is the same for both observers, the time interval Δt for the light to travel the distance ABC is *greater* than the interval $\Delta t'$ for the light to travel the distance $2l_0$. In other words, the time interval between two events (light flash leaves and light flash returns) as observed from S is greater than the same interval as observed in S'. But if the time interval between "ticks" of the clock in S' is greater for observer S than for observer S', S must conclude that *the clock in S' is running slow* compared to the performance of a similar clock in his own frame of reference.

Similarly, if observer S' were observing a clock in S and comparing it with his own clock in S', he would see the path of the light flash in S as greater than $2l_0$ and, believing that the speed of light is always c, would conclude that the clock in S was running slow. Each observer would come to the same conclusion: a clock in the other's frame of reference, moving with a speed close to that of light, runs slow. His own clock behaves in "normal" everyday fashion.

With this qualitative understanding of time dilation ("time stretching"), let us calculate how Δt is related quantitatively to $\Delta t'$. For observer S'

$$\Delta t' = \frac{2l_0}{c} \tag{1}$$

Now observer S sees the light flash following the path ABC, so for him the time interval Δt for one up-and-down trip is

$$\Delta t = \frac{\overline{ABC}}{c} \tag{2}$$

But

$$\overline{ABC} = \overline{AB} + \overline{BC} = 2\,\overline{AB} \tag{3}$$

and, by the Pythagorean theorem,

$$\overline{AB}^2 = \overline{BD}^2 + \overline{AD}^2 \tag{4}$$

where

$$\overline{BD} = l_0 \tag{5}$$

and

$$\overline{AD} = \frac{v\,\Delta t}{2} \tag{6}$$

Therefore,

$$\overline{AB}^2 = l_0{}^2 + (\tfrac{1}{4})v^2(\Delta t)^2 \tag{7}$$

and

$$\overline{AB} = \sqrt{l_0{}^2 + (\tfrac{1}{4})v^2(\Delta t)^2}$$

We have, then,

$$\Delta t = \frac{2\sqrt{l_0{}^2 + (\tfrac{1}{4})v^2(\Delta t)^2}}{c}$$

$$= \frac{\sqrt{(2l_0)^2 + v^2(\Delta t)^2}}{c} \tag{9}$$

But substituting $(\Delta t')c$ for $2l_0$, we have

$$\Delta t = \frac{\sqrt{(\Delta t')^2 c^2 + v^2(\Delta t)^2}}{c} \tag{10}$$

Squaring both sides,

$$(\Delta t)^2 = \frac{(\Delta t')^2 c^2 + v^2(\Delta t)^2}{c^2} \tag{11}$$

and solving for Δt, we finally obtain

$$\Delta t = \frac{\Delta t'}{\sqrt{1 - v^2/c^2}} \tag{12}$$

This is the relationship we sought. It gives Δt, a time interval between events in S' as observed by an observer in S, in terms of $\Delta t'$, the time interval between the *same* two events in S' as observed by an observer in S'. Although we derived this result for a special pair of events (light flash leaves and light flash returns), the result applies to *any* two events at the same location in S'. Note that since $\sqrt{1 - v^2/c^2}$ is always less than unity, Δt must always be greater than $\Delta t'$, and this is what we mean by time dilation. Notice also that when *v* is small compared to *c*, $\sqrt{1 - v^2/c^2}$ is practically equal to unity, and we have Δt *equal* to $\Delta t'$—the situation that exists in classical mechanics. Thus this result of STR is in agreement with classical mechanics when the relative speed is small compared to the speed of light. Finally, be careful when you seek to apply this equation to the time interval between two events in S as observed from S'. How would you write the equation to describe that situation?

Example. If the separation between the mirrors of a light-flash clock is 3.00 m, what is the interval between "ticks" as measured (*a*) in the clock's frame of reference S' and (*b*) in a frame of reference S moving relative to S' with a velocity of 1.50×10^8 m/sec ($\frac{1}{2}$ *c*)? (*c*) What is the interval as measured in a frame S" whose relative velocity is 300 m/sec (that of a jet plane)?

(*a*) $\quad \Delta t' = \dfrac{2l_0}{c} = \dfrac{6.00 \text{ m}}{3.00 \times 10^8 \text{ m/sec}}$

$\quad\quad = 2.00 \times 10^{-8}$ sec

$\quad\quad = 20.0$ nanoseconds

(*b*) $\Delta t = \dfrac{\Delta t'}{\sqrt{1 - v^2/c^2}} = \dfrac{20.0 \text{ nanoseconds}}{\sqrt{1 - (\frac{1}{2})^2}}$

$\quad\quad = 23.3$ nanoseconds

(*c*) $\Delta t = \dfrac{20.0 \text{ nanoseconds}}{\sqrt{1 - (300 \text{ m/sec}/3.00 \times 10^8 \text{ m/sec})^2}}$

$\quad\quad = \dfrac{20.0 \text{ nanoseconds}}{\sqrt{1 - 1.0 \times 10^{-12}}}$

$\quad\quad = 20.0$ nanoseconds

Thus time dilation is present at all relative velocities, but becomes appreciable only when the speed approaches that of light.

Contraction of Lengths

Although time intervals expand, lengths contract according to STR, when observations are made from one frame of reference into another. To see why this must be so, consider once again our two observers in frames S and S'. Now their task is to compare measurements of the length of a steel bar lying *parallel* to the X axis in frame S (Fig. 10-8). Each observer has a clock as before. Observer S measures the length of the bar in the usual way with a meter stick and reports the result as l_0. Observer S', however, has to measure the length in some indirect fashion as his frame S' passes the bar. He starts recording a time interval when he passes the left-hand end of the bar at velocity *v*, and he closes the interval $\Delta t'$ when he passes the right-hand end. S' concludes that the length of the bar is *l*, where

$$l = v \, \Delta t' \quad\quad (1)$$

Observer S sees observer S' travel at a velocity *v* from the left-hand end of the bar to the right-hand end, a distance l_0, in a time Δt measured on *his* clock. Thus,

$$l_0 = v \, \Delta t \quad\quad (2)$$

Dividing the first of these equations by the second, we obtain

$$\frac{l}{l_0} = \frac{\Delta t'}{\Delta t} \qu\quad (3)$$

Now the time interval $\Delta t'$ as measured on the clock in S' by observer S' is *less* than the time interval Δt as observed on the clock in S because of time dilation. The time interval $\Delta t'$ between these two happenings (observer S' at left-hand end of rod in S, observer S' at right-hand end) as observed in S' is related to the time interval Δt between the same two happenings as observed in S by

$$\frac{\Delta t'}{\Delta t} = \sqrt{1 - v^2/c^2} \quad\quad (4)$$

according to what we learned in the last sec-

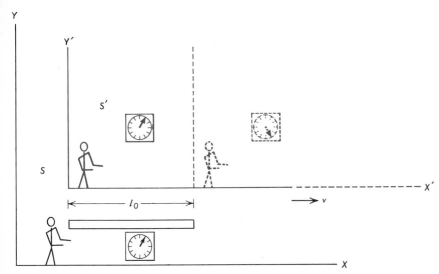

Fig. 10-8. *Contraction of length. The length* l *as observed by* S′ *is less than* l_0 *as observed by* S.

tion. Therefore, we can write equation (3) as

$$\frac{l}{l_0} = \sqrt{1 - v^2/c^2} \qquad (5)$$

and

$$l = l_0\sqrt{1 - v^2/c^2} \qquad (6)$$

The last equation gives the length observed in the direction of motion by an observer moving parallel to the object at a velocity *v*. Since $\sqrt{1 - v^2/c^2}$ is less than unity, *l* must be less than l_0. The equation thus expresses an observed shortening or contraction in the direction of motion.* The difference $l_0 - l$ is called the *Lorentz* contraction, after H. A. Lorentz, one of the pioneers in this field of physics. Furthermore, as was true of time dilation, the Lorentz contraction is appreciably large only at speeds comparable to that of light.

Example. A cosmic-ray particle, traveling vertically downward at a speed of 2.00×10^8 m/sec, goes from the top to the bottom of the

---

* One can show that dimensions *at right angles* to the direction of motion are unchanged.

Empire State Building, whose height with respect to an observer at rest is 380 m. What is the height of the building as observed in the frame of reference of the cosmic-ray particle? Relative to the cosmic-ray particle, the building moves *up* with a velocity of 2.00×10^8 m/sec. To an observer riding along with the particle, the height of the building *l* would be

$$l = l_0\sqrt{1 - v^2/c^2}$$

$$= 380\ \text{m}\ \sqrt{1 - \left(\frac{2.00 \times 10^8\ \text{m/sec}}{3.00 \times 10^8\ \text{m/sec}}\right)^2}$$

$$= 380\ \text{m}\sqrt{1 - (\tfrac{2}{3})^2} = 380\ \text{m}\sqrt{\tfrac{5}{9}}$$

$$= 283\ \text{m}$$

An Experimental Test

How do we subject these strange· ideas of time dilation and length contraction to experimental verification? We cannot make significant observations involving human observers who travel relative to other human observers at speeds comparable to that of light. Such speeds have not been achieved. We can,

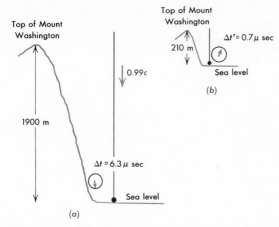

Fig. 10-9. The meson experiment. (a) *The situation as viewed in the frame of reference of an observer at rest with respect to the mountain.* (b) *The situation as viewed by mesons moving at 99% of the speed of light.*

however, use the results of observations on rapidly moving particles of atomic size or smaller to test the special theory of relativity.

A well-known experiment* on the cosmic-ray particle called the *mu meson* showed that the behavior of these fast-moving particles was consistent with STR. The mu meson is an unstable or radioactive particle that breaks up or decays into several other particles. In about 2.2×10^{-6} sec—the half-life—one half of a given number of mu mesons will decay, according to observations made on mesons that have been stopped in a detector. In a longer time, a greater number of mesons will decay. One can compute from the law of radioactive decay that we shall discuss in Chapter 45 how many mesons would survive by the end of a stated time interval.

The experiment consisted of counting the number of mesons arriving in one hour at the top of Mount Washington (altitude 1900 m) in New Hampshire and finding out how many of them decayed and what their "lifetimes" were

* See the article by D. H. Frisch and J. H. Smith in the *American Journal of Physics,* **31,** 342–355 (1963).

when they were stopped in the detector on the top of the mountain. From this information it was possible to predict how many mesons would survive the trip to sea level, 1900 m below the mountain top. Since the mesons were traveling at 99% of the speed of light, the time to travel this distance was 1900 m/3.00×10^8 m/sec or 6.3×10^{-6} sec or 6.3 microseconds. This was almost three times the half-life, so, according to classical mechanics, one would expect only about *one-eighth* ($\frac{1}{2} \times \frac{1}{2} \times \frac{1}{2}$) as many mesons to arrive each hour at sea level as on top of the mountain; the rest would have decayed on the way down. However, when the detector was moved down to sea level, it detected almost *two-thirds* as many mesons per hour as upon the mountain top. Why were the mesons not decaying as expected?

The explanation was given by STR (Fig. 10-9). Although the time interval for the trip down to sea level was 6.3 microseconds *to an observer at rest with respect to the mountain,* to the mesons moving with 99% of the speed of light the time was only 0.7 microsecond. Their "behavior" was in accord with local time in their frame of reference and not with that of the observer. They kept time at one ninth the rate of mesons at rest with respect to the observer. When correction was made for time dilation from the mesons' to the observer's frame of reference, the calculated number of surviving mesons at sea level agreed with the measured number.

Another way of looking at the meson experiment is to consider the Lorentz contraction of the distance from the top of the mountain to sea level. Although that distance was 1900 m to the observer at rest with respect to the mountain, to the mesons moving at 99% of the speed of light the height of the mountain was contracted to only $\frac{1}{9}$ of 1900 m. Thus the mesons saw the mountain flash past in just 0.7 microsecond instead of 6.3 microseconds. The mesons had only 0.7 microsecond to decay during the trip, and more of them survived than would have been expected to survive if the elapsed time in their frame of reference had been 6.3 microseconds.

The meson experiment showed unequivocally the correctness of STR at these high velocities. Other experiments on atomic particles also confirm STR, and that theory has become a part of the everyday mode of operation of physicists and engineers who deal with high-energy particles and accelerators.

Simultaneity

Our ideas of simultaneity have also been affected by STR. Two events that may be simultaneous to an observer in one frame of reference may not be simultaneous to an observer in another.

Suppose that two events take place in frame S' just as observer S' passes observer S at a speed v comparable to that of light (Fig. 10-10). At points a' and b', which are equally distant from observer S', flashbulbs are set off by an electrical circuit that makes them explode at the same time. Light flashes travel outward from the bulbs and move toward observer S' from both a' and b' and also move toward observer S in the other frame of reference. They reach observer S at the same time, and S judges the two events to be simultaneous. If the frame S' were not moving, observer S' would also receive the light flashes simultaneously. However, because of the fact that light is not infinitely fast it will take the light flash from a' somewhat longer to reach S' than that from b'. The reason is that while the light was traveling toward him, observer S' moved to the right a distance r, increasing the distance *from where the light started at a' and decreasing that from where the light started at b'.* The velocity of light, of course, is unchanged by the motion of S'. Observer S' who receives the light flashes at different times can only conclude that the events that produced them could not have occurred

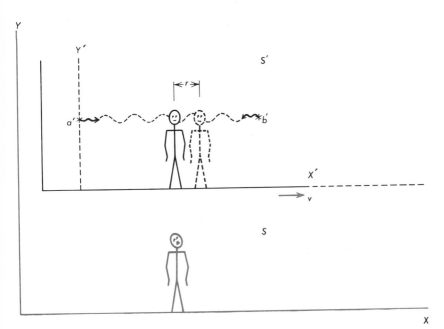

Fig. 10-10. Events that are simultaneous to observer S are not simultaneous to observer S'.

simultaneously. His reasoning is as follows: "I saw the light from *b'* before that of *a'*. I know the distance to me from *b'* is equal to that from *a'* and that the speed of light is the same for all observers. It may be that I am moving through space with a high velocity, but there is no experiment that I can perform in my frame of reference that will demonstrate this. The only conclusion that I can come to from these observations, therefore, is that the event at *b'* must have taken place before the event at *a'* — even though my friend in frame *S* insists that the events were simultaneous."

Thus events that are simultaneous to observer *S* are not simultaneous to observer *S'*. We see once again that the interpretation of events in space and time depends upon the frame of reference in which they are viewed. Of course, if the two events we have been talking about took place in frame *S* instead of *S'*, observer *S'* would conclude they were simultaneous and *S* that they were not.

Events in Space and Time

Earlier in this chapter we saw that in classical mechanics time was independent of the frame of reference of the observer. The time *t* of an event observed by observer *S* either in his own frame of reference or in *S'* was the same as the time *t'* of the same event as observed by observer *S'*, provided the two observers had previously synchronized their clocks. The transformation of the coordinates of the event from *S* to *S'* could be carried out by the equations of Galilean relativity:

$$x = x' + vt' \qquad (1)$$

$$y = y' \qquad (2)$$

$$z = z' \qquad (3)$$

$$t = t' \qquad (4)$$

The special theory of relativity, however, shows that the time and space coordinates of an event are *inseparable*. In the transformation equations that must be used in relativistic

situations (when *v* is comparable to *c*), time and space coordinates enter *each* equation. The *Lorentz transformation equations*[*] are

$$x = \frac{x' + vt'}{\sqrt{1 - v^2/c^2}} \qquad (5)$$

$$y = y' \qquad (6)$$

$$z = z' \qquad (7)$$

$$t = \frac{t' + vx'/c^2}{\sqrt{1 - v^2/c^2}} \qquad (8)$$

S and *S'* frames of reference move with their *X-X'* axes parallel to each other.

Just as the laws of physics could be preserved in classical mechanics by transforming the coordinates by means of the Galilean equations, the laws can be made independent of the observer at relativistic velocities by using the Lorentz transformation equations. The latter allow for the "point of view" of the observer when the two frames of reference are moving relative to each other at a speed comparable to that of light. Notice that when *v* is small compared to *c*, the Lorentz equations reduce to the Galilean equations.

Relativistic Mass

In classical mechanics the mass of a body was constant. In Chapter 5, we found that when a body was acted upon by a resultant force, the ratio of the resultant force *F* at any time to the acceleration *a* was a constant. This constant we called the mass *m*:

$$\frac{F}{a} = m$$

The mass was independent of temperature, velocity, location, and any other conceivable physical variable. In the collisions studied in Chapter 7, we saw that the masses of two colliding bodies determined the ratios of the

---

[*] These equations can be derived from the two postulates of special relativity. See the references at the end of the chapter.

velocities before and after the collision; again, the masses of the bodies did not change just because the velocities were different.

The special theory of relativity, however, demonstrates that the mass of a body is *dependent upon its speed relative to the frame of reference of the observer* and, in fact, is given by

$$m = \frac{m_0}{\sqrt{1 - v^2/c^2}}$$

where m_0 is the observed mass when the body is at rest with respect to the observer and m is its mass when the body moves with a speed v relative to the observer. We notice that when v becomes comparable to c, the speed of light, the quantity $1/\sqrt{1 - v^2/c^2}$ is appreciably greater than unity and m is greater than m_0. When the velocity v is zero, the body is at rest with respect to the observer and its mass m_0 is unchanging; the quantity m_0 is called the *rest mass* of the body.

This prediction has been verified experimentally in the atomic domain by measurements on particles moving at velocities comparable to that of light. For example, measurements of the mass of the electron at high velocities (Chapter 27), show that the mass varies with speed as shown in Fig. 10-11. At approximately one-tenth of the speed of light the mass of the electron begins to be noticeably greater than its rest mass. It increases rapidly, according to the relativistic mass formula, as v increases and becomes extremely large as v approaches c.

Example. The rest mass of an electron is approximately 9.11×10^{-31} kg. What is its mass at a velocity of 1.00×10^8 m/sec (one-third of the velocity of light)?

$$m = \frac{m_0}{\sqrt{1 - v^2/c^2}}$$

$$= \frac{9.11 \times 10^{-31} \text{ kg}}{\sqrt{1 - \frac{1.00 \times 10^8 \text{ m/sec}^2}{(3.00 \times 10^8 \text{ m/sec})^2}}}$$

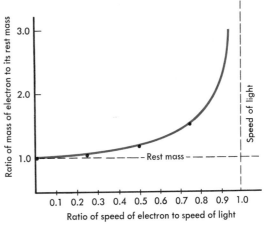

Fig. 10-11. *The mass of an electron varies with the speed according to the relation* $m = m_0/\sqrt{1 - v^2/c^2}$.

$$= \frac{9.11 \times 10^{-31} \text{ kg}}{\sqrt{1 - \frac{1}{9}}} = \frac{9.11 \times 10^{-31} \text{ kg}}{\sqrt{\frac{8}{9}}}$$

$$= 9.66 \times 10^{-31} \text{ kg}$$

Like measurements of time intervals and lengths, measurements of mass are dependent upon the frame of reference of the observer. A proof of the relativistic-mass formula, similar to proofs given above for time dilation and the Lorentz contraction, can be found in the references at the end of the chapter. It is a rather lengthy proof, and we shall not consider it here. For our purposes, it is sufficient to note that mass is a variable in STR, but like all relativistic effects, this one becomes noticeable only at relative velocities comparable to that of light.

Mass-energy

Perhaps the most famous result of Einstein's special theory of relativity is the formula

$$E = mc^2 \qquad (1)$$

which links together mass and energy.* Here

* A simple proof of this equation will be given in Chapter 40 after we have discussed the concept of the photon of light.

E is the *total* energy of the body, m the body's relativistic mass, and c the velocity of light. This equation says that whenever the mass of a body increases by an amount Δm, its total energy increases by an amount ΔE:

$$\Delta E = (\Delta m)c^2 \qquad (2)$$

To understand the relation between mass and energy more completely, let us write the last equation in the form

$$E - E_0 = (m - m_0)c^2 \qquad (3)$$

where m_0 is the rest mass of the body, E_0 the rest energy, and m its mass when it possesses total energy E. E may include kinetic energy, heat energy, energy of excitation (for atoms), and other forms of energy.

If $E - E_0$ is the kinetic energy E_K that the body acquires when it is accelerated from rest to velocity v relative to the observer, then

$$E_K = (m - m_0)c^2 \qquad (4)$$

or

$$E_K + m_0c^2 = mc^2 = E \qquad (5)$$

which tells us that the total energy E equals the kinetic energy E_K plus m_0c^2, the *rest energy*.

$$m = \frac{m_0}{\sqrt{1 - v^2/c^2}}$$

Equation (4) becomes

$$E_K = \left[\frac{m_0}{\sqrt{1 - v^2/c^2}} - m_0 \right] c^2 \qquad (6)$$

This expresses the relation—in the special theory of relativity—among kinetic energy, rest mass, and velocity. Let us ask two questions about it: (1) how is it related to our earlier formula $E_K = \frac{1}{2}m_0v^2$, and (2) is there any limit to the velocity that can be given to a body if we are willing to give it greater and greater amounts of kinetic energy?

To answer the first question we shall write equation (6) as

$$E_K = m_0c^2 \left[\left(1 - \frac{v^2}{c^2} \right)^{-1/2} - 1 \right] \qquad (7)$$

Now according to the binomial theorem of algebra, we can write the first term inside the brackets as

$$\left(1 - \frac{v^2}{c^2} \right)^{-1/2} \cong 1 + \frac{1}{2}\frac{v^2}{c^2}$$

to a good approximation *if v/c is quite small*. In the "everyday" situations of classical mechanics v/c is indeed a very tiny fraction, as we have seen, so equation (7) can be written

$$E_K = m_0c^2 \left[1 + \frac{1}{2}\frac{v^2}{c^2} - 1 \right] = m_0c^2 \left[\frac{1}{2}\frac{v^2}{c^2} \right]$$
$$= \frac{1}{2}m_0v^2 \qquad (8)$$

We see, therefore, that at nonrelativistic velocities, the STR formula for kinetic energy reduces to the one we found for classical mechanics.

We can get a clue to the answer to the second question by noting that as v approaches c in size, according to equation (7), the quantity $(1 - v^2/c^2)^{-1/2}$ grows without limit and E_K approaches infinity. This suggests that v can never equal c and that the speed of light in vacuum is an upper limit to the speed that any material body (not light itself) can achieve.

Experiments with electrons have shown that the speed of light is indeed the "ultimate speed."[*] Electrons were accelerated in a linear accelerator and their final speeds measured. As the speed approached that of light, greater and greater energies were required to produce small increases in velocity. It was not possible to reach the velocity of light within the limits of available energy. The speed of light is apparently the upper limit for material bodies; it is approached, but not attained.

Our final example of the relationship

[*] For a complete description of these experiments see the article by W. Bertozzi in the *American Journal of Physics*, **32**, 551–555 (1964).

between mass and energy concerns the release of energy from the nuclei of atoms—"atomic energy," which is really nuclear energy. Near the end of the nineteenth century, scientists discovered that the nuclei of atoms of certain heavy elements such as uranium, radium, and thorium are unstable and slowly break up of their own accord. These nuclei, in exploding, throw out particles—alpha particles—having great kinetic energies. Since then, it has been demonstrated that huge amounts of energy are stored in the nuclei of *all kinds* of atoms. This energy can be liberated in two ways: first by the joining together or *fusion* of light atoms like hydrogen to form heavier atoms; second, by the splitting or *fission* of heavy atoms such as uranium. When uranium atoms undergo fission into barium, krypton, and other elements, the total mass before and after fission is constant, but the total *rest* mass of the resulting atoms is less than the *rest* mass of the uranium. Thus the fissioning liberates thermal energy in that the fragments of the split nuclei fly apart with great velocities.

Example. When 0.23500 kg of uranium are fissioned, the rest mass of the resulting atoms is 0.23477 kg. How much kinetic energy is liberated?

$$\Delta m = 2.3 \times 10^{-4} \text{ kg}$$
$$\Delta E = (\Delta m)c^2 = 2.3 \times 10^{-4} \text{ kg}$$
$$\times (3.0 \times 10^8 \text{ m/sec})^2$$
$$= 2.1 \times 10^{13} \text{ kg-m}^2/\text{sec}^2$$
$$= 2.1 \times 10^{13} \text{ joules}$$

This is approximately 6 million kilowatt-hours of energy, an amount equal to the electrical energy output of Grand Coulee Dam in 3 hours. At 3¢ per kilowatt-hour, this energy would cost $180,000!

Much evidence leads us to believe that the energy of sunlight is liberated by the fusion of hydrogen into helium. When 4.032 gm of hydrogen are converted into 4.003 gm of helium, 0.029 gm is converted into kinetic energy. The energy thus liberated would heat your house or run your automobile for several years.

Summary

In Newtonian physics, the time coordinate of a specified event is the same for all observers, regardless of their frames of reference or their relative motions. Space coordinates of the event differ, depending on the frame of reference, but can be brought into agreement by using the Galilean transformation equations.

Transformation equations connect the time and space coordinates of an event, as referred to one frame of reference, with those of the same event referred to another.

The principle of relativity in mechanics states that the laws of mechanics are independent of the inertial (uniformly moving) frame of reference in which they are observed. Any such frame of reference is as suitable as any other for observing the laws of mechanics. By the same token, there is no mechanical experiment that one can perform inside a closed system that will detect its state of motion at constant velocity.

The speed of light in a vacuum is found experimentally to be a constant and independent of the inertial frame of reference in which it is measured. When an observer measures the speed of light in vacuum he obtains the same value regardless of his motion relative to the source of light.

Einstein's special theory of relativity (STR) is concerned with frames of reference that move relative to each other at constant velocity. It is based on two postulates or assumptions:

1. *All* the laws of physics (not just those of mechanics) are the same for observers in all inertial frames of reference.
2. The speed of light in vacuum c is the same for all observers.

Although STR applies to all moving frames of reference, its most interesting predictions deal with relative velocities that are comparable to the speed of light. Among these are time dilation, contraction of lengths, the relativity of simultaneity, and increase in mass.

Time dilation means that a time interval Δt between two events as observed in frame S is *greater* than the time interval $\Delta t'$ as observed in frame S' in which the two events take place, when S' moves relative to S with velocity v:

$$\Delta t = \frac{\Delta t'}{\sqrt{1 - v^2/c^2}}$$

Lorentz contraction: the length l of an object as observed from S is related to l_0, its length as measured by an observer at rest with respect to the object in S', by:

$$l = l_0\sqrt{1 - v^2/c^2}$$

Events that are simultaneous to observer S are not simultaneous to observer S'—and vice versa.

The Lorentz transformation equations hold at speeds comparable to that of light and reduce to the Galilean equations at speeds small compared to that of light.

The mass m of an object as determined by an observer with respect to whom the object is moving with a velocity v is related to m_0, the rest mass of the object, by:

$$m = \frac{m_0}{\sqrt{1 - v^2/c^2}}$$

The total energy E of a body is related to its mass m by

$$E = mc^2$$

The kinetic energy E_K is given by

$$E_K = (m - m_0)c^2 = m_0c^2\left[\left(1 - \frac{v^2}{c^2}\right)^{-1/2} - 1\right]$$

In nuclear fission, the total rest mass of the fragments of the split nucleus is less than the rest mass of the original nucleus; the change in rest mass is accompanied by an increase in the kinetic energy of the fragments.

Questions

1. What is a frame of reference?

2. Describe one or more frames of reference that would be appropriate in each of the following situations: (a) attaching a kitchen clock to the wall, (b) locating an office in a multistory building with many parallel corridors on each floor, (c) crossing the United States on the interstate highways, (d) approaching the South Pole, and (e) cruising toward the moon in a space vehicle.

3. Discuss Newton's concepts of absolute and relative time and compare them with the concept of time in the special theory of relativity.

4. Discuss Newton's concept of absolute space. Why was such a concept needed? Is it still needed?

5. Passenger A on a jet airplane in level flight with a ground velocity of 600 mi/hr tosses a pillow to a friend B standing in the aisle 10 ft nearer the front of the plane. Describe the motion of the pillow as viewed (a) by a third passenger C seated halfway between A and B, (b) by an observer on the ground, and (c) by an observer in a military plane that overtakes and passes the jet at a ground velocity of 800 mi/hr.

6. How could the observers in question 5 reconcile their observations?

7. What are transformation equations? Give an example.

8. What is the principle of relativity in mechanics?

9. Discuss the possibility of conducting an experiment that would allow one to determine the velocity of a frame of reference from measurements made entirely within that frame.

10. What are the two postulates of the special theory of relativity? Discuss the significance of each.

11. What was the significance of the Michelson-Morley experiment?

12. Discuss qualitatively what is meant by (a) time dilation, (b) contraction of lengths, and (c) increase in mass. State carefully the kind of measurement involved and the frames of reference being used.

13. Estimate the importance of the relativistic effect in each of the following as observed by an observer at rest on the earth and state whether the principles of STR do or do not apply: (a) the length of a car moving on the open highway, (b) the mass of a jet airplane passing overhead, (c) the lifetime of a soap bubble, (d) the mass of a decelerating fission fragment, (e) the paths of colliding billiard balls, (f) the path of a high-energy proton colliding elastically with a hydrogen atom, and (g) your mass when you run to class.

14. What experimental evidence is there for (a) the Lorentz contraction, (b) the increase of mass with velocity?

15. If the velocity of light were only 100 mi/hr, describe how the following would appear to you if you were in a car traveling 50 mi/hr: (a) the diameter of the steering wheel, (b) the rate of blinking of a "flashing" traffic signal, (c) the width of the highway, and (d) the length of a "measured mile" on the highway.

16. Explain why two events that appear to be simultaneous to an observer in one frame of reference may not be simultaneous to an observer in another frame. State clearly how this result follows from the two postulates of STR.

17. Is the special theory of relativity consistent with classical mechanics? Discuss.

Problems

A

1. Estimate the increase in mass of a 2000-kg car, as determined by an observer at rest on the ground, when the car accelerates from rest to a speed of 72 km/hr. Is the change of mass experimentally detectable?

2. Calculate (a) the kinetic energy and (b) the mass of an electron moving with a speed of 2.0×10^8 m/sec with respect to the observer. Rest mass of the electron is 9.1×10^{-31} kg.

3. At what speed would a meter stick have to move past an observer at rest on the ground to appear to have the length of a yardstick?

4. What is the observed percentage change in length of a jet airplane traveling 300 m/sec as observed from the ground?

5. What would be the time interval between ticks of a seconds clock as observed by an observer passing the clock at a speed of $0.1c$?

B

6. How much heat energy would be liberated if 8.064 kg of hydrogen were fused into 8.006 kg of helium?

7. At what speed with respect to an observer at rest in the laboratory would the mass of an electron be 100 times its rest mass of 9.1×10^{-31} kg?

8. At what speed must the relativistic equations be used if the desired accuracy in the calculated quantities is (a) 1%, (b) 0.1%?

9. At what speed does the kinetic energy calculated by the relativistic equation differ from that calculated by the classical equation by 1%?

10. Calculate the ratio of m/m_0 for speeds relative to the observer of $0.2c$, $0.4c$, $0.6c$, $0.8c$, and $0.98c$. Plot the results against v.

C

11. In a race, plane A flies 100 mi to the east from the starting point and returns. Plane B flies 100 mi north and back. Each plane has an air velocity of 500 mi/hr, and the wind velocity is 50 mi/hr from west to east. Determine which plane will win the race by calculating the time for a round trip for A and for B. (Remember that B must head upwind.)

12. (a) What energy is required to give a 1000-kg spaceship a velocity of $0.5c$? (b) To what mass is this energy equivalent?

For Further Reading

Bondi, Hermann, *Relativity and Common Sense,* Science Study Series, Doubleday and Company, Garden City, 1964.

Born, Max, *Einstein's Theory of Relativity,* Dover Publications, Inc., New York, 1962.

Einstein, Albert and Leopold Infeld, *The Evolution of Physics,* Simon and Schuster, New York, 1938.

Frank, Philipp, *Einstein, His Life and Times,* Alfred A. Knopf, New York, 1953.

French, A. P., *Special Relativity,* W. W. Norton and Company, New York, 1968.

Gamow, George, *Mr. Tompkins in Wonderland,* MacMillan Company, New York, 1942.

Katz, Robert, *An Introduction to the Special Theory of Relativity,* Momentum Books, D. Van Nostrand Company, Princeton, 1964.

CHAPTER ELEVEN

Gravitation

We have seen how the first correct quantitative description of motion, due principally to Galileo, and the three fundamental laws of Newton led to consistent explanations of many mechanical phenomena. Classical mechanics, as the logical structure built by Newton and his followers was called, dominated the thought of physicists for almost two centuries—until the dawn of modern physics early in the twentieth century. The special theory of relativity, discussed in the last chapter, broke away from many of the ideas of classical physics but agreed with Newtonian mechanics at low relative velocities.

One important link must be added now to complete our discussion of mechanics. What does the gravitational force depend upon? We have seen that gravity enters many problems. Although we no longer believe—as some of the followers of Newton did—that gravitational attraction explains all phenomena in physics, an understanding of the gravitational force was and is one of the chief goals of physics.

Newton's discovery of the law of universal gravitation and the demonstration by him and his followers that the principles of physics apply just as strongly to the motion of astronomical bodies as to terrestrial ones was probably the greatest triumph of classical mechanics. That motion in the universe could be calculated by the same principles that govern the fall of an apple produced a revolution in thought about man's place in the universe.

Having discussed the law of universal gravitation and some of its applications, we shall conclude this chapter by considering what relativity physics has to say about gravitation.

The Universe of the Greeks

Aristotle and his followers developed a description of the earth, sun, planets, and stars that was a part of a system of the universe that attempted to explain all phenomena. In their view, the earth was a globe, and the sun was mounted on the surface of a gigantic crystal sphere which revolved around the earth once every 24 hours. On other concentric spheres, revolving with different speeds, whirled the moon, the planets, and the fixed stars (Fig. 11-1). The heavenly bodies were unchanging and perfect. They moved on circles because to the Greeks only circular motion was eternal.

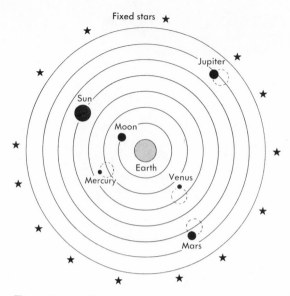

Fig. 11-1. The ancient Greeks assumed that the sun and planets revolved around the earth.

fixed stars in about 1.9 years. However, it does not progress uniformly from west to east. Periodically Mars reverses its motion and travels westward for a while (Fig. 11-2). To explain this motion, Ptolemy assumed that Mars and the other planets traveled on their own, smaller whirling spheres (Fig. 11-3). He explained the motions of the heavenly bodies so cleverly that his theory survived for 13 centuries.

Copernicus Proposed a New Theory

A few years after Columbus discovered America, a young Polish monk named Copernicus, troubled by the complexity and the increasing discrepancies in the Ptolemaic theory, proposed a new theory in which the earth and the planets revolved around the sun (Fig. 11-4). His theory was simpler than Ptolemy's, for it did away with some of the smaller spheres. Copernicus showed that the apparent reversed motion of Mars, for example, might be due to the motion of the earth around the sun. To understand how this can be, hold a pencil between your eyes and a distant building and wag your head sidewise. The pencil will seem to move to and fro, oppositely with reference to the more distant building. In the same manner, as one looks past Mars at the more distant fixed stars and is carried to and fro by the motion of the earth

The theory that the stars revolved around the earth was improved by the Greek scholar Ptolemy who lived in Alexandria, Egypt, in the second century A.D. In order to explain irregularities in the motions of the planets, he assumed that some of them traveled in eccentric circular paths. Ptolemy also had to explain the reversed motions of the planets. For example, Mars journeys eastward and returns to its initial position with reference to the

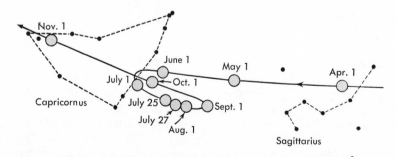

Fig. 11-2. The planet Mars reverses its apparent motion as seen from the earth. (After R. H. Baker, Astronomy, *fifth edition, D. Van Nostrand Company, 1950.)*

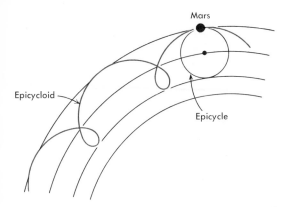

Fig. 11-3. *To explain the motion of Mars and other planets, Ptolemy assumed that each planet moved on its own whirling sphere.*

around the sun, Mars seems to move to and fro in the opposite direction. (This effect is called *parallax*.)

Copernicus, advocating his new doctrine, discovered that the scholars of his day were well satisfied with the theory of Aristotle and Ptolemy. It was a comfortable view of the universe, for it made the earth the center of creation, and the sun a lamp to guide man's footsteps. Copernicus, they thought, would have the earth be a mere grain of sand, revolving as an insignificant planet around a central sun. Copernicus protested that his new theory explained the facts *more simply* than the old one. However, he avoided a fight and spent the rest of his life writing a book—*On the Revolutions of the Heavenly Spheres*—that explained his theory. It was placed in his hands in 1543, newly printed, when he was dying.

Galileo, Tycho Brahe, and Kepler

About 60 years after Copernicus died, a Dutch spectacle maker happened to look through two spectacle lenses, held a certain distance apart, and found that distant ships seemed larger. Thus he accidentally invented the telescope and received a small prize for

his invention. When Galileo heard the news, he constructed a crude telescope. One memorable night he observed that the moon had mountains like those of the earth and that it was not the perfect body of Aristotelian theory. Looking at the planet Jupiter, he saw four moons which revolve about it with different velocities, a small solar system behaving as Copernicus said such a system should. Later he discovered the rings of Saturn and, on the sun, dark patches or sunspots which, moving across its disk, proved that it rotates.

Most important of all, Galileo found that the planet Venus showed phases like the moon (Fig. 11-5). Also, Venus seemed larger at some seasons of the year than at others. Galileo pointed out that, if Venus revolved around the earth, as it should according to Ptolemy's theory, it would remain at a constant distance from the earth and its apparent size would not vary. However, if the earth and Venus both revolved around the sun, then the distance from earth to Venus *should* vary, as it apparently does.

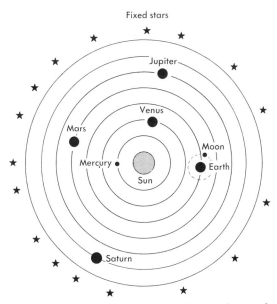

Fig. 11-4. *Copernicus assumed that the earth and planets revolved around the sun.*

Fig. 11-5. The various phases and true relative sizes of the planet Venus. (Courtesy of E. C. Slipher, Lowell Observatory.)

Galileo actively took up the defense of the Copernican system. In a book entitled *Dialogue on the Great World Systems* (the Aristotelian-Ptolemaic system and the Copernican system), written in his native Italian instead of the more scholarly and more exclusive Latin, Galileo patiently and wittily took up the arguments against the Copernican theory and refuted them. He appealed both to experiment and to observation and directed his arguments to the intelligent layman as much as to the scientists of his day. He quickly ran into trouble. His books were burned and he was forced to "detest the error and heresy of the movement of the earth." After spending several years in forced seclusion, Galileo was released. His efforts seemed wasted, but new generations accepted his teachings.

Shortly before Galileo first peered through his telescope and argued in favor of the Copernican theory, a Danish astronomer named Tycho Brahe (Ty'ko Bra'hey) was patiently and skillfully observing the positions of the planets and compiling careful records. After he died, his pupil, Johannes Kepler, studied the records. From his analysis of Tycho's data, Kepler deduced three laws of planetary motion:

First law. The planets move in elliptical paths with the sun at one focus of the ellipse (Fig. 11-6).

Second law. A line drawn from the sun to any planet sweeps over equal areas in equal time intervals (Fig. 11-7).

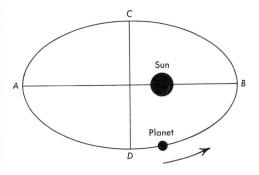

Fig. 11-6. Kepler's first law. The planets move on elliptical, rather than circular, paths with the sun at one focus. However, the ellipticity is less than that shown here.

Third law. The square of the period of revolution of a planet is proportional to the cube of the mean radius of the planet's orbit. If two planets have periods (the time for a complete revolution about the sun) T_1 and T_2 and the mean radii of their orbits are R_1 and R_2, Kepler's third law states that

$$\frac{T_1{}^2}{T_2{}^2} = \frac{R_1{}^3}{R_2{}^3}$$

Kepler obtained his laws by painstakingly fitting curves to the data, guided by a mystical belief that simple numerical relationships govern the universe. Although the

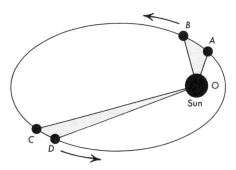

Fig. 11-7. Kepler's second law. If the planet goes from A to B in a given time interval (e.g., one week) and from C to D in the same time interval, the area AOB equals the area DOC.

elliptical orbits of the planets differ very slightly from circular ones, Kepler pursued his studies unwaveringly until he was convinced beyond doubt that the orbits could only be represented by ellipses. However, he was not able to deduce his laws from a general theory of the solar system. He had shown, nevertheless, that it was necessary to abandon the crystal spheres. This fact gave rise to a great difficulty, namely, how to support the heavens. The answer to this question was given by Isaac Newton, who was born a few months after Galileo died.

Newton and the Law of Universal Gravitation

When Newton was twenty-three years old and a student at Cambridge University, he was forced to leave school because of the bubonic plague which was ravaging England. During this enforced vacation of more than a year, he made three important contributions. First, he proved that a prism can disperse a beam of sunlight into rainbow colors; second, he invented the branch of mathematics called calculus; and third, he explained how the attraction of the earth keeps the moon in its orbit and was led to the universal significance of gravitation. A contemporary tells us that one day Newton saw an apple fall and, in his own words, he "began to think of gravitation as extending to the orb of the moon." He wondered whether the earth did not attract the moon even as it did the apple, and whether this attraction might not provide the necessary force to hold the moon in its elliptical orbit.

In order to test the hypothesis, Newton computed the force with which the earth attracts each pound of the moon's substance. He assumed that the attractive force varies inversely as the square of the distance from the earth's center.

Suppose that a 1-kilogram ball is weighed at the earth's surface with a spring balance (Fig. 11-8). In this position it is 6400 km from

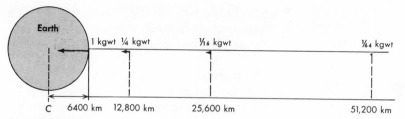

Fig. 11-8. *Weight of a 1-kilogram body at different distances from the earth's center.*

the earth's center of gravity, and its weight is 1 kilogram-weight. Next, imagine that the ball is carried 6400 km away from the surface, to a position 12,800 km from the earth's center. By thus doubling the distance of the ball *from the center,* the pull of the earth—according to the inverse square hypothesis—would be diminished to one-fourth of its previous value. Even though the mass of the ball remained constant, its weight would decrease to $\frac{1}{4}$ kgwt. At a distance of 25,600 km from the earth's center, twice as great as before, the attraction should be $\frac{1}{16}$ kgwt.

The distance to the moon is about 60 times the earth's radius; hence if the inverse square law is true, each kilogram of the moon's substance should be attracted by the earth with a force of $\frac{1}{3600}$ kgwt. Thus the moon's acceleration toward the earth should be correspondingly $\frac{1}{3600}$ g, that is, 9.8 (m/min)/min, instead of 9.8 (m/sec)/sec.* Next if the radius r of the moon's orbit is known and also the moon's velocity v, we can compute its acceleration toward the earth by the equation

$$a = \frac{v^2}{r}$$

The velocity v may be found by dividing the circumference of the moon's orbit by the time required for a "round trip" (27.5 days). When correct values are used, the computed acceleration agrees closely with 9.8 (m/min)/min

---

* Newton, of course, computed the acceleration in feet per minute per minute, but we shall use metric units.

($\frac{1}{3600}$ g), the value predicted by the inverse square law.

Unfortunately, in Newton's day the accepted value of the earth's radius was in error by about 10%, throwing off the value of the radius of the moon's orbit, and the two values of the acceleration did not agree. The twenty-three-year-old student did not publish his work, but shoved his computations into a drawer in his desk. Twenty years later, a new determination of the earth's radius was made, and when the corrected value was used, excellent agreement was achieved. Perhaps you wonder why Newton did not publish his results despite the 10% error. It was partly because he always hesitated to publish anything and had to be prodded by his friends. Also, he doubted the correctness of his assumption that the earth attracted the ball as though the mass of the earth were concentrated at its center. During the twenty years, he justified this assumption mathematically by using the calculus.

Newton's universal law of gravitation, which was first tested in this manner, is as follows:

Every particle in the universe attracts every other particle with a force directly proportional to the product of their masses and inversely proportional to the square of the distance between them.

The Gravitational Attraction Between Small Bodies can be Measured

According to Newton's law of gravitation, a ball of mass m_1 should attract a ball of mass

m_2, at a distance r, with a force F given by

$$F = G \frac{m_1 m_2}{r^2} \qquad (1)$$

The gravitational constant G may be measured by any of several methods, including the method used by Cavendish in 1798. He measured the gravitational attraction of two lead spheres for each other by means of a highly sensitive device called a *torsion balance* (Fig. 11-9). It consisted of a thin quartz fiber supporting a horizontal rod, at each end of which a small lead ball was mounted. When the rod was turned slightly in a horizontal plane, the fiber was twisted. A restoring torque proportional to the angular displacement of the rod was set up in the fiber. By measuring the angular displacement produced by a known torque, Cavendish could calibrate the torsion balance. Thereafter he could determine an unknown torque acting on the rod — and hence the force exerted on the small lead spheres — by measuring the angular displacement produced by that torque. The angular position of the rod and fiber

could be recorded on the scale by a beam of light reflected from the mirror.

Cavendish's experiment was to place two large lead spheres beside the smaller lead spheres as shown. The gravitational attraction to be measured was that between pairs consisting of one large lead sphere and one small one; using two pairs of spheres in a symmetrical arrangement not only balanced the apparatus, but doubled the effect and made the measurements more precise. The torque caused by gravitational attraction twisted the quartz fiber until equilibrium was reached. The angular position of the fiber was noted. Then the large spheres were quickly moved to the alternative positions shown as dotted circles. In this new configuration, the gravitational attraction was different, and the fiber experienced a different torque. After the system had stopped its slow oscillation about the new equilibrium position, the angular displacement could be determined, the torque computed, and the gravitational force F between *one* pair of large and small spheres calculated. G could then be determined from

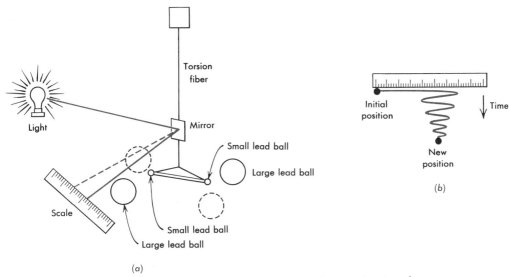

(a)

(b)

Fig. 11-9. *The Cavendish apparatus for measuring the gravitational constant.*

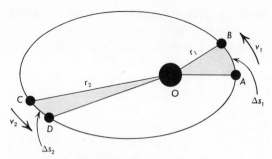

Fig. 11-10. Kepler's law of equal areas.

F, m_1 and m_2, and r the distance between centers of the large and small spheres.

The Cavendish method was later improved, and still other methods devised for measuring G. The value accepted today is

$$G = 6.670 \times 10^{-8} \text{ dyne-cm}^2/\text{gm}^2$$
$$= 6.670 \times 10^{-11} \text{ newton-m}^2/\text{kg}^2$$

Two balls, each having a mass of 1 gm, placed with their centers 1 cm apart, attract each other with a force of about 68×10^{-12} gwt (sixty-eight trillionths of a gram-weight). Clearly, the gravitational force is not large unless one of the bodies is a large mass—a planet or a star.

Motion in Orbits
Under Gravitational Attraction

Newton and those who followed him were able to derive Kepler's laws and go beyond them to explain small irregularities in the motion of planets due to planet-planet attractions. To derive Kepler's laws generally for elliptical orbits requires mathematical techniques that go beyond the scope of this book. Let us show at least that Kepler's laws are consistent with the principles of mechanics that we have discussed so far.

Consider a planet of mass m moving on an elliptical orbit about the sun (Fig. 11-10). At one time in its revolution, the planet moves a distance Δs_1 from A to B in a time Δt at an average velocity v_1 at a distance r_1 from the sun. At a later time, it moves a distance Δs_2

from C to D in the same time interval Δt at an average velocity v_2 at a distance r_2 from the sun. Since the resultant external torque on the planet is negligibly small for our purposes, the angular momentum of the planet in its orbit is constant, and

$$I_1 \omega_1 = I_2 \omega_2$$

where I and ω are the moment of inertia and the angular velocity, respectively. Using the expressions given in Chapter 8 for I and ω

$$mr_1{}^2 \left(\frac{v_1}{r_1}\right) = mr_2{}^2 \left(\frac{v_2}{r_2}\right)$$

or

$$mr_1 v_1 = mr_2 v_2$$

Multiplying both sides by $\frac{1}{2}\Delta t$

$$\tfrac{1}{2} mr_1 (v_1 \, \Delta t) = \tfrac{1}{2} mr_2 (v_2 \, \Delta t)$$

Canceling the mass and noting that $v_1 \, \Delta t = \Delta s_1$ and $v_2 \, \Delta t = \Delta s_2$, we have

$$\tfrac{1}{2} r_1 \, (\Delta s)_1 = \tfrac{1}{2} r_2 \, (\Delta s)_2$$

But $\frac{1}{2} r_1 (\Delta s)_1$ is approximately the area of the triangle *BOA*, and $\frac{1}{2} r_2 (\Delta s)_2$ the area of the triangle *DOC*. Thus area *BOA* equals area *DOC*. Hence, from the law of conservation of angular momentum, one of our fundamental conservation laws, we have shown that Kepler's second law holds. Notice that one consequence of the law is that as the planet moves nearer to the sun—toward its perihelion—its velocity increases.

Next, consider a planet of mass m moving on a *circular* orbit (a fairly good approximation to the actual orbit) about the (stationary) sun whose mass is M_S. Let the radius of the orbit be R and the magnitude of the velocity of the planet—constant in this instance—be v. The gravitational attraction between the planet and the sun is

$$F = \frac{GmM_S}{R^2}$$

The centripetal acceleration of the planet is

v^2/R. By Newton's second law,

$$F = ma$$

or

$$\frac{GmM_S}{R^2} = m\frac{v^2}{R} \qquad (2)$$

But the velocity of the planet is $2\pi R/T$ where T is the period of its motion. Hence

$$\frac{GmM_S}{R^2} = m\frac{4\pi^2R^2/T^2}{R}$$

or

$$T^2 = \left(\frac{4\pi^2}{GM_S}\right)R^3 \qquad (3)$$

Since the quantity in parentheses is a constant, this is Kepler's third law. Table 11-1 contains data on the motion of the planets. Note the constancy of R^3/T^2.

Table 11-1. The Planets

	Mean Distance from Sun (in millions of miles)	Period (in years)	(Distance)^3 (in mi^3)	(Period)^2 (in yr^2)	$\dfrac{(Distance)^3}{(Period)^2}$ (mi^3/yr^2)
Mercury	36.02	0.241	46.73×10^{21}	58.08×10^{-3}	0.805×10^{24}
Venus	67.27	0.615	30.44×10^{22}	37.82×10^{-2}	0.805×10^{24}
Earth	93.00	1.00	80.44×10^{22}	10.00×10^{-1}	0.805×10^{24}
Mars	141.7	1.88	28.48×10^{23}	35.34×10^{-1}	0.805×10^{24}
Jupiter	483.9	11.86	11.33×10^{25}	14.07×10	0.805×10^{24}
Saturn	887.1	29.46	69.81×10^{25}	86.79×10	0.805×10^{24}
Uranus	1784.8	84.01	56.85×10^{26}	70.58×10^2	0.805×10^{24}
Neptune	2796.7	164.79	21.87×10^{27}	27.16×10^3	0.805×10^{24}
Pluto	3669.8	247.70	49.42×10^{27}	61.36×10^3	0.805×10^{24}

Periods in days: Mercury 88, Venus 225, Earth 365.24, Mars 687.

Spacecraft in orbit around the earth and other artificial satellites of the earth, as well as planets, obey Kepler's laws. The orbits of such bodies are sometimes called Keplerian orbits. For a spacecraft to remain in a stable circular orbit at a distance R from the *center* of the earth, equation (2) requires that the velocity of the spacecraft be

$$v = \sqrt{GM_E/R} \qquad (4)$$

where M_E is the mass of the earth. This velocity cannot be given to the spacecraft in one installment at the earth's surface because of the heating effects of air friction. A multistage

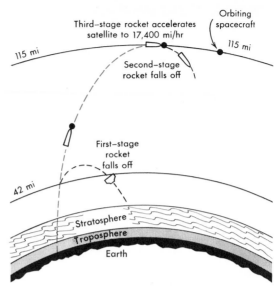

Fig. 11-11. Launching a spacecraft into earth orbit or parking orbit preparatory to going to the moon. Data are from the Apollo 11 mission.

rocket carries the spacecraft up to its orbit, each stage of the rocket increasing the velocity and then dropping off, until the velocity required for a stable orbit at the desired height is reached (Table 11-2). The last stage of the rocket "tilts" the spacecraft into its orbit and gives it a final push (Fig. 11-11).

The Energy of an Object in Orbit

A planet, orbiting spacecraft, or other satellite has (1) kinetic energy E_K due to its orbital

Table 11-2. Characteristics of Stable Circular Orbits of Satellites

Distance from Center of Earth, mi	Distance from Surface, mi	Orbital Speed, mi/sec	Time for one Revolution, hr
4.1×10^3	100	4.83	1.48
4.2×10^3	200	4.77	1.54
4.3×10^3	300	4.71	1.59
4.5×10^3	500	4.61	1.70
5.0×10^3	1.0×10^3	4.37	2.00
6.0×10^3	2.0×10^3	3.99	2.62
9.0×10^3	5.0×10^3	3.26	4.82

velocity and (2) potential energy E_P due to gravitational attraction. The first of these is given by

$$E_K = \tfrac{1}{2}mv^2 \qquad (1)$$

where m is the mass of the satellite and v its velocity. Since, according to the last section, the requirement for a stable circular orbit about the earth is

$$v = \sqrt{GM_E/r} \qquad (2)$$

where M_E is the mass of the earth and r is the radius of the orbit, we can write equation (1) as

$$E_K = \tfrac{1}{2}\frac{GmM_e}{r} \qquad (3)$$

In Chapter 22 we shall show that the potential energy of a body that is acted upon by an inverse-square-law attractive force is inversely proportional to the distance r to the body from the center of the attracting body. For an object under the gravitational attraction of the earth we have

$$E_P = -\frac{GM_E m}{r} \qquad (4)$$

The negative sign tells us that work must be done *on* the satellite to pull it away from the earth. When r is very large, E_P is numerically very small; as r approaches infinity, E_P becomes *less negative* and approaches zero. We are taking the zero point of our scale of potential energy at *infinity* (that is, at a distance far enough from the earth that its gravitational attraction is negligibly small). Thus a smaller *negative* potential energy corresponds to a larger energy on this scale. When r is small, E_P is numerically large, but this corresponds to a greater *negative* potential energy, which, of course, is the same as a smaller energy. For example, if you are in debt by ten dollars at the beginning of the week and in debt by twenty dollars by the weekend, your wealth has decreased by ten dollars.

The total energy E of the orbiting object is

$$E = E_K + E_P = \tfrac{1}{2}\frac{GmM_E}{r} - \frac{GmM_E}{r}$$

$$E = -\tfrac{1}{2}\frac{GmM_E}{r} \qquad (5)$$

This tells us that the *total* energy of a satellite bound by gravitational attraction to another body is negative and is inversely proportional to the distance between their centers. As a satellite moves closer to the earth, for example, its total energy decreases. Note, however, that its period of revolution decreases because the satellite is moving faster. This seems paradoxical at first, but we must remember that although the kinetic energy is increasing, the potential energy is decreasing at a much greater rate, causing the total energy to decrease.

When a satellite loses energy by brushing the outer layers of the atmosphere and moves closer to the earth (decreasing r), its period of revolution decreases, and the satellite moves faster. Air friction increases, and the satellite spirals down into the atmosphere. If it is not equipped with a heat shield, it burns when it enters the denser air.

Escape Velocity

What velocity would a spacecraft have to have just barely to escape from the attraction of the earth? To answer this question we apply the law of conservation of mechanical energy to the situation that exists when the rocket engines have been shut off. At two different distances from the center of the earth the energies $(E_P)_1$, $(E_K)_1$, $(E_P)_2$, and $(E_K)_2$ will be related by

$$(E_P)_1 + (E_K)_1 = (E_P)_2 + (E_K)_2 \qquad (1)$$

Let us now assume that the rocket engines shut off — outside the denser part of the atmosphere — when the rocket has reached the escape velocity v_E at a distance from the center of the earth equal to r_1. The rocket "coasts" outward from then on until at "infinity" its potential energy has *increased* to zero and its kinetic energy has *decreased* to zero. Substituting into equation (1) we have

$$-\frac{GM_E m}{r_1} + \frac{1}{2}mv_E^2 = 0 + 0$$

or

$$v_E = \sqrt{2GM_E/r_1} \qquad (2)$$

where v_E is the "velocity of escape."

Docking in Orbit

In space travel, two spacecraft may have to join one another while both are in orbit so that astronauts can be transferred from one space vehicle to the other. The maneuver is not as simple as it might first seem.

Consider first how a tugboat would catch up with an ocean liner as the liner entered a harbor. The liner would reduce its speed and maintain its course, two steps that would be independent of each other. Coming up from behind, let us say, the tugboat would increase its speed to exceed that of the liner and would follow an intersecting course. Eventually the tug would catch up with the liner.

Suppose now that an astronaut A in one spacecraft is pursuing astronaut B who is ahead of him in another spacecraft in the same circular orbit about the earth. Assume first that A turns on his rocket engine in order to increase his velocity so that he can overtake B. The result is that A increases his total energy. According to equation (5) on page 176, an increase in total energy means that the radius r of the orbit must increase (so that the energy will be *less negative*). But according to Kepler's third law, the consequence of an increase in r is an increase in the period T. So the result of firing the rocket is to move A into an orbit of greater radius and larger period. A would see B apparently drop below him and move ahead.

Suppose, however, that A should fire his rocket in retrograde (backward-firing) fashion to slow himself down. His total energy would decrease, his orbit radius decrease, and his period decrease. Now A would see B apparently move above him and fall behind.

The correct maneuver, if A is to catch B, is evidently for A to move to a higher and "slower" orbit, wait for B to come around on a lower and "faster" orbit, and then fire his retro rocket at just the right time and for just the right interval so that he drops behind B and can carry out the required docking.

Travel Between Earth and Moon

Landings on the moon and lunar exploration began with the historic Apollo 11 mission by the United States in July, 1969. Moon flights are a triumph of applied technology resting on many discoveries in basic science.

At the start of a typical lunar mission, the first stage of the launching rocket exerts a thrust of about 7.6×10^6 lb on the space vehicle, which at liftoff may weigh 6.4×10^6 lb, propels it to an altitude of 42 miles, and gives it a velocity of 6200 mi/hr. This stage then shuts down and separates from the rest of the spacecraft, and the second stage ignites, providing a thrust of about 1.0×10^6 lb and taking the space vehicle to an altitude of 115 miles and to a velocity of 15,500 mi/hr. The second stage separates, and the third-stage engine is turned on for a precisely determined interval (145 seconds for Apollo 11) in order to insert the space vehicle into a circular earth orbit at an altitude of 115 miles and a velocity of 17,400 mi/hr.

The departure from the earth takes place after about one and a half revolutions around the earth. The third-stage engine ignites again, and the space vehicle is directed into its trajectory to the moon (Fig. 11-12) and given its escape velocity of about 24,000 mi/hr. The third stage is then detached, and the astronauts are on their way to the moon, coasting "uphill" against the gravitational at-

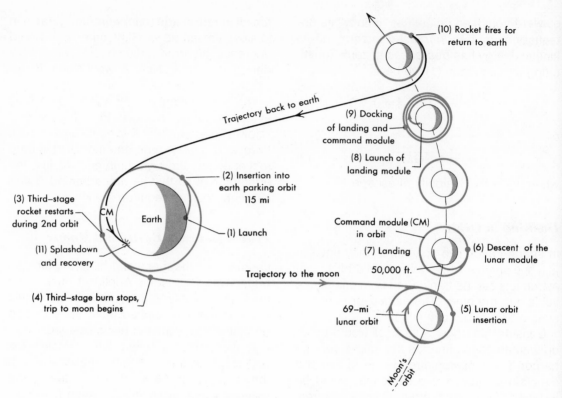

(10) Rocket fires for return to earth

Trajectory back to earth

(9) Docking of landing and command module

(8) Launch of landing module

(2) Insertion into earth parking orbit 115 mi

(3) Third–stage rocket restarts during 2nd orbit

CM

Earth

(1) Launch

Command module (CM) in orbit

(6) Descent of the lunar module

(7) Landing 50,000 ft.

(11) Splashdown and recovery

(4) Third–stage burn stops, trip to moon begins

Trajectory to the moon

69–mi lunar orbit

(5) Lunar orbit insertion

Moon's orbit

Fig. 11-12. *Trip to the moon. NASA diagram for the Apollo 11 mission.*

traction of the earth and then "downhill" as they approach the moon. At an altitude of about 90 miles above the moon, the rocket engine, which is in the forward direction (retrograde position) for this maneuver, is turned on to *decelerate* the spacecraft and allow it to swing into a lunar orbit at an altitude of about 70 miles above the moon's surface.

Lunar landing is accomplished by first separating (undocking) the spacecraft into two parts: a lunar module, in which several of the astronauts make the descent, and a command module, which remains in lunar orbit, carrying at least one other astronaut. The rocket engine in the lunar module then fires in retrograde fashion, slowing down the lunar module and causing it to descend to the moon's surface along a trajectory such as that

shown in Fig. 11-12. As the lunar module approaches the moon's surface, the rocket engine is firing straight down, lowering the module gently to the moon's surface. After lunar exploration, the astronauts reenter the lunar module's ascent stage and take off to rejoin the command module, leaving behind the landing portion of the lunar module. The rocket engine in the ascent stage propels it straight up for a short distance and then tips it into a trajectory that takes it into a lunar orbit below that of the command module. The docking maneuver, described in the last section, follows, and the lunar module and the command module are joined again. After the crew of the lunar module have entered the command module, the lunar module is jettisoned.

Return to the earth begins with the firing of the rocket engine in the command module when the spacecraft is on the far side of the moon. This gives the spacecraft its escape velocity from the moon and inserts it into the trajectory shown, which will return it to the earth in about 60 hours with a velocity of about 24,000 mi/hr at reentry into the earth's atmosphere. Friction with the atmosphere decelerates the command capsule (the rest of the spacecraft is jettisoned just before reentry) as it approaches the surface of the earth. The forward surface of the capsule is coated with a special heat shield (ablative coating) that burns off to keep the temperature of the rest of the command module low. Near the end of the reentry, parachutes open in order further to decelerate the capsule and finally lower it to the surface of the ocean — splashdown.

Determining the Mass of the Earth and the Sun

The attraction of two spheres of unit mass, 1 m apart, being known, that of any two spheres at a known distance apart can be found by equation (1) on page 173. Moreover, we can compute the mass of the earth itself. Our problem is to find how massive the earth must be to exert a force of 9.8 newtons on a body of mass 1 kg at its surface. The radius of the earth is about 4000 mi, or 6.37×10^6 m. Let M_E be the mass of the earth. Then

$$9.8 \text{ newtons} = 6.670 \times 10^{-11} \frac{\text{newton-m}^2}{\text{kg}^2}$$
$$\times \frac{(1.00 \text{ kg})M_E}{(6.37 \times 10^6 \text{ m})^2}$$

$M_E = 5.98 \times 10^{24}$ kg

The mass M_S of the sun can be calculated by equation (3) from a knowledge of the period of the earth and the average radius of its orbit.

$$M_S = \frac{4\pi^2}{G} \frac{R^3}{T^2}$$

Variations of the Acceleration Due to Gravity

The acceleration due to gravity varies from place to place on the earth's surface because of variations in *elevation, latitude,* and *the composition of the earth's crust.*

The pull of the earth on a body above the earth's surface decreases as it moves farther from the earth's center. Hence, *g* is smaller at Denver, Colorado (elevation 5280 ft), than at New York City (Table 11-3). (*Inside the earth,* the gravitational force and *g* are both proportional to the distance to the center of the earth, decreasing to 0 at the center.)

Table 11-3. Values of the Acceleration Due to Gravity

Place	Latitude	Elevation	m/sec²
North Pole	90°	0	9.832
Equator	0°	0	9.780
New Orleans	29° 56'	0	9.793
New York City	40° 48'	0	9.802
Pittsburgh	40° 27'	535 ft	9.801
San Francisco	37° 47'	350 ft	9.799
Chicago	41° 50'	600 ft	9.802
Denver	39° 40'	5280 ft	9.796
Pikes Peak	38° 50'	14,100 ft	9.789

The variation with latitude has two causes. First, the distance from sea level to the earth's center is about $12\frac{1}{2}$ mi greater at the equator than at the poles, for the earth is not a perfect sphere; second, a body at the equator travels in a circle and forever falls toward the earth's center (Chapter 6). Being accelerated downward, it seems to weigh less than if it were at rest. Part of the earth's attraction causes the body to move on its circular path. This apparent decrease in weight is exactly like that which you experience when you are a passenger on an elevator that is accelerated downward. The decrease depends upon the acceleration and hence on the velocity of the body. If the earth were to rotate 18 times as fast as it does, bodies at the equator would seem to have zero weight. At the poles their weight would be unchanged.

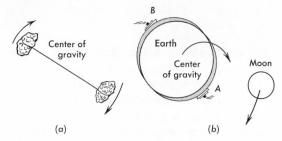

Fig. 11-13. Explanation of tides. (a) *Water accumulates at the outer edge of the whirling sponges.* (b) *Tide at* B *is due to the same cause. That at* A *is caused by the greater attraction of the moon.*

The variations of gravity at different places on the earth's surface are neglected in ordinary affairs, but astronauts walking on the moon's surface must adjust themselves to a much smaller acceleration due to gravity than they are accustomed to. On the moon, a 180-lb astronaut would weigh about 30 pound-weights. Notice that the *mass* of the astronaut, his opposition to being accelerated, would remain constant, but his *weight* would vary.

The Tides

Everyone who lives by the seashore is familiar with the tides, which generally rise every $12\frac{1}{2}$ hours, and knows that the rise of the water is caused chiefly by the attraction of the moon. Few people can explain why there are two tides. It is easy to understand why the water should pile up on the side of the earth which is nearer to the moon, but why should it also accumulate on the opposite side? An analogy will be helpful. Imagine that two water-soaked sponges are tied together by a long cord and tossed into the air. They revolve about their common center of gravity. Then the water will gather at the outer side of each sponge and fly off tangentially (Fig. 11-13a). Now think of the earth and the moon, which revolve around their common center of

gravity, which is located inside the earth (Fig. 11-13b). Water accumulates at A, both because of the moon's attraction and because of the rotation. At B, the two influences oppose each other. However, because of its greater distance, the moon's attraction is *smaller* than at A. The rotational effect is larger; hence the water accumulates at B.

You might expect the tides to be highest when the moon is directly overhead, or somewhat more than 12 hours later. However, the friction of the rotating earth holds the tides back so that, generally, they occur several hours behind schedule. At many locations, high tide occurs when the moon is near the horizon.

A block of wood against the rim of a spinning grindstone will slow down the stone and eventually stop it. Similarly, the tides exert friction against the surface of the earth and decelerate it. The days are becoming slightly longer. The effect is very small, only a few thousandths of a second per century. Perhaps the earth will slow down until finally one side will always face the moon. The moon, having a smaller amount of inertia and less initial kinetic energy than the earth, has long since lost most of its rotational motion, and we always see the same face.

The moon has no oceans or lakes on its surface. Its tides occur only inside its solid crust. Tides also occur in the earth's crust; the Washington Monument, for example, rises and falls more than a foot because of this tide. Tides also occur in the earth's atmosphere.

The sun, like the moon, exerts gravitational attractions on the earth. The tides which it produces are about one-half as great as those caused by the moon. The resultant tides due to the combined attractions of the sun and the moon are abnormally high at times of full moon and new moon, when the sun and moon both "pull together" along the same straight line. The tides are unusually low at the times of the half moon when they pull at right angles to each other.

The General Theory of Relativity

In the last chapter we discussed the special theory of relativity, which deals with frames of reference that move relative to each other with constant velocity. Starting from the assumptions that the laws of physics were the same for all observers and that the velocity of light was independent of the observer's frame of reference, we arrived at a number of important results—time dilation, contraction of lengths, and change of mass, for example.

Einstein developed the general theory of relativity as an extension of the special theory to *accelerating* frames of reference and, especially, to gravitational phenomena. He wanted to explain gravitation by relating it to something simpler and more basic.

One of the earliest experiments performed in physics has had a profound influence on general relativity. This was Galileo's demonstration that bodies falling freely near the surface of the earth had the same acceleration *g*, regardless of their masses. Let us relate Newton's law of gravitation to his second law of motion and examine the conditions under which the acceleration due to gravity is constant. We write the gravitational law for the attraction that one body exerts on a second as

$$F_{1 \text{ on } 2} = \frac{G(m_1)_g(m_2)_g}{r^2} \qquad (1)$$

where the subscript g attached to each mass denotes the *gravitational* mass—the mass defined by the law of gravitation. Newton's second law states that, as a result of this force, body 2 undergoes an acceleration a_2:

$$F_{1 \text{ on } 2} = (m_2)_i a_2 \qquad (2)$$

where $(m_2)_i$ is the *inertial* mass of body 2—its resistance to being accelerated. Combining these two equations we have

$$(m_2)_i a_2 = \frac{G(m_1)_g(m_2)_g}{r^2} \qquad (3)$$

If body 1 is the earth, $(m_1)_g$ is m_E, the mass of the earth, and r is R_E, its radius. Both m_E and

R_E are constants. Body 2 is any other body near the surface of the earth. If $(m_2)_i$ is the same as $(m_2)_g$—that is, if the inertial mass of a body is indeed the same as its gravitational mass—equation (3) reduces to

$$a_2 = \frac{Gm_E}{R_E^2} = \text{constant} \qquad (4)$$

By inserting values of G, m_E, and R_E into the equation we can show that the right-hand side of the equation equals 9.8 m/sec². This tells us that the acceleration of a falling body is independent of its mass and equal to *g*. Theory and experiment are thus in agreement—*if* inertial mass (resistance to acceleration) and gravitational mass are identical. We have a clue here to a possible relationship between acceleration and gravitation—one that Einstein found extremely significant.

Next consider an observer who is inside an elevator without windows. He can observe what is happening inside the elevator, but he cannot see outside. First, he holds a book three feet above the floor, releases it, and observes that it accelerates downward 9.8 m/sec²—normal everyday behavior. A few seconds later he repeats the experiment and is surprised to find that the book does not accelerate to the floor but remains three feet above it. What hypotheses can the man reasonably make to explain this? First, he may decide that gravity has somehow been turned off and the book is no longer being attracted by the earth. If he is a reader of science fiction, for example, he may believe that someone has invented an anti-gravity shield and placed it below the floor of the elevator. But another hypothesis is possible: that the cable supporting the elevator has broken and the cage, man, and book are *all* falling freely down the shaft with an acceleration of 9.8 m/sec². If that were so, the elevator and its contents would be in the "weightless" state, like that of astronauts in orbit around the earth and for the same reason. The point is that no experiments that the man could perform *inside* the elevator could tell him which of these hypotheses was correct. Until the crash came—or the safety devices stopped the elevator—the man could not tell whether gravity had been turned off or whether he were in a frame of reference accelerating downward 9.8 m/sec^2. The two possibilities would be equivalent. To be sure, an observer *outside* the elevator could distinguish between these possibilities, but the man inside could not.

Let us take another example. Suppose a blacked-out spaceship is coasting in space far from any massive bodies of astronomical size. The astronauts and the equipment of the spaceship are in a gravity-free situation. Released objects float in space inside the cabin. Suppose now that such objects were observed to start accelerating in one direction. Again two hypotheses would be possible: first, that the spacecraft was approaching a source of gravitational attraction and, second, that the spacecraft was accelerating. Either one would be defensible, and the astronauts could not choose one over the other.

Consideration of thought experiments of this kind led Einstein to base the theory of general relativity upon a postulate called the *principle of equivalence,* which stated that *it is impossible to detect a difference between the effect of gravitational attraction and the effect of a suitably chosen acceleration of a frame of reference.* Thus we can think of a gravitational force as acting on an object near the surface of the earth or we can think of the frame of reference of the body as being one which is accelerating *upward* 9.8 m/sec^2. Similarly, objects near any "attracting" body can be viewed as being in an accelerating frame of reference whose acceleration can be chosen to have an effect on the objects equivalent to that describable by the gravitational law.

Einstein's strategy was to replace the "force" description of gravitation by a geometrical description using the fundamental coordinates of space and time. His general theory thereby not only provided a description of gravitational phenomena, but was consistent with the special theory of relativity. How this was done and how the predictions of the general theory were investigated experimentally go beyond the scope of this brief introduction but can be studied in some of the books listed at the end of the chapter.

Summary

According to Ptolemy's theory, the earth was a sphere around which the sun, the moon, the fixed stars, and the planets revolved. Copernicus argued that the sun, rather than the earth, was the center of the solar system, because this assumption explained the apparent motions of the planets more simply. Galileo's observations with the telescope yielded new facts which could not be readily explained by Ptolemy's theory, but could be explained by that of Copernicus.

Kepler showed that the planets move around the sun in elliptical paths, that a line drawn from the sun to any planet sweeps over equal areas in equal time intervals, and that the square of the period of revolution of a planet is proportional to the cube of the mean radius of the planet's orbit.

Newton demonstrated that, if gravitational attraction varies inversely with the square of the distance between two bodies, the earth's attraction for the moon provides the centripetal force necessary to hold the moon in its orbit. He was able to derive Kepler's laws from this hypothesis and his laws of motion.

The law of universal gravitation states that the attraction *F* for each other of two spheres of masses m_1 and m_2, their centers being a distance *r* apart, is given by

$$F = \frac{Gm_1m_2}{r^2}$$

in which *G* is the gravitational constant.

The velocity of an earth satellite in a stable circular orbit is inversely proportional to the square root of the radius of the orbit. The satellite's *total* energy *E* (kinetic plus potential) is negative and is inversely proportional to the radius of the orbit. As the radius decreases, the total energy decreases (becomes more negative), the kinetic energy increases, and the period decreases. The velocity of escape from an orbit is inversely proportional to the square root of the radius of the orbit.

Travel from the earth to the moon involves the following steps: (1) launching of the space vehicle into a parking orbit around the earth, (2) escape from the earth orbit into a trajectory that intersects the path of the moon, (3) entrance into a lunar orbit, and (4) descent to the moon's surface. The spacecraft is a multistage rocket, each stage of which is jettisoned as soon as it has accomplished its purpose of accelerating or decelerating the space vehicle.

The mass of the earth may be found by computing the mass of a sphere that will attract a 1-kg sphere with a force of 9.8 newtons, the distance between the centers of the two bodies being equal to the radius of the earth.

The general theory of relativity extends the special theory to accelerating frames of reference and gravitational phenomena. The assumption that inertial mass and gravitational mass are the same leads to the experimentally verified result that the acceleration of freely falling bodies near the earth's surface is independent of their masses. The principle of equivalence states that it is impossible to detect a difference between the effect of gravitational attraction and that of a suitably chosen acceleration of a frame of reference.

Questions

1. How can the mass of the earth be determined?

2. What is the value of the gravitational constant *G* on the moon?

3. Distinguish carefully between mass and weight. How can you compare the masses of two bodies without weighing them? Has your method determined inertial mass or gravitational mass?

4. How much would a kilogram of matter weigh at the center of the earth?

5. Describe Cavendish's method for measuring the gravitational attractions between relatively small bodies.

6. Explain how to compute the mass of the sun, given the mass of the earth. Can you compute the mass of the moon in the same way?

7. Why are bodies said to be "weightless" in a spaceship moving in a circular satellite orbit? Would a coin sink in water in a satellite? Could you crack a nut with a "weightless" hammer?

8. Can an object be "dropped" from a satellite?

9. Sketch the path of the moon around the sun.

10. Discuss the possibilities of changing the period of rotation of the earth.

11. At what phase is the planet Venus brightest as observed from earth?

12. In the cabin of a spaceship, can astronauts drink water through a straw? If not, how do they drink?

13. Cite several kinds of evidence for (a) the rotation of the earth, (b) its revolution around the sun.

14. When an earth satellite moves from one orbit to another of twice the radius, what change occurs in (a) its period, (b) its kinetic energy, (c) its potential energy, and (d) its total energy?

15. Compare the following athletic events as held on the moon (in a lunar city under a heated dome supplied with oxygen) and on the earth: (a) the pole vault, (b) the 100-yard dash, (c) a baseball game.

16. What would happen if a man in a spaceship in orbit around the earth fired a cannon directly at another spaceship ahead?

17. If a spaceship landed on an unknown

planet, how could the astronauts determine its mass relative to that of the earth?

18. Why are spacecraft launched from the earth in an easterly direction? In what direction should a spacecraft to Mars be launched with respect to the direction of the earth's orbital motion?

19. Just after a spacecraft has reached its escape velocity from the earth and begun its trajectory to the moon, the last stage of the launching rocket (stage III) separates from the spacecraft and is turned away from the translunar trajectory. The fuel remaining in stage III is then dumped through its nozzle. What effect does this have on the motion of stage III? What will happen to stage III?

20. Explain how docking is carried out in orbit. Why would an expert hot-rod driver or interceptor pilot have to learn new reflexes to become a good pilot of a spacecraft?

21. What was significant about the fact that *g* is independent of mass for the development of the general theory of relativity?

22. Several patrons of a "fun ride" in an amusement park in the twenty-first century enter a sealed blacked-out chamber and seat themselves on light cardboard chairs. One of them holds a bottle half filled with a soft drink. The machinery outside starts, and the passengers have the following successive experiences: (a) it suddenly becomes impossible to drink from the bottle though a straw, and a coin floats in mid-air: (b) the passengers accelerate toward the ceiling; and (c) the chairs collapse under them, and the passengers are pressed against the floor. What interpretations can the passengers give to these observations? Relate your answers to the principle of equivalence.

Problems

(*Radius of earth* ≅ *6400 km.*)

A

1. What is the acceleration due to gravity (a) 6400 km above the earth's surface, (b) at the distance of the moon, about 384,000 km from the earth's center, (c) at the center of the earth?

2. At what altitude above the earth's surface would the acceleration due to gravity be 4.9 m/sec^2?

3. With what velocity would an object hit the moon if dropped from a height of 3.0 m above the lunar surface?

4. How much would an 80-kg man weigh at an elevation (a) 6400 km, (b) 12,800 km above the earth's surface?

5. What gravitational force exists between two 70-kg men 1.0 m apart?

6. The centers of two lead spheres A and B, each having a mass of 1 megagram (10^6 gm), are 2.0 m apart. (a) Find the force of attraction exerted by A on B. (b) How far would one of the balls travel from rest in 1000 sec if subject to this force and free to move without friction?

B

7. What is the gravitational force between a proton (mass 1.67×10^{-27} kg) and an electron (mass 9.11×10^{-31} kg) which are separated — as in the normal hydrogen atom — by 5.29×10^{-11} m?

8. A spacecraft is to move in a circular parking orbit 1000 km from the surface of the earth. What must the velocity of the spacecraft be?

9. Assuming that the mass of the moon is 1/80 that of the earth, at what distance from the earth's center, on a line from earth to moon, would a man's weight be zero? Assume that the distance from the center of the earth to the center of the moon is 384,000 km.

10. If the mass of the moon is 1/80 that of the earth and its radius is 1/4 as great, (a) how much would an 80-kg astronaut weigh on the moon, and (b) how high could this man jump if he could jump 2 ft while wearing his spacesuit on the earth's surface?

11. Approximately how much would a 70-kg man weigh if the earth were twice as large in diameter as it is and had the same average density?

12. Calculate the velocity of escape from the surface of (a) the sun, (b) the moon, (c) the planet Jupiter.

13. The mass of a spacecraft in circular parking orbit 200 km above the earth's surface is 7.0×10^5 kg. Calculate (a) the period of the spacecraft, (b) its velocity, and (c) its total energy.

C

14. (a) What is the velocity of escape from the surface of Phobos, one of the moons of Mars? It has a radius of 16 km and a mass equal to 10^{-8} that of the earth. (b) What is the acceleration due to gravity at the surface of Phobos?

15. The sun's diameter is 110 times that of the earth and its mass is 333,000 times as great. How many newtons would a kilogram of matter weigh at the sun's surface?

16. (a) Calculate from the following data the gravitational attraction exerted by the earth on the moon: m_1 (earth) $= 6.0 \times 10^{24}$ kg, m_2 (moon) $= 7.5 \times 10^{22}$ kg, distance between centers $= 4.0 \times 10^8$ m. (b) What is the diameter of a steel cable that would barely exert this force without breaking if each square centimeter can exert a force of 7.5×10^4 newtons?

17. The distance from the earth to the sun is 1.50×10^8 km. If the time required for Jupiter to make one trip around the sun is 12 years, calculate the distance of Jupiter from the sun. Assume that Jupiter's orbit is circular.

18. What is the velocity of a spacecraft in a lunar orbit at a height of 860 km above the moon's surface? Radius of the moon is 1720 km.

19. If the velocity of escape for the earth is 7.0 mi/sec, what is the velocity of escape for Mars? Mars has a mass 0.11 times that of the earth and a radius 0.53 times the radius of the earth.

20. A weather satellite moves at a height of 1600 km with a velocity of 25,600 km/hr. Approximately how long is it visible as it travels from horizon to zenith to horizon?

21. Given that the period of a satellite 161 km above the earth's surface is 1.48 hr, calculate the mass of the earth.

22. The gas shell of a nebula is expanding at 16 km/sec. Its diameter is 10×10^{12} km. How long has it been expanding? Assume that 1 year $= 3.0 \times 10^7$ sec.

23. How much time will elapse, according to a time-measuring device in his spacecraft, for an astronaut who travels for 10 earth-years at a velocity that is 0.9 times that of light?

24. What is the orbital angular momentum of a 1500-kg spaceship orbiting the earth at an altitude of 160 km?

25. Plot the gravitational force acting between two 10-kg spherical masses versus distance as the separation between their centers is reduced from 2.0 m to 0.20 m.

26. Show that the velocity v_h with which a body falling freely from a height h hits the surface of the earth is given by

$$v_h = \sqrt{2g_0 R \left(1 - \frac{1}{1 + h/R}\right)}$$

where g_0 is the acceleration due to gravity at the surface of the earth and R is the radius of the earth.

27. The planet Mars has two satellites: Deimos (period 30.2 hours, mean radius of orbit 2.36×10^4 km) and Phobos (period 7.5 hours, mean radius of orbit 9.42×10^3 km). Calculate the mass of Mars.

28. If the sun (mass 2.0×10^{30} kg, radius 7.0×10^8 m) were to undergo gravitational collapse, shrinking to a radius of, say, 1.0 km, (a) what would be the change in the gravitational potential energy of a solar proton (mass 1.67×10^{-27} kg)? (b) What kinetic energy would it so acquire? (c) What would be the density of the sun?

29. The mass of the sun is 2.0×10^{30} kg, its diameter is 1.4×10^9 m, and its period of rotation is 25 days. The masses of the planets are as follows (in units of the earth's mass): Mercury 0.0543, Venus 0.814, Mars 0.107, Jupiter 318, Saturn 95.3, Uranus 14.5, Neptune 17.2,

Pluto 0.83. From these data and those in Table 11-1, calculate (a) the total angular momentum of the giant planets Jupiter and Saturn, (b) the total angular momentum of all of the other planets, and (c) the angular momentum of the sun. (d) What percentage of the total angular momentum of the whole solar system does the sun possess? (e) What percentage of the total mass of the solar system does the sun possess?

30. If a satellite traveling eastward in an orbit 160 km above the equator crosses the meridian at the same time as one traveling west in an orbit 160 km above the equator, when and where will they meet again?

31. Suppose that the mass of the earth became eight times as great (and its radius twice as great) and the masses of water, air, and so on were also to become eight times as great as they now are. What would then be the atmospheric pressure at sea level?

32. Suppose that suddenly gravity ceased to act upon you, your clothing, and anything else attached to you. Nothing else was so affected. At the time this happened, you were standing, motionless, out-of-doors. The air was motionless at the time. You did not have time to kick anything or take hold of anything. Calculate where, with respect to the spot on which you were standing, you would be 6 min later.

33. Discuss the effects of a high-altitude location for the Olympic Games upon the possibility of setting new records. Reduced availability of oxygen is an important factor, but reduced gravitational attraction and reduced buoyancy of the air should also be considered. How would performance be affected by the latter two in the shot put, pole vault, and running long jump? Quantitatively, compare performance at Mexico City (Olympic Games of 1968) and at sea level, noting the following data: density of air at sea level and 20°C, 0.001205 gm/cm³; density at 7349 ft above sea level, same temperature, 0.00091 gm/cm³; radius of the earth (polar), 3950 miles, (equatorial) 3963 miles. The 1968 winners of the three men's events mentioned above made the following records: 67 ft 4¾ in, 17 ft 8½ in, and 29 ft 2½ in, respectively.

For Further Reading

Bondi, Hermann, *The Universe at Large,* Doubleday and Company, Garden City, 1960.

Born, Max, *Einstein's Theory of Relativity,* Dover Publications, Inc., New York, 1962.

Clancy, Edward P., *The Tides,* Doubleday and Company, Garden City, 1969.

Cohen, I. B., *The Birth of a New Physics,* Doubleday and Company, Garden City, 1960.

Einstein, Albert and Leopold Infeld, *The Evolution of Physics,* Simon and Schuster, New York, 1938.

Galilei, Galileo, *Dialogue on the Great World Systems,* University of Chicago Press, Chicago, 1953.

Gamow, George, *Gravity,* Doubleday and Company, Garden City, 1962.

Hess, Wilmot et al., "The Exploration of the Moon," *Scientific American,* October, 1969, p. 54.

James, J. N., "The Voyage of Mariner IV," *Scientific American,* March, 1966.

Knedler, John W., Jr., ed., *Masterworks of Science* (Copernicus, "On the Revolutions of the Heavenly Spheres"), Doubleday and Company, Garden City, 1947.

Koestler, Arthur, *The Watershed,* Doubleday and Company, Garden City, 1960.

Sciama, D. W., *The Physical Foundations of General Relativity,* Doubleday and Company, Garden City, 1969.

Molecular Physics
and
Heat

CHAPTER TWELVE

Molecular Forces in Solids and Liquids

Thus far we have considered mostly the motions and energies of bodies large enough to be seen—baseballs, automobiles, and planets. Now we begin our discussion of molecular physics and heat. In order to understand what we observe, we shall deal with particles too small to be glimpsed through an optical microscope—*atoms* and *molecules.* The principles of mechanics will be of great help to us in understanding the large-scale forces that arise from the interactions of these small particles.

The question of whether or not matter is built up of ultimate particles or atoms has puzzled thinkers since the days of Socrates and Plato. Democritus (400 B.C.) wrote as follows:

"Atoms are infinite in number and infinitely varied in form. They strike together and their lateral motions and whirlings are the beginnings of worlds."

"The varieties of all things depend upon the varieties of their atoms in number, size, and aggregation."

For twenty-three hundred years after Democritus, the view that matter is composed of particles was advocated by some thinkers and opposed by others. On the one hand, some philosophers—and a few scientists—claimed that the existence of atoms explained economically and succinctly the bewildering variety of forms that occur in nature—the varieties of butterflies, the different metallic ores, rain and snow, the species of animals, for example. All of these forms were due to more or less temporary combinations of permanent and indestructible atoms which, upon dissolution of one form, could combine in other forms. To understand nature, one must identify the atoms and learn how they combine. The opposing view, held by most philosophers and scientists before the seventeenth century, was that forms in nature represent stages in processes by which matter moves to its designated place in the universe: iron is of the earth, for example, and the rusting of iron is a process by which it returns to the earth (not the combination of iron atoms with oxygen atoms, as we now believe). According to this view, one must know the "natural homes" of matter to understand the processes by which matter changes. Only the heavens were unchanging. Galileo described the Aristotelian point of view—to which he himself, an atomist, did not subscribe — in his *Dialogue on the Great World Systems.*

"Simplicio: I see in the Earth plants and animals continually generating and decaying; winds, rains, tempests, storms arising; and, in a word, the aspect of the Earth to be perpetually metamorphosing; none of which mutations is to be discerned in the celestial bodies, the constitution and the appearance of which are most exactly the same as what they were time out of mind, without the generation of anything new or the corruption of anything old."

The triumph of the atomic theory over the opposing theory followed the careful *quantitative* investigations of chemical reactions by Lavoisier, Dalton, and other chemists at the end of the eighteenth and the beginning of the nineteenth century. The results of their painstaking measurements of the relative masses of substances entering into chemical reactions and the relative volumes of reacting gases could be explained adequately only on the basis of an atomic theory. Later in the nineteenth century, physicists began the study of the structure of atoms and molecules that is still going on and that brings added conviction that the atomic theory is correct. Although space does not permit us in this book to review the chemical evidence for the atomic theory, we will consider later in some detail the developments in physics that led to present knowledge of the structure of atoms and molecules.

Today no reputable scientist doubts that atoms are real. This faith is justified because the atomic theory is useful. It enables us to interrelate many experimental facts that can be explained in no other way and it predicts new facts for investigation. Also, it affords a unity which pleases the mind and delights the imagination.

Atoms and Molecules

The word atom comes from a Greek expression, *atomos,* meaning uncut. The term is well chosen. Atoms are very stable bodies in chemical reactions. They cannot be broken up by the hottest furnace or by any other means ordinarily used by chemists. Under nuclear bombardment, to be discussed later, atoms can be disintegrated. *An atom is a particle indivisible in chemical changes.*

From a variety of kinds of evidence from physics and chemistry, scientists have identified over one hundred different elements or species of atom. (See Appendix I) Ranked according to atomic number Z, the atoms found in nature start with hydrogen ($Z = 1$) and end with uranium ($Z = 92$). Elements with atomic numbers greater than 92 have been made in small quantities by bombarding other elements with high-energy particles from accelerators. Each element has an average atomic mass given in terms of carbon-12 as 12.0000 units (grams per gram-atom). Typical values are: hydrogen, 1.0080 units; helium, 4.003 units; carbon, 12.011 units; oxygen, 15.999 units; silver, 107.870 units; uranium, 238.03 units.

Atoms are extremely small. The diameter of an atom is roughly one hundred-millionth of a centimeter (10^{-8} cm). The masses of individual atoms have been measured by physicists and found to range from about 10^{-24} gm to 10^{-22} gm. Suppose you were to cut a silver dime into two equal pieces, putting aside half and again dividing the other half, and so on. Imagine that you could repeat the sectioning process about 73 times. Then you would have a silver atom which could be disrupted only by very violent means.

The different kinds of atoms combine into many different kinds of molecules. Chemists have identified hundreds of thousands of different molecules and studied their properties. The molecules of some substances, such as helium, consist of single atoms. Others, such as hydrogen, have two atoms of a single element. The hydrogen chloride molecule consists of an atom of hydrogen joined to one of chlorine. If hydrogen chloride is broken up, hydrogen and chlorine result. Their properties

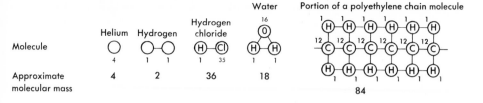

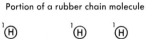

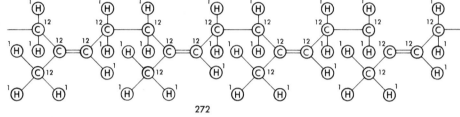

Fig. 12-1. *A few of the structures of molecules.*

are quite different from those of hydrogen chloride. Still other molecules are built up of three, four, and even thousands of atoms. Some molecules have simple geometric shapes: the hydrogen molecule has a dumb-bell shape. Others are, for example, long chains or spiral chains. Molecular masses of molecules that have been studied range from 2.00 units for hydrogen to tens of thousands times that for the large and complex molecules of organic chemistry (Fig. 12-1). *A molecule is the smallest particle of a substance that has the properties of that substance.*

The average distance between the centers of the neighboring molecules in a glass of water is about three hundred-millionths of a centimeter. Suppose that the glass of water were magnified until it became as large as the moon. Then one of its molecules would be about as big as a baseball. Since molecules are exceedingly small, the number of them in a glassful of water is very great. Suppose that you poured a glassful of water into the Atlantic and allowed sufficient time so that this water became equally dispersed through all the oceans. If you then dipped up a second

glassful of water, it is probable that you would recover about two thousand of the selfsame molecules.

Molecular Forces and the Three Phases of Matter

In the last chapter, you learned that two spheres exert gravitational attractions on each other that depend inversely on the square of the distance between their centers. Molecules have also been found to exert forces on each other, due not to gravitation but to the interaction of electric charges in the molecules. (Gravitational forces also exist, but are much weaker than the other forces.) At small distances of separation, the molecular forces are attractive, but decrease rapidly with distance (inversely as the distance to the eighth power) as the molecules separate. At distances of about 10^{-5} cm or so, the force between molecules is practically zero. When molecules collide, on the other hand, approaching each other very closely, the forces become repulsive ones: molecules resist being "superimposed" on each other. The qualitative picture of molecular forces that we will use in our discussion of molecular physics and heat is that molecules act like hard spheres that attract each other appreciably only when the distances between molecules are less than a few tens or hundreds times the diameter of the spheres.

Observations in everyday life remind us of the short range of molecular forces. Break a piece of chalk into two pieces. If you carefully fit the broken pieces together and push them against each other, they still will not adhere to each other. The molecules of the two surfaces are not sufficiently close to attract one another appreciably. Two pieces of ordinary glass placed in contact with each other do not attract each other strongly. If the surfaces of the plates are unusually flat and free from microscopic hills and valleys, they adhere strongly and are separated with difficulty.

Matter exists in three different phases: the gaseous, the solid, and the liquid. Gases expand indefinitely. They have neither fixed volume nor shape. Remove the stopper of a bottle of ammonia in a closed room and soon you will detect the odor of the vapor in any part of the room. Open a door into a hall, and after a short period the odor will be noticed there also. The molecules of a gas are not tightly joined together, but, like a swarm of flies, they dart about on zigzag paths, bumping into one another and the walls of their enclosure. Forces among gas molecules are small except when molecules collide or when the density of the gas is high and many molecules are crowded into a small space. At low densities, gas molecules go their individual ways, disorganized and uncooperative. Although the velocities of individual gas molecules are widely distributed, the average velocity is high: at 0°C the average velocity of oxygen molecules is 425 m/sec.

In solids the molecules are held together by comparatively strong molecular forces so that a solid body has a definite size and shape. A steel spring can be distorted by stretching; but, when the distorting forces are removed, it springs back to its original configuration. A definite crystal structure, as described below, is characteristic of solids. With certain exceptions, the motion of molecules in a solid is limited to vibration about a fixed point.

Liquids have definite volumes but indefinite shapes. Ten gallons of gasoline occupy the same volume whether contained in a cylindrical tank or a rectangular one. A liquid conforms to the shape of the container; it is not rigid. The structure of liquids is not nearly so well understood as that of a solid or the lack of structure of a gas. Liquids are in many ways intermediate between gases and solids. Since a substance in liquid form usually occupies a somewhat greater volume than that of the same mass of solid, it is clear that liquid molecules have more room to move around, somewhat like gas molecules, and that molecular forces in liquids must be weaker than in solids. On the other hand, there is some evidence that groupings of molecules, somewhat like the arrangements in crystals, form and dissolve in liquids. Gases mix very readily; solids, hardly at all. Some liquids mix readily; others, like oil and water, do not mix.

Finally, some substances have the properties of both solids and liquids. A lump of pitch is brittle and rigid so that it breaks into small pieces when it is struck by a hammer; yet, if it is placed in a cup and is left there for several weeks, it gradually conforms to the shape of the container like a liquid. Silicone putty can be rolled into a solid ball, yet the ball flows into a smooth flat "pool" in a few minutes. A piece of glass tubing supported horizontally at its ends gradually sags under its own weight. Substances that have no definite, permanent shape are regarded as extremely viscous liquids. They are said to be "amorphous" (structureless) because their molecules are not arranged in definite patterns.

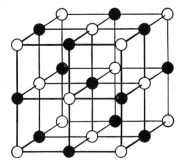

Fig. 12-2. In a sodium chloride crystal, the ions of sodium (black) and chlorine (white) are at the corners of cubes.

Crystal Structure

True solids are said to be *crystalline,* and their particles are arranged regularly like blocks built into a wall. In some crystals, the building blocks are single atoms; in others, they are groups of atoms; and in still others, they are electrically charged atoms or groups of atoms, called ions. Examine ordinary table salt through a magnifying glass. You will see that the individual grains are cubical. Sometimes, in salt mines, cubical crystals are found which are several inches on each side.

In each sodium chloride crystal, large or small, the ions (Chapter 23) of sodium and chlorine are arranged at the corners of tiny cubes (Fig. 12-2), held in place by electrical forces between the ions. The whole crystal is composed of a three-dimensional array of such tiny cubes, the model resembling the "jungle gym" found in children's playgrounds. Iron solidifies into a cubic structure. Graphite has a hexagonal structure (Fig. 12-3a). Quartz crystals also show a hexagonal structure.

There are seven crystal systems in all. The cubic structure of sodium chloride and the hexagonal structure of graphite are two of the most common. Which crystal structure a substance has when it solidifies depends upon many things including the nature of the forces and the size of the atoms or groups of atoms forming the building units. The same substance can have different crystal structures above and below a certain temperature. X-ray studies, to be described later, give us some of our most useful information about crystals.

The physical properties of a solid—its mechanical strength and its conduction of heat and electricity, for example—depend upon the kind of crystal structure and the degree to

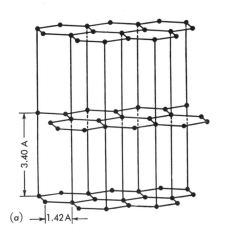

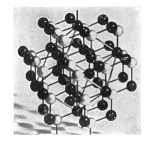

(a)   →|1.42A|←   (b)

Fig. 12-3. (a) In a graphite crystal, the carbon atoms are in layers, relatively far apart [1 A (angstrom) = 10^{-8} cm]. (b) In diamond, the atoms close together and interlock so that diamond is hard.

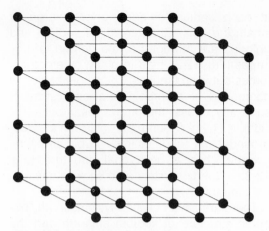

Fig. 12-4. *An ideal crystal.*

which the crystal is perfectly organized. Diamond, a form of pure carbon, is one of the hardest substances known. Graphite, also pure carbon, a greasy material, is so soft that it can be smeared over a surface like paint and is used as a lubricant. The carbon atoms in graphite are arranged in parallel planes which are a considerable distance apart (Fig. 12-3*a*). The attractive forces between adjacent planes are relatively small, like those between the sheets of paper in a book. The layers slip readily, hence graphite is greasy. The atoms in a diamond are arranged in pyramid fashion. They are closer together than in graphite, and they interlock so that diamond is very hard (Fig. 12-3*b*). Diamonds were made millions of years ago in nature's laboratory. Carbon at very high temperatures was

subjected to tremendous pressures which compressed it, and the atoms remained interlocked when the material was cooled. It is now possible to make synthetic industrial diamonds in the laboratory by establishing these conditions. The atoms in a piece of mica, which has a hexagonal crystal structure, are arranged in parallel layers. The attractive forces between atoms within a layer are large, but the forces between adjacent layers are small. Hence, mica can be split into thin sheets.

An ideal crystal would be one in which a perfectly uniform structure prevailed throughout the substance (Fig. 12-4). Real crystals lack the perfect regularity of an ideal crystal. Although single crystals, in which the unit structures are in orderly array, can be grown in the laboratory, they invariably have defects within them. One of the atoms may be missing at a point where the crystal structure requires it (Fig. 12-5*a*); one of the atoms may move from its regular position into the space between crystal layers (Fig. 12-5*b*); or the crystal may have a dislocation caused by the presence of an extra layer of atoms slipped into the crystal (Fig. 12-5*c*). Finally, we should note that single crystals are comparatively rare. Most solids are composed of small crystallites oriented in various directions (Fig. 12-6). Hammering, rolling, and rapid changes of temperature break up the regularity of single crystals and produce many-crystallite solids.

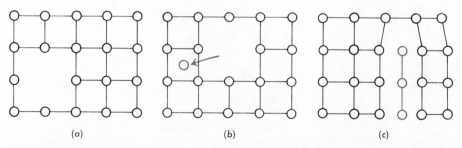

(a)   (b)   (c)

Fig. 12-5. *Imperfections in crystals:* (a) *a vacancy,* (b) *a displaced atom,* (c) *an edge dislocation.*

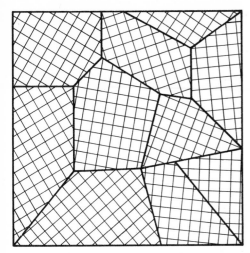

Fig. 12-6. Crystallites within a solid.

The degree of organization of a solid—the extent to which it resembles an ideal crystal—determines many of its physical properties. Melted copper cooled very slowly, without jarring, may form a single crystal with all the atoms in an orderly array. If the melted metal is cooled suddenly, or if it is jarred while it solidifies, a jumble of microscopic crystals results. Similarly, when a housewife freezes ice cream in a refrigerator and does not stir the liquid, fairly large ice crystals form so that the mixture is gritty. Stirring prevents large crystals from forming. The mechanical properties of large single crystals differ from those of solids composed of many smaller crystals. A copper rod as thick as a lead pencil, in the "single crystal" state, can be bent with the fingers of one hand. This bending makes adjacent crystal layers slip past one another. The bending introduces dislocations so that you need a hammer and anvil to straighten the rod again. It is as rigid as ordinary copper. Copper and zinc, intimately mixed, form *brass*, an alloy which is harder than either of its constituents. Brass and many other metals may be hardened by cold working. In cold working, the metal is pounded or rolled into sheets so that the

microscopic crystals merge with one another. However, when the metal is annealed by heating, its crystals grow and the metal becomes softer (Fig. 12-7). We shall refer to the crystal structure repeatedly as we discuss the various physical properties of solids.

Hooke's Law of Stretching

When an engineer designs a suspension bridge or a dentist fashions a bridge of gold to span the gap where a tooth has been extracted, each is concerned with the cohesive properties of the materials. Let us discuss some of these properties, beginning with the resistance that a solid offers to a stretching force.

Often we use the words *stress* and *strain* interchangeably, but in physics they differ in meaning. Suppose that you stretch a rubber band. Its molecules, across any cross section, will attract each other and exert a restoring force. We define the stretching stress as *the restoring force per unit area.* Thus if the restoring force is 10 pound-weights and if the cross-sectional area of the rubber band is 0.010 in.², the stretching stress is 1000 pound-weights/in.²

$$\text{Stretching stress} = \frac{\text{Restoring force}}{\text{Cross-sectional area}} = \frac{F}{A}$$

Strain means distortion. Suppose that a rubber band is 6 in. long and that you stretch

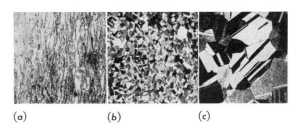

(a)   (b)   (c)

Fig. 12-7. Heating brass promotes the growth of its crystals: (a) cold-worked, (b) heated 30 min at 300°C, (c) heated 30 min at 750°C. (Courtesy of Dr. E. W. Skinner.)

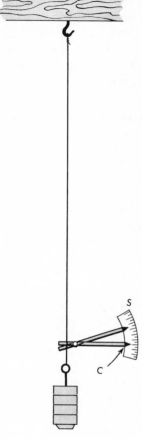

Fig. 12-8. Measuring the elongation of a wire.

it 0.12 in. Then the stretching strain, or fractional elongation, is 0.12 in./6 in. = 0.020. Notice that strain is a pure number. It has no units.

$$\text{Stretching strain} = \frac{\text{Elongation}}{\text{Length}} = \frac{e}{L}$$

If you stretch a piece of chewing gum or taffy and release it, it will not contract. It has been deformed plastically. Stretch a rubber band, and it will snap back almost to its original length. The rubber is more elastic than the chewing gum. *Elasticity is the property of a body which tends to restore its original dimensions when the deforming stress is removed.*

The elastic properties of the metal in a wire can be studied by using simple apparatus. Attach one end of the wire to a rigid support and the other to a weight carrier. Let one end of a lever be fastened to the wire at a point near its lower end (Fig. 12-8). The motion of the free end of the lever will measure the elongation of the wire. When you suspend several equal weights, one at a time, the successive elongations are equal. When all the weights are removed, the pointer *C* returns to its original position. However, if the load is increased sufficiently, the successive elongations become greater and we say that the *elastic limit* has been exceeded. Then the wire stretches plastically like chewing gum, and eventually it breaks.

The results of such an experiment are shown in Fig. 12-9. From this graph it is clear that, until the stress reached 50,000 lb/in.^{2}, the elongations were directly proportional to the loads. After this elastic limit was exceeded, the elongations began to increase and the wire broke when the stress reached 64,000 lb/in.^{2} Figure 12-10 shows similar curves for other materials.

In the experiment just considered, the strain was proportional to the stress until the elastic limit was reached. That is, when the stretching force was doubled, the elongation was also doubled. This is a special case of an important law, first stated by the English physicist, Robert Hooke, nearly three centuries ago:

Until the elastic limit is reached, strain is directly proportional to stress.

Three centuries ago, scientists were more jealous and suspicious of one another than they are today. Hooke wished to secure the credit for discovering this law without telling anyone what he had discovered. Then he would be able to continue his investigations without competition, expecting to make a full disclosure later. Hooke published his law first as an array of letters arranged in alphabetical

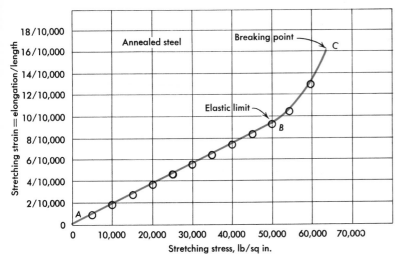

Fig. 12-9. *Stress and strain in a wire. What is the relation between them?*

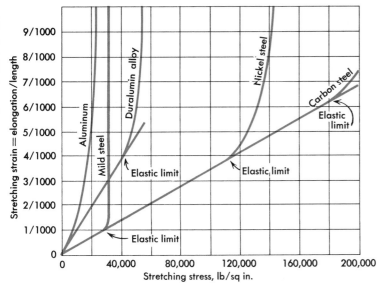

Fig. 12-10. *Stress-strain curves. In what respect do mild steel and carbon steel differ? (Courtesy of A. Karpov.)*

order: *ceiiinosssttuv* which can be rearranged to give the Latin expression *Ut tensio, sic vis* (As the stretching, so is the force).

When an engineer designs a suspension bridge, he must know how much each wire will stretch when subjected to a given load. In calculating the strain, he uses an experimentally determined elastic constant called Young's modulus. We define it as *the ratio of the stretching stress to the strain which it produces.*

$$\text{Young's modulus} = \frac{\text{Stretching stress}}{\text{Stretching strain}}$$

$$Y = \frac{F/A}{e/L}$$

Hooke's law then can be written

$$F = \frac{A \cdot Y}{L} e = ke$$

where the constant $k = AY/L$ holds for a given piece of wire. Springs also obey Hooke's law of stretching below their elastic limits. (See Table 12-1.)

Example. In a demonstration of the strength of steel a 2000-pound automobile was suspended by a steel wire. Find the elongation produced by the load, assuming that the wire was 10 in. long and 0.040 in.² in cross section. Young's modulus of the steel = 30,000,000 lb/in.²

Let the elongation be *e*.

$$\text{Stress} = \frac{2000 \text{ lb}}{0.040 \text{ in.}^2} = 50,000 \text{ lb/in.}^2$$

$$\text{Strain} = \frac{e}{10 \text{ in.}}$$

$$Y = \frac{50,000 \text{ lb/in.}^2}{e/10 \text{ in.}} = 30 \times 10^6 \text{ lb/in.}^2$$

$$e = 0.017 \text{ in.}$$

The restoring force that opposes the stretching of a wire is due to the molecular attractions within the crystal acting across any cross section of the wire. Hence the elastic force has the same origin as the forces that led to the organization of the crystal when the substance solidified. If the layers of the crystal are separated slightly by a distorting force, the molecular forces act to pull the layers back to their original positions. Why then do the molecular forces cease to be as effective at the elastic limit? We must lay the blame on the imperfections called dislocations within the crystal. Under a sufficiently large stretching force, the edge dislocation shown in Fig. 12-11 can move in the direction of the force, causing a slip of one section of a crystal layer. Since the force required to do this is much less than the force to pull an entire layer forward the same distance, the wire no longer obeys Hooke's law. It has passed its elastic limit; further increases of force produce further slipping of crystal layers due to the motion of the many dislocations. The wire acquires a permanent elongation or set

Table 12-1. Values of Young's Modulus

Substance	Young's Modulus		Stress at Elastic Limit lb/in.²	Breaking Stress lb/in.²
	lb/in.²	dynes/cm²		
Aluminum, rolled	10,000,000	7.0×10^{11}	25,000	29,000
Aluminum alloy (nickel 20%)	9,400,000	6.47×10^{11}	23,000	60,000
Gold	11,380,000	7.85×10^{11}		25,000
Gold alloy (10% copper)				65,000
Iron, wrought	26–29,000,000	$18.3–20.4 \times 10^{11}$	21–26,000	42–52,000
Lead, rolled	2–2,400,000	$1.47–1.67 \times 10^{11}$		3000
Phosphor bronze			60,000	80,000
Rubber		0.05×10^{11}		
Steel, annealed	29,000,000	20×10^{11}	40,000	64,000

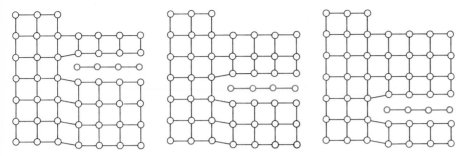

Fig. 12-11. The motion of a dislocation carries a part of the crystal plane downward.

and eventually reaches the breaking point.

According to calculations, materials would have higher elastic limits if dislocations were less numerous or if the motion of dislocations were impeded in some way. "Whiskers" are tiny metallic filaments (Fig. 12-12) that were discovered as an unwanted growth on the metal coatings in electronics equipment. Physicists in the Bell Laboratories and in other laboratories investigated the whiskers in an effort to eliminate them and found to their surprise that tin whiskers were almost 1000 times stronger than tin wires. They had discovered an almost perfect crystal. Because of their small diameters (around 10^{-5} cm), whiskers do not permit the growth of

Fig. 12-12. "Whiskers" as seen on a piece of tin-plated metal under a microscope. The whiskers, which are single crystals of tin, are up to 0.25 in. long and about 0.0001 in. in diameter. (Courtesy Bell Telephone Laboratories.)

dislocations as other materials do, and hence their mechanical strengths remain high. Further study of the properties of whiskers promises to produce stronger and lighter materials for the engineer to use in construction.

If dislocations must be present in materials, their weakening effects can be minimized by blocking their motion. Working materials by hammering or rolling, as we have seen, breaks up the crystal into smaller crystallites or grains. Dislocations cannot roll over the grain boundaries very rapidly because of the change of direction involved. Hence polycrystalline materials are stronger than single crystals. Adding impurities, such as carbon, to metals produces stronger materials. In carbon steel, the carbon atoms diffuse between the layers of iron atoms and tend to prevent dislocations from moving. Carbon steel has a higher elastic limit than mild steel which contains less carbon (Fig. 12-10). In case hardening of steel, the heated steel is placed in an atmosphere of a carbon-containing gas. The carbon diffuses a short distance into the steel and produces a hard shell on the surface.

Other Kinds of Strain

Bodies can be not only stretched but also compressed, bent, and twisted. In all these deformations elastic forces may arise which tend to restore the body to its original size and shape. Besides stretching strain, there are two other kinds, *volume* strain and *shearing* strain. In general, the strain in a body is the fractional deformation produced in it by a stress.

Suppose that a rubber ball is placed in a liquid confined in a vessel and that the hydrostatic pressure is increased, compressing the ball. Then the change of stress is the increase in pressure, Δp, and the volume strain is the fractional change in volume, $\Delta V/V$, that is produced. Hooke's law of volume elasticity states that, up to the elastic limit,

$$\Delta p = - B \frac{\Delta V}{V}$$

where B is the modulus of volume elasticity.

Suppose that you push horizontally, with a small force, on the upper cover of a book and attempt to force it sideways. The cover will exert an opposite, restoring force. A metal block, subjected to a similar shearing force, will resist also. The shearing stress is the ratio F/A of the restoring force to the area of the top of the block over which the force acts. Note that the force acts parallel to the surface in shear. The shearing strain is the ratio of the displacement a of the top of the block to its thickness b (Fig. 12-13).

The ratio of shearing stress to shearing strain is called the *rigidity coefficient* (or modulus). By Hooke's law

$$\frac{F}{A} = n \frac{a}{b}$$

where n is the rigidity coefficient.

Twisting is related to shearing. When one end of a rod is twisted, the angular displacement of the rod is proportional to the torque acting on it. The rod may be thought of as a large number of disks which slide past each other when there is a shearing strain. The strain is greatest at the outer edge of the disk and it is zero along the axis of rotation. The torque τ required to produce an angular deflection θ (in radians) is given by

$$\tau = \frac{n \pi r^4 \theta}{2l}$$

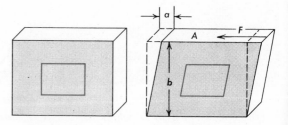

Fig. 12-13. A shearing stress F/A *results from a shearing strain* a/b.

in which n is the shear modulus of elasticity; r, the radius of the rod; and l is its length.

Solids have stretching elasticity, shear elasticity, and volume elasticity; liquids and gases have volume elasticity only (Table 12-2).

Table 12-2. Shear and Volume Moduli of Elasticity in dynes/cm²

	Shear	Volume
Aluminum	2.5×10^{11}	7.0×10^{11}
Copper	4.2×10^{11}	12×10^{11}
Glass	2.5×10^{11}	45×10^{11}
Iron, cast	5.5×10^{11}	10×10^{11}
Mercury	0	2.8×10^{11}
Steel, hot-rolled	8.2×10^{11}	16×10^{11}
Water	0	0.22×10^{11}

Molecular Forces in Liquids

If you dip a clean glass rod into a vessel of water and then remove it, some of the water will cling to the glass. The attractive forces between the water molecules and those of the glass are greater than those between adjacent water molecules. If you dip the rod into a vessel of clean mercury and then remove it, none of the mercury clings to the glass. Here the attraction of the mercury molecules for one another, that is, their *cohesion*, is greater than the *adhesion* of the mercury molecules to the glass.

If you dip a block of glass into water (Fig. 12-14a), neighboring water molecules exert a cohesive force *OA* on a molecule at *O*. The glass exerts a *greater* adhesive force *OB*. The resultant force *OC* must be normal to the surface of the liquid. (Why?) Hence we expect the water to creep up the glass—as we find that it does. The *angle of contact* α is small. The cohesion of mercury is greater than its adhesion to glass. Dip a glass block into mercury (Fig. 12-14b). A molecule at *O* is strongly attracted by the nearby mercury molecules and less strongly attracted by the glass. The resultant force *OC* pulls the mercury away from the glass. The angle of contact is large. For *pure* water and *clean* glass, the angle of contact between the film and the glass is nearly zero. For impure water and ordinary glass, it is about 25°. For mercury after exposure to air, α is about 130°. When the angle of contact is small, we say the liquid "wets" the glass. Wetting agents, used in dyeing fabrics, make the angle of contact small between

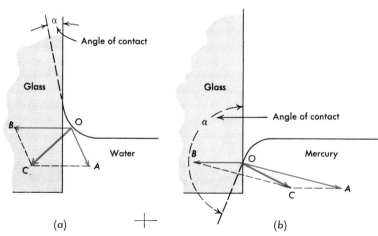

Fig. 12-14. (a) *The adhesive force* OB *is greater than the cohesive force* OA; *hence the liquid is pulled toward the glass.* (b) *Cohesion is greater than adhesion.*

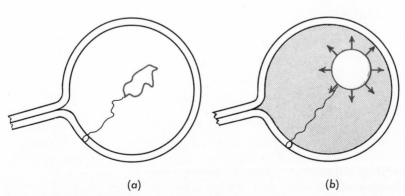

Fig. 12-15. *Cohesion pulls the mercury into droplets.*

the fibers and the liquid dye so that the dye penetrates the fabric. Waterproofing coatings, used on raincoats, create large angles of contact between the water and the fibers.

Surface Films on Liquids

Mercury spilled on a table top breaks up into globules of different sizes (Fig. 12-15). The smaller droplets are nearly spherical, and the larger ones are flattened. If gravity did not act, each droplet would be a perfect sphere. This tendency to form spherical droplets manifests itself in all liquids. Gunshot is made by pouring melted lead through a sieve at the top of a tower. The falling liquid breaks up into spherical drops, which solidify before striking a pool of water at the bottom of the tower.

Touch the surface of a mercury globule with a needle. A dimple forms at the point of contact and disappears when you remove the needle. The drop behaves like an inflated rubber balloon, as though it were enveloped in an elastic membrane.

Elastic films appear at the surfaces of all liquids—strictly speaking at the interfaces between liquids and a gas, such as air, or between two different liquids that do not mix. You can float a safety razor blade on a water surface. The blade quickly sinks to the bottom when the film is ruptured. Some insects are not wet by water, hence they can walk on the surfaces of water pools. A slight depression or dimple is formed where each foot pushes against the water surface. Most insects are wet by water; hence they are gripped by the film and cannot escape. For a fly, the forces of the surface films are much stronger than gravitational forces.

If you dip a ring of wire into a soap solution and then remove it (Fig. 12-16a), the film will behave like an elastic sheet of stretched rubber. Rupture the film inside a loop of thread. The film will pull the thread out on a circle (Fig. 12-16b). The fact that the loop is uniformly curved shows that the tension of the film is the same in different parts of the surface.

Float a toothpick or a match on water. The floating object remains at rest, because the pull of the film on one side is balanced by that on the other. A little alcohol dropped on the water at one side of the toothpick will weaken the film on that side so that it yields and the toothpick is pulled away by the stronger film on the other side.

If you put little flecks of camphor onto the

(a)   (b)

Fig. 12-16. *Soap film. In* b *the film pulls the cord into a circle.*

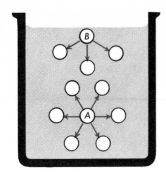

Fig. 12-17. Molecular forces explain surface tension.

surface of water, the camphor will dissolve more at one point than another, reduce the surface tension, and be pulled in the opposite direction. Thus the camphor flecks dart about like small motor boats. Touching the water with a clean toothpick or a match will not stop the camphor. However, if the wood is first rubbed through your hair, it will collect a trace of oil which will spread over the water surface, and the motion of the camphor will cease. The oil film required to do this is only one molecule thick. Near the end of the last century, the British physicist Lord Rayleigh showed that the behavior of an oil film on water could be used to measure the diameter of a single oil molecule. He allowed olive oil to fall on water and spread into a circular layer just thick enough to keep camphor from moving. Knowing the mass of the oil added and its density, he calculated the volume of the oil layer. He measured the area of the layer and, by dividing the volume by the area, found its thickness. The thickness of the oil layer—one of the first measurements of the diameter of a single molecule—was about 10^{-7} cm. Later the American physicist Irving Langmuir performed similar experiments in which he allowed the oil layer to spread out freely and measured the diameters of a number of different molecules.

Surface tension can be explained in terms of the attractive forces that the molecules of a liquid exert on each other. Near the center of a beaker of water (Fig. 12-17), molecule *A* has an average force of zero acting upon it because the attractions of its neighbors tend to cancel each other. Near the surface of the water, molecule *B* experiences attractive forces in the downward direction only. If *B* should move slightly above the surface, there would be a resultant force acting downward to prevent its escape. A liquid surface thus resists deformation like an elastic sheet and reduces its area to a minimum. A liquid drop falling in a vacuum takes the shape of a sphere because the sphere has the smallest surface area for a given volume.

The tension of a stretched drumhead or a sheet of rubber may be measured by the pull that it exerts on a unit length of its boundary. Similarly, we define the coefficient of surface tension σ (Greek *sigma*) of a liquid as the *force exerted per unit length of boundary*. Usually, surface tension coefficients are expressed in dynes per centimeter (Table 12-3). The force *F* exerted by a film of length *L* is then

$$F = \sigma L$$

One way of measuring the coefficient of surface tension of a liquid is to use a frame of thin wire dipped into the liquid (Fig. 12-18). The wire is pulled upward by means of a delicate spring balance, and the force *F* required to rupture the film is measured. There are *two* surface films, front and back; hence, if the width of the frame is *w*, the total width of the two films is 2*w*. The coefficient of surface ten-

Table 12-3. Values of Coefficients of Surface Tension of Liquids in Contact with Air

	dynes/cm or ergs/cm²
Water at 15°C (59°F)	73.5
Water at 100°C (212°F)	58.9
Methyl alcohol at 15°C	24.7
Mercury at 15°C	487
Platinum at 2000°C	1819

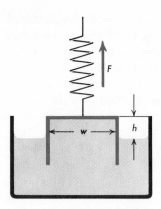

Fig. 12-18. Measuring the pull of a (double) surface film.

sion of the liquid is given by

$$\sigma = \frac{\text{Force}}{\text{Width of both films}} = \frac{F}{2w}$$

When we blow a soap bubble, we do work in creating the surface film and we store up potential energy in it. In Fig. 12-18, suppose that the height of the film formed in the frame is *h*. The work done in creating the film is *Fh* or $(2\sigma w)h$, and this is also the potential energy stored in it. The area of the film is $2wh$. Hence, the energy per unit area is $2\sigma wh/2wh$ and is equal to the coefficient of surface tension σ. Thus a water film whose coefficient of surface tension is 75 dynes/cm has 75 ergs/cm^2 of energy per unit area. A surface film tends to contract so as to minimize its surface and therefore its potential energy.

The pressure in a soap bubble depends upon the surface tension coefficient σ of the film and also upon the radius of the bubble. Suppose a hemispheric bubble of radius *R* rests on a glass plate. Let the excess pressure in the bubble be *p*. The bubble has two surface films. The glass exerts on each film a force $2\pi R\sigma$; hence the total down force acting on the half-bubble is $2(2\pi R\sigma)$. The excess pressure *p* of the confined air exerts an up force $p(\pi R^2)$. Thus

$$2(2\pi R\sigma) = p(\pi R^2)$$

$$p = \frac{4\sigma}{R}$$

Notice that the smaller the bubble is (the smaller *R* is), the greater is the pressure. In Fig. 12-19, which soap bubble will increase in size?

Capillarity

The tendency of liquids to be forced into minute pores and tiny openings is called *capillarity* (Latin *capillus*, a hair). Examples are the rise of oil in a lamp wick, and the absorption of ink by blotting paper. Capillarity is useful in agriculture, for it raises water from depths of several feet toward the surface of a field where the water may be utilized by growing plants. If the soil near the surface contains many fine capillaries, the moisture will escape by evaporation. In order to prevent this escape, the farmer cultivates his fields frequently. Thus he breaks up the long fine capillaries so that moisture does not rise to the surface and escape. Concrete pavements are laid on cinders in which the openings are so large that water does not rise in them. If the pavements were laid on the

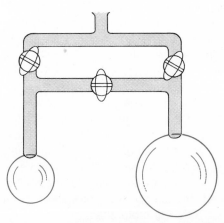

Fig. 12-19. The excess pressure in the small bubble is greater than that in the large bubble.

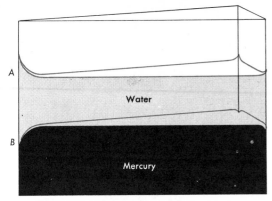

Fig. 12-20. *Cohesion of mercury is greater than adhesion of mercury and glass. Water, however, adheres to clean glass.*

ground, capillary action would tend to form soft mud, incapable of supporting the weight and likely to freeze in winter.

When a board is placed in water, the liquid penetrates the microscopic pores, and the wood swells. If the board has been bent, it will tend to straighten out as a garden hose does when water is admitted to it at high pressure.

Capillarity is explained by taking into account the cohesion of the molecules of the liquid and their adhesion to other kinds of molecules. Consider some mercury and water in a wedge-shaped glass vessel (Fig. 12-20). As

we have seen, the angle of contact between water and clean glass is small, and that between mercury and glass is large. Both walls attract the water near the narrow corner of the wedge; hence the water is drawn rather far up above the level surface of the liquid. Strong cohesive forces pull the mercury away from the narrow corner so that the boundary there is lower than in other parts of the surface. Similarly, if one dips a glass tube with a capillary bore into water, the liquid will rise in the bore because the adhesion of water to glass is greater than the cohesion of the water (Fig. 12-21*a*). If he dips the tube into mercury, the down pull of the mercury on the molecules at the surface is greater than the adhesive force between the mercury and the glass; hence the film inside the tube descends below the level of the surface outside (Fig. 12-21*b*).

We can determine the coefficient of surface tension by measuring the elevation of a column of the liquid in a capillary tube. Let h be the elevation of the liquid surface in a capillary bore above that in the vessel; let r be the radius of the bore and D_m the mass density of the liquid. The film pulls upward along the circular boundary of the column in the bore, the circumference of which is $2\pi r$. The upward pull of the film is $2\pi r\sigma$ (Fig. 12-22). The

(a)   (b)

Fig. 12-21. (a) *Strong adhesion pulls the water up inside the tube with a capillary bore (diameter exaggerated in the drawing).* (b) *Strong cohesion pulls the mercury down, away from the glass.*

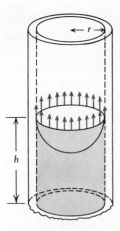

Fig. 12-22. *The up pull $2\pi r\sigma$ on the film is equal to the weight $\pi r^2 hD_m g$ of the liquid.*

volume of the column is $\pi r^2 h$, its mass is $\pi r^2 hD_m$, and its weight is $\pi r^2 hD_m g$. In equilibrium:

Upward pull of film = Weight of liquid column

$$2\pi r\sigma = \pi r^2 hD_m g$$

and

$$\sigma = \frac{rhD_m g}{2} \qquad (1)$$

We assume that the surface tension forces are directed vertically upward at the edge of the film. When the angle of contact α of the film with the wall is not zero, equation (1) should be written

$$\sigma = \frac{rhD_m g}{2 \cos \alpha}$$

Summary

Atoms are particles that cannot be divided by chemical methods. A molecule of a substance is the smallest particle that has the properties of that substance.

Intermolecular forces are negligibly small at distances greater than a few hundred-thousandths of a centimeter.

Gases have neither intrinsic shapes nor volumes, liquids have definite volumes but not definite shapes, and solids have both definite shapes and volumes. In amorphous substances the molecules are arranged at random. In crystalline substances they are arranged in regular patterns which are repeated throughout each single crystal. The nature of the crystal structure and the defects in it determine many of the physical properties of the solid.

The stress in a body is the restoring force per unit area. The strain of a body is its fractional deformation: the stretching strain, for example, is the elongation of a body per unit length. Elasticity is the tendency of a body to resume its original shape or size after it is deformed or distorted. Hooke's law: The strain in a body is proportional to the stress, until the elastic limit is reached. After the elastic limit is exceeded, the ratio of stress to strain decreases. Young's modulus for a substance is the ratio of the stretching stress to the stretching strain.

Cohesion is the attraction of molecules of the same kind for one another; adhesion is the attraction between molecules of different kinds. Cohesive forces tend to draw bodies of liquid into spherical drops. These attractive forces are manifest at the surfaces only of the liquids, which behave as though they were surrounded by elastic films. The coefficient of surface tension of a liquid is the force per unit length of film. The energy per unit area of a film is equal to its coefficient of surface tension.

Capillarity is defined as the tendency of a liquid to be forced into capillary tubes or, in certain liquids, to be expelled. The formula for the rise of a liquid of surface tension coefficient σ in a tube of radius r is $h = 2\sigma/rD_m g$, D_m being the mass density of the liquid.

Questions

1. How many kinds of atoms are there? How many kinds of molecules are there?

2. Describe a method for determining the diameter of a molecule.

3. What is a single crystal? Are solids usually single crystals? Discuss.

4. Explain why diamond is harder than graphite.

5. How can the sizes of the crystallites in a solid be controlled?

6. Why is a lever-type indicator of increase of length used in Fig. 12-8? How could you calculate the calibration factor of the indicator in order to determine extensions of the wire from changes in pointer readings? How could you determine the calibration factor experimentally?

7. Explain the action of glues, cements, and solders in terms of molecular forces.

8. Why does quick freezing preserve the flavor of foods?

9. Explain (a) the malleting of metal into a cavity in a tooth, and (b) why a dental plate (denture) clings to the roof of the mouth.

10. Why do the bristles of a paint brush stay apart under water but draw together when the brush is removed?

11. Mercury spilled on the clean surface of a glass plate gathers into droplets, but water does not. If the surface is greasy, the water also forms droplets. Explain.

12. Two toothpicks, floating on a water surface, are parallel to each other and a small distance apart. A hot needle is touched to the water surface between them, and the two fly apart. Explain.

13. (a) If a glass rod is coated with paraffin or with oil, water will not cling to it. Explain. (b) If the fat and grease are removed from cotton, it will soak up much more water than untreated cotton. Why?

14. Why does the end of a glass rod become rounded when it is heated to the softening point in a Bunsen flame?

15. A cartoon showed a bricklayer standing on a sagging plank loaded with bricks so that, as he laid bricks and the wall rose, the plank rose, keeping him even with the top of the brick wall. Would this be possible?

16. A glass plate lifted from a water surface comes off wet. When the plate is lifted from a mercury surface, none of the liquid clings to it. Compare the forces required to lift the plate in the two cases.

17. Several fog particles merge to form a single particle. Is the total potential energy of the surfaces increased or diminished?

18. Water will rise in a long capillary tube to an elevation h. If the tube is shortened so that its top is at an elevation less than h above the water surface, will water run out of the tube, creating a "fountain"? Explain.

19. Discuss the forces involved in each of the following: (a) a piece of thin aluminum foil rests on a surface of water, (b) the same foil folded to make a boat floats on water, (c) the foil rolled into a ball sinks.

Problems

(*See tables for constants.*)

A

1. A strip of rubber 10.0 cm long and 0.50 cm² in cross section is stretched 0.20 cm by a force of 4.0 kgwt. What are (a) the stretching stress and (b) the stretching strain?

2. How much will an annealed steel wire 10 ft long and 0.050 in.² in cross section be stretched by a force of 1000 lb?

3. How large a load can be supported by the wire of the preceding problem (a) without permanently deforming it, and (b) without breaking it?

4. The smallest force that will break a rope 1.00 cm in diameter is 314 kgwt. What is the breaking stress of the rope?

5. What force does a surface film exert on one side of a toothpick 4.0 cm long, floating on water at 59°F?

6. What is the radius of the bore of a capillary in which water at 59°F rises to elevations of (a) 1.00 mm, (b) 1.00 cm, (c) 1.00 m? From

these results, do you think that surface tension is sufficient to raise the sap to the top of a tree?

7. Given that the surface tension of ethyl alcohol is 22.3 dynes/cm and its mass density is 0.79 gm/cm^3, calculate the height to which alcohol will be raised in a vertical tube, the bore radius of which is 0.040 cm.

8. How many grams of water at 15°C will be lifted by a glass capillary tube, the bore diameter of which is (a) 1.2 mm, (b) 0.60 mm?

9. How high will water at 15°C rise in a glass capillary tube if the radius of the bore is 0.60 mm?

B

10. An inverted U-shaped platinum wire is dipped into water at 15°C and then raised, forming a water film. The width of the U at the water surface is 3.0 cm. What force is required to break the two-surface film formed in it?

11. A wire 200 cm long and 4.0 mm^2 in cross section is stretched 2.00 mm by a force of 2000 gwt. What are (a) the stretching stress, (b) the stretching strain, (c) Young's modulus of the material?

12. A force of 7.0 kgwt is required to break a piece of cord. How much force is required for a cord made of the same material that is (a) twice as long and of the same diameter, (b) twice as large in diameter and of the same length?

13. One end of a wire made of annealed steel is attached to a beam and the other end welded to a wire made of aluminum-nickel alloy (20% nickel). The two have the same length and diameter. A load attached to the free end of the aluminum-nickel wire stretches it 5.0 mm. How much is the steel wire stretched?

14. A solid rubber ball 6.00 cm in diameter is submerged in a lake at such depth that the pressure exerted by the water is 1.00 kgwt/cm^2. Find the change in volume. The volume coefficient of elasticity is 1.0×10^7 dynes/cm^2.

15. What force is required to lift a ring made of thin wire, 4.00 cm in diameter, from the surface of water at 59°F?

16. How much would the tendon of a man's leg be stretched by a force of 15 kgwt if the tendon is 10 cm long and 0.50 cm in diameter? Young's modulus is 1.63×10^6 gwt/cm^2.

17. What torque will produce an angular deflection of 22.5° for a steel rod 10 cm long and of radius 4.0 mm?

C

18. A soap film, whose surface tension coefficient is 30.0 dynes/cm, is formed on a "hairpin" having a crosspiece 8.0 cm long. Compute (a) the force opposing outward motion of the crosspiece, (b) the work done in pulling it out 4.0 cm, (c) the resulting increase in area of the film, (d) the increase in energy per unit area (in ergs per square centimeter) of new film surface. (Compare this value with the surface tension coefficient.)

19. A toothpick 5.0 cm long floats on water; the water film on one side has a surface tension coefficient of 50 dynes/cm. On the other side, camphor reduces the surface tension to 30 dynes/cm. What resultant force acts on the toothpick?

20. A hemispherical water bubble of radius 2.50 cm is formed on the surface of a pool. The surface tension coefficient is 75 dynes/cm. (a) Find the force tending to prevent the bubble from escaping. (b) What is the force per unit area of water surface inside the bubble?

21. A load of 70 tons is carried by a steel column having a length of 24 ft and a cross-sectional area of 10.8 in.^{2} What decrease in length will this load produce if Young's modulus is 30×10^6 lb/in.^{2}?

22. Compute the pressure due to surface tension inside a mercury droplet 0.012 cm in diameter.

23. An aluminum rod, 1.5 m long and 0.60

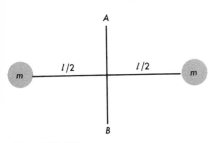

Fig. 12-23.

cm in diameter, is clamped at one end. At the other end, a disk 15 cm in radius is attached to the rod and supported by a bearing. A cable wound around the rim of the disk supports a weight-hanger. What total load is required to twist the rod through 45°?

24. Prove that the edge of the film of water that rises between the glass plates in Fig. 12-20 is an equilateral hyperbola whose equation is $xy = k$.

25. Estimate the volume of water in the top layer of molecules on Lake Erie. (Area of the lake is 9910 square miles. Assume that the molecules are spheres with a diameter of 3×10^{-8} cm.)

26. Derive an expression for the increase in length of a light rod of initial length l that is rotated horizontally with angular velocity ω about a vertical axis *AB*. (See Fig. 12-23.) Small bodies of mass m are attached at the ends of the rod.

27. Nylon rope is used in mountain climbing because of its elasticity—it gradually slows a fall that would break a man's back if he were arrested by a rope having little stretch. Suppose that you were required to select a nylon rope that would provide protection for a climber falling freely 20 meters. He and his equipment weigh 90 kgwt, and a maximum deceleration of 3 *g* is tolerable. You are free to measure the stretch versus tension of several sizes of rope, and the breaking strength is known. Describe how you would make your choice and justify it by some calculations. State your assumptions clearly.

For Further Reading

Andrade, E. N. daC., "Robert Hooke," *Scientific American,* December, 1964, p. 94.

Bernal, J. D., "Structure of Liquids," *Scientific American,* August, 1960, p. 124.

Boys, C. V., *Soap Bubbles and the Forces Which Mould Them,* Doubleday and Company, Garden City, 1959.

Davis, Kenneth S., and John Arthur Day, *Water: The Mirror of Science,* Doubleday and Company, Garden City, 1961.

Derjaguin, B. V., "The Force Between Molecules," *Scientific American,* July, 1960, p. 47.

Holden, Alan, and Phylis Singer, *Crystals and Crystal Growing,* Doubleday and Company, Garden City, 1960.

Reiner, Marcus, "The Flow of Matter," *Scientific American,* December, 1959, p. 122.

Rogers, Eric M., *Physics for the Inquiring Mind,* Chapter 6, Princeton University Press, Princeton, 1960.

Thomson, Sir George, *The Forseeable Future,* The Viking Press, New York, 1960.

Wannier, Gregory, "The Nature of Solids," *Scientific American,* December, 1952, pp. 39–48.

CHAPTER THIRTEEN

Temperature and Expansion

In earlier chapters we have considered principally mechanical energy and its transformations. Now we shall deal with *thermal energy* or *heat energy,* whose transfer is associated with changes in the irregular disorderly motion of molecules. One of the most important ideas that we shall use in the study of heat is *temperature.* We shall see later that the average kinetic energy of gas molecules, for example, is proportional to the temperature of the gas and is increased by the addition of heat to the gas under constant volume.

A baby gets his first ideas of temperature by means of sensations of hot and cold. However, our senses are not reliable enough for establishing a temperature scale. In zero weather, a block of iron seems colder than a piece of wood at the same temperature. It is better to relate temperature to the direction of heat flow. If you dip a hot poker into a bucket of water, heat will pass from the hot metal to the colder liquid. In general, heat flows from hotter bodies to colder ones. *Temperature is that which determines the direction of heat flow.* Bodies at the same temperature are said to be in thermal equilibrium with each other; the net exchange of heat between them is zero.

Thermometers

A thermometer is a device for measuring temperatures. Galileo invented the first thermometer in 1593. It consisted of a bulb attached to a glass tube, the lower end of which was submerged in a vessel of water. Initially, when the bulb was heated, some of the air was expelled from it, and as the air cooled again, water was forced upward into the tube (Fig. 13-1). Thereafter changes of temperature of the bulb caused the level of liquid to rise and fall in the tube. This thermometer was the ancestor of the clinical thermometers of today. To use it, the physician first placed the bulb in his own mouth. As the confined air was heated, the water level in the tube moved downward to a point which was marked. He then placed the thermometer in the patient's mouth and, if the water level moved to the same position as before, the doctor reasoned, "This man's temperature is the same as mine, so it must be normal!" Galileo's description of his thermometer shows how scientists were limited by the lack of standard units of measure. He described it as "a glass bottle about the size of a hen's egg, the neck of which was two palm's lengths and as narrow as a straw."

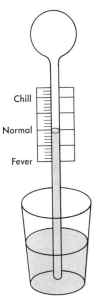

Chill

Normal

Fever

Fig. 13-1. Galileo's primitive thermometer. Why was it less accurate than the cheapest thermometer of today?

Galileo attached a scale to his thermometer in order to measure the height of the water column. He could not trust the accuracy of its readings from one day to the next, for variations in barometric pressure caused the water to rise and fall in the tube. About 50 years later the Grand Duke of Tuscany, in Italy, needed a reliable thermometer for experiments on the artificial hatching of eggs. He wanted an instrument which would not be affected by varying barometric pressures. He inverted Galileo's bulb and tube, filled it with alcohol to a certain level of the tube, and sealed its upper end. Temperature variations were indicated by the rise and fall of the liquid surface due to the expansion or contraction of the alcohol. The stem was marked off in

degrees, forming the thermometer essentially as we know it today.

Any physical property that depends upon temperature can be used as the basis of a thermometer. Modern liquid-in-glass thermometers rely upon the expansion of mercury or alcohol with rise of temperature. The mercury thermometer (Fig. 13-2) contains a relatively large thin-walled reservoir of mercury connected to a glass tube with a capillary bore. Nitrogen is often introduced into the thermometer above the mercury in order that its pressure will tend to prevent the boiling of the mercury at the higher temperatures. The liquid-in-glass thermometer is limited in range, because at lower temperatures the liquid freezes, and at higher temperatures, its vapor pressure may burst the thermometer bulb. The *resistance thermometer* measures temperatures by the varying resistance of a coil of platinum wire (Chapter 23). The *thermocouple* generates an electric voltage due to temperature difference (Chapter 25).

The *optical pyrometer* is used to measure the temperatures of red-hot or white-hot bodies only. It consists (Fig. 13-3) essentially of a telescope with a small incandescent electric lamp mounted in front of the eyepiece. An electric battery causes the current to heat the filament. This current is regulated by a variable resistor (rheostat). The observer looking through the telescope sees an image of the hot body whose temperature is to be measured. He also sees an image of the hot filament of the electric lamp. He varies the resistance until the filament is just as bright as the image of the hot body, so that the filament becomes invisible. Then he measures the electric current by means of the ammeter and reads the corresponding temperature from a calibration chart. The range of the op-

Fig. 13-2. The mercury thermometer.

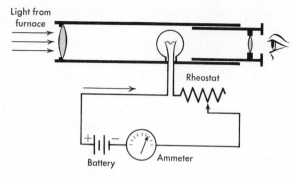

Fig. 13-3. *Optical pyrometer. The operator varies the electric current until the filament of the lamp seems just as bright as the furnace. Then he reads the ammeter and next the corresponding temperature from a chart.*

tical pyrometer is from 1500°F to many thousands of degrees. It can be used to measure the surface temperature of the sun.

When the temperature of a gas is raised, the kinetic energy of the molecules increases and the pressure of the gas, at constant volume, increases. The pressure of the confined gas in a *gas thermometer* (Fig. 13-4) is $p_B + p$, in which p_B is the barometric pressure and p is the pressure due to the mercury column of height h. When the temperature of the gas is increased, the gas expands. Then the open tube is raised so as to restore the volume to its previous value and the pressure is noted. From the calibration curve of temperature versus pressure, the temperature can be obtained. Constant-volume gas thermometers, although cumbersome and inconvenient, are used in research work for the accurate measurement of temperatures. They may be employed to measure high temperatures such as those of furnaces, and temperatures far below the freezing point of mercury.

Fixed Points and Temperature Scales

About a century after Galileo's invention, Fahrenheit, of Danzig, devised the thermomet-

ric scale which is in everyday use in English-speaking countries. Fahrenheit wished to avoid negative readings and therefore chose for the zero point a temperature colder than the ordinary winter temperatures of his locality. He found that he could produce a conveniently low temperature by using a mixture of a salt (ammonium chloride) and ice in certain proportions and chose this point as the zero of the new scale. As a second fixed point on the scale, Fahrenheit used the normal temperature of the human body. On the thermometers of today, the value is 98.6°F. The temperature of the body is no longer used in standardizing thermometers, since it varies considerably even in healthy people. The upper fixed point is the temperature of boiling water, at "normal" pressure (76 cm-Hg), which by definition is 212°F. For the lower point, the melting point of ice is taken as 32°F.

Scientists generally use the *Celsius* (formerly centigrade) *temperature scale.* The normal freezing point of water on the Celsius scale is taken as 0° and the normal boiling point as 100° (Fig. 13-5). A number of other fixed points—temperatures which can be conveniently and reproducibly established in the laboratory—have been set by interna-

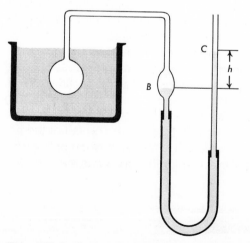

Fig. 13-4. *Constant-volume gas thermometer.*

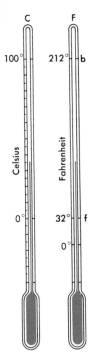

Celsius

Fahrenheit

100°   212° b

0°   32° f

0°

Fig. 13-5. The fixed points of the Fahrenheit and Celsius thermometers: $212°F = 100°C$; $32°F = 0°C$.

tional agreement (Table 13-1) and are used in calibrating thermometers. A third temperature scale, one of fundamental importance, is the **absolute Kelvin** temperature scale, on which the boiling point of water is about 373°K, the freezing point of water is about 273°K, and the zero is approximately −273°C (more accurately −273.15°C). Temperatures within a few millionths of a degree of the Kelvin zero have been reached. One of the advantages of this scale is that the pressure of a gas at constant volume is proportional to its Kelvin temperature. Figure 13-6 shows the readings of a constant-volume gas thermometer. If the origin of temperatures were shifted from 0°C to −273.15°C, the pressure of the gas in the region of the plotted points would be proportional to temperature *on this new scale.* This rule holds closely for gases at ordinary temperatures and pressures. By definition, it

would hold for an ideal gas (Chapter 15) at *any* temperature. The Kelvin temperature scale, of course, can be used in discussing the behavior of any physical system.

Table 13-1. Some Fixed Points on the International Temperature Scale of 1948

	Temperature (°C)
Primary fixed points (normal pressure)	
Oxygen point: equilibrium between liquid oxygen and its vapor	−182.970
Ice point[a]	0.0000
Steam point	100.0000
Sulfur point: equilibrium between liquid sulfur and its vapor	444.600
Gold point: equilibrium between solid and liquid gold	1063.0
Secondary fixed points (normal pressure)	
Freezing point of mercury	−38.87
Naphthalene point: equilibrium between naphthalene and its vapor	218.0
Freezing point of platinum	1769
Melting point of tungsten	3380

[a] By international agreement, the zero of the Celsius scale is now set in relation to the triple point of water (the temperature at which liquid water, ice, and water vapor are in equilibrium) which is taken as 0.0100°C.

In converting Fahrenheit temperature readings to Celsius, and vice versa, keep in mind that the 180 degrees on the Fahrenheit scale from the freezing point to the normal boiling point of water are equal to 100 degrees on the Celsius scale (Fig. 13-7). Hence

$$\frac{\text{Fahrenheit degrees above freezing}}{180 \text{ F}°}$$

$$= \frac{\text{Celsius degrees above freezing}}{100 \text{ C}°}$$

or

$$\frac{\text{Fahrenheit temperature} - 32°\text{F}}{180 \text{ F}°}$$

$$= \frac{\text{Celsius temperature} - 0°\text{C}}{100 \text{ C}°}$$

That is,

$$\frac{5 \text{ C}°}{9 \text{ F}°} = \frac{\text{Celsius temperature} - 0°\text{C}}{\text{Fahrenheit temperature} - 32°\text{F}}$$

For
ideal gas
−273.15°C
Pressure, cm-Hg
70
60
50
40
30
20
10
−250
−200
−150
−100
−50
0
+50
+100

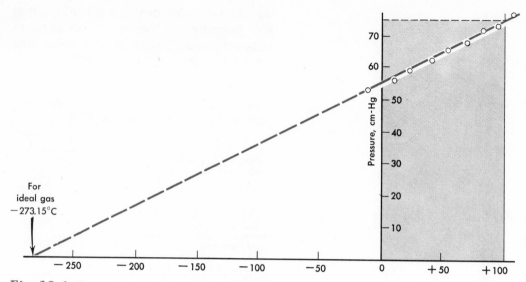

Fig. 13-6. *Variation of pressure with temperature at constant volume. The pressure of a perfect gas would become zero at about −273°C or 0° Kelvin.*

and

$$t = \tfrac{5}{9}(t_F - 32°)$$

where t is the Celsius temperature and t_F the Fahrenheit temperature. The relation between the Celsius temperature t and the corresponding Kelvin temperature T is particularly simple:

$$T = t + 273°$$

Example. The temperature of a room is 68°F. (*a*) What is the equivalent reading on a Celsius thermometer? (*b*) What is the corresponding temperature on the Kelvin scale?

(*a*)
$$\frac{5\ C°}{9\ F°} = \frac{t - 0°C}{68°F - 32°F}$$

$$t = 20°C$$

(*b*)
$$T = t + 273°$$
$$T = 293°K$$

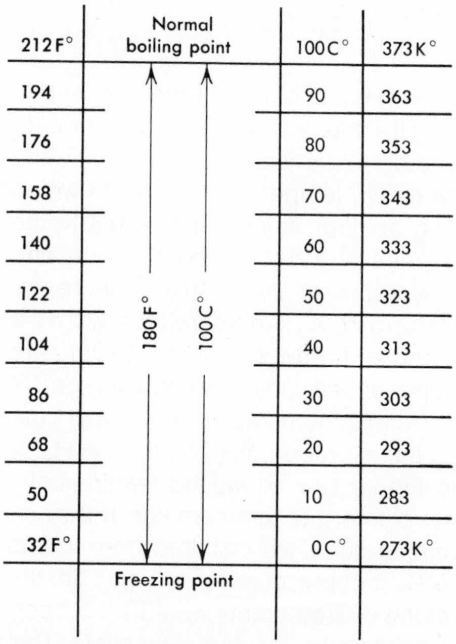

212 F°
Normal
boiling point
100 C°
373 K°
194
90
363
176
80
353
158
70
343
140
60
333
122
180 F°
100 C°
50
323
104
40
313
86
30
303
68
20
293
50
10
283
32 F°
0 C°
273 K°
Freezing point

Fig. 13-7. *Corresponding Fahrenheit, Celsius, and Kelvin temperatures.*

The Linear Expansion of Solids

One of the consequences of a change in the temperature of a body is that the body expands or contracts. When a solid body is heated, the kinetic energy of its atoms increases, they vibrate or oscillate within the crystal with greater average speeds, and the average distance between neighboring atoms increases so that the body expands. There are numerous familiar examples of expansion and contraction caused by variation in temperature. The length of the roadway of the Golden Gate Bridge at San Francisco varies more than 5 ft during the year. Hot water poured into a thick glass tumbler sometimes fractures the glass because the surface layers near the water expand before the outer layers become heated. Thin-walled tumblers are less likely to break because all parts of the glass are heated more uniformly. The rivets of the steel frame of a building are hammered while hot; in cooling they draw the steel plates tightly together. When concrete highways are laid, gaps are left between adjacent sections in order to permit expansion.

When a metal rod is heated, the expansion $l - l_0$ is proportional to the original length l_0 of the rod at temperature t_0, and to the change of temperature $(t - t_0)$, where l is its length at temperature t. Thus:

$$l - l_0 = \alpha l_0 (t - t_0)$$

$$\Delta l = \alpha l_0 \, \Delta t$$

$$\alpha = \frac{l - l_0}{l_0 (t - t_0)} = \frac{\Delta l}{l_0 \, \Delta t}$$

In this equation, α, the *coefficient of linear expansion* or *linear thermal expansivity,* of the material, equals the *fractional expansion per degree change in temperature.* For example, when an iron wire 1 km long is heated 1 C°, it expands 1.2 cm. The coefficient of linear expansion is 1.2 cm/(100,000 cm × 1 C°) = 0.000012/C° (see Table 13-2).

Table 13-2. Coefficients of Expansion (Thermal Expansivities)

Material	1/C°	1/C°		1/F°	
		Fractional Expansion per Degree at 20°C (68°F)			
A. Linear					
Aluminum	0.000 024	24	$\times 10^{-6}$	13	$\times 10^{-6}$
Brass	0.000 019	19	$\times 10^{-6}$	11	$\times 10^{-6}$
Copper	0.000 017	17	$\times 10^{-6}$	9.4	$\times 10^{-6}$
Glass, ordinary	0.000 009 0	9.0	$\times 10^{-6}$	5.0	$\times 10^{-6}$
Glass, Pyrex	0.000 004 0	4.0	$\times 10^{-6}$	2.2	$\times 10^{-6}$
Invar alloy (nickel-steel)	0.000 000 9	0.9	$\times 10^{-6}$	0.5	$\times 10^{-6}$
Iron	0.000 012	12	$\times 10^{-6}$	6.7	$\times 10^{-6}$
Platinum	0.000 009 0	9.0	$\times 10^{-6}$	5.0	$\times 10^{-6}$
Quartz, fused	0.000 000 59	0.59	$\times 10^{-6}$	0.33	$\times 10^{-6}$
Steel	0.000 013	13	$\times 10^{-6}$	7.2	$\times 10^{-6}$
Tungsten	0.000 004 3	4.3	$\times 10^{-6}$	2.4	$\times 10^{-6}$
B. Volume					
Alcohol, ethyl	0.001 01	10.1	$\times 10^{-4}$	5.5	$\times 10^{-4}$
Mercury	0.000 182	1.82	$\times 10^{-4}$	1.00	$\times 10^{-4}$

Example. One of the steel cables supporting a suspension bridge is 1000 m long. Find its expansion when the temperature changes from 0°C to 40°C.

$$\Delta l = \alpha l_0 \, \Delta t$$

$$\Delta l = 0.000013/\text{C}° \times 100{,}000 \text{ cm} \times 40 \text{ C}°$$
$$= 52 \text{ cm}$$

Applications of Thermal Expansion

If a surveyor uses a steel tape which is correct at 70°F and the actual temperature is 100°F, the tape will be longer than the length marked upon it, and the measured value will be too small. When tapes are made of Invar, a nickel-steel alloy, the errors are about $\frac{1}{14}$ as great as when steel tapes are used; such errors may be neglected in ordinary work. The pendulum rod of a clock expands when the temperature rises, and the clock loses time. To correct this, a vessel of mercury is used as a bob (Fig. 13-8). The expansion of the rod increases the length of the pendulum, but the expansion of the mercury, relatively much greater, raises the center of gravity and decreases the effective length. The two changes can be made practically to neutralize each other so that the clock keeps time more accurately.

Molecular Physics and Heat 215

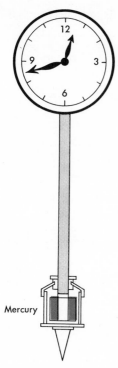

Fig. 13-8. *A compensated pendulum. Increase of temperature lowers the center of gravity of the bob and raises that of the mercury.*

If a bar of iron and one of brass are riveted together as in Fig. 13-9 and the bar is heated in a flame, the brass expands more than the iron, and the bar bends. Many thermostats for controlling furnaces and electrical heating devices utilize such bimetallic bars. In Fig. 13-10, the bar is used to operate an electrical switch. If the temperature of the room falls, end *A* of the bar moves sidewise until it touches the contact at the left, closing an electric circuit and either starting the motor in an oil burner or energizing an electromagnet which opens the drafts or gas supply valve in other kinds of heaters. As the room heats again, the bar moves backward until it breaks the contact and turns off the furnace.

Volume Expansion

When a block of iron is heated, not only does the length increase, but also the width, thickness, and volume. The equation for computing volume expansion resembles that for linear expansion. We define the volume expansivity of a substance as *the fractional change of volume per degree change of temperature.* For example, if 100 cm³ of iron expands 0.0036 cm³ when its temperature is raised 1 C°, its volume expansivity is 0.000036/C°.

Let the initial volume of a substance be V_0, the final volume V, the rise of temperature Δt, and the volume expansivity β (Greek *beta*). Then

$$V - V_0 = \beta V_0 \, \Delta t$$

The volume expansivity of a solid substance that expands equally in all directions can be found by multiplying its linear coefficient by 3. Suppose that when a cubical block of material is heated 1 C°, the edge *AB* expands 1/10,000% (Fig. 13-11). If the height and depth of the cube did not change, the volume would also expand 1/10,000%. However, the change of each of these dimensions is equal to that along *AB*, so that the volume change is 3/10,000%.

A more mathematical proof is the following: Let the initial volume of a cube be V_0, the ini-

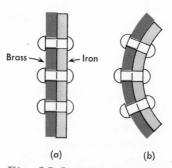

Fig. 13-9. *Differential expansion. Which expanded more, the brass or the iron?*

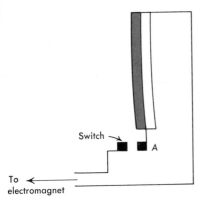

Switch

To electromagnet

Fig. 13-10. A thermostat.

tial length of one of its edges l_0, and the length after the expansion $(l_0 + l_0\alpha \, \Delta t)$. The volume after expansion is

$$V = (l_0 + l_0\alpha \, \Delta t)^3$$
$$= l_0^3 + 3l_0^3\alpha \, \Delta t + 3l_0^3\alpha^2 \, \Delta t^2 + l_0^3\alpha^3 \, \Delta t^3$$

$\alpha \, \Delta t$ is small compared with l_0; hence its square and cube are so small that the third and fourth terms may be neglected. (For example, if $\alpha \, \Delta t = 0.01$, $\alpha^2 \, \Delta t^2$ is 0.0001.)

$$V \cong l_0^3(1 + 3\alpha \, \Delta t) = V_0(1 + 3\alpha \, \Delta t)$$

Hence, the volume expansivity is approximately *three times* the linear expansivity.

Liquids and gases undergo volume expansion. Put water into a glass bulb (Fig. 13-12), and plunge the bulb into hot water. At first the liquid surface at *A* moves downward, and afterward it rises. The initial fall of the surface occurs because the expansion of the glass walls increases the volume of the bulb. Later, when the liquid itself is heated, it expands more than the volume of the enclosure does; hence the level at *A* rises. The apparent increase equals the *difference* between the increase in volume of the liquid and that of the container.

Example. The brass gasoline tank of a car has a volume of 15 gallons. It is filled to the brim with gasoline, the average volume ex-

pansivity of which is 0.00096/C°. What volume of gasoline will overflow if the temperature rises 20 C°?

Expansion of gasoline
$$= 15 \text{ gal} \times 0.00096/\text{C}° \times 20 \text{ C}°$$
$$= 0.288 \text{ gal}$$

Linear expansivity of brass $= 0.19 \times 10^{-4}/\text{C}°$

Volume expansivity of brass
$$= 0.57 \times 10^{-4}/\text{C}°$$

Expansion of brass container
$$= 15 \text{ gal} \times 0.57 \times 10^{-4}/\text{C}° \times 20 \text{ C}°$$
$$= 0.017 \text{ gal}$$

Hence

Overflow $= 0.288 \text{ gal} - 0.017 \text{ gal} = 0.271 \text{ gal}$

Water has a maximum density at 4°C. Suppose that you put ice-cold water into the bulb (Fig. 13-12), and heat the water very slowly to prevent sudden expansion of the glass. The water will *contract* as you heat it from 0°C to 4°C, and then it will expand like most liquids. The crystal structure of ice can be thought of as made of a number of tetrahedra—an oxygen atom at the apex with four hydrogen atoms (two belonging to the molecule and two belonging to a neighboring molecule) at

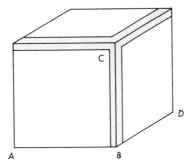

C

D

A   B

Fig. 13-11. Volume expansion. When heated, the cube expands equally along AB, BC, BD. Its volume expansion is three times its linear expansion.

Fig. 13-12. *Apparatus to demonstrate relative expansion.*

the base. It takes more space to stack these tetrahedra in ice than the same number of water molecules required in the liquid. Hence, the volume of the water is a minimum at 4°C and increases from 4°C to 0°C. Its density is a maximum at 4°C (Fig. 13-13). Moreover, just at 4°C, the density of water varies little with change of temperature. For this reason the scientists who originally devised the metric system defined the kilogram as the mass of 1000 cm³ of water at 4°C. Later on, the definition of the kilogram was changed to the one given in Chapter 1. Later we shall discuss some geological consequences of the physical fact that water expands when its temperature decreases from 4°C to 0°C.

Summary

Heat tends to flow from points of higher temperature to those of lower temperature. Temperature difference determines the direction of heat flow.

To transform Celsius temperatures into Fahrenheit, or vice versa, use the relation

$$\frac{t_F - 32°F}{180\ F°} = \frac{t - 0°C}{100\ C°}$$

The Kelvin temperature T corresponding to a Celsius temperature t is given by

$$T = t + 273°$$

The normal boiling point of water (at 76 cm-Hg) is 212°F or 100°C, and the freezing point is 32°F or 0°C. The zero of the Kelvin scale is approximately −273°C.

Most substances expand on heating. The linear expansivity of a substance is the fractional expansion in length per degree rise of temperature. The volume expansivity is the fractional change in volume per degree change in temperature. For solids, it equals three times the linear expansivity.

The density of water is a maximum at 4°C.

Questions

1. Discuss the choice of fixed points for the Celsius and the Fahrenheit scales.

2. Derive the equations for converting a Fahrenheit temperature to the equivalent

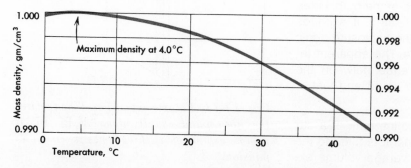

Fig. 13-13. *Density of water at different temperatures.*

Celsius temperature and from a Celsius temperature to the equivalent Fahrenheit temperature.

3. Compare the properties of mercury with those of water as a thermometric substance. State advantages and disadvantages of each.

4. Why is the maximum density of water 1.000 gm/cm^3?

5. A thermometer is immersed suddenly in hot water. What precautions must be taken in using it to measure the temperature of the water?

6. Prove that twice the linear expansivity of a substance equals its areal expansivity.

7. Discuss the change of density of water with temperature near its freezing point. What effect does this have when a lake freezes?

8. An iron rod connects the opposite sides of a circular iron hoop. If the system is equally heated, will the hoop remain circular? Explain.

9. Answer question 8 if the iron rod is replaced by (a) an aluminum rod, (b) a tungsten rod.

10. After studying the table of linear expansivities, explain why a glass beaker is more likely to crack when cooled quickly than a beaker made of quartz.

11. What are the advantages of an ideal gas thermometer over a mercury-in-glass thermometer?

12. What kind of wire would be suitable for sealing into Pyrex glass and what kind into ordinary glass?

13. Two steel rings are almost concentric. When the larger one is heated, it will just slip onto the smaller one. How could they be separated some time later?

14. How could the temperature of each of these be measured: liquid air, the surface of the moon, the surface of the sun, the interior of a boiling teakettle, a glowing tungsten filament?

15. What factors place a practical limit on the temperature range of (a) an optical pyrometer, (b) a constant-volume gas thermometer, (c) a mercury-in-glass thermometer?

16. How would you measure the temperature of one of the moons of Jupiter?

17. In using a mercury barometer to determine atmospheric pressure, why is it necessary to correct for room temperature?

18. Estimate how far the top of an obelisk 100 feet tall is deflected from the vertical on a hot summer day.

Problems

A

1. A spherical cavity in a block of copper has a volume of 2.000 cm^3 when the block is at 15°C. Find the volume of the cavity at 80°C.

2. When a steel rail was first rolled at a steel mill, its length was 14 m and its temperature was 520°C. Find its contraction in cooling to 20°C.

3. If steel rails 14 m long are laid, just touching each other, at 30°C, what will be their separation at −10°C?

4. An aluminum rod is 56.0 cm long at 20°C. At what temperature will it be 1.0 mm longer?

5. A cable of a steel suspension bridge is 1.60 km long. How much will the cable contract when the temperature decreases from 10°C to −40°C?

B

6. What is the reading of a Fahrenheit thermometer when the reading on the Celsius scale is one-half as great?

7. A glass flask of volume 1200 cm^3 is full of mercury at 20°C. How many cubic centimeters will overflow when the temperature is raised to 40°C? (The linear expansivity of glass is 9.0 × 10^{-6}/C°. The volume expansivity of mercury is 1.82 × 10^{-4}/C°.)

8. A surveyor finds that the apparent length of a lot is 160 ft when measured at 100°F with a steel measuring tape that is correct at 70°F. What is the correct length of the lot?

9. A metal rod was 52.342 cm long at 20°C and 52.396 cm long at 98°C. What is the

linear expansivity of the metal? (Keep the correct number of significant figures.)

10. A 2.0-ft³ brass sphere at 19°C has its temperature raised 10°C. Find the increase in volume.

11. If a 2000-cm³ solid expands 1.00 cm³ when heated 100 C°, what is the volume expansivity?

C

12. A steel rod 30.0 cm long and 2.00 cm² in area is cooled from 340°C to 20°C. What force would prevent it from contracting at all?

13. If the thin brass rod in a simple compensated clock pendulum is 100 cm long and the mercury reservoir has a cross-sectional area of 5.0 cm², how many cubic centimeters of mercury must be placed in the reservoir to keep the period of the clock constant?

14. A round hole 8.000 cm in diameter is cut from a plate of iron, of which the linear expansivity is 1.2×10^{-5}/C°, at 0°C. What will be the diameter of the hole at 40°C?

15. A round brass plug has a diameter of 8.0010 cm at 10°C. At what temperature will it fit snugly into the hole of problem 14? (The plug and the plate are to have the same temperature.)

16. The temperature of an iron rod 8.0 ft long and 0.20 in.² in cross section is lowered 20 F°. What force will be required to stretch it to its original length? (Young's modulus is 30×10^6 lb/in.²)

17. Suppose that the earth at the equator were encircled by an iron band that fitted snugly. If the temperature of the iron were raised 1/600 C° and the earth did not expand, what would be the mean distance of the band above the earth's surface after the expansion? Assume that the earth's radius is 6×10^8 cm.

18. The top of the column of a mercury barometer is 74.9 cm above the level of the mercury in the reservoir at a temperature of 20°C. What is the reading of the barometer at 0°C if the atmospheric pressure does not change?

19. A metal strip A of length l, thickness d, and coefficient of linear expansion α_1 is riveted to a metal strip B of the same length l and thickness d, but different coefficient of linear expansion α_2, so that their ends coincide, forming a bimetallic bar (Fig. 13-9). Show that if the temperature of the bar is changed by Δt, the bar bends into the arc of a circle of radius R given by

$$R = \frac{d}{(\alpha_1 - \alpha_2)\, \Delta t}$$

20. A glass bulb is partially filled with mercury and is then evacuated and sealed off. When the temperature is changed by an amount Δt, the volume of the evacuated part of the bulb does not change. Calculate what part of the initial volume of the bulb the mercury originally occupied.

21. Dumet wire can be sealed into common glass, because its linear expansivity is equal to that of glass, 9×10^{-6}/C°. The core of the wire, of nickel steel (Invar), is surrounded by a sleeve of copper. If the diameter of the wire is 1.00 mm, what is the diameter of the nickel-steel core?

For Further Reading

Emiliani, Cesare, "Ancient Temperatures," *Scientific American,* February, 1958, p. 54.

Hoyle, Fred, "Ultrahigh Temperatures," *Scientific American,* September, 1954, pp. 144–154.

Kantrowitz, A., "Very High Temperatures," *Scientific American,* September, 1954, p. 132.

MacDonald, D.K.C., *Near Zero, the Physics of Low Temperature,* Science Study Series, Doubleday and Company, Garden City, N. Y., 1961.

Zemansky, Mark W., *Temperatures, Very Low and Very High,* Momentum Book, D. Van Nostrand Company, Princeton, 1964.

CHAPTER FOURTEEN

Heat and Its Transfer

In the last chapter we discussed temperature and the expansions that occur when the temperatures of bodies are raised. Let us now consider the quantity of *heat* that is required to produce a given change of temperature in a body and the ways in which heat is transferred from one body to another.

Quantity of Heat

Heat is energy in transit from one body to another, and therefore it may be expressed in energy units—joules or ergs, for example. Special units for heat, however, were introduced into physics during the latter part of the eighteenth century when the relation of heat to energy was not understood. These special units—such as the *calorie* (cal)—are still used, although we no longer believe that heat is a substance, as most eighteenth-century scientists did. In a later chapter, we shall state the relation between the calorie and the joule.

One calorie is the heat required to raise the temperature of one gram of water 1 C°. Since the amount of heat required to cause unit temperature rise in unit mass of water varies slightly with temperature, for exactness

it is necessary to specify the temperature range. The calorie is usually defined as the heat required to raise the temperature of 1 gram of water from 14.5°C to 15.5°C. (Often in biology and in physiology the *kilocalorie*—the heat required to raise the temperature of one kilogram of water 1 C°—is used. It equals 1000 calories, and is abbreviated kcal.)

The heat Q required to raise the temperature of a body of mass m from a temperature t_1 to a temperature t_2 is given by

$$Q = mc(t_2 - t_1)$$

where c, the specific heat of the substance, is the heat absorbed per unit mass per unit increase in temperature or is the heat given out per unit mass per unit decrease in temperature. Experimentally determined values of the specific heats of a number of substances are given in Table 14-1.

Note that water has one of the highest specific heats. If you heat a pound of iron and a pound of water (in a suitable vessel) separately over two equal Bunsen flames, the iron becomes so hot in a few minutes that water boils when a few drops are sprinkled on the upper surface of the block. Meanwhile the water in the vessel is heated so slightly that you can dip your finger into it without discomfort. A given mass of water requires more heat per degree rise of temperature than does an equal mass of iron.

Example. How much heat is required to raise the temperature of 800 gm of copper, specific heat = 0.092 cal/gm-C°, from 20°C to 80°C?

$$Q = 800 \text{ gm } (0.092 \text{ cal/gm-C°})(80°C - 20°C)$$
$$= 4420 \text{ cal}$$

Table 14-1. Specific Heats

	cal/gm-C°
Solids	
Aluminum	0.212
Brass	0.090
Carbon (graphite) at −50°C	0.114
at 11°C	0.160
Copper	0.092
Glass (soda)	0.16
Gold	0.0316
Ice at 0°C	0.51
Iron	0.115–0.119
Lead	0.030
Silver	0.056
Zinc	0.093
Liquids	
Alcohol, ethyl	0.60
Mercury	0.033
Water (by definition) at 15°C	1.00

Gases	At constant volume	At constant pressure
Air	0.171	0.240
Hydrogen	2.44	3.43
Steam at 100°C	0.364	0.482

Calorimeters

Heat measurements in the laboratory are usually made by means of *calorimeters* (Latin, *calor* = heat). The principle of the calorimeter is illustrated by the simple form (Fig. 14-1) sometimes used in the introductory physics laboratory. A shiny metal cup *K*,

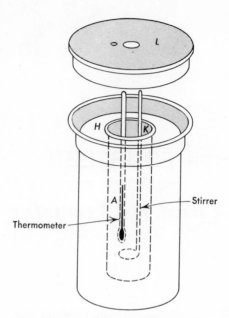

Fig. 14-1. A simple calorimeter. How is heat loss prevented?

containing the calorimetric liquid—often water—is supported inside a larger vessel by means of a fiber disk in order to insulate the inner vessel. A stirrer keeps the calorimetric liquid circulating until thermal equilibrium is attained. The thermometer measures the temperature of the inner cup before, during, and after, the addition of heat. When heat is added, part goes to the water contained in the cup and part to the container.

In the *method of mixtures* it is assumed that the heat given up by a hot body placed in the calorimeter is equal to that absorbed by the calorimeter and its contents. Precautions must be taken to ensure that an exchange of heat between the calorimeter and its surroundings will not destroy the correctness of this assumption. In the student laboratory, one can plan the experiment so that the initial temperature of the calorimeter and its contents will be about as much below room temperature as the final temperature of the calorimeter will be above room temperature after the hot body has been placed in the calorim-

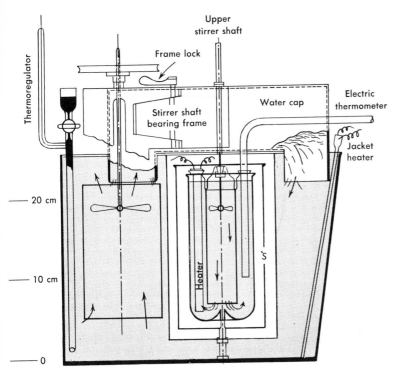

Labels in figure:
Upper stirrer shaft
Frame lock
Thermoregulator
Stirrer shaft bearing frame
Water cap
Electric thermometer
Jacket heater
Heater
S
20 cm
10 cm
0

Fig. 14-2. A compensated-loss calorimeter.

eter. The heat gained *from* the room thus approximately cancels the heat lost *to* the room. In research calorimeters (Fig. 14-2) elaborate care is devoted to insulating the calorimeter from its surroundings either by compensating for the heat exchange with the surroundings or by correcting for it. Unless these precautions are taken, heat measurements are far from precise.

The *water equivalent* of a calorimeter or other body is the mass of water that will absorb the same amount of heat as the calorimeter for an equal rise of temperature. For example, since the specific heat of copper is 0.092 times that of water, the water equivalent of a 100-gm copper calorimeter is

$$100 \text{ gm} \times \frac{0.092 \text{ cal/gm-C}°}{1 \text{ cal/gm-C}°} = 9.2 \text{ gm}$$

Example. A lead block of mass 300 gm at 100°C is placed in 100 gm of water at 15.0°C,

contained in a brass calorimeter of mass 200 gm. What is the final temperature of the mixture?

(*a*) Heat lost by hot body
= 300 gm × 0.030 cal/gm-C°(100°C − t)

(*b*) Heat gained by cold body
= (100 gm × 1 cal/gm-C°)(t − 15°C)
+ (200 gm × 0.090 cal/gm-C°)(t − 15°C)

Assuming that (*a*) and (*b*) are equal,

$$9.0(100°C − t) = 118(t − 15°C)$$
$$t = 21.0°C$$

A bomb calorimeter, a type of calorimeter used for determining the *heat of combustion* of a fuel or a food, is represented in Fig. 14-3. The heat of combustion is the heat evolved per unit mass of substance burned. The bomb calorimeter consists of a massive steel "bomb" containing a cup filled with a known quantity of test substance—coal, for example. A thin iron wire, suspended from two elec-

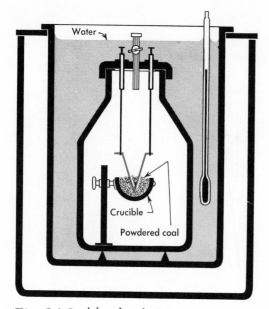

Water

Crucible

Powdered coal

Fig. 14-3. A bomb calorimeter.

trically insulated rods, is buried in the coal. After the cover is screwed on tightly, oxygen from a tank is added until the pressure is about 15 atmospheres. This bomb is then immersed in a large calorimeter containing water, and the initial temperature is carefully read. An electric current, sent through the fine iron wire, heats it red hot and ignites the material, which burns violently until all the carbon is oxidized. The heat evolved during

the combustion causes a rise of the temperature of the calorimeter. This rise is noted, and the heat evolved is computed in the usual manner. Typical heats of combustion are given in Table 14-2.

Respiration calorimeters are used to study metabolism, the life processes that convert the energy of foods into energy for the performance of work and for the maintenance of the temperature of the body. The subject is placed in a small chamber with double walls which insulate him from the outside. Oxygen is provided from a tank, and the exhaled gases are collected and their water-vapor content measured. Heat given off by the subject's body is absorbed by water flowing through continuous piping inside the chamber. By measuring the temperature of the incoming water, that of the outgoing water, and the mass of water passing through the piping per unit time, the rate at which heat enters the circulating water can be calculated. About one-fourth of the heat evolved from the subject is contained in the water vapor that he exhales; this can be determined. His total production of heat is that carried off by the circulating water plus the heat contained in the exhaled water vapor.

Conduction

We turn now to the processes by which heat flows from one body to another because of a temperature difference. Heat may be transferred by *conduction,* by *convection,* and by *radiation.* Let us first discuss conduction.

If you push one end of a solid rod into burning coal, the other end will eventually become warm, showing that heat has been transmitted along the rod. You recall that the forces are relatively strong between the particles that comprise a solid. The atoms behave as if they were connected to their neighbors by springs. The atoms of the rod at the end nearer to the flame are bombarded by

Table 14-2. Heats of Combustion of Fuels and Foods

Fuels	Calories per gram	Foods	Calories per gram
Alcohol, ethyl	10,000	Apples	640
methyl	5300	Beans, navy	3540
Coal, anthracite	6500	Bread, white	2660
bituminous	6000–8000	Butter	7950
Coke	8000	Buttermilk	365
Gas, coke-oven	4000–6000	Cream, 40%	3810
natural	8000–12,000	Lard	9300
water	3000–6000	Milk	715
Gasoline	11,000	Potatoes,	
Hydrogen	34,000	boiled	970
Wood (oak)	4000	Spinach,	
		cooked	575
		Sugar,	
		granulated	3940

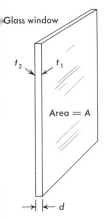

Glass window

t_2 t_1

Area = A

d

Fig. 14-4. *Heat transferred per second through a flat slab such as a window depends upon its area, its thickness, the temperature difference and the material.*

the energetic gas molecules in the flame, absorb energy, and oscillate more violently than before. They transfer kinetic energy to their neighbors, and this energy is passed onward as heat toward the cooler end of the rod. (If the rod is made of metal, the process is especially rapid because free electrons within the metal assist in carrying kinetic energy down the rod.) In liquids and gases, the molecules move more freely than in solids, but the conduction process still consists in a transfer of kinetic energy from the more rapidly moving molecules to the slower ones. *In conduction, heat alone is transferred; atoms are not.*

In winter weather, considerable heat escapes from a heated room by conduction through the glass windows. In general, if the direction of heat flow is perpendicular to a flat slab of material and no heat is conducted outward through the edges, the heat transmitted, Q, is found experimentally to depend upon the *cross-sectional area A* of the slab, its *thickness d,* the *temperature difference* $t_2 - t_1$ between the two surfaces, the *time interval* τ (Greek, *tau*), and the *thermal conductivity k* of the substance (Fig. 14-4):

$$Q = kA\left(\frac{t_2 - t_1}{d}\right)\tau$$

The thermal conductivity k is the *heat that flows per unit time through unit area of a slab of unit thickness, the temperature difference between the faces being* 1° The quantity $(t_2 - t_1)/d$ is called the *temperature gradient.*

The metals are the best conductors (Fig. 14-5); then come the non-metallic solids and liquids. The gases generally are the poorest conductors of all. Hydrogen is an exception. Its conductivity is greater than that of asbestos paper or of cork. (See Table 14-3.)

Table 14-3. Thermal conductivities in

$$\frac{\text{cal/sec}}{\text{cm}^2 \times \text{C°/cm}}$$

A. Metals	k	B. Non-metallic solids	k	C. Liquids and gases	k
Aluminum	0.50	Asbestos paper	0.0004	Air at 0°C	0.000057
Copper	0.99	Brick, machine	0.000091	Alcohol, ethyl at	0.00057
Iron	0.163	made, dry		20°C	
Lead	0.083	Concrete	0.0020	Hydrogen at 0°C	0.00041
Silver	1.01	Cork board	0.00011	Water at 20°C	0.0014
		Glass, window	0.0025		
		Ice	0.0050		
		Snow	0.00026		
		Soil, moist	0.0037		
		Wood	0.0004		
		Wool felt	0.00010		
		Wool, glass	0.00015		
		Wool, rock	0.00010		

Conduction of heat through cylinders, through spheres, and through objects of still other geometrical shapes is of interest in applications of physics. The heating engineer must calculate the rate at which heat is lost by

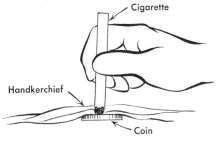

Cigarette

Handkerchief

Coin

Fig. 14-5. *Why doesn't the handkerchief get scorched?*

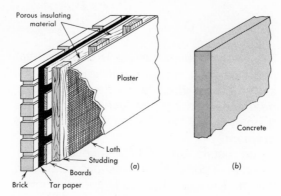

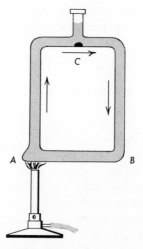

Fig. 14-6. *Insulation of homes. Why is the wall in* (a) *a better insulator than the wall in* (b)?

conduction from a steam pipe in the ground; the geophysicist, the rate at which heat escapes from the hot interior of the earth. The method of calculation in such instances is similar to that used above for a flat slab, but the mathematics involved is somewhat more complex.

Woolen cloth and the fur coat of the fox have low thermal conductivities because of the air that is trapped between adjacent fibers. Cotton, silk, and linen cloth conduct heat better because their fibers do not imprison as much air. The walls of refrigerators are heat-insulated by successive layers of cork or by layers of mineral wool. Some refrigerators are wasteful because householders pay more attention to chrome and enamel than to heat insulation! Residences are insulated in order to oppose both the escape of heat in winter and its entrance in summer. Fortunately, both aims may be attained by the same means. Usually heat insulation is achieved by using several layers of material, sometimes porous, separated by air spaces. In Fig. 14-6a, there are six layers. Such a wall conducts less than one-fourth as much heat per unit time as a wall of solid concrete of equal size (Fig. 14-6b). The insulation may be further improved by filling the

spaces between the studding with asbestos or with mineral wool

Convection

Heat can be transferred within a fluid by the process called *convection.* If a rectangular glass tube filled with water is held in a flame as shown in Fig. 14-7, the liquid in the column above *A* will expand and become less dense than that above *B*. The pressure at *B* will be greater than that at *A*, and the hot water will be forced upward from *A*, causing circulation. The circulation can be made visible by dropping a few crystals of potassium permanganate into the water at *C*. *In convection, heat is transferred by the transfer of matter.*

Convection is a more complicated process than the process of conduction through geometrically simple shapes. We shall not go into the methods of calculation used in the investigation of convection, but limit ourselves to giving some examples. In the hot-air "gravity" furnace, the cold air in the return pipe is more dense than the air in the pipes leading from the furnace up to the registers. Thus the cold air descends and forces the hot air to rise.

Fig. 14-7. *Causing convection in a liquid.*

Similarly, in a hot-water gravity-circulated heating system, the cool water descends through the return pipes, forcing warm water to rise to the radiators. (Many modern hot-air heating systems include an electrically driven blower to circulate the warm air more rapidly, and hot-water systems often have a water pump for a similar purpose.)

In heat-insulating a dwelling, give attention to preventing convection losses. Close small openings in window frames by weatherstripping, and provide storm windows. Put a layer of glass wool or the like in the attic in order to prevent convection losses from the ceiling. It may be advisable to put layers of glass wool in the hollow spaces between the plaster and the outside wall of the dwelling. Pulling down the window shades of a room reduces heat loss by forming a layer of still air near the window. Finally, a vapor barrier—sheets of metal foil or plastic-coated paper—should be installed on the side of the insulation material nearer the living areas. This will help to prevent water vapor from passing through the insulation, condensing, and soaking the building materials on the lower-temperature side of the insulation.

A wind is a convection current in the atmosphere caused by unequal heating. The air at the earth's surface near the equator gets strongly heated, expands, rises and streams away from the equator. For this reason, the equator is a region of low barometric pressure. At the poles, the air in the upper atmosphere gets chilled, descends, and streams outward. Thus the poles are regions of high pressure. If the earth did not rotate, the polar winds would stream directly toward the equator, then rise and return to the poles. Actually, the earth's rotation and the interaction of cold air masses from the poles with warm air masses greatly modify these winds by breaking the single pole-to-equator cycle into several cycles (Fig. 14-8). Air rising at the equator and tending to move northward gains an eastward component of velocity relative to

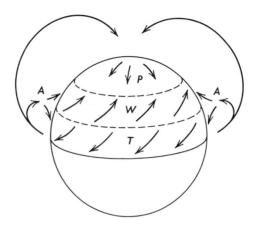

P Polar easterlies
A Antitrades
W Prevailing westerlies
T Trade winds

Fig. 14-8. The planetary winds.

the earth. This is because the linear velocity of a point on the earth decreases from about 1000 mi/hr at the equator to zero at the poles. Thus, the warm air, originating at the equator, will have both north and east components of velocity as it moves away from the equator. The air descends to the earth at latitudes of about 30° and returns to the equator, forming a region of high pressure at these latitudes. Thus at latitudes lower than 30°, the steady surface winds blow toward the equator from the northeast. They are called the "trade winds."

Consider the air streaming southward from the north pole. This air is moving toward regions where the earth's surface moves faster. Thus the winds are directed from the northeast. The air stream eventually rises at latitudes of about 60° and returns to the poles. Thus a second region of low pressure is formed. Notice the third circulating air stream. At the earth's surface, it is directed from latitudes of about 30° toward the poles. This stream forms the "westerlies," so important in the United States.

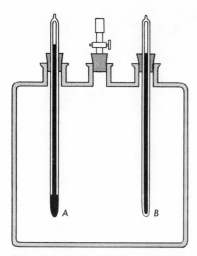

Fig. 14-9. *The blackened bulb* A *absorbs more radiation than the silver-coated bulb* B.

Important factors in climate are the land and sea breezes. The westerly air currents from the Pacific Ocean, traveling over California, Oregon, and Washington, account for the relatively mild and uniform climates in these sections. Near the eastern coast, the climate is much more variable, because the westerlies have passed over great land areas where the temperatures fluctuate widely. The breezes at the seashore often vary with the time of day. The land heats rapidly during daylight, and the heated air rises like the gases in a chimney. Then the sea breeze blows in from the ocean. At night the land surface cools quickly, the cold air descends, and a land breeze blows out over the water.

Radiation

The third important mode of heat transfer is by *radiation*. Your hand held under the bulb of an incandescent lamp feels warm, showing that the hand absorbs radiant energy. This energy is not carried by conduction, for air is a poor conductor of heat. Neither is it transferred by convection, for warm air rises. When a thermometer placed in direct sunlight is

heated, it is even more evident that conduction and convection are not important factors. In fact, there are relatively few molecules in the depths of space, hence no energy can come to us from the sun by conduction or by convection. The energy is carried at the velocity of light solely by electromagnetic radiation, which is of the same nature as light or radio waves. This radiation, incident on a piece of stone or any other body, agitates its molecules and thus effects a transfer of heat.

In conduction, energy is transferred from molecule to molecule. In convection, the energy and the molecules are transferred together. But radiation is of the same nature as light and can travel through a vacuum. We shall have much more to say in Chapter 32 about electromagnetic radiation. Here we discuss its role in the process of heat transfer.

Mount two thermometers *A* and *B* (Fig. 14-9) in a highly evacuated flask, and place the flask in direct sunlight. There is no gas in the bottle; hence only radiation heats the thermometer bulbs. If the thermometers are of the same dimensions and structure, they will be heated at the same rate. However, if the bulb of *A* is coated with lampblack or soot, and *B* with silver, *A* will absorb more radiation than *B* and hence its temperature will rise faster than that of *B*. Lampblack absorbs more than 97% of the incident radiation and reflects less than 3% of it; an untarnished silver surface absorbs less than 10% and reflects the rest. A body which would absorb *all* the incident radiation and reflect none of it we call a perfectly black body or, more briefly, a *black body.* If you transfer the two thermometers mounted in the evacuated bottle to a refrigerator, the temperature of the blackened thermometer will fall more rapidly than that of the other one. Experiments prove that the blacker body—the better absorber—is the better emitter of radiation and the poorer reflector.

A small hole in a box is an almost perfect absorber of radiation—because radiation en-

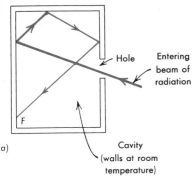

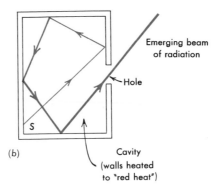

Fig. 14-10. *A black-body cavity is an almost perfect* (a) *absorber of radiation,* (b) *emitter of radiation.*

tering the box is almost completely absorbed by the walls as it is reflected back and forth inside (Fig. 14-10a). Similarly, radiation inside a heated cavity (Fig. 14-10b) is strengthened by internal reflections, in each of which emitted radiation from the walls is added to the radiation reflected from that point. Therefore, we would find the hole to be a maximum emitter.

The *emittance e* of a surface is the ratio of its radiation output to the radiation output of a perfectly black body—such as a small hole in a cavity—of the same dimensions at the same temperature. The emittance of a black body is 100% by definition; that of lampblack is 97%, and that of polished silver about 6%.

Table 14-4. Typical Emittances and Reflectances of Surfaces at 100°C

	e	r
"Black body" (by definition)	1.00	0.00
Lampblack	0.97	0.03
Asbestos paper	0.93	0.07
Gray lusterless paint	0.91	0.09
Black lusterless paint	0.91	0.09
White canvas	0.88	0.12
Window glass	0.88	0.12
Galvanized iron, tarnished	0.50	0.50
Aluminum bronze paint	0.28	0.72
Galvanized iron, bright	0.15	0.85
Copper, polished	0.07	0.93
Aluminum foil, polished silver	0.06	0.94

Many experiments prove that the *absorptance a* of a surface, the ratio of the radiant energy absorbed by it to the total radiation striking it, is equal to its emittance *e*. Suppose that a surface absorbs one-fourth of the radiation striking it. This means that it absorbs one-fourth as much radiation per unit area per unit time as a black body would. Then it also emits one-fourth as much at the same temperature. For any surface

Absorptance = Emittance   (Kirchhoff's law)

$$a = e$$

Kirchhoff's law can readily be proved from the following considerations. Imagine two equal spheres *A* and *B* suspended in a closed space whose walls are at the same temperature as *A* and *B*. Suppose that *A*, a perfectly black body, absorbs 100 cal/sec. It must emit radiation at the same rate in order that its temperature remain equal to that of the walls. Suppose that the surface of *B* emits one-fourth as much radiation as the black-body surface of *A*. Then it must absorb one-fourth as much. Its emittance, 0.25, is equal to its absorptance (see Table 14-4). Otherwise the spheres could not be in thermal equilibrium.

The *reflectance r* of a surface is the ratio of the radiation that it reflects to the total radiation incident upon it. Since incident energy must be *either* absorbed or reflected,

$$a + r = 1$$

and, by Kirchhoff's law,

$$e + r = 1$$

Some Examples of Radiation

The *Dewar flask* or *thermos bottle* was invented by the British physicist James Dewar to contain liquid air. As shown in Fig. 14-11, the flask is double walled. The space between the walls is evacuated to minimize convection and conduction. The walls are coated with silver to decrease radiation. The inner cups of calorimeters are covered with a highly reflecting coating for the same reason. Instrumentation satellites are often coated with a thin layer of gold to reflect the sun's radiation and prevent violent swings of temperature in the interior of the satellite as it moves from the earth's shadow into sunlight.

Hot water or steam radiators should be painted with a coating which is a good emitter. A chromium-plated radiator would be much less effective than one painted black.

The housewife is justified, from a scientific standpoint, in using pots and kettles with

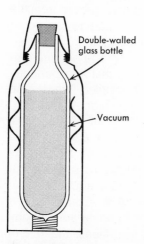

Double-walled glass bottle

Vacuum

Fig. 14-11. Dewar or thermos flask. How does it minimize loss by conduction and by radiation?

shining walls, for they are poor emitters. However, the bottoms of the kettles should be blackened to increase radiant energy absorption from the flame.

The rate at which radiation from the sun strikes the earth has a great effect upon climate. The average annual temperature in any region depends partly upon the rate at which it receives energy from the sun. A screen 1 cm² in cross section, placed perpendicular to the sun's rays at the outer borders of the atmosphere, would receive nearly 2 cal of radiant energy per minute. Much of this energy is reflected by clouds in the upper atmosphere so that even on clear days less than 50% reaches the earth's surface. A little of the energy is stored by growing plants, but most of it is reradiated into space. Century by century, the earth receives energy and radiates it at nearly the same rate so that its temperature remains nearly constant.

The absorption of the atmosphere is due principally to water vapor, carbon dioxide, ozone, fog, and dust. Of these, dust has been in the past by far the most variable constituent. Much of it is thrown into the atmosphere by volcanoes, where it remains suspended for months and even years. Some scientists believe that a volcanic eruption is followed by a period of cooler weather due to absorption of the sun's radiation by ejected dust. Cooler periods have followed eruptions of Tamboro in 1815, Krakatau in 1883, Mont Pelee in 1902, and Katmai in 1912.

Recently, carbon dioxide, produced by technological civilizations all over the world, has been increasing in the atmosphere. There is speculation that the increased absorption of the sun's radiation by carbon dioxide may be causing the gradual warming of the earth's climate. This is one of the many reasons for concern about pollution of the air.

The radiation that a surface receives depends upon the angle of incidence of the sun's rays. The surface *DC* (Fig. 14-12), on which the rays are incident normally, receives

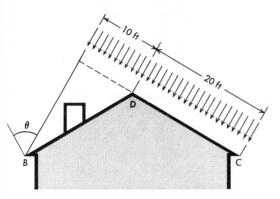

Fig. 14-12. *The surface* DC *receives twice as much radiant energy per unit time as* DB.

energy at a rate twice that at which the equal surface receives it. Let radiation be incident normally on a surface of area A at a rate R_0A, where R_0 is the incident energy per unit area perpendicular to the sun's rays per unit time. $\mathscr{R}$, the rate at which energy strikes a surface of equal area, if the angle of incidence between the rays and the perpendicular to the surface is θ, is

$$\mathscr{R} = R_0A \cos \theta$$

The Arctic regions are cooler than the tropics for two reasons. First, sunlight incident near the North Pole must penetrate a greater depth of atmosphere than an equal beam incident near the equator. Second, in the Arctic, the sun's rays are incident obliquely and hence are distributed over a larger area than in the tropics (Fig. 14-13).

Radiation and Temperature

The exact relationship between temperature and rate of radiation is given by the *Stefan-Boltzmann law of total radiation*. The derivation of the law from the principles of thermodynamics goes beyond the scope of this book. We shall state the law and note that it has been amply verified by experiment:

The rate at which a black body radiates

energy is proportional to the fourth power of its Kelvin temperature. That is, a black body of area A and Kelvin temperature T emits energy at a rate Q/τ given by

$$\frac{Q}{\tau} = e\sigma AT^4$$

where e is the emittance and σ (Greek *sigma*) is the Stefan-Boltzmann constant and equals 1.35×10^{-12} cal/cm²-sec-K°⁴ or 5.67×10^{-8} joule/m²-sec-K°⁴. For a black body, e is 1. For other surfaces, values of e are given in Table 14-4.

Example. How much energy is radiated per second from the filament of an incandescent electric lamp if the surface area is 0.300 cm² and the temperature is 3000°K? Assume (*a*) that the filament is a black body and (*b*) that its emittance is 0.40.

(*a*) $Q/\tau = 1.35 \times 10^{-12} \dfrac{\text{cal}}{\text{cm}^2\text{-sec-K}^{\circ 4}}$

$\times$ 0.300 cm² $\times$ (3000 K°)⁴ = 33.0 cal/sec

(*b*) $Q/\tau = 0.40 \times 33.0$ cal/sec = 13.2 cal/sec

Not only does a body radiate energy to surrounding objects, but also it absorbs radiant

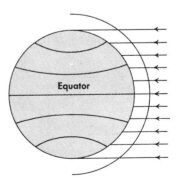

Fig. 14-13. *The Arctic regions receive less energy per acre than the tropics because the radiations must pass through a thicker atmosphere and because they are incident more obliquely.*

energy which is incident upon it. When you stand near a hot stove you feel warm because you absorb more energy from the stove than you emit to it. If you stand near a cake of ice, you feel chilly because you radiate more energy to the ice than you absorb from it. Suppose that a black body of surface area A at temperature T_b is in an enclosure whose walls are at a temperature T. Then the body will absorb radiant energy at a net rate given by

$$\frac{Q}{\tau} = \sigma A(T^4 - T_b^4) \qquad \text{(Prevost's law)}$$

Summary

One calorie is the amount of heat required to raise the temperature of one gram of water 1 C°.

The specific heat of a substance is the heat absorbed per unit mass, per degree rise of temperature.

The heat of combustion of a substance is the heat per unit mass evolved when the substance is completely burned.

In conduction, heat energy is passed from molecule to molecule without transfer of matter.

The heat conducted per unit time through a flat slab equals $kA(t_2 - t_1)/d$, k being the thermal conductivity of the material, A the area of the slab, $(t_2 - t_1)$ the temperature difference between the faces, and d the distance between them.

In convection, matter and heat are transferred together.

In radiation, energy may be transferred without matter.

The percentages of incident radiations which are absorbed by different surfaces vary widely. A black body absorbs all the radiant energy. The absorptance of a surface is equal to its emittance. A black body therefore is a maximum emitter of radiation.

The rate of energy emission by a black body of area A is expressed by

$$\frac{Q}{\tau} = e\sigma A T^4$$

where $\sigma = 1.35 \times 10^{-12}$ cal/cm²-sec-K°⁴

Questions

1. Why is snow a better heat insulator than ice?

2. Which is the better heat insulator: (a) ice or glass, (b) wood or concrete?

3. Which contains more heat energy at the same mass and temperature: a hot-water bottle filled with water, or a brick?

4. What change in temperature will 100 cal of heat produce in 1.00 gm of (a) mercury, (b) lead, (c) aluminum?

5. Although air is a poorer conductor of heat than wool, why does a cloth covering decrease the loss of heat from a hot body?

6. A piece of paper wrapped on a brass rod may be held in a gas flame without being scorched. If wrapped on a wooden rod, it scorches quickly. Why?

7. Why should the coating of ice be removed from the cooling unit of a refrigerator at frequent intervals?

8. Are several sheets of newspaper a good heat insulator?

9. Why do travelers in the desert wear heavy clothing?

10. In winter weather, why will a wet glove freeze to an iron bar more quickly than to a wooden one?

11. Would you choose aluminum paint for (a) a radiator, (b) a spaceship?

12. Why does frost form on the inside of windows in a heated house in winter?

13. What is the best experimental approximation we have to a black-body radiator?

14. What effect on climate does the high specific heat of water have?

15. Why can water be boiled in a paper cup?

16. How does the amount of heat received at the North Pole on June 21 compare with

that received at the equator on the same date?

17. What purpose do storm windows serve in cold climates?

18. Which melts faster, ice covered with a layer of soot or clean ice? How could you apply this to opening a channel for ships on a frozen lake?

19. In winter, when the outside temperature is 30°F and the inside temperature is 70°F, what would be typical temperatures at the interior and exterior surface of a window pane?

Problems

A

1. A copper calorimeter of mass 200 gm contains 100 gm of water. How many calories are required to raise the temperature from 16°C to 30°C?

2. The glass in the 25 windows of a house has a total area of 16 m². The glass is 0.30 cm thick, and the temperature difference between the two surfaces is 0.50 C°. (*a*) How much heat is transmitted in 12.0 hr? (*b*) How much bituminous coal (heat of combustion 7.0×10^3 cal/gm) must be burned to supply this heat?

3. A pond is covered by a sheet of ice 3.00 cm thick. The temperature of the upper surface of the ice is −20°C, and that of the lower surface is 0°C. At what rate is heat conducted through each square centimeter of ice at the surface?

4. A layer of snow 10.0 cm thick covers a field. If the surface of the ground is at 0°C and the upper surface of the snow is at −20°C, how much heat is transferred per hour, per square meter of surface?

5. The rate of heat flow into an aluminum cooking vessel 1.00 mm thick is 8.50 (cal/sec)/cm². If the temperature of the inside surface is 100°C, what is the temperature of the outside surface?

6. The hot-water pipes of a furnace in a cellar are coated with a black paint having an emittance of 0.85. The heat loss by radiation is 7.5×10^6 cal/day. What would be the loss if the pipes were painted with aluminum paint, the thermal emittance of which is 0.27?

7. The temperature inside a furnace is 1727°C. At what rate, expressed in (cal/sec)/cm², is energy radiated through a small hole in the furnace wall?

B

8. A 30-gm nickel ball is removed from a furnace and dropped into a copper calorimeter of mass 200 gm containing 200 gm of water at 18°C. If afterward the temperature of the water rises to 23°C, find the temperature of the furnace. (Specific heat of nickel, 0.113 cal/gm-C°.)

9. Two hundred grams of lead shot at 80°C are poured into 150 gm of water at 10°C contained in a calorimeter of mass 200 gm and specific heat 0.10 cal/gm-C°. Find the temperature of the mixture. Try to estimate the answer before solving.

10. (*a*) If the average density of the air is 1200 gm/m³, how much heat is required to heat 200 m³ of air at constant pressure from 15°C to 25°C? (*b*) How much coal, heat of combustion 7.0×10^3 cal/gm, would supply the energy if the stove were 40% efficient?

11. How much lead shot at 100°C must be put into a liter of water at 20°C so that the final temperature may be 45°C?

12. A glass test tube, having a mass of 26 gm and a specific heat of 0.20 cal/gm-C°, contains 25 gm of water at 40°C. A thermometer with a water equivalent of 2 gm, at 18°C, is inserted into the water. How many degrees is the temperature of the water lowered?

13. When 30 gm of a test fuel are burned in a bomb calorimeter whose water equivalent is 3000 gm, the temperature of the calorimeter increases from 25.0°C to 65.0°C. Calculate the heat of combustion of the fuel.

14. A slab of stone 75 cm by 40 cm and 10 cm thick is exposed on the lower surface to steam at 100°C. A cake of ice rests on the upper surface. In 40 min, 4800 gm of ice are melted, absorbing 80 cal/gm in melting. What is the thermal conductivity of the stone?

c

15. (a) The walls of a building are of solid concrete. Their thickness is 28.0 cm, and the total surface area is 450 m². If the average temperature difference in winter is 10 C°, how much heat is transferred per day of 86,400 sec? (b) How much coal, heat of combustion 7.0×10^3 cal/gm, would supply this heat? (c) What would the coal cost for 100 days at 1.2 cents/kg?

16. The temperature of the water flowing into a car heater is 75°C, and that of the water as it leaves is 35°C. If 40 liters of water flow through the heater per second, how much heat is liberated in 100 sec?

17. A liquid A of specific heat 0.50 cal/gm-C° at 25°C is mixed with another liquid B of specific heat 0.25 cal/gm-C° at 50°C. The final temperature of the mixture is 30°C. What per cent of the total mass is the mass of A?

18. At what rate does a tungsten lamp filament (emittance 0.40) radiate energy if its surface area is 0.80 cm² and its temperature is 2400°K?

19. A ball, the surface area of which is 150 cm², radiates energy at a rate of 10 cal/sec. (a) What is its temperature if it is a black body? (b) At what rate would it radiate at that temperature if the emittance of the surface were 0.40? (c) At what rate would it radiate energy as a black body if the temperature were 2600°K?

20. A sheet of iron 3.0 mm thick is coated with a sheet of lead 1.0 mm thick. If the outer surface of the lead is at 80°C and the outer surface of the iron is at 20°C, what is the temperature of the boundary surface between the lead and the iron?

21. The brick walls of a house have a total area of 350 m² and are 20 cm thick. The density of the walls is 2.0 gm/cm³. The specific heat of the material is 0.20 cal/gm-C°. (a) In order to heat the walls 20 C°, how much heat is required? (b) To supply this heat, how much coal, heat of combustion 7.0×10^3 cal/gm, must be burned in a furnace that is 35% efficient?

22. A chemical engineer wants to remove 12.0×10^6 cal/min from a liquid. The liquid flows through a heat exchanger in which heat is given to water which enters at a temperature of 35°C and leaves at a temperature of 80°C. How rapidly (kilograms/minute) must the water circulate through the heat exchanger?

23. The area of a ribbon filament in an experimental lamp is 10.0 cm² and the emittance of the filament is 0.40. When the air has been removed from the lamp, 740 cal/sec (3000 watts) of electrical energy input power are required to raise the filament to a temperature of 2000°K. With the lamp filled with nitrogen, 950 cal/sec (4000 watts) are required to raise the filament to this temperature. Calculate (a) the rate of radiation by the filament, (b) the rate of conduction of heat by the electric wires or leads, and (c) the rate at which heat is removed by convection.

24. Thermal energy flows out of the interior of the earth at a rate of about 4×10^{-8} erg/gm-sec. Estimate the rate at which the earth radiates energy into space, assuming that the surface temperature of the earth remains constant.

25. The sun radiates energy at a rate of 67,000 kw/m². (a) Estimate its surface temperature, assuming it to be a black body. (b) Assuming that the sun radiates equally in all directions, estimate the rate at which energy is received per square meter of the earth's surface in the tropics at noon if 50 per cent of the radiation is absorbed in the atmosphere. (Radius of the sun, 6.96×10^5 km. Distance of the sun from earth, 1.50×10^8 km.)

26. A factory chimney is 100 m high, and

the cross-sectional area of the opening is 1.00 m². The average density of the gases is 0.0011 gwt/cm³; that of the air outside is 0.0012 gwt/cm³. (*a*) Estimate the difference in pressure at the base of the chimney, inside and outside. (*b*) If the flue at the bottom of the chimney is closed by a barrier 1.00 m² in area, estimate the difference between the forces pushing against it from the two sides.

27. Heat is conducted across a slab of area A consisting of two plates in contact, one of thickness l_1 and thermal conductivity k_1, and the other of thickness l_2 and conductivity k_2. The exposed surface of the plate of thickness l_1 is at a temperature t_1 and that of the other plate at t_2. Derive expressions for (*a*) the temperature of the two inner surfaces that are in contact and (*b*) the rate of conduction of heat across the slab.

28. Estimate the amount of sugar a bottle of soft drink would have to contain merely to provide the body with enough energy to raise the temperature of the soft drink from the temperature of melting ice to body temperature after it has been drunk. The heat of combustion of sugar is 3900 cal/gm.

For Further Reading

Barker, M. E., "Warm Clothes," *Scientific American,* March, 1951, pp. 55–60.

Dyson, F. J., "What Is Heat?" *Scientific American,* September, 1954, p. 58.

Wilson, Carroll L. and William H. Matthews, eds., *Man's Impact on the Global Environment,* MIT Press, Cambridge, 1970, pp. 170f.

CHAPTER FIFTEEN

The Kinetic Theory of an Ideal Gas

So far, in our discussion of heat and temperature, we have dealt mainly with large-scale effects. When two bodies are at different temperatures, heat may be transferred from one to the other by conduction, convection, or radiation. Thermometers are used to measure the changes in temperature, and calorimeters to measure the quantities of heat exchanged. An increase in the temperature of a solid or liquid is usually accompanied by expansion. We referred to molecules in a qualitative way to explain what happens in several heat processes, but we made little *quantitative* use of the notion that matter consists of molecules.

In this chapter we shall use the kinetic theory of molecular behavior to explain quantitatively the properties of an ideal gas and hence to throw some light on the behavior of real gases at pressures not exceeding a few atmospheres. Connections between the principles of mechanics studied earlier and those of heat will serve to remind you that the underlying ideas of physics are closely tied together. Knowledge of one part of physics helps one in gaining an understanding of other parts. The relatively simple theory of an ideal gas will help you to understand what a

theory is, why it is useful, and what its limitations are.

What is a Theory?

In the natural sciences, a theory is an hypothesis devised to account for known experimental facts, to show relationships among them, and to predict new relationships. A successful theory explains the known facts satisfactorily within the limits of existing experimental precision and predicts where the experimenter should look for new results and what those results are likely to be. A good theory does this rather economically: it makes assumptions sparingly, relies usually, although not always, on well-established concepts, and contains as few new concepts as the theorist can get along with. Successful theories are general, relating many areas of knowledge within the science to each other; they are fertilely predictive of new results; and they give the scientist a feeling of pleasure at the unity of nature. Theories fail also. They may make predictions that are incorrect, or they may simply become ponderous and complicated. They are then usually modified r, in the great revolutions of science, entirely supplanted by new theories.

We have already met several theories in his book, and we shall meet many others. The Ptolemaic theory of the universe provides a good illustration of the life and death of a theory. From its formulation in the second century A.D. until the publication of Copernicus' volume *The Revolutions of the Heavenly Spheres* near the end of the sixteenth century, the geocentric theory of the universe ruled supreme. With its complicated system of spheres and epicycles, it was able to account for the apparent motions of the planets within the limits of the rough astronomical measurement of those times. Moreover, the Ptolemaic theory pleased many people in that it made the earth—the home of man—the center of the universe. However, as we have seen, growing precision of measurement by such astronomers as Tycho Brahe in the sixteenth century brought the Ptolemaic theory into difficulties: it became increasingly difficult to adjust the epicycles and other geometrical devices of the theory to explain astronomical data. Copernicus made his break with the tradition of fourteen centuries by developing the heliocentric theory, which explained the known facts at least as well as the Ptolemaic theory and did it more economically, with a smaller number of epicycles. Galileo seized upon this new theory and found it in better accord with his studies of motion on the earth. Galileo—like most scientists since his day—believed that the same principles that govern the motion of bodies on the earth govern the motion of distant planets. These principles of celestial motion, of course, remained for Kepler to formulate, Newton to explain, and the followers of Newton to work out in precise and satisfying detail. The Ptolemaic theory was officially dead by the beginning of the eighteenth century, but its successor, the detailed Copernican theory, had already met at the hands of Kepler and Newton the fate of all theories:

modification to accommodate new knowledge and to do it simply.

One of the devices much used in physical theories is the model. A model is a kind of mental framework on which scientists hang their ideas in forming new theories. The model contains simplifications from the real-life situation to make it easier to apply mathematics to the physical system being studied. The Newtonian model of the solar system, for example, was that of mass points—the planets—traveling about the massive sun, held in their elliptical orbits by a gravitational force described by the law of universal gravitation. Our model of a perfect crystal, discussed in Chapter 12, is that of a uniform geometrical lattice of one of seven kinds of symmetry with atoms or groups of atoms occupying the lattice points. In this chapter we shall discuss the kinetic model of an ideal gas. Later we shall encounter other models. In the simpler situations, models may be represented by physical devices: pingpong balls mounted on toothpicks or springs in a lattice arrangement, for example, are often useful in thinking about crystals. But physics deals with increasingly abstract ideas, and some of the models may not be readily represented in this way. To be useful in physics, however, all models must serve to hold together ideas and must lend themselves readily to calculation.

Gas Gauges and Pumps

Let us first note some of the experimental equipment used in handling gases in the laboratory and measuring their pressures. Our theory—as we shall see—rests upon the data obtained in the laboratory with pressure gauges, thermometers, and volume-measuring cylinders fitted with pistons.

The pressure of a confined body of gas may be measured by an *open-tube manometer* (Fig. 15-1). This is a U-shaped tube containing a liquid of known density, usually mercury, but occasionally water or oil. One end of

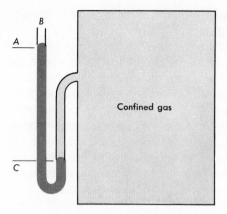

Fig. 15-1. Open-tube manometer. The excess pressure in the reservoir is the pressure due to the liquid column of height AC.

the tube is connected to the gas container; the other is open to the atmosphere. The pressure at A in the open tube is the atmospheric pressure p_B. At C there is an additional pressure p' due to the weight of the liquid in the vertical column between A and C. The total pressure p is the sum of the two, that is,

$$p = p_B + p'$$

Example. An open-tube manometer connected to a gas system reads 10 cm-water, and the barometric pressure is 74 cm-Hg. What is the total pressure?

$$p = 74 \text{ cm-Hg} + 10 \text{ cm-water}$$
$$= 74 \times 13.6 \text{ gwt/cm}^2 + 10 \text{ gwt/cm}^2$$
$$= 1016 \text{ gwt/cm}^2$$

The open-tube manometer is inconvenient for accurate measurements because the barometric pressure must be determined at frequent intervals. The *closed-tube manometer* indicates the total pressure directly. In this device (Fig. 15-2), the end is sealed, and the space above the column of mercury is highly evacuated, like that in the tube of a barometer. The liquid column DE is supported entirely by the pressure of the gas confined in the tank. Thus the pressure is equal to

that produced by this column of liquid of height $h = DE$.

Higher pressures—for example, those of gases stored in commercial cylinders—are measured by the Bourdon type of pressure gauge, represented in Fig. 15-3. The thin-walled metal tube is bent so as to lie along the arc of a circle. When the gauge is attached to a container filled with a gas whose pressure is greater than the atmospheric pressure, the walls bulge and the tube straightens a little. A gear attached to the tube rotates the pointer so that its position measures the pressure. Most Bourdon gauges indicate "gauge pressure," the excess gas pressure above that of the atmosphere. Occasionally, Bourdon gauges are calibrated as rough vacuum gauges for pressures less than atmospheric. When the bent tube is being evacuated, it gradually curls into a smaller circle as the push of the atmosphere squeezes the walls together.

The rotary or mechanical pump, used in laboratories for producing moderately high vacua (Fig. 15-4), consists essentially of an iron cylinder, fitted with four vanes sliding in slots, which rotates in a larger cylindrical opening. Each vane is pushed outward from

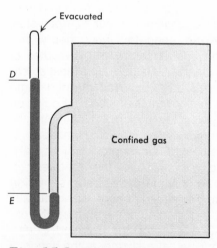

Fig. 15-2. Closed-tube manometer. The column DE *measures the total gas pressure.*

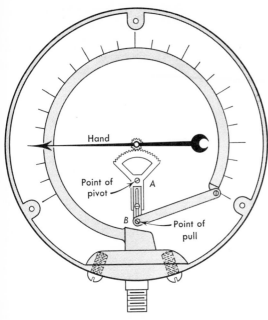

Fig. 15-3. Bourdon gauge. Increase of pressure in the tube makes it bulge and uncurl.

of the kind described in the preceding paragraph is required to lower the pressure to a millimeter-of-mercury or so. Then the diffusion pump can reduce the pressure still lower.

At low pressures, gauges for measuring the pressures of gases must be much more refined than the simple manometers discussed above. At pressures from a few hundredths of a millimeter-of-mercury to a few ten-thousandths, for example, a *thermocouple gauge* is often used. It contains a heated filament mounted in a glass or metal enclosure that is sealed to the vacuum system whose residual pressure is to be measured. A thermocouple is sealed within the same enclosure and measures the temperature of the heated filament. The more gas that remains in the vacuum system, the more the filament is cooled by convection and conduction, and the lower is its temperature. As the pressure drops during the pumping-out process, the temperature of the filament rises and the reading of the thermocouple gauge in-

the axis by a spring. Gas entering at the intake is trapped in the space between two vanes, is carried around, and is forced out at the outlet. Two such pumps operated in series give pressures as low as three-millionths of an atmosphere.

The diffusion pump (Fig. 15-5) has no pistons or other mechanically moving parts. Combined with a rotary pump, it can reduce the pressure to a billionth of an atmosphere. An oil, such as silicone oil, is boiled in the boiler at the bottom, and the vapor rushes upward through the vertical tube to emerge from the ducts as umbrella-shaped jets. Molecules from the vessel to be evacuated enter the oil-vapor jets and are driven downward through the inner tube. Then the oil vapor strikes the air-cooled wall of the pump, liquefies, and returns to the boiler. The gas remaining behind is pumped out through a side tube. A diffusion pump will not operate at ordinary pressures. An auxiliary "fore" pump

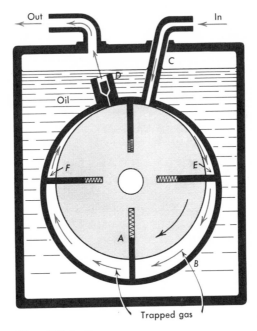

Fig. 15-4. Rotary pump.

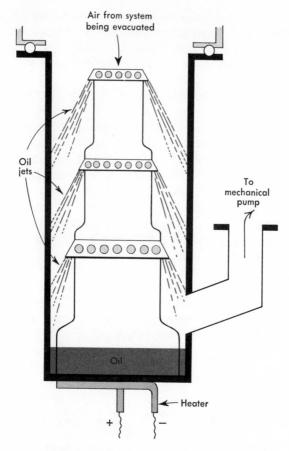

Air from system
being evacuated

Oil
jets

To
mechanical
pump

Oil

← Heater

+   −

Fig. 15-5. *Diffusion pump. The oil-vapor jet drives air downward, reducing the pressure in the evacuated vessel.*

creases. Thus, the gauge can be calibrated to read gas pressure.

Some Experimental Facts about Gases

The **state** of a given mass of gas is specified by giving corresponding values of its pressure p, volume V, and temperature T. As the gas undergoes various processes, its state changes from that described by one set of values of p, V, and T to that described by another set. One important kind of information about gases is the equations that describe how the values of p, V, and T change from one

state to another. We shall use "ideal gas" in this section to refer to a real gas at pressures of not more than a few atmospheres, stating later more precisely what our model of an ideal gas is.

1. In the discussion of the gas thermometer in Chapter 13, you learned that the pressure of an ideal gas is proportional to the Kelvin temperature at constant volume:

$$\left.\frac{p_1}{p_2} = \frac{T_1}{T_2}\right\} \quad \text{Isochoric (constant volume) process}$$

2. The volume of a gas at a constant pressure of not more than a few atmospheres is also proportional to the Kelvin temperature. Suppose that 273 cm³ of dry air at 0°C is confined in a tube (Fig. 15-6) at atmospheric pressure by means of a piston. Let the atmospheric pressure be constant. If the temperature of the confined air is raised 1 C°, its volume will increase 1 cm³, or $\frac{1}{273}$ of its initial volume at 0°C. When the temperature is increased further, the expansion per degree change of temperature is constant, so that at 100°C the volume of the gas is 373 cm³. If the experiment is performed using a different gas, such as hydrogen or helium, the expan-

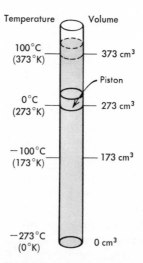

Temperature   Volume

100°C
(373°K)   373 cm³

                   Piston

0°C
(273°K)   273 cm³

−100°C
(173°K)   173 cm³

−273°C
(0°K)   0 cm³

Fig. 15-6. *Volume of a gas at different temperatures.*

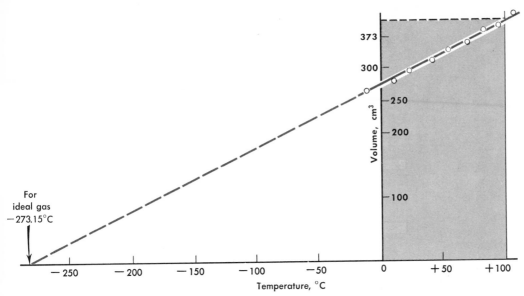

Fig. 15-7. *Variation of volume with temperature at constant pressure. The volume of an ideal gas would become zero at about −273°C or 0° Kelvin.*

sion per degree is very nearly the same as before. *All gases undergo approximately the same fractional change in volume per unit change of temperature, namely, a change of $\frac{1}{273}$ of the volume at 0°C per C°.* Figure 15-7 shows the graph of volume versus Celsius temperature at one pressure. If the zero of temperatures is shifted from 0°C to −273.15°C, the zero of the Kelvin scale, the volume and the Kelvin temperature are seen to be proportional to each other at constant pressure. Similar curves can be plotted for other pressures, giving a family of straight lines which can all be extrapolated to pass through the Kelvin zero. In general,

$$\left.\frac{V_1}{V_2}=\frac{T_1}{T_2}\right\} \quad \text{Isobaric process (Charles' law)}$$

3. At constant temperature, the pressure of a given mass of gas is directly proportional to its density and inversely proportional to its pressure. This relationship was discovered by Robert Boyle more than three centuries ago. He used a J-shaped tube (Fig. 15-8) into

which mercury was poured, trapping air in the closed, shorter end. When the mercury surfaces in both sides were at the same level, the pressure of the confined air equaled that of the atmosphere, for example, 76 cm-Hg. Boyle then poured mercury into the open arm, increasing the pressure until the volume of the confined air was reduced to one-half of its previous value. In this condition, with the temperature unchanged, the mercury surface in the open tube was 76 cm above that in the closed tube, thus making the total pressure of the confined gas 152 cm-Hg. In other words, the pressure of the confined gas was doubled, as was its density. Boyle found that when the pressure was tripled, the density was tripled, and the volume was reduced to one-third. He verified the law by making numerous other measurements, using different pressures (Fig. 15-9). Similar curves can be plotted for other temperatures giving a family of *isothermal* curves. We have

$$\left.\frac{p_1}{p_2}=\frac{V_2}{V_1}\right\} \quad \text{Isothermal process (Boyle's law)}$$

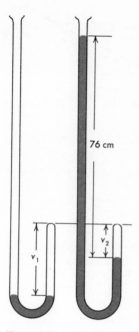

76 cm

v_1

v_2

Fig. 15-8. *Boyle's apparatus.*

These three relationships connecting p, V, and T can be combined into the *equation of state* of an ideal gas. We have $p_1V_1/T_1 = p_2V_2/T_2 = p_3V_3/T_3$ or, in general,

$$\frac{pV}{T} = nR \qquad \text{(General gas law)}$$

In this equation n is the number of moles (gram-molecular masses) of the gas, and R is the gas constant which has the value 8.32 joules per mole-Kelvin degree. If the temperature is constant, it can be combined with nR, and the equation of state yields Boyle's law; if the pressure is constant, the equation expresses the law of Charles.

Example. When a balloon left the earth in a cosmic ray flight, the volume of the gas was 10,000 m³, the pressure was 74 cm-Hg, and the temperature was 27°C or 300°K. Find the volume of the gas at an elevation where $p = 14.8$ cm-Hg and $t = -73$°C = 200°K.

$$\frac{p_1V_1}{T_1} = \frac{p_2V_2}{T_2}$$

$$\frac{74 \text{ cm-Hg} \times 10,000 \text{ m}^3}{300°\text{K}} = \frac{14.8 \text{ cm-Hg} \times V}{200°\text{K}}$$

$$V = 33,300 \text{ m}^3$$

A three-dimensional graph called a *p-V-T* surface connects the possible values of pressure, volume, and temperature for the ideal gas being studied (Fig. 15-10). Curve *ab* is an isothermal curve showing how the pressure and volume change at a constant temperature T_1. Similarly curve *cd* is an isochoric curve at constant volume V_1, and *ef* is an isobaric at constant pressure p_1. Any process that the perfect gas undergoes must be on the *p-V-T* surface; it need not be an isochoric, isobaric, or isothermal process, however. For example, curve *ag* is an *adiabatic curve* showing the changes of state that occur if the gas is expanded without allowing heat to enter. Notice that the adiabatic curve comes down somewhat more steeply than the isothermal curve that starts at the same point on the surface.

Such are the "gas laws" for the states of an ideal gas. In addition to explaining the gas laws, however, a theory of an ideal gas should account for other observations, such as the diffusion of gases. If you remove the stopper

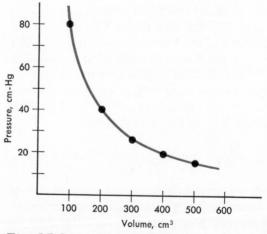

Fig. 15-9. *At constant temperature, the pressure of an ideal gas is inversely proportional to the volume.*

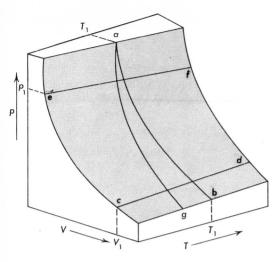

Fig. 15-10. A p-V-T *surface for an ideal gas.*

of an ammonia bottle, after a few minutes the odor of ammonia may be noticed in all parts of the room. If a pellet containing bromine gas—a dark brown gas—is broken within an evacuated glass vessel, the bromine fills the space almost immediately. These facts suggest that molecules must have high velocities, but that one gas diffusing through another does so rather slowly. Moreover, gases and vapors which are composed of relatively large and massive molecules diffuse more slowly than gases whose molecules are smaller and less massive. Hydrogen or illuminating gas diffuses more rapidly than air. To demonstrate this, connect a glass U-shaped manometer tube to the opening in an unglazed porous cup (Fig. 15-11). Air will diffuse into and out of the porous cup at equal rates and the pressure in the cup will remain equal to that of the atmosphere. Place an inverted glass jar over the cup and flood the space with hydrogen. The hydrogen molecules, having higher speeds, will diffuse into the cup faster than the air molecules diffuse outward, hence the pressure in the cup will increase. Remove the jar. Now the hydrogen will escape more rapidly than the air enters, so that the pressure in the cup will diminish.

Finally, the specific heat of a gas is a thermal property of interest to us. Can the kinetic theory of an ideal gas account for the measured specific heats? Before attempting to answer this question, let us turn to the kinetic theory and learn to what extent the theory accounts for the experimental facts reviewed in this section and where it fails.

Kinetic Theory of an Ideal Gas

By an ideal gas we mean one whose molecules move at random within the container, have the same mass, and are negligibly small compared to the distances they travel between collisions. We shall further assume that collisions between the molecules and between the molecules and the walls of the container are elastic ones: mechanical energy is conserved. Finally, we shall imagine that the molecules do not exert forces on each other except during collisions. Note that these

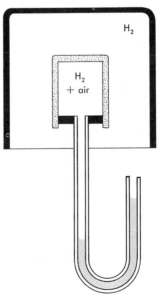

Fig. 15-11. Differential diffusion. Hydrogen diffuses into the porous cup faster than air diffuses out of it. Therefore the pressure in the cup increases.

assumptions, which were introduced to simplify the model of an ideal gas, will limit to ideal-gas conditions the application of the results of our theory.

The pressure of a gas is produced by the bombardment of the molecules. Consider a box that contains many gas molecules. These molecules, moving with average speeds of hundreds of meters per second, bump against each other and against the walls of the enclosure. They hit one wall so often that they exert against it a nearly constant force. If the speed of the molecules increases, they will make more impacts per second and will hit the wall harder at each impact. If the volume of the enclosure is diminished so as to crowd the molecules into closer quarters, they will hit the wall more often and will exert a still greater force. Thus the force per unit area exerted against the wall—the pressure—depends upon both the *speeds* of the molecules and the *size* of the enclosure.

Let us now derive a formula for the pressure of an ideal gas. Suppose that a cubical box (Fig. 15-12), the length of each side of which is l, contains N molecules of a perfect gas, each of which has a mass m. Consider one molecule that moves to and fro, striking the face A at right angles. When the molecule moves to the left, its velocity is negative. When it hits the wall A in an elastic collision, the molecule of velocity $-v$ and momentum $-mv$ is brought to rest and sent back in the opposite direction with momentum $+mv$. The

total change of its momentum during the impact is $+mv - (-mv) = 2mv$.

Since the width of the cubical box is l, the molecule must travel, to and fro, a distance $2l$ between successive impacts against A. The number of impacts per unit time is velocity/distance $= v/2l$. Thus the average change of momentum at A per unit time is the change of momentum per impact times the number of impacts per second, or

$$\frac{\Delta(mv)}{\Delta t} = 2mv \times \frac{v}{2l} = \frac{mv^2}{l}$$

In accord with Newton's laws of motion, the force acting on a body may be defined either as mass times acceleration or as the rate of change of its momentum. By the second definition, the average force F exerted by the molecule on wall A is

$$F = \frac{mv^2}{l}$$

Thus far we have assumed that only one molecule is in the box and that it continues to move to and fro, striking A normally. Let us go a step further and assume that there are N molecules in the box and that because of their random motions one-third of them move sidewise striking A, one-third move up and down, and the remainder strike the front and back faces of the box. (More rigorous procedures here can remove some of the artificiality of this assumption, but the final result is the same.) We would expect the molecules to have a wide range of velocities: some at any instant moving with great velocities, others momentarily at rest. However, let the average speed of the molecules—that is, the average of the magnitude of the velocity—be $\bar{v}$. One third of the molecules pound against A; hence the average force F exerted against it is

$$F = \frac{1}{3}\frac{Nm\bar{v}^2}{l}$$

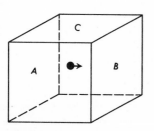

Fig. 15-12. *A molecule in a box.*

The area of the surface A is l^2; hence the pressure is given by

$$p = \frac{F}{A} = \frac{1}{3}\frac{Nm\bar{v}^2}{(l^2)l} = \frac{1}{3}\frac{Nm\bar{v}^2}{V}$$

where V is the volume of the box. We have thus obtained by applying the principles of mechanics to the molecular model of an ideal gas a formula for the pressure of the gas

$$p = \frac{1}{3}\frac{Nm\bar{v}^2}{V} \qquad (1)$$

No one has ever seen a molecule or can hope to see one; hence it is interesting to see how simply average molecular speeds may be determined. In equation (1) Nm/V is the mass of the gas per unit volume, that is, its mass density D_m. Hence

$$p = \frac{1}{3}(D_m\bar{v}^2)$$

Using this expression we now compute the mean molecular speed* of oxygen at 32°F or 0°C. Let the pressure be 76 cm-Hg or 1,010,000 dynes/cm², The density D_m under these conditions is the mass of one mole of oxygen, 32.0 gm, divided by the volume that a mole occupies at 76 cm-Hg and 0°C, 22,400 cm³, equaling 0.00143 gm/cm³.

$$p = \frac{1}{3}(D_m\bar{v}^2)$$
$$\text{1,010,000 dynes/cm}^2 = \frac{1}{3}(0.00143 \text{ gm/cm}^3)\bar{v}^2$$
$$\bar{v} = 46,100 \text{ cm/sec}$$
$$= 461 \text{ m/sec}$$
$$\cong 0.3 \text{ mi/sec}$$

To determine how the average energy and average speed of the gas molecules depend upon temperature, we must bring in the experimentally determined equation of state of an ideal gas

$$pV = nRT$$

* More precisely, the speed that enters the pressure formula is the square root of the average of the squared speeds of the molecules—the "root-mean-square," or rms, speed—which is about 8% larger than the average or mean speed.

where p is the pressure, V the volume, n the number of moles or gram-molecular masses, and T the Kelvin temperature. From the pressure formula we have

$$pV = \frac{1}{3}Nm\bar{v}^2$$

Hence,

$$nRT = \frac{1}{3}Nm\bar{v}^2 = \frac{2}{3}(\frac{1}{2}Nm\bar{v}^2)$$

Nm is the total mass of the gas and $\frac{1}{2}Nm\bar{v}^2$ is its total translational molecular kinetic energy. Hence *the total translational kinetic energy of the molecules of an ideal gas is proportional to its Kelvin temperature,* or

$$(E_K)_{\text{total}} = \frac{3}{2}nRT$$

The average translational kinetic energy per single molecule E_K is the total kinetic energy divided by the number of molecules N.

$$E_K = \frac{(E_K)_{\text{total}}}{N} = \frac{3}{2}\frac{nRT}{N} = \frac{3}{2}\frac{R}{N/n}T$$

The number of molecules per mole N/n is *Avogadro's number* N_0, whose value, a constant for *all* gases (Chapter 23), is 6.025×10^{23} molecules/mole. We can compute the average kinetic energy of one molecule of *any* gas at 0°C:

$$E_K = \frac{3}{2}\frac{8.32 \text{ joules/mole-K}°}{6.02 \times 10^{23} \text{ molecules/mole}}273°\text{K}$$

$$= 5.66 \times 10^{-21} \text{ joule/molecule}$$

This is the average translational kinetic energy of an oxygen molecule, a hydrogen molecule, or the molecule of any other gas at 0°C. Incidentally, it is also the average translational kinetic energy of a mammoth colloidal particle that is visible through a microscope. However, molecules can possess additional energy of rotation and of vibration.

Comparison of the Results of Kinetic Theory with Experiment

A real gas should behave like an ideal gas when the density is low and intermolecular at-

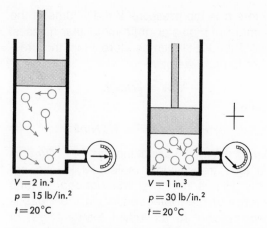

V = 2 in.³
p = 15 lb/in.²
t = 20°C

V = 1 in.³
p = 30 lb/in.²
t = 20°C

Fig. 15-13. When the volume decreases, the pressure exerted by the molecules increases (temperature constant).

tractions are negligible. We find that the agreement between theory and experiment is rather good and moreover, that the model of an ideal gas helps us to organize our thoughts about gases.

The theory relates the experimental gas laws to each other. For example, Boyle's law can be obtained from the pressure formula. Let the pressure of a gas be changed from p_1 to p_2 at constant temperature so that the average molecular velocity is constant.

$$\frac{p_1}{p_2} = \frac{\frac{1}{3}Nm\bar{v}^2/V_1}{\frac{1}{3}Nm\bar{v}^2/V_2} = \frac{V_2}{V_1}$$

Hence

$$\frac{p_1}{p_2} = \frac{V_2}{V_1} \quad \text{(Temperature is constant)}$$

Suppose that 2 in.³ of air at a pressure of 15 lb/in.² is confined in a cylinder closed by a sliding piston (Fig. 15-13). When the piston is pushed inward, reducing the volume to 1 in.³, the density of the gas is doubled. According to the kinetic theory, if the temperature of the gas is the same as before, the molecules have the same average speed, there are twice as many hits per second against the piston, and the pressure is doubled.

Similarly, if the volume of a gas is constant, increasing the Kelvin temperature causes a proportional increase in the pressure. Suppose that the Kelvin temperature is quadrupled. Since the *square* of the average velocity is proportional to the Kelvin temperature according to the kinetic theory, the average velocity is doubled. If the average velocity of the molecules is doubled, without change of density, the force exerted against each wall of the container is quadrupled. At the doubled speed, each molecule hits the wall twice as often per unit time. Moreover, at each impact it delivers twice as much momentum. Since both factors are doubled, the force against the wall and the pressure of the gas are quadrupled. Thus, quadrupling the Kelvin temperature at constant volume quadruples the pressure.

The kinetic theory correctly predicts the *distribution of molecular speeds* at a given temperature. The calculations go beyond the scope of this book, but the Maxwell-Boltzmann velocity distribution formula, derived

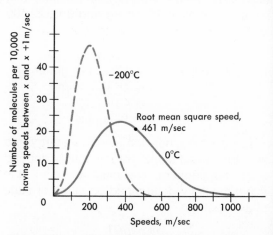

Fig. 15-14. The Maxwell-Boltzmann distribution curve for the speeds of oxygen molecules at 0°C and −200°C. (After O. H. Blackwood, T. H. Osgood, A. E. Ruark, et al., An Outline of Atomic Physics, third edition, John Wiley and Sons, 1955.)

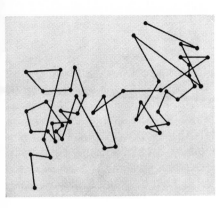

Fig. 15-15. Path of a molecule through a gas. (Magnification 50,000.)

from the kinetic theory, is in good agreement with experimental results (Fig. 15-14).

The kinetic theory explains the *diffusion* of gases. As mentioned earlier, ammonia diffuses through the air of a room. The ammonia molecules dart hither and thither through the air with speeds of several hundred meters per second. In view of their high speed, the relatively low rate of diffusion or migration is surprising. It is easily understood, however, when we consider that a molecule does not make a beeline or travel in a straight line, but collides with other molecules at irregular intervals and moves about in random directions (Fig. 15-15). Like a man idly wandering in a crowd, it progresses slowly. The average distance that an oxygen molecule goes between collisions, called the *mean free path*, is about 9.9×10^{-6} cm at 0°C and 1 atmosphere (atm) pressure. We have also noted that gases and vapors which are composed of relatively large and massive molecules diffuse more slowly than gases whose molecules are smaller and less massive. There are two reasons for this difference according to the kinetic theory: first, the larger molecules have smaller average speeds (Why?); and second, being more bulky, they collide more frequently with other molecules. The rates at which molecules or colloidal particles in a fluid diffuse are proportional to their average

velocities. But the average velocities are inversely proportional to the square roots of the molecular weights *M*. Thus, treating the molecules of both gases as point masses, we have

$$\frac{\text{Diffusion rate of } A}{\text{Diffusion rate of } B} = \frac{\bar{v}_A}{\bar{v}_B} = \sqrt{\frac{M_B}{M_A}}$$

For example, the molecular weight of hydrogen is about one-sixteenth as great as that of oxygen. Therefore, at a given temperature and pressure, we expect hydrogen to diffuse four times as fast as oxygen.

As we have noted previously, the equation $E_K = 3RT/2(N/n)$ applies not only to the molecules of a gas but also to molecules in solution and to colloidal particles having masses billions of times as great as that of the hydrogen molecule.

Brownian movements of microscopic particles are caused by bombardment of the particles by neighboring molecules. The translational kinetic energy of a Brownian particle was first determined in 1908 by the French chemical physicist Perrin. He focused his microscope on a certain particle of a wax called gamboge, suspended in a liquid, and watched it as it wandered about. By measuring its rate of wandering, he was able to estimate its average speed. After determining the mass of the particle, he computed the kinetic energy ($\frac{1}{2}mv^2$) and found that his value agreed with that for the hydrogen molecule at the same temperature as computed by the pressure equation.

The kinetic theory equation for gas pressure may be modified so as to predict the *specific heat* at constant volume of a monatomic gas such as helium. Earlier we saw that the pressure formula and the equation of state for an ideal gas led to

$$\frac{1}{3}Nm\bar{v}^2 = nRT$$

In this equation, *m* is the mass per gas molecule, *N* is the number of molecules, and *n* is the number of moles of gas. We then saw that

$$(E_K)_{\text{total}} = \tfrac{1}{2}Nm\bar{v}^2 = \tfrac{3}{2}(\tfrac{1}{3}Nm\bar{v}^2) = \tfrac{3}{2}nRT$$

The left side of the equation is the total translational kinetic energy of the gas and

$$R = 8.32 \text{ joules/mole-K}° \cong 2 \text{ cal/mole-K}°$$

If the volume of the ideal gas is constant, the change in the translational kinetic energy $\Delta(E_K)_{\text{total}}$ is equal to the heat added, ΔQ. By definition, the heat added per mole per degree change of temperature is the *molecular heat* at constant volume c_V. Hence,

$$c_V = \frac{\Delta Q}{n\,\Delta T} = \frac{\Delta(E_K)_{\text{total}}}{n\,\Delta T} = \tfrac{3}{2}R \cong 3 \text{ cal/mole-K}°$$

This quantity, divided by the molecular weight of any *monatomic* gas such as helium, gives its specific heat at constant volume within 1%.

The computed value of c_V is much too low for any *diatomic* gas. We explain the molecular heat of a diatomic gas, such as hydrogen, by assuming that the dumbbell molecules of such a gas can tumble freely. Assuming that kinetic energy is equally distributed among the different motions—the principle of equipartition of energy—we deduce that the hydrogen molecules can have extra kinetic energies of rotation. Whereas a monatomic "point" molecule, such as that of helium, can move independently in only three ways—parallel to the X, Y, and Z axes—a diatomic molecule can also rotate around two axes. A monatomic molecule has 3 degrees of freedom and its molecular heat c_V at constant volume is 3 cal/mole-K°. A diatomic molecule, with 5 degrees of freedom, might be expected to have a molecular heat c_V of 5 cal/mole-K°. The specific heat of hydrogen, accordingly, should be 5 cal/mole-K° divided by 2 gm/mole or around 2.5 cal/gm-K°. The experimentally determined specific heat of hydrogen at 20°C is 2.44 cal/gm-K°.

In Chapter 17, we shall discuss the molecular heat of a gas at constant pressure.

Failures of the Kinetic Theory of an Ideal Gas

Earlier in this chapter we remarked that most theories fail sooner or later and must be modified or replaced. As an explanation of the properties of real gases at relatively low densities, the kinetic theory of a perfect gas, as we have seen, has been quite successful. At pressures of several hundred atmospheres and above, it fails to predict the observed behavior. Figure 15-16 shows how the product of pressure and volume of two real gases—hydrogen and oxygen—behaves at 0°C with increasing pressure. Note that if the two gases were ideal gases, pV would be a constant, according to Boyle's law. As you see, it is not. The ratio of pV for hydrogen to $(pV)_0$ for a perfect gas increases with increasing pressure. As the hydrogen molecules are crowded closer and closer together at higher pressures, they no longer act like mass points. The molecular force of *repulsion* becomes effective at small molecular separations and prevents decreases in volume proportional to the increases in pressure so that the product pV increases. Oxygen behaves somewhat differently. As the pressure increases, pV first becomes less than $(pV)_0$ because the intermolecular *attraction* adds to the effect of the external pressure and reduces the volume more than one would expect. However, as the pressure continues to increase, the molecular forces of repulsion come into play and the product pV increases. Extensions of kinetic theory have been developed to explain this behavior.

If the temperature of real gases is lowered as they are compressed, they soon cease to resemble an ideal gas. At sufficiently low temperatures, the molecules move so sluggishly that they form clusters and liquefy. Some gases are more "permanent" than others. Under normal pressure, oxygen liquefies at −183°C, hydrogen at −253°C,

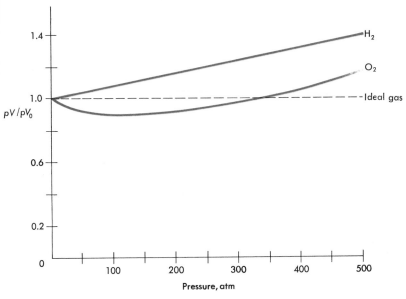

Fig. **15-16.** *The ratio of* pV *for hydrogen and oxygen at 0°C to* (pV)₀ *for an ideal gas.* (*After A. G. Worthing and D. Halliday,* Heat, *John Wiley and Sons, 1948.*)

and helium, the most permanent gas, at −269°C.

Finally, we should note that the simple kinetic theory of molecular heats of gases does not explain the experimental fact that the molecular heats of gases such as hydrogen vary markedly with temperature. The explanation of this temperature variation comes from quantum theory.

Summary

A theory is an hypothesis devised (1) to account for known experimental facts, (2) to show relationships among them, and (3) to predict new relationships. Theories that fail to do these things are modified or, in some instances, replaced by other theories.

A model is a framework around which scientists build theories. The model of an ideal gas is that of a swarm of molecules moving at random at average distances of separation that are large compared with their diameters. Collisions are elastic, and intermolecular attractive forces are negligible.

Among the many experimental facts that must be accommodated by a theory of a gas are p-V-T data, the distribution of molecular velocities, the facts of diffusion, and the specific heats of gases.

The pressure of a gas, produced by molecular bombardment, is given by $p = \frac{1}{3}Nm\bar{v}^2/V$. The equation of state of a perfect gas is $pV = nRT$. From these two relations, one can deduce that the average total kinetic energy of an ideal gas is proportional to the Kelvin temperature, compute the average molecular speed and energy at a given temperature, compare the diffusion rates of two gases, compute the specific heat of a monatomic gas at constant volume, and show the following p-V-T relationships:

$$\frac{p_1}{p_2} = \frac{V_2}{V_1} \qquad \text{(Constant temperature)}$$

$$\frac{p_1}{p_2} = \frac{T_1}{T_2} \qquad \text{(Constant volume)}$$

$$\frac{V_1}{V_2} = \frac{T_1}{T_2} \qquad \text{(Constant pressure)}$$

The kinetic theory of a perfect gas explains the behavior of real gases adequately at low pressures, but fails at high pressures and low temperatures.

Questions

1. Define root-mean-square velocity. How does it differ from mean velocity? (Use velocities of 1, 2, and 3 m/sec as examples.)

2. What is the significance of Avogadro's number?

3. Why is Kelvin temperature, rather than Celsius temperature, used in the general gas law?

4. Discuss the fundamental assumptions involved in the kinetic theory of an ideal gas.

5. Compare the advantages and disadvantages of a closed-tube manometer with those of an open-tube manometer.

6. Upon what does the rate of diffusion of a gas depend?

7. Show by the kinetic theory model of an ideal gas why doubling the mean molecular speed of a gas quadruples its pressure.

8. How can the mean molecular speed of a gas be determined? How is it related to density?

9. Show how Boyle's law may be derived from the equation $p = \frac{1}{3}Nm\bar{v}^2/V$.

10. Prove that when the two gases are at the same temperature, the mean (translational) kinetic energy of a hydrogen molecule equals that of an oxygen molecule.

11. (a) What is Brownian motion? (b) Describe Perrin's experiment and state what it proved.

12. Why is the molecular heat of a diatomic gas at constant volume greater than that of a monatomic gas?

13. What conclusions can you draw from these data:

	rms velocity (cm/sec)	mean free path (cm)
H_2	18.4×10^4	18.3×10^{-6}
N_2	4.93×10^4	9.44×10^{-6}
O_2	4.61×10^4	9.95×10^{-6}

14. By what fraction of its initial volume would an ideal gas expand when the temperature increased 1 C° if the initial temperature were (a) 0°C, (b) 127°C (c) −273°C? Assume constant pressure.

15. Under what conditions will raising the temperature of a gas 1 C° make it expand 1/373 of its initial volume?

16. In an experiment at constant temperature, these data were obtained for a confined gas:

Pressure (cm-Hg)	102.5	96.3	87.3	79.4	68.9	58.7
Volume (cm³)	31.3	33.3	35.9	40.4	46.3	54.3

Plot p versus V and p versus $1/V$. Interpret these results. What type of curve results from each plot?

17. How do you account for the high diffusion rate of H_2?

18. Estimate the number of gas molecules in an inflated automobile tire.

19. How would you expect the mean free path of a gas molecule to depend on the density of the gas?

20. If the pressure of an ideal gas is plotted against $1/V$ at constant temperature, what is the significance of the slope of the straight line that results?

21. Design a hot-air balloon with the construction shown in Fig. 15-17. The passenger, gondola, nylon cords, stove, etc. weigh 100 kg all together. The fabric of which the bag is made weighs 0.05 kgwt/m². The outdoor temperature is 25°C. Make a graph of the radius of the inflated bag versus its inside air temperature. Select the final size of the balloon on an economic basis, stating your assumptions about the cost of materials, etc.

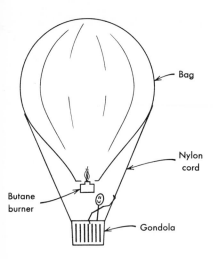

Fig. 15-17.

Problems

A

1. The volume of a steel hydrogen tank is 0.25 m³. What mass of hydrogen will it contain at 10 atm pressure if the density of hydrogen at 1 atm is 90 gm/m³? The temperature is constant.

2. How many cubic meters of helium at a pressure of 40 atm are required to fill a balloon of volume 2.5×10^3 m³ at a pressure of 1 atm without change of temperature?

3. The volume of the inner tube of an automobile tire is 3.0×10^4 cm³. If the pressure, above atmospheric, is 150 cm-Hg and the barometric pressure is 75 cm-Hg, what volume will the air occupy if the tire bursts? The temperature is constant.

4. How many cubic centimeters of air at atmospheric pressure must be pumped into the tire described in the preceding problem in order to raise the pressure above atmospheric from 150 cm-Hg to 225 cm-Hg at constant temperature?

5. If Boyle's law held, at what depth in the sea would the density of air in a bubble equal that of salt water, density 1.03 gwt/cm³? The temperature is 10°C.

6. A cylinder with a sliding piston contains 3000 cm³ of air at 27°C. If the pressure remains constant, at what Celsius temperature will the volume be 4000 cm³?

B

7. A cylindrical tank having a volume of 0.50 m³ and a depth of 1.0 m contains air at normal pressure. Water is forced into the tank from a faucet until the excess or gauge pressure is 4.0 atm. What is (*a*) the volume of the confined air, (*b*) the depth of the water?

8. How many grams of oxygen, molecular mass 32 gm/mole, are contained in a tank having a volume of 20 liters if the pressure is 10 atm? The temperature is 0°C. (At 0°C, a mole of gas occupies 22.4 liters at a pressure of 1 atm.)

9. At 0°C, the pressure of 1 mole (28 gm) of nitrogen is 76 cm-Hg and its volume is 22,400 cm³ Calculate the average velocity of a nitrogen molecule under these conditions.

10. An air bubble rose from a point 10 m below the surface of a pond. When the bubble reached the surface, its volume was twice that at the bottom without change of temperature. What was the atmospheric pressure in cm-Hg?

11. A short, heavy metal cylinder is lowered, open end downward, into fresh water to a depth of 30 m without change of temperature. The atmospheric pressure is 74 cm-Hg. (*a*) What is the pressure of the air trapped in the cylinder? (*b*) What fraction of the cylinder will be filled with water?

12. If the temperature of 1.0 cm³ of an ideal gas were raised from −272°C to −267°C, how much would it expand if the pressure were constant?

13. The density of air at 0°C and 76 cm-Hg is 1.29 gwt/liter. Find its density at 47°C and 100 cm-Hg.

14. The volume of the gas contained in a

stratosphere balloon was 7.2×10^3 m³ at 75 cm-Hg and 27°C. What was the volume when the balloon rose to a region where the pressure was 5.00 cm-Hg and the temperature was −33°C?

C

15. A thin-walled cylinder, closed at one end, is thrust open end downward into a fresh-water lake to such a depth that water rises halfway up the cylinder. The barometer reads 76 cm. (a) How far below the surface of the lake is the level of the water in the cylinder? (Density of mercury is 13.6 gwt/cm³.) (b) If the cylinder weighs 200 gwt in air and its volume is 300 cm³, at what depth will it tend neither to rise nor to sink?

16. In 3.00 min, 300 bullets strike and embed themselves in a vertical wall. The mass of each bullet is 6.00 gm, and its horizontal velocity is 0.80 km/sec. Find (a) the momentum imparted to the wall per second, (b) the average force exerted against the wall. Assume that the firings are evenly spaced.

17. Air is trapped in the closed end of a glass tube by a column of mercury 40 cm long. When the tube is vertical, open end up, the length of the column of trapped air is 10 cm. What is the length of the column after the tube is rotated in a vertical plane (a) 30°, (b) 60°, (c) 90°, (d) 120°, (e) 150°, (f) 180° from its initial position? Atmospheric pressure is 76 cm-Hg.

18. Calculate the value of the gas constant R in joules per mole-Kelvin degree. The volume of one mole of an ideal gas at 76 cm-Hg pressure and 0°C is 22,400 cm³.

19. There are about 20 molecules per cubic centimeter in space. Find the mean free path of the molecules, assuming that the diameter of a molecule is 1.0×10^{-8} cm.

20. (a) If a bullet of mass 8.00 gm and horizontal velocity 1.20 km/sec strikes and embeds itself in a wooden block of mass 3920 gm suspended like a pendulum bob by a cord 1.0 m long, to what angle with the vertical will the cord swing back? (b) What will the angle be if the bullet recoils elastically?

21. (a) Using the temperature data in Fig. 2-17, page 26, calculate the average velocity of a hydrogen molecule at an altitude of 400 miles above the earth's surface. (b) Calculate the velocity of escape at this altitude. Is it likely there is much molecular hydrogen in the upper atmosphere?

22. Four grams of hydrogen and 32.0 grams of oxygen are placed in a 10.0-liter container at a temperature of 0°C. (a) What is the total pressure in the container? (b) A spark is created in the chamber, and the two gases react. What is the pressure if the temperature is then 100°C?

For Further Reading

Alder, B. J. and T. E. Wainright, "Molecular Motions," *Scientific American,* October, 1959, p. 113.

Conant, James B., ed., *Robert Boyle's Experiments in Pneumatics,* Harvard Case Histories in Experimental Science, Harvard University Press, Cambridge, 1950.

Morrison, Philip and Emily, "High Vacuum," *Scientific American,* May, 1950, p. 20.

Rogers, Eric M., *Physics for the Inquiring Mind,* Chapter 25, Princeton University Press, Princeton, 1960.

CHAPTER SIXTEEN

Change of Phase

When a substance is heated, usually the speeds of its molecules increase and its temperature rises. There are three familiar exceptions: first, when a solid melts; second, when a solid sublimes; and third, when a liquid evaporates. In these changes of phase, heat energy is utilized in separating the molecules from one another, thus increasing their potential energies.

Melting and Freezing

Suppose that we place some crushed ice, at a temperature below the freezing point, in a vessel on a stove. The temperature will rise at first, showing that the molecular kinetic energy of the ice is increasing. When the melting point is reached, the temperature will remain constant until all the ice is melted. Then the temperature of the water will rise (Fig. 16-1).

In ice, the molecules are neatly arranged in crystals. As the ice melts, absorbing heat, its molecules are torn away one by one, and equal energy is required to dislodge each molecule. Thus the temperature of the melting ice remains constant. Similarly, when water freezes, giving out heat, its temperature does not change. *During the melting or freezing of water the temperature remains constant at 0°C under standard pressure.* Normal melting temperatures of several substances are given in Table 16-1.

The molecules of a non-crystalline substance such as butter are not arranged regularly, and some molecules are more tightly bound than others. Hence a non-crystalline substance gradually softens as the temperature rises; it has no definite melting point.

Ice in melting absorbs heat; when water freezes, heat is evolved. The heat of fusion L_f of a substance is defined as *the amount of heat required per unit mass to melt the substance without change of temperature.*

$$L_f = \frac{\text{Heat absorbed}}{\text{Mass of substance melted}} = \frac{Q}{m}$$

The heat of fusion of water is approximately 80 cal/gm.

Example. If 200 gm of water at 40°C is poured onto several kilograms of cracked ice at 0°C, how much ice is melted? The final temperature is 0°C.

The heat lost by the water is

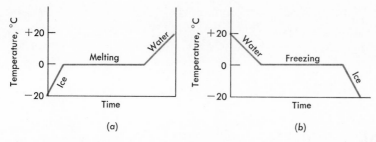

Fig. 16-1. (a) *The temperature of the ice remains constant while it melts, and* (b) *that of the water remains constant while it freezes.*

$$Q = cm(t_2 - t_1)$$
$$= 1 \text{ cal/gm-C}° \times 200 \text{ gm} \times 40 \text{ C}°$$
$$= 8000 \text{ cal}$$

The mass *m* of ice melted is

$$m = \frac{Q}{L_f} = \frac{8000 \text{ cal}}{80 \text{ cal/gm}} = 100 \text{ gm}$$

Behavior of Substances near the Freezing Point

Most substances contract when they solidify, but water, bismuth, cast iron, type metal used in printing, and a few others expand. Some-

Table 16-1. Melting Points and Heats of Fusion of Crystalline Substances at Normal Pressure

	Melting Point, °C	Heat of Fusion, cal/gm
Alcohol, ethyl	−117.3	24.9
Aluminum	658	76.8
Ammonia	−75	108.1
Carbon dioxide (at 5.2 atm)	−57.0	45
Copper	1083	42
Gold	1063.6	15.8
Helium	−271.5	
Hydrogen	−259.14	
Mercury	−38.7	2.8
Oxygen	−218.4	3.3
Platinum	1773.5	27.2
Rose metal, alloy	98.0	
Tin	232	14.0
Water	0	79.7
Wood's metal	62	

times the expansion of water in freezing does damage—for example, in bursting the radiator of a car. Often it is beneficial, as in the disintegration of rocks to form soil. One important effect of the expansion of water on freezing is that ice, being less dense than water, floats. If the water of a lake contracted on freezing, the ice formed in winter would go to the bottom, where, heat-insulated by the water, it would remain unmelted during the following summer. The effects on climate and marine life would be profound.

Increase of pressure raises the freezing point of most substances. Water is an exception. Increased pressure increases opposition to expansion, and water cannot freeze without expanding. A pressure of about 1 ton/in.² is required to lower the freezing point of water by 1 C°. When you skate on ice, the sharp edges of your skate blades exert great pressure so as to lower the melting point of the ice. It melts and forms a slippery film under the runner. By squeezing soft snow, you exert great pressure at the contact points of the crystals so that the snow melts. When you release the pressure, the water freezes, forming a hard, compact ball. In order to demonstrate the melting of ice due to high pressure, suspend a load of 10 kilograms or so from a thin wire looped over a block of ice (Fig. 16-2). At the increased pressure, the ice melts. As the process—called *regelation*—continues, the water refreezes. The ice below the wire melts and absorbs heat from

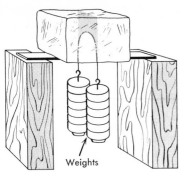

Fig. 16-2. Regelation. The ice beneath the wire melts and absorbs heat from the water above it so as to freeze it and close the gap.

the water above it. Hence the water film freezes again and closes the gap.

Water, freed of dissolved air by boiling and not jarred or shaken, may be cooled several degrees below the normal freezing point. If jarred, it solidifies suddenly (Fig. 16-3). The absence of dissolved air and other impurities favors this "undercooling." Plumbers know that hot-water pipes of a vacant unheated residence are more likely to burst in winter than the cold-water pipes. The liquid in the cold-water pipes freezes so gradually that some of it creeps along the pipe back to the water main. Thus the pressure does not rise sufficiently to do damage. The water in the hot-water pipes has been freed of dissolved air and impurities by previous heating. It therefore undercools and then freezes suddenly, often bursting the pipes.

The addition of impurities lowers the melting points of most substances. In making ice cream, salt is added to the ice to lower its melting point. The mixture of the two solids—ice and salt—is unstable at 0°C. A solution of salt in liquid water is stable, however, so the ice and salt dissolve to form salt water. The heat required to melt the ice comes from the contents of the freezer and from the salt water that has already formed. Hence, the temperature of the freezing mix-

ture drops below 0°C. In general, the freezing points of liquids are lowered when foreign substances are dissolved in them. Antifreeze solutions for the radiators of automobiles are made by dissolving alcohol, ethylene glycol, etc., in the water.

Sublimation

There is abundant evidence, as we have seen, that the molecules of gases and liquids move about at high speeds, bumping into one another and the walls of the container. If a vessel of water is left uncovered, the water gradually evaporates.

In most solids, however, the binding forces are so great that few molecules escape. In others, such as solid iodine and camphor, the transition from the solid phase to the gas phase—*sublimation,* as it is termed—is appreciable. Common ice vaporizes or sublimes. If wet clothing is hung out to dry in zero weather, the water soon freezes, but in the course of a few hours, practically all the ice evaporates and the clothing is ready to be ironed. "Dry ice" (solid carbon dioxide) sublimes readily at room temperatures. Many solid substances sublime so slowly that the only evidence of evaporation is the odor. A few grains of the solid perfume base called

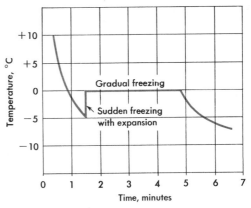

Fig. 16-3. Undercooling. Water can be cooled below 0°C. Then it freezes suddenly.

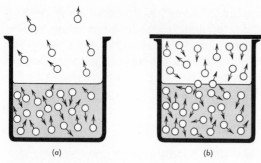

Fig. 16-4. (a) *Evaporation of a liquid. Its speedier molecules escape.* (b) *When the space is saturated, the rates of evaporation and condensation are equal.*

musk will give a perceptible odor for several years. Heat must be added for a solid to sublime. The heat of sublimation of solid iodine, for example, is about 58 cal/gm at 25°C. The pressure of the iodine gas in equilibrium with the solid at that temperature is 0.031 cm-Hg. In metals, the pressure of the metallic gas is much smaller; the pressure of gaseous cesium at 29°C is only 1.2×10^{-7} cm-Hg.

Evaporation and Vapor Pressure

When water evaporates from an open vessel, the more energetic molecules plunge through the elastic surface film and escape. Then they disperse by diffusion and convection so that the quantity of liquid gradually diminishes. Close the vessel so that the vapor cannot escape (Fig. 16-4). As evaporation proceeds, the vapor above the liquid becomes more dense. Not only do molecules escape from the liquid, but also others from the vapor return to it. Soon the two rates become equal, and *dynamic equilibrium* is established. Then we say that the space above the liquid is *saturated,* and the vapor density remains constant. *When a vapor in contact with its liquid is saturated, the rates of evaporation and of condensation are equal.* The satu-

rated vapor exerts a pressure called the *saturated vapor pressure* and has a density called the *saturated vapor density.*

Dissolved substances lower vapor pressures. Place salt water and pure water in separate containers *A* and *B* under cover and reduce the pressure somewhat with a vacuum pump (Fig. 16-5). Then the amount of water in *B* will slowly diminish and that in *A* will increase. Each salt molecule in *A* attracts surrounding water molecules and opposes their evaporation. Thus the vapor pressure near the salt solution is less than that near the water, and vapor diffuses from *B* to *A*. In general, any solid substance, dissolved in a liquid, lowers the vapor pressure.

The pressure of a saturated vapor is independent of volume. To study the effect of volume variations on vapor pressure, suppose that a few drops of ether are placed in a long cylinder closed by a piston. The liquid will evaporate until the space becomes saturated. Push the piston downward (Fig. 16-6*a*), thus compressing the vapor and raising its pressure and density. Then the rate of condensation will be greater than that of evaporation. Vapor will condense until its density reaches the initial saturation value. Pull the piston upward to its first position (Fig. 16-6*b*), and the liquid ether will evaporate until the pressure and density will be the same as before. *At constant temperature the pressure of a saturated vapor is independent of its volume.* To produce an unsaturated condi-

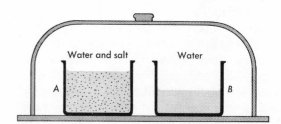

Fig. 16-5. *The vapor pressure near the salt water in* A *is less than that near the water in* B. *Therefore water diffuses from* B *to* A.

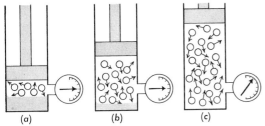

(a)   (b)   (c)

Fig. 16-6. *The vapor pressure is the same in* (a) *and* (b), *although the vapor occupies a greater volume in* (b). *In* (c), *all the liquid has evaporated and the pressure of the gas decreases as the volume increases.*

tion, pull the piston upward so far that *all* the liquid evaporates. Then on further expansion (Fig. 16-6c) the space will become unsaturated, and the pressure will decrease. In the unsaturated condition, the vapor obeys the gas laws of the last chapter as does an ordinary gas. *A gas is merely a vapor far removed from saturation.*

The vapor pressures of liquids may be studied by means of the apparatus diagramed in Fig. 16-7. *A* is a mercury barometer. In *B*, the space above the mercury contains ethyl ether at 20°C. The ether vapor in *B* pushes the mercury column down 44 cm; hence the vapor pressure of ether at 20°C is 44 cm-Hg. Suppose that we tilt the tube containing ether sideways as in *B'*. Immediately afterward, the space above the mercury will be supersaturated. However, sufficient ether vapor will condense so that finally, the vapor pressure will again be 44 cm-Hg. If the tube *B* had contained air at 44 cm-Hg pressure, tipping it to *B''* would have increased the pressure. (Why?)

Rise of temperature of a liquid increases its saturated vapor pressure and saturated vapor density (Fig. 16-8 and Table 16-2). There are two reasons for this. First, the number of molecules in the vapor becomes greater; and second, they move faster. They exert greater forces at the walls of the enclosure both because they make more impacts per unit

time and also because each molecule has a greater change of momentum per impact. At ordinary room temperatures, the vapor pressure of water is a few millimeters-of-mercury; in a power-plant boiler, it is many atmospheres.

Boiling

A liquid usually evaporates at its upper surfaces only, but above a certain temperature bubbles of vapor are produced inside the liquid, and it *boils*. Thereafter, the temperature remains constant, regardless of the rate of heating. When the temperature is below the boiling point, bubbles of vapor cannot be produced below the surface. The vapor pressure is less than that of the atmosphere. If a bubble were formed, it would quickly collapse. *The boiling point of a liquid is the temperature at which the saturated vapor pressure of the liquid equals the pressure at the surface.*

The boiling point of a liquid depends upon the applied pressure. At the summit of Pikes Peak in Colorado, water boils at about 180°F, and boiling water is not hot enough to cook

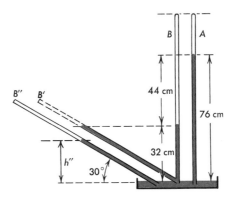

Fig. 16-7. *The pressure of the gas or vapor in* B *is 44 cm-Hg. If the space contains a saturated vapor, tilting the tube to* B' *will not change the pressure. If it contains air, the pressure will change. Why?*

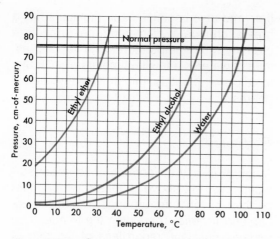

Fig. 16-8. Saturated vapor pressure versus temperature: curves for different liquids.

boiling point at normal pressure of a liquid is the temperature at which its saturated vapor pressure is 76 cm-Hg (Table 16-3).

The influence of pressure on the boiling point explains the operation of the ordinary coffee percolator and the geyser. Near the bottom of the funnel of a percolator, the boiling point is higher than that at the surface of the water. (Why?) Bubbles of steam form there and rise, pushing water ahead of them in the narrow vertical tube. This liquid, rising to regions of lower pressure, is above its boiling point, and it bubbles so violently that a spray is thrown over into the coffee contained in the upper vessel. Geysers are gigantic percolators. The temperature of the water, deep in the earth, is far above the normal boiling point (Fig. 16-10). When this water begins to boil and rises toward the surface, the pressure decreases and the superheated liquid boils so violently that a jet is thrown into the air. After a jet is expelled, the

beans and some other foods. We can determine the boiling points of water at reduced pressures by means of the apparatus sketched in Fig. 16-9. Water is boiled in the flask, the flame is removed, and the stopper is inserted at once. Then cold water is poured over the flask to lower the vapor pressure. The steam above the liquid is condensed, and the water continues to boil vigorously until the temperature falls almost to that of the room. The temperature of the vapor is measured by means of the thermometer, and the pressure by means of the open-tube manometer. The

Table 16-2. Boiling Points and Heats of Vaporization of Liquids at 76 cm-Hg

	Boiling Point, °C	Heat of Vaporization, cal/gm
Alcohol, ethyl	78.5	204
Ammonia	−33.3	327
Ether	34.5	83.9
Helium	−268.8	6.0
Hydrogen	−252.8	108
Mercury	357	70.6
Oxygen	−183	50.9
Sulfuric acid	338	122
Sulfur dioxide	−10.0	95
Water	100.0	540
Water (at 0.46 cm-Hg)	0.01	597

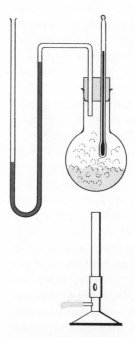

Fig. 16-9. Apparatus to measure the boiling points of water at reduced pressures.

Table 16-3. Saturation Vapor Densities and Pressures of Water

Temperature		Vapor Density, gm/m³	Vapor Pressure, mm-Hg
°C	°F		
−10	14.0	2.16	2.15
0	32.0	4.85	4.58
5	41.0	6.80	6.54
10	50.0	9.41	9.21
11	51.8	10.02	9.84
12	53.6	10.67	10.52
13	55.4	11.35	11.23
14	57.2	12.06	11.99
15	59.0	12.83	12.79
20	68.0	17.30	17.54
40	104.0	51.1	55.3
60	140.0	130.5	149.4
80	176.0	293.8	355.1
95	203.0	505	634
96	204.8	523	658
97	206.6	541	682
98	208.4	560	707
99	210.2	579	733
100	212.0	598	760
101	213.8	618	787
200	392.0	7840	11,659

dry, drinking water is kept in porous vessels so that evaporation cools the contents.

To demonstrate the cooling effect of evaporation, place a small vessel of water under the bell jar of an air pump with a dish containing sulfuric acid to absorb the water vapor. When the pressure is lowered sufficiently, the water boils. The rapid evaporation cools the water until it begins to freeze. While this is occurring, ice, boiling water, and steam are in contact with each other. The pressure and temperature—called the *triple point*—at which the ice, liquid water, and its vapor are in equilibrium are 0.46 cm-Hg and 0.0100°C, respectively.

Since evaporation is a cooling process, heat must be supplied in order to keep the temperature of a liquid constant when it vaporizes. The heat required is proportional to the mass of liquid which is evaporated. *The heat of vaporization of a liquid L_v is the heat per unit mass required to vaporize the liq-*

pressure and temperature at the bottom of the column fall nearly to the normal values. Then the water accumulates again, and a considerable period is required to heat it sufficiently to boil at the enhanced pressure caused by the column that fills the tube. "Old Faithful," a geyser in Yellowstone Park, merits its name because it erupts regularly at intervals of about 65 minutes.

Evaporation and Heat

When a liquid evaporates, its speedier molecules are more likely to escape than the slower ones. These escaping molecules carry away more than their average share of kinetic energy. Hence evaporation cools the liquid. On a hot day, one fans himself in order to increase evaporation from the skin. Water vaporizing from the leaves of trees cools the surrounding air. In regions where the air is

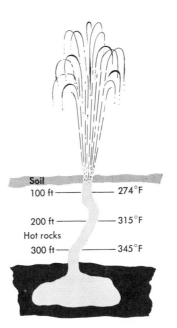

Fig. 16-10. A geyser. The water, deep underground, gets heated far above its normal boiling point.

uid *if the temperature is constant.* Heats of vaporization are expressed in calories per gram and depend upon the pressure (see Table 16-3).

$$L_v = \frac{\text{Heat absorbed}}{\text{Mass of substance vaporized}} = \frac{Q}{m}$$

Figure 16-11 shows the heat required to convert 1 gm of ice at −20°C into steam at 120°C. Notice that the heat of vaporization of the water depends upon its pressure and therefore its boiling point. At lower temperatures the molecules have less kinetic energy; hence more heat energy must be added to make them evaporate. For a similar reason, the surface tension of a liquid increases as its temperature is lowered.

Example. One hundred grams of steam is condensed at 100°C in a vessel containing ice and water at 0°C. How much ice will be melted if the final temperature is 0°C? To condense the steam:

$$Q_1 = mL_v = 100 \text{ gm} \times 540 \text{ cal/gm} = 54,000 \text{ cal}$$

To cool the condensed steam to 0°C:

$$\begin{aligned} Q_2 &= mc \; \Delta t \\ &= 1 \text{ cal/gm-C}° \times 100 \text{ gm} \times 100 \text{ C}° \\ &= 10,000 \text{ cal} \end{aligned}$$

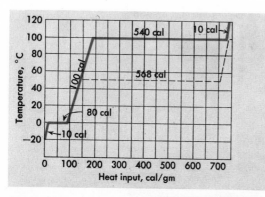

Fig. 16-11. *Heat input per gram to raise the temperature of ice 20 C°, melt it, heat the water to the boiling point, evaporate it, and to "superheat" the vapor.*

$$Q_1 + Q_2 = 64,000 \text{ cal} = m_i \times 80 \text{ cal/gm}$$

$$m_i = 800 \text{ gm, mass of ice melted}$$

Vaporization and Condensation in Weather Processes

One of the quantities used to describe the weather is dampness or *relative humidity.* By this expression we mean the ratio of the water vapor in a given space to the amount that the space would contain if it were saturated at that temperature (Table 16-2). That is

Relative humidity

$$= \frac{\text{Density of water vapor}}{\text{Saturation density of water vapor}}$$

For example, if the actual vapor density is 10 gm/m³ and the saturation vapor density is 20 gm/m³ at that temperature, the relative humidity is 50%.

The temperature at which the water vapor in a space is saturated is called the **dew point.** Put a little warm water in a polished metal cup and add ice until a film of dew first appears on the surface of the vessel. The temperature of the water is the dew point. Dew-point observations are used in determining relative humidities. The method will be understood from the following example.

Example. What is the relative humidity out-of-doors if the temperature is 20°C and the dew point is 15°C?

From Table 16-3, the saturation vapor density of water at 15°C is found to be 12.83 gm/m³. This is the amount of vapor *actually present* per unit volume. From the table, note how much water vapor would be present if the space were saturated at the actual temperature, 20° C, namely, 17.30 gm/m³. Hence the relative humidity is (12.83 gm/m³)/(17.30 gm/m³) = 74.2%.

In determining relative humidities, weather observers use the wet-and-dry-bulb hygrometer. It consists of two thermometers, the bulb of one being covered by a cloth moistened with water. If the space were saturated, water would not evaporate from the cloth, and the two thermometers would indicate the same temperature. When the air is unsaturated, water evaporates from the bulb and the wet bulb is cooled. The drier the air, the greater is the rate of evaporation, and hence the greater the temperature difference of the two thermometers. A chart provided with the instrument indicates the relative humidity corresponding to any pair of temperature readings.

Since discomfort in hot weather is related to both temperature and humidity, some weather bureaus issue a temperature-humidity index reading. This is computed by adding the dry-bulb and wet-bulb thermometer readings on the Fahrenheit scale, multiplying the sum by 0.4 and adding 15. For example, when the temperature is 82°F and the relative humidity is 90%, the temperature-humidity index is 80. Half the populace is said to be uncomfortable when the index reaches 75.

Dew forms when moisture-laden air comes into contact with objects which are at temperatures below the dew point. When this temperature is below freezing, the vapor is deposited as frost.

Fog and clouds form when the atmosphere is cooled below the dew point. This cooling may be caused by the mixture of colder air with moist, warm air. A second method of cooling is by expansion. As air rises, its temperature falls about 1 C° for each 185-ft increase in elevation. Eventually, at a certain level, the air gets cooled to the dew point and becomes saturated, so that fog or clouds form. On calm, humid summer days we see small puffy "wool pack" or cumulus clouds. Each cloud is formed by the condensation of water vapor from a vertical air stream. Soaring birds rise on these currents. Pilots flying gliders travel scores of miles, soaring upward in the successive "air fountains," and down

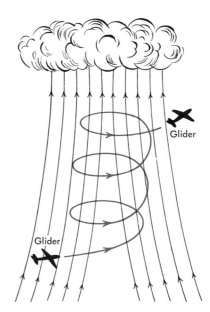

Fig. 16-12. Cumulus clouds. Rising air, cooled by expansion, becomes saturated at a certain elevation.

again from each column to the next (Fig. 16-12). Traveling by airplane, you feel a bump whenever the plane strikes updraft air. You notice a "hole in the air" when the plane leaves the rising stream.

A Vapor-Liquid Equation

In the last chapter we discussed the kinetic theory of an ideal gas. We saw that the gas equation $pV = nRT$ would apply to an *ideal* gas at any pressure, volume, and temperature, but that it is not valid for real gases at very high pressures and low temperatures, or for liquids. Modifications must be made to the equation of state of an ideal gas if it is to apply to real gases.

First, the molecules of a real gas are not mere points. They have finite volumes. Thus the actual space in which the molecules can move is less than the space of the container. We can improve the gas equation by writing $(V - b)$ instead of V, in which b is an experi-

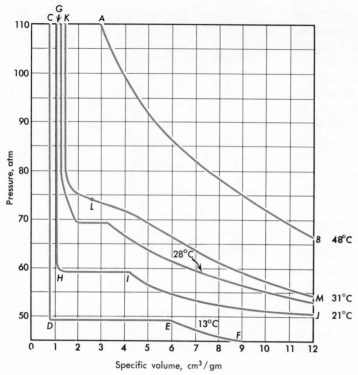

Fig. 16-13. Isothermal curves for carbon dioxide.

mentally (empirically) determined constant for a given gas, accounting for the volume of the molecules themselves. In carbon dioxide at normal pressure, b is about $\frac{1}{50}$ of V. This means that in CO_2 at 50 atm pressure, b is about equal to V. At that pressure, the molecules of CO_2 are packed together, like oranges rather loosely packed in a crate. They cannot dart about freely.

The second reason why real gases do not obey the ideal gas law is that the molecules attract one another and therefore tend to cohere in clusters. This effect is very great in a highly compressed gas. The correction factor is a/V^2, where a is an empirically determined constant for a given gas. In CO_2, under normal conditions, a/V^2 is about 7 thousandths of an atmosphere. We can take account of the decreased volume and the clustering effect by using an equation developed

by van der Waals, one of several different equations proposed by different investigators as equations of state of a real gas:

$$(p + a/V^2)(V - b) = nRT$$

The van der Waals equation fits the experimental data for a real gas in the vapor region and even to some extent in the liquid region. Neither it nor any other proposed equation of state provides complete agreement with experiment along isotherms similar to those in Fig. 16-13.

In Fig. 16-13, *BA* represents values of the pressure of a mole of CO_2 at 48°C as the volume per unit mass changes. The molecules have kinetic energies so great along *BA* that they do not form clusters. The curve resembles that for an ideal gas. *FEDC* shows the pressures at 13°C. As the vapor is compressed from *F* toward *E*, it begins to liquefy at **E** and is totally liquefied at **D**. Then the molecules are packed so tightly that further increase of pressure compresses them but little. Along *JIHG*, at a higher temperature, the vapor begins to liquefy at **I**. The curve *MLK* is most interesting. The vapor liquefies at the *critical temperature* of 31.1°C and at 73 atm pressure, but fails to liquefy at temperatures greater than this. At higher temperatures, the gas cannot be liquefied by any pressure, however great.

Liquefaction of Gases

The vapors of many substances such as water, alcohol, and gasoline can be liquefied at ordinary temperatures by compressing them. The so-called "permanent" gases, such as oxygen, nitrogen, hydrogen, and helium, must first be cooled below their *critical temperatures,* which are far below room temperature. *The critical temperature of a substance is that temperature above which it cannot be liquefied by any pressure, however great* (Table 16-4).

A thick-walled glass tube half filled with liquid carbon dioxide can be used to determine the critical temperature of that substance (Fig. 16-14).* If the tube is immersed in a beaker of water and warmed slowly, as the temperature rises, the vapor becomes more dense, the liquid less dense, and the film of the surface between the two less sharply defined. The film fades away entirely at 31.1°C because the vapor and the liquid are equally dense. At higher temperatures, the entire space is filled with vapor (or gas), and the carbon dioxide cannot be liquefied by any pressure, however great. As the tube is

* In order to avoid injury if the tube explodes, the observers should be protected—by a plate-glass shield, for example. In a safer experiment developed by Professor Harald Jensen of Lake Forest College, liquid propane (critical temperature 96.8°C and critical pressure 42 atm) in a tube immersed in a beaker of water is warmed through the critical point.

Table 16-4. Normal Boiling Points, Critical Temperatures, and Critical Pressures

Substance	Boiling Points at Normal Pressure, °C	Critical Temperatures, °C	Critical Pressures, atm
Air	−194	−141	37
Ammonia	−33.3	132	111
Carbon dioxide	−80	31.1	73
Ether	34.5	194	35.5
Helium	−268.8	−268	2.26
Hydrogen	−252.8	−240	12.8
Oxygen	−183	−119	50
Water	100.0	374	218

cooled again below its critical temperature, the film between liquid and vapor reappears.

Does a tank of carbon dioxide contain a liquid? The answer depends upon the conditions. At any temperature above 31.1°C, the tank could contain vapor only. At temperatures between this value and the freezing point, −56.6°C, both liquid and vapor will be

Fig. 16-14. Observation of critical temperature. Raising the temperature decreases the density of the liquid and increases that of the vapor. At the critical temperature, the two become equal and the film vanishes.

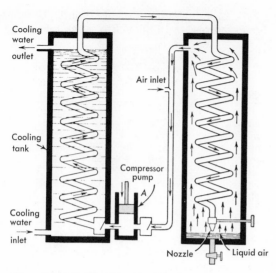

Fig. 16-15. *Making liquid air.*

present, separated by a surface film. Below the freezing point, the tank will contain only solid CO_2 and its vapor.

In order to liquefy air, it must be cooled below its critical temperature ($-141°C$). In one often-used method, the pump A (Fig. 16-15) compresses air to a pressure of 200 atm. Heat, resulting from the compression, is absorbed in the cooling tank. Then the air enters the second coil and expands as it escapes through the nozzle. In the highly compressed gas, the neighboring molecules attract each other. When the gas expands, work must be done in separating them. Thus part of their kinetic energy is used in the separation, and the gas is slightly cooled (Joule-Thomson effect). After escaping from the nozzle, the chilled gas travels upward in the heat-insulated chamber and cools the compressed gas traveling downward through the coiled pipe. As the process continues, the escaping gas becomes progressively colder until, finally, part of it liquefies and may be drawn off like water.

Hydrogen is liquefied by a similar process. However, at ordinary temperatures the molecules do not attract each other. The hydrogen

gas must first be cooled by passing it through coils immersed in liquid air so as to chill it sufficiently that the molecules may attract one another strongly. Also, the hydrogen is allowed to do work in a small expansion engine, similar to a steam engine, at the expense of its store of kinetic energy, thus further undergoing a decrease of the temperature toward the critical point.

p-V-T Surfaces for Real Substances

In Chapter 15, we considered the *p-V-T* surface for an ideal gas, a surface that passed through all points representing allowed values of the pressure, volume, and temperature of an ideal gas. Planes passed through the surface parallel to the plane containing the *p* and *V* axes, for example, cut the surface in isothermal curves. Constant-pressure and constant-volume processes could be similarly represented on the surface. The entire *p-V-T* surface was a three-dimensional graph that summed up all of the available information about the possible states of an ideal gas.

Figure 16-16 shows the *p-V-T* surface summarizing the experimental data for a real substance, such as carbon dioxide, CO_2. It is more complicated than that for a perfect gas because real substances, as we have seen, undergo changes of phase. Notice first the regions for the solid (solid shading), liquid (cross-barred), vapor (dotted), and gas (dotted). A set of values of *p*, *V*, and *T* corresponding to a point lying within one of these regions represents a stable state of the substance in that phase. For example, if the pressure, volume, and temperature of CO_2 in a cylinder (fitted with a movable piston and provided with means of changing the temperature of the gas) are adjusted experimentally to correspond to a point lying within the region *degf*, the CO_2 will remain in the vapor phase. If the *p-V-T* point lies within *abih*, the CO_2 will be a solid, and if within *jkfc*, a liquid.

Next notice the regions separating the

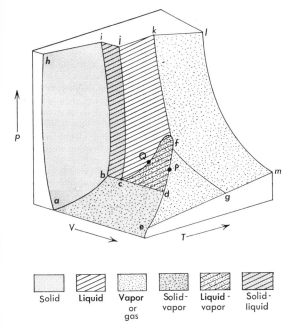

Fig. 16-16. A p-V-T *surface for carbon dioxide.*

solid, liquid, and vapor or gas phases. These are transition regions for change of phase. Suppose that the CO_2 is at the point P on the line df separating the vapor from the liquid-vapor region dcf. If the pressure is increased without change of temperature, the CO_2 will condense; it will gradually liquefy with decrease in volume until its state reaches a point Q on the curve cf. At Q, the CO_2 is entirely liquid. Further increase of pressure will produce only a slight change of volume because liquids are highly incompressible. Similarly, the region $bcji$ is the transition region for melting of a solid or freezing of a liquid, and the region $abde$ for the sublimation of a solid or the transition from a vapor to a solid. The line cd is the triple line: along this line solid, vapor, and liquid are in equilibrium with each other. Note further that curve gfk is the critical isotherm and point f the critical point. At temperatures above the temperature at the critical point, the substance cannot be lique-

fied, regardless of what changes in pressure and volume one makes.

The gas, vapor, liquid-vapor, and solid-vapor regions have comparatively gentle slopes, corresponding to the ease with which gases and vapors can be compressed. The solid, solid-liquid, and liquid regions are "precipices": huge changes of pressure are required in these regions to produce small changes in volume. Note that crossing the solid-liquid region from the solid to the liquid involves an increase in volume in CO_2 corresponding to expansion of the substance. Water contracts when it melts, and on its *p-V-T* surface the solid region would jut out beyond the liquid region.

Note finally that if planes parallel to the plane containing the p and V axes (constant-temperature planes) are passed through the surface between d and f, isothermal curves similar to the three lowest curves in Fig. 16-13 are traced. A constant-temperature plane through point f traces the critical isotherm of Fig. 16-13, and constant-temperature planes at higher temperatures trace isotherms, such as *AB* in Fig. 16-13, which resembles the isotherms of a perfect gas. Thus, much information can be represented by a *p-V-T* surface.

Summary

Heat is evolved when a liquid freezes; when the liquid evaporates, heat is absorbed if the temperature remains constant. Dissolved substances lower the freezing point of any liquid. Water, bismuth, cast iron, and a few other substances expand on solidification; most substances contract. The heat of fusion of a substance is the heat required per unit mass to melt the substance without change of temperature. It is expressed in calories per gram.

Sublimation is a change of phase from the solid to the vapor.

The more energetic molecules are more

likely to escape when a liquid evaporates; hence vaporization cools a liquid. The heat of vaporization of a liquid is the heat required per unit mass to change it from a liquid to a vapor, without change of temperature. A vapor in contact with its liquid is saturated if the rates of evaporation and of condensation are equal. The density and the vapor pressure of a saturated vapor increase with temperature and are independent of its volume. Dissolved substances decrease the vapor pressure of a liquid.

Vapor bubbles are produced inside a liquid when it boils. At the boiling point, the vapor pressure equals the pressure to which the liquid is subjected. The boiling point of a liquid depends on the applied pressure.

The relative humidity of a space is the ratio of the actual water-vapor density to the density if the space were saturated.

A liquid and its saturated vapor have the same density at the critical temperature, and the surface film between them disappears. A gas cannot be liquefied when above its critical temperature.

Questions

1. Explain why the addition of salt (sodium chloride) to water lowers the freezing point.

2. Discuss the influence of pressure on the melting point of a substance.

3. How can the heats of fusion and of vaporization of water be determined experimentally?

4. Why does evaporation cool a liquid?

5. How is the phenomenon of regelation involved in (a) ice skating, (b) making a snowball?

6. Discuss the experimental evidence that the pressure and the density of a saturated vapor are independent of its volume.

7. What becomes of the heat energy utilized in melting ice, and what becomes of the heat liberated when water freezes?

8. How would the presence of air in the space above a liquid affect (a) the saturated vapor pressure of the liquid, and (b) its rate of vaporization?

9. Carbon dioxide freezes when it escapes from a tank through a cloth bag, forming "snow." Discuss in terms of a p-V-T surface for CO_2.

10. Using the concept of critical temperature, distinguish between a vapor and a gas.

11. On a winter day, the dew points indoors and outdoors were the same, yet the relative humidities were different. Explain.

12. Why is it difficult to make snowballs in very cold weather?

13. Why cannot water be heated above $100°C$ at normal pressure?

14. Steam enters a radiator at $100°C$ and water leaves it at the same temperature. Is the room heated thereby? Explain.

15. A vessel of pure water and one of salt water are placed in a closed container. In which vessel does the liquid level decrease? Why?

16. Would water boil at $100°C$ on Mars? At $50°C$ on earth? At more than $100°C$ on earth?

17. Why is a steam burn more harmful than a burn caused by hot water? Give numerical examples to support your answer.

18. Does an electric fan cool the air in a room? Why does it make you feel more comfortable?

19. If a glassful of water has ice cubes floating in it, what will happen to the water level when the ice melts?

20. Why is it better to give a person with a high fever a sponge bath with alcohol and water rather than just with water?

21. Why do the windows of some houses have frost on the inside in winter, while others do not?

22. Why is the dimension of time absent from some of the heat equations?

23. Make a list of the unusual properties of water that you have encountered so far. Show how each one can affect you.

24. On an extremely precise balance a kil-

ogram of ice is compared with a kilogram of water. Which has the greater mass?

25. Would you expect the heat of vaporization of water to be greater at 20°C or at 100°C? Why?

26. If there is a little water at the top of the mercury column in a barometer tube, will the barometer read too high or too low? How will the reading be affected by temperature change?

27. It has been suggested that liquid ammonia might serve as a substitute for water as the basis of life elsewhere in the universe. Some of the properties of liquid ammonia are these: its boiling and freezing points at 1 atmosphere pressure are −33°C and −78°C, respectively; solid ammonia sinks in liquid ammonia; a process similar to photosynthesis involving ammonia would yield nitrogen; and nitrogen does not support combustion-like processes of respiration. Discuss the suitability of liquid ammonia as a basis for life processes.

28. Consider question 19 again. What effect would each of the following, not considered there, have on the final level of the liquid: (a) the ice cubes contained some grains of sand; (b) the ice cubes contained some air bubbles; (c) the water (and the glass) were not at 0° to start with, but were at room temperature; (d) the liquid was not water at all and had a density less than that of water.

Problems

A

1. At what rate must heat be supplied to melt 2.5 kg of ice per hour at 0°C? The final temperature of the water is 10°C.

2. How much heat is required to change 40 gm of ice at −30°C into steam at 140°C? The pressure is 76 cm-Hg. (Specific heat of steam is 0.48 cal/gm-C°; specific heat of ice is 0.51 cal/gm-C°.)

3. What mass of water at 80°C must be added to 300 gm of ice at 0°C so that all the ice may be melted?

4. A copper ball of mass 200 gm and specific heat 0.092 cal/gm-C° at 170°C is placed in a cavity in a block of ice at 0°C. How much ice will be melted?

5. If the dew point is 10°C, what is the relative humidity (a) in a room where the temperature is 20°C, (b) outdoors where it is 15°C?

6. Suppose that outdoors on a summer day the temperature is 30°C and the dew point is 10°C. If the air as it rises cools 0.6 C° for each 100 m, at what elevation will clouds begin to form?

7. (a) When the outdoor air temperature is 10°C and the relative humidity is 100%, what is the vapor density? (b) What would the relative humidity be if this air, entering a room, were heated to 20°C?

B

8. When 3500 cm³ of air from a room are drawn through a drying tube containing calcium chloride, the mass of the tube increases from 54.2350 gm to 54.2632 gm. What are (a) the vapor density, (b) the dew point, and (c) the relative humidity if the temperature is 15.0°C; and (d) the relative humidity if the temperature is 10.0°C?

9. Two hundred grams of water are undercooled to −10°C. When a chip of ice is dropped into the water, ice is suddenly formed, and the temperature rises to 0°C. How much ice is produced? Specific heat of ice is 0.51 cal/gm-C°.

10. What temperature results if 150 gm of ice at 0°C are placed in 100 gm of water at 50°C?

11. One hundred grams of ice are put into 400 gm of water at 20°C. What is the final temperature of the mixture?

12. Thirty grams of steam at 100°C are condensed in 360 gm of water at 20°C which is contained in a copper calorimeter of mass 200 gm and specific heat 0.090 cal/gm-C°. What is the final temperature?

13. How many grams of water at 20°C can be heated to 100°C by condensing 15.0 gm of steam at 100°C?

14. How much ammonia must be evaporated in a refrigerator to absorb sufficient heat to freeze 1700 gm of water at 0°C? (Neglect the heat required to warm the vapor.)

15. If 20 gm of steam at 200°C are mixed with 80 gm of ice at −40°C and condense at 100°C, what is the resulting temperature of the mixture?

16. How much heat is required to change 40.0 gm of tin at 132°C into liquid tin at its melting point, 232°C? (Specific heat of tin is 0.054 cal/gm-C°; heat of fusion, 13 cal/gm.)

C

17. A copper calorimeter of mass 220 gm and specific heat 0.10 cal/gm-C° contains 398 gm of water at 45°C. If 200 gm of ice are melted in it, the resulting temperature of the mixture is 5.0°C. Compute the heat of fusion of ice.

18. Five hundred grams of water at 20°C, 50 gm of ice at 0°C, and 20 gm of steam at 100°C are mixed in a 300-gm copper calorimeter which was initially at 20°C. Compute the resulting temperature. (Specific heat of copper is 0.090 cal/gm-C°.)

19. One thousand grams of mercury (density 13.6 gwt/cm³) at 100°C are poured into a 100-gm soda-glass beaker containing 100 gm of water and 20 gm of ice, all at 0°C. Compute the final temperature.

20. The cap of a jar is screwed on when the jar is at 100°C. The jar contains steam and a few grams of boiling water. What is the pressure when the jar cools to 40°C?

21. The cap of a jar containing a few grams of ethyl alcohol, boiling at 78°C, is screwed on tightly. What is the pressure in the jar when it cools to 19°C?

22. Each day 28,000 gm (7 gal) of water are evaporated at 100°C to humidify a building. Find the cost per 30 days if natural gas is used costing $15.00 per billion calories. (As-

sume that the initial temperature of the water is 20°C and that the efficiency is 50%.)

23. Air is trapped above a mercury column in an arrangement similar to that in Fig. 16-7. The tube is 100 cm long, the barometer reads 76 cm, and the mercury column is 36 cm high. What will be the pressure in the tube if the tube is tilted to an angle of 30° with the horizontal?

24. What is the equivalent of 540 calories in mass? What per cent would it be of one gram?

25. If the heat of vaporization of air is 52 cal/gm and its mean specific heat at constant pressure is 0.24 cal/gm-C°, compare liquid air with ice as a refrigerant if the material to be cooled is brought to 0°C. Liquid air boils at −182°C.

26. On a certain day the dew point is 8°C and the relative humidity is 35%. The saturated vapor pressure of water at 8°C is 0.80 cm-Hg. If the air were saturated, what would the pressure of the water vapor be?

27. When the relative humidity is 80%, the temperature 25°C, and the atmospheric pressure 732.6 mm-Hg, what is the pressure due to dry air?

28. Taking into account vapor pressure, calculate the maximum height of a wall over which water can be siphoned at (a) 100°C, (b) 75°C. The barometric pressure is 76 cm-Hg.

29. If, in the van der Waals equation, p is in newtons/m² and V is in m³, what are the proper units for a and b?

30. Plot p and $p \times V$ against V for carbon dioxide at 0°C by using the van der Waals equation for the following values of V, in liters: (a) 0.02, (b) 0.01, (c) 0.008, (d) 0.007, (e) 0.006, (f) 0.005, (g) 0.004, (h) 0.0033, (i) 0.003. For carbon dioxide, $a = 0.007$ atm-liter², $b = 0.002$ liter.

31. Air is removed from a cylinder fitted with a piston, and water vapor (steam) allowed to enter until the pressure is 10 cm-Hg, the volume is 50 cm³, and the temperature is 60°C. If the piston is pushed down slowly,

keeping the temperature at 60°C until the volume is 25 cm³, what is the final pressure?

32. Estimate whether the addition of one ice cube can cool a cup of boiling coffee to drinking temperature.

For Further Reading

Battan, Louis J., *Harvesting the Clouds,* Doubleday and Company, Garden City, 1969.

Battan, Louis J., *The Nature of Violent Storms,* Doubleday and Company, Garden City, 1961.

Chalmers, Bruce, "How Water Freezes," *Scientific American,* February, 1959, p. 114.

Davis, Kenneth S. and John Arthur Day, *Water, the Mirror of Science,* Doubleday and Company, Garden City, 1961.

Edinger, James G., *Watching for the Wind,* Doubleday and Company, Garden City, 1967.

Hall, H. Tracy, "Ultrahigh Pressures," *Scientific American,* November, 1959, p. 61.

MacDonald, D. K. C., *Near Zero, The Physics of Low Temperature,* Doubleday and Company, Garden City, 1961.

Reiter, Elmar R., *Jet Streams,* Doubleday and Company, Garden City, 1967.

Westwater, J. W., "The Boiling of Liquids," *Scientific American,* June, 1954, p. 64.

CHAPTER SEVENTEEN

Heat, Work, and Engines

In the preceding chapters, we discussed heat as "disorderly" energy that is transferred from one body to another because of a temperature difference between them. We used a special unit of heat—the calorie. Now we shall consider the relation between heat and work and the equivalence of the special units of heat and the units of mechanical energy. Two important laws will appear in this chapter: one that states the conservation of energy in thermal processes and one that states the terms laid down by nature when thermal energy is used.

Heat and Work

Before 1845 many scientists believed that heat was a substance called *caloric* that could be squeezed out of matter by friction, like water from a sponge. Other scientists favored the view that heat is a kind of energy. Nearly half a century earlier, Benjamin Thompson,* who later became Count Rumford, performed an experiment that should have settled the question. While superintendent of an arsenal at Munich, Germany, he noticed that a great deal of heat was produced while a cannon was being bored. He

wished to see if the supply of "caloric" in the iron could be exhausted by long-continued boring. He purposely blunted the tool so that it would not cut the metal and found that heat flowed from the iron hour after hour, as long as the drill was turned. Rumford reasoned that, since heat was evolved from the iron almost without limit, it could not be a form of matter. Further, he argued, to get heat from the iron, one had to put in energy, so that heat is a form of energy. He concluded: "It is hardly necessary to add that anything which an isolated body or system of bodies can continue to furnish without limitation cannot possibly be a material substance."

Though Rumford's work undermined the foundations of the caloric theory, the theory was not abandoned until about 45 years later, when the British physicist Joule (jool) determined with considerable precision the work

---

* Benjamin Thompson (1753–1814) was born in Woburn, Massachusetts. As a loyalist in the American Revolution, he left his home and went to England. There he helped to found the Royal Institution. Later, in Germany, he organized poor relief, got the beggars off the streets, and studied dietetics to find how to feed them cheaply. Thompson did outstanding work in research on heat. He and Benjamin Franklin were among the few colonial scientists whose abilities were recognized in Europe.

required to produce the same change of temperature in a system as that produced by a unit quantity of heat. His apparatus was a churn, the paddles of which were driven by descending weights (Fig. 17-1). The stirring of the water produced an increase in temperature of the water. By noting the rise in temperature of a known mass of water (taking into account the heat absorbed by the churn), Joule calculated the amount of heat Q that would have to be added from outside the apparatus to produce the *same* change of temperature. The work done in producing the increase in temperature was the product of the weight w of the descending body and the total distance h through which it moved. The ratio of the work done W to the heat Q that would be required to produce the same temperature change in a system is J, the *mechanical equivalent of heat.*

$$J = \frac{W}{Q}$$

James Joule, the son of a prosperous brewer, first measured the mechanical equivalent of heat in 1840 when he was 22 years old. He spent most of his life making better determinations. Joule's final value for the mechanical equivalent of heat J was within 1% of the value accepted today (Table 17-1), which is

$$J = 4.1858 \text{ joules/cal}$$
$$\cong 42,700 \text{ gwt-cm/cal}$$

Joule's experiments and many others have shown that heat is energy in transit.

Table 17-1. Determinations of the Mechanical Equivalent of Heat[a]

Year	Investigator	Method	Value, joules/cal	% Deviation
1879	Rowland	Churn: water	4.189	+0.1
1899	Barnes and Callendar	Electric heating: continuous flow	4.182	−0.1
1921	Jaeger and Steinwehr	Electric heating	4.1850	−0.02
1939	Osborne, Stevenson, and Ginnings	Electric heating	4.1858	0

[a] See M. W. Zemansky, *Heat and Thermodynamics*, McGraw-Hill Book Company, New York, 1943.

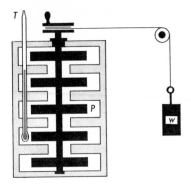

Fig. 17-1. *Joule's churn. The descending weight did work in stirring the water and thus generated heat. One calorie of heat is produced by doing 4.19 joules of work.*

Example 1. The falls at Niagara are 50 m high. How much is each kilogram of water heated when it strikes at the bottom of the falls?

Energy transformed
$$= mgh$$
$$= (1 \text{ kg})(9.8 \text{ m/sec}^2)(50 \text{ m})$$
$$= 490 \text{ joules}/(4.19 \text{ joules/cal})$$
$$= 117 \text{ cal}$$

Rise of temperature $= 0.12 \text{ C}°$

Example 2. A 4.90-gm lead rifle bullet has a velocity of 10^5 cm/sec (1 km/sec). How many degrees does the temperature of the bullet rise when it strikes a target if one-twentieth of the energy remains in the bullet?

$$E_K = \tfrac{1}{2}mv^2 = \tfrac{1}{2}(4.90 \text{ gm})(10^5 \text{ cm/sec})^2$$
$$= 2.45 \times 10^{10} \text{ dyne-cm or ergs}$$
$$= \frac{2.45 \times 10^{10} \text{ ergs}}{4.19 \times 10^7 \text{ ergs/cal}} = 585 \text{ cal}$$

Energy remaining in bullet $= 29.2$ cal

Let $\Delta t =$ rise of temperature.

$$29.2 \text{ cal} = 0.030 \text{ cal/gm-C}° \times 4.90 \text{ gm} \times \Delta t$$

$$\Delta t = 200 \text{ C}°$$

The First Law of Thermodynamics

Thermodynamics is the branch of physics that describes thermal properties in terms of such quantities as pressure, volume, and temperature. These quantities, as we have seen, depend upon the internal state of the system. However, thermodynamics does not attempt a molecular description of thermal behavior, leaving that to the branch of physics called *statistical mechanics.*

The first law of thermodynamics is essentially an application of the law of conservation of energy to situations in which heat is transferred. The first law relates changes in the internal energy of a system to the heat transferred and to work done. Let us consider the role of each of these.

The internal energy of a system is, in molecular terms, the sum of the kinetic energies of the molecules in their random disorderly motions and the potential energies due to the mutual attractions of the molecules. (Previously we have referred to internal energy as thermal energy.) In thermodynamics, however, one simply accepts the internal energy as a way of describing the system and does not attempt to relate it to such quantities as the molecular speeds. The internal energy depends upon the state of the system. For the kind of system that we shall discuss, this means that the internal energy depends upon the pressure, volume, and temperature. *Changes* in the internal energy in a thermal process are of greatest interest to us. They can be produced by (1) the transfer of heat to the system from another system at a different temperature, or (2) the performance of work by the surroundings on the system or by the system on its surroundings, or (3) by both heat transfer and the performance of work. Take a simple example: The internal energy (the kinetic energy of random molecular motion) of a block of iron can be increased by doing work on it: one could squeeze the iron in a heat-tight press, causing a rise of temperature of the iron. Or, one could transfer heat to the iron from another body at a higher temperature by leaving the iron in the sunlight for an hour, again causing a rise in its temperature.

You recall that in mechanics we compared energy to the balance in a bank account and the performance of work to deposits or withdrawals. Thermal processes follow a similar bookkeeping except that there can be two kinds of deposits or withdrawals against the balance of internal energy: (1) heat transfers to or from another system at a different temperature and (2) the performance of mechanical work by or on the system being considered. That there is such a quantity as internal energy and that nature keeps accurate books on what happens to it is expressed in the *first law of thermodynamics.* If heat Q is transferred *to* a system and the system does mechanical work W *on* its surroundings, the change in the internal energy U stored in the system is $U_2 - U_1$; and the first law of thermodynamics states that:

$$Q = (U_2 - U_1) + W$$

Let us apply the first law to two processes that occur at constant pressure. First consider the heat of vaporization of water at 100°C and normal pressure. We have seen that the heat of vaporization is 540 cal/gm. How much of the 540 cal added to 1 gm of the water in vaporizing it is retained as internal energy? To answer this question, we must compute the work that the water does in expanding from the liquid to the vapor. Suppose that 1 gm of water is confined in a cylinder fitted with a frictionless piston of area A that can expand against a constant pressure of 1 atm. If the piston is pushed back a distance Δl when the water is heated and completely vaporized, the work W done by the system *on* its surroundings (the atmosphere) is

$$W = F \cdot s = (pA)(\Delta l) = p(A \, \Delta l) = p \, \Delta V$$

where ΔV is the change in volume of the water. As a liquid, 1 gm of water occupies approximately 1 cm³; as a vapor at normal pressure, 1672 cm³. A pressure of 1 atm is 76 cm-Hg or 1.01×10^6 dynes/cm².

$$W = p\,\Delta V$$

$$= 1.01 \times 10^6 \frac{\text{dynes}}{\text{cm}^2}\,(1672\ \text{cm}^3 - 1\ \text{cm}^3)$$

$$= 168.7 \times 10^7\ \text{dyne-cm or ergs}$$

$$= 168.7\ \text{joules}$$

$$= 168.7\ \text{joules} \times 1\ \text{cal}/4.19\ \text{joules}$$

$$\cong 40\ \text{cal, work done } \textit{by} \text{ the water in expanding}$$

By the first law of thermodynamics

$$Q = (U_2 - U_1) + W$$
$$540\ \text{cal} = (U_2 - U_1) + 40\ \text{cal}$$
$$U_2 - U_1 = 500\ \text{cal}$$

Thus, of the 540 cal of heat transferred to 1 gm of water to vaporize it at the normal boiling point, 40 cal of work are done in pushing back the atmosphere in expanding, and the remaining 500 cal are stored as increased internal energy of the water.

Next let us use the first law of thermodynamics in the calculation of the molecular heat of a gas at constant pressure. Suppose that we confine 1 mole (32 gm) of oxygen in a cylinder having a piston (Fig. 17-2). Let the initial pressure be 1 atm, the volume 22,400 cm³, and the temperature 0°C or 273°K. First, lock the piston so that the volume of the gas will remain constant. Raise the temperature 1 K°. In Chapter 15, we saw that the heat input to the gas is 5 cal so that the molecular heat of the oxygen at constant volume is 5 cal/mole-K°. The gas does no work because its volume is constant. Hence, all of the heat transferred is stored as internal energy. Now release the piston. The compressed gas will expand and will do work in pushing up the piston. The gas will do work at the expense of its internal energy and be cooled. It turns out that we must add 2 cal to the mole of gas in order to restore the internal energy used in doing work. Thus, while the molecular heat of oxygen at *constant volume* is 5 cal/mole-K°, that at *constant pressure* is 7 cal/mole-K°. To

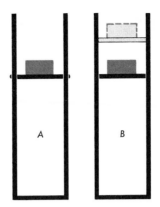

Fig. 17-2. *The gas in A, at constant volume, has a smaller specific heat than the gas in B.*

understand this, let us first write the first law of thermodynamics

$$Q = \Delta U + W$$

and divide each term by the number of moles n of the gas and the temperature change ΔT

$$\frac{Q}{n\,\Delta T} = \frac{\Delta U}{n\,\Delta T} + \frac{W}{n\,\Delta T}$$

As we saw in the preceding example, the work done at constant pressure is $p\,\Delta V$. Thus,

$$\frac{Q}{n\,\Delta T} = \frac{\Delta U}{n\,\Delta T} + \frac{p\,\Delta V}{n\,\Delta T}$$

Now $Q/(n\,\Delta T)$ is defined as the molecular heat c_p at constant pressure. $\Delta U/(n\,\Delta T)$ is the molecular heat c_V at constant volume, since heat added at constant volume increases the internal energy and no work is done. Therefore,

$$c_p = c_V + \frac{p\,\Delta V}{n\,\Delta T}$$

The pressure p is 1 atm or 1,010,000 dynes/cm², ΔV is 22,400 cm³/273, and ΔT is 1 K°.

$$\frac{p\,\Delta V}{n\,\Delta T} = \frac{1,010,000\ \text{dynes/cm}^2\,(22,400\ \text{cm}^3)}{(1\ \text{mole})\,273\,(1\ \text{K}°)}$$

$$= 8.32 \times 10^7\ \text{ergs/mole-K}°$$
$$\cong 2\ \text{cal/mole-K}°$$

Hence

$$c_p = 5 \text{ cal/mole-K}° + 2 \text{ cal/mole-K}°$$
$$= 7 \text{ cal/mole-K}°$$

Heat Engines

To convert other forms of energy completely into internal energy requires no special equipment. Friction *always* does this. An electric current also produces a rise of temperature in a conductor. The inverse process—the conversion of thermal energy into other kinds of energy—is wasteful and usually requires specialized equipment.

One of the devices much used in the past to derive work from thermal energy is the *piston-type steam engine.* Although steam locomotives are used only rarely by railroads these days, the piston-type steam engine has an honorable record of service to mankind. Moreover, attempts to understand and improve the operation of the steam engine led to many of the principles of physics that we are discussing. In Fig. 17-3, steam from the boiler at high pressure and temperature—and hence high internal energy—enters the cylinder and pushes the piston to the left. Meanwhile, spent steam at lowered pressure and

temperature on the other side of the piston is forced out through *E* into the exhaust pipe. After the piston has moved to the left sufficiently, the slide valve is automatically shifted so as to make the steam enter the space at the other side of the piston, thus reversing the motion. Note that only a part of the heat that enters the engine is used in doing work; the rest is exhausted with the spent steam.

Modern heat engines using steam employ a rotational motion rather than the reciprocating motion of the piston steam engine. The *turbine* steam engine resembles a water wheel or a windmill. Jets of steam directed against the blades or "buckets" of the turbine exert forces on the blades so as to keep the wheel in motion. Turbines having several sets of blades mounted on the same shaft are efficient in performing work at the cost of the internal energy of the steam (Fig. 17-4). Fixed blades are mounted between adjacent moving wheels (Fig. 17-5). The steam jet strikes against the first set of moving "buckets" and rebounds, giving up some of its energy. Then the jet, reversed by the fixed blades mounted on the turbine casing, strikes the next revolving wheel, and again rebounds with further diminished energy. In this manner, it worms its way through the turbine, gradually gives up energy, and emerges with greatly reduced velocity, energy, and pressure. Because of diminishing pressure, the steam expands as it travels through the turbine, and so the last sets of wheels are larger than the first. The turbine does work in turning the shaft of an electric generator. Notice again that heat flows into the turbine in the high-pressure, high-temperature steam, work is done by the steam, and some heat is exhausted in the spent steam.

The internal-combustion engine differs from the steam engine in that the fuel is burned in the cylinder instead of in the firebox of a boiler. The gasoline automobile engine is the most familiar type of internal-combustion

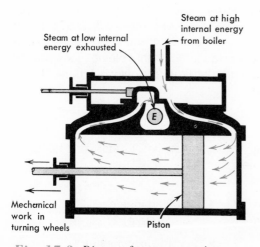

Fig. 17-3. Piston of a steam engine.

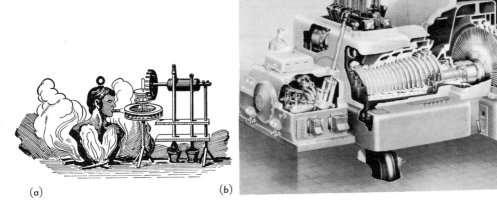

(a)   (b)

Fig. 17-4. (a) *Steam turbine of long ago.* (b) *A modern steam turbine* (General Electric).

engine (Fig. 17-6). The four strokes of its complete cycle are:

1. *The intake stroke.* The piston moves downward (*B*), and a stream of air is forced past the nozzle of the carburetor at *A*, through the intake valve, into the cylinder. Because of the high velocity of the air, its pressure is reduced, and gasoline from the carburetor jet is sprayed into the stream to produce the explosive mixture.

2. *The compression stroke.* The intake valve is closed, and the piston is forced upward, compressing the gas (*C*). During the compression stroke the explosive mixture is heated to a degree depending upon the ratio of the initial and final volumes. In high-compression engines, the reduction in volume is more than eight-fold. Higher compression gives greater efficiency. However, excessive compression makes the mixture burn explosively so that the engine "knocks." This knocking can be prevented by adding tetraethyl lead to the fuel.

3. *The power stroke.* Just before the piston completes its upward stroke, the mixture is ignited by an electric spark. The heat pro-

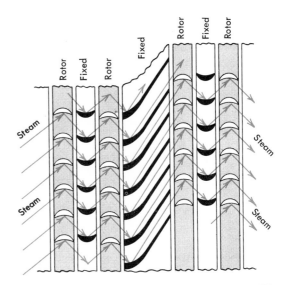

Fig. 17-5. *Rotors and stators of a turbine. The steam strikes a moving "bucket," then a fixed bucket, and so on, gradually losing momentum and energy.*

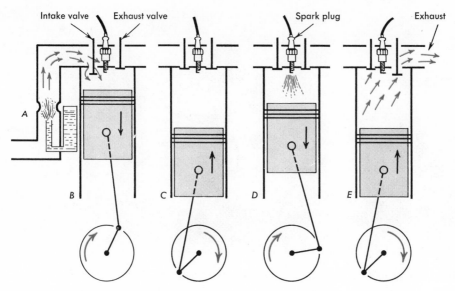

Fig. 17-6. *Four-cycle engine:* A, *carburetor nozzle;* B, *intake stroke;* C, *compression;* D, *power;* E, *exhaust.*

duced by the combustion raises the temperature of the gas and increases its pressure to over 25 atm. Then the piston is driven downward, work is done, and the gas is cooled by the expansion (*D*).

4. *The exhaust stroke.* The exhaust valve opens (*E*), the piston moves upward, and the warm gases are expelled, carrying away heat. The four strokes of the cycle are then repeated.

Power-lawn-mower gasoline engines have only one cylinder; the energy is delivered during only one stroke out of each four, that is, once for each two revolutions of the shaft. A flywheel keeps the system going during the three stages when no energy is delivered. In the automobile engine, several cylinders are mounted side by side. Their pistons are connected to the same shaft so that the power strokes occur more often and the motion is more uniform.

The Diesel engine has the same four stages of operation as the ordinary gasoline engine, but it has neither carburetor nor electric-spark

ignition. At the input stroke, air only, instead of a mixture, is forced into the cylinder. During the next stroke, the air is compressed about sixteenfold. The hot air ignites a jet of oil forced into the cylinder. The Diesel engine is more efficient than the gasoline engine. Other examples of heat engines could be given—the gas turbine, the jet engine, and the rocket engine. Let us return, however, to the discussion of the principles underlying all of these heat engines.

The Efficiency of Heat Engines

The word efficiency has several meanings. For example, the *boiler* efficiency tells us what fraction of the energy of the fuel is delivered as energy of the steam. The *engine* efficiency tells what fraction of the energy supplied to the engine is converted into work. The *overall* efficiency means the ratio of the useful mechanical energy to that of the fuel. The energy losses for a certain automobile traveling at 30 mi/hr on a level road are

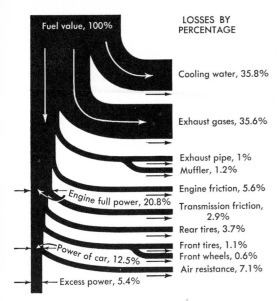

Fuel value, 100%

LOSSES BY PERCENTAGE

Cooling water, 35.8%

Exhaust gases, 35.6%

Exhaust pipe, 1%
Muffler, 1.2%

Engine friction, 5.6%

Engine full power, 20.8%

Transmission friction, 2.9%

Rear tires, 3.7%

Front tires, 1.1%
Front wheels, 0.6%

Power of car, 12.5%

Air resistance, 7.1%

Excess power, 5.4%

Fig. 17-7. Energy losses of an automobile engine.

shown in Fig. 17-7. Notice that most of the energy is wasted in the cooling system and in the exhaust gases. Less than 5% is used in overcoming road friction.

At the moment, we are interested in the thermal efficiency η (Greek *eta*) of the heat engine. By definition, η is the ratio of the work W done by the engine during one cycle to Q_1, the heat supplied to it in that cycle. Since the work done is the heat supplied Q_1 minus the heat Q_2 exhausted to the sink with the spent gases, we have (Fig. 17-8)

$$\eta = \frac{W}{Q_1} = \frac{Q_1 - Q_2}{Q_1}$$

If, for example, 1000 cal of heat are supplied to a heat engine and 900 cal are exhausted, the mechanical equivalent of 100 cal of work has been done and the thermal efficiency of the engine is 10%. Under what conditions will the thermal efficiency be a maximum? Can it ever be 100%? Table 17-2 gives experimentally determined efficiencies for several heat engines. However, the greatest progress in increasing engine efficiencies has come

not from small improvements in the design of actual engines, but, strangely enough, from consideration of a theoretical engine that never ran—the Carnot engine.

Table 17-2. Typical Thermal Efficiencies

	%
Steam engine, reciprocating-type	12–27
Steam engine, turbine-type	32–50
Automobile engine	15–20
Diesel engine	30–40
Turbojet airplane engine	14

The Carnot Ideal Heat Engine

Sadi Carnot (Sah' di Kahr no'), a brilliant young French engineer, answered many questions about thermal efficiencies in 1824. He imagined an ideal heat engine of maximum efficiency which operated without friction. He assumed that the heat entered the cylinder from the *source* at the temperature T_1 and that it gave up heat to the *sink* at a lower temperature T_2. He also assumed that no wasted heat escaped through the walls of the cylinder. Suppose that the cylinder and piston contain an ideal gas as the working substance. Suppose initially, that the engine is thermally connected to the hot source and the gas is at the source temperature (Fig. 17-9a). Let the piston move up so that the gas expands at constant temperature, doing work.

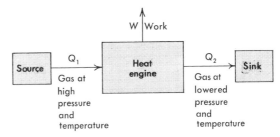

Fig. 17-8. The thermal efficiency of a heat engine is the ratio of work done to the heat supplied. Some heat is always exhausted to the sink.

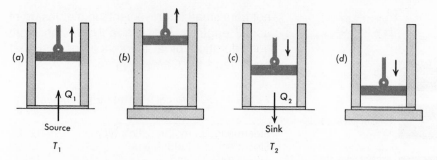

Fig. 17-9. *Ideal Carnot engine—its four cycles. (a) Does work and receives heat from source (isothermal expansion). (b) Does work, receiving no heat (adiabatic expansion). (c) Isothermal compression. (d) Adiabatic compression.*

It absorbs heat Q_1 from the source so that the gas temperature remains constant. Next, the cylinder is transferred to the insulating pad (Fig. 17-9*b*), and the gas is allowed to expand further *adiabatically,* that is, *without absorbing or emitting any heat.* During this adiabatic expansion the gas does work and is cooled to a temperature which is exactly that of the cool sink. Then the cylinder is placed on the sink and the gas is compressed at constant temperature (Fig. 17-9*c*). During the compression, heat Q_2 flows into the sink. Finally, the cylinder is placed on the insulating pad and the gas is compressed (Fig. 17-9*d*) until it reaches its initial volume and pressure, and the cycle has been completed.

Figure 17-10 indicates the volume and pressure changes during the four stages of the Carnot cycle. During the first, isothermal expansion, the pressure decreases along *AB*, and heat flows into the gas. During the second, adiabatic expansion *BC*, the pressure and the temperature both decrease. Then the gas is compressed isothermally along *CD* and heat flows into the sink. Finally, the gas is compressed adiabatically along *DA*, and the system goes back to its initial condition.

During the isothermal expansion, the engine received heat Q_1 from the boiler; during the compression it expelled heat Q_2

into the sink. Thus the thermal efficiency of the system is

$$\eta = \frac{Q_1 - Q_2}{Q_1} = 1 - \frac{Q_2}{Q_1}$$

But the heat Q_1 received by a Carnot engine during one stroke and the heat Q_2 given out at the other stroke may be shown to be proportional to the temperatures T_1 and T_2.

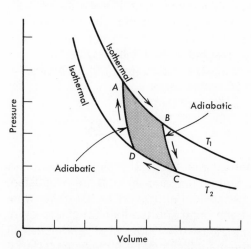

Fig. 17-10. *Carnot cycle. AB, isothermal expansion; BC, adiabatic expansion; CD, isothermal compression; DA, adiabatic compression.*

$$\frac{Q_1}{Q_2} = \frac{T_1}{T_2}$$

Therefore, the efficiency of the Carnot engine operating between temperatures T_1 and T_2 is

$$\eta_C = \frac{T_1 - T_2}{T_1}$$

All practical heat engines waste energy in friction and other ways, so that the efficiency of a Carnot ideal engine is a maximum for given temperatures of the source and the sink.

Suppose that a steam engine receives steam from a boiler at 400°K, uses part of this heat to do work, and ejects the remainder to a sink at a temperature of 300°K. Then the efficiency of the engine must be less than that for a Carnot engine, i.e.,

$$\eta < \eta_C = \frac{400°K - 300°K}{400°K} = 0.25 = 25\%$$

Consideration of the Carnot engine tells us that, in order to secure high efficiency, the source, or boiler, temperature of an engine should be high and the exhaust temperature should be low. Thus, because of its higher boiling point, water is a better working substance than alcohol. In recent years, turbine engines have been devised to use mercury as a working substance. The hot liquid mercury from the exhaust of the mercury vapor turbine is used in a steam boiler to produce steam to operate a steam turbine engine. Thus the total drop from the temperature of the mercury vapor to the exhaust temperature of the steam turbine is relatively great. In the internal-combustion engine, the temperature of the burning gases is higher than that attained in steam boilers; hence these engines are correspondingly more efficient.

The Refrigerator

We may think of a refrigerator as a heat engine operated in reverse. A heat engine receives heat from a hot source, uses part of this heat to do work, and delivers the remainder to a sink. A refrigerator receives heat from a cold enclosure, does work and increases the quantity of heat, and delivers the heat to a warmer body.

Usually the working substance of a refrigerator is a liquid of low boiling point, such as ammonia, sulfur dioxide, or Freon-12 (CCl_2F_2). In Fig. 17-11 Freon-12 evaporates at A, absorbing heat from the cooling chamber. The electric motor-driven pump compresses the vapor and heats it. The warm, compressed vapor is cooled by the fins and liquefies. The liquid passes through the expansion valve and returns to the cooling chamber where it again evaporates. Then the cycle is repeated.

Many buildings, especially in the southern states, are cooled by "heat pumps" during the summer and heated by them in winter. In summer, such a system pumps heat out of the building into the air. In winter, the refrigerator is reversed and pumps heat into the building, refrigerating the outdoors.

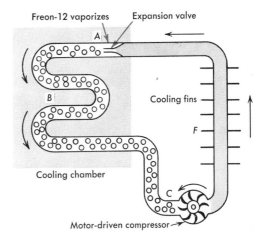

Fig. 17-11. Mechanical refrigerator. Liquid, vaporizing at A, absorbs heat. The vapor, warmed by compression at C, is cooled by the fins and liquefies.

The Second Law of Thermodynamics

If 1000 joules of mechanical work is done on a system without the passage of heat to or from it, the internal energy stored in the system increases by 1000 joules. However, if 1000 joules of heat is supplied to a heat engine, it is *not* possible to convert it entirely into1000 joules of work unless the sink is at 0°K—an unrealizable condition; some of the heat must be exhausted to a sink. Many experiments have convinced us of the truth of this statement, *the second law of thermodynamics: It is impossible to devise a heat engine which, working in a cycle, will convert heat from a hot body entirely into work without exhausting heat to a colder body.*

Heat flows readily from hot to cold bodies. However, to take heat from a cold body and exhaust it to a hot body requires the performance of work. Another statement of the second law, applicable to refrigerators, which can be shown to be equivalent to the one given above, is: *It is impossible to devise a refrigerator which, working in a cycle, will take heat from a cold body and exhaust it to a hot body without the performance of work.*

The theoretical Carnot engine has the highest efficiency of conversion of heat into work of any heat engine operating between two specified temperatures. Since even the Carnot engine could not convert heat entirely into work, practical engines must be expected to fall far short of accomplishing this. Notice that the *first* law of thermodynamics does not forbid us to convert heat entirely into work: energy would be conserved in such a process. It is the *second* law of thermodynamics, an entirely independent principle, that sets the terms on which nature allows us to use heat.

The Degradation of Energy

The energy of the universe constantly becomes more degraded and less available. According to the principle of the degradation of energy, it is easy to let energy run "downhill" and hard to get it "uphill" again. Suppose that you ignite a spoonful of gasoline. You start a chain reaction, setting free the readily available chemical energy of the highly organized gasoline molecules and converting it into the less available energy of randomly moving air molecules, thus "wasting" the energy. Suppose that you burn the gasoline in the cylinders of your car. This time the energy may do some useful work in overcoming friction, but the final result is the same. Chemical energy has been converted into energy of random molecular motion. Many experiments and observations have convinced us that all other kinds of energy tend to be converted into the energy of random molecular motion and to become less available. This, in a sense, is the most lowdown, degraded kind of energy.

In the physical universe, disorder and uniformity continually increase. Natural processes tend to move toward an increase in the disorder and the uniformity of the universe. Biological systems, through evolution, become more highly organized. However, this is not an exception to the law of degradation. A living organism is not an isolated system. It must receive energy from and give energy to its surroundings. When we take this energy exchange into account, the disorder of the entire system, including the environment of the living organisms, *does* increase with time.

Other kinds of energy continually decrease, and the energy of random molecular motion in the universe constantly increases. We can slow down or speed up this process, but we cannot prevent it entirely. Moreover, heat itself flows downhill from hot to cold. According to a widely accepted theory, as time goes on, we approach the day when the sun and other stars will be cold unchanging cinders, all at the same temperature. This dismal condition has been called the *death of heat.* It is comforting to learn that, probably, the death of heat is billions of years away and that, if other catastrophes are averted, life on this little earth may persist for millions of years. Also, the principle of the degradation of energy may not always hold. It is a statistical law, justified only by experience.

The Sun Supplies Most of our Energy

On a clear day at noon, each acre of the earth's surface at the equator receives energy at the rate of more than 5000 horsepower. If purchased at three cents per kilowatt-hour, this energy would cost more than one hundred dollars per hour. New devices, such as the solar battery, the solar stove, and the solar space heater, can convert a small fraction of the sun's radiation on the earth to usable energy. However, the earth reradiates most of the solar energy into space. A little is used in evaporating water which, falling as rain, supplies our rivers, which can turn water wheels. Some of the energy generates winds which drive windmills. Most important of all is the energy stored in growing plants, providing food to maintain animal life. Part of the energy received millions of years ago was stored in coal and oil. This energy we expend at an ever-increasing rate to maintain our industrial civilization.

In America we are spending our petroleum resources rapidly. Although we are learning how to extract more of the oil in known oil fields and are discovering new fields with greater accuracy, our needs for petroleum are increasing too. The mining of coal is more difficult, but the reserves are larger, and a century or so will elapse before our supplies are seriously depleted. When these carbon deposits are gone, we can use water power, but if all the available falls were utilized, they would take care of only a small fraction of our requirements. As the other, cheaper sources of energy are depleted, atomic energy will become more important in our economy.

Both the availability of fuels and the effect of their use on the environment must be considered. Unlimited energy resources would not greatly benefit mankind if their use led to constantly increasing pollution of the atmosphere, land, and water. An ecological view must be taken in assessing the appropriateness of uses of fuel that may be quite feasible from a technological standpoint.

Summary

Rumford, Joule, and others have proved that heat is energy in transit. Approximately 42 million ergs or 427 gramweight-meters of work are required to produce the same change of temperature as would be produced by the addition of one calorie of heat. $W = JQ$.

The first law of thermodynamics states that the heat added to a body equals the increase in its internal energy plus the work done by the body on its surroundings. The specific heat of a gas at constant pressure is greater than that at constant volume because of the work done when the gas expands.

Heat engines take in heat, convert part of it into mechanical energy, and exhaust the rest. The maximum efficiency of a heat engine supplied with heat from a source at absolute temperature T_1 and delivering heat at a sink temperature T_2 is that of a Carnot engine:

$$\eta_C = \frac{T_1 - T_2}{T_1}$$

The second law of thermodynamics asserts that it is impossible to transfer heat from a colder to a warmer body without doing work and that it is impossible to operate a heat engine without exhausting heat to a colder body.

Questions

1. When a bottle containing a carbonated drink is uncapped suddenly, a dense fog appears above the surface of the liquid. Explain.

2. A toy rocket car is propelled by gas from a small CO_2 cartridge. Explain the following:

(*a*) an accelerating force acts on the car, (*b*) the escaping gas does work, (*c*) at the end of the trip, the cartridge is cold to the touch.

3. State several advantages of turbines over reciprocating-type steam engines as means of propulsion of ships.

4. What are advantages and disadvantages of high-compression engines?

5. Discuss the dependence of the efficiency of a heat engine upon the temperature of the source.

6. Discuss the possibility of converting all the heat energy of steam into mechanical energy.

7. Why do most heat engines have a relatively low efficiency?

8. How can a theoretical concept guide technological development? Use the Carnot ideal heat engine as an example.

9. What was the caloric theory? Was it a foolish theory? Why did it survive so long?

10. State the first law of thermodynamics, carefully defining "heat," "internal energy," and "work."

11. State the second law of thermodynamics as applied to (*a*) a heat engine, (*b*) a refrigerator.

12. Discuss the degradation of energy. Give several examples.

13. An ice cube is floating in a glass of water. Would you expect to observe the water suddenly begin to boil and the temperature of the ice to decrease? Is such behavior forbidden by the law of conservation of energy? Discuss.

14. Is the change in the internal energy of ice more or less than 80 cal/gm when ice melts? Discuss.

15. Why is the specific heat of a gas at constant pressure greater than that at constant volume?

16. If all the gravitational potential energy were converted into heat, estimate the increase in temperature of water that falls 108 meters over Victoria Falls.

17. At what rate could a car be propelled

by utilization of solar energy, assuming that the efficiency of conversion from solar into electrical energy is 10%? State your other assumptions.

18. Will keeping the door of an electric refrigerator open warm or cool a room? Explain.

19. A bullet traveling at a speed of 800 m/sec strikes a target and is heated by the impact. In what respect does the kinetic energy after the impact differ from that before?

20. Which quantity is the greater energy: (*a*) 1 cal or 1 kilowatt-sec, (*b*) 1 cal or 4 joules?

21. What is meant by the "death of heat"?

Problems

A

1. A boy weighing 50 kgwt slides down a rope 5.0 m long and reaches the ground with zero speed. (*a*) How much work is done in overcoming friction? (*b*) How much heat is produced?

2. A box weighing 60 kgwt is dragged 200 m along a level street. How much heat is developed if the coefficient of friction is 0.20?

3. A meteor of mass 2.0×10^7 gm has a velocity of 12 km/sec. How many calories of kinetic energy of translation has it?

4. A 70-kg block of ice is dragged 20.0 m across a level floor. How much ice is melted due to friction if the coefficient of friction is 0.10? (Assume that all of the heat is used to melt ice.)

5. From what elevation must 10 gm of ice fall to develop sufficient heat to melt it completely? Assume that one-half of the heat remains in the ice.

6. What is the efficiency of a heat engine that operates between 30°C and 100°C?

B

7. At noon on a sunny day, sunlight imparts 1.20 cal/min-cm² to the roof of an automobile,

the area of which is 2.5 m². (a) Find the power in kilowatts and in standard horsepower. (b) If the roof is covered with solar-battery elements which convert solar energy into electrical energy at an efficiency of 10%, what electrical power is produced?

8. A 16-gm bullet at 0°C, traveling 600 m/sec, strikes and embeds itself in a block of ice at 0°C. How much ice is melted?

9. If a man works with a total efficiency of 30% and he weighs 100 kgwt, how far can he climb vertically by expending the energy liberated by the combustion of 30 gm of bread?

10. The weight of an automobile is 1600 kgwt and its speed is 30 m/sec. It is stopped in 5.00 sec. (a) What was its initial kinetic energy? (b) What was the power of the brakes (i.e., at what rate did they absorb energy)? (c) How much heat was generated?

11. A soda biscuit has a mass of 8.0 gm, and the heat of combustion is 3000 cal/gm. How many such biscuits must a man whose mass is 90 kg eat to supply sufficient energy for him to climb to the top of a building 80 m high? Assume that (a) 100% and (b) 20% of the energy is utilized.

12. The burning of 1 gal of gasoline liberates about 3.0×10^7 cal of heat. To what elevation would this amount of heat lift a car weighing 1600 kgwt if 18% of the heat were converted into mechanical energy?

13. An automobile engine wastes 80% of the energy of the gasoline (see problem 12) in heat carried away by the radiator water and the exhaust. The car travels 24 miles per gallon of gasoline. How much heat is carried away per minute when the car travels 60 ft/sec?

14. The engine of the car in problems 12 and 13 has a mass of 150 kg, the copper radiator 20 kg, and the cooling system contains 15 kg of water. What time is required for the engine to heat up from 20°C to 70°C? Assume no heat losses.

15. The powder used to fire a 20-gm lead

bullet generates 7000 cal of heat. If the muzzle velocity of the bullet is 1000 m/sec, what is the efficiency?

16. A steam engine receives 16,000 cal at a boiler temperature of 427°C and exhausts 12,000 cal at a temperature of 227°C. Calculate (a) the actual efficiency of the engine, (b) its Carnot efficiency.

C

17. What is the velocity of a lead bullet at 30°C that is just melted by the heat of impact when the bullet strikes the target? Assume that all the heat remains in the lead. (Melting point of lead: 327°C; heat of fusion: 5.9 cal/gm.)

18. Ten per cent of the energy of combustion of gasoline is utilized in accelerating a 1500-kg car. (a) How much gasoline (heat of combustion: 11,000 cal/gm) is needed to increase the velocity from 0 to 40 km/hr? (b) If the mass density of the gasoline is 0.75 gm/cm³ and gasoline costs 10 cents per liter, what is the cost of this acceleration?

19. One gram of helium contained in a cylinder fitted with a sliding piston is heated from 0°C to 273°C. The volume changes from 5600 cm³ to 11,200 cm³, but the pressure remains 76 cm-Hg. If the molecular heat at constant pressure is 5 cal/mole-K°, find (a) the amount of heat added, (b) the work done by the expanding gas, (c) the increase in internal energy of the gas, (d) the molecular heat at constant volume, (e) its specific heat at constant volume.

20. An iron spring having an elastic constant of 5.0 kgwt/cm is compressed 6.0 cm, then held in acid and completely dissolved. (a) What average force opposed the compression? (b) How much heat is evolved by the energy due to compression? (c) How much would this heat raise the temperature of the spring if its mass were 0.50 kg and all the heat were used for this purpose?

21. Suppose an ideal Carnot engine were operated backward so as to deliver heat from

a source at 7.0°C to a room at 27.0°C. How much heat would be taken from the source to supply 2.5×10^5 cal?

22. A liter of air initially under standard conditions is heated to 1.00°C, keeping the pressure constant. (a) How much work does the gas do in expanding? (b) If the specific heat of the air at constant pressure is 0.237 cal/gm-C°, what is the specific heat of air at constant volume?

23. A jet aircraft burns 5000 kg of kerosene in a flight of 600 mi. The efficiency of the engine is 14%, and the heat of combustion is 1.10×10^4 cal/gm. Find the work done by the engine.

24. Assuming that 0.80 cm of rain fell during a storm and that the distance through which the water descended was 1000 m, estimate (a) the amount of heat evolved in condensing the water vapor that falls on 1.0 km² of the earth's surface, and (b) the average rise of temperature of the air through which the droplets fall. Density of the air is 1200 gm/m³, and the heat of vaporization is 500 cal/gm.

25. A gram of ice at 0°C occupies a volume of 1.087 cm³ at normal pressure, and a gram of liquid water at 0°C, 0.999 cm³. (a) How much work is done on the ice by the atmosphere when the ice melts at 0°C? (b) How much energy is given to the water molecules in the melting process?

For Further Reading

Brown, Sanborn C., *Count Rumford, Physicist Extraordinary,* Doubleday and Company, Garden City, 1962.

Hart, Ivor B., *James Watt and the History of Steam Power,* Henry Schuman, New York, 1949.

Heilbrunn, L. V., "Heat Death," *Scientific American,* April, 1954, p. 70.

Sandfort, John F., *Heat Engines,* Doubleday and Company, Garden City, 1962.

Schrödinger, Erwin, *What is Life?,* Cambridge University Press, Cambridge, 1946.

Ubbelohde, A. R., *Man and Energy,* Penguin Books, Baltimore, 1963.

Wilson, Carroll L. and William H. Matthews, eds., *Man's Impact on the Global Environment,* MIT Press, Cambridge, 1970.

Vibrations,
Wave Motion,
and
Sound

CHAPTER EIGHTEEN

Vibrations

In earlier chapters, we considered several types of motion, beginning with the simplest—that of a particle moving in a straight line with constant velocity. Rectilinear motion with constant acceleration was discussed next, and after that, the motion of a particle moving in a circular path with constant speed but changing direction. Now you will study the to-and-fro or vibratory type of motion that is more complex than the others, yet very necessary to an understanding of sound, X-rays, light, radio, and other kinds of wave motions.

What is a Vibration?

Familiar examples of vibrations are the motion of pendulums, the swaying of trees, and the quiverings of the strings of musical instruments. *A vibration is a motion that repeats at regular intervals of time.*

Consider the ball and spring represented in Fig. 18-1. When you displace the ball to your left a distance *r*, the spring exerts an opposite, elastic *restoring force* which acts toward the mid-position of the ball. When you release the ball, the restoring force accelerates it until it reaches its mid-position. Then

the force decelerates the ball until it reaches its extreme right position. The inertia of the ball opposes the change of its velocity and keeps it moving after it passes the midpoint—or equilibrium position.

Suppose that you surrounded the ball and spring with glycerine or tar. Then friction would quickly stop the vibration. In order that a body may vibrate, three conditions must be fulfilled: (1) It must have inertia—this is provided by its mass—to keep it moving past the midpoint of its path. (2) There must be an elastic restoring force to accelerate the body toward the midpoint. (3) The friction opposing the motion must be small.

Three important terms used to describe a vibration are its *period, frequency,* and *amplitude.* The *period T* of a simple vibration is the time required for *one complete to-and-fro motion* or cycle. For example, the period of the vibrator in Fig. 18-1 might be $\frac{1}{3}$ sec ($\frac{1}{3}$ sec per cycle). The *frequency f* is the *number of complete vibrations (or cycles) per unit time.* If the period of a vibrator is $\frac{1}{3}$ sec, its frequency is 3 vibrations (cycles) per second or 3 hertz, where 1 hertz = 1 cycle/sec. The *amplitude* is the *maximum*

Fig. 18-1. A ball vibrating with simple harmonic motion. The restoring force is proportional to the displacement.

displacement of the vibrating body from the midpoint. The amplitude of the vibrator in Fig. 18-1 is r.

Simple Harmonic Motion

The most familiar type of vibration, called *simple harmonic motion,* is illustrated by the vibrator which we have just described. When you displace the ball slightly from its midpoint or position of equilibrium, the restoring force is directly proportional to the displacement. Doubling the displacement, for example, doubles the restoring force. This fact distinguishes simple harmonic motion from all other types.

Simple harmonic motion is that type of vibration in which the restoring force acting on the vibrating body is proportional to its displacement.

According to Hooke's law, when an elastic body is distorted, the restoring force is directly proportional to the displacement. Thus, if Hooke's law is obeyed, the body will vibrate with simple harmonic motion.*

* Not all elastic bodies obey Hooke's law, however. For example, the restoring force in an O-shaped loop made of spring steel is a constant independent of the displacement for small displacements.

The period of a simple harmonic vibrator depends upon the stiffness of the spring (or other agency) that pulls the body back to its position of equilibrium. If one spring is replaced by a stiffer one that exerts a greater restoring force at a given displacement, then the force urging the ball toward the midpoint will be greater, the acceleration will be greater, and a smaller time will be required for a vibration to occur. The stiffness of the spring is expressed by the *spring-constant k,* which is the *restoring force per unit displacement.* For example, if a restoring force F of 30,000 dynes acts on a ball when the displacement s is 1 cm, and a restoring force of 60,000 dynes acts on it when the displacement is 2 cm, its spring-constant F/s is 30,000 dynes/cm.

According to Hooke's law

$$F = -ks$$

where the negative sign means that the direction of the force is opposite to that of the displacement. By Newton's second law, we have the force $F = ma$ where a is the acceleration of the ball at any instant. Therefore

$$ma = -ks$$

or

$$a = -\left(\frac{k}{m}\right)s$$

The acceleration of a particle in simple harmonic motion is thus proportional to the displacement at any instant. Since the acceleration is therefore *not* constant, we cannot use such equations as $s = v_0 t + \frac{1}{2}at^2$ (Chapter 4) to describe the motion. We shall see shortly how the motion can be described.

If the spring is replaced by one that is one-fourth as stiff, so that the spring-constant is one-fourth as great, we find experimentally that the period of vibration will be doubled. *The period of a simple harmonic vibrator is inversely proportional to the square root of the spring-constant.*

The period of vibration depends not only upon the spring-constant, but also upon the mass of the vibrator. Suppose that the ball, Fig. 18-1, is replaced by one having a greater mass. Then the inertia of the system is increased, it accelerates more slowly at a given displacement, and the period increases. Quadrupling the mass, experiment shows, doubles the period. In general, *the period of a simple harmonic vibrator is directly proportional to the square root of the mass of the vibrator.* These experimental facts are expressed by the following equation for the period:

$$T = 2\pi \sqrt{\frac{\text{Mass}}{\text{Spring-constant}}} = 2\pi \sqrt{\frac{m}{k}} \quad (1)$$

Note that the period is *independent of the amplitude.* We shall derive this equation in the next section.

Example. A 2.45-kg ball is suspended by a light spiral spring. A force of 0.10 kgwt is required to pull the weight downward 0.10 m. Find (a) the spring-constant of the spring and (b) the period of vibration.

(a) $k = 0.10$ kgwt/0.10 m
 $= 1.0$ kgwt/m $= 9.8$ newtons/m

(b) $T = 2\pi \sqrt{\dfrac{2.45 \text{ kg}}{9.8 \text{ newtons/m}}}$

$= 2\pi \sqrt{\dfrac{2.45 \text{ kg}}{9.8 (\text{kg-m/sec}^2)/\text{m}}} = \pi$ sec

Describing Simple Harmonic Motion

Since the acceleration varies in simple harmonic motion, we must adopt a new mathematical approach to calculate the displacement and velocity at any time. It is convenient to make use of an imaginary *reference circle.* For example, let Fig. 18-2 represent successive positions of the pedal of an upended bicycle, whose rear wheel revolves at constant angular velocity. Light from the setting sun casts a shadow of the pedal on the wall. The pedal moves steadily from position *a* to *b* to *c* and so on. Its shadow accompanies it, moving from *A* to *B* to *C* along the wall. The point *B* is the *projection* or "shadow" of *b* on the line *DD'*, that is, the point where a line drawn normal to *DD'* from *b* intersects it.

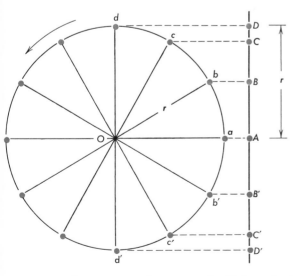

Fig. 18-2. *Imaginary reference circle. If the point* b *moves with uniform circular motion, its projection* B *vibrates with simple harmonic motion.*

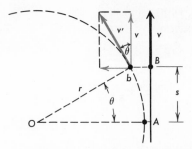

Fig. 18-3. *If the reference point* b *moves with velocity* v', *the velocity* v *of the vibrating particle at* B *is* v' cos θ. *Its displacement at* B *is* r sin θ.

Similarly, *A* and *C* are projections of *a* and *c*. If a point moves on the circle with constant velocity, its projection on *DD'* will vibrate with simple harmonic motion. Thus *simple harmonic motion may be regarded as the motion of a point which is the projection of a point moving with uniform circular motion; the line upon which projection occurs lies in the plane of the circle.* A mathematical check on the correctness of this statement will be given below. Introducing the circle of reference is advantageous because it allows us to draw information about simple harmonic motion from the simpler uniform circular motion.

The *phase* of a vibration tells us what fraction of a vibration has been completed. Usually phase is expressed in terms of the angle through which the reference particle has moved from some chosen position. For example, in Fig. 18-2, *b* has moved through 30° since it left the zero point, the position marked *a*. The phase of *b* is therefore 30°. When the phase is 90°, the imaginary point will have made one-fourth of a complete vibration. After *b* has made one complete vibration, its phase will be 360° or 2π radians.

In Fig. 18-3, let the angular velocity of the reference point be ω, let its period be *T*, and let its phase be θ at a time *t*. Then

$$\theta = \omega t = \frac{2\pi}{T}t$$

In Figs. 18-2 and 18-3, when the vibrating particle is at *B* and the reference particle is at *b*, the phase of the vibrating particle is 30°. Its displacement is the vertical projection *AB* of the line *bO* and is one-half (that is, sin 30°) times the amplitude *r*. In general, for any simple harmonic motion whose amplitude is *r* and whose displacement is zero at zero time, the displacement *s* when the phase angle is θ is given by

$$s = r \sin \theta = r \sin \omega t = r \sin \frac{2\pi t}{T}$$

In Fig. 18-3, the vector *v'* represents the velocity of the imaginary particle at *b*. The vector *v*, the vertical component of *v'*, is the velocity of the simple harmonic vibrator. At a time *t* when the phase angle is θ

$$v = v' \cos \theta = v' \cos \omega t = \frac{2\pi r}{T} \cos \frac{2\pi t}{T}$$

To obtain the acceleration of the vibrating particle at any time, recall that the centripetal acceleration of the reference particle moving on a circle with constant speed *v'* is

$$a_C = -\frac{(v')^2}{r} = -\frac{(2\pi r/T)^2}{r} = -\frac{4\pi^2 r}{T^2}$$

where the negative sign tells us that the acceleration is directed toward the center of the circle. The acceleration *a* of the vibrating particle (Fig. 18-4) is the vertical projection, that is, the vertical component, of a_C or

$$a = a_C \sin \theta = -\frac{4\pi^2 r}{T^2} \sin \frac{2\pi}{T} t$$

Since the displacement $s = r \sin (2\pi/T)t$, we have

$$a = -\frac{4\pi^2}{T^2} s = -\omega^2 s \qquad (2)$$

where ω, the angular velocity, equals $2\pi/T$. If the acceleration of the vibrating particle is *a*,

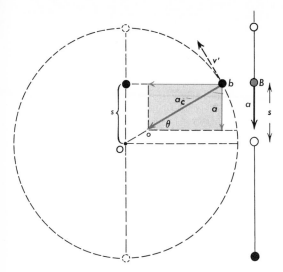

Fig. 18-4. The simple harmonic vibrator.

we can substitute for it F/m, according to Newton's second law, where F is the restoring force and m the mass of the vibrating particle. Hence

$$\frac{F}{m} = -\frac{4\pi^2}{T^2}\,s$$

or

$$\frac{F}{s} = -\frac{4\pi^2 m}{T^2}$$

But $F/s = -k$, the spring-constant of the vibrator, and we have

$$k = \frac{4\pi^2 m}{T^2}$$

or

$$T = 2\pi\sqrt{\frac{m}{k}} \qquad (3)$$

By substituting k/m for $4\pi^2/T^2$ in equation (2) above, we obtain

$$a = -\frac{k}{m}\,s \qquad (4)$$

Finally, note that

$$F = -\left(\frac{4\pi^2}{T^2}\,m\right)s$$

$$= -ks$$

in accord with the definition of simple harmonic motion given above.

By this derivation* we have shown, first, that the acceleration of projected uniform circular motion is simple harmonic motion because our calculation of the acceleration led consistently to equations (2) and (4) and second, that the period of simple harmonic motion is indeed $T = 2\pi\sqrt{m/k}$.

We can summarize the discussion of simple harmonic motion as follows:

1. If the amplitude of a simple harmonic vibrator is r and its period T, the circle of reference has a radius r and the reference particle has a period T. $T = 2\pi\sqrt{m/k}$. The phase of the vibrating particle is the central angle $\theta = (2\pi/T)t$ of the reference particle, and the angular velocity $\omega = 2\pi/T$.
2. The displacement of the vibrating particle at any time t is $s = r \sin \omega t$.
3. Its velocity is $v = r\omega \cos \omega t$.
4. Its acceleration is $a = -\omega^2 r \sin \omega t = -\omega^2 s$.
5. The restoring force is $F = -m\omega^2 s$.

---

* *A calculus proof.* If a body performs simple harmonic motion, which is described by $s = r \sin \omega t$, we can readily use calculus to show that it obeys Hooke's law. Let us first calculate its velocity v:

$$v = \frac{ds}{dt} = \frac{d}{dt}(r \sin \omega t) = r\omega \cos \omega t$$

Next we find the acceleration a:

$$a = \frac{dv}{dt} = \frac{d}{dt}(r\omega \cos \omega t) = -r\omega^2 \sin \omega t$$

$$a = -\omega^2 (r \sin \omega t) = -\omega^2 s$$

For a, we write F/m by Newton's second law:

$$\frac{F}{m} = -\omega^2 s$$

and

$$F = -\omega^2 m s = -ks$$

where k is the spring-constant. This last equation is just Hooke's law.

Example. In Fig. 18-3, suppose that the pedal of the bicycle moves on a circle of radius 20 cm and that the pedal revolves once per second. What is (a) the velocity of the pedal, (b) the velocity of its shadow when (1) $\theta = 0°$, (2) $\theta = 30°$, (3) $\theta = 60°$, (4) $\theta = 90°$, (5) $\theta = 180°$?

(a) $v' = \dfrac{2\pi r}{T} = \dfrac{2\pi \times 20 \text{ cm}}{1 \text{ sec}} = 40\pi \text{ cm/sec}$

(b) $v = v' \cos \theta$: (1) $v = 40\pi$ cm/sec, (2) $v = 0.86 \times 40\pi$ cm/sec, (3) $v = 20\pi$ cm/sec, (4) $v = 0$ cm/sec, (5) $v = -40\pi$ cm/sec.

The Energy of a Vibrating Body

Suppose that the ball mounted on the spring (Fig. 18-1) is gradually pulled sideways. Then the magnitude of the restoring force increases from zero to a value ks when the displacement is s, where k is the spring-constant. The *average* force during this displacement is $\frac{1}{2} ks$.

The total work done in causing the displacement, equal to the potential energy E_P of the spring at the displacement, is given by*

$$E_P = \text{Average force} \times \text{Displacement}$$
$$= (\tfrac{1}{2}ks) \cdot s = \tfrac{1}{2}ks^2$$

If friction is negligibly small, the ball when released will continue to vibrate with constant amplitude and constant energy. If the amplitude of vibration—the maximum displace-

---

* *A calculus proof.* The element of work dW done *on* the spring in a small displacement ds is

$$dW = -Fds$$

where we are exerting a force $-F$ to oppose the restoring force F. According to Hooke's law, $F = -ks$, and therefore

$$dW = -(-ks)ds = ksds$$

Let us integrate this expression from $s = 0$ to $s = s$.

$$W = \int_0^W dW = k \int_0^s sds = \tfrac{1}{2}ks^2$$

Since the energy E_P equals the work done on the spring in this displacement, we have

$$E_P = \tfrac{1}{2}ks^2$$

ment—is r, then $\frac{1}{2}kr^2$ is the *potential* energy at maximum displacement, the *kinetic* energy at zero displacement, and the *sum* of the two energies at *any* displacement.

Example. A ball has a mass of 1.0 kg and the spring-constant of the spring to which it is attached is 100 newtons/m. It vibrates with an amplitude of 0.10 m. What are (a) the potential energy when the displacement is maximum and (b) the velocity v of the ball at the midpoint?

(a) $E_P = \tfrac{1}{2}kr^2$
$= \tfrac{1}{2} \times 100 \text{ newtons/m} \times 0.01 \text{ m}^2$
$= 0.50$ joule

(b) E_K (at midpoint) $= \tfrac{1}{2}mv^2$
$= \tfrac{1}{2} \times 1.0 \text{ kg} \times v^2$
$= 0.50$ joule
$v = 1.0$ m/sec

We summarize the variations of velocity, acceleration, force, and potential and kinetic energy for a vibrator in Fig. 18-5.

The Pendulum

The motion of a pendulum was first studied by Galileo when he was a young student just beginning his scientific career. Galileo observed that the period of the vibrations of a pendulum seemed to be constant as the motion gradually died down, and he verified this conclusion by timing them.

The period of a "simple" pendulum consisting of a small ball suspended by a long cord of length l, swinging through a small angle, is given by

$$T = 2\pi \sqrt{\dfrac{l}{g}} \qquad (5)$$

g being the acceleration of gravity.

Notice (a) that the period is proportional to the square root of the length l of the pendulum, (b) that it is inversely proportional to the square root of the acceleration due to gravity, g, and (c) that it does *not* depend

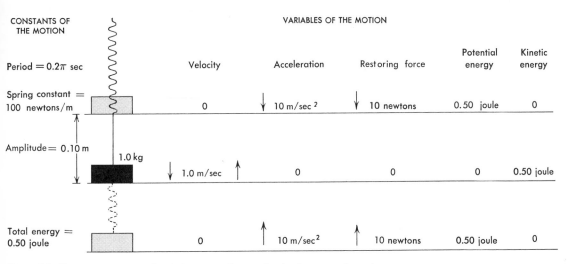

CONSTANTS OF THE MOTION	VARIABLES OF THE MOTION					
		Velocity	Acceleration	Restoring force	Potential energy	Kinetic energy
Period $= 0.2\pi$ sec						
Spring constant $=$ 100 newtons/m		0	$\downarrow$ 10 m/sec^2	$\downarrow$ 10 newtons	0.50 joule	0
Amplitude $= 0.10$ m	1.0 kg	$\downarrow$ 1.0 m/sec $\uparrow$	0	0	0	0.50 joule
Total energy $=$ 0.50 joule		0	$\uparrow$ 10 m/sec^2	$\uparrow$ 10 newtons	0.50 joule	0

Fig. 18-5. *Maximum and minimum values of velocity, acceleration, restoring force, potential energy, and kinetic energy for a vibrator with the characteristics given to the left of the drawing.*

upon the mass of the pendulum nor upon its amplitude.

Galileo's discovery, that the period of a pendulum was constant, was quickly exploited by physicians, who used pendulums to measure the pulse rates of their patients. The doctor suspended a ball from a cord, set it into motion, and counted the number of vibrations it made during a chosen number of pulse beats. Soon clocks were invented in which pendulums regulated the rate of ticking. Improvements were devised by Christian Huygens, by Galileo's son, by Robert Hooke, and by many others. Accurate clocks enabled scientists for the first time to measure small time intervals accurately. Thus accelerations could be determined.

The pendulum is useful in determining the value of g, the acceleration of gravity. First the period T_0 is observed at some location where g_0 has been precisely determined. Then the period T is observed at some other location and the local value of g is computed from equation (5) by

$$\frac{T_0}{T} = \sqrt{\frac{g}{g_0}}$$

The simple pendulum moves with simple harmonic motion and therefore can be described by the equations given earlier in this chapter. Let us calculate the period of a simple pendulum. Figure 18-6 represents a simple pendulum of length l, weight w, and mass m. The weight $w = mg$ can be replaced by two components, F' parallel to the cord and F at right angles to it. The angle θ is assumed to be small so that the triangles ABC and CEG are nearly similar. Hence (approximately)

$$\frac{F}{mg} = \frac{s}{l}$$

$$\frac{F}{s} = \frac{mg}{l}$$

Now F is the restoring force and s the displacement, so that F/s is the spring-constant k. From equation (1),

$$T = 2\pi \sqrt{\frac{m}{k}} = 2\pi \sqrt{\frac{m}{mg/l}} = 2\pi \sqrt{\frac{l}{g}}$$

When the mass of the pendulum is doubled, the spring-constant is also doubled; hence the period is unchanged.

Vibrations, Wave Motion, and Sound 293

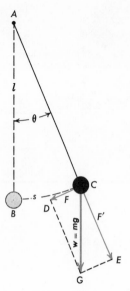

Fig. 18-6. The simple pendulum.

(The error arising from the assumption that the triangles are similar is only about 0.02% if the pendulum swings through an angle θ of 3° from the position of equilibrium. If the angle is 60°, the error is 6%.) For larger angles of swing, we use the equation

$$T = 2\pi\sqrt{l/g}\left[1 + \tfrac{1}{4}\sin^2\left(\frac{\theta}{2}\right) + \tfrac{9}{64}\sin^4\left(\frac{\theta}{2}\right)\right]$$

The equation $T = 2\pi\sqrt{l/g}$ gives the period of a *simple* pendulum, the mass of which is concentrated at a point. The mass of the suspending cord is zero. The mass of a *physical* pendulum, such as a meter stick pivoted at one end, is distributed. The period of a physical pendulum is given by

$$T = 2\pi\sqrt{\frac{\text{Moment of inertia}}{\text{Torque/Angular displacement}}}$$

$$= 2\pi\sqrt{\frac{I}{\tau/\theta}}$$

Notice as before, that in rotational motion, moment of inertia corresponds to mass in linear mechanics. Torque corresponds to force, and angle to distance.

Resonance

A child who pushes a playmate seated in a swing soon learns to make the frequency of his pushes equal to the natural frequency of the swing. By so doing, he adds energy to the vibrating system, a little at a time, until the amplitude of the vibration is very great. Similarly, in Fig. 18-7, the pendulum *B*, connected to *A* and *C* by springs, readily makes *A* vibrate. Pendulum *C*, which is not in resonance ("in tune") with *A*, vibrates scarcely at all.

Vibrator A is in resonance with vibrator B if they have the same natural frequency

The sharpness of "tuning" between two vibrators depends upon the friction that opposes the motion of the "receiver" vibrator. In Fig. 18-8, curve *a* shows how the response of a steel spring, clamped in a vise, depends upon the frequency of the impulses acting on it. In *b*, the spring is partly submerged in water so that fluid friction opposes the vibration.

You will meet other illustrations of resonance in sound, light, and radio. Sometimes resonance is called *sympathetic vibration*.

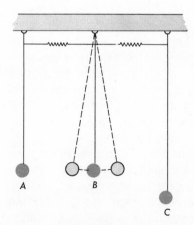

Fig. 18-7. Resonance. The vibrating pendulum B, *connected (coupled) to* A *and* C, *readily sets* A *into vibration. The two, having the same frequency, are in resonance.*

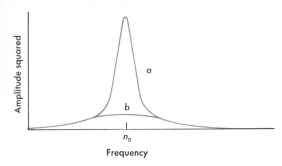

Amplitude squared

a

b

n_0

Frequency

Fig. 18-8. Resonance of a vibrating spring. The amplitude of vibration becomes large when the applied force reaches the natural frequency f_0 of the spring. (a) Little damping. (b) Large amount of damping.

Summary

A vibration is a to-and-fro motion that repeats at equal time intervals. The displacement of a vibrator is its distance from its position of equilibrium. The amplitude of a vibration is its maximum displacement. The restoring force acting on a vibrator is the force urging it toward its position of equilibrium.

A body vibrates with simple harmonic motion if the restoring force F is directly proportional to the displacement s, i.e., $F = -ks$, where k is the spring-constant, and the negative sign reminds us that the restoring force and the displacement are oppositely directed.

Simple harmonic motion may also be described as the motion of the projection on a straight line of a point moving with constant speed on a circle, the circle and the line being in the same plane.

The phase of a vibration is the angle through which the reference particle has moved since the particle left a chosen starting position.

The period of a vibration is the time required for a complete to-and-fro motion. The frequency is the number of vibrations per unit time. The period of a simple harmonic

vibrator is given by

$$T = 2\pi \sqrt{\frac{m}{F/s}}$$

The period of a simple pendulum of length l vibrating through a small angle is given by

$$T = 2\pi \sqrt{\frac{l}{g}}$$

When vibrators A and B have the same natural frequency, they are in resonance and A can readily make B vibrate, that is, can transfer energy to B.

Questions

1. How is simple harmonic motion defined? How is it related to uniform circular motion? List several examples of simple harmonic motion.

2. If the period of vibration of a pendulum is 0.5 sec, what is its frequency? To what is the product of frequency times period equal?

3. Given a spring and a meter stick, how could you secure data to enable you to calculate the period of a body of known mass suspended from the spring?

4. How is the period of a mechanical vibrator affected (a) if both the mass and the spring-constant are doubled, (b) if the constant of the spring is quadrupled but the mass is unchanged?

5. When an additional passenger enters an automobile, does the period of vibration of the car on its springs increase, decrease, or remain constant?

6. What is the period of the pendulum in a grandfather clock?

7. In what way could a pendulum be used for determining standards of mass, length, and time?

8. Discuss the meaning of the minus sign in $F = -ks$. Define clearly each of the quantities in the equation.

9. In simple harmonic motion, when is the speed one-half of the maximum speed?

10. Why do the wheels of a moving vehicle sometimes appear reversed in motion in a motion picture?

11. What is the maximum length possible for a simple pendulum?

12. If a grandfather clock with a brass pendulum were moved from the first to the third floor of a building, what non-mechanical action could be taken so that it would still keep correct time?

13. As the length of a simple pendulum was changed, the following data were obtained. Plot period T versus length l and T^2 versus l. From the curves determine the acceleration of gravity g.

Period (sec)
| 1.98 | 1.83 | 1.67 | 1.58 | 1.50 | 1.40 |

Length (cm)
| 98 | 84 | 70 | 63 | 56 | 49 |

14. Derive the equation for the period of vibration of a meter stick vibrating with negligible friction about an axis at one end of the stick, perpendicular to the stick.

15. How would a pendulum clock behave in a spacecraft in an orbit around the earth (a) if the spacecraft spins about a longitudinal axis, (b) if the spin velocity is zero?

16. Four different clocks—a grandfather clock, a wristwatch (balance-wheel type), an hourglass, and a sundial—are all constructed and regulated to keep correct time. Suppose now that another set of similar clocks, each four times as large in each dimension as the first set, is constructed and adjusted to the correct time. What times will they show 30 minutes later? Discuss the effect of the change in dimensions upon the rate of each clock—as quantitatively as you can.

17. What is the effect on the rate of a watch of taking it to the top of a high mountain? Consider the action of the air that surrounds the balance wheel.

Problems

A

1. A child's 60-gm rubber ball is suspended from a long rubber band. If an additional force of 2000 dynes will pull it down 10 cm, find (a) the period of vibration of the system and (b) the frequency.

2. (a) When a 50-kg boy entered an automobile, its center of gravity descended 0.30 cm. What was the spring-constant of the springs? (b) The weight of the boy and car body was 400 kg. What was the period of vibration of the boy and the car on its springs?

3. The crank of a grindstone is 40 cm long. The crank is turned uniformly, making 1.0 rev/sec. The shadow cast by the handle on the level ground at noon vibrates with simple harmonic motion. Find (a) the period, (b) the frequency, (c) the amplitude of the vibration.

4. What is the period of a simple pendulum 80 cm long at a place where $g = 981$ cm/sec^2?

5. When a boy sits on a swing, his center of gravity is 2.5 m below the beam to which the ropes are tied. Find the approximate period of vibration.

B

6. A simple pendulum of mass 80 gm, 100 cm long, is pulled sidewise so that its displacement is 5.0 cm. What restoring force acts on it?

7. A 6.0-kg ball is suspended by a light spiral spring. A force of 2.0 kgwt is required to pull the ball downward 20 cm. Find its period of vibration.

8. The acceleration due to gravity on the moon's surface is $\frac{1}{6}$ that on the earth. That on the sun is 28 times that on the earth. A pendulum beating seconds on the earth is 25 cm long. How long a pendulum would beat seconds (a) on the moon, (b) on the sun?

9. When a 300-gm bird is on its perch in a

cage suspended by a spring, the cage is pulled 0.25 cm below its level when empty. The mass of the cage is 900 gm. Compute the period of vibration (a) of the empty cage, (b) of the cage when the bird is on its perch.

10. A swing has a period of 4.0 sec, and the amplitude is 1.0 m. Find (a) the speed of the imaginary particle moving in the reference circle for this vibration, (b) the maximum speed of the swing, (c) the length of the ropes.

11. In problem 3, find the speed of the shadow (a) when the crank is horizontal and (b) when it makes angles of 90°, 60°, and 30° with the horizontal.

12. A platform vibrates up and down with simple harmonic motion of amplitude 80 cm. What is its maximum frequency if a box resting on it remains in contact?

13. A particle is executing simple harmonic motion of amplitude 20 cm and period 12 sec. Find the displacement at a time (a) 2 sec, (b) 3 sec, (c) 6 sec, and (d) 9 sec after the particle reaches its maximum displacement.

C

14. An automobile travels along a road having a wavy surface. A passenger vibrates up and down with simple harmonic motion, the amplitude being 4.90 cm. What is the greatest frequency at which he will remain in contact with the seat?

15. A tuning fork makes 100 vibrations per second and the amplitude of vibration of a particle on the end of the fork is 1.00 mm. Find (a) the speed of the particle at the midpoint of the vibration, and (b) its speed 1/600 sec later.

16. Imagine a hole bored through the earth from pole to pole. The weight of a ball dropped into the hole can be shown to be proportional to its displacement from the center of the earth. (a) What would be the period of oscillation of a 10-kg ball? Assume

that $r = 6.4 \times 10^6$ m. (b) What would be the maximum velocity?

17. Find the period of vibration of a 100-gm uniform meter stick supported vertically at one end. Neglect friction and assume that the amplitude of vibration is small.

18. A block of iron of mass 50 kg attached to a spring vibrates with simple harmonic motion. If the amplitude of vibration is 25 cm and the time of a complete vibration is 0.80 sec, find the maximum kinetic energy of the block.

19. How much work is required to stretch a spring 10 cm if a 2.0-kgwt force will stretch it 1.5 cm?

20. Plot the ratio T/T_0 versus the angular amplitude. T_0 is the period for a simple pendulum as calculated from the approximate formula and T is the period calculated from the more exact formula. Cover the range of angular amplitude from 1° to 10° in steps of 1°.

21. A ball oscillates with simple harmonic motion, the period being π sec and the amplitude 12 cm. What is (a) the maximum speed of the ball, (b) its speed when its phase is 60°? (Assume 0° phase at maximum speed.) (c) What restoring force acts on the ball of mass 25 gm when its phase is 0°, 30°, 90°?

22. A 50-kg boy jumps from the top of a wall 1.0 m to the ground while wearing a pair of shoes, each shoe having two springs attached to it. The springs have a compression constant of 30 kgwt/cm. Assuming each spring does one-fourth of the work of stopping the boy, calculate how much each spring is compressed at maximum compression.

23. A 20-kg body falls freely and strikes a spring whose spring-constant is 60 kgwt/cm. The spring is compressed 3.0 cm. (a) Through what total vertical distance does the body move before it is brought momentarily to rest? (b) Describe the subsequent motion of the body.

24. Find the period of oscillation of a thin

ring (diameter 1.0 m, mass 1.0 kg) hanging on a knife edge.

25. A stick loaded at one end weighs 21.13 gwt and has a uniform cross-section area of 1.0 cm². It is dropped vertically into a liquid and oscillates up and down with a period of vibration of 0.25 sec. What is the mass density of the liquid? Neglect the damping of the motion.

For Further Reading

Feather, Norman, *An Introduction to the Physics of Vibrations and Waves,* Edinburgh University Press, (Aldine Publishing Company, Chicago, Ill.) 1961.

Press, Frank, "Resonant Vibrations of the Earth," *Scientific American,* November, 1965.

CHAPTER NINETEEN

Wave Motion

Some Definitions

If you drop a stone into a quiet pool and watch the waves which travel out in ever-widening circles, you will notice that small twigs and leaves, floating on the surface, are tossed about as the crests (high points) and troughs (low points) of the waves pass them. They are not carried forward but vibrate about fixed positions of equilibrium (Fig. 19-1). *A wave motion is a vibratory disturbance traveling through a medium.*

If you tie one end of a long rope to a hook fastened to the ceiling and suddenly jerk the other end sidewise, a single disturbance or "pulse" will travel up the rope. Each particle will move to and fro as the pulse passes (Figs. 19-2*A* and 19-2*B*). Vibrate the end, and a series of equally spaced waves will travel upward (Fig. 19-2*C*). Each particle of the rope moves *at right angles* to the direction of travel, hence we say that the waves are *transverse.*

Imagine that you vibrate one end of a long horizontal cord up and down with simple harmonic motion. Thus you will cause a series or train of waves which will travel along the cord. Each particle in it will vibrate with simple har-

monic motion (Fig. 19-3). Notice the space from *p* to *d*. It includes one complete hill and valley—one complete wave. The particles *p* and *d* are at the crests of waves. They are in the same condition of vibration. The distance between them is one wavelength. *One wavelength is the shortest distance between particles that are in the same condition of vibration.*

Consider the phases of different points on a train of waves. In Fig. 19-3, when the first wave reaches *a*, the phase of *a* is zero and it is moving upward. At this instant, the phase of *b* is 30°, that of *c* is 60°, and of *d* 90°. The phase of *m* is 360°. Notice that the particles *a* and *m* are in the same condition of vibration. (The displacement of each is zero and each is moving upward.) The distance between *a* and *m* is one wavelength. Notice also that the phase of *m* is 360° greater than that of *a*. Similarly, the phase difference between *b* and *n* is 360°. In general, particles which are one wavelength apart differ in phase by 360°.

Suppose that the speed of the waves along the cord pictured in Fig. 19-3 is 2 m/sec and that you vibrate the end once per second. Then, each wave will have traveled 2 m when the next one is started, so that this distance is

Fig. 19-1. *Water waves. Each particle vibrates. The particles do not advance. (Courtesy of Servel.)*

the wavelength. Vibrate the rope twice per second, and the wavelength will be one-half as great as before. If 10 waves are generated per second, each will travel $\frac{2}{10}$ m before the next one is started; this is the wavelength. In general, if f is the frequency of the waves (f waves are produced per unit time) and the speed is v, the wavelength λ (Greek *lambda*) is given by

$$\lambda = \frac{v}{f}$$

so that

$$v = f\lambda$$

This relationship is illustrated by the data in Table 19-1

Table 19-1. Frequencies, Speed, and Wavelengths

Frequency of Vibration, vib/sec	Speed of Waves, m/sec	Wavelength, m
1	2	2
2	2	1
10	2	0.2
20	2	0.1
f	v	v/f

Example. The wavelength of the waves on a lake is 6 m, and 30 waves pass a floating

barrel per minute. What is the speed of the waves?

$$v = f\lambda = 30/\text{min} \times 6\,\text{m} \times 1\,\text{min}/60\,\text{sec} = 3\,\text{m/sec}$$

Common Properties of Waves

Waves of many kinds occur in nature: ripples, deep-water waves on the ocean, waves on stretched strings, sound waves, and light waves, to name a few. All obey the equation $v = f\lambda$. They have other common properties. If you understand these similarities you will be able to predict the behavior of many apparently different kinds of waves.

This section contains a brief introduction to common wave properties—illustrated by the behavior of ripples. Later we shall discuss these properties in more detail as we discuss the various kinds of waves.

Water ripples that spread from a small vibrating source (Fig. 19-1) have circular wave fronts. *A wave front is a line connecting points that have the same phase.* The wave front is circular because the velocity of the disturbance is the same in all direc-

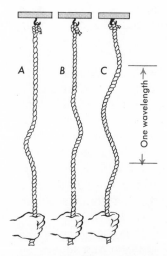

Fig. 19-2. *Transverse waves. Each particle vibrates at right angles to the direction of travel.*

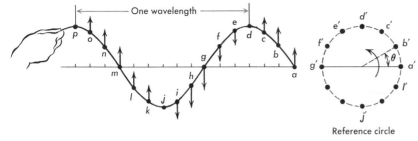

Fig. 19-3. *Phase difference in a wave train. The points* a *and* d *differ in phase by 90°;* d *and* p, *one wavelength apart, differ in phase by 360° or 2π radians.*

Fig. 19-4. *Ripples traveling to the right strike a plane surface and are reflected upward. (Courtesy Educational Services Incorporated.)*

tions in this case. In other kinds of waves this is not necessarily so; the velocity may be greater in one direction than another. The frequency with which the crests of the ripples pass a given point is equal to the frequency of the vibrator. However, if the source is moving, the frequency of the ripples is not that of the source; we shall learn later how to calculate the frequency of the waves from a moving source when we study the Doppler effect.

The ripples expand outward and would travel forever if it were not for (1) viscous forces that damp out the waves and (2) obstructions of one kind or another. Ripples in a viscous oil are soon suppressed; those on a lake may travel far over the surface of the water before dying out. Obstructions have various effects on waves:

Reflection. If the ripples from a vibrating line source like the vibrating bar in Fig. 19-4 — plane waves — hit a vertical wall they are reflected. The reflected ripples are also plane waves, but they travel in a different direction at the same velocity. In Fig. 19-7a, ripples with circular wave fronts are reflected backward.

Refraction. Let the ripples move from deeper to shallower water. In shallow water the velocity of the ripples is less. The wave fronts change their direction (Fig. 19-5) —

Fig. 19-5. Ripples traveling to the right pass over a glass plate, undergo a decrease in velocity, and are refracted downward. (Courtesy Educational Services Incorporated.)

unless they hit the shallow water at right angles—bending toward the perpendicular to the boundary between deeper and shallower water. The waves are *refracted* as their velocity decreases; their frequency is the same, but their wavelength is less. On going from shallower to deeper water, refraction again occurs, but the bending is in the other direction.

Diffraction. When ripples meet obstacles with openings in them, the wave behavior depends upon the size of the opening. If it is large, the wave fronts of the ripples that pass through are only slightly changed (Fig. 19-6a). If the opening is small—about as large as the wavelength—the ripples bend into the "shadow" of the obstacle. The bending is called *diffraction.*

Interference. When ripples cross each other, they go their separate ways (Fig. 19-7a); the crossing does not affect what becomes of the waves later. However, in special situations (Fig. 19-7b), ripples come together at certain points in such a way that the crest of one ripple always meets the crest of another; the displacement of the water at that point is the sum of the separate displacements and a doubly high crest results. The ripples *interfere constructively* at those points. At other points, crest always meets trough, and the ripples cancel each other—*destructive interference* occurs.

Polarization. All of the particles of the surface of the water move up and down always—not at one angle to the surface at one time and at some other angle at another—as the ripples go by. The ripples are *polarized.* Not all waves are polarized. Vibrate the end of the rope in Fig. 19-3 in a random fashion—not just up and down. The particles of the rope will move in random directions also—and the waves are unpolarized.

Waves on Stretched Cords

Experiments with stretched cords demonstrate that, the more tightly the cords are

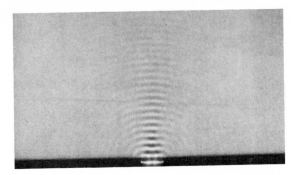

Fig. 19-6. Diffraction of waves. (a) *The wavelength is less than the width of the opening and the diffraction effects are relatively slight.* (b) *The wavelengths are greater; diffraction is more pronounced.*

stretched, the greater are the speeds of the waves. If a rope is replaced by one of greater mass per unit length, the tension being kept constant, the waves travel more slowly. In general, for a rope or cord of mass per unit length m/l, the tension being F, the speed v is given by

$$v = \sqrt{\frac{F}{m/l}}$$

Example. What is the speed of transverse waves in a cord which is 2.0 m long and has a mass of 0.1 kg? The tension is 5.0 newtons.

$$v = \sqrt{\frac{5.0 \text{ kg-m/sec}^2}{0.1 \text{ kg/2.0 m}}} = 10 \text{ m/sec}$$

Let us now derive the equation for the velocity of waves on a stretched cord. The

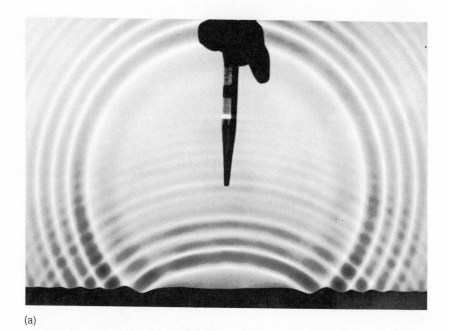

(a)

(b)

Fig. 19-7. (a) *Ripples cross each other.* (b) *Ripples interfere, producing maximum disturbance at some points and minimum disturbance at others.* (*From* PSSC Physics, *D. C. Heath and Company, 1960.*)

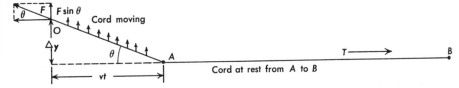

Fig. 19-8. *Deriving an equation for the velocity of a wave.*

derivation will help you to understand how the familiar laws of mechanics apply to waves. Suppose that in time t a vibrator attached to the cord at the left of Fig. 19-8 raises the end of the cord a *small* distance Δy, starting a pulse down the cord. The tension in the cord is F, and the upward force exerted by the vibrator is $F \sin \theta$ where θ is the angle between the moving part of the cord and the part at rest. The total length of the cord is l and its mass is m.

Consider the work done by the vibrator in raising the end of the cord. The vibrator works against an elastic force $F \sin \theta$, so, using the equation for work done against a spring ($W = \frac{1}{2}ks^2$), we have

$$W = \frac{1}{2}\left(\frac{F \sin \theta}{\Delta y}\right)(\Delta y)^2 = \frac{1}{2}F (\sin \theta) \, \Delta y$$

where $(F \sin \theta)/\Delta y$ is the spring-constant k. The disturbance travels a distance vt, where v is the velocity of the pulse in the time that the end of the cord is raised a distance Δy. Hence

$$\sin \theta = \frac{\Delta y}{\overline{OA}}$$

But for small pulses, $\overline{OA} \cong vt$, and

$$\sin \theta = \frac{\Delta y}{vt}$$

Therefore

$$W = \frac{1}{2}F\frac{\Delta y}{vt}\,\Delta y = \frac{1}{2}F\frac{(\Delta y)^2}{vt}$$

The work done by the vibrator in time t increases the energy of the cord from O to A. The energy of each bit of the cord at any time

is part kinetic and part potential. *On the average* the total energy of each bit equals its maximum kinetic energy. Thus we can treat the cord from O to A as if each bit had the same velocity v' where $v' = \Delta y/t$. The total energy of the cord is therefore the sum of the maximum kinetic energies:

$$E = \frac{1}{2}\left(\frac{m}{l}\right)vt(v')^2$$

where $(m/l)(vt)$ is the mass of the cord from O to A. The work W done in raising the end O must equal the energy E of OA. Therefore

$$\frac{1}{2}F\frac{(\Delta y)^2}{vt} = \frac{1}{2}\left(\frac{m}{l}\right)vt(v')^2$$

But $v' = \Delta y/t$, and

$$\frac{1}{2}F\frac{(\Delta y)^2}{v} = \frac{1}{2}\left(\frac{m}{l}\right)vt^2\left(\frac{\Delta y}{t}\right)^2$$

or

$$F = \left(\frac{m}{l}\right)v^2$$

Hence

$$v = \sqrt{\frac{F}{m/l}}$$

Progressive Waves in Cords

Progressive waves are ones in which the disturbance travels continually outward from the source. Waves in a long rope or ripples on a large pond are progressive waves. When waves are reflected upon themselves, interference results and *standing waves* are formed. We shall discuss the difference

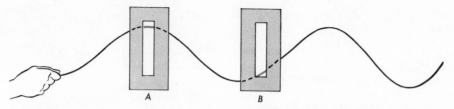

Fig. 19-9. A progressive wave in a long rope. Each bit of rope moves up and down with simple harmonic motion. The bit of rope at A is ahead in phase of that at B.

between progressive waves and standing waves in the next section.

In a progressive wave train, each particle undergoing the wave motion performs simple harmonic motion. If one end of a long rope is vibrated (Fig. 19-9), progressive waves move down the rope. Since the waves are eventually damped out, very little energy is reflected. A piece of cardboard with a slit cut in it allows one bit of the rope to be observed. The bit of rope moves up and down as the waves pass it—as if it were attached to a spring. Its motion is simple harmonic. All of the other particles of the rope are in simple harmonic motion also with the same frequency and (approximately) the same amplitude. The difference among them is one of phase. Bits of rope near the source of the waves are ahead in phase of the ones farther from the source. A crest reaches *A* (Fig. 19-9) before it reaches *B*, although both *A* and *B* are in simple harmonic motion. *In a progressive wave there is a progressive decrease in phase with increasing distance from the source.*

The displacement y_0 of a particle of the rope at the source of the waves (Chapter 18) is

$$y_0 = r \sin\left(\frac{2\pi}{T}\right)t$$

where r is the amplitude with which the end of the rope is vibrated and T is the period of the vibration (Fig. 19-10). The phase of this vibration at any time t is $(2\pi/T)t$. The equation for the displacement of a particle at a distance x from the source has the same form as that for y_0 except that the phase is less. It is less by the time for a disturbance to go from O to x multiplied by $2\pi/T$. If the wave velocity is v, this time is x/v. Thus

$$y = r \sin\left(\frac{2\pi t}{T} - \frac{2\pi x}{T v}\right)$$

or

$$y = r \sin \frac{2\pi}{T}\left(t - \frac{x}{v}\right)$$

This is the wave equation for a progressive wave traveling toward larger values of *x*. Notice that it describes (1) the motion for a given

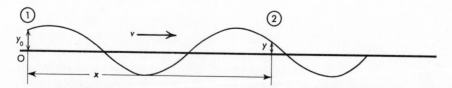

Fig. 19-10. The displacement of particles in a progressive wave. y = r sin(2π/T)(t − x/v).

particle (x specified) at all times and (2) the motion of all particles at a given time (t specified). In other words, the equation describes (1) how one particle moves up and down when seen through a slit (Fig. 19-9) and (2) what the whole wave profile is at any one time (Fig. 19-10).

Example. One end of a long cord is vibrated with a period of 2 sec and an amplitude of 4 cm. The velocity of the waves is 50 cm/sec. (a) Calculate the displacements of a particle 100 cm from the source at times $t = 4$ sec, 4.5 sec, 5 sec, and 5.5 sec. (b) Calculate the displacements of particles at 0, 25 cm, 75 cm, and 100 cm from the source at a time $t = 2$ sec.

(a) For a particle at $x = 100$ cm,

$$y = 4 \text{ cm sin} \frac{2\pi}{2 \text{ sec}} \left(t - \frac{100 \text{ cm}}{50 \text{ cm/sec}} \right)$$

When $t = 4$ sec,

$$y = 4 \text{ cm sin} \frac{2\pi}{2 \text{ sec}} (4 \text{ sec} - 2 \text{ sec})$$

$$= 4 \text{ cm sin } 2\pi \text{ radians}$$
$$= 4 \text{ cm sin } 360° = 0$$

Similarly, when

$$t = 4.5 \text{ sec}, \quad y = 4 \text{ cm}$$
$$t = 5 \text{ sec}, \quad y = 0$$
$$t = 5.5 \text{ sec}, \quad y = -4 \text{ cm}$$

(b) At a time $t = 2$ sec,

$$y = 4 \text{ cm sin} \frac{2\pi}{2 \text{ sec}} \left(2 \text{ sec} - \frac{x}{50 \text{ cm/sec}} \right)$$

For a particle at $x = 0$

$$y = 4 \text{ cm sin} \frac{2\pi}{2 \text{ sec}} \left(2 \text{ sec} - \frac{0}{50 \text{ cm/sec}} \right)$$

$$= 4 \text{ cm sin } 2\pi \text{ radians} = 0$$

Similarly, for

$$x = 25 \text{ cm}, \quad y = -4 \text{ cm}$$
$$x = 50 \text{ cm}, \quad y = 0$$
$$x = 75 \text{ cm}, \quad y = 4 \text{ cm}$$

Interference of Waves

Interference occurs when two or more wave trains, having traveled along different paths, recombine to produce a resultant effect. Suppose that you fasten two ropes to a single rope tied to a beam. If you vibrate the ends of the two ropes *in phase,* the two wave trains arriving at *F* will combine to produce waves of greater amplitude (Fig. 19-11a). If you vibrate the ends oppositely, the waves will oppose each other at *F*. The energy will be reflected (Fig. 19-11b).

Interference is the combination of waves that have traveled different paths to produce larger or smaller waves.

A stretched rope provides an excellent example of interference. To demonstrate its behavior, let two students hold the opposite ends of a rope and let each student jerk his

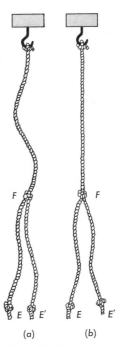

(a)   (b)

Fig. 19-11. Interference. (a) *Two wave trains combine at* F *to produce larger waves (constructive interference).* (b) *The two waves oppose each other at* F *(destructive interference).*

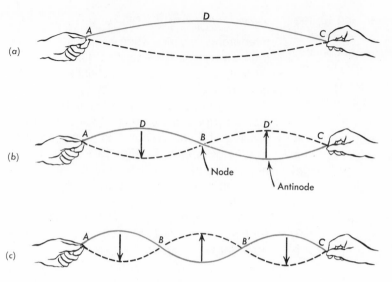

(a)

(b)

Node

Antinode

(c)

Fig. 19-12. Standing waves. (a) *Crests meet crests at* D, *producing maximum motion. Crests meet troughs at* A *and* C. (b) *The frequency is doubled. Antinodes occur at* D *and* D'. *Nodes are at* A, B, *and* C. (c) *Nodes are at* A, B, B', *and* C. *The cord vibrates with tripled frequency.*

hand upward at the same instant, so as to cause two pulses. They will meet at the midpoint and cause a great displacement there. Then let the ends move downward, and two downward pulses will cause a great downward displacement at the midpoint. Continued vibrations can be timed so that the rope vibrates in one segment (Fig. 19-12a). This is called a *standing wave* because, unlike a progressive wave, it does not seem to advance. (Actually, two trains of waves continually travel to and fro in the rope.) Vibrate the ends of the rope oppositely so that an upward pulse starts from *A* when a downward one starts from *C*. The two will interfere and counteract each other at *B*, which will be a *node* or point of minimum motion (Fig. 19-12b). Maximum motions occur in the *antinodes D* and *D'*, so that the rope vibrates in two segments. Each is one-half wavelength long. By vibrating the ends more rapidly, the rope may be made to oscillate in 3, 4, 5, or

more segments. (At each antinode, constructive interference occurs; at each node, destructive interference.)

A wire stretched between two pegs will oscillate in a similar manner. Pluck it upward at the midpoint, and two upward pulses will travel toward the ends. Since the ends are clamped, the pulses are reflected as downward pulses for a reason that we shall discuss later in this chapter. Then the pulses move to the opposite ends, where they are reflected as upward pulses. This process continues; the pulses run to and fro so that the wire vibrates in one segment. Hold a card at the midpoint to damp out the vibration of lowest frequency. The wire will vibrate in two segments with a frequency twice as great as before. By touching it at other places, you can make it vibrate in 3, 4, 5, or more segments.

The vibration of lowest frequency, when the string vibrates in one segment, we call its *fundamental* or *first harmonic.* The other modes of vibration in 2, 3, 4, . . . loops, of frequencies 2, 3, 4, . . . times that of the fundamental, are the second, third, fourth, etc., harmonics. The word harmonics originated from the study of stringed musical instruments. However, it is also applied to vibrations unrelated to music.

In a standing wave—unlike a progressive wave—the particles within any loop are all in phase with each other. They are 180° out of phase with particles in neighboring loops. The nodes in a standing wave are one-half wavelength apart. Hence, standing wave patterns are often used to measure wavelengths: λ equals twice the distance between successive nodes.

Not all bodies vibrate harmonically. The various modes of vibration of a bell, for example, have frequencies that are not integral multiples of a fundamental frequency.

Rods can also vibrate. Suppose that you support a long thin rod at its ends and vibrate them up and down in phase. Then two sets of transverse waves will travel to and fro along the rod. At the midpoint they always reinforce each other; hence the amplitude of the resulting wave is a maximum, and an antinode exists there. At each end the two waves oppose each other so that minimum amplitude results and a node is produced (Fig. 19-13*a*). If the two ends are vibrated in opposite phase, the two waves will counteract each other at the midpoint so as to produce a node (Fig. 19-13*b*). If you hold the rod at its midpoint, a node is formed there. The incident wave and the reflected wave reinforce each other at the ends, producing antinodes (Fig. 19-13*c*). The rod, if supported at one end, can vibrate with a node at this end and an antinode at the other (Fig. 19-13*d*). In general, there will be a node wherever the vibration of the rod is restricted.

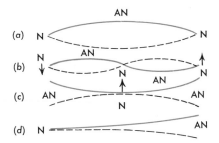

Fig. 19-13. *A vibrating bar can emit several harmonics.*

Compressional Waves

So far we have discussed transverse waves. Compressional waves are also important. Suspend a long massive spring horizontally by cords (Fig. 19-14). Push the end *A* toward *B* to compress the first section of the spring. Then *B* will compress the next section and exert a force on the loop at *C*, accelerating it to the right and compressing the next spring. In this manner, the compression will pass from each section to the next one beyond, and a pulse of compression will travel to the right through the system. Pull end *A* to the left to stretch the first section of spring. Then the second section will be stretched. Thus the impulse will be passed on to the next section, and a pulse of expansion will progress toward the right. Vibrate end *A* so as to compress and stretch the first section of spring periodically. A series or train of equally spaced disturbances will travel along the system. In this experiment, the wavelength is the distance from one point of maximum compression or expansion to the next (the distance between points whose phase differs by 360°). The loops of the spring vibrate in a direction *parallel to the line of travel,* and the waves are called *compressional* (or *longitudinal*). Compressional waves, like transverse ones, have velocities that depend upon the properties of the medium in which they are traveling. Compressional waves also

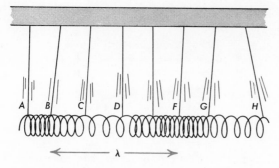

Fig. 19-14. Compressional waves can travel through this long spring.

show interference effects. We shall discuss the velocities and interference of compressional waves in the next chapter.

Change of Phase on Reflection

When a wave travels from one medium to another, in which the speed is different, part of the energy is reflected and part is transmitted. The reflected wave may undergo a 180° change of phase. Transverse waves as well as compressional waves show these effects. We shall illustrate them for the latter.

Suppose that a compression travels from *A* to the right along the line of cars in Fig. 19-15 in which all the springs are equally stiff. When the spring between *C* and *D* is compressed, the car *D* will move to the right, and part of the energy will travel onward. However, the car *D* is more massive than *C*; hence it will not move sufficiently to relieve the compression of the spring. Part of the energy will be re-

flected, and a compression will travel backward toward *A*.

Suppose that a compression travels from *F* to the left. When the spring between *D* and *C* is compressed, the car *C* will move to the left, and part of the energy will be transmitted as before. The car *C*, having smaller mass than *D*, will move so readily that the spring behind it will be *stretched*. Thus *D* will be pulled to the left, and a wave of *expansion* will be reflected.

In general, *when waves travel from a medium in which the speed is higher to one in which it is lower, a compression is reflected as a compression and the phase is changed by* 180°. *When waves travel from a medium in which the speed is lower to one in which it is higher, compressions are reflected as rarefactions and the phase is unchanged.*

Summary

A wave motion is a vibratory disturbance traveling through a medium. Waves include ripples on water, deep-water waves, sound waves, light waves, radio waves, waves on strings, and many others.

A transverse wave is one in which the particles vibrate at right angles to the direction in which the wave travels. In a compressional wave, the directions of vibration and of travel are parallel.

The wavelength is the smallest distance between particles that are in the same condition of vibration. The velocity of a wave is

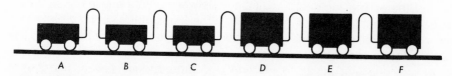

Fig. 19-15. Phase change on reflection. A compression traveling to the right is reflected at D *as a compression with 180° change of phase.*

A compression moving to the left will be reflected at D *as an expansion with no change of phase.*

given by $v = f\lambda$. The velocity of a wave on a stretched cord is $v = \sqrt{F/(m/l)}$.

In a progressive wave each particle executes simple harmonic motion. Particles nearer the source are ahead in phase of particles farther from the source. The displacement of a progressive wave is given by

$$y = r \sin \frac{2\pi}{T}\left(t - \frac{x}{v}\right)$$

When waves travel from one medium into another in which the velocity is less, part of the energy is generally reflected with a change of phase of 180°. If the velocity in the new medium is greater, the phase of the reflected wave is unchanged.

In interference, waves traveling along different paths combine to produce resultant waves.

In standing waves, there are nodes, or points of minimum motion, and antinodes, or points of maximum motion. A stretched string or a thin strip of metal vibrates in one segment in its fundamental vibration of lowest frequency. It can also vibrate at higher frequencies, called harmonics, which are exact integral multiples of the fundamental.

Questions

1. Distinguish between compressional (longitudinal) and transverse waves.

2. How do progressive and standing waves differ?

3. Describe what happens when waves are reflected, refracted, or diffracted. Refer to the ripple-tank pictures in Figs. 19-4 to 19-7.

4. What causes damping of waves?

5. Show that $(2\pi x/Tv)$ has the dimensions of an angle.

6. Give a familiar example of standing waves not discussed in this chapter.

7. What determines whether there will be a change of phase when a wave is reflected?

8. How will the speed of waves in a cord be changed if you double both the tension and the mass per unit length?

9. Plot the sine function $y = \sin \theta$ by looking up values for y for values of θ at intervals of 5° from 0° to 360° in a table of sines. (See Appendix D.)

10. If a piece of chalk squeaks when you write on the blackboard, how can you raise the sound frequency so that it is inaudible (i.e., ultrasonic)?

11. Prove that Fl/m has the dimensions of (velocity)^2.

12. What range of wavelengths is possible in a vibrating string?

13. Draw on your own experience to describe examples of the occurrence of the following in light or television waves: reflection, refraction, diffraction, interference, polarization?

14. When a string vibrates with n loops, prove that the frequency f of the waves is given by

$$f = \frac{n}{2l}\sqrt{\frac{F}{m/l}}$$

l = length of the string, m = mass of the string, F = tension of the string.

15. Show by sketches the fundamental wave and the second and third harmonics for (a) a vibrating string, (b) a vibrating rod clamped at the middle, (c) a vibrating rod clamped at the end.

16. Sketch the resultant waves when two waves, A and B, of equal wavelength and amplitude travel in opposite directions in a stretched string. Make a series of drawings as each wave advances in steps of 45°, where CD represents 360°. Interpret your drawings. (See Fig. 19-16.)

17. Show that the equation for a progressive wave on a stretched string can be written as $y = r \sin 2\pi(t/T - x/\lambda)$.

Problems

A

1. What is the wavelength of a broadcasting station that transmits waves at a

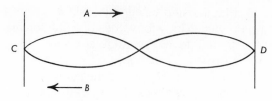

Fig. 19-16.

frequency of 600 kilocycles (6.00 × 10⁵ cycles) per second? (Speed of waves = 3.0 × 10⁸ m/sec.)

2. The speed of the transverse waves in a stretched string is 200 m/sec and the length of the string is 1.00 m. (*a*) How many to-and-fro journeys does a wave make per second? (*b*) What is the frequency of the fundamental?

3. If the velocity of water waves of wavelength 2.0 m is 4.0 m/sec, how many per second will pass a given point?

4. What is the frequency of the third harmonic for a rod clamped at one end that has for its fundamental a frequency of 20 vib/sec?

5. What is the wavelength of the third harmonic of a string that has a fundamental of wavelength 60 cm?

B

6. If the speed of transverse waves in a cord is 20 m/sec, what is the tension in newtons when the cord is 3.0 m long and its mass is 0.15 kg?

7. Find the speed of transverse waves in a cord if the tension is 4.08 kgwt and the mass per unit length is 25 gm/cm.

8. Transverse waves in a certain rope travel at a speed of 12.0 m/sec. The tension is 1.02 kgwt. What is the mass per unit length of the rope?

9. A load weighing 20 kgwt is attached to the lower end of a 1.0-kg rope 10 m long, tied to a beam. If the lower end of the rope is struck with a stick, a pulse will travel up the rope, be reflected, and return. What time is required for the to-and-fro journey? (Neglect tension due to the weight of the rope.)

10. A 20-m cable weighing 4.0 kgwt is stretched between two trees with a tension of 100 kgwt. How long will it take a transverse wave to make a round trip if the cable is struck at one end?

11. Two stretched wires, *A* and *B*, have equal diameters, lengths, and tensions. The ratio of linear densities of the wires is *A/B* = 11/7. What is the ratio of their fundamental frequencies?

C

12. Find the mass per unit length of a 1.0-m vibrating string with a second harmonic of 300 vib/sec. The tension is 100 newtons.

13. One end of a cord is vibrated by a tuning fork at 60 vib/sec and an amplitude of 4.0 cm. The velocity is 2.0 m/sec. (*a*) Calculate the displacements of a particle 4.0 cm from the source at times *t* = 0.040 sec, 0.045 sec, 0.050 sec, 0.055 sec. (*b*) Calculate the displacements of particles 0, 1.0 cm, 2.0 cm, 3.0 cm, 4.0 cm from the source at a time *t* = 0.02 sec.

14. Draw on a sheet of graph paper a sine curve *A* having an amplitude of 4 cm and a wavelength of 6 cm. Also, on the same axes, draw a sine curve *B* of amplitude 2 cm and wavelength 3 cm. Finally, draw the resultant curve starting with both waves in phase and continuing until the waves are back in phase. Interpret these graphs physically.

15. Draw a sine-wave train *A* consisting of 11 waves of amplitude 1.0 cm and wavelength 2.0 cm. Below this graph, draw a train *B* consisting of 10 waves of amplitude 1.0 cm, but of wavelength 2.2 cm. Let *A* and *B* be initially in phase with each other. Then plot a combination wave train, in which the displacement at each point is the sum of those of *A* and *B*. Interpret these graphs physically.

16. A progressive wave traveling to the right on a stretched string is described by $y = r\sin(2\pi/T)(t - x/v)$, and the reflected wave by $y = r\sin(2\pi/T)(t + x/v)$. Derive the algebraic sum of these two displacements and show

analytically that the resulting equation represents a wave pattern characterized by nodes (points at which the string is always at rest) and antinodes (points at which the string vibrates with maximum amplitude).

17. A man whose mass is 98 kg sits on a seat supported by springs. If the seat oscillates up and down, making 1.00 vibration per second, the amplitude being 8.0 cm, what are (a) the greatest force, and (b) the least force, exerted on him by the springs?

18. A compressed wave travels along a long spring, the displacement of the particles at a point on the spring from their equilibrium position at a given instant being represented by $d = A \sin 2\pi(t/T - x/\lambda)$. Sketch the wave patterns at times $t = 0$, 0.1 sec, 0.2 sec, 0.3 sec, etc. Let $T = 1.20$ sec, $A = 5$ cm, and $\lambda = 20$ cm.

For Further Reading

Bascom, Willard, "Ocean Waves," *Scientific American,* August, 1959, p. 74.

Bascom, Willard, *Waves and Beaches,* Doubleday and Company, Garden City, 1964.

Bernstein, J., "Tsunamis," *Scientific American,* August, 1954, p. 60.

Defant, Albert, *Ebb and Flow, The Tides of Earth, Air, and Water,* University of Michigan Press, Ann Arbor, 1958.

Oliver, Jack, "Long Earthquake Waves," *Scientific American,* March, 1959, p. 87.

CHAPTER TWENTY

Sound

Consider a rubber balloon, connected to a cylinder and piston (Fig. 20-1). Moving the piston downward causes the balloon to expand so that the walls move outward, compressing the surrounding air. Then the air molecules rush outward, crowding against their neighbors. These are accelerated in turn, and a pulse of compression travels outward. Raising the piston causes the balloon to contract. The surrounding air rushes inward, forming a rarefaction. Then the disturbance progresses outward from the balloon just as the compression did. Moving the piston up and down makes the balloon expand and contract periodically, so that a train of alternate compressions and rarefactions is set up. The wavelength is the distance between successive points of maximum compression. The frequency of the compressional waves thus produced is too low for our ears to hear them. They are of the same nature, however, as sound waves and have the velocity of sound waves.

The prongs of a tuning fork vibrate in simple harmonic motion, creating alternate compressions and rarefactions in the air (Fig. 20-2). Thus a train of waves is produced—a part of which is shown in Fig. 20-2—and each air molecule vibrates with simple harmonic motion in a direction parallel to the line of travel. The waves can be heard, and we call them *sound waves. Sound is a compressional wave train that can be heard.* Psychologists define sound as a sensation. Therefore they would justly claim that no sound can exist on a planet where there are no living beings. Many arguments arise from confusion as to the meanings of words.

Sound waves can be produced not only in air, but in water, in iron, or in any medium having volume elasticity. To demonstrate the transmission of sound through wood, scratch one end of a table with a pencil. A student who listens with his ear close to the other end of the table will hear the sound.

In order for sounds to be transmitted, there must be an elastic medium. If an electric bell is suspended in a bottle containing air, the vibrations of the bell cause sound waves. They vibrate the glass walls, and the walls in turn generate waves outside the bottle. If you pump air out of the bottle, the sound will become nearly inaudible. Sound cannot travel through a vacuum.

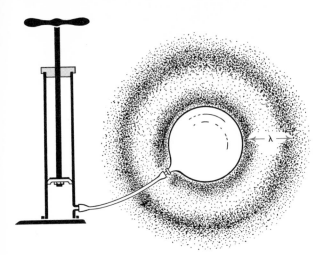

Fig. 20-1. Producing a compressional wave in air.

The Speed of Sound

The fact that sound travels more slowly than light is well known. The flash of a lightning discharge is seen before the sound of the thunder is heard. The jet of steam from a distant factory whistle is observed before the sound is perceived. The speed of sound in air may be determined approximately by the following simple experiment. Two groups of students are provided with signal rockets and stop watches. One group travels a mile or so by automobile along a straight highway, measuring the distance by means of the odometer of the car. This group discharges a rocket, and the first group observes the time elapsing between seeing the flash of the exploding rocket and hearing the report. Distance divided by time gives the speed of sound. Then the experiment is repeated, but the first group of students fires off the rocket and the second group observes the time. Averaging the two values reduces error if a wind, blowing from the one group to the other, tends to increase one speed and to decrease the other.

More sophisticated laboratory experiments for measuring the speed of sound use two microphones and a precision timer. A sharp pulse of sound emitted by a whistle travels from one microphone to another a known distance away. The time interval between the electrical signals produced by the microphones when the sound pulse strikes them can be measured accurately by means of an oscilloscope (Chapter 31). The separation of the microphones divided by this time is the

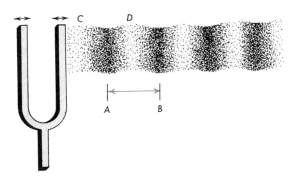

Fig. 20-2. Vibrating tuning fork. Its prongs alternately compress and rarefy the air, producing compressional waves that can be heard. The distance from compression to compression is one wavelength. Only a part of the wave fronts from the two prongs of the tuning fork is shown.

speed of sound. Still other methods for measuring the speed of sound depend upon resonance; we shall consider them later in this chapter.

Table 20-1. Speeds of Sound at 0°C

	m/sec
Air	332
Oxygen	317
Hydrogen	1270
Water	1432
Iron	5000

Two factors determine the speed with which a compressional wave travels through a medium: (1) the massiveness of the particles that vibrate, and (2) the elastic forces between them. Suppose that the suspended balls represented in Fig. 20-3 were replaced by other more massive balls, the same set of springs being retained. Then the inertia would be increased, a longer time would be required to set each ball in motion, and the speed of the wave would diminish. Suppose, next, that stiffer springs were used. The balls would respond more quickly because the forces exerted on them would be greater; the speed of the waves would increase. The speed of sound in a substance of mass density D_m is given by

$$v = \sqrt{\frac{E}{D_m}}$$

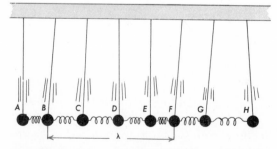

Fig. 20-3. The velocity of the compressional waves in this system depends upon the mass of the balls and the stiffness of the springs.

in which E is the elastic constant of that substance. This equation might be used to compute the speed of sound in iron, and E would be its Young's modulus. For a liquid or a gas, E is the modulus of volume elasticity. The moduli of volume elasticity of gases are proportional to their pressures p. The value for diatomic gases such as air, hydrogen, oxygen, and nitrogen is approximately

$$E = 1.40p$$

In general, for gases, $E = \gamma p$ in which γ (Greek *gamma*) is the ratio of c_p, the specific heat of a gas at constant pressure, to c_V, that at constant volume. For diatomic gases at room temperature (Chapter 17), c_p is approximately 7 cal/mole-K° and c_V (Chapter 15) is about 5 cal/mole-K°; their ratio γ is $\frac{7}{5}$ or 1.40.

The speed of sound in any diatomic gas is given by

$$v = \sqrt{\frac{\gamma p}{D_m}} = \sqrt{\frac{1.40p}{D_m}} \qquad (1)$$

The proofs of the equations for the speed of sound are similar to that for the speed of waves in a stretched cord.

Example. At 0°C and 76 cm-Hg (1.013×10^5 newtons/m²), the density of air is 1.293 kg/m³. Compute the speed of sound.

$$v = \sqrt{\frac{1.40 \times 1.013 \times 10^5 \text{ newtons/m}^2}{1.293 \text{ kg/m}^3}}$$

$$= 332 \text{ m/sec}$$

In accord with equation (1):

1. *The speed of sound in a gas is independent of the atmospheric pressure.* Doubling the pressure of a gas doubles the density (Boyle's law), and the ratio p/D_m is unchanged.
2. *The speed is independent of the frequency of the sound.*
3. *The speed is directly proportional to the square root of the absolute temperature.*

To show that the speed of sound is directly proportional to the square root of the Kelvin temperature T, suppose that the Kelvin temperature of a gas is quadrupled while the pressure is kept constant. Then the density will be one-fourth of its previous value, and by equation (1) the speed of sound will be doubled.

In more mathematical terms,

$$v \text{ varies as } \sqrt{\frac{p}{D_m}}$$

According to the general gas law,

$$pV \text{ varies as } T$$

and

$$\frac{p}{D_m} \text{ varies as } T$$

$$D_m \text{ varies as } \frac{p}{T}$$

Therefore

$$v \text{ varies as } \sqrt{T}$$

$$\frac{v_1}{v_2} = \sqrt{\frac{T_1}{T_2}}$$

Applications Making Use of the Speed of Sound

Submarines may be located by means of sound detectors. Sonar—SOund Navigation And Ranging—devices are carried by destroyers and other naval ships to detect and locate submarines by underwater sounds that they emit or reflect. In *passive* sonar devices, a transducer—a device like a microphone for converting the energy of underwater sound waves into electrical energy—"listens" for sound emitted by the engines of a submarine. Incident sound waves set up electrical signals in the transducer; these signals are amplified, producing "pinging" sounds in a loudspeaker. The sonar is highly directional; it not only reports that a submarine is nearby, but can tell the approximate direction.

Active sonar systems (much like radar devices of which you have undoubtedly heard) measure the distance to a target. A pulse of sound waves with a frequency between 5000 and 50,000 cycles/sec is emitted by a transducer—which can act as a source of sound as well as a detector—and travels under water until it strikes a submarine. Some of the sound energy is reflected back to the destroyer and produces an electrical signal in the transducer. The electrical signal makes a "ping" in a loudspeaker and also marks the end of the time interval required for sound to travel under water from the destroyer to the submarine and return. Knowing the speed of underwater sound—approximately 1500 m/sec—one can compute from the electronically measured time interval the distance from the destroyer to the submarine.

Active sonar is also used for peacetime purposes. Navigators use it in the form of a depth gauge to determine the depth of water under a ship. The transducer of a depth gauge sends sound pulses downward and they are reflected from the floor of the ocean. Fishermen use sonar to determine the depth of water and also to locate large schools of fish. The nuclear submarines that sail across the North Pole use sonar to map the ice overhead and to find clear places for surfacing.

By a method similar to sonar, petroleum deposits are located by measuring the distances from the earth's surface down to oil-bearing layers of rock. Dynamite is exploded at the surface, the time for the waves to descend and return is noted, and then the depth is computed from the known speed of sound in the earth. Antarctic explorers use this method also for determining the thickness of ice layers and the geological composition of the underlying land.

Sound travels better with the wind than against it. Figure 20-4 represents sound waves emitted by the source S. In still air, each wave travels the same distance, in all

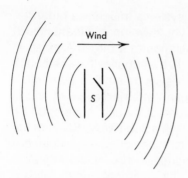

Fig. 20-4. *Sound waves in the wind.*

directions, in a given time. Thus each wave front is spherical. In the figure, a wind blows to the right and its speed increases with elevation. The sound waves are distorted by the wind. Because of this distortion, the waves traveling toward the right are deflected downward toward the earth. Those traveling to the left, against the wind, are deflected upward. Thus sound travels better with the wind than against it.

Interference of Sound Waves

Suppose that sounds from a tuning fork enter the tubes at *A* (Fig. 20-5) and travel along different paths. The intensity of the resultant waves at *D* depends upon the difference of path. If this difference is $\frac{1}{2}$, $\frac{3}{2}$, $\frac{5}{2}$, etc., wavelengths, compression will meet rarefaction, and the two sounds will interfere destructively and tend to cancel each other. If the difference is 0, 1, 2, 3, etc., wavelengths, compression will meet compression and rarefaction will meet rarefaction, and the sounds will interfere constructively and tend to reinforce each other.

When you pull the cork out of an empty bottle, you hear a popping sound. If you repeat the experiment after partly filling the vessel with water, the pitch will be higher than before. The frequency of the sound reinforced by an air column depends upon its length. Hold a vibrating tuning fork near the opening

of a vertical tube filled with water and gradually lower the water level in the tube. When the length of the air column is about one-fourth of the wavelength, the column will reinforce the sound from the fork. The phenomenon is one of resonance: the air column can be set into vibration by sound from a tuning fork of the proper frequency (Fig. 20-6).

Fix your attention on the lower prong of the fork. When it moves downward, it sends a *compression* into the tube, and a *rarefaction* outward, away from it. While the prong is moving downward, the compressional wave travels down to the water—a distance of $\frac{1}{4}$ wavelength. The compression is reflected as a compression at the water surface and travels an equal distance back to the opening. The total distance, down and up, is $\frac{1}{2}$ wavelength. Thus the compression gets back to the opening just when the fork begins to move upward from its equilibrium position. The prong, moving upward, will send out the next compression. The two compressions will join and go out together, producing a sound wave of increased amplitude.

This statement is so important that we re-

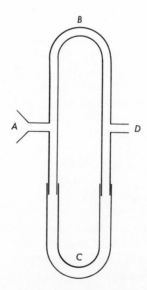

Fig. 20-5. *Interference of sound waves.*

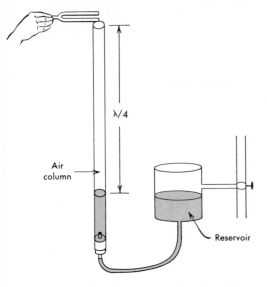

Fig. 20-6. *Reinforcement by interference. When the length of the narrow air column is $\frac{1}{4}\lambda$, each reflected compression reinforces a distance $\frac{1}{2}\lambda$ while the fork makes $\frac{1}{2}$ vibration.*

peat it. *When the length of the air column is one-fourth of the wavelength, a compression travels down and back while the fork makes one-half vibration. Thus the reflected compression reinforces the next one sent out by the fork.*

We can also explain reinforcement by the air column by considering standing waves. When such a narrow air column vibrates with its fundamental frequency (first harmonic), there is a velocity node at its closed end (air molecules are prevented from moving longitudinally there) and an antinode at its open end (Fig. 20-7a). When it emits its third harmonic, there are two nodes and two antinodes (Fig. 20-7b). For the fifth harmonic, there are three nodes and three antinodes (Fig. 20-7c). Like a rod clamped at one end, the column can vibrate when its length is $\frac{1}{4}$, $\frac{3}{4}$, $\frac{5}{4}$, . . . times the wavelength.

The existence of nodes and antinodes in a tube can be demonstrated by means of the apparatus diagramed in Fig. 20-8. It consists

of a piece of pipe, one end of which is closed, the other covered by a sheet of cellophane or paper. Holes, drilled at equal intervals along the upper edge of the pipe, serve as jets for burning gas. When no sounds are incident on the cellophane, all the gas jets are equally high. When an organ pipe is sounded nearby, the sound waves travel to and fro in the tube. At points where they interfere to cause nodes, the jets are of constant height. At the antinodes, where the motion of the molecules is greater, the pressure varies, and the jets dance up and down. If an organ pipe of variable pitch is used, the nodes and antinodes, moving to and fro as the wavelength is varied, give a fascinating spectacle.

So far, we have seen that sounds of constant wavelength λ are reinforced by tubes closed at one end of lengths $\frac{1}{4}\lambda$, $\frac{3}{4}\lambda$, $\frac{5}{4}\lambda$, etc. A complex sound consists of a mixture of

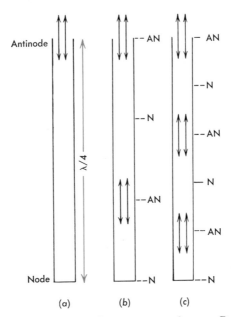

Fig. 20-7. *Vibrating air columns. Reinforcement occurs when the length of the air column is $\frac{1}{4}$, $\frac{3}{4}$, $\frac{5}{4}$, . . . times λ. The arrows suggest the amplitudes of longitudinal vibrations. These vibrations are greatest at velocity antinodes.*

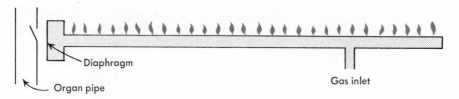

Fig. 20-8. *Dancing gas jets. At certain points, the gas pressure varies, making the jets jump up and down. At other points, the pressure is constant.*

sounds of different frequencies. Such a sound can be reinforced by a tube of *fixed* length. A blast of air, blown past the edge of such a tube (as in the whistle or organ pipe), is a complex sound. The tube will pick out and reinforce sounds of certain frequencies within the mixture. These frequencies are 1, 3, 5, etc., times that of the fundamental, whose wavelength is approximately 4 times the length of the tube.

Example. A boy blows across the top of an empty bottle whose height is 20 cm. What is the frequency of the third harmonic produced if the speed of sound is 34,400 cm/sec?

The wavelength of the fundamental note is approximately 4 times the height of the bottle or 80 cm. The frequency of the *fundamental* is

$$f_{\text{fund}} = \frac{v}{\lambda_{\text{fund}}} = \frac{34{,}400 \text{ cm/sec}}{80 \text{ cm}} = 430 \text{ vib/sec}$$

The frequency of the third harmonic,

$$f_3 = 3f_{\text{fund}} = 1290 \text{ vib/sec}$$

A narrow tube open at both ends can also emit sounds. When the fundamental tone is generated, there is a velocity node at the midpoint and an antinode near each end. A sudden compression at one end of the tube will travel through the tube and escape. As it leaves, the air behind it expands suddenly, forming a rarefaction. This will travel back through the tube, escape, cause another compression, and so on. Thus, a single disturbance gives rise to a series of compressions

and rarefactions which travel past each other. These waves interfere and produce a node at the middle of the tube. The air is least restricted near the ends; hence these must be regions of maximum vibration, or antinodes (Fig. 20-9). The length of the tube is approximately equal to the distance between two antinodes in the fundamental, hence the wavelength of the fundamental or first harmonic approximately equals twice the length of the tube. All of the harmonics—not just the odd-numbered ones as in the closed-end organ pipe—are present; their frequencies, as always, are proportional to the integers 1, 2, 3, ...*

---

* A harmonic is a vibration whose frequency is an integral multiple of the frequency of the fundamental, which has the lowest frequency in the system being studied. The fundamental itself is the first harmonic; the other harmonics have frequencies 2, 3, 4, 5, . . . times that of the fundamental. *Not all harmonics are present in certain systems, however.* The word *overtone,* on the other hand, is used to mean *other sounds* of frequencies greater than that of the fundamental. The frequencies of overtones are not necessarily integral multiples of that of the fundamental although they may be so in some systems. We shall stress "fundamental" and "harmonics" in this book, but an example of the other term "overtone" is in order: A closed-tube organ pipe produces sounds whose frequencies are proportional to 1, 3, 5, 7, ... The fundamental, or first harmonic, has the lowest frequency; the third, fifth, seventh, etc., harmonics are present and have frequencies 3, 5, 7, . . . times that of the fundamental. The sounds whose frequencies are greater than the fundamental are the overtones. Thus the first overtone—*for this system*—is the third harmonic; the second overtone is the fifth harmonic; the third overtone is the seventh harmonic; and so on. For a vibrating bell, the first overtone might have a frequency 2.73 times that of the fundamental; this would not be a harmonic.

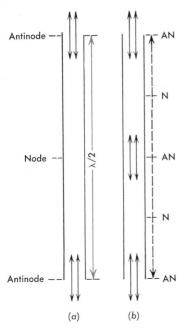

Antinode --

Node --

$\lambda/2$

Antinode --

AN

N

AN

N

AN

(a)   (b)

Fig. 20-9. A tube open at both ends. It emits: (a) first harmonic — one node and two antinodes, (b) second harmonic — two nodes and three antinodes.

Sounds of **different** frequencies interfere to produce **beats.** When two tuning forks of slightly different frequency are sounded together, a rising and falling of sound is heard. At a certain instant, compressions from both forks reach the ear, and the resultant sound is very loud. Later, when the two sounds are out of phase, the compressions from one fork counteract the rarefactions from the other. Thus, as the forks get into and out of step or phase, the sound increases and decreases in intensity. The periodic reinforcements of sounds are called beats, and the frequency with which this occurs is the beat frequency.

Suppose that two men walk side by side and that one takes 60 steps per minute, the other, 59. They will be in step **once** per minute. If the second man takes 58 steps per minute, he will take 29 steps while the first

takes 30, and they will be in step **twice** per minute. The number of times per minute that the two men get into step equals the difference between the frequencies of the steps. Similarly, the beat frequency f of the beats produced when two tuning forks of frequencies f_1 and f_2 are sounded together is given by

$$f = f_2 - f_1$$

The phenomenon of beats is utilized in tuning pianos. Suppose that a tuning fork of frequency 514 vib/sec is sounded along with a certain piano wire and that 8 beats/sec are heard. Suppose, further, that as the string is tightened, the number of beats gradually decreases to 2/sec. In this condition, the wire has a frequency 514 vib/sec − 2 vib/sec = 512 vib/sec.

The Doppler Effect

As we noted earlier, the frequency of a wave, usually equal to that of the source, can be changed by motion of the source or the observer or both. When an observer travels toward the source of sound, more waves enter his ear per second, and the frequency is raised. When he travels away from the source, fewer waves reach him per second, and the frequency is lowered. This phenomenon is called the **Doppler effect** and holds for all waves. We shall consider here the Doppler effect in sound.

When your automobile passes another, the horn of which is sounding, you may notice a pronounced drop in the frequency of the sound. Before passing the car, the frequency of the sound was raised above normal. Afterward, it was lowered below normal — hence the sudden drop in frequency at the instant of passing. If both cars were at rest, the frequency of the sound you hear would be given by the usual formula

$$f = \frac{\text{Wave speed}}{\text{Wavelength}} = \frac{v}{\lambda} \qquad (2)$$

When you travel toward the source of sound with a velocity v', the apparent wave speed is $v + v'$; hence the frequency is given by

$$f' = \frac{v + v'}{\lambda} \qquad (3)$$

Dividing equation (3) by equation (2),

$$\frac{f'}{f} = \frac{v + v'}{v} \qquad (4)$$

If you move away from the source,

$$\frac{f'}{f} = \frac{v - v'}{v} \qquad (5)$$

Suppose next that the observer is at rest and the source of sound is moving with a velocity v'. Let the moving source send out a compression and then—at a time $1/f$ later—send out the next compression. The distance between the compressions will be $v(1/f) - v'(1/f)$ because the moving source will have caught up with the first compression by the distance $v'(1/f)$ during the time $1/f$ between the sending out of successive compressions. But this distance between compressions is the *apparent* wavelength λ':

$$\lambda' = v\frac{1}{f} - v'\frac{1}{f}$$

or

$$\lambda' = \frac{v - v'}{f} \qquad (6)$$

and the apparent frequency f' will be given by

$$f' = \frac{v}{\lambda'} = \frac{fv}{v - v'}$$

or

$$\frac{f'}{f} = \frac{v}{v - v'}$$

In general, for a moving source when the observer is at rest,

$$\frac{f'}{f} = \frac{v}{v \pm v'}$$

where the positive sign in the denominator on the right is used when the source moves away

from the observer and the minus sign is used when the source moves toward the observer.

The Intensity of Sound Waves

The *intensity* of a sound means its *power per unit area.* Sound waves that deliver energy at a rate of one-millionth of a watt per square centimeter thus have an intensity of 1 microwatt per square centimeter. The intensity can be shown to be proportional to the square of the amplitude of the sound wave and to the square of the frequency.

The intensities of sound waves are astonishingly small. A barely audible sound has an intensity of 10^{-16} (a ten million-billionth of a) watt per square centimeter. The energy delivered to your ear by such a sound would lift a mosquito vertically about 1 ft in a year! The deafening shriek of a nearby factory whistle would lift the mosquito about 30 ft/sec! (Table 20-2).

Table 20-2. Intensities and Intensity Levels of Familiar Sounds

Source of Sound	Intensity Level, Decibels	Intensity Level, Bels	Relative Intensity	Intensity, watts/cm²
Jet planes	130–170	13–17	10^{13}–10^{17}	10^{-3}–10
Beginning of painful hearing	140	14	10^{14}	10^{-2}
Boiler factory	100	10	10^{10}	10^{-6}
Machine shop	70	7	10^{7}	10^{-9}
Ordinary conversation	60	6	10^{6}	10^{-10}
Whisper at 4 ft	20	2	10^{2}	10^{-14}
Empty church, cave	10	1	10^{1}	10^{-15}
Faintest perceptible sound	0	0	10^{0}	10^{-16}

The intensities of sound waves differ so greatly that we express their ratios by exponents. If sound *A* is 100 times (10^2) as intense as sound *B*, we say that its *intensity level* is 2 *bels* higher. If *C*'s intensity is $\frac{1}{10}$ (10^{-1}) as great as *A*, its intensity level is 1 bel lower. The following table will help you to understand:

Relative intensity I/I_0	10^0	10^1	10^2	10^3	10^6
Intensity level Bels	0	1	2	3	6

The difference between the intensity levels B_1 and B_2 of two sounds is the logarithm to base 10 of the ratio of their intensities. $B_2 - B_1 = \log I_2/I_1$. A smaller unit, the decibel, is commonly used. When two sounds differ in level by 1 decibel, the ratio of their intensities is $10^{0.1} \cong \frac{5}{4}$.

The *period of reverberation* of a room, a term used in architectural acoustics, is the time required for a sound to die down to one-millionth of its original intensity, that is, for the intensity level to fall by 60 decibels. This period, for a large hall, should be between 1 and 2 sec so that echoes will not be troublesome. The period of reverberation for a hall can be controlled by covering the walls with porous acoustic plaster or acoustic tile to reduce the period to a desired value. In designing an auditorium, the architect must know the *absorptivities* of the walls, floor, and ceiling. The sound absorptivity of a surface (Table 20-3) is the fraction of the incident sound energy that is absorbed by or transmitted through that surface. All the energy incident on an open window is transmitted, and none of it is reflected; therefore it is considered to be a standard, perfect absorber.

Table 20-3. Sound Absorptives (at 500 cycles/sec)

Open window	1.00
Compact audience	0.95
Heavy curtains in thick folds	0.50
Acoustical plaster	0.50
Unvarnished wood	0.10
Ordinary plaster, varnished woods, glass, linoleum	0.05

Hearing

The human ear is a marvelously intricate device for detecting sound. Figure 20-10 is a simplified diagram showing parts of the ear that are most important in hearing. Waves entering the outer ear cause the drum to vibrate.

Three tiny interlocked bones—the *hammer, anvil,* and *stirrup*—transmit the motion to a flexible membrane covering a "window" into the cochlea (Latin, *a snail*). This tube, about an inch long and coiled like a snail's shell, is divided into halves by a long, flexible separating wall (the basilar membrane) on which there are thousands of nerve endings. Compressional waves, set up in the liquid filling the cochlea, cause the basilar membrane to vibrate. When the sounds are of low frequency, the compressional waves penetrate to the far end of the tube and affect the nerve endings located there. At the highest audible frequencies, the waves do not penetrate so far and only the nerve endings nearest to the ear drum are affected. Thus the sensation of pitch depends upon the excitation of the different nerve endings by sounds of different frequencies.

Loudness is a psychological term and means the amount of *sensation* that a sound produces. It depends upon the intensity of the sound waves, their frequency, and the sensitiveness of the ear. To a totally deaf person, all sounds have zero loudness.

Figure 20-11 shows how the sensitivity of the normal ear depends upon frequency. The upper curve $A'BD$ marks the intensity at which sounds become painfully loud. Below $AB'C$ they are inaudible. Notice that the normal ear cannot detect sound waves of frequency less than 16 per second or "ultrasonic" waves of frequency greater than about 20,000 per second. The ear is most sensitive to sounds of frequency between 1000 and 2000 vib/sec.

The curve $A'C$ represents the sensitivity curve of a certain person whose hearing is 25% impaired. His deafness is most pronounced for tones of low frequency.

Music

Almost everyone loves music. Unfortunately, the scientific aspects of music to most people

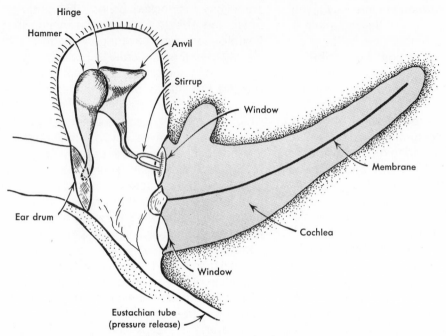

Labels on figure: Hinge, Hammer, Anvil, Stirrup, Window, Membrane, Ear drum, Cochlea, Window, Eustachian tube (pressure release)

Fig. 20-10. The human ear. Simplified diagram. Vibrations of the ear drum, transmitted through the lever system to the window, set up waves in the liquid in the cochlea. High frequency waves affect nerves near the "window"; those of lower frequency travel farther along the tube.

today are as they were to Sir Francis Bacon in the days of Shakespeare, "something that is ill-understood, something that may truthfully be reckoned as one of the subtlest pieces of nature."

In musical sounds, the waves are spaced in an orderly manner. Noises have no regularity; a noise is a jumble of sound waves. Hold a piece of cardboard against the edge of a revolving saw with uniformly spaced teeth. The emitted sound will have a definite pitch and will be musical. If the teeth are spaced unevenly, the sound waves will be irregular, and a noise will be heard.

The musician and the psychologist distinguish three characteristics of a musical sound, *loudness, pitch,* and *quality.* The loudness depends upon the sensitiveness of the ear. It also depends upon the *power* of the

sound or the time rate at which sound energy enters the ear, and hence upon the amplitude of the vibration. The pitch is the position of the sound on a musical scale. When two tuning forks are of the same frequency, the musician judges that the sounds that they emit have the same pitch. (However, in complex blends of frequencies, it is easy to fool the ear.) The third characteristic, the quality, depends upon the combination of harmonics. We distinguish the music of the violin from that of the saxophone by the qualities of their sounds.

In most musical instruments, the sound-producing agencies are strings, stretched membranes, rods, reeds, or air columns. In the piano and violin, the frequencies of the strings are varied by altering their lengths, tensions, and diameters. An important part of these instruments is the sounding board, set

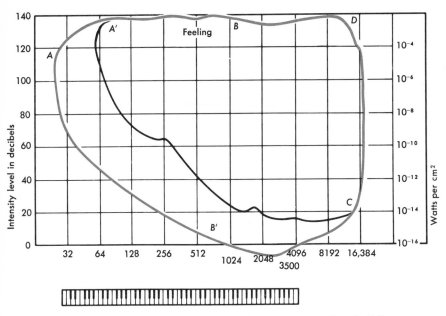

Fig. 20-11. *Sensitiveness of the normal ear to sounds of different frequencies. A barely audible sound of frequency 32/sec is 10^7 times more intense (70 decibels higher in level) than one of frequency 1000/sec. The curve A'C marks the lower limit for a certain person whose hearing is impaired for sounds of low frequency.*

into vibration by the strings, to increase the area of contact with the air. In the xylophone the frequencies of the rods or bars are varied by varying their lengths. The harmonica and accordion have reeds fixed at one end and set into vibration by an escaping stream of air. Most organ pipes have no reeds. Instead, a jet of air strikes against a sharp edge, causing a disturbance which travels up and down the tube, forcing the jet itself to move in and out periodically with frequencies determined by the dimensions of the tube. The flute operates on the same principle.

In the clarinet, the frequency of the vibrating reed is determined not only by its own stiffness and dimensions, but partly by standing waves set up in the tube, and partly by the pressure exerted upon it by the player. The clarinet responds like a closed pipe in which the odd-numbered harmonics predom-

inate. In the trumpet, cornet, and French horn, the lips of the player vibrate. The pitch depends principally upon the length of the air column, but the musician, by changing the tension of his lips, can determine whether the fundamental or some higher harmonic is most prominent. The bugler cannot change the length of the resonant column of his instrument, but must rely solely upon blowing the various harmonics. Most bugle calls use only the third, fourth, fifth, and sixth harmonics.

The human vocal cords are a musical instrument! In the human voice mechanism, the two vocal cords form lips at the end of the windpipe. When the cords are loose and flabby, they vibrate as a whole with low frequency. As the tension increases, a smaller portion of each cord vibrates, and the frequency is increased both because of the

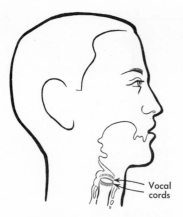

Fig. 20-12. Air passing between the vocal cords makes them vibrate and produce sounds which form the basis of speech.

increased tension and because of the **decreased mass** of the vibrators (Fig. 20-12). The sounds produced by the vocal cords are reinforced by the cavities and air columns in the throat and mouth, giving each voice its own quality.

When two tones are sounded together and the result is pleasing, they are said to be *in harmony.* The frequency ratio (interval) of two harmonious tones is often expressible by small integers. For example, a note and its octave have a frequency ratio 2/1 and are most harmonious. The notes *do, mi, sol,* sounded simultaneously, constitute a harmonious cord. Their frequencies are proportional to 4, 5, 6, and the ratio of any two of these numbers is expressed in small integers, 4/5, 4/6, 5/6. The notes *mi* and *fa* sounded together are less harmonious; their frequency ratio is 15/16. Actually, what is harmonious depends much upon a person's musical background. Many musical compositions accepted today would have sounded outlandish to an ancient Greek.

Sound waves can be studied by means of a cathode-ray oscilloscope (Fig. 20-13). When no sound waves strike the microphone, the luminous spot of the oscilloscope (Chapter 31) travels horizontally across the screen.

Sound waves striking the microphone set up an electrical wave that is the counterpart of the compressional wave. The electrical wave deflects the spot of the oscilloscope, causing it to trace the characteristic wave form of the sound. When a tuning fork is struck with a wooden stick, a harsh, metallic clang is heard, which rapidly dies out. Afterward, the tone is almost pure; its graphic representation is a sine wave. The oscilloscope trace, Fig. 20-14*a*, tells us that the clang of the fork when it is first struck is due to a harmonic of frequency greater than that of the fundamental of the fork. Figure 20-14*b* reveals harmonics in the sound from a good violin, played by an expert. In Fig. 20-14*c*, sounds of voice and orchestra combine a tumult of harmonics. Nevertheless, the ear can distinguish the violins from the flutes. Beats produced by two tuning forks of different frequencies are recorded in Fig. 20-15. It is possible to blend pure tones of different frequencies to reproduce the quality of sound from a given source

Fig. 20-13. A cathode-ray oscilloscope is used to study the wave form of sound produced by an electric organ. (Courtesy Hugh Lineback.)

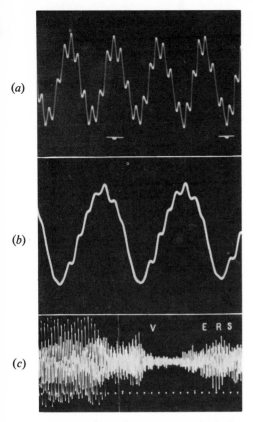

(a)

(b)

(c)

Fig. 20-14. Sound wave traces: (a) the clang from a tuning fork just after it has been struck, (b) a violin, (c) the sextette from Lucia.

such as a human voice, the violin, or almost any other musical instrument (Fig. 20-16). Similarly a complex wave form can be analyzed into its component sine waves by a mathematical procedure called *Fourier analysis.*

Summary

Sound is a compressional wave train that can be heard. The speed of sound in a gas is independent of the pressure and directly proportional to the square root of the Kelvin temperature. The speed of sound in diatomic gases, such as oxygen or nitrogen, is given by $v = \sqrt{E/D_m} = \sqrt{1.40p/D_m}$.

The sounds emitted by a tube closed at one end are the odd-numbered harmonics whose frequencies are 1, 3, 5, 7, . . . times that of the fundamental tone, or first harmonic. Those emitted by a tube open at both ends have frequencies 1, 2, 3, 4, 5, . . . times that of the fundamental.

Two sound sources of different frequencies combine to cause a sound which rises and falls in intensity, producing beats. The number of beats per unit time is given by $f = f_2 - f_1$.

When an observer moves toward or away from a stationary source of sound at a speed v' the frequency f' of the sound that is heard is given by

$$\frac{f'}{f} = \frac{v \pm v'}{v}$$

The intensity of a sound is the power per unit area. It may be expressed in microwatts per square centimeter. The difference in intensity levels $B_2 - B_1$ between two sounds, in bels, is the logarithm of the ratio of their intensities. $B_2 - B_1 = \log I_2/I_1$. One bel = 10 decibels.

The psychological characteristics of musical sounds are loudness, pitch, and quality. The loudness depends upon the amplitude of the sound vibrations and the sensitiveness of the ear, the pitch upon the frequency, and the quality upon the blend of frequencies that are present.

Questions

1. Why can sound be heard more easily under water in a swimming pool than through the air?

Fig. 20-15. Trace caused by sound from two forks of different frequencies showing "beats."

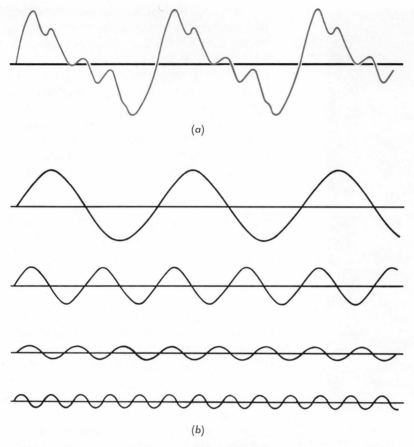

(a)

(b)

Fig. 20-16. (a) *The note* G$_4$ *(392 vib/sec) as produced by a violin.* (b) *The sound analyzed into its four most prominent harmonics, each of which is a pure sine wave. Frequencies are proportional to 1, 2, 3, 5.*

2. Discuss the vibration of air columns in a tube closed at one end and also in a tube open at both ends. Sketch the standing wave patterns for several of the harmonics present.

3. Is resonance desirable in (*a*) the sounding box of a violin, (*b*) a radio loudspeaker?

4. Why do most children have high-pitched voices?

5. If a cornet and a piano are in tune at 20°C, will they be in tune at 30°C? Why?

6. Would a plucked violin string vibrate for a longer or shorter time if the violin had no sounding board? Why?

7. What is meant by high-fidelity (hi-fi) reproduction of sound?

8. Are there sound waves on Mars? Discuss.

9. Does the reverberation time in an empty classroom increase or decrease when a class enters?

10. Which travels faster, sound or a bullet, (*a*) through air, (*b*) through water? Discuss.

11. How could you determine your distance from a cliff by the echoes?

12. A tuning fork is sounded first in air and then in hydrogen. Compare (a) the frequencies, (b) the wavelengths, of the sounds produced in the two gases.

13. Show that $(E/D_m)^{1/2}$ has the dimensions of velocity.

14. What are beats? How can they be produced?

15. Compare the three physical characteristics of sound waves and the three psychological characteristics.

16. Describe an observation from your personal experience or an experiment that shows that the velocity of sound is independent of frequency.

17. What is the difference between the intensity of a sound and its loudness? How much does the intensity of a sound change when the intensity level is (a) doubled, (b) tripled? Discuss the advantage of describing the loudness of sounds by "pianissimo" and by "15 db."

18. What is the Doppler effect? Describe a situation where you have observed it. Would it apply to music from a car radio in a car moving toward the radio station to which you were listening?

Problems

Speed of sound in air: 0°C, 332 m/sec; 20°C, 344 m/sec.

A

1. A tuning fork emits a sound of frequency 800 vib/sec. Find the wavelength (a) in air when the speed is 340 m/sec, (b) in hydrogen when the speed is 1360 m/sec.

2. At 27°C, sound travels in air 2000 km/hr. How fast does it travel in the stratosphere if the temperature is −80°C?

3. The speed of sound in oxygen at 0°C is 317 m/sec. Find the speed in hydrogen at this temperature. The density of hydrogen is $\frac{1}{16}$ that of oxygen. (Assume normal pressures.)

4. Thunder was heard 4.0 sec after a lightning stroke. How far away was the lightning? (The temperature was 20°C.)

5. A circular saw has 60 evenly spaced teeth. How many revolutions per minute must it make in order to emit a note of frequency 600 vib/sec when a stick of wood is pressed against its edge?

6. A sound is emitted under water by a fathometer mounted on a steamship. The echo from the bottom is heard 2.5 sec later. How deep is the ocean at that point if the speed of sound in the water is 1430 m/sec?

7. The frequency of the tone of the lowest pitch that is audible to most people is 20 vib/sec and that of the highest audible pitch is approximately 20,000 vib/sec. Find the wavelengths of these sounds in air at 20°C.

8. An empty cartridge shell is 2.5 cm deep. (a) What is the fundamental frequency of the note that it emits when a person blows across the edge? (b) What is that of the third harmonic? (The temperature is 20°C.)

B

9. Young's modulus of a piece of metal is 16.2×10^{11} dynes/cm², and the density of the metal is 7.8 gm/cm³. What is the speed of a compressional wave in it?

10. The intensity of a sound wave is 10 microwatts/cm². To what should this be increased in order to produce a 5.0-db rise in intensity level?

11. What is the frequency at 20°C (a) of the fundamental and (b) of the fifth harmonic of a 1.0-m organ pipe, open at one end?

12. The speed of sound at 27°C is 350 m/sec. How long is a closed tube that emits a fundamental tone of frequency 600 vib/sec at this temperature? At what temperature would the same tube emit a tone of frequency 300 vib/sec?

13. In tuning a piano string, a fork of frequency 500 vib/sec is used. As the wire is progressively tightened, the number of beats decreases until the value is 6/sec. What then is the frequency of the string?

14. The frequency of the fundamental tone of a gong is 1000 vib/sec. What is the frequency of the sound that an observer hears (a) if he approaches the source at a speed of 30 m/sec, (b) if he moves away from the source at this speed? (Speed of sound is 330 m/sec.)

15. If the speed of sound is 330 m/sec, how long must four tubes closed at one end be in order to emit the chord (do, mi, sol, do') of frequencies proportional to 4, 5, 6, and 8? Let the lowest note have a frequency of 256 vib/sec.

16. Two organ pipes, each closed at one end, are, respectively, 0.80 m and 0.85 m long. Find the frequency of the beats of the two fundamental tones if the speed of sound is 330 m/sec.

C

17. If a phonograph record turning at 33 rpm plays an orchestral composition in tune for $A = 440$ vib/sec, what will be the frequency of A if the disk turns at 35 rpm? Will it still sound like music at 35 rpm?

18. (a) How many beats per second are there when notes of fundamental frequency 30/sec (mi) and 32/sec (fa) are sounded together? (b) What is the number of beats between the eleventh harmonic of the mi and the tenth harmonic of the fa when they are sounded together? Discord due to some harmonics is more pronounced than that due to the fundamental tones.

19. A tube closed at one end emits a fundamental tone of frequency 340 vib/sec. What is the frequency of the harmonic produced when a hole located 8.33 cm from the closed end is uncovered? (Speed of sound is 340 m/sec.)

20. The closing of a double window decreases the intensity level of the street noises entering through that window from 90 decibels to 70 decibels. What percentage of the sound was prevented from entering?

21. A tuning fork emits sound of the same frequency as that of the fundamental tone of a closed organ pipe 60 cm long at 20°C. When the organ pipe is at a lower temperature, 4 beats per second are heard. What is the approximate temperature, assuming that the frequency of the fork is constant?

22. A sound, the intensity level of which is 2.0 bels, is reflected from a wall that has an absorptivity of 90%. What is the intensity level of the reflected sound?

23. When one automobile A passed another one B at rest, the frequency of the sound emitted by the horn of A as heard by a passenger in B changed from 1200 vib/sec to 1150 vib/sec. What was the speed of A? (The temperature was 20°C.)

24. An automobile approached a cliff, and the driver observed that the sound of his own car horn reflected from the cliff changed in pitch from do (256 vib/sec) to re (288 vib/sec). What was the speed of the car? (The temperature was 20°C.)

25. A piece of chalk (held at one end) ordinarily squeaks when one writes on a blackboard. It does not produce an audible squeak if its length is less than about 2.5 cm. What is the speed of sound in the chalk?

26. The fundamental frequency of an open organ pipe 2.0 m long filled with air is the same as the frequency of the third harmonic of a closed pipe filled with hydrogen. Calculate the length of the closed pipe.

27. Two sources of sound, each of frequency 600 vib/sec, are separated by a distance of 3.0 m. If an observer moves along the line connecting the two sources at a velocity of 15 cm/sec, calculate the beat frequency that he hears.

28. Would it be possible for an observer to travel between two fixed tuning forks, having frequencies of 430 vib/sec and 440 vib/sec, at such a velocity that they would appear to have the same pitch? If so, calculate the velocity.

For Further Reading

Benade, Arthur H., *Horns, Strings, and Harmony,* Doubleday and Company, Garden City, 1960.

Griffin, Donald R., *Echoes of Bats and Men,* Doubleday and Company, Garden City, 1959.

Hutchins, Carleen, "The Physics of Violins," *Scientific American,* November, 1962, p. 78.

Knudsen, Vern O., "Architectural Acoustics," *Scientific American,* November, 1963, p. 78.

Van Bergeijk, W. A., J. R. Pierce, and E. E. David, Jr., *Waves and the Ear,* Doubleday and Company, Garden City, 1960.

Von Bekesy, Georg, "The Ear," *Scientific American,* August, 1957, p. 66.

Electricity
and
Magnetism

CHAPTER TWENTY-ONE

Electric Charges

We begin the discussion of electricity and magnetism with *electrostatics*—the study of electric charges at rest. It may seem strange to you that we do not go immediately to a consideration of electric currents—charges in motion—which are so important in our technological society. The principles that govern electric charges at rest, however, underlie the study of electric currents. The concepts of this chapter and the next will help you to understand electric currents and other electric phenomena.

Electrostatics is important in its own right. Many of the forces responsible for the structure of atoms and molecules are electrostatic in origin. To understand the structure of matter, you must know the principles of electrostatics. Also, an increasing number of devices—electric precipitators, electrostatic loudspeakers, and high-energy accelerators, for example—are based on the principles of electrostatics.

The ancient Greeks recorded some of the first experimental facts about electrical charges over 2600 years ago. They observed that a piece of amber rubbed with fur attracted light objects such as feathers and bits of straw. The Greeks loved to speculate, but

they did not experiment systematically. Twenty-two centuries elapsed, and empires rose and fell, before there was further progress. Then, in the days of Shakespeare and Galileo, came the great awakening. That was a golden age not only of geographical exploration and literary creativeness, but also of science. Men threw off the old belief in tradition and authority and asked questions of nature by experiments.

About the time that Galileo discovered the laws of the pendulum and of accelerated motion, Sir William Gilbert, physician to Queen Elizabeth I, became interested in the study of magnetism and, incidentally, of electrical phenomena. He showed that many substances after being rubbed behave as amber does. Gilbert described this condition by saying that substances are *electrified,* from the Greek word for amber, *elektron.*

Positive and Negative Charges

Rub two pieces of sealing wax with flannel, and suspend one of them by a cord. Bring the other piece close to the first one (Fig. 21-1a), and the two will repel each other. Rub a glass rod with silk and bring it near to the sus-

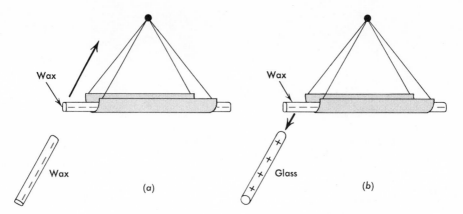

Fig. 21-1. Electric charges. (a) *Two wax rods, rubbed with flannel, repel each other.* (b) *A glass rod, rubbed with silk, attracts a wax rod rubbed with flannel.*

pended piece of wax (Fig. 21-1*b*). The two will attract each other. These experiments demonstrate that there are *two kinds* of electricity. The kind appearing on the glass we call *positive;* that on the wax is *negative. Any positively charged body behaves like glass that has been rubbed with silk. A negatively charged body behaves like wax rubbed with flannel. Like charges repel each other, but unlike charges attract.*

Any two dissimilar materials brought into contact become electrically charged to a greater or lesser extent. Your shoes scuffing over the pile of a rug or a plastic tile floor, the tires of a moving car, a comb passing through your hair—all acquire electrical charges. Rubbing forces the two substances into close contact.

The attractive forces between unlike charges and the repulsive forces between like charges depend upon—among other things—the distance between the charges, becoming noticeably greater as the distance becomes less. We shall investigate the exact law of force between small charged bodies later in this chapter.

Where do electric charges come from? According to an extremely simplified model of

the atom—one that we shall modify considerably later—the ordinary hydrogen atom (H^1) is a tiny earth-moon system made up of two particles, a *proton* and an *electron.* The proton is the nucleus or kernel of the system and is positively charged. It comprises almost all the mass of the atom. The electron, the moon of the system, is negatively charged and is about 1/1840 as massive as the proton.

All heavier atoms are believed to be built up of protons, electrons, and other particles called *neutrons.* The neutron, a trifle more massive than the proton, is uncharged. The helium atom (He^4) has two protons and two neutrons in the nucleus and two satellite electrons. Oxygen (O^{16}) has eight protons and eight neutrons in the nucleus and eight satellite electrons (Fig. 21-2).

The number of protons in the nucleus is called the *atomic number Z* of the element. The atomic numbers of hydrogen, helium, and oxygen are respectively 1, 2, and 8. The total number of protons and neutrons in the nucleus is called the *mass number M* of the element. Thus the atomic number of ordinary oxygen is 8 and its mass number is 16. *Isotopes* are atoms of an element that have the same numbers of protons, but different numbers of

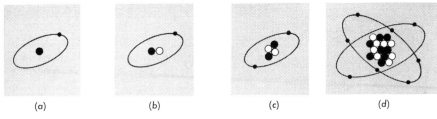

Fig. 21-2. *Structure of* (a) *ordinary hydrogen,* (b) *heavy hydrogen,* (c) *helium,* (d) *oxygen.*

neutrons. They have different mass numbers. Hydrogen has a "heavy" isotope, deuterium, of mass number 2, and another, tritium, of mass number 3. Uranium has an isotope of mass number 238, one of mass number 235, and a third of mass number 234.

The outermost electrons may be knocked off from the atoms or molecules of any material, but more work is required to remove electrons from some substances than from others. In the following list, the substances are arranged in order of increasing work required to remove an electron: the metals, fur, glass, flannel, silk, sealing wax, ebonite (hard rubber), sulfur.

Electrons may be removed more readily from fur than from sealing wax. When the two are rubbed together, electrons are transferred from the fur to the wax. The fur, having lost some of its normal supply of electrons, becomes positively charged. The wax gets excess electrons and becomes charged negatively. In some substances, each atom is ordinarily neutral; it has as many electrons as protons. In others, one kind of atom may have acquired excess electrons at the expense of other atoms. In table salt—sodium chloride—each chlorine atom has an excess electron and each sodium atom a deficiency of one electron.

An uncharged body is neutral; hence whenever one body acquires a positive charge by losing electrons, some other body or bodies must acquire an equal negative charge by gaining them. According to this model of the atom, *a negatively charged body has an excess number of electrons. A positively charged body has a deficiency of electrons.* Notice that by electrifying a body you do not *create* electricity, you merely *transfer* charged particles—electrons—from that body to other bodies.

The Electroscope

A negatively charged wax rod suspended by a cord may be used to detect charges on other bodies. If you move a positively charged body close to one end of the rod, the two attract each other. The approach of a negatively charged object causes repulsion. A more sensitive device to detect charges, called the *electroscope,* was invented in 1787, nearly two centuries after Gilbert's pioneer work. You can construct an electroscope by fastening one end of a strip of gold foil to a metal rod which, for insulation, is mounted in an amber or ebonite plug (Fig. 21-3). Charge the electroscope negatively by touching the metal rod with a stick of wax that you have rubbed with flannel. The electrons transferred to the insulated rod will distribute themselves over it and the gold leaf. The electrons on the rod will repel those on the leaf, causing it to diverge. Increasing the charge on the system will make the leaf diverge further. A further refinement—discussed later in this chapter—is needed to make the electroscope an instrument for quantitative measurement of electric charges.

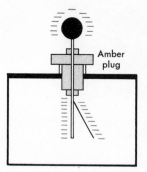

Fig. 21-3. A gold-leaf electroscope.

An electroscope can be used to tell whether a body is charged positively or negatively. Suppose that you move a charged body *A* near the negatively charged electroscope and that the leaf diverges. The charge on the body *A* repels electrons from the knob and forces them down to the gold leaf; hence the charge on *A* must be negative. What would have happened if the charge had been positive?

The *dosimeter,* used for measuring the intensity of energetic radiations, is a special kind of electroscope (Fig. 21-4). When gamma rays—high-energy electromagnetic waves that are emitted by some radioactive nuclei—enter the chamber of the charged dosimeter, they strip electrons from the air molecules, separating negative and positive

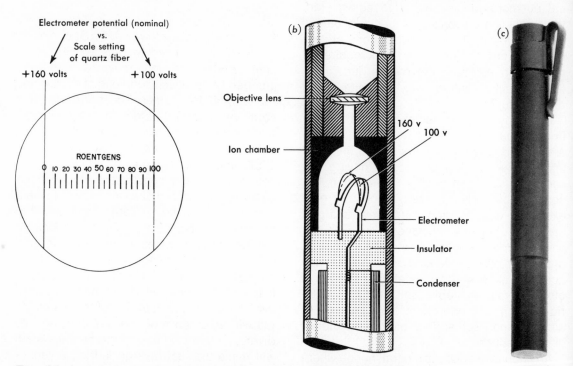

Fig. 21-4. A dosimeter for measuring the intensity of energetic radiations: (a) a view of the reticle of a dosimeter, (b) the construction of the dosimeter showing its resemblance to an electroscope; (c) the outer case, resembling a fountain pen. (Courtesy of Bendix Aviation Corporation.)

charges. The negatively charged particles are attracted to the positively charged fiber, discharging it. The rate at which the fiber returns to its uncharged position is proportional to the intensity of the gamma rays. Dosimeters are used by nuclear workers to safeguard their health by avoiding overexposure to energetic radiations.

Conductors and Non-conductors

For a century after Gilbert's experiments on electricity, electrical science progressed slowly. Then Stephen Gray, in London, found that he could transmit electric charges through a hemp cord suspended by silk threads. He replaced the silk threads by copper wires and found that the electrical charges escaped, and that metals are good conductors of electricity.

Charge an electroscope negatively and touch its knob with wire which is "grounded"—by being connected to a pipe running to damp earth, for example (Fig. 21-5). The gold leaf will move downward, showing that the excess electrons have escaped from the electroscope through the wire to the earth. When electric charges travel in a definite direction in this manner, they constitute an *electric current.* When they do not travel in a definite direction, they are termed *static*, even when they move about at random.

Substances that offer great resistance to the passage of charges are called *insulators.* There is no sharp boundary between conductors and insulators. The metals and carbon are good conductors. Pure water, amber, rubber, and glass are insulators; impure water containing dissolved acids, salts, or bases is intermediate in conductivity. Some materials, called *semiconductors,* conduct well only under certain conditions, as when light falls on them or when their temperature is raised.

In any metallic body some of the electrons are loosely attached to the atoms, and they

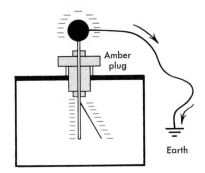

Fig. 21-5. *Discharging an electroscope. Excess electrons on the knob repel others on the wire so that they escape to the earth.*

are free to wander among neighboring atoms, just as gas molecules of one kind diffuse through molecules of another. Mobile electrons are called *conduction* charges. When you connect a negatively charged metallic body to earth by a wire, some of the excess free electrons in it, repelled by those nearby, move over onto the wire. These excess electrons repel other electrons, driving them along the wire. These, in turn, act on others, so that finally the body is discharged. What would have happened if the body had been charged positively (for example, by contact with a glass rod which had been rubbed with silk)?

Charges are also present in insulators although these are not free charges as in metals. The electrical charges in insulators are bound to the atoms and molecules of the substance. As we shall see in the next chapter, bound charges can appear on the surface of an insulator under certain conditions. These bound charges—called *polarization charges*—exert forces on free charges near them.

Finally, in some liquids and gases, electricity is conducted by the motion of charged atoms or groups of atoms called ions. The air contains ions, produced, for example, when energetic cosmic ray particles tear electrons away from air atoms. Each positively charged

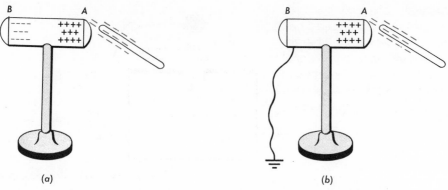

B   A   B   A

(a)   (b)

Fig. 21-6. Charging by induction. (a) Electrons on the rod repel free electrons in the metal, driving them to the far end of the cylinder. (b) Then electrons escape to the earth, leaving the cylinder charged positively.

air ion can take an electron from a negatively charged electroscope, thus slowly discharging the electroscope. However, electrons are the particles whose motion we shall consider in this first part of our study of electricity. The positively charged protons in the nuclei of atoms in a metal wire, for example, are "locked in" and do not take part in the conduction process.

The Behavior of Free Charges on Metallic Conductors

A conductor may be electrified by touching it against another charged body; this is charging by *conduction*. A second important method, which helps us understand how free charges behave on metallic conductors, is *induction.* Hold a negatively charged rod near one end of an insulated metal conductor, which for convenience is a cylinder (Fig. 21-6). (It can just as well be a sphere, a flat sheet, or a conductor of any other shape.) Free electrons of the metal will be repelled and will crowd toward the far end of the cylinder. This end is negatively charged and the near end positively charged. In order to test the nature of the charges, touch the cylinder

at *B* with a metal ball mounted on a glass rod for insulation. Then touch the ball against a negatively charged electroscope, and the divergence of the gold leaf will increase. This indicates that the charge at *B* is negative. Touch the ball at *A*, and some of its electrons will move over onto the cylinder so that the ball will be positive. When it is touched against the knob of the negatively charged electroscope, the gold leaf will move down. Next, touch the cylinder with a wire connected to earth. Electrons will escape, for a path is provided by which they may be repelled still farther from the negatively charged wax rod. Disconnect the wire, and *then* remove the wax rod. The charges on the cylinder will distribute themselves symmetrically (by flow of electrons). *When you charge the cylinder by induction, no electricity escapes from the wax rod, and the induced charge is of the opposite sign. A free charge induces charges on conductors in its neighborhood.*

Why does a charged body attract an uncharged one? Bring a positively charged body *A* near an uncharged pith ball *B* that has been sprayed with a metallic paint and suspended by a cord (Fig. 21-7a). *A* will induce a

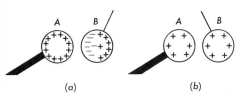

(a)   (b)

Fig. 21-7. B *is initially neutral.* (a) *The plus charges on* A *pull electrons over to the near side of* B *so that it is attracted.* (b) *Like charges repel each other.*

charge on **B**. Electrons, attracted by the charges on **A**, will move to the nearer side of the pith ball. The negative charge on **B** is nearer to **A** than the positive charge; hence **B** will be attracted by **A**. If the two bodies touch each other, electrons will be transferred from **B** to **A**. Then both bodies will be charged positively and they will repel each other (Fig. 21-7*b*).

Excess charges reside on the outside of a conductor. To show this experimentally, charge an insulated, hollow metal vessel negatively, and touch a metal ball suspended by a silk cord against the outer surface (Fig. 21-8). The ball will be charged by conduction. Its charge may be tested by means of the electroscope and will be found to be negative. Next, ground the ball to remove excess charges and then lower it into the vessel so as to touch the bottom. Remove it and touch it against the knob of an uncharged electroscope. The leaf will not diverge, for the ball is uncharged. Charge the ball negatively, touch it to the inside of the vessel, and test the ball for excess charge: it has lost its charge. Repeat the experiment as often as you wish, and the result will always be the same. Electrons will not remain on the ball when it touches the inside surface of the vessel. They move to the outside of the vessel. Experiments like this with either positively or negatively charged bodies show that *excess electrostatic charges reside only on the outside of a conductor.*

Charges are most dense where the surface is most curved. The charges on an insulated sphere that is far from other bodies are distributed uniformly over the surface but those on the egg-shaped conductor crowd together at the regions where the surface is most curved (Fig. 21-8). To test this statement, touch a coin, mounted on an insulator, against the flat side of the charged vessel, then against the knob of an electroscope, and note the deflection of the gold leaf. Repeat the experiment, touching the coin against the curved end of the vessel. More of the electricity will be transferred to the coin, and the gold leaf will diverge more than before. Thus you can demonstrate that the static charges on a conductor are most dense in regions where the surface is most curved.

Charges escape readily from points. Since charges are more dense at curved portions of a conducting surface than in flat regions, it follows that the accumulation of electricity at the point of a needle which is attached to a charged body will be very great. Electric charges readily escape from needle points. To demonstrate this effect, attach a sewing

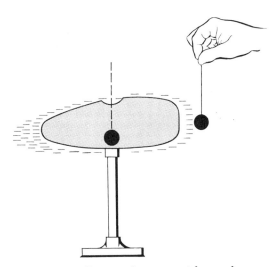

Fig. 21-8. Excess charges reside on the outer surface of the conductor. They are most dense in regions where the surface is most curved.

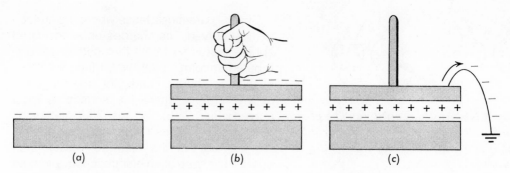

Fig. 21-9. An electrophorus.

needle to the knob of an electroscope and charge it so that the leaf stands out nearly at right angles. The gold leaf will move down, showing that the charge is escaping, carried away by air ions whose charge is opposite to that on the needle.

Devices for Generating Static Charges

Small electrostatic charges, as we have seen, can be produced by rubbing a wax rod with flannel or a glass rod with silk. The electrophorus (e lek trof' e rus) is a convenient device for generating somewhat larger electrostatic charges by induction. It consists of a flat plate made of wax, or some other insulator, and a metal plate with an insulating handle. Rub the wax with flannel so as to pepper its surface with excess electrons (Fig. 21-9). Place the metal plate on the wax. The surfaces are rough and they touch each other at only a few points, so that only a little of the charge on the wax escapes. The remaining excess electrons on the wax repel others in the metal and drive them to the upper surface. Touch the plate with your finger or a "grounded" wire. Electrons will escape from the plate, and it will be charged positively. Lift the plate from the wax, then move your finger close to it. You will see and feel a spark. You can repeat the operation many times without

appreciably decreasing the charge on the wax.

When you lift the positively-charged plate from the negatively-charged wax, you do work in overcoming the electrical attraction between them. Thus you store electrical energy in the system, and this energy causes the heat and light of the electric spark. If you lift the metal plate only a millimeter or so, the

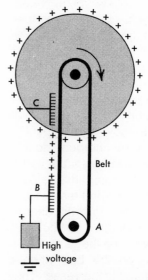

Fig. 21-10. Van de Graaff generator. The belt carries positive charges into the metal shell, where they attract electrons from the comb C, thus charging the sphere positively.

spark will be less intense than if you lift the plate higher. Why is this?

A Van de Graaff generator can generate electric charges continuously. Figure 21-10 represents an insulated metal shell with a rubberized cotton belt driven by an electric motor. The metal comb *B* is so highly charged that its teeth attract electrons from the belt, charging it positively. The belt carries the positive charges up into the metal shell, where they attract electrons from the teeth of a second metal comb *C*. Thus the shell becomes charged positively. Large Van de Graaff generators, based on this principle, produce voltages of 10-12 million volts. These devices are used to bombard the nuclei of atoms (Chapter 46). Small Van de Graaff generators are used for classroom demonstrations.

Electric batteries also furnish electric charges. The charges on the terminals of a battery are exactly the same as those on the electrostatic generator. Batteries are used for producing steady currents of charges, as we shall see.

Measuring Charge Quantitatively

An electroscope can be used to measure electrostatic charge quantitatively, if it is suitably modified to screen the charge being measured. In Fig. 21-11 the knob of the electroscope has been replaced by a cylindrical metal container with a cover.

Suppose that you wish to compare the negative charges on two metal spheres. One of the spheres is lowered on an insulating thread into the container without touching the walls, the cover is replaced, and the deflection of the leaf noted. The positive charge induced on the inside of the container (held there by the charge on the sphere) equals the charge on the sphere. (We shall discuss why this is so in Chapter 22.) It also equals the *negative* charge on the outside of the container and on the electroscope leaf. Thus the

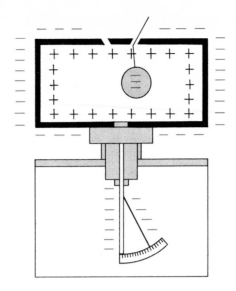

Fig. 21-11. *Calibrating an electroscope.*

induced charge on the electroscope *equals* the charge on the sphere. (The charge on the electroscope would *not* equal the charge on the sphere if the sphere were merely touched to the knob of an ordinary electroscope; the sphere's charge would be shared between it and the electroscope.) Suppose that the first sphere is then removed, the second one inserted, and the deflection of the leaf noted. If the deflection equals that produced by the first sphere, we conclude that the two spheres were equally charged. *Equal charges produce equal deflection.*

To calibrate the electroscope, you could, in principle, prepare a series of equally charged spheres and lower them in groups of one, two, three . . . at a time into the container, noting the deflection produced by total charges equal to one, two, three . . . times an arbitrary unit charge. You would find that the deflections increase as the total charge increases, and you could plot a charge-deflection graph. This procedure is an idealized one—it would be tedious to carry it out—but it suggests a way of experimentally obtaining equal charges and of finding the ratio of any two charges by comparing their deflections.

The Law of Force Between Point Charges

Now we are ready to discuss the law of force between point charges. Mathematically, a point charge is one which, like the mass points we discussed earlier, is a geometric point. In practice, we can treat as a point charge the charge on a body whose dimensions are negligibly small compared to the distance to another charged body.

Nine years after the signing of the Declaration of Independence, the French physicist Augustine Coulomb showed experimentally that the force exerted by an electric point charge q_1 on another point charge q_2 varies directly as the product of the two charges and inversely as the square of the distance r between them. This law resembles Newton's law of gravitation. It may be expressed mathematically as

$$F \sim \frac{q_1 q_2}{r^2}$$

where the symbol $\sim$ means "varies as."

Coulomb's apparatus was similar to that used by Cavendish to measure the gravitational attraction between two bodies (Chapter 11). It consisted of a metal sphere mounted at one end of a light rod and balanced by a load at the other end. The rod was suspended from a knob by a torsion fiber in which the restoring torque was proportional to the angle of twist θ (Fig. 21-12). First he charged the ball A, and brought a second, equally charged, ball B near it. Electrical repulsion pushed the suspended ball away. By turning the knob through an angle θ, he twisted the fiber sufficiently to force the suspended system back to its initial position. When he doubled the charge on the second ball, he found that, to bring the system back, the knob must be turned through an angle 2θ. Doubling the charge on one ball doubled the force of repulsion. The force was proportional to the charge on the ball.

Coulomb showed further that if the fiber had to be twisted through an angle θ when the centers of the two balls were at a distance r apart, a twist of 4θ was required to reduce the distance to $r/2$. Thus the repulsive force varied inversely as the square of the distance.

We must be careful to specify the conditions that underlie Coulomb's law. As stated, the law holds for point charges in vacuum. In practice, the force between two point charges is almost the same in air as it is in vacuum so that force measurements in air differ only slightly from those made in empty space. With a suitable modification, the law accounts for the force between point charges in an insulating fluid if the electrical properties of the fluid are the same in all directions and the surfaces of the fluid are far from the point charges. The law does *not* describe the force between two charged bodies of finite size in the presence of other conducting or non-conducting bodies. The reason for this is that induced charges appear on nearby conductors and polarization charges appear on non-conductors; these additional charges also exert forces on the charged bodies under consideration. The general problem of calculating the forces between charged bodies is a difficult one and requires advanced mathematics for its solution. We shall limit ourselves to relatively simple cases.

To convert Coulomb's law from a proportionality into an equation, we introduce the constant k_0 whose value depends upon what we choose as our system of units and upon our definition of unit charge:

$$F = k_0 \frac{q_1 q_2}{r^2}$$

Several systems of electrical units have been developed and are in use. The rationalized mks system that we shall stress is being increasingly used by scientists and engi-

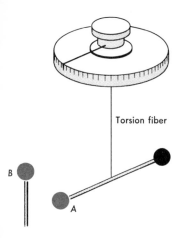

Torsion fiber

B

A

Fig. 21-12. *Coulomb's apparatus.*

neers. The units in it were chosen to simplify the equations of electricity and magnetism and to give as much prominence to electric charge as a fundamental quantity as to mass, length, and time. The unit of charge in the mks system is the *coulomb.* It will be defined later (Chapter 28) in terms of electric current. Anticipating this more fundamental definition of the coulomb by an independent laboratory experiment, we can define it now in terms of Coulomb's law:

One coulomb is that point charge which at a distance of one meter in vacuum from an equal, like point charge repels it with a force of approximately 9 × 10⁹ newtons. Thus in the mks system k_0 equals 9×10^9 newton-meter²/coulomb².

More precisely k_0 equals $1/(4\pi\epsilon_0)$ where ϵ_0 (Greek *epsilon*), called the *permittivity* of a vacuum, equals 8.85415×10^{-12} coulomb²/newton-meter². The factor 4π is introduced arbitrarily for reasons of convenience. Inserting it into Coulomb's law permits cancellation of a factor of 4π in equations that we shall encounter later. The permittivity ϵ_0 is an *experimentally* determined constant. It is *not* arbitrary, but follows from Coulomb's law under the specification that

the force be in newtons, the distance in meters, and the charge in coulombs. Rounding off the factor $4\pi\epsilon_0$ to $\frac{1}{9} \times 10^{-9}$ coul²/newton-m², we can write Coulomb's law for point charges in a vacuum as

$$F = \frac{q_1 q_2}{4\pi\epsilon_0 r^2} = 9 \times 10^9 \text{ newton-m}^2/\text{coul}^2 \frac{q_1 q_2}{r^2}$$

Example. What force does a positive point charge of 2×10^{-6} coul exert on a negative point charge of -5×10^{-6} coul at a distance of 0.10 m in a vacuum?

$$F = 9 \times 10^9 \text{ newton-m}^2/\text{coul}^2$$

$$\times \frac{(2 \times 10^{-6} \text{ coul})(-5 \times 10^{-6} \text{ coul})}{0.010 \text{ m}^2}$$

$$F = -9 \text{ newtons}$$

The negative sign indicates that the force between the charges is an *attractive* one. A positive force is a repulsive one.

In the *electrostatic* system of units, still used in scientific articles, unit charge, sometimes called the statcoulomb, is defined as that point charge which, acting in a vacuum, exerts a force of 1 dyne on an equal charge at a distance of 1 cm. The constant k_0 in the electrostatic system has the value 1 dyne-cm²/statcoulomb². Three billion statcoulombs equal one coulomb.

When two or more other point charges exert forces on a point charge q_1, the total force acting on the charge q_1 is the resultant of the separate forces that the other charges would exert on q_1 if acting alone.

Example. Three point charges ($q_1 = 2.0 \times 10^{-6}$ coul, $q_2 = 5.0 \times 10^{-6}$ coul, and $q_3 = 6.7 \times 10^{-6}$ coul) are placed in vacuum at the corners of a right triangle as shown in Fig. 21-13. What is the resultant force on q_1?

The force exerted by q_2 on q_1 is

$$F_{1,2} = 9 \times 10^9 \text{ newton-m}^2/\text{coul}^2$$

$$\times \frac{(2.0 \times 10^{-6} \text{ coul})(5.0 \times 10^{-6} \text{ coul})}{0.010 \text{ m}^2}$$

$$= 9 \text{ newtons}$$

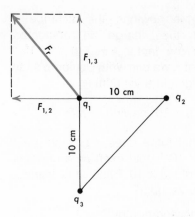

Fig. 21-13. *Two point charges exert forces on a third.*

Similarly, the force exerted by q_3 on q_1 is

$F_{1,3} = 9 \times 10^9$ newton-m^2/coul^2

$$\times \frac{(2.0 \times 10^{-6} \text{ coul})(6.7 \times 10^{-6} \text{ coul})}{0.010 \text{ m}^2}$$

$= 12$ newtons

$F_{1,2}$ and $F_{1,3}$ are at right angles. Their resultant F_r is

$$F_r = \sqrt{(9 \text{ newtons})^2 + (12 \text{ newtons})^2}$$
$$= 15 \text{ newtons}$$

Rutherford Scattering of Alpha Particles

One of the best experimental tests of Coulomb's law in the atomic world was carried through in 1911 by Sir Ernest Rutherford in England. Coulomb's law was used to account for the scattering of alpha particles—helium atoms stripped of their two electrons—by the atoms in thin metal foils. Not only were the experimental results in accord with the predictions of Coulomb's law for point charges, but they also suggested and strongly supported Rutherford's nuclear model of the atom.

Before Rutherford performed his experiments, it was believed that an atom was a tiny "pudding" of positive charge with electrons embedded in it like raisins. If such an atom were bombarded by alpha particles, the alpha particles would change their directions *only slightly* in passing through because of the repulsion between the positively charged alpha particles and the positive charge in the atom. Rutherford decided to bombard atoms with alpha particles and, as a test of this model, measure the angles through which the alpha particles were scattered.

Rutherford's apparatus is shown in Fig. 21-14. Alpha particles—a combination of two protons and two neutrons with a total positive charge twice that of an electron—are emitted at high energy by radioactive nuclei of radium and other heavy elements. S is a radioactive source from which alpha particles are emitted in all directions. Some pass through the slit and form a narrow beam on the other side. Alpha particles in the beam strike a thin metal foil. Most of them pass through the foil without change of direction. Some are scattered. Rutherford proceeded to count the number scattered through various angles θ by rotating a fluorescent screen about the foil as center and observing with a microscope the tiny flashes of light that the impinging alpha particles made on the fluorescent screen.

The results of the experiment were not at all what the "pudding" model of the atom had predicted. Not only were a considerable number of alpha particles scattered through large angles θ on the other side of the foil, but an appreciable number were scattered backward on the side of the foil toward the slit. A new model of the atom was needed.

Rutherford developed a model of the atom in which the positive charge of the atom was concentrated at its center in a relatively tiny, but massive "nucleus." The heavy, positively charged nucleus behaved like an almost stationary point charge, and Rutherford could calculate the repelling force exerted by the nucleus on an approaching alpha particle.

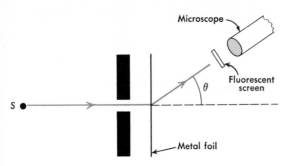

Fig. 21-14. Rutherford's apparatus for study-ing the scattering of alpha particles by the nuclei of a metal foil. (From O. H. Blackwood, T. H. Osgood, A. E. Ruark, et al., An Outline of Atomic Physics, third edition, John Wiley and Sons, 1955.)

When the alpha particle passed relatively far from the nucleus (Fig. 21-15a) the inverse square force was small, and the alpha par-ticle underwent a small change of direction. When the impact parameter p was small (Fig. 21-15b), the alpha particle approached nearer the nucleus, the repelling force was large, and the alpha particle was hurled back, scattered through a large angle.

The experimental results of the scattering experiments were in good agreement with the quantitative predictions of the nuclear model of the atom and firmly established the nuclear model. Rutherford was able to show that the Coulomb scattering by nuclei accounted for all except a few scattering events. He used the scattering data to calculate that the diam-eter of the nucleus could not be greater than about 3×10^{-12} cm as compared with atomic diameters of about 10^{-8} cm. Thus the outer diameter of the atom is 10,000 times that of its nucleus.

The few scattering events that could not be explained by Rutherford's theory were those at very large angles at which the alpha par-ticles come extremely close to the nucleus. Later, as we shall see, it was found that a spe-cific nuclear force—not electrostatic in

origin—operates within the nucleus and is responsible for these failures of Rutherford's theory.

Lightning

During violent storms, the clouds become highly charged with electricity, and gigantic sparks or lightning strokes jump from one cloud to another or to the earth. These sparks heat the air through which they pass, causing it to expand suddenly and producing thunder. The reason for the charging of the clouds is not well understood. It is known that, when water drops are broken up by a blast of air, the small droplets are negatively charged and the larger droplets are positive. One theory of thunderstorms assumes that when raindrops are broken up by violent gusts of wind, positively charged drops rise more slowly than the lighter negatively charged ones so that the lower clouds acquire positive charges.

Lightning discharges often occur from cloud to cloud and, sometimes, between a cloud and the earth. Also, a phenomenon called **return shock** is observed. When a strongly charged cloud is overhead, trees, buildings, and other objects become oppo-sitely charged by induction. After the light-ning occurs or the charged cloud moves

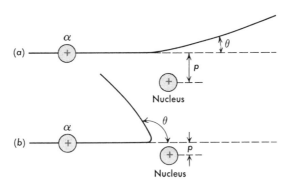

Fig. 21-15. Scattering of alpha particles from a positively charged nucleus.

away, they lose their charge. If a person comes in contact with them before they are completely discharged, he may be hurt even though lightning does not strike him.

The first American scientist who became famous was Benjamin Franklin (1706–1790), printer, business man, statesman, and philosopher. At the age of seventeen, he ran away from his Boston home to Philadelphia, where he arrived friendless and hungry. Only 26 years later, he retired from business with enough wealth to be a gentleman of leisure. For several happy years, he amused himself by making electrical experiments. Using a few pieces of cloth, needles from the sewing room, and some pans and bits of wax from the kitchen, he made a discovery and an invention that released men from one of their greatest fears, so that he became the best-known scientist of his day.

Franklin discovered that electric charges escape more readily from sharp points than from blunt ones, and he applied his discovery to the invention of the lightning rod. At that time there was considerable doubt as to the nature of lightning. Franklin suggested that if, as some supposed, it was an electrical phenomenon, the charges might be collected from the clouds by means of a long, pointed rod erected on the roof of a tall building. He sent this suggestion to a friend in England, who presented it to the Royal Society of London. The British scientists gave courteous attention to the message from the wilds of America, but they did not act. Franklin's friend sent it on to Paris, where it was welcomed.

One stormy day, King Louis XV and all his court watched while a soldier climbed to the top of a tower and raised a long pole which supported a wire. Its sharp point collected charges from the air so that a spark jumped from the lower end of the wire to his hand. Experiment showed that charges collected from the wire had the same properties as those secured from glass or wax. Lightning is nothing but an electric spark. Meanwhile, Franklin, in far-off Pennsylvania, weary of waiting, decided to experiment for himself. There were no high buildings in Philadelphia, and so he flew a silk kite during a thunderstorm. The lower end of the kite string was attached to a silk ribbon held by Franklin, who stood in a dry place, so that the electrical charges could not escape. He was thrilled to see the loose fibers of the kite string bristle out like hairs of an angry cat. When he touched a door key tied to the wet string he received a painful shock. Thus, again, the identity of lightning and the electric spark were proved. Franklin's experiment was dangerous. Shortly after, a Russian investigator was killed while repeating it.

Though Franklin performed the experiments mainly to satisfy his curiosity, his discovery quickly proved to have a practical value. He suggested that houses might be protected from lightning by mounting on their roofs grounded metal rods. Soon these lightning rods appeared on buildings all over Europe. Formerly the lightning stroke had been regarded as a means of punishment by an offended deity; thereafter it was considered as a natural phenomenon, like rain or hail, which might be guarded against by well-understood means.

A building can be perfectly shielded from lightning by completely surrounding it with metal screens. Excellent protection is secured more simply by mounting sharp metal rods, 3 ft high, at intervals of 15 ft or less, along the highest parts of a building and connecting them carefully to damp earth. The efficacy of lightning rods is proved by the fact that steeples, factory chimneys, and skyscrapers are struck scores of times without being damaged. During a storm, the Empire State Building in New York City was struck 15 times in 15 minutes. Lightning *does* strike twice in the same place!

Remember that a structure fitted with lightning rods gives nearly complete protection within a large cone about it. At street level the radius of this cone is about one and one half times the height of the building. The crowded houses of a city discharge the clouds harmlessly; hence there is little danger from lightning. In the open country or on golf courses, keep away from isolated trees and small sheds, for they are likely to "draw" the lightning. In a building that is not equipped with lightning rods, stay away from fireplaces and chimneys. Their sooty linings afford good conductors for lightning. If you are in an open field, crouch down with your feet close together and do not touch the ground or any object on the ground except with your two feet. It is better to do this and be drenched rather than stand up and be "burned."

Dangers from Static Electricity

In cotton mills and paper mills, electrical charges are generated when the cloth or paper passes over metal rollers. Sometimes electrical sparks cause disastrous fires. This danger is avoided by humidifying the air so that water films conduct away the electricity from the surfaces. The friction of gasoline flowing through a hose at a filling station may electrify the metal nozzle so that a spark is produced when the nozzle is touched against the gasoline tank of a car. The danger is decreased if the accumulation of charges on the nozzle is prevented by "grounding" it. Similarly, mechanics are careful to ground an airplane before filling its tanks with fuel. The friction of a gasoline truck's tires against the road may generate a charge so that, later, when someone opens a valve, the gasoline vapors ignite. Often a metal chain is attached to the truck. One end, dangling against the ground, carries away the electrostatic charges. Never rub clothing with a gasoline-soaked cloth; the friction may generate an electrical charge, cause a spark, and ignite the vapor.

Sometimes explosions occur in operating rooms of hospitals when the vapor of anesthetics is ignited by static sparks. In order to minimize this danger, the doctors and nurses wear cotton clothing. The operating table is grounded to conduct away electricity.

Summary

Positive charges appear on glass when rubbed with silk; negative charges, on sealing wax when rubbed with flannel. Like charges repel each other; unlike charges attract.

An atom of ordinary hydrogen is composed of one positively charged proton as a nucleus and one negatively charged electron. The proton is about 1840 times more massive than the electron. Heavier atoms are believed to be built up of protons, neutrons, and electrons. When a body is negatively charged, it has excess electrons; if positively charged, there is a deficiency of electrons.

In metallic conductors many of the electrons are free to travel among the atoms like molecules of a gas.

When electric charges are "static" they do not progress in any definite direction. Excess electrostatic charges reside on the outer surface of a conductor; their density is greatest in regions of greatest curvature.

In charging a body A negatively by induction, a body B, positively charged, is brought near to it. Then A is grounded, electrons travel from the earth to it, and the connection to the earth is broken. Meanwhile the charge of B does not change.

Coulomb's law states that a point charge q_1 exerts on a point charge q_2 at a distance r in vacuum a force F given by

$$F = \frac{k_0 q_1 q_2}{r^2}$$

where k_0 is approximately 9×10^9 newton-m^2/coul^2.

Questions

1. Illustrate the meaning of atomic number, mass number, and isotope for (a) hydrogen, (b) uranium.

2. Describe the physical process of charging a body (a) by contact (conduction) and (b) by induction.

3. What are the units and numerical values for the constants k_0 and ϵ_0 we have adopted in Coulomb's equation?

4. When are the electrical units we have adopted said to be (a) rationalized, (b) mks units?

5. How could you construct an electroscope out of materials available in your kitchen?

6. What is the significance of Rutherford's work on alpha scattering (a) as a test of Coulomb's law, (b) in the development of the nuclear model of the atom?

7. Discuss the differences between atoms of He^3 and H^3 in regard to (a) the composition of their nuclei, (b) the number of planetary electrons in each atom.

8. According to the simple theory presented thus far, why do metals conduct electricity better than insulators? What are semiconductors?

9. Why do electric charges escape from sharp points on conductors more readily than from blunt ones?

10. How could you determine experimentally whether the electrostatic charge built up on your car because of friction between the tires and the road is positive or negative?

11. If you charge a fountain pen by rubbing it on a coat sleeve, why will it pick up small pieces of paper?

12. When you unroll friction tape with a quick pull in a darkened room, why do you often see a glow?

13. You can pick up an electrostatic charge by walking on a rug. Why is this effect usually greater in winter than in summer?

14. During a thunderstorm, long sparks jumped from an ungrounded sink to a nearby water faucet in the kitchen of an isolated farmhouse. Explain.

15. Does charging a body change its mass?

16. Why does an electrostatic machine not work well in summer when the humidity is high? What would be the best season of the year to demonstrate this machine? What steps could be taken to improve its performance in a humid climate?

17. An electrophorus can be charged and discharged repeatedly, producing spark after spark. Estimate the electrical power that could be developed this way.

18. Sketch curves showing how the electrostatic force between two charged bodies varies with separation (a) if the distance dependence is $1/r$, (b) if it is $1/r^2$, (c) if it is $1/r^3$.

Problems

(1 μcoul = 1 microcoulomb)
Assume that the charged bodies are point charges in a vacuum.

A

1. What force does a charge of 10 μcoul exert on a charge of 40 μcoul at a distance of (a) 1.0 cm, (b) 2.0 cm, (c) 0.10 cm?

2. A pith ball A carrying a charge of 20 μcoul is suspended 6.0 cm above another charged ball B. The charge on B exerts a force of 0.50 newton on the charge on A. What is the charge on B?

3. The mass of a pith ball is 0.204 gm. It carries a negative charge of 0.020 μcoul. A small ball is located 2.00 cm above it. How great a positive charge must the upper ball carry in order to lift the lower one?

4. How far must a charge A of 10.0 μcoul be from a charge B of 5.0 μcoul in order that A may exert on B a force of 0.50 newton?

B

5. Two balls, *A* and *B*, each having a charge of 25 μcoul, are 6.0 cm apart. (*a*) What force does the charge on *A* exert on the charge on *B*? (*b*) At what distance of separation will the force be halved?

6. Three positive charges, each 200 μcoul, are located, *A* at the zero mark, *B* at the 0.10-m mark, and *C* at the 0.20-m mark of a meter stick. What is the force exerted (*a*) by the charges *A* and *B* on *C*, (*b*) by the charges *A* and *C* on *B*?

7. A pith ball *A* of mass 0.102 gm and a positive charge of 0.10 μcoul is located on a smooth horizontal insulating surface at a point 50 cm from a ball *B* carrying a positive charge of 0.20 μcoul. (*a*) What force is exerted by *B* on *A*? (*b*) What will be the acceleration of *A* when it is first released? Neglect the pull of gravity.

8. At what distance above *B* (problem 7) must *A* be located so that the force of electrical repulsion on *A* may equal its weight?

9. What would be the magnitude of equal and unlike charges, separated by a distance equal to that from the moon to the earth, so that the attractive electrical force between them would be equal to the attractive gravitational force between moon and earth?

C

10. Three charges of 50 μcoul each are located at the corners of an equilateral triangle 2.00 cm on each side. (*a*) What force is exerted on *A* by *B* and *C*? (*b*) What would be the force if *B* were negative instead of positive?

11. Two 1.0-gm pith balls are each given a positive charge of 1.0 μcoul and suspended by 1.0-m silk cords from the same point. What will be the equilibrium distance between the pith balls?

12. Two pith balls, each having a mass of 0.102 gm, are suspended from the same point by means of silk threads each 0.20 m long. The balls are equally charged, and the two threads make an angle of 90°. What is the charge on each ball?

13. A proton located 1.0×10^{-8} cm from another proton is free to accelerate. What will be its acceleration at the instant of release? The charge on a proton is 1.60×10^{-13} μcoul, and its mass is 1.67×10^{-27} kg.

14. The hydrogen ion of a hydrogen-chloride molecule has a positive charge of 1.6×10^{-13} μcoul. The chlorine ion has an equal negative charge. The two ions are 1.0×10^{-8} cm apart. What is the electrostatic force exerted on the chlorine ion by the hydrogen?

15. An electron of charge -1.6×10^{-13} μcoul revolves around a proton of equal positive charge. The mass of the electron is about 9.0×10^{-28} gm. The distance of the proton from the electron is 0.50×10^{-8} cm. Find (*a*) the centripetal force acting on the electron, (*b*) the speed of the electron, (*c*) its frequency of revolution. (*Hint:* equate the electrical attraction to the centripetal force.)

16. According to the theory of Rutherford scattering, the number of monoenergetic alpha particles per unit cone (solid angle) scattered at an angle ϕ with the initial direction of the beam is proportional to $\dfrac{1}{\sin^4 \phi/2}$. If 33 particles per hour are scattered into a unit solid angle at an angle of 120° in a certain scattering experiment, calculate the number to be expected at angles of 90°, 60°, and 30°.

For Further Reading

Battan, Louis J., *Harvesting the Clouds,* Doubleday and Company, Garden City, 1969.

Battan, Louis J., *The Nature of Violent Storms,* Doubleday and Company, Garden City, 1961.

Franklin, Benjamin, *Experiments and Observations on Electricity,* ed. by I. B. Cohen, Harvard University Press, Cambridge, 1941.

Lewis, Harold W., "Ball Lightning," *Scientific American,* March, 1963, p. 106.

Moore, A. D., *Electrostatics,* Doubleday and Company, Garden City, 1968.

CHAPTER TWENTY-TWO

Electric Fields and Potential Difference

What Is a Field?

An electric charge q will exert a force on any other nearby charge. We say that q sets up an *electric field* in the surrounding space and that this field acts on any other charge in it. Thus *an electric field is a region in which forces act on electric charges.*

The idea of an electric field seems mysterious, for it is something that you cannot feel or taste or hear or see. It is a concept that enables us to make our knowledge of the physical world more orderly. It seems advisable, before discussing this subject further, to point out that all of us live in another, more familiar, kind of field — that produced by the earth's attraction. A *gravitational* field is a region in which gravitational forces act on bodies. We define the gravitational *field strength* at any point as the force per unit mass acting on a body at that point. For example, if the acceleration due to gravity at a point is 980 cm/sec^2, then a force of 980 dynes acts on a 1-gm body at that point. Therefore, the strength of the earth's gravitational field at that point is 980 dynes/gm.

A body dropped from rest experiences a force directed through the earth's center. A line to the center, with an arrow pointing in the direction in which the force acts on the body, we call a *gravitational line of force.* We can represent the strength of a gravitational field by the number of lines of force drawn through unit area normal to the field. For example, to represent a field strength of 980 dynes/gm, we draw 980 lines, with arrow points, perpendicularly through each square centimeter of surface (Fig. 22-1). Near the earth's surface, the gravitational lines of force are nearly parallel.

Electric Lines of Force

A gravitational line of force indicates by its direction (the direction of the tangent) at each point on it the direction of the gravitational force on a body at that point. Similarly, in electricity, we define an electric line of force as *the line whose direction at each point is the direction of the force on a positively charged body.*

The force that is exerted on a body located in an electric field depends on the quantity of charge on the body and also on the *strength* of the electric field. The strength E of an electric field at a point is defined as the *force per*

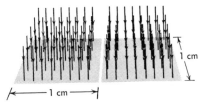

Fig. 22-1. *Representation of a gravitational field. If 980 lines pass through each square centimeter, the gravitational field strength is 980 dynes/gm.*

unit charge acting on a positively charged body placed at that point. *E* is a vector quantity, and an electric field is a vector field.

$$\text{Electric field strength} = \frac{\text{Force}}{\text{Quantity of charge}}$$

$$E = \frac{F}{q}$$

The mks unit of electric field strength is the newton per coulomb; the cgs unit is the dyne per statcoulomb.

If a small pith ball carrying a charge of 4×10^{-9} coul experiences a force of 12×10^{-5} newton, then the electric field strength at the point where the pith ball is located is

$$E = \frac{12 \times 10^{-5} \text{ newton}}{4 \times 10^{-9} \text{ coul}}$$

$$= 3 \times 10^4 \text{ newtons/coul}$$

In computing the electric field strength at a point, imagine that a small known charge *q* on an object of small size is located at that point, calculate the force acting on *q*, and then find the force per unit charge. Experimentally, we can also find *E* by measuring the force acting on a *small* charged body and dividing the force by the charge. We specify that the test charge be small in order to minimize the inducing of charges on nearby conductors and that the body be small so that polarization charges on nearby insulators will be small. In the early sections of this chapter, we shall discuss point charges in vacuum, introducing insulators later.

Example. Two point charges $q_1 = +10 \times 10^{-9}$ coul and $q_2 = -10 \times 10^{-9}$ coul are each located 0.10 m from point *P* in vacuum (Fig. 22-2). What is the electric field strength at *P*?

The electric field strength *E* at *P* is the *resultant* force per unit charge due to both q_1 and q_2. Imagine that a small charge *q* is located at *P* and let the force exerted on it by the positive charge be F_1, and that due to the negative charge F_2. By Coulomb's law

$$F_1 = 9 \times 10^9 \frac{\text{newton-m}^2}{\text{coul}^2} \left[\frac{q(10 \times 10^{-9} \text{ coul})}{0.010 \text{ m}^2} \right]$$

$$= (9.0 \times 10^3 \text{ newtons/coul})q$$

$$F_2 = 9 \times 10^9 \frac{\text{newton-m}^2}{\text{coul}^2}$$

$$\left[\frac{q(-10 \times 10^{-9} \text{ coul})}{0.10 \text{ m}^2} \right]$$

$$= -(9.0 \times 10^3 \text{ newtons/coul})q$$

The directions of F_1 and F_2 are shown in Fig. 22-2. F_1, a repulsive force, is along the line between q_1 and *P*, pointing *away* from q_1; F_2 points *toward* q_2. The resultant force *F* on *q* is

$$F = \sqrt{F_1{}^2 + F_2{}^2}$$

$$= (12.7 \times 10^3 \text{ newtons/coul})q$$

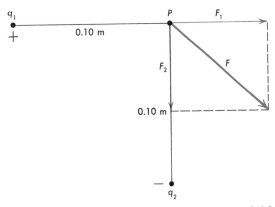

Fig. 22-2. *Calculating the electric field strength at* P *due to* q_1 *and* q_2.

$$E = \frac{F}{q} = 12{,}700 \text{ newtons/coul}$$

We can easily write an equation for the electric field strength E at a point at a distance r in a vacuum from a single point charge q. Imagine placing a test charge q_1 at that point. The force F on q_1 is

$$F = \frac{k_0 q q_1}{r^2}$$

and

$$E = \frac{F}{q_1} = \frac{k_0 q}{r^2}$$

where k_0 as before approximately equals 9×10^9 newton-m²/coul².

Gauss's Law

Figure 22-3 represents a point charge q of 1 coul in vacuum at the center of two concentric spheres of radii 1 m and 2 m. Notice that near

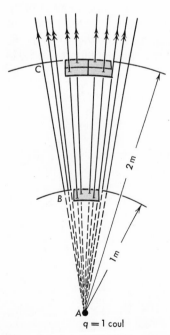

Fig. 22-3. *The field strength at* C *is* $\frac{1}{4}$ *that at* B. q/ϵ_0 *lines originate at* A.

q, where the field is strong, the lines are closely crowded together. What is the field strength E at point B on the surface of the first sphere? E is equal to the force per unit charge placed at that point; that is,

$$E = \frac{k_0 q}{r^2} = \frac{q}{4\pi\epsilon_0 r^2} = 9 \times 10^9 \text{ newtons/coul}$$

We represent this field strength by imagining the drawing of 9×10^9 (9 billion, or more precisely $|k_0|$) barbed lines through each square meter of the sphere's surface. *The number of lines per unit area, normal to the surface, is defined to be equal to the magnitude of E at that point.* At C, the field strength is one-quarter of that at B, hence we draw $\frac{9}{4} \times 10^9$ lines per square meter.

We can readily see that the same number of lines passes through each sphere. The total area of the first sphere is 4π m²; hence, multiplying this area by the density of the lines, we find that the total number of lines piercing the surface of the first sphere is $4\pi \times 9 \times 10^9$ lines. The number of lines passing through the second sphere is $4\pi \times 4$ m² $\times \frac{9}{4} \times 10^9$ lines/m² or $4\pi \times 9 \times 10^9$ lines, the same number that passes through the first sphere. In general, since the area of a sphere with radius r is $4\pi r^2$ and since the number of lines per unit area at a distance r is $q/4\pi\epsilon_0 r^2$, the total number of electric lines N that pass through *any* sphere around q is

$$N = (4\pi r^2)\frac{q}{4\pi\epsilon_0 r^2} = \frac{q}{\epsilon_0} = 4\pi k_0 q$$

The number of electric lines of force that pass through a closed surface surrounding a charge q is q/ϵ_0 (Gauss's law).

Although we have derived Gauss's law for a point charge and spherical geometry, it holds for *any* collection of static charges of total charge q surrounded by a *closed* surface of *any* shape.

Electric lines of force due to static charges originate on positive charges and end on negative charges. Figure 22-4a represents

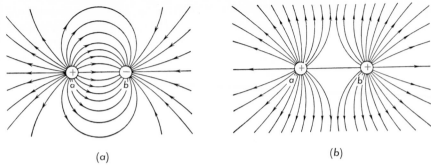

(a)   (b)

Fig. 22-4. *Electric fields due to* (a) *equal unlike charges,* (b) *equal like charges.*

the resultant electric field due to equal, unlike charges. Experimentally one can show this by scattering gypsum crystals on a glass plate before bringing up the charged pith balls *a* and *b*. The gypsum crystals align themselves along the electric lines of force. In Fig. 22-4*b*, the field is due to equal, like charges. Notice that the lines of force are more concentrated in regions where the field is stronger.

Gauss's law is an important and useful generalization. Let us use it to find the electric field strength near a uniformly charged flat metal plate. Suppose that a large plane conductor carries a free charge per unit area σ (Fig. 22-5). The charge resides on its surface, as we saw in the last chapter.

The electric field strength E near the surface of the conductor is *perpendicular* to the surface. Why is this so? If it were not perpendicular, it would have a component parallel to the surface. Because of this component, forces would act on free conduction charges on the surface, causing them to move in a current. Since the charges are at rest, no parallel component of field strength can be present, and the field strength is perpendicular to the surface. Hence the electric lines of force are also perpendicular to the surface.

Now consider an imaginary cylindrical closed surface that passes through the surface of the conductor and extends a little above and a little below it. The cross-sec-

tional area of the cylinder is A. The charge q enclosed by the cylinder is the charge per unit area σ times the area A or σA. The number of electric lines that pass through the closed surface, according to Gauss's law, is

$$N = \frac{q}{\epsilon_0} = \frac{\sigma A}{\epsilon_0}$$

All of these lines go through the upper base of the cylinder: first, no lines can pass through the sides because the lines are vertical; and second, no lines penetrate into the metal at the bottom of the cylinder because a field there would set up a current in the metal. At

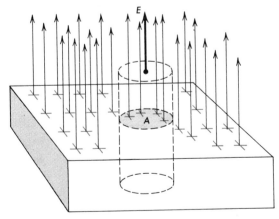

Fig. 22-5. *A portion of a large plane conductor with a free charge per unit area σ.*

the top of the cylinder the number of lines per unit area is

$$\frac{N}{A} = \frac{\sigma A}{\epsilon_0 A} = \frac{\sigma}{\epsilon_0} = E$$

Thus the electric field strength near a very large plane conductor with free charge per unit area σ is σ/ϵ_0, is directed perpendicular to the surface, and does not penetrate into the conductor.

The Conservation of Electric Charge

The generalization that whenever one body acquires a positive charge, some other body must acquire an equal negative charge was mentioned in the preceding chapter. This law, called the law of conservation of electric charge, may be demonstrated by an experiment — called the ice-pail experiment — devised more than a century ago by Michael Faraday. Connect an insulated metal vessel to an electroscope — as in Fig. 21-11 of Chapter 21 — and lower a positively charged metal ball into the vessel. Then electrons are drawn to the inner walls, the outside acquires

a positive charge, and the leaf of the electroscope diverges (Fig. 22-6a). Conservation of electric charge requires that the positive charge induced on the outside of the vessel equal the negative charge induced on the inside.

Notice that the number of electric lines leaving the ball is equal to the number reaching the inner surface of the vessel by Gauss's law; this is so because the electric field inside the metal is zero, and hence *no* lines exist inside the conducting vessel. Thus the negative charge inside the container must equal the positive charge on the ball because otherwise there would be a net charge within the vessel and some lines would enter or leave its interior. If the ball is removed, the leaf collapses. Instead, touch the charged ball against the bottom of the vessel. The divergence of the leaf does not change. Consider what this means. If the induced charge on the outside of the container had been less than that on the inside or on the ball, the leaf would have moved down somewhat. If the induced charge had been greater, the leaf would have diverged. The leaf does not move; hence we reason that the induced charge on the inside and the induced charge on the outside were exactly equal. *Electric charge — like energy, linear momentum, and angular momentum — is conserved* (Fig. 22-6b)

Potential Difference

Electricity plays an important part in our lives because electrical energy can be transferred from place to place silently, conveniently, and economically. In a great electrical energy generating plant, coal is fed into the furnaces of the steam boiler. There the chemical energy of the coal is converted into the energy of the steam and then into the mechanical energy of the whirling turbines. These drive giant generators, in which mechanical energy is converted into electrical energy to light our

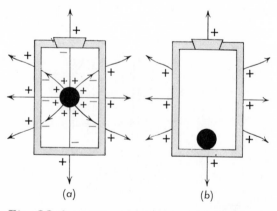

(a)   (b)

Fig. 22-6. (a) *The negative charge induced on the inside equals the inducing charge on the ball.* (b) *The positive charge on the outside of the vessel must have equaled the negative charge on the inside. Electric charge is conserved.*

buildings, to lift elevators, and to operate telephones, radios, and refrigerators.

What determines the amount of energy delivered by the transfer of electric charge? Before answering this question, let us review a few facts as to energy relationships in gravitation and in the flow of fluids.

The amount of work that water can do after passing over a waterfall depends upon the *height* of the falls and the *quantity of water* that is transferred. The height of Niagara Falls is 50 m; hence each kilogram of water that goes over loses 50 kgwt-m or 490 joules of potential energy. The total work done is given by

Work done = Weight of water × Distance of fall

When a liquid flows through a horizontal pipe from one reservoir to another, the work done is equal to the volume of liquid transferred times the difference in pressure between the two ends of the pipe. For example, if 2 m³ of water is transferred and the pressure difference is 100 newtons/m², the work done is 200 newton-m or 200 joules.

Work done = Volume of water
$\qquad\qquad$ × Decrease of pressure

In electricity, difference of *potential* corresponds to difference of *level* in gravitation and to difference of *pressure* in liquid flow in the sense that all three are the work done per unit quantity transferred. Consider the two oppositely charged spheres A and C (Fig. 22-7). A tiny positively charged pith ball B will be attracted by the charges on A and repelled by those on C. If you force the pith ball from A toward C, you must do work and you will increase the energy of the system. We define the *electrical potential difference V* between two points as the *work required, per unit charge, to transfer positive charge from one point to the other.*

Potential difference
$$= \frac{\text{Work done}}{\text{Quantity of charge transferred}}$$

Fig. 22-7. *The potential difference between* A *and* C *is the work per unit charge required to force electricity from one body to the other.*

$$V = \frac{W}{q}$$

Potential difference, like work, is a scalar quantity. Two or more potential differences are combined algebraically.

Usually, in gravitational problems, we are more interested in the *difference* of level between points than in their elevations above sea level. Similarly, in electrical theory, we are often concerned with the potential *difference* between two points rather than with their potentials. The electrical potential of the earth is commonly taken as zero, and the potentials of other bodies are measured with reference to the earth. (However, in some important problems, the potential of any point which is infinitely distant from any charged body is assumed to be zero.)

The most widely used unit of potential difference, the *volt*, named for the Italian physicist Volta, is defined as the *potential difference between two points such that one joule of work is required to force one coulomb of charge from one point to the other.* One volt equals one joule per coulomb. The units of potential difference in other systems of electrical units are given in Appendix H.

Example. In starting the engine of an automobile, 200 coul of electricity were transferred through the motor. If 2400 joules of work was done, what was the potential difference between the terminals?

$$V = \frac{W}{q}$$

$$= \frac{2400 \text{ joules}}{200 \text{ coul}}$$

$$= 12.0 \text{ volts}$$

Electric potential differences are produced by such devices as Van de Graaff generators—already referred to—and also by chemical batteries, electric generators, and photovoltaic cells, which we shall discuss in later chapters. Instruments used to measure potential differences are called *voltmeters.* A gold-leaf electroscope can serve as a voltmeter if it is fitted with a scale to measure the angle of deflection of the gold leaf. A potential difference of several thousand volts may be measured by determining the largest gap across which a spark will jump in air. The width of this largest spark gap depends upon the applied potential difference, upon the air pressure, and upon the shapes of the two terminals. The sparking potential for two needle points 1 cm apart in air at normal pressure is about 8500 v. For two polished spheres 1 cm in diameter and 1 cm apart, it is 25,000 v. Moving-coil instruments are more commonly used for measuring potential differences; they will be described later in Chapter 27.

Electric field strength is related to electric potential difference. A parallel-plate capacitor consists of two parallel plane conductors separated by a vacuum or an insulator. The charges on the plates are equal and of opposite sign. Let the distance between the two horizontal metal plates of a parallel-plate vacuum capacitor be s, and let the potential

difference between them be V (Fig. 22-8). If the plates of the capacitor are large or if the separation between them is small, the electric lines of force will be uniformly dense except at the edges. Thus, near the center of the capacitor, the electric field strength and the force on a given charge are constant. Suppose that a small particle carrying a positive charge q is transferred from the lower plate to the upper one. Then the work done against electrical forces is given by

$$W = qV \qquad (1)$$

But work equals the external force F_1 times the displacement s; hence

$$W = F_1 s \qquad (2)$$

From equations (1) and (2)

$$qV = F_1 s$$

$$\frac{V}{s} = \frac{F_1}{q}$$

Now the force F exerted by the field on the charged particle is equal and opposite to F_1. Thus

$$F = -F_1$$

and

$$\frac{V}{s} = -\frac{F}{q} = -E$$

Hence

$$E = -\frac{V}{s}$$

where E is the field strength. Therefore, the electric field strength at any point equals the negative rate of change of potential difference with displacement parallel to the field; in other words, the electric field strength is the negative *gradient* of the electric potential. This result, which was derived for the uniform field of a parallel-plate capacitor, can be shown to be true for non-uniform electric fields also. You can readily show that one newton per coulomb equals one volt per

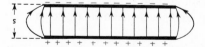

Fig. 22-8. A parallel-plate capacitor. The field strength between the plates can be expressed in newtons per coulomb or in volts per meter.

meter and that both are proper units of electric field strength.

Measuring the Charge of the Electron

Electricity is atomic in nature. Electric charge is not continuous, but is an aggregate of small elementary charges. Measuring electric charge is really a process of counting the number of tiny elementary charges in it.

The charge on a single electron was first precisely determined by R. A. Millikan during the years 1910–1918. His method was essentially as follows: By means of an atomizer or sprayer, a mist of microscopic oil droplets was produced above the upper plate of a parallel-plate capacitor whose plates were separated by air (Fig. 22-9). A few of these droplets drift downward through a hole in the upper plate. Suppose that one of them, because of the friction of spraying, acquires an excess charge of one electron. If the upper plate is charged positively with respect to the lower one, the electric field between the plates opposes the downward motion of the droplet. The strength E of the electric field may be so adjusted that the up force Eq of the electric field is exactly equal to the down pull mg of the earth's gravitational field. That is,

$$Eq = mg$$

or

$$q = \frac{mg}{E} = \frac{mg}{(-V/s)} \qquad (3)$$

where V is the potential difference between the plates and s is their separation. From this equation the charge q of the droplet may be determined if mg, the weight of the droplet, is known. Millikan determined the weights of the microscopic droplets by observing how fast they drifted downward in the earth's gravitational field when no electric field acted upon them. Such droplets fall with constant velocity, for the down pull of gravity is balanced by the friction of the air and the small buoyant

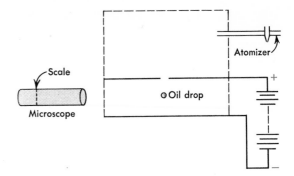

Fig. 22-9. The scheme of Millikan's apparatus for measuring the charge on the electron.

force. Moreover, the smaller the droplet, the smaller will be its velocity. The law of fall in air (Stokes' law) had previously been established experimentally for spheres of different known weights and diameters.

Let the constant terminal velocity with which the droplet drifts downward in the gravitational field be v_t, the radius of the oil droplet be a, and the coefficient of viscosity of the air be η. According to Stokes' law, the backdrag of air friction F_f on the droplet is

$$F_f = 6\pi\eta a v_t$$

The weight of the droplet w is its volume times the density of the oil D times the acceleration due to gravity

$$w = mg = VDg = \tfrac{4}{3}\pi a^3 Dg$$

The buoyant force F_b acting on the droplet equals the weight of the displaced air (Archimedes' principle), whose density is d:

$$F_b = \tfrac{4}{3}\pi a^3\, dg$$

As the droplet falls, it is in equilibrium, and the sum of the up forces equals the sum of the down forces:

$$F_f + F_b = w$$

$$6\pi\eta a v_t + \tfrac{4}{3}\pi a^3\, dg = \tfrac{4}{3}\pi a^3\, Dg \qquad (4)$$

By measuring the rate of fall (v_t) over the scale of the microscope and knowing the quantities η and d for air and D for the oil used, one can calculate the radius and hence the mass of a given droplet. Substituting for m in equation (3) and knowing the potential difference required to hold the same droplet stationary, he can calculate q.

The charge of one electron was the smallest value that Millikan ever found on any droplet. Moreover, the other charges on the droplet were always exactly twice, three times, or n times this value. Therefore, he assumed that the droplet had 2, 3, or n excess electrons, each electron having exactly the same charge.

By experiments of this nature, Millikan and others proved that the charge of the electron is 1.60×10^{-19} coulomb; there are 6.25 billion billion electrons in one coulomb of negative charge. If all the inhabitants of the United States were to count this tremendous number at the rate of one per second, working day and night without stopping, several thousand years would be required to complete the task.

Potential Due to a Point Charge

Forces that depend upon $1/r^2$, where r is the distance from the source, are called inverse-square forces. They can be written $F = K/r^2$. For gravitational forces, $K = -Gm_1m_2{}^*$; for electrostatic forces, $K = k_0q_1q_2$. Inverse-square law forces lead to $1/r$ potentials: the gravitational potential, which is the work per unit mass, and the electric potential, which is the work per unit charge, both vary inversely with the first power of the distance from the source. In this section we shall derive the equation $V = k_0q/r$ for a point charge. The same proof could be followed to show that

---

* In Chapter 11, the law of universal gravitation was stated as $F = Gm_1m_2/r^2$ where the sign of the right-hand side of the equation was positive. We must now modify this equation to express the fact that the force between two point masses is always an *attractive* one. We can do this by inserting a negative sign on the right: $F = -Gm_1m_2/r^2$.

$V_G = -Gm/r$ for the gravitational potential of a point mass.

Suppose that we move a charge q_1 from the point b to the point a, slightly nearer the charge q (Fig. 22-10). We must exert a force to overcome the electric repulsion and must do work equal to the product of the average force and the distance moved. By doing this work, we store up electrical potential energy in this system. When the charge is at b, its distance from q is r_b, and the force of repulsion is k_0qq_1/r_b^2. At a, the force is k_0qq_1/r_a^2. Assume that a is very close to b. Then the average force acting on q_1 is nearly $k_0qq_1/(r_a \times r_b)$. Thus the work input, or energy increase, is

$$\text{Work done} = \Delta W \cong \frac{k_0qq_1}{r_ar_b}(r_b - r_a)$$

$$= k_0qq_1\left(\frac{1}{r_a} - \frac{1}{r_b}\right)$$

The total energy increase to move q_1 from an infinitely distant point to a is equal to the sum of a very large number of terms:

$$\Sigma(\Delta W) \cong k_0qq_1\left[\left(\frac{1}{r_a} - \frac{1}{r_b}\right) + \left(\frac{1}{r_b} - \frac{1}{r_c}\right) \right.$$
$$\left. + \left(\frac{1}{r_c} - \frac{1}{r_d}\right) + \left(\frac{1}{r_d} - \cdots - \frac{1}{r_\infty}\right)\right]$$

where r_∞ means a very large distance from q so that the electric force there is zero. Notice that all the terms cancel out except the first and last. Hence,

$$W = k_0qq_1\left(\frac{1}{r_a} - \frac{1}{r_\infty}\right) = \frac{k_0qq_1}{r_a}$$

since $1/r_\infty$ is zero. The potential V at r_a is the work done per unit charge in bringing charge from a great distance ("infinity") to point a.

$$V = \frac{W}{q_1} = \frac{k_0q}{r_a}$$

In the rationalized mks system, $K_0 = \dfrac{1}{4\pi\epsilon_0} \cong 9 \times 10^9$ newton-m²/coul². Thus the potential at a point 1 m from a charge of

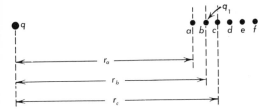

Fig. 22-10. The potential at a due to the charge q is $k_0 q / r_a$.

$1/(9 \times 10^9)$ coul is

$$V = \left(\frac{9 \times 10^9 \text{ newton-m}^2}{\text{coul}^2}\right)\left(\frac{1}{9 \times 10^9}\text{coul}\right)\frac{1}{1 \text{ m}}$$

$$= 1 \text{ joule/coul} = 1 \text{ v}$$

We have used an approximate method. More exact methods yield the same result.*

Capacitance

A free charge induces charges on other conductors in its neighborhood. The general problem of finding relationships between the charge on one conductor and the charges on neighboring conductors is quite complicated. We shall consider a simple case — that of the parallel-plate capacitor.

---

* *A calculus derivation.* The element of work dW done when q_1 is moved from a distance r to $r - dr$ is given by

$$dW = F(-dr) = \left(\frac{k_0 q q_1}{r^2}\right)(-dr)$$

$$dW = -k_0 q q_1 \frac{dr}{r^2}$$

To find the work W done when q_1 is moved from some great distance ($r \to \infty$) to r_a, we integrate the last expression:

$$\int_0^W dW = -k_0 q q_1 \int_\infty^{r_a} \frac{dr}{r^2}$$

$$W = -k_0 q q_1 \left[-\frac{1}{r}\right]_\infty^{r_a}$$

$$W = k_0 \frac{q q_1}{r_a}$$

The potential V at r_a, as before, is

$$V = \frac{W}{q_1} = \frac{k_0 q}{r_a}$$

The amount of energy that a tank of compressed air contains depends jointly upon the pressure and the volume of the tank, and might, therefore, be specified in terms of the mass of air the tank contains at 1 atm pressure. In electricity, the amount of charge that a system contains depends upon the electrical potential of the body and also upon its electrical *capacitance*. The greater the capacitance of a body, the greater is the quantity of charge that is required to produce a given change in its potential.

The capacitance C of a capacitor is the quantity of charge on either plate per unit potential difference between the plates.

$$\text{Capacitance} = \frac{\text{Quantity of charge}}{\text{Potential difference}}$$

$$C = \frac{q}{V}$$

The unit of capacitance, the *farad*, named after Michael Faraday, equals one coulomb per volt. The microfarad, more commonly used, is one-millionth of a farad.

The two plates of a parallel-plate capacitor, when close together, can contain a greater electrical charge per unit potential difference than when they are further apart. The capacitance of such a parallel-plate capacitor is given by

$$C = \frac{\epsilon_0 A}{s}$$

in which s is the distance between the plates and A is the area of either plate.

To understand why this is so, suppose that each of the two parallel plates of a capacitor has an area A. Let the distance between them be s and the charge on the plates be $+q$ and $-q$. Then the charge per unit area σ is q/A. Earlier we saw that the electric field strength near a plane conductor with a charge per unit area σ is

$$E = \frac{\sigma}{\epsilon_0}$$

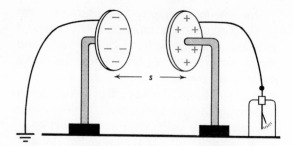

Fig. 22-11. *Increasing the separation* s *of the plates increases the potential difference between them. The charge on either plate is constant.*

The electric field near the center of a parallel-plate capacitor is uniform, and *E* is constant except near the edges. Thus the work done per unit charge in transferring charge from one plate to the other is

$$V = Es = \frac{\sigma}{\epsilon_0}s$$

$$V = \frac{qs}{\epsilon_0 A} \qquad (5)$$

Hence

$$C = \frac{q}{V} = \frac{\epsilon_0 A}{s}$$

Charge one plate of a parallel-plate capacitor positively and connect it to a gold-leaf electroscope (Fig. 22-11). Ground the other plate. A negative charge will be induced on it so that the plates have equal and opposite charges. Bring the grounded plate near the other. The nearness of the negative charge reduces the potential of the positive plate: less work would be done in taking a unit charge from one plate to the other. Thus, decreasing the separation of the plates *decreases* the potential difference between them, and the leaf of the electroscope collapses. At the same time, the capacitance *increases* because a given charge is being held at a smaller potential difference. If you increase the separation, the potential dif-

ference increases and the leaf diverges. The capacitance decreases.

We can readily calculate the energy of a charged capacitor. The energy required to fill a water tank of height *h* is equal to the weight of the water times the *average vertical distance* $\frac{1}{2}h$, through which it must be lifted. Similarly, the work required to charge a capacitor from zero potential difference to a potential difference *V* is given by

$$W = q(\tfrac{1}{2}V)$$

But

$$q = CV$$

Therefore

$$W = \tfrac{1}{2}CV^2$$

and this is also the energy *U* of the charged capacitor.

$$U = \tfrac{1}{2}CV^2$$

Example. How much energy has a capacitor of capacitance 1 microfarad when charged to a potential difference of 1 v?

$$U = \tfrac{1}{2} \times 10^{-6} \text{ farad} \times \frac{1 \text{ coul/v}}{1 \text{ farad}} \times (1 \text{ volt})^2$$

$$\times \frac{1 \text{ joule/coul}}{1 \text{ v}}$$

$$= 5 \times 10^{-7} \text{ joule}$$

Combinations of Capacitors

Capacitors can be connected "in parallel" or "in series." Figure 22-12*a* represents a battery which causes a potential difference *V* across three capacitors 1, 2, and 3, connected in parallel. What is their combined capacitance, that is, what would be the capacitance *C* of some capacitor 4 which would contain the same charge *q* at the same potential difference? The charge on 4 is

$$q = q_1 + q_2 + q_3$$

But $q = CV$; hence

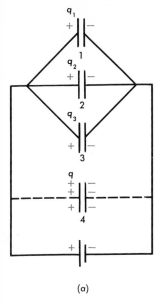

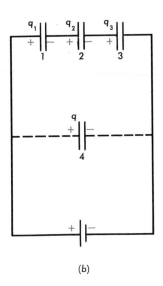

(a)   (b)

Fig. 22-12. (a) *Capacitors in parallel:* $C = C_1 + C_2 + C_3 + \cdots$ (b)
Capacitors in series: $1/C = 1/c_1 + 1/c_2 + 1/c_3 + \cdots$

$$CV = C_1V_1 + C_2V_2 + C_3V_3$$

and since

$$V = V_1 = V_2 = V_3$$
$$C = C_1 + C_2 + C_3$$

Figure 22-12*b* represents the three capacitors connected in series with one another to the battery. 4 is an equivalent single capacitor. The fall of potential *V* across the capacitor 4 is equal to the sum of the potential differences across the three capacitors. That is,

$$V = V_1 + V_2 + V_3$$

For any capacitor, $V = q/C$. Hence

$$\frac{q}{C} = \frac{q_1}{C_1} + \frac{q_2}{C_2} + \frac{q_3}{C_3}$$

But the charges q_1, q_2, and q_3 imparted to the positive plates of each of the three capacitors are equal. The charge *q* imparted to the single equivalent capacitor must also have the same value.

Therefore:

$$q = q_1 = q_2 = q_3$$

$$\frac{1}{C} = \frac{1}{C_1} + \frac{1}{C_2} + \frac{1}{C_3} \quad \text{or} \quad C = \frac{1}{\dfrac{1}{C_1} + \dfrac{1}{C_2} + \dfrac{1}{C_3}}$$

Dielectrics

So far we have discussed free conduction charges and induced conduction charges, both in vacuum. The former are transferred to one body by contact with another charged body, and the latter appear on conductors when free charges are nearby. There is a third way in which charges can appear—as bound charges on the surfaces of insulators due to the polarization of the insulators. Insulators are referred to as *dielectrics* in this regard.

What do we mean by polarization? Ordinarily, matter is electrically neutral, as we have seen. This means that the center of positive charge in the molecules of most dielectrics, such as glass, coincides with the center of negative charge (Fig. 22-13*a*). However, in an

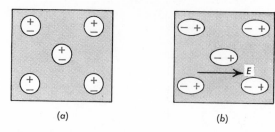

Fig. 22-13. The polarization of molecules in a dielectric in an electric field.

electric field, the electrons are somewhat displaced from their normal positions because of the electric forces that act on them. The molecule becomes an induced electric *dipole* with one end positively charged and the other negatively charged (Fig. 22-13*b*). At the same time, the tiny dipoles tend to line up with their positive ends pointing in the direction of the field. If the electric field is turned off, the molecules cease to be dipoles. Some molecules—such as those of gaseous hydrogen chloride—are permanent electric dipoles. An electric field applied to them merely swings them into line with the field, their positively charged ends pointing in the direction of the field.

Whether the molecules are permanent electric dipoles or induced electric dipoles, the application of an electric field causes bound charges to appear at the surfaces of dielectrics as the dipoles align themselves with the field. In the interior of the dielectric, the positive and negative ends of neighboring dipoles cancel. At one surface, however, the positive ends are exposed, giving a positive surface charge, and at the other surface the negative ends produce a negative surface charge (Fig. 22-14).

The polarization *P* of a dielectric slab between the plates of a charged parallel-plate capacitor is *the surface charge per unit area of exposed surface.* For some dielectric materials—but not all—*P* is directly proportional to the electric field strength *E*.

Surface charges must be considered as well as free charges and induced charges in computing the electric field strength. We see that the effect of the polarization charges is to *oppose* the electric field set up by the charges on the plates. The electric field between the plates is thus less than it would be in the absence of the dielectric. Therefore, the work done per unit charge in taking charge through the dielectric from one plate to another—the potential difference—is *decreased,* and the capacitance is *increased.*

To illustrate these effects, suppose that a flat container of mercury, a conductor, is placed between the oppositely charged plates *A* and *B* (Fig. 22-15). Then electrons will move to the left and accumulate at the face next to *A*, and positive charges will appear next to *B*. The total induced charges at the two mercury surfaces will equal the charges on *A* and *B*, and the field strength inside the mercury will be *zero*. Now suppose that the mercury is replaced by oil—a dielectric. The electrons in each oil molecule are slightly displaced to the left in the field so that there is a negative surface charge at the surface of the oil near *A* and a positive charge near *B*. Because of these polarization

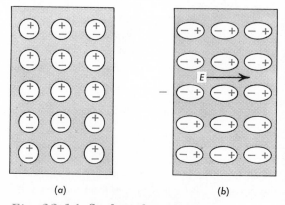

Fig. 22-14. Surface charges due to polarization. (a) *Molecules of a dielectric before an electric field is applied.* (b) *Bound charges appear on the surfaces in an electric field.*

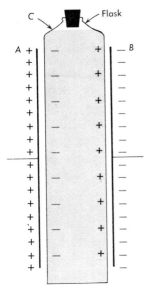

Fig. 22-15. *A capacitor. A dielectric material between the plates becomes electrically polarized so as to weaken the field and increase the capacitance.*

charges the field strength inside the oil will be *less* than if the medium were air. (It will not be zero, however.) Thus less work will be required to transfer unit charge across the liquid than would be required if the medium were air, the potential difference between the two plates of the capacitor is diminished, and the capacitance of the system is increased.

The dielectric constant κ (Greek *kappa*) of a material is the ratio of the capacitance C of a capacitor with the material between the plates to the capacitance C_0 when the plates are separated by a vacuum (Table 22-1).

$$\kappa = \frac{C}{C_0}$$

Thus the capacitance of a parallel-plate capacitor whose plates are separated by a material of dielectric constant κ is

$$C = \frac{\kappa \epsilon_0 A}{s}$$

Point charges exert smaller forces on each other when surrounded by a dielectric because of polarization charges. Coulomb's law for point charges in a dielectric of dielectric constant κ is

$$F = \frac{q_1 q_2}{4 \pi \epsilon_0 \kappa r^2}$$

However, this equation can be used *only when the dielectric has the same properties in all directions and has no exposed surfaces, that is, it has an infinite extent.*

The atoms of some chemical compounds are held together mainly by electrostatic (ionic) attractions. For example, sodium atoms yield electrons to nearby chlorine atoms to form oppositely charged ions. These cling together in sodium chloride crystals. However, sodium chloride readily dissolves in water. This is because the water, having a very high dielectric constant, weakens the forces of attraction between neighboring sodium and chlorine ions.

Table 22-1. Dielectric Constants

Air at 0°C (normal pressure)	1.0006
Alcohol, ethyl, at 0°C	28.4
Barium titanate	approx. 5000
Ebonite	2.7
Glass	5.4–9.9
Hydrogen at 0°C (normal pressure)	1.0003
Mica	5.66–5.97
Paraffin	2.10
Petroleum	2.13
Vacuum (by definition)	1.0000
Water	78.5

Summary

An electric field is a region in which a charged body would experience a force.

The strength of the electric field at a point is the force per unit charge acting on a body located at that point. Units of field strength are the newton per coulomb and the volt per meter.

The number of electric lines of force passing through a closed surface that surrounds a charge q is q/ϵ_0 (Gauss's law). The electric field strength near a charged plane conductor equals σ, the charge per unit area divided by ϵ_0.

The potential difference between two points is defined as the work required per unit charge to transfer positive charge from one point to the other. One volt equals one joule per coulomb. The electric field strength is the negative gradient of the electric potential.

Electric charge is conserved. If one body gains charge, another loses an equal amount. Millikan showed that the elementary charge, that on an electron, is approximately 1.6×10^{-19} coul. Measuring charge is basically a process of counting elementary charges.

The capacitance C of a body or system is the quantity of charge per unit difference of potential. One farad equals one coulomb per volt.

The capacitance C of a parallel-plate vacuum capacitor is given by $C = \epsilon_0 A/s$; that of a parallel-plate capacitor whose plates are separated by a dielectric material, by $C = \kappa\epsilon_0 A/s$, in which κ is the dielectric constant of the dielectric.

The total electric force on a free charge is the sum of the forces exerted by other free charges, by charges induced on nearby conductors, and by polarization charges on nearby insulators.

The capacitance of several capacitors connected in parallel is given by $C = C_1 + C_2 + \cdots$; if connected in series, by

$$\frac{1}{C} = \frac{1}{C_1} + \frac{1}{C_2} + \cdots \quad \text{or}$$

$$C = \frac{1}{\dfrac{1}{C_1} + \dfrac{1}{C_2} + \cdots}$$

Questions

1. Compare a gravitational field with an electric field, discussing them in terms of such concepts as field strength, lines of force, and potential.

2. Cite several differences between gravitational fields and electric fields.

3. By what test could you determine whether the electrical potential difference between two points is one volt?

4. Can an electric field ever exist inside a hollow metal vessel?

5. What is the relationship between electric potential and electric field strength? Are they both vector quantities?

6. How could you use Gauss's law to derive the electric field strength due to a long straight charged wire?

7. What is the charge on (a) an electron, (b) a proton, (c) an alpha particle?

8. (a) How can an electric field be represented graphically? (b) How can the strength of the electric field be represented? Illustrate for a uniform field of 10 newtons/coulomb.

9. Describe how a book-type capacitor could be made and used. Let two plates open and close like stiff pages in a book.

10. Prove that a farad per meter is equal to a coulomb^2 per newton-meter^2.

11. How is the polarization of the dielectric in a parallel-plate capacitor related to the surface charge density? What is an electric dipole?

12. Why does the electric field strength between two charged bodies initially in vacuum diminish when a material substance is inserted between them?

13. If a metal sheet is inserted midway between the plates of a parallel-plate capacitor, does the capacitance increase, decrease, or stay the same? Discuss.

14. Why is capacitance generally given in microfarads?

15. Estimate the dimensions needed for a 1-farad parallel-plate capacitor with air as a dielectric.

16. The plates of a capacitor (in air) are charged so that their potential difference is 100 v. Oil is poured into the space between

the plates. How does this affect (a) the potential difference, (b) the electric field strength between the two plates, (c) the electrical energy of the system?

17. Three sheets of copper and three of glass, all 1.0 mm thick and 10 cm × 10 cm in area, are to be used to make a parallel-plate capacitor. What combination will give (a) the largest capacitance, (b) the smallest capacitance?

Problems

A

1. A force of 0.04 newton is required to hold a charge of 15 μcoul at a certain point. What is the electric field strength at that point?

2. What is the electric field strength at a point 0.10 m away from a charge of 800 μcoul?

3. How much work is required to take a charge of 12 coul from one point to another one when the potential difference between them is 250 v?

4. The potential difference between the two terminals of a storage battery is 12.6 v. How much work is required to transfer 8.0 coul of electricity from one terminal to the other?

5. Six hundred joules of work are required to transfer 50 coul of charge from one terminal of a battery to the other. What is the potential difference between the terminals?

B

6. (a) What is the electric field strength at a point midway between a positive charge of 200 μcoul and a negative charge of 50 μcoul if the charges are 10 cm apart? (b) What would be the field strength if both charges were negative?

7. Three capacitors, 2.0, 3.0, and 6.0 microfarads, are connected in series to a 12-v source. Find the charge on each capacitor.

8. What is the capacitance in microfarads of a capacitor that acquires a charge of 360 μcoul when connected to a 60-v battery?

9. What is the capacitance in farads of a capacitor made by pasting sheets of tinfoil to opposite faces of a sheet of paraffin paper? The area of the paper between the foils is 500 cm², the thickness of the paper is 0.0010 cm, and its dielectric constant is 2.6.

10. The capacitance in farads of an isolated sphere is equal to $1/(9 \times 10^9)$ times its radius in meters. (a) Compute the capacitance of the earth. (b) The addition of how many coulombs of charge from cosmic radiation would increase the potential of the earth by 1.0 v? ($r = 6400$ km.)

C

11. Charges $A = +400$ μcoul and $B = -400$ μcoul are located 10 cm apart at the ends of a diameter of a circle. What are the electric field strengths at points on the circle that are (a) 6.0 cm, (b) 4.0 cm, (c) 2.0 cm from A?

12. A glass jar, 10 cm in diameter, has metal coatings inside and outside extending 15 cm up the sides, with none on the bottom. What is its capacitance in farads and in microfarads if the glass is 0.30 cm thick and has a dielectric constant of 5?

13. The electric potential difference between two parallel plates 0.0010 m apart is 200 v. (a) What is the electric field strength between them in volts per centimeter and newtons per coulomb? (b) What acceleration will a hydrogen ion experience in this field? Its mass is 3.32×10^{-27} kg and its charge is 1.60×10^{-13} μcoul.

14. In Millikan's experiment, an oil droplet of radius 0.400×10^{-4} cm and mass density 0.80 gm/cm³ is in equilibrium between the plates when it carries a charge of 1 excess electron. What are (a) the electric field strength and (b) the potential difference in volts between the plates if they are 0.600 cm apart?

15. How many electric lines pass through a sphere of radius 2.00 cm if a positive charge

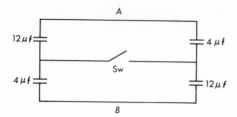

— figure diagram with labels: A, B, 12 μf, 4 μf, 4 μf, 12 μf, Sw

Fig. 22-16.

of 4.00 μcoul is located at its center?

16. Each of the two plates of a parallel-plate capacitor in air has a charge density of 4.0 μcoul/m². (a) What is the field strength between the plates? (b) What is the potential difference between them if they are 1.00 cm apart?

17. If a 1-microfarad capacitor at a potential difference of 2 v is connected in parallel with a 2-microfarad capacitor at a potential difference of 1 v, (a) what will be the resulting potential difference across each capacitor? (b) Calculate the total energy before and after the capacitors are connected. Explain how energy is conserved in this example.

18. (a) What is the total kinetic energy of 10^{10} electrons that have been accelerated by passing across a potential difference of 1.0×10^6 volts? (b) What is the velocity of the electrons?

19. An alpha particle of mass m, velocity v_0, and charge $+2e$ approaches a nucleus of charge $+Ze$, where Z is the atomic number of the nucleus. Assuming that the nucleus is very massive and that energy is conserved,

derive an expression for the distance of closest approach of the alpha particle to the nucleus.

20. Two point charges of 100 μcoul and 300 μcoul are 0.25 m apart in a vacuum. How much work is done in reducing their separation to 0.10 m?

21. A beam of electrons, each of mass m, is fired into the uniform electric field between the large plates of a parallel-plate capacitor at right angles to the direction of the electric field. The separation of the plates is d, and the potential difference between them is V_0. If the electrons enter the field at a point halfway between the plates with an initial velocity v_0, derive expressions for (a) the distance from the edge to the point at which the beam strikes one of the plates and (b) the velocity of the electrons just before they strike.

22. (a) If the voltage across AB (Fig. 22-16) is 240 v, find the charge on each capacitor. (b) If switch Sw is closed, find the change in charge on each capacitor.

For Further Reading

Anderson, David L., *The Discovery of the Electron,* D. Van Nostrand, Princeton, 1964.

Millikan, Robert A., *Electrons (+ and −),* University of Chicago Press, Chicago, 1947.

Millikan, Robert A., *The Autobiography of Robert A. Millikan,* Prentice-Hall, New York, 1950.

CHAPTER TWENTY-THREE

Current and Resistance

In the preceding two chapters, we discussed static electric charges. If you charge an insulated metal sphere positively, the positive charge will be equally distributed over the sphere and will remain at rest. However, if you connect a wire from the sphere to the ground, charge will flow to the ground. The positive charge flowing from the sphere to the ground constitutes an *electric current.* The concept of a current of charge plays a fundamental role in the theory of electricity and magnetism. In addition, most of the practical uses of electricity deal with currents. We must understand what they are.

Most of what we discuss in this section of the book can be called the "classical" treatment of current and resistance, going back to the work of the early nineteenth-century physicists. However, modern physics—especially the branches of it called quantum mechanics, statistical mechanics, solid state physics, and plasma physics—has greatly added to our understanding of the fundamental processes by which electric charges move through conducting media; we shall consider some of these developments in a later chapter.

Direction of Current

The positive charges in a wire cannot flow because they are tightly bound in the nuclei of the atoms. It is the electrons that travel, and they move from points of *lower* (more negative) to those of *higher* (less negative) potentials. Before the discovery of the electron, however, people assumed that an electric current in a wire was directed from positive to negative, as the positive charges would, in fact, flow *if they were free.* In solving practical problems of electrical circuits, the older convention of regarding an electric current as a flow of positive charges is often followed. However, this imaginary, assumed, positive current is opposite to the actual electron flow in a wire.

Either convention can be consistently followed in discussing circuits. We shall stress *electric* currents of positive charge in discussing currents in metallic conductors except when the motion of individual particles is concerned. Then we shall usually refer to *electron* currents.

Current and Potential Difference

A positively charged insulated metal sphere is at a positive potential difference with respect to the ground. The potential difference is — as always — the work done per unit charge to take positive charge from the ground to the sphere. It is also the work done per unit charge *by* positive charge as it flows to the ground. Connected to the ground by a wire, the sphere will lose both its positive charge and its potential difference as the positive charge flows to the ground against the resistance offered by the grounding wire. The electric current will be a *transient* current, lasting a small fraction of a second.

To provide a steady potential difference, connect an electrostatic generator to the sphere (Fig. 23-1) so that the generator charges the sphere to a potential of, say, +1200 v. Connect the sphere to one end of the insulated wooden meter stick. (Such a stick is slightly conducting. Spraying it with a sodium chloride solution and drying it will increase its ability to conduct.) Then, after an initial charge has been transferred, no charges flow along the stick, and the current

is zero. Each point on the stick has the same electric potential. Connect the right end of the meter stick to the earth through a wire. Now the left end *A* is at +1200 v, *B* is at +600 v, and *C* at zero potential. Moreover, these potentials are maintained because the electrostatic generator replenishes the charge on the sphere as charge flows to the ground over the meter stick. The difference of potential between any two points on the stick is associated with a current through it.

The current *I* is the *charge transferred per unit time* through any cross section of the stick.

$$I = \frac{q}{t}$$

In the mks system of units the charge *q* is measured in coulombs, the time *t* in seconds, and the current *I* in *amperes.* The ampere as the unit of current will be defined independently in Chapter 28. For the moment, we shall merely relate the ampere to the coulomb: *one ampere is the current which transfers one coulomb per second through any cross section of a conductor.*

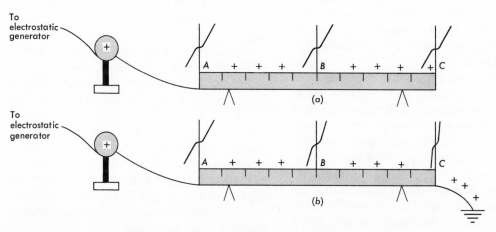

Fig. 23-1. (a) *An insulated meter stick is connected to an insulated sphere that is kept charged by an electrostatic generator. Each point on the stick is at the same electric potential.* (b) *One end of the meter stick is grounded.*

Our simple experiment with the electrostatic generator indicates that positive charge in a current flows from points of higher concentration of positive charge (points of higher potential) to points of lower concentration of positive charge (lower potential).

There are interesting resemblances between the flow of liquids and the flow of electricity. Suppose that water flows from a reservoir through a horizontal pipe into a basin. From there it is pumped back into the reservoir, so that a continuous water current flows in the circuit. The water current is measured in gallons per second just as the electric current is measured in coulombs per second. The pressure of the water as it enters the pipe from the reservoir is the work done per unit volume as water flows through the pipe, just as the potential difference of charge on the sphere is the work done per unit charge as charge flows to the ground over the meter stick. The pressures at points along the pipe decrease in the direction of flow as do the electric potentials along the meter stick.

The laws of the flow of electricity in a metallic conductor can be studied conveniently by using the circuit represented in Fig. 23-2a. It consists of a dry cell D, an ammeter or current-measuring device C, and a voltmeter, together with interconnecting wires. The dry cell "pumps" the electric charge around the circuit, the ammeter (a moving-coil instrument that we will discuss in Chapter 27) measures the current, and the voltmeter indicates the difference of potential between the ends of the wire. We represent the cell by the symbol ⊣|⊢ and the electrical resistance offered by the wire by ⌁⌁⌁ (Fig. 23-2b). (The connecting wires are made of an alloy whose resistance varies little with temperature.) The electric current is directed from the positive terminal of the cell, through the connecting wires, to the negative terminal of the cell. We shall discuss the way charge is conducted *through* the cell at the end of this chapter.

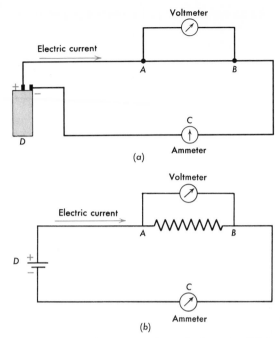

Fig. 23-2. *The current through the wire* AB *is proportional to the potential difference between its ends.* (a) *The physical circuit.* (b) *The circuit diagram.*

We measure the current and the difference of potential when one dry cell is in the circuit. Next we connect two dry cells at D; the positive terminal of one cell is connected to the negative terminal of the other. The voltmeter reading and the ammeter reading will both be doubled. Further experiments like this indicate that *the current through a metallic conductor is directly proportional to the potential difference between its ends* over a remarkably wide range of experimental conditions. The results are stated in Ohm's* law:

* Georg Ohm (1787–1854) was a German high-school teacher. His enunciation of this law in 1827 aroused such bitter antagonism that he lost his position. Years later, when his work was corroborated by other scientists, he was honored by a professorship in physics at the University of Munich. Ohm stated his law only eight years after Oersted discovered the magnetic effect of a current (Chapter 26). Ohm had no reliable voltmeters, ammeters, or batteries. He employed thermocouples to generate currents.

The current I *through a metallic conductor is directly proportional to the potential difference* V *between its ends.*

Or

$$I = \frac{V}{R}$$

where R is an experimentally determined constant—called the *electrical resistance*—for a given conductor. We shall define the *ohm*, the unit of electrical resistance, in the next section.

Before concluding this discussion of the relation between current and potential difference, we must point out that not all conductors obey Ohm's law. The current is *not* proportional to potential difference in many devices that are important applications of physics—electronic tubes, neon signs, and transistors used in radio receivers, to name just a few. We shall discuss these devices in later chapters. However, metallic conductors in circuits obey Ohm's law over a wide range of conditions. Our further discussion of it will be well worthwhile.

Electrical Resistance

Every metallic conductor opposes the passage of electrical charges through it. This opposition arises because the moving charges collide with vibrating atoms in the crystal structure of solids and other particles of the conductor. In so doing, the moving charges give up energy, which appears as heat.

In a perfect single crystal—no dislocations, no impurities, and no lattice vacancies—at absolute zero, electrical resistance would be at a minimum. Conductors, called superconductors, cooled to within a few degrees of absolute zero, carry currents without diminution for weeks. Lead, for example, becomes a superconductor below the temperature of 7.2°K. Electrons can slip through crystal lattices at low temperatures with little loss of energy. At room temperatures in ordinary polycrystalline material, the vibrations of the atoms and the irregularities in the crystal lattice cause resistance. The electrons drift slowly—at velocities of about one ten thousandth of a meter per second—through the wire in the direction *opposite* to that of the electrical field strength in the wire.

According to Ohm's law, *electrical resistance is the ratio of the potential difference to the current for a conductor at a given temperature.* The *ohm*, the unit of resistance, is defined in terms of the ampere and the volt, as follows:

One ohm is the resistance of a conductor through which the current is one ampere when the potential difference between the ends of the conductor is one volt.

One ohm equals one volt per ampere.

Example. The potential difference between the terminals of an electric iron is 120 v, and the current is 5.0 amp. What is the resistance of the wire?

$$R = \frac{V}{I} = \frac{120 \text{ v}}{5.0 \text{ amp}} = 24.0 \text{ ohms (written 24.0 } \Omega)$$

The resistance of a metallic conductor depends upon its length. Doubling the length of a wire doubles the potential difference for a given current, indicating that the resistance has doubled. Stated another way, doubling the length doubles the potential difference required to produce a given current in the wire. This is in accord with our model of a current in an electrical conductor because doubling the length should double the average number of collisions that an electron makes in traveling through the wire. *The resistance is directly proportional to the length of the wire.*

The resistance of a wire depends also upon its cross-sectional area. Suppose that first one high-resistance wire is connected across the terminals of a dry cell and that afterwards a second, equal wire is connected beside the first one. Then equal currents will be established in the two wires and the total current will be doubled. Thus the resistance of the two wires together is one-half that of the single wire. Moreover, the combined cross section of the two wires is twice that of the single wire. In general:

The resistance of a conductor is inversely proportional to its cross-sectional area.

We can summarize in a single equation the dependence of the resistance R at a given temperature upon length l, cross-sectional area A, and the electrical properties of the material:

$$R = \rho \frac{l}{A}$$

in which ρ (Greek *rho*), the *resistivity* of the substance, is RA/l and is numerically equal to the resistance of an imaginary conductor, made of the substance, 1 m long and 1 m² in cross section (Table 23-1).

Finally, the resistance of a conductor depends upon its temperature. Wrap several feet of thin iron wire onto a piece of asbestos, and connect the wire, through an ammeter, to a dry cell. Heat the coil in the flame of a Bunsen burner, and the current through the wire will diminish because the resistance increases. The resistance of all pure metallic conductors increases with temperature, but that of most forms of carbon and of some alloys diminishes.

The variation of the resistance of a pure metal with temperature resembles that of the pressure variation of a gas with temperature. This suggests that the free electrons in a metal behave like a gaseous atmosphere. When the temperature rises, the atoms in the crystal vibrate with greater amplitudes; the forward motion of the electrons is impeded; they drift along the wire more slowly; and the resistance increases.

The resistance R of a conductor at a Celsius temperature t can be computed from the resistance R_0 at 0°C by means of the empirical equation

$$R = R_0(1 + \alpha t)$$

in which α is the *temperature coefficient of resistance* of the substance. (See Table 23-1.)

Table 23-1. Resistivities and Temperature Coefficients of Resistance

Substance	Resistivities at 0°C, ohm-meters	Temperature Coefficient Per C°
Aluminum	2.8×10^{-8}	0.0039
Carbon	$3500 \ \times 10^{-8}$	−0.0005
Copper	1.7×10^{-8}	0.0040
Glass, plate	$9 \ \times 10^{17}$	
Iron	$10 \ \times 10^{-8}$	0.0062
Manganin alloy	$44 \ \times 10^{-8}$	0.00001
Mercury	$94 \ \times 10^{-8}$	0.00088
Nichrome alloy	$100 \ \times 10^{-8}$	0.0004
Quartz, fused (an excellent insulator)	$5 \ \times 10^{22}$	
Silver	1.5×10^{-8}	0.0040
Slate	$2 \ \times 10^{12}$	
Sodium chloride-water solution (5%)	$15 \ \times 10^{-8}$	
Tungsten	5.5×10^{-8}	0.0045

The electron-gas model explains the increase of resistivity of metals with increase of temperature by the greater disorder introduced into a crystal as the temperature is raised. The vibrational amplitudes of the atoms are greater, and the electrons are unable to slip through the crystal lattice readily. The effect of this is a greater resistivity.

The actual situation is not this simple, however. Some materials—semiconductors in particular—show a *decrease* in resistivity with increase in temperature. Also, even for metals, as we noted above, spectacular decreases of resistivity occur at temperatures near 0°K in the phenomenon of superconductivity. These changes cannot be explained by a simple electron-gas theory.

The resistivity of a metal is also sensitive to pressure, impurities, and elastic strains. Increase of pressure distorts the crystal lattice and generally leads to an increase in resistivity. Impurity atoms—indium in lead, for ex-

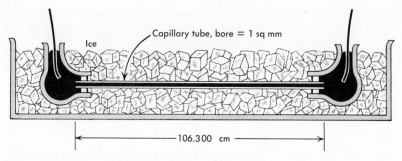

Fig. 23-3. A standard 1-Ω resistor. The column of mercury is 106.300 cm long. It is 1 mm² in cross section. The temperature is 0°C. The resistance is 1.000495 Ω.

Capillary tube, bore = 1 sq mm

Ice

106.300 cm

Some Resistors

ample — scattered through the crystal lattice, either substituting for the atoms of the metal in lattice sites or occupying interstitial positions, affect the "orderliness" of the crystal, and in metals, cause an increase in resistivity. In semiconductors, as we shall see, impurity atoms can have an opposite effect, actually decreasing the resistivity. Strains produced by shearing the metal, working it, stretching it, and so on, usually increase the resistivity.

Some Resistors

In setting up practical standards of resistance, it is necessary to specify the length of the conductor, its cross-sectional area, the temperature, and the substance of which it is composed.

The resistance at 0°C of a column of mercury 106.300 cm long and 1 mm² in cross-sectional area is approximately 1 Ω (more precisely it is 1.000495 Ω). Figure 23-3 represents the structure of a resistance standard of these dimensions. The ends of the glass tube are inserted into bulbs filled with mercury to make contact with the wires through which the electricity flows.

It is convenient to remember that the resistance of a copper wire 1000 ft long and 0.1 in. in diameter is about 1 Ω. Then the resistances of other copper wires can be found by proportion. For example, the resistance of a copper wire 4000 ft long and 0.2 in. in diameter would also be 1 Ω.

Coils of wire of known resistance are useful in making electrical measurements. The wires are made of an alloy, the resistivity of which varies but little with temperature. Several of the resistance coils are mounted in a "resistance box," so that different coils may be connected into a circuit (Fig. 23-4a). The ends of the wires are soldered into brass blocks mounted side by side. Tapered plugs are inserted between adjacent blocks. When all the plugs are in place, most of the charges flow through the massive bar, and the resistance is practically zero. Removing any plug forces the current to traverse one of the coils. For example, if the 1-Ω and the 3-Ω plugs are out, the resistance is 4 Ω. In electronic circuits, a wide variety of resistors is used. Some are made of carbon and some of resistance wire. They are usually cylindrical in shape, but come in a wide range of resistances and heat-dissipating capacities (Fig. 23-4b).

The current in a circuit may be varied by changing the amount of resistance in the circuit. One way to do this is by means of a slide-wire rheostat. It consists of a spiral coil of wire wound around a tube made of porcelain or some other insulating material (Fig. 23-5a). The current enters the coil at the left-hand terminal, say, and leaves it at the movable contact or slider, whence it travels along

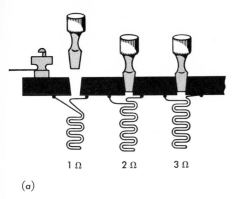

(a)

The resistance will therefore be very small, and the current large. In electronics, variable resistors are often smaller (Fig. 23-5*b*) than the rheostats we have just discussed. Some are toroidal (doughnut-shaped); the moving contact travels along a circular path. Measurements of resistance will be discussed in the next chapter.

An Alternative Statement of Ohm's Law

We can readily put Ohm's law into another mathematical form which is independent of

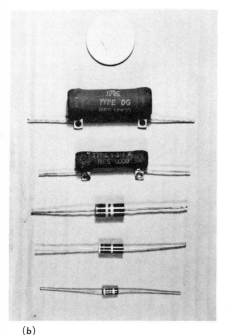

(b)

Fig. 23-4. Resistors. (a) *Standard resistance coils.* (b) *Resistors used in electronic circuits.*

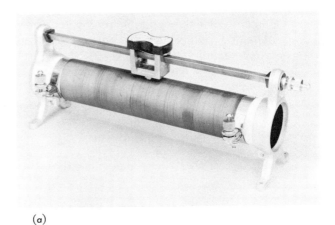

(a)

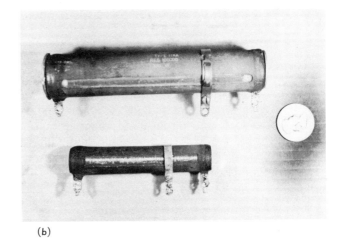

(b)

the massive bar and leaves the rheostat. When the slider is at the midpoint as shown, the charges flow through the left-hand part of the coil only. When the slider is pushed to the right as far as possible, the current must traverse the entire coil; the resistance is great and the current small. When the slider is at the extreme left, the current does not enter the coil but goes directly along the massive bar.

Fig. 23-5. Variable resistors. (a) *A slide-wire rheostat.* (b) *A variable resistor used in electronics.*

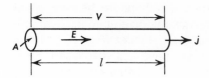

Fig. 23-6. *Deriving the relation* $E = j\rho$.

the shape of the conductor and is more useful for many purposes. Consider a regular cylindrical conductor of uniform resistivity ρ, length l, and cross-sectional area A. A potential difference V is maintained between the ends of the conductor by a battery (Fig. 23-6). The current in the conductor is I. The *current density j*, defined as the *current per unit cross-sectional area,* is given by $j = I/A$. The magnitude of the electric field strength E within the conductor is expressed by $E = V/l$. Thus, substituting in Ohm's law El for V and jA for I, we have

$$jA = \frac{El}{R}$$

or

$$j\frac{RA}{l} = E$$

Note that RA/l equals the resistivity ρ. Therefore,

$$E = j\rho$$

The electrical field strength in the conductor equals the product of the current density and the resistivity. Although we have derived this result for a very simple case, it can be shown to be true more generally for conducting substances and to apply to E, j, and ρ at a point in a conductor.

The Wiedemann-Franz Law

The conduction of charge in metallic conductors and the conduction of heat by them are closely connected. We found in Chapter 14 that the "free" electrons in a metal take a

prominent part in the heat conduction process, diffusing away from higher-temperature regions and carrying kinetic energy with them. The same free electrons we have just seen are the electrical charge carriers in metallic conductors.

The law of Wiedemann and Franz, first discovered experimentally and later derived theoretically from the electron-gas model, states that *the ratio of the thermal conductivity (κ) of a substance to its electrical conductivity (σ) (this is defined as $1/\rho$) at a given temperature is the same for all metals.* Furthermore, the ratio of κ to σ is directly proportional to the Kelvin temperature T:

$$\frac{\kappa}{\sigma} = TL$$

L, the Lorentz constant, can be derived from the theory; it equals approximately 2.2×10^{-8} joule^2-coul^2-°K^2.

The law of Wiedemann and Franz agrees quite well with experimental results, although not exactly. We mention the law, which provides a connection between the physics of heat and the physics of electrical conduction, to remind you that (1) the different areas of physics are intimately related to each other and (2) physical theory based on experiment is increasingly able to explain quantitatively and relate to each other experimental results in several areas of physics.

Conduction of Electricity by Liquids

We think of an electric current in a metallic conductor as a swarm of electrons migrating slowly from the negative pole of a battery toward the positive pole, in a direction *opposite* to that of the electric field strength. An electric current in a solution consists of two oppositely moving streams of charged atoms or groups of atoms called *ions* (Greek, *wanderers*). Moreover, an electric current through a solution does not merely produce heat as in a metal. It decomposes the solu-

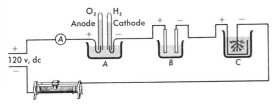

Fig. 23-7. *Chemical effects of an electric current. Gases are liberated in* A, *copper in* B, *and lead in* C.

tion. *Any solution that conducts electricity is called an electrolyte.*

To demonstrate some effects produced by an electric current in electrolytes, connect a battery in series with a slide-wire rheostat and three cells, *A*, *B*, and *C*, containing solutions of sulfuric acid (H_2SO_4), copper sulfate ($CuSO_4$), and lead acetate together with a little acetic acid (Fig. 23-7). We call the positive electrode of each cell its *anode* (Greek *away-road*), the negative electrode, the *cathode* (*toward-road*). After you close the switch, bubbles will appear at both electrodes in *A*, hydrogen being set free at the cathode and oxygen at the anode. Copper is deposited on the cathode in *B*, while it slowly dissolves from the copper anode. In the third cell, lead is deposited on the cathode, forming a treelike structure.

As another example, when hydrogen chloride (HCl) is dissolved in water, its molecules dissociate to form singly charged hydrogen ions H^+ and chlorine ions Cl^-. Each hydrogen ion, lacking one electron, is positively charged, and each chlorine ion, having one excess electron, is negatively charged. When inert platinum electrodes are dipped into a cell containing hydrogen chloride in solution and connected to an outside source of potential difference, the hydrogen ions are driven toward the negative electrode, and the chlorine ions toward the positive electrode (Fig. 23-8). The current consists of these two ion streams. When each hydrogen ion reaches the cathode, it captures an electron from the metal and becomes a hydrogen atom. Two

hydrogen atoms then join to become a hydrogen molecule, and hydrogen gas escapes at the cathode. Each chlorine ion reaching the anode gives up its excess electron, and chlorine gas escapes at the positive electrode.

Electrolysis is widely used to coat metals with tin, gold, silver, nickel, chromium, and copper. Spoons to be electroplated are connected to the negative terminal of a battery and suspended in a water solution of silver salt. Bars of silver, dipping into the electrolyte, are connected to the positive electrode. Positive silver ions, forced into solution from the positively charged anode, are driven through the solution to the negatively charged cathode. Arriving there, each ion receives one electron and is deposited as ordinary silver. One atom dissolves from the anode for every atom going out of the liquid at the cathode. The concentration of silver ions in the solution remains constant.

All the energy required to cause an ion to go into solution from the anode is recovered when the ion goes out of solution at the cathode. Thus the sole energy expended in operating electrolytic cells is that wasted in heating the electrolyte and the wires of the circuit.

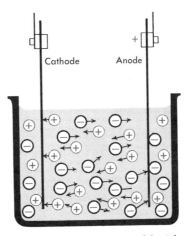

Fig. 23-8. *Hydrogen chloride electrolyte. The H^+ ions go to the cathode and the Cl^- ions to the anode.*

Electrolysis is used to decompose water into hydrogen and oxygen. This is accomplished by sending a current through cells containing solutions of sulfuric acid in water. Inert platinum electrodes are used. The acid is completely ionized into H^+ and SO_4^{--}. The water is slightly ionized into H^+ and OH^-.* Hydrogen ions receive electrons from the cathode and are liberated as gaseous hydrogen. Sulfate ions give up their excess electrons to the anode and change into SO_3 and O so that gaseous oxygen is liberated from the solution. The SO_3 and water react to form sulfuric acid.

Measuring the Mass of an Atom by Electrolysis

The masses of atoms of some of the metallic elements can be readily measured by electrolysis. To understand how this is done, you must first know Faraday's law of electrolysis.

Suppose that three electrolytic cells containing solutions of a silver salt, a copper salt, and an aluminum salt are connected in series. Let a current through the cells continue until 96,500 coul of charge has passed through each cell. Then 107.9 gm of silver (1 gram-atom of silver), $\frac{1}{2} \times 64$ gm of copper ($\frac{1}{2}$ gram-atom of copper), and $\frac{1}{3} \times 27$ gm of aluminum ($\frac{1}{3}$ gram-atom of aluminum) will be driven out of solution. (One gram-atom of an element has a mass in grams numerically equal to its atomic mass.) The quantity of charge (96,500 coul) that will liberate one gram-atom of a univalent element we call a *faraday.* A univalent atom gains or loses one electron in ionization, a bivalent atom gains or loses two electrons, and so on.

The mass m of any element liberated from an electrolytic solution by the transfer of a charge $q = It$ is given by

---

* In fact, each H^+ ion is captured by a water molecule, forming a hydronium (H_3O^+) ion.

$$\frac{m}{M/z} = \frac{It}{96,500 \text{ coul}}$$

where M is the atomic mass of the element and z is the valence. This is Faraday's law of electrolysis. Thus

$$m = \left[\frac{M/z}{96,500 \text{ coul}} \right] It$$

The quantity in brackets we call the *electrochemical* equivalent of the element (Table 23-2).

Table 23-2. Electrochemical equivalents

Element	Atomic mass, gm/gram-atom	Valence	Electrochemical Equivalent, gm/coul
Hydrogen	1.008	1	$1.008/96,500 = 0.000\ 010\ 45$
Silver	107.87	1	$107.87/96,500 = 0.001\ 118$
Copper	63.54	1	$63.54/96,500 = 0.000\ 658\ 4$
Copper	63.54	2	$63.54/(96,500 \times 2) = 0.000\ 329\ 2$
Lead	207.19	2	$207.19/(96,500 \times 2) = 0.001\ 073$
Oxygen	16.00	2	$16.00/(96,500 \times 2) = 0.000\ 082\ 9$
Nickel	58.71	2	$58.71/(96,500 \times 2) = 0.000\ 304\ 4$
Zinc	65.37	2	$65.37/(96,500 \times 2) = 0.000\ 338\ 7$
Aluminum	26.98	3	$26.98/(96,500 \times 3) = 0.000\ 093\ 2$

In Chapter 22 we discussed how Millikan measured the charge of the electron. He found that one coulomb or one ampere-second of negative charge contains about 6.25 billion billion electrons. It follows that one faraday of negative charge should contain 96,500 coulombs $\times 6.25 \times 10^{18}$ electrons per coulomb $= 6.03 \times 10^{23}$ electrons. Suppose that one faraday of charge flows through an electrolyte such as a solution of a silver salt whose ions are singly charged. For each electron entering at the cathode, one atom will be deposited. It follows that there are 6.03×10^{23} (more precisely, 6.025×10^{23}) atoms in a gram-atom of *any* element. This is the *Avogadro number.*

When we know the Avogadro number, we can compute the mass per atom of an element from its atomic mass. For example, 107.9 grams of silver make a gram-atom; hence the mass m_a of a silver atom is

$$m_a = \frac{107.9 \text{ gm}}{\text{gm-atom}} \left(\frac{1 \text{ gm-atom}}{6.03 \times 10^{23} \text{ atoms}} \right)$$

$$= 178.9 \times 10^{-24} \text{ gm}$$

So small is this mass that, if the atoms in a silver dime were dispersed uniformly through all the oceans, there would be several hundred of them for each spoonful of water.

A more convenient method for measuring the ratio of the mass to the charge of ions of any isotope of any element relies upon the deflection of a beam of charged atoms in a magnetic field in vacuum (Chapter 27).

Summary

Positive charge flows from points of greater positive potential to those of smaller positive potential; negative charge in the opposite direction. The current is the charge transferred per unit time through any cross section of a conductor.

One ampere transfers one coulomb of charge per second through any cross section of a conductor.

The current in a metallic conductor is directly proportional to the potential difference between its ends. Ohm's law:

$$I = \frac{V}{R}$$

where R is the resistance.

The resistance of a conductor is the ratio of the potential difference to the current at a given temperature.

One ohm is the resistance of a conductor through which the current is one ampere when the potential difference between the ends of the conductor is one volt.

The resistance R at a Celsius temperature t is given by $R = R_0(1 + \alpha t)$ where R_0 is the resistance at 0°C and α is the temperature coefficient of resistance.

The resistance of a conductor is directly proportional to its length and inversely proportional to its cross-sectional area. The resistivity of a substance is RA/l and is numerically equal to the resistance of an imaginary conductor made of that substance 1 m long and 1 m² in cross-sectional area.

Ohm's law can also be written as $E = j\rho$, where E is the electric field strength in the conductor, j the current density, and ρ the resistivity of the conductor.

Solutions that conduct electricity are called electrolytes. In an electrolyte, the metallic and hydrogen ions are positive. The other ions are negative. Currents in an electrolyte consist of positive ions moving to the cathode (negative terminal) and negative ions moving to the anode (positive terminal).

Univalent atoms gain or lose one electron each in ionization, and bivalent atoms gain or lose two electrons. Faraday's law of electrolysis states that the mass of any element liberated in electrolysis is proportional to its atomic mass M and inversely proportional to its valence z. Thus

$$m = \frac{M}{z} \times \frac{1}{96{,}500 \text{ coul}} It$$

Questions

1. What are other names for (a) a coulomb per second, (b) a joule per coulomb, (c) a volt per ampere?

2. State Faraday's law of electrolysis in words and in algebraic form.

3. Does resistance always increase with increase in temperature? Cite evidence in support of your answer.

4. How do electric current and electron current differ? Why are they both used?

5. How can the charge on a proton be determined by using Faraday's law? How can the mass of a silver atom be found?

6. What is a faraday? How does it differ from a farad?

7. A wire is connected across the terminals of a small battery. How is the current affected if (a) the length of the wire is doubled, (b) its cross-sectional area is doubled, (c) both the length and the cross-sectional area are dou-

bled? Assume that the potential difference between the ends of the wire is constant.

8. How can electrochemical equivalents be (a) calculated, (b) measured experimentally?

9. What happens to the resistance of a conductor (a) if its radius is doubled and its length quadrupled, (b) if its radius is doubled and its length halved?

10. Determine the units in the mks system for the product of current density and resistivity. What physical quantity has these units?

11. Show that the units on *both* sides of the equation describing the Weidemann-Franz law reduce to volt^2/K°.

Problems

A

1. How many coulombs of charge are transferred by a current of 0.60 amp in 1.5 hr?

2. An incandescent lamp is connected to a 120-v source; the current is 0.50 amp. What is the resistance of the lamp?

3. An electric iron of resistance 30.0 Ω is connected to a 120-v line. What is the current through the iron?

4. A piece of wire has a resistance of 24.040 Ω at 0°C and 24.242 Ω at 100°C. What is its temperature coefficient of resistance?

5. The resistance of the copper windings of a generator at 15°C is 30.0 Ω. After the generator operates for several hours, the resistance of the coils is 40.0 Ω. What is the temperature of the wires?

6. A piece of uniform wire 5.0 m long and 1.00 mm in diameter has a resistance of 1.00 Ω. Find the resistance of a wire of this material 10.0 m long and 0.50 mm in diameter.

7. A copper wire of resistance 20 Ω is drawn out so that its length is doubled and its cross section is reduced one-half. What is its resistance afterward?

8. A wire made of copper is 1200 cm long and 0.20 cm in diameter. What is its resistance at 0°C?

9. The resistance of a wire at 0°C is 273 Ω. At 80°C it is 313 Ω. What is its temperature coefficient of resistance?

B

10. When a coil of copper wire is in an ice bath at 0°C, its resistance is 184 Ω. When it is immersed in a hot liquid, the resistance is 268 Ω. What is the temperature of the liquid?

11. A motion-picture projector lamp requires a current of 5.00 amp. Its resistance, when operating, is 8.0 Ω. What additional resistance is required (a) if the potential difference available at the wall outlet is 120 v, (b) if it is 110 v?

12. When a tungsten lamp is connected to a 120-v source, the current is 0.83 amp and the temperature is 2800°C. (a) What is the resistance of the filament? (b) What is its resistance at 0°C?

13. How many kilograms per hour of oxygen will be liberated by a current of 500 amp in an electrolytic cell containing acidulated water?

14. What time is required to deposit 80 kg of copper from a $CuSO_4$ solution in an electrolytic purification process if the current is 40 amp?

15. A current passes through two electrolytic cells in series. One contains a solution of silver nitrate, the second a solution of copper sulfate. It is found that 2.78 gm of silver are deposited. What mass of copper is deposited?

C

16. A current of 2.0 amp electroplates silver onto a spoon for 8.0 min. (a) How much silver is deposited, and what is its cost if the price is 8.0 cents/gm? (b) How many atoms of silver are deposited?

17. One gram-molecule of a gas at 0°C and 76 cm-Hg occupies 22.4 liters. How many faradays of charge must flow to liberate 44.8 liters (a) of hydrogen and (b) of oxygen at normal temperature and pressure?

18. One gram-molecule of a gas at 0°C and 76 cm-Hg occupies 22.4 liters. (a) How many grams of water must be electrolyzed in order to produce 1.5 m³ of hydrogen at 27°C and 114 cm-Hg? (b) How many coulombs of charge must be transferred through the electrolytic cell?

19. How much wire of diameter 0.50 mm and resistivity 20×10^{-7} Ω-cm at 0°C would be needed to make a 10.0-Ω coil at 80°C? The temperature coefficient of resistance is 0.0040/C°.

20. A platinum resistance thermometer has resistance of 300 Ω at 20°C, and a temperature coefficient of resistance of 0.00365/C°. What is the temperature of a furnace in which it has a resistance of 1200 Ω?

21. A generator in a camp supplies 5.00 amp at 120 v to a distant lamp. The two wires leading to the lamp are each 0.25 mi long, and the resistance of each wire is 0.60 Ω/mi. (a) Find the fall of potential in each wire. (b) Find the reading of a voltmeter connected across the lamp terminals. (c) Make a sketch of the connections.

22. A heating coil through which the current is 10 amp when the voltage difference is 120 v is to be made of nichrome wire 0.020 cm in diameter and is to operate at 800°C. (a) Calculate the length of wire required. (b) Describe a convenient configuration of the coil if it is not to be excessively long.

For Further Reading

Magie, W. F., *A Source Book in Physics,* "Ohm," p. 465, McGraw-Hill, New York, 1935. (Reprinted by Harvard University Press.)

CHAPTER TWENTY-FOUR

Electrical Circuits

In the preceding chapter, we considered how the current through a metallic conductor depends upon its resistance and the difference of potential across its terminals. Now we shall see how Ohm's law may be applied to complex combinations of resistors to find, first, the current in each resistor; second, the difference of potential (voltage) across its terminals. Applications of this kind will strengthen your understanding of the basic principles of electricity. Furthermore, we need this information about currents and voltages to be sure that electrical devices—electric lamps, radio and television receivers, and the like—will operate properly. Some basic electrical measurements will be discussed: for example, certain measurements of current, voltage, and resistance.

In this chapter we shall deal with direct-current circuits, in which the current at any point is in one direction. Much of what we learn can be applied later to alternating-current circuits.

Resistors in Series

When the same electric charges must flow through several conductors, we say that the conductors are connected *in series.* For example, the ammeter and the two automobile lamps represented in Fig. 24-1 are in series. Remember that *in a series combination the current is the same in every part of the circuit.* We insert an ammeter in *series* with the resistor in order to measure the current through it, and a voltmeter *across* (in parallel with) the resistor in order to measure the potential difference across it.

Many people think that electricity is consumed in a circuit. This is a mistake. The electrons are not destroyed, but, like air molecules in a series of pipes, are driven forward. We pay the electric company to force electrons through the lamps, motors, etc., of our homes and factories. We pay for the *energy* that we use.

In Fig. 24-2, let the current through B be I_B, let that through C be I_C, and let the current through the battery be I. Since the resistors and the battery are connected in series, the sum of the potential differences across the separate resistors equals the potential difference V across the terminals of the battery

$$V = V_B + V_C$$

As we have just seen, the current is every

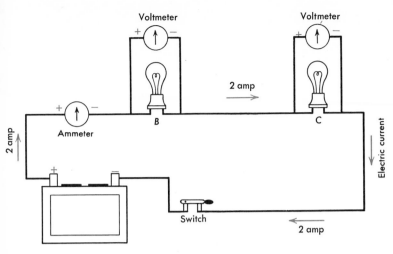

Fig. 24-1. *An electric circuit. The lamps are connected in series. The electric current in each part of the circuit is the same.*

where the same in a series circuit:

$$I = I_B = I_C$$

Applying Ohm's law to the entire circuit,

$$V = IR$$

where R is the total resistance of B and C. Applying Ohm's law to B and C separately, we have

$$V_B = I_B R_B \quad \text{and} \quad V_C = I_C R_C$$

Therefore

$$IR = I_B R_B + I_C R_C$$

and

$$R = R_B + R_C$$

When several conductors are connected in series, *the resistance of the combination R equals the sum of their individual resistances*, $R_1, R_2, R_3 \ldots$

$$R = R_1 + R_2 + R_3 + \cdots$$

The resistances of the two conductors B and C connected to the 6-v battery are 1 Ω

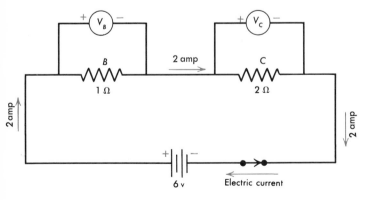

Fig. 24-2. *Resistors in series. The main current is equal to that through each resistor.* $I = I_B = I_C.$ $V = V_B + V_C.$ $R = R_B + R_C.$

and 2 Ω, respectively (Fig. 24-2). The total resistance of the circuit is therefore 3 Ω. By Ohm's law, the current everywhere in this series circuit is V/R, the *total* voltage drop (the potential difference across the terminals of the battery) divided by the *total* resistance, 6 v/3 Ω, or 2 amp.

Applying Ohm's law to B and C separately, the readings of the voltmeters across B and C are:

Across B: $V_B = 1 \ \Omega \times 2 \ \text{amp} = 2 \ \text{v}$
Across C: $V_C = 2 \ \Omega \times 2 \ \text{amp} = 4 \ \text{v}$

Suppose that we should replace the two conductors B and C by a single, equivalent conductor. Its resistance must be such that a potential difference of 6 v across its terminals will cause a current of 2 amp. The resistance R of this equivalent conductor would be

$$R = \frac{V}{I} = \frac{6 \ \text{v}}{2 \ \text{amp}} = 3 \ \Omega$$

This is the total resistance of the series circuit; it is equal to the sum of the two resistances.

Why does the sum of the potential differences across the separate resistors in a series circuit equal the potential difference across the terminals of the battery? Recall that potential difference is work done per unit charge. Consider the work done against electrical resistance by 1 coul of charge as it drifts completely around the circuit of Fig. 24-2. At the positive terminal of the battery the potential (referred to the negative terminal) is +6 v or +6 joules/coul. The energy of 1 coul of charge there is 6 joules/coul × 1 coul or 6 joules. As the charge passes through resistor B, it does 2 joules of work because the potential difference across B is 2 joules/coul. As the charge goes through C it gives up the remaining 4 joules of energy by doing work. The battery increases the energy of the charge to 6 joules, as the charge passes through it, by converting chemical energy into electrical energy. We see that the work done

per unit charge *by* the charge against resistance in the external circuit equals the work done per unit charge *on* the charge by the battery.

Resistors in Parallel

Suppose that we connect a copper wire having a resistance of 60 Ω across supply lines from an electrical generator having a potential difference of 120 v. Then the current will be 2 amp. Now we connect a second, equal, copper wire across the lines. Two paths are available for current, and the current in the lines will be twice as great, that is, 4 amp. Imagine the two wires laid side by side and pushed together to form a single conductor. It has the same length as each of the two, and twice the cross section of one; hence its resistance is one-half as great, that is, 30 Ω. Evidently, then, two conductors of equal resistance, when connected "in parallel" with each other, have a combined resistance which is one-half that of either conductor alone. Now replace one of the wires by another one which is one-half as large in cross section and has twice as great a resistance, that is, 120 Ω. The currents through the two wires are

$$I_1 = \frac{120 \ \text{v}}{60 \ \Omega} = 2 \ \text{amp}$$

$$I_2 = \frac{120 \ \text{v}}{120 \ \Omega} = 1 \ \text{amp}$$

The total current is 3.0 amp; hence the resistance of the single conductor having the same resistance as the two in parallel is:

$$R = \frac{120 \ \text{v}}{3.0 \ \text{amp}} = 40 \ \Omega$$

The resistance R of any parallel combination of conductors or resistance R_1, R_2, R_3, etc. (Fig. 24-3), may always be found in this manner by imagining that each of the elements is connected to a 120-v source or one having *any assumed value*. However, the

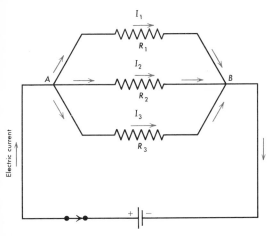

Fig. 24-3. *Resistors in parallel.* $I = I_1 + I_2 + I_3$. $V = V_1 = V_2 = V_3$.

total resistance can be found more directly.

Let the total potential difference from A to B in Fig. 24-3 be V and the total current I; let the total resistance of the circuit be R. The total current is the current anywhere before it divides at A or after it recombines at B. The current I must equal the sum of those through the three resistors.

$$I = I_1 + I_2 + I_3 = \frac{V}{R}$$

The total potential difference V across the resistors in parallel equals the potential difference across any one of them. The reason for this is that the work per unit charge must be the same from A to B, regardless of whether the charge goes through R_1, R_2, or R_3. Thus

$$V = V_1 = V_2 = V_3$$

Applying Ohm's law to each resistor:

$$I_1 = \frac{V_1}{R_1}; \quad I_2 = \frac{V_2}{R_2}; \quad I_3 = \frac{V_3}{R_3}$$

and

$$I = \frac{V_1}{R_1} + \frac{V_2}{R_2} + \frac{V_3}{R_3} = \frac{V}{R}$$

Hence

$$\frac{1}{R} = \frac{1}{R_1} + \frac{1}{R_2} + \frac{1}{R_3} \quad \text{or} \quad R = \frac{1}{\dfrac{1}{R_1} + \dfrac{1}{R_2} + \dfrac{1}{R_3}}$$

R is the total resistance of R_1, R_2, and R_3 in parallel or, stated in another way, it is the resistance of an equivalent resistor that would have the same total current as R_1, R_2, and R_3 in parallel.

Example. An electric lamp of resistance 100 Ω, a toaster of resistance 50 Ω, and a percolator of resistance 50 Ω are connected in parallel to a 120-v source. What is the resistance of an electric iron connected to the same supply wires that takes as much current as all three, and what is the current through it?

$$\frac{1}{R} = \frac{1}{100 \ \Omega} + \frac{1}{50 \ \Omega} + \frac{1}{50 \ \Omega} = \frac{5}{100 \ \Omega}$$

$$R = 20 \ \Omega$$

$$I = \frac{V}{R} = \frac{120 \ v}{20 \ \Omega} = 6.00 \ \text{amp}$$

Measuring Resistances

One of the simplest ways to measure the resistance of a conductor is to connect it in series with a battery and an ammeter A which measures the current I (Fig. 24-4). A voltmeter V connected in parallel with the conductor in-

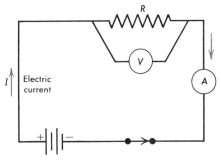

Fig. 24-4. *Measuring the resistance of a conductor by the ammeter-voltmeter method.*

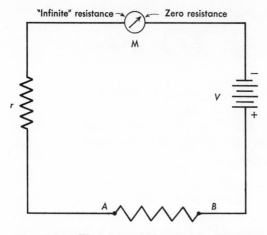

M

r

V

A   B

Fig. 24-5. *The circuit of a simple ohmmeter.*

between the terminals, one can determine its resistance to an accuracy of a few per cent.

The *Wheatstone bridge* is used for more accurate measurements of resistances. The student-laboratory type of Wheatstone bridge is represented in Fig. 24-6; *a* is the resistor of unknown resistance R_x and *b* a standard resistor whose resistance R_{st} is accurately known. *DF* is a uniform resistance wire. By shifting the contact *S*, we can adjust the ratio R_1/R_2 of the resistances of the two parts of the wire *DF*. A battery causes currents through *AC* and *DF*. A galvanometer *G*, a device for measuring small currents, is connected as indicated.

The contact *S* is shifted until the galvanometer current is zero. (The bridge is then said to be balanced.) Then we know that the potential difference across the galvanometer is zero. Thus the potential drop $I_x R_x$ across *a* is equal to the drop $I_1 R_1$ across the top section of the wire. Similarly, the drop $I_{st} R_{st}$ across *b* is equal to that across the bottom section.

dicates the potential difference across it. The resistance *R* of the coil is then computed by Ohm's law from the ratio of *V* to *I*. (The resistance of the voltmeter should be very great compared with *R* so that it carries a negligible fraction of the current. Otherwise the current through the voltmeter will have to be taken into account.)

An *ohmmeter* is a convenient instrument for measuring the resistance of a resistor quickly. It contains (Fig. 24-5) within a carrying case an electrical meter *M*, a battery, and an internal resistor *r*. Suppose the terminals *A* and *B* of the ohmmeter are connected by a short copper wire of practically zero resistance. The battery will establish a current in the circuit equal to *V/r*. The resistance *r* is so chosen that the needle is deflected to the last division at the right-hand end of the scale. This scale division is marked 0. If the terminals are not connected at all—open circuit or very large resistance—the needle will not be deflected at all; the first mark on the scale at the left is thus the "infinite resistance" mark. By inserting various resistors of known resistance *R*, the ohmmeter scale can be calibrated for intermediate deflections of the needle, and the scale marked in ohms. Thereafter, by inserting a unknown resistor

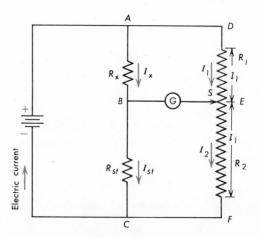

Fig. 24-6. *Wheatstone bridge circuit. When the bridge is "balanced," the potential drop* $I_x R_x$ *across* AB *is equal to that across* DE. *The current through* AB *is equal to that through* BC. $R_x/R_{st} = R_1/R_2 = l_1/l_2$.

In consequence

$$\frac{I_x R_x}{I_{st} R_{st}} = \frac{I_1 R_1}{I_2 R_2}$$

But $I_x = I_{st}$ and $I_1 = I_2$. (Why?)
Therefore,

$$\frac{R_x}{R_{st}} = \frac{R_1}{R_2}$$

Also, since the wire is uniform,

$$\frac{R_x}{R_{st}} = \frac{l_1}{l_2}$$

In the Wheatstone bridge, one potential difference is adjusted relative to another until the galvanometer current is *zero*. This method — a common one in experimental physics — is called a *null* method.

A network problem. The network represented in Fig. 24-7 affords an exercise which will help you to understand how Ohm's law is applied to more complicated circuits.

(a) What is the total current if the potential difference causing it is 120 v? First, find what single resistance is equivalent to the parallel combination *BCD*.

$$\frac{1}{R} = \frac{1}{4.0\ \Omega} + \frac{1}{6.0\ \Omega} + \frac{1}{12\ \Omega} = \frac{6}{12\ \Omega}$$

$$R = 2.0\ \Omega$$

The resistance of *A* is 3.0 Ω; that of the parallel combination is 2.0 Ω; therefore, the total resistance of the network is 5.0 Ω.
The total current I is given by

$$I = \frac{120\ \text{v}}{5.0\ \Omega} = 24\ \text{amp}$$

(b) What is the voltage across the parallel combination?
The resistance of the parallel combination is 2.0 Ω, and the current is 24 amp; hence the voltage V_2 is given by

$$V_2 = 24\ \text{amp} \times 2.0\ \Omega = 48\ \text{v}$$

(The voltage across *A* is 24 amp $\times$ 3 Ω = 72 v. We check our work by noting that 72 v + 48 v = 120 v.)
(c) What is the current I_1 through *D*?
Divide the fall of potential V_2 across the parallel combination by the resistance of *D*.

$$I_1 = \frac{V_2}{R} = \frac{48\ \text{v}}{12\ \Omega} = 4.0\ \text{amp}$$

You can find the currents through *B* and *C* in the same manner.

Voltaic Cells

Electric currents may be produced in circuits by several devices, such as electric genera-

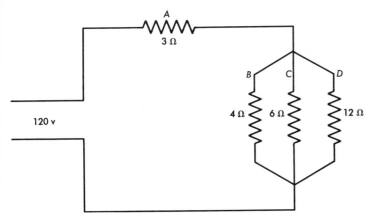

Fig. 24-7. A network.

tors and voltaic cells or batteries. We shall discuss voltaic cells in this section, reserving the discussion of generators for a later chapter.

In 1790, the year after the American Constituition went into effect, Luigi Galvani, an Italian biologist, made a discovery leading to the invention of the voltaic cell. He noticed that sparks from an electrostatic machine caused the legs of a newly killed frog to contract. He then wanted to see if a thunderstorm would have the same effect so he attached freshly prepared frogs' legs to a lightning rod and found that they contracted whenever there was a distant flash of lightning. In a further experiment, Galvani hung frogs' legs against an iron lattice by brass hooks. He observed that the legs contracted whenever they touched the iron; an electrostatic machine or connection to a lightning rod in a thunderstorm was not needed. By further experiments, he found that pairs of different kinds of metals could cause the twitching.

Scientists were greatly interested in the connection between electricity and living tissue. Physicians hoped, by electric shocks, to make the dead live again. Galvani claimed that to produce the currents, *living tissue* must touch the metals. Other workers opposed this view. A bitter controversy raged for ten years. Then another Italian, Alessandro Volta, for whom the volt is named, settled the dispute. He found that currents could be produced without living tissue by means of disks of copper and zinc, separated by pieces of cloth moistened with a salt solution. He made the first *voltaic cell* by dipping a strip of zinc and one of copper into a vessel of salt water. He found that the voltage could be increased by connecting several cells in series, positive to negative. Such a combination of cells he called a *battery* because of the analogy to a group of cannon.

Let us review the action of a zinc-copper cell—called a Daniell cell—in producing a continuing potential difference. Zinc—im-

mersed in a zinc sulfate solution—dissolves from the rod A (Fig. 24-8). Each atom leaves two electrons on the rod and thus becomes a zinc ion. The concentration of zinc ions increases until equilibrium is established; the rate of further dissolving equals the rate at which zinc ions leave the solution. The negative potential of the zinc rod then becomes constant. Copper in a copper sulfate solution also dissolves, forming positive ions and establishing a negative potential on the copper rod. However, copper has a smaller solution pressure than zinc; hence zinc ions diffuse through the porous barrier and force copper ions out of solution onto the copper rod. When equilibrium is established, the potential difference between B and A is about 1.08 v with B positive relative to A. When the two rods are connected by a wire, there is a current, and 1 gram-atom of zinc dissolves for each gram-atom of copper forced out of solution. The energy thus liberated maintains the electric current against the energy dissipated due to resistance.

An electric cell may be constructed by using any two different conductors and an electrolyte. It is essential that the two con-

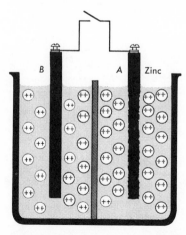

Fig. 24-8. A zinc-copper voltaic cell. Zinc has a greater solution pressure than copper. Zinc ions diffuse through the wall and force copper ions out of solution.

ductors have different solution pressures. Place a clean copper cent and a dime in your mouth, separating them by your tongue. Permit the two to touch, and you will experience the nasty taste of the ions. Dentists are careful not to permit a gold filling to touch one made of a silver amalgam, for the two in contact with each other and with the saliva would form a closed circuit, causing the silver to dissolve. Never use salt water as antifreeze in the radiator of your car. Electrolytic action would eat away the soldered seams.

Substances that have smaller solution pressures than hydrogen are said to be *electropositive,* and those having greater pressures are *electronegative.* Below are indicated the potentials of several metals, when each is combined with a hydrogen electrode at zero potential and an electrolyte to form a voltaic cell.

Lithium$^-$ −2.96 v, sodium$^+$ −2.71 v, aluminum^{+++} −1.70 v, zinc^{++} −0.76 v, iron^{++} −0.44 v, lead^{++} −0.12 v, hydrogen$^+$ 0.00 v, copper^{++} +0.34 v, silver$^+$ +0.789 v, mercury^{++} +0.800 v, gold^{+++} +1.5 v.

When a voltaic cell having an acid electrolyte operates, hydrogen ions are forced over to the positive electrode, where they pick up electrons, are liberated, and form a gaseous film. The accumulation of the gas around the electrode is called *polarization.* It has undesirable effects. First, the film has a high electrical resistance and it opposes the flow of electricity. Second, the hydrogen film effectively changes the nature of the electrode. For example, a polarized zinc-copper cell acts like a zinc-hydrogen cell, which has a smaller voltage. Polarization is avoided or diminished in several ways in different types of voltaic cell.

Several Kinds of Voltaic Cells

The best-known type of voltaic cell is the *dry cell.* Its negative electrode is a zinc can which contains a paste of manganese

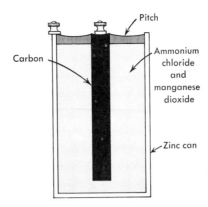

Fig. 24-9. A dry cell. Its cathode is zinc; its anode carbon. The electrolyte is ammonium chloride.

dioxide, graphite, and ammonium chloride solution (Fig. 24-9). The positive electrode is a carbon rod. When the cell generates a current, zinc ions, dissolving from the cathode, force hydrogen ions out of solution ($Zn + 2NH_4Cl \rightarrow ZnCl_2 \cdot 2NH_3 + 2H$). When the cell generates a large current, hydrogen accumulates and causes polarization. (When the circuit is opened, this hydrogen slowly combines with manganese dioxide to form manganic oxide and water.)

Sometimes voltaic cells are used as standards of potential difference in laboratory experiments. Such a cell must be made of pure, carefully chosen chemicals and materials, and it must not be greatly affected by temperature variations. The *Weston normal cadmium cell* is best adapted for this purpose. Its positive electrode is mercury; its negative electrode is cadmium amalgamated with mercury. The electrolyte is a saturated solution of cadmium sulfate. The depolarizer is mercurous sulfate. The voltage of a Weston cell is 1.0183 v at 20°C. Its rate of change of voltage with temperature is about 40 microvolts per Celsius degree.

A voltaic cell that can be restored to its original condition by forcing a current through it in a reversed direction is called a *storage cell.* In the lead cell of an automobile battery, the positive electrode is a lead grid containing lead dioxide (PbO_2). The negative electrode is pure lead; the electrolyte is a solution of sulfuric acid in water.

When a lead cell is discharging, each PbO_2 molecule that is reduced at the positive plate liberates two doubly charged oxygen ions. Meanwhile, one sulfate ion is forced onto the plate. Thus this plate loses two electrons. The negative plate receives one sulfate ion and thus gains two electrons from the solution. In recharging the cell, these actions are reversed.

The chemical equation for the reversible reaction is $PbO_2 + Pb + 2H_2SO_4 \rightleftharpoons 2PbSO_4 + 2H_2O$ + Electrical energy. It should be emphasized that chemical energy, not electricity, is stored in the cell.

In a completely discharged lead storage cell, the positive and negative plates are coated with lead sulfate—a poor conductor of electricity. The internal resistance of the cell is then so great that it cannot be readily recharged. A lead storage battery should never be allowed to discharge completely.

As a lead cell is discharged, the concentration of sulfate ions in the solution decreases and that of oxygen ions increases. These oxygen ions combine with hydrogen ions to form water. Since the sulfate ions that are removed from the solution are more massive than the oxygen ions that replace them, the decrease in density of the solution, measured by means of a hydrometer, indicates the degree of discharge of the cell.

The storage battery of a car must deliver large currents, and its internal resistance must be low. The plates or electrodes are placed close together, separated by thin wooden or rubber strips. All the positive plates are connected in parallel to the positive terminal and the others to the negative one (Fig. 24-10). The maximum voltage of a newly charged cell is about 2.2 v. Six cells connected in series comprise the 12-v battery used in cars.

Electromotive Force and Terminal Voltage of a Voltaic Cell

A generator or voltaic cell is said to be the source of an *electromotive force* (emf). Such

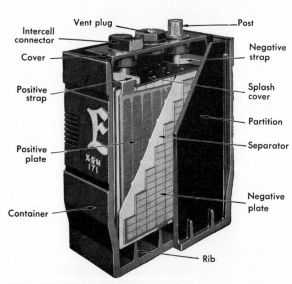

Fig. 24-10. *A storage cell. (Courtesy of Electric Storage Battery Company.)*

a source converts some other kind of energy into electrical energy. A voltaic cell converts *chemical* energy into *electrical* energy. Electromotive force is not a force! An electromotive force $\mathscr{E}$ is the *energy* per unit charge U/q given by the source to charge passing through it.

$$\mathscr{E} = \frac{U}{q}$$

The unit of emf, the volt, is the same as the unit of potential difference. *A source has an emf of one volt if each coulomb of charge passing through it acquires one joule of energy.*

$$1 \text{ volt (emf)} = \frac{1 \text{ joule of electrical energy}}{1 \text{ coulomb}}$$

Figure 24-11 represents a battery having an emf of 4 v and a negligible internal resistance, connected in series with a 1-Ω resistor and a 3-Ω resistor. The current is 1 amp so that 1 coul of charge flows through any cross section of the circuit every second. Since the emf of the battery is 4 v, the battery generates 4 joules of electrical energy per coulomb. This electrical energy is utilized in heating the two resistors: 1 joule/coul in heating *A*, and 3 joules/coul in heating *B*. The total *IR* drop in potential across the two resistors is

equal to the emf $\mathscr{E}$ of the battery. That is:

$$\mathscr{E} = I_1R_1 + I_2R_2$$
$$4 \text{ v} = 1 \text{ amp} \times 1 \text{ Ω} + 1 \text{ amp} \times 3 \text{ Ω}$$

However, when the battery is producing a current, the potential difference *V* across its terminals — the terminal voltage — is *less* than the emf. This is because of the loss of energy per unit charge due to *internal resistance r* inside the battery itself. Suppose that an automobile storage battery has an emf of 12 v and that the internal resistance is 0.020 Ω. On open circuit, the reading across its terminals is 12 v. When the battery causes a current of 50 amp in driving the starting motor, there is a fall of potential inside the battery and the terminal voltage is reduced. Thus

Emf − Loss of potential in battery
= Terminal voltage

$$\mathscr{E} - Ir = V \quad \text{(Discharging)}$$

$$12 \text{ v} - 50 \text{ amp} \times 0.020 \text{ Ω} = 11 \text{ v}$$

When a battery is being charged, both its emf and the *Ir* voltage drop due to internal resistance oppose the current.

$$\mathscr{E} + Ir = V \quad \text{(Charging)}$$

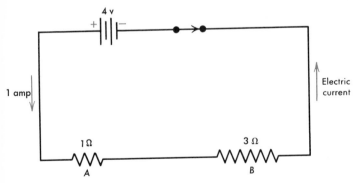

Fig. 24-11. *The total* IR *drop in potential across the two resistors is equal to the emf of the battery. The internal resistance is assumed to be negligible.*

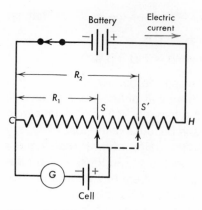

Fig. 24-12. *Potentiometer circuit. When the galvanometer current is zero, the emf of the cell is exactly equal to the fall of potential across CS.*

Notice that the potential difference across the terminals of the battery is *greater* than its emf when it is being charged.

When a cell or battery is on open circuit—causing no currents—there is no current and no drop of potential inside the battery. Thus *the voltage across the terminals of a battery on open circuit is equal to its emf.*

The Potentiometer

Potential differences and electromotive forces can be measured by means of voltmeters, or, more accurately, by means of *potentiometers.* In Fig. 24-12 a battery causes a steady current I in the circuit, and therefore a potential difference across the terminals of the uniform resistor wire *CH.* The test cell whose emf is to be determined and a galvanometer are connected in the lower branch circuit, and the sliding contact S is adjusted until the galvanometer current is zero. Since this current is zero, the emf of the test cell $\mathscr{E}_x$ must be exactly equal to the potential difference IR_1 between C and S. Then the test cell is replaced by a standard cell of precisely known emf, and the slider is again

adjusted to a position S' such that the galvanometer current again is zero. The emf of the standard cell $\mathscr{E}_{st}$ equals the potential difference IR_2 between C and S'. Then

$$\frac{\mathscr{E}_x}{\mathscr{E}_{st}} = \frac{IR_1}{IR_2} = \frac{R_1}{R_2} = \frac{l_1}{l_2}$$

where $l_1 = CS$ and $l_2 = CS'$.

Kirchhoff's Laws

Kirchhoff's laws may be used to deal with more complicated networks having batteries in different parts or "loops."

We wish to find the currents I_1, I_2, and I_3 in Fig. 24-13. Arbitrarily assume that these currents are directed as shown by the arrows.

Kirchhoff's first law states that *the algebraic sum of the currents to any point in the circuit is zero.* This means that the sum of the currents toward a point is equal to the sum of those away from it. The first law is thus merely a statement of the *law of conservation of electric charge*: at a point of branching, all charge must be accounted for; none can be created or destroyed. Thus at the point *A,*

$$I_1 - I_2 - I_3 = 0 \qquad (1)$$

The second law states that *around any loop or circuit in the network, the algebraic sum of the emf's and the IR drops in potential is zero.* The agreement as to the sign of an emf or an electron current is as follows: If we go through the source of emf from its negative to its positive pole, its emf is considered positive. If we go through a resistor in the same direction as the current, the potential difference (*IR*) is negative. The second law is thus a statement that the electric forces acting on charges in a circuit with steady currents are conservative ones: the net work done on a unit charge around a closed loop is zero.

Since we do not know the directions of the currents to begin with, we *arbitrarily* assume

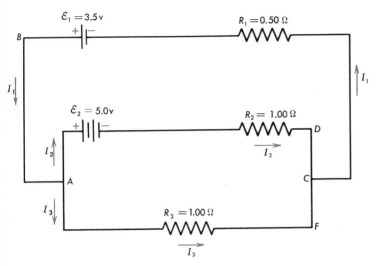

$\mathcal{E}_1 = 3.5\text{v}$ $R_1 = 0.50\ \Omega$

$\mathcal{E}_2 = 5.0\text{v}$ $R_2 = 1.00\ \Omega$

$R_3 = 1.00\ \Omega$

Fig. 24-13. A more complicated network. The algebraic sum of the currents to any point is zero. The sum of the emf's and the IR drops around any branch is zero.

directions as shown by the red arrows. Let us then consider the loop *AFCBA*. The sum of the emf's and the *IR* drops is

$$-1\ \Omega\ I_3 - 0.5\ \Omega\ I_1 + 3.5\text{ v} = 0 \qquad (2)$$

Similarly, for *ADCBA*

$$-5.0\text{ v} - 1\ \Omega\ I_2 - 0.5\ \Omega\ I_1 + 3.5\text{ v} = 0 \qquad (3)$$

Solve equation (1) for I_1 and substitute into equations (2) and (3).

$$3.5\text{ v} - 0.5\ \Omega\ I_2 - 1.5\ \Omega\ I_3 = 0 \qquad (4)$$

$$-1.5\text{ v} - 1.5\ \Omega\ I_2 - 0.5\ \Omega\ I_3 = 0 \qquad (5)$$

Multiply equation (4) by three and subtract.

$$10.5\text{ v} - 1.5\ \Omega\ I_2 - 4.5\ \Omega\ I_3 = 0$$

$$-1.5\text{ v} - 1.5\ \Omega\ I_2 - 0.5\ \Omega\ I_3 = 0$$

$$12\text{ v} = 4\ \Omega\ I_3; \quad I_3 = 3\text{ amp}; \quad I_2 = -2\text{ amp}$$

(What does the minus sign tell us?)

$$I_1 = 1\text{ amp}$$

Electrophysiology

Galvani's experiments suggested that electrical phenomena played important parts in living bodies. In the centuries since his discovery, physiologists and biophysicists have learned a great deal about the way such phenomena explain the action of nerves in conducting electrical signals throughout the body.

The basic unit of a nerve is the cell, which is a complicated entity indeed. For our present purpose of describing nerve conduction, we can consider the cell as consisting of a gelatinous cytoplasm contained in a semipermeable membrane. Many cells are strung together, one after the other, to form nerve fibers, and the fibers are bunched together in parallel strings to form the nerves themselves. The cytoplasm in the interior of the cell con-

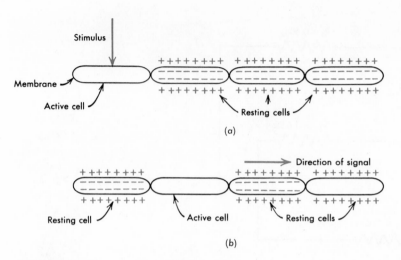

Stimulus

Membrane

Active cell

+ + + + + + +   + + + + + + +   + + + + + + +
- - - - - - -   - - - - - - -   - - - - - - -
+ + + + + + +   + + + + + + +   + + + + + + +

Resting cells

(a)

Direction of signal

+ + + + + + +   + + + + + + +   + + + + + + +
- - - - - - -   - - - - - - -   - - - - - - -
+ + + + + + +   + + + + + + +   + + + + + + +

Resting cell

Active cell

Resting cells

(b)

Fig. 24-14. The conduction of an electrical signal by a nerve fiber.

tains potassium ions (K$^+$) at a relatively high concentration, sodium ions (Na$^+$) at a much lower concentration, and negatively charged ions (Cl$^-$). The fluid that bathes the cell, outside the membrane, has a lower concentration of potassium ions and a somewhat higher concentration of sodium ions. The net effect is that *in its resting state* the nerve cell is electrically polarized with the interior of the membrane *negatively* charged at a potential about 70 millivolts lower than that of the exterior. The membrane separates the two ionic mixtures and maintains the potential difference when the cell is resting. The equilibrium, however, is a delicate one, and, if disturbed, will lead to a mixing of the two ionic populations, interior and exterior, and a sharp change in the potential difference.

This is what happens when a suitably large stimulus is applied to the resting nerve cell. The stimulus—delivered by a signal from another nerve, for example, or by one from a sensitive nerve ending in the ear—causes the membrane to become permeable to ions (Fig. 24-14a); sodium ions suddenly enter the cell and potassium ions leave it. The result is that the cell in its active state is first depolarized and then actually given a slight polarization

in the reverse direction. The cell recovers, however. The membrane allows the concentrations of potassium ions inside the cell and of sodium ions outside to build up again, and the cell returns to its resting state. The whole event takes only about 1 millisecond.

How does the "firing" of one nerve cell in this way cause the conduction of a signal along the nerve? The action is somewhat analogous to the spreading of a flame along a burning string: the high temperature of the flame at one point kindles the adjacent portion of the string. Similarly, the ionic motion through the membrane walls of the active cell constitutes a current that disturbs the equilibrium of the neighboring cell (Fig. 24-14b), and the neighboring cell fires. That causes the next cell to fire, and so on. The nerve impulse—a depolarization signal—travels down the line of cells at a speed of between 1 meter/sec (for small nerves) and 100 meters/sec (for large nerves).

When the nerve impulse reaches a muscle, it can make the muscle contract, and vice versa—the contraction of a muscle can cause an "action potential" or electromotive force. By clamping one electrode to your ankle and another to your wrist, a physician can mea-

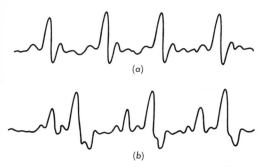

Fig. 24-15. *Electrocardiograms.* (a) *The normal heart.* (b) *The malfunctioning heart.*

sure the emf's generated by your heart and record an electrocardiogram (Fig. 24-15).

Summary

In a series circuit, the current is everywhere the same, the total potential difference is the sum of the individual potential differences, and the total resistance R is given by

$$R = R_1 + R_2 + R_3 + \cdots$$

If the conductors are connected in parallel, the potential difference across each is the same, the total current is the sum of the currents through the branches, and the combined resistance is given by

$$\frac{1}{R} = \frac{1}{R_1} + \frac{1}{R_2} + \frac{1}{R_3} + \cdots$$

or

$$R = \cfrac{1}{\cfrac{1}{R_1} + \cfrac{1}{R_2} + \cfrac{1}{R_3} + \cdots}$$

The Wheatstone bridge formula is as follows (see Fig. 24-6):

$$\frac{R_x}{R_{st}} = \frac{R_1}{R_2}$$

A voltaic cell consists of two electrodes, made of dissimilar substances, in contact with an electrolyte. The substance forming the negative electrode has a greater tendency to dissolve than that forming the positive electrode. Polarization is the accumulation of hydrogen around the positive electrode. A storage cell is a voltaic cell that can be restored to its initial condition or recharged by forcing current through it in a reversed direction.

The electromotive force $\mathscr{E}$ of a source of electrical energy is the electrical energy U it liberates per unit quantity of charge. $\mathscr{E} = U/q$.

The terminal voltage of a source having internal resistance r when delivering a current I is $V = \mathscr{E} - Ir$. When a cell is on open circuit, its emf equals its terminal voltage.

Kirchhoff's laws: (1) the algebraic sum of the currents to any point in a circuit is zero; and (2) around any loop in a network, the algebraic sum of the emf's and the IR drops in potential is zero.

Questions

1. What are the advantages of connecting electrical devices in parallel with the generator?

2. What is stored in a "storage" battery? Why are the plates of a storage cell made large and placed close together?

3. Why must the internal resistance of a lead storage cell be low?

4. Why should the emf of a cell be measured with a potentiometer rather than with a voltmeter?

5. How could you make a simple voltaic cell? What is the highest voltage obtainable from a single cell?

6. Compare the rules of combination of capacitors with those for resistors in (a) series circuits and (b) parallel circuits.

7. Why is the total potential difference across three resistors connected in series to a battery equal to the sum of the potential differences across each?

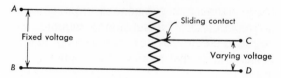

Fixed voltage

Sliding contact

• C

Varying voltage

• D

Fig. 24-16.

8. Discuss the accuracy of the following methods for determining the resistance of a resistor: (*a*) ammeter-voltmeter method, (*b*) use of an ohmmeter, (*c*) use of a Wheatstone bridge.

9. A Daniell cell is usually stored with a 20-Ω resistor connected between its terminals. Discuss why this is desirable.

10. What is the advantage of using a sensitive galvanometer in a Wheatstone bridge? When the bridge is balanced, will interchanging the galvanometer and the battery affect the balance?

11. The emf of a storage battery is 6.0 v. Under what conditions is the voltage across its terminals (*a*) greater than, (*b*) less than 6.0 v? Why?

12. How should a battery be connected to a battery charger when it is being charged: plus to plus, or plus to minus? Explain.

13. State and derive the equations for the total electromotive force of three cells in series and in parallel.

14. How can three resistors of resistances 2 Ω, 3 Ω, and 6 Ω be connected to give a total resistance of (*a*) 4 Ω, (*b*) 1 Ω?

15. Estimate the total resistance when the following are connected in parallel: (*a*) 1 Ω and 10^6 Ω; (*b*) 1 Ω and 10^{-3} Ω; (*c*) 1 Ω, 10^3 Ω, and 10^6 Ω.

16. Explain how the circuit in Fig. 24-16 could be used to obtain varying voltages. (Such a device is called a potentiometer, but it has a different purpose than the potentiometer shown in Fig. 24-12.)

17. Does the emf of a voltaic cell increase, decrease, or remain constant when (*a*) the size of the electrodes is increased, and (*b*) the distance between them is decreased?

Problems

Draw circuit diagrams for all problems.

A

1. A 2.0-Ω coil is connected in a circuit in parallel with a 4.0-Ω coil, and the latter carries a current of 4.0 amp. (*a*) What is the potential difference across the combination? (*b*) What is the current through the 2.0-Ω coil?

2. How many lamps, each having a resistance of 144 Ω, can be connected in parallel with each other across the two wires of a 120-v lighting system if the maximum allowable current is 10 amp?

3. Conductors of 6.0-Ω, 3.0-Ω, and 4.0-Ω resistance are connected in series with a 6.0-Ω storage battery. Find (*a*) the current through each resistor and (*b*) the voltage across each resistor.

4. What is (*a*) the greatest, (*b*) the smallest total resistance that can be secured by combinations of four resistors of resistance 4 Ω, 8 Ω, 12 Ω, and 20 Ω?

5. Two wires of total resistance 0.20 Ω lead from a 120-v generator to a set of lamps. If 10 lamps, each of resistance 50 Ω, are connected in parallel, find (*a*) the total current, (*b*) the voltage drop in the wires, and (*c*) the voltage across the lamps. Neglect the internal resistance of the generator.

6. A 12.0-v battery of negligibly small internal resistance is connected in series with a 3.0-Ω coil and a 4.0-Ω coil. What is (*a*) the cur-

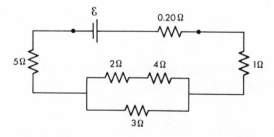

Fig. 24-17.

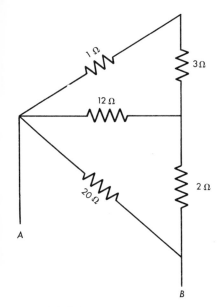

1 Ω
3 Ω
12 Ω
20 Ω
2 Ω
A
B

Fig. 24-18.

rent in the circuit, and (*b*) the voltage across the 4.0-Ω coil?

7. Coils of resistance 2.0 Ω and 6.0 Ω are connected in parallel with each other to a battery. The current through the smaller resistance is 6.0 amp. Find (*a*) the potential difference across the combination, and (*b*) the total current.

B

8. The emf of a cell as measured by a potentiometer is 1.5 v. When the current is 1.0 amp, the voltage across the cell is 1.3 v. What is the internal resistance of the cell?

9. What is the voltmeter reading when a voltmeter is placed across a 10-v cell with an internal resistance of 0.20 Ω, which is being charged at 2.0 amp?

10. Conductors of resistance 2.0 Ω, 3.0 Ω, and 4.0 Ω are connected in parallel with each other and in series with a wire of resistance 1.0 Ω. The current through this wire is 10.0 amp. What is the current through the 2.0-Ω conductor?

11. What is the terminal potential dif-

ference of a battery of emf 20.0 v and internal resistance 0.050 Ω when connected to an external resistance of 4.5 Ω?

12. A uniform wire, 2.0 m long, with a resistance of 11 Ω, is connected in series with a 6.0-v battery and a rheostat having a resistance of 1.0 ohm. If a voltmeter is connected across 50 cm of the wire, what will it read?

13. The current through the 2-Ω resistor in Fig. 24-17 is 1 amp. Find the emf of the battery, whose internal resistance is 0.20 Ω.

C

14. Five dry cells, each of emf 1.5 v and resistance 0.50 Ω, are connected in series in a circuit with 1.5-Ω external resistance. One of the cells is reversed so as to oppose the current. What are the current through the reversed cell and the voltage across it?

15. Conductors of resistance 3.0 Ω, 5.0 Ω, 12.0 Ω, and 20.0 Ω are connected in parallel with each other, and the four are connected through a 4.5-Ω resistor to a 6.0-v battery of negligible internal resistance. Find (*a*) the current through the battery, and (*b*) that through the 3.0-Ω resistor.

16. In the ammeter-voltmeter method of determining resistance, it is assumed that the current through the voltmeter is negligibly small. Find the error arising from this assumption if the resistance of the voltmeter is

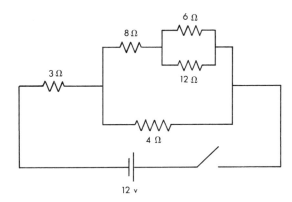

6 Ω
8 Ω
3 Ω
12 Ω
4 Ω
12 v

Fig. 24-19.

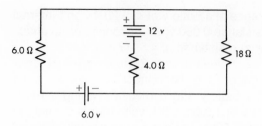

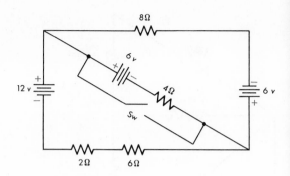

Fig. 24-20.

Fig. 24-21.

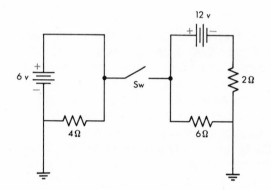

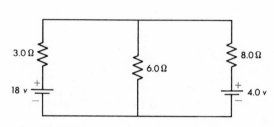

Fig. 24-22.

Fig. 24-23.

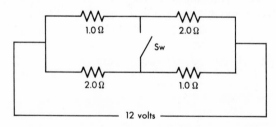

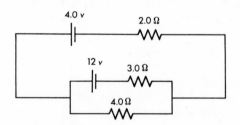

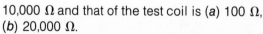

Fig. 24-24.

Fig. 24-25.

10,000 Ω and that of the test coil is (*a*) 100 Ω, (*b*) 20,000 Ω.

17. Coils of resistances 1.0 Ω, 2.0 Ω, 3.0 Ω, and 4.0 Ω are connected in parallel with one another, and the group is connected in series with a coil of resistance 2.0 Ω and a dry cell of internal resistance 0.52 Ω and emf 1.50 v. Compute (*a*) the current through the cell, (*b*)

the voltage across the 3.0-Ω resistor, and (*c*) the current through the 4.0-Ω resistor.

18. A network consists of a 10-Ω resistor *A* connected in series with a pair of resistors, *B* and *C*, in parallel. The network is connected to a 120-v source. The current through *A* is 3.0 amp. The resistance of *B* is 120 Ω. What is the resistance of *C*?

19. Find the total resistance between *A* and *B* in the circuit shown in Fig. 24-18.

20. Calculate the current in the 6-Ω resistor in the circuit shown in Fig. 24-19.

21. (*a*) In how many ways can battery *A*, of emf 6.0 v and internal resistance 0.60 Ω, and battery *B*, of emf 12.0 v and internal resistance 1.20 Ω, be connected to a 10-Ω resistor? (*b*) What is the current in the 10-Ω resistor in each case?

22. Calculate (*a*) the current in the 18-Ω resistor and (*b*) the voltage across the 6.0-Ω resistor in Fig. 24-20.

23. What is the change in the current in the 6.0-Ω resistor in Fig. 24-21 when the switch Sw is closed?

24. What is the change in the current in the 4.0-Ω resistor in Fig. 24-22 when the switch Sw is closed?

25. Find the current through the 6.0-Ω resistor in Fig. 24-23.

26. (*a*) Find the current in each resistor in Fig. 24-24 before the switch Sw is closed. (*b*) Find the change in the current in each resistor when the switch is closed.

27. Calculate the currents in the circuit in Fig. 24-25.

For Further Reading

Galambos, Robert, *Nerves and Muscles,* Doubleday and Company, Garden City, 1962.

CHAPTER TWENTY-FIVE

Electric Power and Heating Effects

Electrical Power and Energy

In mechanics, you learned that power is the time rate of doing work and may be expressed in such units as horsepower, watts (joules per second), and kilowatts. The power of a waterfall depends upon the height of the fall and upon the number of kilogram-weights of water transferred per unit time. Similarly, in electric circuits the power expended in heating a resistor, charging a storage battery, or turning a motor depends upon the difference of potential between the terminals of the device and the electric current through it. As you learned in Chapter 22, the work W done in transferring a charge q between two points is given by

$$W = qV$$

in which V is the potential difference between them. Therefore the power P is given by

$$P = \frac{W}{t} = \frac{Vq}{t} = VI$$

1 joule/sec = 1 watt = 1 v × 1 amp

Example. An incandescent lamp is connected to a 110-v generator. The current is 0.50 amp. What is the power?

$P = VI$
= 110 v × 0.50 amp
= 110 joules/coul × 0.50 coul/sec
= 55 joules/sec
= 55 watts

Electrical energy U is the electrical power P times the time t:

$$U = P \cdot t$$

The kilowatt-hour is a unit of energy. You can buy electrical energy at a certain price per kilowatt-hour. Your electric bill depends upon the energy supplied, which, in turn, depends upon the *current*, the *time*, and the *voltage.* Suppose that the current through the heater coils of a stove is 10 amp at 100 v, so that the power is 1000 joules/sec, that is, 1000 watts or 1 kilowatt. Then in 1 hour (3600 sec), the total input energy is 1 kilowatt-hour or 3,600,000 joules.

1 kilowatt-hour (kw-hr) = 3.6 million joules

Example. An electric refrigerator requires 200 watts and operates 8 hr/day. What is the cost of the energy to operate it for 30 days at 6 cents/kw-hr?

200 watts × 8.0 hr/day × 30 days

$$= 48{,}000 \text{ watt-hr} \times \frac{1 \text{ kw-hr}}{1000 \text{ watt-hr}}$$
$$= 48 \text{ kw-hr}$$

$$48 \text{ kw-hr} \times 6 \text{ cents/kw-hr} = \$2.88$$

The *electron volt* is a unit of energy used in atomic and nuclear physics. It is *the energy acquired by a particle whose charge equals that of the electron whenever the particle passes through a potential difference of one volt in vacuum.* What is the relation between the electron volt and the joule or the erg? Suppose that an electron, whose charge is 1.60×10^{-19} coul, goes from the negative plate to the positive plate of a vacuum parallel-plate capacitor whose plates have a potential difference of 1 v. The work done by the electric field in accelerating the electron is

$$\begin{aligned} W = Vq &= (1.60 \times 10^{-19} \text{ coul})(1 \text{ v}) \\ &= (1.60 \times 10^{-19} \text{ coul})(1 \text{ joule/coul}) \\ &= 1.60 \times 10^{-19} \text{ joule} \\ &= 1.60 \times 10^{-12} \text{ erg} \end{aligned}$$

Thus the kinetic energy acquired by the electron just before it strikes the positive plate is 1 electron volt (ev) or 1.60×10^{-19} joule. An alpha particle, whose charge is $2 \times 1.60 \times 10^{-19}$ coul, would acquire an energy of 2 ev, or 3.20×10^{-19} joule, in passing from the positive plate to the negative. *Any* quantity of work or energy can be converted into electron volts by the relation

$$1 \text{ ev} = 1.60 \times 10^{-19} \text{ joule}$$

The Heating Effect of a Current

Electric currents can do work in many ways, such as charging storage batteries, running electric motors, and generating heat. We can always compute the input power of a device by

$$P = VI \qquad (1)$$

in which V is the potential difference between the terminals of the device and I is the current.

Let us first apply this equation to the conversion of electrical energy into thermal energy. Let the resistance of the device be R. Then, by Ohm's law,

$$V = IR \qquad (2)$$

Eliminating V from equations (1) and (2), we have

$$P = (IR)I = I^2R$$

Or, if we eliminate I, we have

$$P = \frac{V^2}{R}$$

This is Joule's law of heating.

To convert joules into calories, remember that 4.19 joules = 1 calorie.

Example. A coffee percolator is rated at 800 watts.

(*a*) How many calories of heat does it generate in 100 sec?

$$\begin{aligned} P &= 800 \text{ watts} \\ &= 800 \text{ joules/sec} \times 1 \text{ cal/4.19 joules} \\ &= 191 \text{ cal/sec} \end{aligned}$$

$$Q_1 = 191 \text{ cal/sec} \times 100 \text{ sec} = 19{,}100 \text{ cal}$$

(*b*) What time would be required for this percolator to heat 1 liter ($\cong 1000$ gm) of water from 20°C to 100°C? Neglect the heat loss to the percolator and its surroundings.

$$\begin{aligned} Q_2 &= 1.00 \text{ cal/gm-C}° \times 1000 \text{ gm} \times 80°C \\ &= 80{,}000 \text{ cal} \end{aligned}$$

Therefore

$$Q_2 = 191 \text{ cal/sec} \times t = 80{,}000 \text{ cal}$$
$$t = 419 \text{ sec} = 6.97 \text{ min}$$

(*c*) If, in fact, 450 sec were required, what was the efficiency of the device?

$$\text{Eff} = \frac{191 \text{ cal/sec} \times 419 \text{ sec}}{191 \text{ cal/sec} \times 450 \text{ sec}} \times 100\% = 91\%$$

When charge flows through two resistors connected in series, the heat produced in each is proportional to its resistance. Sup-

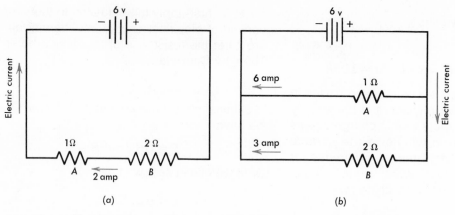

Fig. 25-1. *Heating of resistors connected* (a) *in series, and* (b) *in parallel. In series, the larger resistance develops the greater amount of heat. In parallel, the smaller does so.*

pose that a 1-Ω resistor *A* and a 2-Ω resistor *B* are connected in series to a 6-v battery of negligible internal resistance (Fig. 25-1*a*). Then the current through each resistor will be 2 amp. The I^2R loss in the 1-Ω resistor will be 4 watts; that in the 2-Ω resistor will be 8 watts.

Suppose that the two are connected in parallel (Fig. 25-1*b*). Then the voltage across each will be 6 v. The current through the 1-Ω resistor will be 6 amp and its power will be 36 watts. The current through the 2-Ω resistor will be half as great; its power will be 18 watts. The resistor of smaller resistance develops the greater amount of heat.

In buildings, electrical devices are connected in parallel across the supply lines. The resistance of high-power devices is smaller than that of low-power ones. The resistance of a "30-watt, 120-v" lamp is twice that of a "60-watt, 120-v" lamp.

When electrical supply wires are accidentally short-circuited by being brought into contact with each other, the resistance of the circuit so formed may be only a few hundredths of an ohm. The current becomes very large and heats the wire to dangerously high temperatures. To avoid this danger, *fuses* are connected in series with the supply lines.

Each fuse consists of a wire or strip of lead, aluminum, or an alloy which melts at a low temperature. When the current attains a prescribed value, for instance 15 amp, the metal melts, and the circuit is opened. The fuse must be replaced after the short circuit has been repaired. The wire is mounted in a receptacle, which prevents the melted metal from setting fire to the surroundings (Fig. 25-2). Circuit breakers—electromagnetic devices that open the circuit when the current exceeds a preset value and that can be reset when the overload is removed—are being increasingly used in place of fuses.

Electrical Power and Electromotive Force

Let us calculate the electrical power required to charge a storage battery. In the last chapter you learned that the terminal voltage of a storage battery during the charging process is greater than its electromotive force $\mathscr{E}$ by the amount of the internal voltage drop within the battery. Thus, if the battery charger sets up a terminal voltage *V* in sending a charging current *I* through the internal resistance *r*, we have

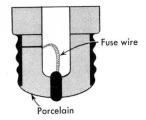

Fuse wire

Porcelain

Fig. 25-2. A screw-base fuse.

$$V = \mathscr{E} + Ir$$

The power delivered to the battery is V times I:

$$P = \mathscr{E}I + I^2 r$$

What do the terms on the right-hand side of the equation mean? The first term, the product of the emf and the current, is the rate at which energy is transformed from electrical energy into stored chemical energy—useful energy. The second term is the rate at which electrical energy is converted into unusable thermal energy in the battery electrolyte and plates.

Example. By means of a battery charger, a potential difference of 7.0 v is applied to the terminals of a storage battery whose emf is 6.0 v and whose internal resistance is 0.50 Ω. (*a*) What is the charging current? (*b*) At what rate is energy being converted into chemical energy, and at what rate is heat produced?

(*a*) On charging, the terminal voltage is

$$V = \mathscr{E} + Ir$$

$$7.0 \text{ v} = 6.0 \text{ v} + I(0.50 \ \Omega)$$

$$I = 2.0 \text{ amp, the charging current}$$

(*b*) $$P = \mathscr{E}I + I^2 r$$

$$P = (6.0 \text{ v})(2.0 \text{ amp}) + (2.0 \text{ amp})^2(0.50 \ \Omega)$$
$$= 12 \text{ watts} + 2 \text{ watts}$$

The rate of converting electrical energy into stored chemical energy is 12 watts and the rate of heating is 2 watts.

Finally, we can calculate the power delivered by a *discharging* battery to an external circuit. Upon discharging, the battery has a terminal voltage V which is also the voltage drop across the external circuit.

$$V = \mathscr{E} - Ir \qquad (3)$$

Or the power delivered by the battery is

$$P = \mathscr{E}I - I^2 r \qquad (4)$$

In this equation, the first term on the right is the rate at which chemical energy is converted into electrical energy and the second term is the rate of heating. The difference between the two is the rate at which useful energy is delivered to the external circuit.

Although we have considered only batteries so far, the foregoing discussion applies to *any* source of emf, including electrical generators, thermocouples, and other devices. All such devices have an internal resistance that must be considered in calculating the useful power that they will deliver. It is worth noting that the power delivered by such a device is a maximum when *the external, or load resistance equals the internal resistance.* The proof of this statement requires the mathematical techniques of calculus and will not be given here. However, we can illustrate its application. From equation (3) above,

$$IR = \mathscr{E} - Ir$$

where R is the load resistance. Hence,

$$I = \frac{\mathscr{E}}{R + r}$$

and equation (4) becomes

$$P = \frac{\mathscr{E}^2}{R + r} - \frac{\mathscr{E}^2 r}{(R + r)^2}$$

where P is the power delivered to the load. You can easily show that when $R = 0$ (no load), the delivered power is zero. When R is very large, the denominators of both terms in the last equation become very large, and the terms themselves become very small; again $P = 0$. But when $R = r$, the delivered power is $\mathscr{E}^2/4r$; this is the largest deliverable power

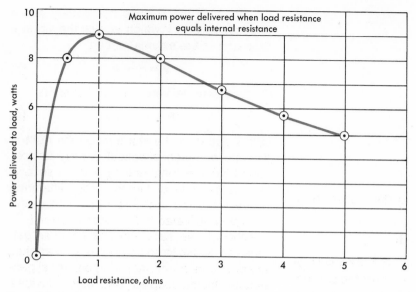

Fig. 25-3. *The power delivered to a load is greatest when the load resistance equals the internal resistance of the source of emf. The emf is 6 v, and the internal resistance is 1 Ω.*

from that battery. Figure 25-3 is a graph of delivered power plotted against load resistance for an emf of 6 v and an internal resistance of 1 Ω.

The Incandescent Lamp

The development of the incandescent lamp illustrates how technological progress accompanies advances in basic science. Discoveries made in the laboratory by physicists who are seeking new knowledge of nature sooner or later prove useful in solving the practical problems of mankind.

Before the development of incandescent lamps, houses were lighted by the flickering flame of the fireside, used by Abraham Lincoln as a boy, by candles, gas flames, and kerosene lamps. The first incandescent lamps, devised more than a century ago, were platinum wires heated red hot by currents from voltaic cells. The lamps had little practical use, both because of their small lumi-

nous efficiencies and because the batteries were expensive and inconvenient. The development of the generator, based on the scientific discovery of electromagnetic induction by Michael Faraday in 1831, provided an economical source of electrical energy and led to a search for filament materials that could be operated at a higher temperature than platinum. In 1859, Sir Joseph Swan of England devised a lamp whose filament was a narrow strip of paper carbonized or charred by heating in an oven. The strip was mounted in an evacuated bulb, but the available air pumps were so ineffective that high vacua could not be produced. The hot carbon filament was soon oxidized. In 1879, twenty years after Swan devised his first lamp, Thomas Edison of the United States made filaments by carbonizing cotton threads. By that time adequate air pumps had been invented, and oxidation caused little trouble. However, the filaments were fragile, and the first lamps operated for only a few hours. Then one was constructed which had a remarkably long life compared with the others. Edison and his assistants anxiously watched it day and night until it "burned out" after 40 hours! Edison was encouraged to seek better materials. He tried hundreds of substances and found that bamboo, split into narrow strips and charred by heating, provided rugged carbon filaments.

George Westinghouse secured the contract for supplying the lamps for the World's Fair at Chicago, in 1893. He was dismayed to find that the Edison patents prevented him from sealing the wires of his lamps into the glass bulbs. He cleverly got around this difficulty by mounting the wires on stoppers which were plugged into the bulbs. A third of a million of these lamps dazzled the visitors at the Fair.

The carbon filament of the Edison lamp could be raised to a higher temperature than a platinum filament; therefore it was a better emitter of light. Tungsten is preferable to carbon because the filament can be heated still hotter. Early investigators found it impossible to draw tungsten into fine wires because the metal was very brittle. This brittleness was due to traces of other substances; ways were found to remove these impurities. By heat treatments and by hammering at high temperatures, the crystal structure of tungsten was so modified that tungsten wires could be drawn to a hairlike thinness. The filaments were mounted in evacuated bulbs, giving the vacuum-type tungsten lamp, which was introduced in 1908.

About seven years later, a young college instructor, Irving Langmuir, spent his summer vacation working at the General Electric Research Laboratories. He decided to study the thermal conductivity of nitrogen by observing how much more power input such a gas-filled lamp required than a vacuum-type lamp of the same dimensions. After many months' work he found the gas impeded evaporation of the filament so that he could heat it hotter and get more light for a given energy input.

Fig. 25-4. An incandescent tungsten-filament lamp.

Despite the increased heat loss due to conduction, the gas-filled lamp was more efficient in producing light than the vacuum-type lamp. Today, most incandescent lamps of power greater than 60 watts are gas-filled (Fig. 25-4).

In planning this study, Langmuir did not expect to improve the efficiencies of lamps. He wanted to do "pure" research to discover nature's laws. Nevertheless, the gas-filled lamp he developed saves us many millions of dollars each year. Pure research often yields big dividends in practical usefulness.

The improvement in lighting during the last century has been so gradual that people are scarcely conscious of it. Today, using a gas-filled incandescent lamp, one can buy light for two cents which would cost two dollars if provided by paraffin candles. To replace the present lighting of your home by candles would cost about $200 per month.

Electrical Heating is Relatively Expensive

Electrical energy may readily be converted in a resistor into heat with 100% efficiency. Nevertheless, electrical heating is expensive

Table 25-1. Approximate Comparative Costs of Fuels Producing Heating Equivalent to that Obtained with $100 Worth of Bituminous Coal

Coal, anthracite (buckwheat)	$ 190
bituminous	100
Electricity (turbines)	1000
Gas, natural	80
Oil, heavy	140

(Table 25-1). This is because elaborate and costly generating and transmission systems are required. Moveover, steam turbines and generators can convert only 20 to 50% of the energy of steam into electrical energy for reasons that we discussed in Chapter 17. Generation of electrical energy by waterfalls is usually more expensive than by steam, for dams are costly to construct and maintain. Nuclear energy plants that derive energy from the fission of uranium are becoming competitive with coal-burning electrical generating stations in localities where "carbon" fuels—coal, oil, and gas—are expensive. As the earth's reserves of fossil carbon fuels are used up, nuclear energy plants will become more economically feasible. They are also less productive of pollutants of the air than other kinds of electrical generating stations but produce heated water and radioactive wastes that must be disposed of in ways that will not adversely affect the environment.

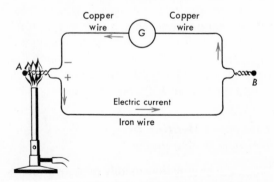

Fig. 25-5. *Seebeck effect. At* **A** *electrons "evaporate" from iron to copper.*

Thermoelectricity

Electrical energy is readily changed into heat, but the inverse process, the production of electrical energy by heat, is difficult. One way to do this on a small scale is by means of a *thermocouple.* Connect two copper wires to a millivoltmeter or galvanometer and fuse the ends of the copper wires to the ends of a piece of iron wire.

Heat one of the iron-copper junctions with a Bunsen burner as shown in Fig. 25-5. The galvanometer needle will be deflected, showing that there is an emf and a flow of electrons. The appearance of an electromotive force in a circuit consisting of two dissimilar metals (or two semiconductors) closely bonded at two junctions, one junction at a higher temperature than the other, is called the *Seebeck effect.* The Seebeck and related effects are called *thermoelectricity.*

To study the Seebeck effect more carefully, we replace the galvanometer by a potentiometer and measure the emf of the thermocouple at several temperature differences of the junctions, maintaining the cold junction at 0°C by immersing it in an ice bath. By measuring the emf's generated by various thermocouples under different conditions, we conclude that (1) the emf depends upon the metals of which the thermocouple is made, being relatively large for iron-constantan thermocouples, for example, and small for copper-iron at a given temperature difference; (2) the emf is not large for most thermocouples, amounting to not more than 40–50 *milli*volts for a temperature difference of 1000°C; (3) the emf is not directly proportional to temperature difference but often rises to a maximum and then decreases (Fig. 25-6).

The Seebeck effect is due to the migration of electrons from one of the metals to the other at the junctions of a thermocouple. In the copper-iron thermocouple (Fig. 25-5), for example, electrons move from iron to copper at

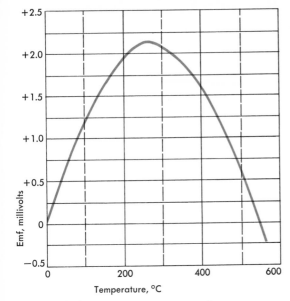

Fig. 25-6. *Thermoelectric emf-temperature curve for a copper-iron couple. One junction is at 0°C.*

both the hot and cold junctions, establishing a potential difference at each junction. However, at the hot junction, *more* electrons migrate from iron to copper than at the cold junction. The potential differences at the junctions oppose each other, like batteries connected + to + and − to −, but the potential difference at the hot junction is the greater. The total emf in the circuit is such as to send an electric current from the hot junction to the cold in the iron.

An analogy may help to make the thermoelectric process clear. Suppose that a ring-shaped glass tube is partly filled with alcohol (Fig. 25-7). Heating the liquid at *A* will increase the vapor pressure there. The liquid will vaporize more rapidly at the warmer surface than at the cooler surface, and the vapor will condense at *B*. A current of alcohol will exist in the tube as long as you maintain the temperature difference between *A* and *B*. The current of alcohol is analogous to the current of electrons. The direction of the electron cur-

rent in a thermoelectric device is opposite to that of the conventional *electric* current.

Several thermocouples connected in series form a *thermopile,* which is more sensitive to temperature changes than a single thermocouple (Fig. 25-8) and can be used as a thermometer after it has been calibrated by being subjected to known temperature differences. Temperature differences as small as a millionth of a Celsius degree can be measured by these devices. Thermopiles have been used in biological research to study the heat evolved by living tissue. For instance, a piece of muscle of a newly killed frog is suspended near one set of junctions of a thermopile mounted in a vessel kept at constant temperature. The heat evolved by the muscle raises the temperature of the nearby junctions and deflects the galvanometer to which the thermopile is connected. When the muscle is stretched by suspending a heavier body from it, the galvanometer deflection increases, showing that more heat is emitted. The generation of heat by the muscle when working and when resting has been studied under various conditions, so that better in-

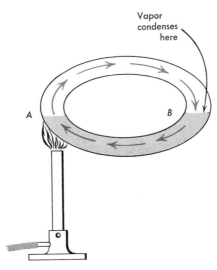

Fig. 25-7. *Temperature difference causes alcohol to circulate.*

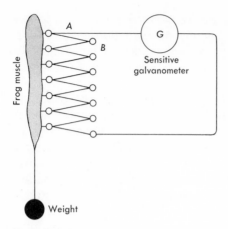

Fig. 25-8. *A thermopile. It can measure the rise of temperature caused by stretching the muscle of a frog.*

Table 25-2. Relative Costs of Energy "

Energy Conversion System	Unit Cost of Energy to Consumer in Dollars Per Kilowatt-Hour
Central-station power plant	0.02
Magnetohydrodynamic generator	0.02
Diesel-electric, railroads	0.03
Gas turbine	0.03
Iron-nickel alkaline storage battery	0.1
Lead-acid storage battery	0.2
Gasoline engine, automotive vehicles	0.2
Draft animals	0.3
Thermoelectric generators	0.3
Thermionic generators	0.3
Thermoelectric gas safety pilots	4
Fuel cells	4
Silicon solar cells	6
Candlelight	10
Leclanche primary battery	40
Man	50
Mercury batteries	500
Electric-watch battery	70 000

" From William T. Reid, *The Physics Teacher*, November, 1971.

sight into the physics of living organisms has been gained.

Suppose that you removed the Bunsen burner, Fig. 25-7, and installed a pump in the glass tube to force vapor from *A* toward *B*. Thus you would cause liquid to vaporize at *A* and vapor to condense at *B*. In consequence, *A* would be cooled and *B* would be heated. Similarly, if you force an electron current around the thermoelectric circuit (Fig. 25-5), electrons will "evaporate" from iron to copper at one junction, cooling that junction, and they will condense from copper to iron at the other junction, heating it. This heating and cooling at the junctions is called the *Peltier* effect. The Peltier effect is the basis of thermoelectric refrigerators.

Recent progress in solid state physics has led to the development of thermoelectric devices that convert thermal energy into electrical energy with much greater efficiencies than were formerly available. Thermoelectric generators and thermoelectric refrigerators are now feasible technological devices. Table 25-2 compares the cost to the consumer of energy delivered by several different energy converters.

Thermoemission

When the temperature of a conductor is raised, its free electrons move faster, like the molecules of a gas. When they have sufficient kinetic energy, they plunge through the surface of the conductor and escape. Thomas Edison accidentally discovered this effect in 1874. He wanted to study the evaporation of carbon from the filaments of his new electric lamps so that he could make them last longer. He thought that the evaporating carbon atoms might be charged, and he tried to capture them on a charged plate. Edison's device is represented in Fig. 25-9. He found that, when the plate was positive, charges would flow through the galvanometer, but not when it was negative.

Today, we explain this behavior by saying that electrons can escape from the hot filament but not from the plate. Edison could see no immediate practical use for this tube. Twenty-five years later, it was applied in radio. We shall consider it again in Chapter 31.

Figure 25-10 shows how the electron flow from filament to plate depends upon the temperature of the filament and upon the voltage in a two-element vacuum tube—a diode.

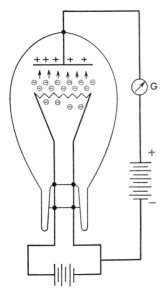

Fig. 25-9. *Edison discovered that an electron current flows from the filament to the plate when the plate is positive.*

When the voltage between filament and plate is zero, electrons that evaporate accumulate near the wire forming a *space charge*. These electrons condense on the filament just as fast as they evaporate. When the plate is positive, some of the electrons are pulled over to it. They flow through the battery and back to the filament. When the voltage of the plate is made more positive, the current increases. Finally, all the electrons that evaporate are pulled to the plate, and the current becomes constant or "saturated." If the filament is raised to a higher temperature, the space charge becomes more dense, and the saturation current becomes greater.

Summary

The energy U converted into heat by a current in a conductor of resistance R can be computed by $U = IVt = I^2Rt$; 4.19 joules = 1 cal; 1 kw-hr = 3,600,000 joules.

The power delivered to a battery of emf $\mathscr{E}$ during charging is $P = \mathscr{E}I + I^2r$ where r is the

internal resistance. The power delivered by the battery to an external load is $P = \mathscr{E}I - I^2r$. The power delivered is a maximum when the internal resistance equals the load resistance.

Tungsten lamp filaments are preferable to carbon filaments, for they can be operated at higher temperatures and therefore are more efficient as emitters of light. In a gas-filled incandescent lamp, the gas acts as a blanket, impeding evaporation of the filament. Such lamps, operated at higher temperatures, are generally more efficient than vacuum-type lamps.

A thermoelectric emf can be generated by establishing a temperature difference between the two junctions of two wires made of different materials.

At high temperatures, electrons leave a metal surface in the process of thermoemission. The emitted electrons form a space charge from which electrons can be attracted by a positively charged plate.

Questions

1. Both a small flashlight lamp and a larger incandescent lamp that is operated at 120 v have currents of 1 ampere. Why does the

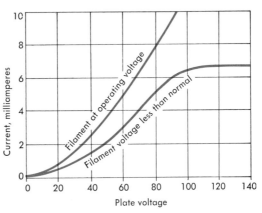

Fig. 25-10. *Plate current versus plate voltage curves for a two-element vacuum tube at two different filament voltages.*

large lamp require more electrical power than the flashlight lamp?

2. The glass bulb of a gas-filled 100-watt incandescent lamp is hotter than the bulb of a vacuum lamp of equal wattage. Why?

3. Which gives off more heat per hour, a 500-watt lamp or a 500-watt electric motor that is turning a drill?

4. Under what conditions is the power a maximum when delivered by a device that is a source of emf?

5. Estimate the cost of operating each of these for a week under normal use: desk lamp, refrigerator, television set, electric iron, electric clock.

6. In a "brown-out" (reduction of power output by an electrical generating station) the voltage supplied to a house is reduced from 115 volts to 105 volts. What effect would this have on the operation of (a) a 115-v, 100-watt lamp, (b) a 115-v, 1600-watt electric heater?

7. Why does the cord of an electric heater not glow when the heating element does?

8. Which uses more energy, a 250-watt TV set in 1 hr, or a 1200-watt toaster in 10 min?

9. Estimate the power that you develop when you run upstairs. Compare this with the power of a 100-watt lamp.

10. Estimate how many persons, each working at the rate of 0.1 horsepower in turning a generator, would be needed to illuminate a classroom.

11. Estimate the number of electrons per second that flow through a 100-watt lamp operated at 120 v.

12. Which is brighter when both are connected in series to a 120-v line: a 100-watt (120-v) lamp or a 10-watt (120-v) lamp? Explain.

13. Compare the power in the 2-Ω resistor in each of the following circuits: (a) a 6-v battery is in series with 1-Ω and 2-Ω resistors, and (b) a 4-v battery is in parallel with 12-Ω and 2-Ω resistors.

14. Prove that, when two lamps A and B are connected first in parallel with a battery and

then in series to the same battery, $1/P_s = (1/P_{A_p}) + (1/P_{B_p})$, where P_s is the total power developed in series, P_{A_p} is the power in A in parallel, and P_{B_p} is the power in B in parallel. Assume no change in resistance of the lamps and a negligible internal resistance of the battery.

Problems

A

1. An automobile battery having an emf of 6.0 v delivers a current of 12.0 amp when the headlights are turned on. If the starter motor also is operated, the total current is 75.0 amp. Find the electrical power in each case.

2. An incandescent lamp is marked "120 v, 75 watts." (a) What is the current in the lamp under normal conditions of operation? (b) How much does it cost per hour to operate the lamp if the price of electrical energy is 6.0 cents/kw-hr?

3. A small electric clock is connected to a 120-v source and the current is 0.060 amp. Find (a) the power, (b) the energy used in 700 hr (approximately 1 month), (c) the cost of this energy at 6.0 cents/kw-hr.

4. A 20-kg elevator motor uses energy at the rate of 50 kw and requires 2.00 min to raise an elevator to the top of a skyscraper. What does the electrical energy for the trip cost at 6.0 cents/kw-hr?

5. What kinetic energy in electron volts is given to an electron when it is accelerated by a potential difference of 3.0 million volts in a vacuum? How many joules is this?

6. In winter, ice forms in a water tank at a rate of 1200 gm/hr. What is the minimum power of an electric lamp that, immersed in the tank, would prevent freezing?

7. A large power plant generates 2.0 kw-hr of electrical energy for each kilogram of bituminous coal burned (heat of combustion 6000 cal/gm). What is the efficiency of the system?

B

8. An electric iron uses energy at a rate of 1320 watts when heating at the maximum rate and 360 watts when the heating is a minimum. The voltage is 120 v. What are the current and the resistance in each case?

9. A 200-watt lamp is totally immersed in 1500 gm of water. How much will the temperature of the water rise in 3 min? Neglect heat losses from the water.

10. Two 6.0-Ω coils are connected first in series, then in parallel across a 12-v battery of negligible internal resistance. What are (a) the total current and (b) the total wattage in each case?

11. An electric heater to be attached to a water faucet uses 6.0 amp at 120 v. It is advertised as "delivering scalding hot water in 1 min." What time would be required for it to heat 2000 gm of water (about 2 quarts) from 20°C to 80°C if no heat escaped?

12. Calculate the "line voltage drop" in a power line having a total resistance of 0.30 Ω if it supplies 1.0 kw at 220 v to an electric motor.

13. An electric ironer requires 11.0 amp at 120 v. (a) What is its power? (b) How much does it cost per hour of operation if the price of electrical energy is 6.0 cents/kw-hr?

14. If no heat escapes, (a) how much electrical energy is required to heat a bathtub of water of mass 50 kg from 20°C to 70°C, and (b) how much will the energy cost at 6.0 cents/kw-hr?

15. A dry cell having an emf of 1.50 v and a negligible internal resistance is connected to a lamp having a resistance of 5.00 Ω. What are (a) the current, (b) the power, and (c) the energy liberated in 1 hr?

C

16. When water is separated into hydrogen and oxygen by electrolysis, the opposing emf is 1.23 v. How much electrical energy, in kilowatt-hours, is required to decompose 1 gram-molecule of water (18 gm)?

17. An electric clock requires 5.00 kw-hr of energy per month (700 hr). (a) How much water per month would have to fall through a vertical distance of 1.00 m over a water wheel to supply this energy? (b) How much water, falling 1.00 m, would supply the energy utilized by a grandfather clock driven by a body of mass 1000 gm, which descends 6.0 m in the 700 hr? Assume that the overall efficiency of the water wheel is 50% in both cases.

18. For how long would the energy released by the conversion of one microgram of matter keep fifty 200-watt lamps operating? Assume 20% efficiency of the generating station.

19. The storage battery of a car when fully charged can generate a current of 8.0 amp for 6.0 hr. Its average emf is 12.6 v. (a) How much energy is liberated? (b) How many kilograms of water would this energy heat from 20°C to 80°C?

20. A dry cell used to operate an electric doorbell costs $1.25. Its average emf is 1.50 v, and it delivers a current of 1.50 amp for 100 hr. What are (a) the power, (b) the total energy, and (c) the cost of the energy per kilowatt-hour?

21. A flashlight battery of negligible internal resistance delivers a current of 1.5 amp for a total life of 6.0 hr. The emf is 3.0 v. (a) How many joules of energy are liberated? (b) How many calories of heat are produced?

22. The results of measurements of voltage (V) and current (I) are given below. Calculate from these data the resistance (R) and the power (P) at each voltage and plot R and P versus V. Why does R vary?

V(v)	I(amp)
120	0.805
100	0.730
80	0.635
60	0.540
40	0.445
20	0.305

23. One junction of an iron-copper thermocouple (Fig. 25-6) is at 0°C. The thermoelectric emf is 1.6 millivolts. (*a*) What possible values can be assigned to the temperature of the hot junction? (*b*) What is the useful range of the thermocouple?

24. A hobbyist who is building a hi-fi set at home needs a resistor whose resistance is 10,000 Ω and which can dissipate 2 watts of power. He has available carbon resistors with the following ratings: two 5 K, 1 w; two 20 K, 1 w; one each of 110 K, 0.5 w; 110 K, 2 w; 11 K, 0.5 w; 11 K, 2 w; 10 K, 0.5 w; and 10 K, 1.5 w, where 5 K means 5000 Ω. What combination will give him the desired resistance/power characteristics and at the same time be economical in the use of resistors (i.e., will not use high-power resistors where others would do)?

For Further Reading

Hogben, Lancelot, *Science for the Citizen,* Chapter 14, W. W. Norton and Company, New York, 1951.

Hogerton, John F., "The Arrival of Nuclear Power," *Scientific American,* February, 1968, p. 21.

Joffe, A. F., "The Revival of Thermoelectricity," *Scientific American,* October, 1958, p. 31.

CHAPTER TWENTY-SIX

Magnetism

The ancient peoples knew of the existence of magnetism. Certain black stones called lodestones (leading stones) attracted pieces of iron. The name magnet is probably derived from Magnesia, a province in Asia Minor where the magnetic iron ore was found. In ancient literature, we find many tall stories about magnetism. We read that Archimedes used a very strong lodestone to draw nails from the hulls of enemy ships, causing them to sink. Some of the stories are true. The Roman historian Pliny quotes the Greek philosopher as saying, "There is a divinity among you moving you like that contained in the stone which Euripedes calls a magnet. . . . This stone not only attracts iron rings, but also imparts to them a similar power to attract other rings."

About three centuries before Columbus discovered America, someone observed that a magnetized needle placed on a bit of wood floating on water would turn to a nearly north-south direction. This invention of the magnetic compass enabled the mariner to venture far out to sea, guided by the magic needle when the skies were obscured by fog.

During the reign of Queen Elizabeth I, William Gilbert, the queen's physician, investigated magnetic phenomena. Gilbert was a careful investigator, testing his ideas by experiments with apparatus. He studied the magnetism of the earth, the forces between bar magnets and the forces between electric charges, and the way in which an iron bar acquired magnetism when suspended in a north-south direction for a long time. He also speculated about the nature of the universe, believing that the universe was infinite in extent and that other stars might have planets as does the sun. Gilbert discovered many of the facts that will be described in this chapter.

Magnets and Magnetic Poles

A permanent magnet is a body that attracts pieces of iron, nickel, and a few other materials. If you dip a magnetized steel rod into a box of iron filings and then remove it, the filings will cling to it at regions near the ends. These regions or seats of attraction where the filings adhere we call the *poles* of the magnet. Suspend a magnetized steel knitting needle at its midpoint by a cord, or place it on a block of wood floating in a basin of water. The magnet will swing around to a nearly north-south position. The end that points to-

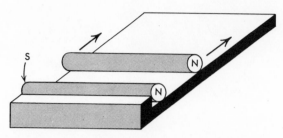

Fig. 26-1. Repulsion of like poles. A cylindrical alnico magnet at rest at the bottom of an inclined plane repels another magnet when like poles are opposite each other. When equilibrium is reached, the two magnets are parallel to each other. What does this indicate about the relative strengths of the two poles of each magnet?

ward the Arctic is called the *north-seeking pole* or more briefly the *north pole.* The other one is the south-seeking pole or the south pole.

If you bring the north pole of a permanent magnet near the north pole of another magnet, the two repel each other (Fig. 26-1); but if a south pole is brought near, attraction occurs. In general, *like poles repel, but un-like poles attract each other.* This law is similar to that for electric charges, but the phenomena are different. Poles exert forces on poles, charges on charges, but magnetic poles do not attract or repel stationary electric charges.

The forces exerted upon each other by two magnetic poles separated by a given distance can be measured by apparatus like that represented in Fig. 26-2. A slender magnet is pivoted like the beam of a balance or trip scale. The poles are approximately "point" poles, concentrated at the ends. Another magnet is supported as shown, separated from the first one by vacuum or air; its north pole repels the lower one downward. A suitable load is attached to the other end of the balance to force the pole up to its initial position. The weight of the added body then equals the force of repulsion of one pole on the other, and the distance between them may be taken as that between the ends of the magnets.

Experimentally we find that two north point poles at a distance *r* apart repel each other with a force *F* that is inversely proportional to the square of the distance of separation.

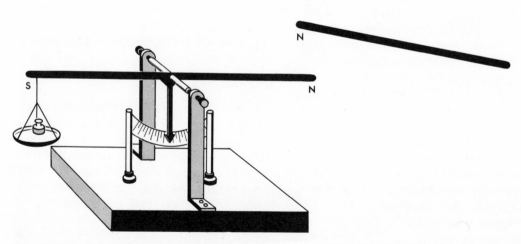

Fig. 26-2. Apparatus to measure the repulsive force between magnet poles.

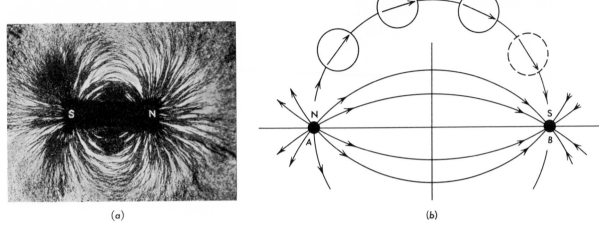

(a)   (b)

Fig. 26-3. (a) *Iron filings reveal the magnetic field of the bar magnet.*
(b) *The magnetic field of a north magnetic pole and an equal south magnetic pole.*

$$F \sim \frac{1}{r^2}$$

If the upper magnet is replaced by another
slender magnet, the force between its north
pole and that of the lower magnet will in general be different at a given separation from
that observed earlier. However, the force will
again be inversely proportional to the square
of the distance between the poles.

The law of force between point magnetic
poles depends upon distance in the same
way as the gravitational force between point
masses and the electrical force between
point charges. It is much more difficult to establish point poles experimentally, however,
and they have much less practical importance than point charges. *Isolated* point
poles have not been discovered. Although a
theory of magnetism can be derived with considerable precision from the idea of point
poles, we shall base our description on a
more fundamental idea—that *magnetism accompanies the motion of electrical
charges.* Before investigating this idea, how-

ever, we shall discuss the magnetic fields of
permanent magnets.

Magnetic Fields and Magnetic Lines

Any region in which a magnetic pole would
experience a force is called a *magnetic field.*
In order to demonstrate the existence of the
field near a magnet, lay a magnetized bar of
steel on a table and cover it with a sheet of
cardboard. Sprinkle iron filings onto the
cardboard and tap it lightly. Then the filings
will arrange themselves from pole to pole
along curved magnetic lines (Fig. 26-3a). Figure 26-3b shows how the magnetic field due
to a north magnetic pole and an equal south
pole can be mapped by the use of small compasses. Each magnetic line indicates, at
every point on it, the direction of the force that
would act on a north pole if it were placed at
that point. Several small compasses are located on one of the lines, and each compass
needle is tangent to this line. Notice that each
magnetic line is directed from the north pole
to the south pole.

Fig. 26-4. *The Oersted medal. Oersted demonstrates that an electric current produces a magnetic field. (Medal awarded by the American Association of Physics Teachers for contributions to the teaching of physics.)*

Electric Currents Produce Magnetic Fields

When Gilbert experimented on static electricity and on magnetism, he noted that the forces exerted by magnet poles on each other resembled those exerted by electric charges on other charges. Gilbert suspected that electricity and magnetism were related, but he could not prove the relationship. At the dawn of the nineteenth century, as you will recall, the Italian physicist Volta showed that currents could be produced by a voltaic cell, consisting of two dissimilar metal plates in an electrolyte. The cell could cause continuous electric currents in conductors and helped greatly in the investigation of electrical phenomena. In 1820, Hans Oersted (pronounced Or'sted), a Danish professor of physics, noticed during a classroom demonstration that an electric current in a wire deflected a nearby compass needle (Fig. 26-4). He had

discovered electromagnetism—that electric currents produced magnetic fields. Oersted's discovery aroused feverish activity among other scientists. Soon it was proved that a current in a long straight wire produces magnetic lines which are circular, in planes at right angles to the wire. To show that this is true, pass a thick, copper wire vertically through a hole in a horizontal piece of cardboard on which there are iron filings. Establish a current of 30 or 40 amp in the wire, tap the cardboard, and the filings line up in concentric circles (Fig. 26-5). Turn off the current, tap the cardboard, and the filings are disarranged.

In Fig. 26-6, several small compasses are at equal distances from a vertical wire. When the electric current is directed upward, the compass needles turn so that the north poles of their needles point counterclockwise as viewed from above. The magnetic field is therefore counterclockwise around the wire. It exerts on a magnet a deflecting force that is

Fig. 26-5. *An electric current in a straight wire produces a magnetic field. The lines of force are circular. (From* PSSC Physics, *D. C. Heath and Company, 1960.)*

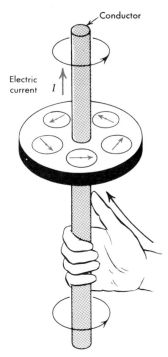

Conductor

Electric current I

Fig. 26-6. Applying the right-hand rule. The thumb indicates the direction of the electric current, the fingers that of the magnetic lines. The magnetic field is directed counterclockwise as viewed from above.

at right angles to the direction of the current. When the current is reversed, each needle turns about and points oppositely. The direction of the magnetic field produced by an electric current can be found by a right-hand rule which you will use in studying generators, motors, and the like.

Imagine that you grasp the wire with your right hand so that your thumb points in the direction of the electric current. Then your fingers will point in the direction of the magnetic field.

Following Oersted's discovery, other physicists investigated the kinds of magnetic fields that accompany currents in conductors of different shapes. A circular loop of wire, a flat circular coil of many turns, a long helical coil or solenoid — all were found to be surrounded

by magnetic fields when charges passed through them (Fig. 26-7). (You can verify the directions of the magnetic fields by applying the right-hand rule.)

Eventually it was discovered that wires were not essential, only the *motion* of charge, wherever charge might be found. Henry Rowland, an American physicist, demonstrated the existence of a magnetic field near a rotating disk that carried an electrostatic charge on its rim. Rowland's apparatus consisted of a glass disk with a metal strip at its edge, a means of rotating the disk at high velocities, and a magnet mounted so that its deflections could be observed. When Rowland placed an electrostatic charge on the metal strip and rotated the disk, the magnet was deflected, indicating the presence of a magnetic field in the expected direction. Reversing the direction of rotation reversed the direction of the magnetic field. Clearly there was no qualitative difference between the magnetic field established by an electric current in a loop of wire and that established by a moving charged body. Later experiments showed that beams of charged particles — such as the electron beams we find in a cathode-ray oscilloscope or a television picture tube — were also accompanied by magnetic fields.

Note particularly that *constant* magnetic fields do not exert forces on electric charges at rest with respect to the field. Likewise, *constant* electric fields do not exert forces on stationary magnetic poles.

Having demonstrated that magnetic fields are caused by the motion of electric charges, we might next attempt to calculate the strength of the magnetic field from the current in various situations. However, we must first establish a quantity by which to describe the magnetic field, just as we employed the electric field strength to describe an electric field. Having defined and illustrated this field variable for magnetic fields and seen how it is related to magnetic forces, we will be in a

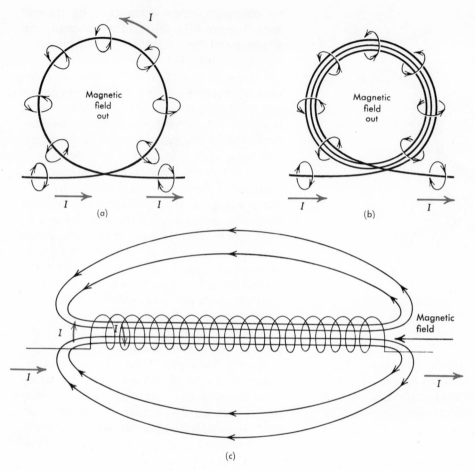

Fig. 26-7. *The directions of the magnetic fields due to electric currents in conductors of different shapes as determined by the right-hand rule:* (a) *a single circular loop,* (b) *a flat coil of many turns,* (c) *a solenoid.*

position to consider how to calculate it from the current.

The Process of Magnetization

So far we have discussed magnetic fields caused by permanent magnets and magnetic fields caused by currents of charged particles. Are the processes by which the magnetic fields are established the same for permanent magnets and for currents in wires, or are they different? We find that the processes

are the same. Basically, magnetic fields are caused by the motion of charges relative to the observer whether these moving charges are conduction currents in a wire or a beam of particles or atomic currents in a permanent magnet. Although it is an oversimplified picture, which we must correct later, individual atoms can be said to contain electric currents because of (1) the motion of electrons around the nucleus, (2) the spinning motion of the electrons, and (3) the spinning motion of many kinds of nuclei. Because of these "cur-

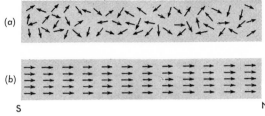

Fig. 26-8. (a) *In unmagnetized iron, the magnetic dipoles are disorganized.* (b) *When all the dipoles are organized, as is suggested in this greatly simplified picture, the magnet is "saturated."*

rents," individual atoms act like tiny electromagnets. They set up magnetic fields that affect other atoms and are in turn affected by magnetic fields.

You will recall that a body may be charged negatively by adding electrons to it, or positively by taking electrons away from it. When a piece of iron is magnetized, no particles are added to it or taken away from it. When we magnetize a piece of iron, we rearrange the atomic magnets, or *magnetic dipoles,* already existing in the metal. In an unmagnetized piece of iron, the dipoles are disorganized. When the iron is magnetized they are regimented so that the individual magnetic fields point in the same general direction. If all the tiny magnets are lined up in this manner, we say that the magnet is saturated (Fig. 26-8).

The following simple experiments help you to visualize the process of magnetization. Fill a long, narrow glass tube with iron filings and stroke it from end to end with the north pole of a bar magnet. This operation causes many of the magnetized filings to swing around so that they point in the same direction. Afterward, one end of the tube will attract the north pole of a compass, and the other end will repel it. Shake the tube so as to rearrange the filings, and you will demagnetize it so that neither end will repel the north pole of the compass. Break a magnetized knitting needle or a hacksaw blade in two, and each part will be a magnet (Fig. 26-9). Then break one of these pieces in two, and again the parts will be magnetized. If you could repeat this process time after time, you would eventually arrive at the ultimate, atomic magnetic dipoles. The magnetism of a permanent magnet is thus distributed throughout its volume. It is most apparent at the poles because there the effects of neighboring atomic dipoles do not cancel each other.

If a piece of iron is in a magnetic field, heating it with a flame or jarring it by hammering will help the dipoles to line up so as to increase the magnetization. If a magnet is not in a field, heating or jarring it will help to demagnetize it.

The similarity between the magnetic field of a bar magnet and that of a current-carrying solenoid reinforces our belief that there is basically no difference between the magnetism of the two (Fig. 26-10). Notice that the external fields of the magnet and the solenoid are almost indistinguishable. We cannot see

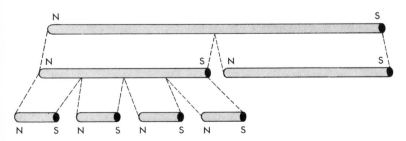

Fig. 26-9. Making little magnets out of a big one.

Fig. 26-10. The magnetic field of (a) a permanent magnet, and (b) a current-carrying solenoid.

the shape of the field inside the magnet, but we can see what apparently happens in the interior of the solenoid. *The magnetic lines are continuous.* As we can easily verify with a small compass, the lines leave one end of the solenoid—its "north pole"—pass around it, and enter the other end, closing upon themselves. Unlike electric lines that begin on positive charges and end on negative charges in an electrostatic field, magnetic lines are closed upon themselves. Next we notice that the external magnetic field is apparently strongest—judging by the concentration of the iron filings—at the poles of the permanent magnet and at the ends of the solenoid. Again, we cannot predict from what we know so far what the field is inside the magnet, but we notice that the internal field of the solenoid seems to be even stronger than the external field at its ends.

Magnetic Forces on Currents

The magnetic field of a current in a straight wire exerts a sideways deflecting force on a

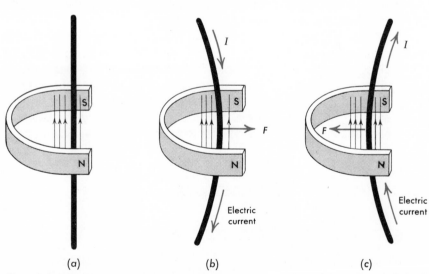

Fig. 26-11. Magnetic deflecting forces act on a current-carrying conductor. The current, the magnetic field of the U-magnet, and the deflecting force are at right angles to each other.

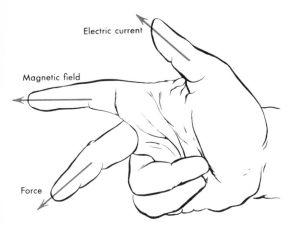

Fig. 26-12. The right-hand rule for magnetic deflecting forces.

nearby magnet. According to Newton's third law, we might expect the magnet to exert by means of *its* magnetic field a deflecting force on the current-carrying wire. Experimentally it is easy to demonstrate that this occurs.

In Fig. 26-11, a flexible vertical wire is placed in the magnetic field between the poles of a U-magnet (Fig. 26-11a). The magnetic lines in the field of the U-magnet are horizontal, running into the page from the north pole of the magnet to the south pole. When an electric current is sent *down* the wire, the wire is deflected to the *right*. When

the current is *upward,* the deflecting force is to the *left.* You can predict the direction of the deflecting force in such a situation by another right-hand rule, often called the motor rule:

Holding the thumb, index finger, and middle finger of your right hand at right angles to each other (Fig. 26-12), point the thumb in the direction of the electric current and the index finger in the direction of the magnetic field. The middle finger will point in the direction of the deflecting force.

If a wire carrying a current is placed in a magnetic field of unknown direction and turned in various directions in the field, for one direction of the wire the deflecting force will be a maximum. For another direction, the force will be *zero.* Using a small magnetic needle, one can verify that the force is a maximum when the current is perpendicular to the magnetic field and that the force is zero when the current is parallel to it. This enables us to define the direction of a magnetic field without reference to magnetic poles: *The direction of a magnetic field is that of the electric current in a conductor so aligned in the field that it experiences* no *magnetic force.*

One can predict the direction of the deflecting force on a current-carrying conductor

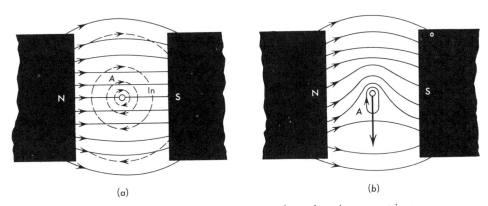

(a)   (b)

Fig. 26-13. Magnetic fields due to a magnet and an electric current in a wire. (a) They oppose each other below the wire and reinforce each other above it. (b) The combined field. It exerts a down force on the wire.

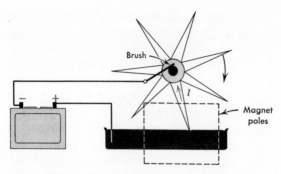

Fig. 26-14. A star-wheel motor. The electric current directed from the mercury to the axle is thrust sidewise by the field of the magnet poles.

in a magnetic field either by the right-hand rule or by a rule stated by Faraday: Imagine that magnetic lines are stretched elastic bands that repel one another. Find where the magnetic lines in the combined field are most concentrated; the force will be directed away from this concentration. Figure 26-13a represents two poles of a magnet and its magnetic field. The dotted circles represent the magnetic lines due to an electric current in the wire *A* directed into the page. Notice that the two fields oppose each other below the wire and strengthen each other above it. Figure 26-13b shows the magnetic lines of the combined fields. According to Faraday's rule, they exert a force on the wire which pushes it downward. The right-hand rule leads to the same direction for the force.

Figure 26-14 represents a simple star-wheel motor. The electric current is directed toward the axle from the pool of mercury. The magnetic deflecting force pushes sidewise on the current and makes the wheel revolve.

In Fig. 26-15, the electric current in each of two parallel wires is in the magnetic field of the other wire. Notice that, according to the right-hand rule, the deflecting forces push the wires together when the currents are in the same direction (Fig. 26-15a) and push them apart when the currents are oppositely directed (Fig. 26-15b).

The Magnetic Induction

We are now ready to define a quantity that describes a magnetic field in the same way that the electric field strength describes an electric field. *The magnetic induction B is the force per unit length per unit current exerted on a current-carrying straight conductor when the conductor is so oriented that the force is a maximum.* B is a vector quantity; its direction is not that of the force, but of the magnetic field as stated in the preceding article. As a defining equation we have:

$$B = \frac{F}{Il}$$

where *F* is the maximum force, *I* the current, and *l* the length of the conductor. If *F* is in newtons, *I* in amperes, and *l* in meters, *B* is in newtons per ampere-meter.* In principle we

* The units of *B* in other systems of units are given in Appendix H.

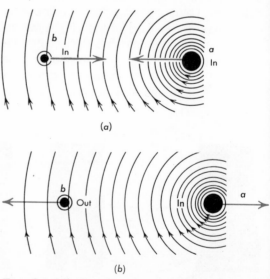

Fig. 26-15. The magnetic fields due to electric currents in two parallel wires. (a) The currents are in the same direction, and the deflecting forces push the wires together. (b) The currents are opposite and the wires are pushed apart.

can measure B at a point by finding the maximum force per unit current per unit length exerted on a small current-carrying conductor at that point. In fact, however, there are more convenient ways of measuring B, as we shall see in Chapter 29.

Example. A 0.2-m straight wire stretched horizontally carries an electric current of 10 amp from east to west in a magnetic field whose magnetic induction is 0.1 newton/amp-m directed vertically downward. What are (a) the magnitude of the magnetic deflecting force on the wire and (b) its direction?

(a) The magnetic field is at right angles to the current. Therefore, the force is a maximum, and

$$F = BIl = (0.1 \text{ newton/amp-m})(10 \text{ amp})(0.2 \text{ m})$$
$$= 0.2 \text{ newton} \cong 0.02 \text{ kgwt}$$

(b) By the right-hand rule: thumb (electric current) points west, index finger (magnetic induction) points vertically downward, and the middle finger (deflecting force) points *south*.

Suppose that a straight conductor of length l carrying a current I is placed in a magnetic field so that the angle between the magnetic induction B and l is θ. Then experimentally we find that it is only the component of B perpendicular to the wire — $B \sin \theta$ — that is effective in causing a deflecting force. Hence

$$F = BIl \sin \theta$$

If $\theta = 90°$, this equation reduces to the one given above in defining B.

Just as the electric field strength indicated the degree of concentration of electric lines of force, the magnetic induction indicates the degree of concentration of magnetic lines or magnetic *flux*. For this reason, B is often called the magnetic flux density. If the magnetic induction at a given point is 10 newtons per ampere-meter we agree to represent the field at that point by drawing 10 magnetic lines of flux per square meter of area that is perpendicular to the field. One magnetic line

is called in the mks system of units one *weber*. Therefore, 10 magnetic lines per square meter, representing a magnetic induction of 10 newtons per ampere-meter, is a magnetic flux density of 10 webers per square meter. Thus, in this example,

$$B = 10 \text{ newtons/amp-m}$$
$$= 10 \text{ magnetic lines/m}^2 = 10 \text{ webers/m}^2$$

The first of these equalities expresses B as a deflecting force per unit current per unit length, and the second and third express B as the density of magnetic lines. In this chapter, we shall stress the first of these properties of B, leaving the others to later chapters.

Because, as you have seen, magnetic lines are continuous, closing upon themselves, the total number of magnetic lines passing through any closed surface is *zero*. That is, just as many magnetic lines must enter as leave the surface. You can readily verify this by imagining the solenoid of Fig. 26-10b completely enclosed by a closed surface. The number of magnetic lines leaving one part of the closed surface must equal the number returning at another part. Notice that magnetic fields are different in this respect from electrostatic fields, in considering which, we found earlier that the number of electric lines leaving a closed surface was generally not zero but proportional to the enclosed charge.

The Earth is a Magnet

Before the experiments of Gilbert, it was believed that some object in the skies, such as the north star, made a compass turn to a north-south position. Gilbert suggested that the earth is a gigantic magnet. He fashioned a piece of lodestone into a sphere and mounted a magnetized needle on a horizontal axis so that its pole could tilt up and down instead of sidewise. He found that a "dipping" needle located near the equator of the small, magnetized sphere became parallel to its axis, that

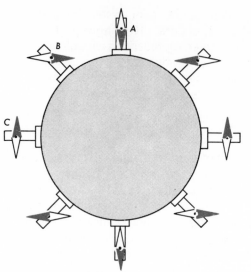

Fig. 26-16. Gilbert's lodestone had a magnetic field like that of the earth.

tized sphere nor a large permanent magnetic dipole, but an electromagnet. According to this theory, the magnetism of the earth is caused by currents of charged particles in the earth's liquid core. Convection within the hot interior of the earth sets up these currents and they are maintained by electrical generator action, according to principles that we shall take up in Chapter 29.

In Fig. 26-19 the vector **AC** represents the magnetic induction B_0 of the earth's field, 5.4×10^{-5} weber/m² at a certain location. The dip at this location is 60°. The vector **AD** represents the *horizontal component* B_h of the field, 2.7×10^{-5} weber/m², and **AE** the *vertical component* B_v, 4.7×10^{-5} weber/m². If θ is the angle of dip, $B_h = B_0 \cos \theta$.

The earth's geographic and magnetic poles do not coincide; hence, at most localities, the compass does not point true north. The angle between the true north and the compass direction is called the *declination*. Figure 26-18 shows lines of equal declination in the United States. At Portland, Oregon, the compass points about 25° east of north, and at Portland, Maine, about 20° west. The change

is, to a line through the two poles (Fig. 26-16). As the compass was moved toward *A*, its north pole dipped downward more and more until at *A* the needle pointed toward the center of the sphere. The earth acts on a dipping compass in the same way. As the compass is carried northward from the equator, the tilt of the dipping needle increases until, at a certain location, the needle is vertical. This location is vertically above one of the earth's two magnetic poles.

The earth's field resembles not only that of a magnetic sphere, such as Gilbert's, but also that of a huge magnetic dipole placed as shown in Fig. 26-17. The axis of the magnet is not parallel to the axis of rotation of the earth but is inclined to it at an angle of about 17°. The magnetic poles are more than a thousand miles from the geographic poles. The angle of inclination, or the dip, is about 60° at the southern boundary of the United States and about 75° at the Canadian border (Fig. 26-18); at the magnetic pole near Hudson Bay, it is 90°.

Recent theories of the earth's magnetism hold that the earth is not a uniformly magne-

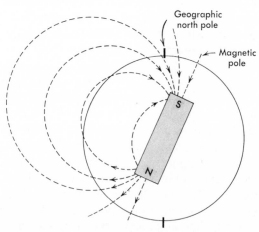

Fig. 26-17. The earth's magnetic field resembles that of a magnetic dipole. The magnetic poles are more than a thousand miles from the geographic poles.

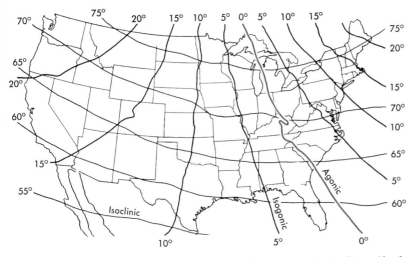

Fig. 26-18. *Magnetic map of the United States, not including Alaska and Hawaii. The "isogonic" lines connect points of equal declination; the east-west "isoclinic" lines connect points of equal dip.*

of declination of the compass was first observed during Columbus's pioneer voyage to America. The sailors were alarmed by this discovery, but today it is well understood by navigators.

Surveyors, mariners, and air pilots, who rely upon the magnetic compass in determining direction, need to know the declinations for their localities. In the United States, the information is supplied by the U.S. Coast and Geodetic Survey. The declination at a given place is not constant, but changes slowly as the years go by. In some places, the change is several degrees in a century. The earth's magnetic field also varies irregularly, hour by hour, during the day, the changes in declination being a few hundredths of a degree. These fluctuations are unusually large during magnetic storms. Probably they are due to changes in the strength of the streams of electrons which the earth receives from the sun. Magnetic storms are most prevalent at times when many spots are present on the sun's surface and the electron currents from sun to earth are most intense.

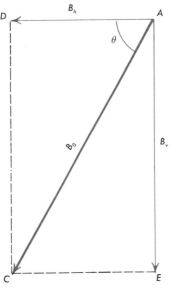

Fig. 26-19. *The magnetic induction of the earth's field AC may be resolved into the horizontal component AD and the vertical component AE.*

Summary

A permanent magnet attracts iron filings, which cling to it at regions called poles. Two magnetic point poles in air or vacuum exert forces on each other that are inversely proportional to the square of the distance of separation.

A magnetic field is a region in which forces act on magnets. A magnetic line indicates by its direction at each point the direction of the force on a magnetic north pole. The magnetic lines of a magnet are directed from its north pole to its south pole in the region surrounding the magnet.

Magnetic fields are caused by the motion of electric charges, whether these moving charges are in a wire or in a beam of charged particles or are atomic currents in a permanent magnet. The magnetic lines caused by an electric current in a long straight wire are a set of circles around the wire in planes at right angles to it. The direction of the field may be found as follows: Grasp the wire with your right hand, with your thumb pointing in the direction of the electric current; then your fingers will point in the direction of the field.

In an unmagnetized piece of iron, the atomic magnetic dipoles are arranged at random. Magnetization causes them to line up in a certain direction. When all the dipoles are thus organized, the magnet is "saturated."

Deflecting forces act on current-carrying conductors in magnetic fields. The deflecting force is a maximum when a straight conductor is perpendicular to the magnetic lines. The direction of the deflecting force is then given by the following rule. Point the thumb of your right hand in the direction of the electric current and the index finger in the direction of the magnetic field; the middle finger, held at right angles to the other two, will point in the direction of the deflecting force.

The magnetic induction is the maximum force per unit length per unit current when a straight conductor is placed in a magnetic field. It is also the number of magnetic lines per unit area.

$$B = \frac{F}{Il \sin \theta}$$

The earth is a great magnet. Each magnetic pole is over a thousand miles from the nearer geographic pole. At the magnetic equator a dipping compass needle is horizontal, and at the magnetic poles it is vertical. The variation of the compass from a north-south direction is called the declination. The declination at any location changes slowly from year to year. Smaller changes occur from hour to hour.

Questions

1. State two or more ways in which magnetostatic phenomena differ from electrostatic phenomena.

2. What are the basic laws for the forces between unlike and like magnetic poles? Compare these with the force laws for electric charges.

3. Design a rotating-disk motor.

4. How can the direction of a magnetic field be defined (a) in terms of the force on a north magnetic pole, (b) in terms of the force on a current-carrying conductor?

5. What happens when a long magnet is broken in half? Explain in terms of the model of a magnet.

6. How do heating and jarring affect a magnet? Why?

7. How is a current-carrying solenoid like a magnet?

8. How could a small compass be used to plot the magnetic field of a current-carrying solenoid?

9. If you were on a desert island and had a compass, could you determine the declination?

10. How has the discovery of magnetism affected the history of the world?

11. Is a magnetic "pole" a point or a region? How can it be studied?

12. A wire carries an electric current down through a magnetic field directed west to east. In what direction is the wire deflected?

13. How could you tell whether a permanent magnet or a current-carrying solenoid was producing a magnetic field if it was not possible to see the source of the field?

14. From Fig. 26-18, estimate the magnetic declination and the magnetic dip (a) at your locality, (b) at New York City, (c) at New Orleans, (d) at Los Angeles.

15. How is the magnetism of a permanent magnet explained in terms of the magnetic fields of currents?

16. Discuss several ways of showing that a moving charge sets up a magnetic field.

17. What is the direction of the force between two parallel power lines both carrying currents from east to west?

18. Where on the earth's surface does the north-seeking end of a magnetic compass point southwest? In what direction would a compass point at the north geographic pole?

19. Would a small magnet on a smooth level shelf at 45° north latitude tend to move? If so, in what direction? (Assume that the magnet is already lined up with the earth's field.)

20. Have the locations of the earth's magnetic poles ever changed substantially? What kinds of evidence might be sought to study this possibility?

Problems

A

1. A wire carrying a current of 5.0 amp perpendicular to a magnetic field of induction 1.0 newton/amp-m experiences a force of 1.0 newton. What is the length of the wire?

2. If the force between two magnetic point poles separated 0.10 m is 0.80 newton, what will be the force if the distance is (a) 0.05 m, (b) 0.20 m?

3. If the magnetic induction of the earth's field is 6.0×10^{-5} weber/m^2 at a given location, what are the (a) horizontal and (b) vertical components for a dip angle of 30°?

B

4. The filament of an incandescent lamp is at right angles to a magnetic field of flux density 0.20 weber/m^2. What sidewise force does a section of the filament 5.0 cm long experience when the current through it is 0.50 amp?

5. What sidewise force acts on an east-west horizontal wire 50 m long in which there is a current of 200 amp? The vertical component of the earth's magnetic induction is 5.0×10^{-5} weber/m^2.

6. What force acts upon a metal bar 60 cm long in a transverse field whose magnetic induction is 1.5 webers/m^2 when the current through the bar is 15 amp?

7. A horizontal straight wire 0.80 m long carries an electric current of 10 amp from south to north in a magnetic field whose magnetic induction is 0.5 newton/amp-m upward. Find (a) the magnitude of the deflecting force, (b) the direction of the force.

C

8. Find the force exerted on the earth by a 1.0-m wire carrying a current of 1.0 amp and perpendicular to the earth's field. $B = 6.0 \times 10^{-5}$ weber/m^2.

9. An experiment to determine the earth's magnetic induction at a certain location gave the vertical component of B as 4.7×10^{-5} weber/m^2 and the horizontal component as 2.5×10^{-5} weber/m^2. Find (a) the dip angle, (b) the total magnetic induction.

10. A 2.0-m wire carrying 60 amp is placed in a magnetic field of flux density 0.5 weber/m^2. What will be the force on the wire if the angle between the wire and the field is 60°?

11. A wire weighing 0.30 kgwt/m, connected to flexible leads, is in a magnetic field of flux density 1.5 webers/m^2. What current will cause the wire to be "weightless"? How must the current be directed?

12. A uniform magnetic field of magnetic induction 10.0 webers/m^2 is applied at right angles to the plane of a rigid circular loop of wire of radius 10.0 cm carrying a current of 100 amp. What force is exerted on an insulating pin that holds the two ends of the loop together?

13. A metal bar of mass m rests on two parallel horizontal metal rails separated by a distance l. The bar carries a current I from one rail to the other, and the coefficient of friction between the bar and the rails is 0.40. Derive an expression for the smallest magnetic induction that would cause the bar to slide. What is the direction of the field?

For Further Reading

Berge, Glenn L. and George A. Seielstad, "The Magnetic Field of the Galaxy," *Scientific American,* June, 1965, p. 46.

Livingston, William C., "Magnetic Fields on the Quiet Sun," *Scientific American,* November, 1966, p. 54.

CHAPTER TWENTY-SEVEN

Magnetic Forces

In the last chapter you learned that magnetic fields accompany moving electric charges. A straight wire carrying a current is surrounded by circular magnetic lines centered on the wire. The magnetism of a permanent magnet is distributed throughout the volume of the magnet, but it, too, is caused by moving charges — currents within the elementary atomic magnets that are aligned during the process of magnetization.

When a current-carrying conductor or a beam of charged particles is placed in a magnetic field, deflecting forces act on the moving charges. In this chapter we shall apply our basic force equation $F = BIl$ to a variety of situations so that you will become familiar with magnetic forces. We shall assume that the required magnetic fields exist; either they are produced by laboratory magnets, or the magnetic field is that of the earth. In the next chapter, we shall look into ways of calculating the magnetic inductions produced by various sources of magnetic fields.

Torque on a Coil Carrying a Current

A magnetic field exerts a sideways deflecting force on a current at right angles to the mag-

netic field. As a consequence, a coil carrying a current in a magnetic field experiences a *torque*. In Fig. 27-1, a rectangular coil that is free to rotate about the axis shown is placed in the magnetic field between two poles. The two sides of the coil that are at right angles to the field are shown in cross section. The electric current in the coil is directed *out* of the page in conductor *A* and *into* the page in conductor *B*. The magnetic field is directed from the north pole to the south. Applying the right-hand rule for magnetic forces, we see that the deflecting force on *B* is *up* and that on *A* is *down.* Each force exerts a clockwise torque on the coil about the axis. The torque of the magnetic forces will cause the coil to rotate in a clockwise sense.

In Fig. 27-2, we show the coil in a view at right angles to the one given in Fig. 27-1. Suppose that the coil consists of a single rectangular loop of length *l* and width *w* with its long sides *AD* and *BC* at right angles to the field of magnetic induction *B*. A force out of the page, $F = BIl$, acts on the side *BC*. The lever arm of this force with reference to the axis is *w*/2; therefore, the force causes a torque $F \times w/2$ on the loop of wire. The other side, *AD*, experiences an equal but opposite force. When the coil is in the position shown, the sides *AB* and

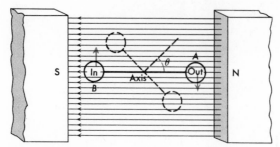

Fig. 27-1.Coil in a magnetic field. The field exerts a torque that tends to turn the coil.

CD experience *no* forces because they are parallel to the field. Therefore the total torque on the coil is

$$\tau = F \times w = (BI)(lw) = BIA$$

in which *A* is the area of the coil when the plane of the coil is parallel to the field.

Suppose that the perpendicular to the plane of the coil is not at right angles to the field but instead makes an angle θ with it (Fig. 27-1). Then the forces on *AD* and *BC* are unchanged, but the lever arms of these forces about the axis become $(w/2) \sin \theta$. Thus the total torque due to forces on *AD* and *BC* is

$$\tau = BIA \sin \theta$$

There are forces on *AB* and *CD*, but these forces exert no torque about the axis shown. (Why?) Finally, if there are *N* turns in the coil instead of one turn, the torque is *N* times greater, and we have

$$\tau = BIAN \sin \theta \qquad (1)$$

Magnetic Moment

In discussing the behavior of a current-carrying coil in a magnetic field, it is often useful to talk about the *magnetic moment* of the coil. *The magnetic moment $\mathcal{M}$ is the torque per unit magnetic induction when the perpendicular to the plane of the coil is at right angles to the magnetic field.* From equation (1) in the last section

$$\mathcal{M} = \frac{\tau_{max}}{B} = \frac{BIAN \sin 90°}{B}$$

$$\mathcal{M} = IAN$$

Thus, the magnetic moment is also the product of the current in the coil, the area of the coil, and the number of turns. This is true even if the coil is not rectangular although we have proven it only for a rectangular coil.

We can also talk about the magnetic moment of a permanent magnet. Placed in a uniform magnetic field—one in which the magnetic induction is constant—a magnet, such as a compass needle, experiences a torque: the field exerts equal and opposite forces on the poles of the magnet (Fig. 27-3) and these forces set up torques about the axis of suspension. The total torque τ is a maximum when the magnet makes an angle of 90° with the magnetic field. The magnetic moment of the magnet is defined as the maximum torque per unit magnetic induction:

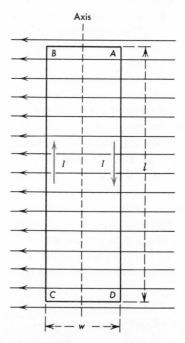

Fig. 27-2.Torque on a coil viewed perpendicularly to the axis of rotation.

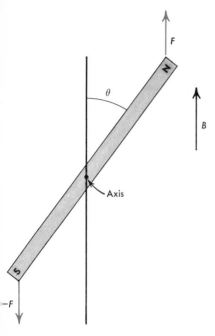

Fig. 27-3. *A magnetic field exerts a torque on a permanent magnet. If the field is uniform, the resultant force is zero.*

$$\mathcal{M} = \frac{\tau_{max}}{B}$$

Once we have determined the magnetic moment of the magnet by measuring the maximum torque in a known field, we can calculate the torque acting on it in any other field when the angle between the magnet and the field is θ:

$$\tau = \mathcal{M}B \sin \theta$$

In a uniform magnetic field, the resultant force acting on the magnet is zero. (Why?) The resultant force on a dipole is *not* zero in a non-uniform field.

Moving-Coil Meters

Measurements of charge, current, potential difference, and other electrical quantities are usually made with moving-coil meters whose principal mechanism is a current-carrying coil in a magnetic field. The most commonly used *galvanometer* consists of a coil pivoted between the poles of a U-shaped magnet (Fig. 27-4). When electrons flow through the coil, it experiences a torque so that the coil is deflected, moving a needle or pointer on a scale. This motion is opposed by a spiral spring, and the angle of deflection is proportional to the torque and hence (equation 1 above) to the current in the coil.

A *voltmeter* is a galvanometer having a suitable resistor connected in series with its coil (Fig. 27-5). The deflection of the coil is directly proportional to the voltage across its terminals and inversely proportional to the total resistance of the coil and resistor. By adjusting the resistance of the resistor, we can make the reading of the pointer on the scale equal to the voltage across the terminals of the instrument. Then it serves as a voltmeter.

Example. A galvanometer has a resistance of 20.0 Ω. A current of 0.100 amp deflects the coil through 10 scale divisions. (a) What resistance must be connected in series with the coil to convert the galvanometer into a volt-

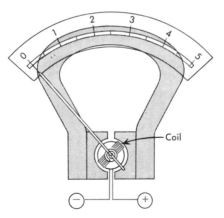

Fig. 27-4. *A galvanometer. A current in the pivoted coil causes a torque which rotates the coil. A spiral spring opposes the displacement. The pole pieces of the magnet are so shaped that the component of the magnetic induction parallel to the plane of the coil is constant.*

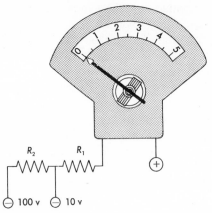

Fig. 27-5. *A two-range voltmeter. To convert the galvanometer into a voltmeter, connect a suitable resistor in series with the coil.*

meter which reads 10 scale divisions when connected to a battery having an emf of 10.0 v? (*b*) What resistance is required to convert it into a 100-v voltmeter?

(*a*) The current required to cause this scale reading is 0.100 amp, hence

$$R = \frac{V}{I} = \frac{10 \text{ v}}{0.100 \text{ amp}} = 100 \text{ } \Omega$$

The galvanometer coil resistance is 20 Ω, hence that of the series coil must be 80 Ω.

(*b*) Applying the same method,

$$R = \frac{100 \text{ v}}{0.100 \text{ amp}} = 1000 \text{ } \Omega$$

The required external resistance is 1000 Ω − 20 Ω = 980 Ω.

A galvanometer may be converted into an *ammeter* by connecting a shunt or low-resistance bypass resistor in parallel with the coil (Fig. 27-6). The resistance of the shunt must be so adjusted that the scale reading indicates the total current through the galvanometer and the shunt. Let this total current be *I* and the current through the galvanometer coil i_c; then that through the shunt is $(I - i_c)$. Let the resistance of the coil be R_c

and that of the shunt r_s. Since the coil and the shunt are connected in parallel with each other, the potential difference across the coil equals the potential difference across the shunt.

$$i_c R_c = (I - i_c) r_s$$

Example. The coil resistance of a galvanometer is 20.0 Ω and a current of 0.100 amp deflects the coil through 10.0 scale divisions. What is the required resistance of the shunt so that the total current shall be 10.00 amp when the scale reading is 10.0 divisions?

$$I = 10.00 \text{ amp}, \quad i_c = 0.100 \text{ amp},$$
$$(I - i_c) = 9.90 \text{ amp}$$

$$0.100 \text{ amp} \times 20.0 \text{ } \Omega = 9.90 \text{ amp} \times r_s$$

$$r_s = 0.202 \text{ } \Omega$$

The resistance of an ammeter is so small that it would be damaged if it were connected across a battery or other source of emf. Ammeters are always connected *in series* in a circuit, and voltmeters are connected *in parallel* with the part of the circuit across which the potential difference is to be measured.

A *multimeter* is a moving-coil instrument that can be used as a voltmeter, ammeter, or

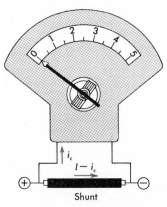

Fig. 27-6. *An ammeter. To convert a galvanometer into a ammeter, connect a suitable shunt (low resistance) in parallel.*

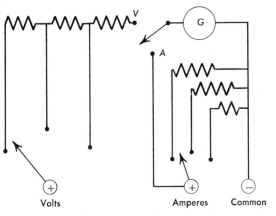

Fig. 27-7. *The circuit diagram of a simple multimeter that can be used as a voltmeter or as an ammeter. The arrows represent the movable contacts of range-selector switches.*

ohmmeter. A range-selector switch permits one to choose the proper series resistor or shunt resistor that is to be connected to the single galvanometer movement *G* (Fig. 27-7).

The Simple Direct-Current Motor

The direct-current motor, like a moving-coil meter, contains a current-carrying coil placed in a magnetic field. A torque acts on the coil. Unlike the coil of a meter, whose rotation is restrained by a spiral spring, the motor is designed to rotate continuously. Electrical energy supplied to the motor is converted into mechanical energy as the motor does work on the device that is attached to the shaft of the motor—the engine of a car, a sewing machine, an electric fan, or a pump for circulating fluids, for example.

In Fig. 27-8, a single loop of wire, placed in a magnetic field, is connected to a battery by slip rings and brushes—so that the connecting wires will not become twisted. Applying the right-hand motor rule, we see that the torque on the loop will cause it to rotate as shown. However, when the loop has turned 90° it will stop rotating because *the torque reverses direction* as the loop passes the 90°

position. We must somehow reverse the current in the loop so that the torque will be always in the same direction.

We can reverse the current automatically each half-revolution by using a split-ring commutator. The split-ring commutator shown in Fig. 27-9 consists of two metal half-rings, each insulated from the other. As the loop rotates, each brush touches first one half-ring, then the other. Thus, the commutator reverses the connections to the battery twice for each revolution. The motor rotates continuously in one direction.

The four-pole motor represented in Fig. 27-10 causes a more constant torque than the one just described. How does it differ from the single-loop motor we have been discussing? First, it contains many turns of wire in each coil instead of a single turn. Second, the rotating coils are wound on an iron armature which strengthens the magnetic field about the coils—for reasons that we shall look into in the next chapter. Third, the poles of the armature fit closely—allowing just enough space for free rotation—within the poles of the field electromagnet; this further strengthens the magnetic field about the coils. Fourth, the armature has two coils mounted at right angles to each other. The split-ring commu-

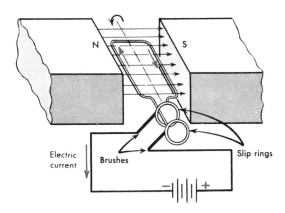

Fig. 27-8. *A single loop carrying a current in a magnetic field. The loop turns 90° from the position shown and then stops.*

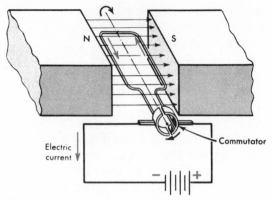

Fig. 27-9. *A single-loop direct current motor. The commutator reverses the electrical connections to the coil every half-revolution — at times when the plane of the coil is perpendicular to the magnetic field — so that the torque acting on the coil is always in the same direction.*

tator is divided into four segments. In the position shown, the vertical coil carries a current so that the armature rotates counterclockwise. After it turns 45°, the brushes will pass from one pair of segments to the other so that the other coil is energized. As the rotation continues, one of the coils is never more than 45° from the position in which the maximum torque acts on it.

Practical motors have drum-type armatures. In the drum armature, numerous coils are wound in the slots of the iron cylinder, and the commutator has many segments. The contact with each coil is made when it is nearly parallel with the field, and the torque exerted on the armature is nearly constant.

If the input current to the motor shown in Fig. 27-10 were reversed, the direction of the current in the top and bottom coils would be reversed. At the same time, the polarity of the field magnet would also reverse, and the armature would rotate in the same direction as before. The reversal of the current does not

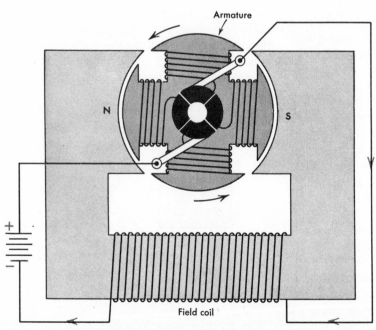

Fig. 27-10. *Four-pole motor. The vertical poles are always energized.*

reverse the direction of rotation; hence motors of this type can be driven either by direct or by alternating current. They are often used in washing machines, vacuum sweepers, and electric shavers, and are called *universal motors.*

In order to reverse the direction of rotation of a d-c motor, one must reverse the direction of the current in the armature coils *without changing the current in the field coils.*

Later we shall discuss motors that operate only on alternating current.

The Motion of Charged Particles in a Magnetic Field

We have seen that, in general, deflecting forces act upon moving charged particles in a magnetic field. In a straight current-carrying conductor placed in a magnetic field, the electrons experience a deflecting force as they drift along the wire in a current. They are confined to the wire, however; hence, the wire as a whole is pushed sidewise.

What is the deflecting force on each electron? Let the number of free electrons in a length l of the wire be n and let them drift forward with a velocity v. Then all the electrons in the length l, having a total charge $q = ne$, where e is the charge on the electron, will move through any cross section of the wire in time $t = l/v$. The current I is q/t or ne/t or nev/l. The n electrons in the length l experience a force F given, as we have seen, by

$$F = BIl = B(nev)$$

if the wire is at right angles to the magnetic field.

Or, the force on *one* electron is

$$F_e = \frac{F}{n} = Bev$$

This equation holds also for charged particles in free space. In general, the deflecting force on a single particle is

$$F = Bqv \sin \theta$$

where q is the charge on the particle and θ is the angle between its velocity and the magnetic field.

One finds the direction of the deflecting force by the right-hand motor rule, treating the moving electrons as an electric current whose direction is *opposite* to that of the electrons. A beam of positively charged particles would be treated as a conventional electric current.

The mass of an electron may be found by first accelerating it by an electric field and then pushing it sideways by a magnetic field. (Television tubes deflect the electron beam in this way.) Figure 27-11 represents a highly evacuated tube. The wire *A* is heated by a battery. The heating increases the speed of the free electrons in the metal, so that some of them plunge through its surface by the process of thermoemission. These electrons are accelerated toward the anode *B*. Some of them pass through the hole in it and strike the fluorescent screen on the glass at *C* so as to

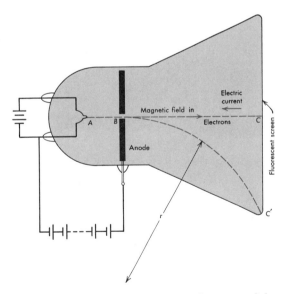

Fig. 27-11. *Device to measure the mass of the electron. An electric field accelerates the electron from* A *to* B. *Then a magnetic field pulls it sidewise along the arc* BC′.

make it emit light. Suppose now that we establish a magnetic field directed into and perpendicular to the page on the right side of the anode. The magnetic deflecting force will act on each moving electron. Since the force is *always* at right angles to the velocity vector, the electrons will be forced sidewise, along a circular path.

Let the charge of the electron be *e* and let the potential difference between the plates *A* and *B* be *V*. The work done in accelerating each electron from *A* to *B* is charge-times-voltage, or *eV* This work imparts kinetic energy to the electron, increasing its velocity from 0 to *v*, so that

$$eV = \tfrac{1}{2}mv^2 \qquad (2)$$

Each electron, after passing through the hole in the anode, is pushed sidewise by the magnetic field and travels on a circle of radius *r*. The centripetal force acting on it is

$$F_e = \frac{mv^2}{r} \qquad (3)$$

The magnetic deflecting force is

$$F_e = Bev \qquad (4)$$

From equations (3) and (4) we have

$$\frac{mv^2}{r} = Bev$$

and, combining this result with equation (2) by eliminating *v*,

$$\frac{e}{m} = \frac{2V}{B^2 r^2} \qquad (5)$$

Knowing the potential difference *V*, the magnetic induction *B*, and the radius of the circle on which the electrons move, we can compute the ratio of the charge to the mass of the electron, obtaining

$$\frac{e}{m} = 1.76 \times 10^{11} \text{ coul/kg}$$

Earlier, we saw that Millikan obtained the value 1.60×10^{-19} coul for *e*, the charge on

the electron. Hence we can calculate the rest mass of the electron

$$m_0 = 9.11 \times 10^{-31} \text{ kg}$$

At velocities near that of light, you will recall, the mass of the electron is appreciably greater than this, increasing with velocity according to the equation

$$m = \frac{m_0}{\sqrt{1 - v^2/c^2}}$$

where *c* is the speed of light (Chapter 10).

Example. An electron is accelerated by passing through a potential difference of 1000 v before entering a magnetic field. What magnetic induction will cause the electron to move on a circular path of radius 1.0 m?

Solving equation (5) for B^2, we obtain

$$B^2 = \frac{2V}{(e/m)r^2}$$

$$B^2 = \frac{2(1.0 \times 10^3 \text{ v})}{(1.76 \times 10^{11} \text{ coul/kg})(1.0 \text{ m}^2)}$$

$$= 1.14 \times 10^{-8} \text{ v/(coul/kg)(m}^2)$$

Let us be careful about units. We expect *B* to be expressed in newtons per ampere-meter. First we substitute 1 joule/coul for 1 v. Next we write the joule as the newton-meter, and the coulomb as the ampere-second. We then have

$$B^2 = 1.14 \times 10^{-8} \quad \frac{\text{newton-m/amp-sec}}{(\text{amp-sec/kg})(\text{m}^2)}$$

or, rearranging the units, we have

$$B^2 = 1.14 \times 10^{-8}$$
$$(\text{newton/amp}^2)(\text{kg/sec}^2)(1/\text{m})$$

Multiplying the numerator and denominator by 1 m gives us

$$B^2 = 1.14 \times 10^{-8}$$
$$(\text{newton/amp}^2)(\text{kg-m/sec}^2)(1/\text{m}^2)$$

But 1 kg-m/sec^2 is 1 newton. Hence

$$B^2 = 1.14 \times 10^{-8} \text{ (newton)}^2/(\text{amp}^2\text{-m}^2)$$

and

$B = 1.06 \times 10^{-4}$ newton/amp-m
$$= 1.06 \times 10^{-4} \text{ weber/m}^2$$

The Cyclotron

The **cyclotron,** invented by E. O. Lawrence, who received the Nobel prize for its invention, is an excellent example of the usefulness of magnetic deflecting fields in the acceleration of particles to high energies. It consists in part of two hollow semicircular electrodes, called dees, which resemble one of the halves of an empty shoe-polish can, cut in two along a diameter (Fig. 27-12). These electrodes, enclosed in an evacuated box between the poles of a large electromagnet, are connected to a radio oscillator having a frequency of about ten million cycles (10 megacycles) per second. The oscillator causes a large potential difference between the dees to vary cyclically with this frequency.

Suppose that at a given instant the left electrode is at a positive potential of 100,000 v above that of the other electrode. Positively charged hydrogen ions released at the center of the dees in the gap between the two electrodes are accelerated toward the right. Each ion gains 100,000 electron volts of energy as it passes across the gap. In the interior of the metallic dees, *no* electric field acts on the ions and the magnitude of the velocity of each remains constant. The magnetic field pulls the ions sidewise and forces them to travel in a semicircular path and to return again to the gap. During this part of its motion, the ions move with a constant speed **v** under the action of a centripetal force provided by the magnetic deflecting force so that

$$Bev = \frac{mv^2}{r}$$

Meanwhile, the potentials of the two electrodes are reversed so that now the right one is positive. When the ions reach the gap, they are again accelerated and their kinetic en-

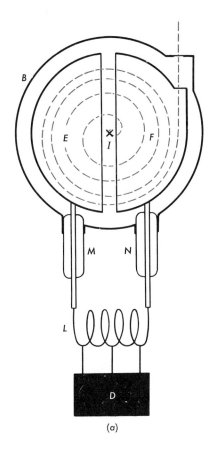

(a)

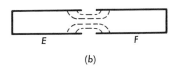

(b)

Fig. 27-12. *The dees of a cyclotron.* (a) *Protons, liberated at* I, *spiral around, acquiring energy each time they cross the gap.* (b) *A cross-sectional view of the dees.*

ergies again increase by 100,000 electron volts. Because of their higher speed, the ions now travel on a semicircle of larger radius, but again return to the gap and again receive energy.

We can readily calculate the frequency with which the oscillator must change the voltage between the dees in order to synchro-

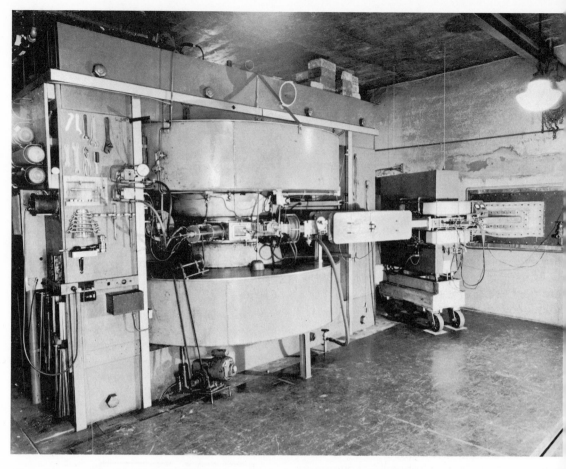

Fig. 27-13. *The University of Pittsburgh cyclotron. The diameter of the pole pieces of the electromagnet is 48 in. This cyclotron produces protons of 8 Mev energy. The apparatus to the right of the picture is part of an analyzing system for reducing the spread in energy of the cyclotron beam so that nuclear reactions can be studied. (Courtesy of Dr. A. J. Allen.)*

nize the voltage with the motion of the ions. Writing $2\pi r/T$ for v, where r is the radius of a particular circular path of an ion and T is the time that would be required for a *complete* revolution at that radius, we can change the previous equation to

$$Be\left(\frac{2\pi r}{T}\right) = \frac{m(2\pi r/T)^2}{r}$$

$$Be = \frac{m2\pi}{T}$$

$$\frac{1}{T} = \frac{1}{2\pi}\left(\frac{e}{m}\right)B$$

But $1/T$ is the frequency f. Hence

$$f = \frac{1}{2\pi}\left(\frac{e}{m}\right)B$$

The frequency depends only upon the ratio of charge to mass and upon the magnetic induction. It does not depend upon the radius of the path at any time or upon the velocity of the ion. It is this principle that makes it possible to use an oscillator of constant frequency.

In this way the ions circle around in increasing circular arcs with increasing speeds and energies until they are allowed to escape from the cyclotron and strike a target. Here they bombard the nuclei of the material of which the target is made and cause nuclear reactions to take place, as we shall see in a later chapter. If an ion receives 100,000 electron volts of energy every time it crosses the gap, and if it crosses the gap 80 times, then its final energy will be 8 million electron volts (Fig. 27-13), written 8 Mev.

The cyclotron is limited to producing particles whose energy is less than about 30 or 40 Mev. This is because the mass of the particle increases relativistically with increasing velocity, and the particle falls behind the changing voltage across the gap until the particle is decelerated at the gap instead of being accelerated.

To overcome this difficulty, other accelerators, such as the *synchrocyclotron* or frequency-modulated cyclotron, were developed. The synchrocyclotron resembles the ordinary cyclotron except that the pole faces of the electromagnet are much bigger—184 in. in diameter, for example, instead of 48 in.—there is only one D-shaped electrode, and the frequency of the oscillator decreases as a batch of ions moves out from the center. As the ions tend to fall behind the changing voltage, the frequency with which the voltage changes also decreases so that the ions always arrive at the gap just in time to be accelerated.

Charged Particles in the Earth's Magnetic Field

In the cyclotron, the ions move in a plane perpendicular to the magnetic field; the path of each ion is a spiral of increasing radius within that plane. If a charged particle enters a uni-

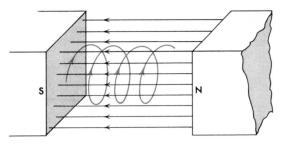

Fig. 27-14. If the initial velocity of a negatively charged particle is not perpendicular to a uniform magnetic field, the particle moves on a helix.

form magnetic field in such a direction that it has a component of velocity *parallel* to the magnetic field, its path is a helix—a spiral of constant radius drawn out in the direction of the magnetic field (Fig. 27-14).

The magnetic field of the earth is not uniform, as we have seen, but somewhat resembles that of a bar magnet. It does, however, exert deflecting forces on charged particles whenever the particles move across magnetic lines. The acceleration produced by the deflecting forces and the initial motion of the charged particles combine to produce rather complicated paths.

Primary cosmic rays are charged particles, mostly protons, that enter the earth's magnetic field from space. They have high energies, occasionally greater than 10^{18} electron volts. Many reach the denser parts of the earth's atmosphere where, as we shall see later, they produce nuclear reactions. The products of these reactions can be detected by suitable instruments at the surface of the earth and by instrument packages carried by rockets to high altitudes. Cosmic ray particles of lesser energies are in general deflected by the earth's magnetic field. Figure 27-15 shows typical paths of protons incident at various latitudes. Near the magnetic pole in Canada, protons (curve *a*) approaching the earth travel almost parallel to the magnetic lines; they move along open spirals that take

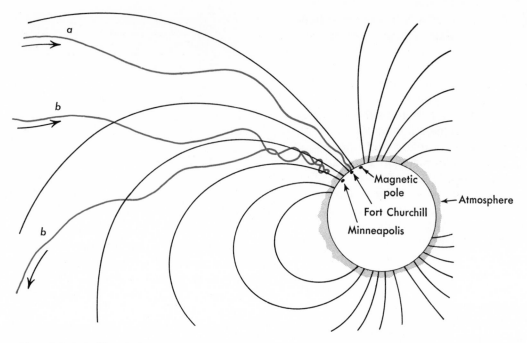

Fig. 27-15. The deflections of protons in the earth's magnetic field. (Adapted from Scientific American.)

them into the earth's atmosphere. At the latitude of Minneapolis, protons of the same energy (curve *b*) travel across the earth's magnetic field and move in tighter and tighter spirals as they enter the stronger portions of the field. The trajectories of these particles can be calculated. The theory predicts that these particles will not reach the earth's atmosphere, but will be essentially "reflected" and sent back out of the earth's magnetic field along an opening spiral. More energetic particles can penetrate the earth's atmosphere, where their energy is dissipated in a variety of nuclear reactions.

Belts of high-energy electrons and protons surround the earth. The belts were discovered by Dr. James Van Allen in one of the earliest flights of satellites bearing instruments for detecting energetic particles. In the regions of greatest concentration in the belts, instruments have detected as many as 10^7 par-

ticles per square centimeter per second. Within the belts, the electrons move back and forth on spiral paths along magnetic lines. The spirals become tighter as the electrons approach the edges of the magnetosphere at the poles, where they are reflected. The electrons trapped by the earth's magnetic field in this fashion spiral back and forth (Fig. 27-16) and also drift eastward. Protons perform a similar motion but, of course, spiral in the opposite sense and drift westward. Finally these particles lose their energies by collisions with air molecules near the poles. They are replaced by other electrons and protons arriving from the sun.

Charged particles—principally protons—streaming out from the sun constitute a *solar wind* which interacts with the earth's magnetic field. Figure 27-17 shows the earth's magnetic cavity—a region of the interplanetary magnetic field within which the

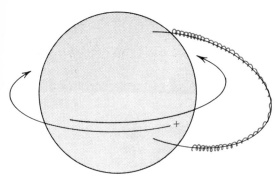

Fig. 27-16. *Electron paths in the magne-*
tosphere.

earth's field predominates. On the side toward the sun, the earth's cavity extends outward to a distance of about 10 times the radius of the earth. At right angles to this direction it extends to 14 earth radii. In the direction opposite to the sun, the cavity has a long cylindrical tail (only part of which is shown in the diagram) extending to several earth-moon distances from the earth. The belts of trapped charged particles lie within the magnetic cavity. Where the solar wind impinges on the cavity and sweeps around it, a shock wave is set up, analogous to that which forms when a plane flies at supersonic velocities. The solar wind, incidentally, helps to explain why the tail of a comet always streams out in a direction opposite to that of the sun.

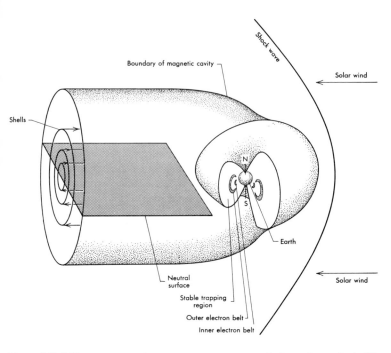

Fig. 27-17. *The earth's magnetic cavity and the solar wind. The neu-*
tral surface separates the upper half-cylinder where the magnetic lines
point toward the earth from the lower half-cylinder where they point
away from the earth.

Summary

The torque on a current-carrying coil of N turns placed in a magnetic field is

$$\tau = BIAN \sin \theta$$

The magnetic moment of such a coil is the maximum torque per unit magnetic induction, or IAN.

Electrical moving-coil meters contain a current-carrying coil that is placed in the field of a permanent magnet. The rotation of the coil is opposed by a spiral spring. A galvanometer is a current-indicating device. To transform a coil-type galvanometer into a voltmeter, connect a resistor of suitable resistance in series with the coil. To convert it into an ammeter, connect a resistor (shunt) of suitable resistance in parallel. An ammeter should be connected in series in a circuit.

A direct-current motor also contains one or more current-carrying coils placed in a magnetic field. The split-ring commutator permits the motor to turn continuously by periodically reversing the current in the coils. Electrical energy supplied to the motor is used in doing mechanical work.

A moving charged particle in general experiences a deflecting force in a magnetic field

$$F = Bqv$$

The mass of the electron can be measured by accelerating an electron beam by means of an electric field and then deflecting it by a magnetic field. The electron's rest mass is 9.11×10^{-31} kg.

In the cyclotron, charged particles move on an expanding spiral path in a perpendicular magnetic field. Whenever an ion passes from one electrode to another, its energy increases by a fixed amount. The period of the ion's motion is constant.

Charged particles are trapped in the Van Allen belts in the earth's magnetic field. The solar wind, a flux of charged particles from the sun, interacts with the earth's magnetic field, setting up a magnetic cavity.

Questions

1. Describe how one might calibrate an ammeter.

2. A stream of positive ions travels east across a magnetic field directed south. In what direction is the ion stream deflected?

3. How could you determine the velocity of an electron in a discharge tube if its charge and mass were known? Describe both the measurements and the calculations.

4. Why must an ammeter always be connected in series?

5. How should a voltmeter be connected to measure the potential difference across a resistor? What would happen if it were connected in series with the resistor?

6. How could you make a galvanometer into (a) a voltmeter, (b) an ammeter? Can the same galvanometer be used for both?

7. What is the significance of the belts of trapped particles in (a) understanding the nature of the earth's magnetic field, (b) understanding the nature of cosmic rays, (c) space travel?

8. How would you interpret the readings of two identical voltmeters connected in series across a 10-Ω coil in which the current was 5 amp?

9. Why is a cyclotron limited to energies not greater than about 40 Mev (million electron volts)? How must cyclotron design be modified for higher energies?

10. Discuss the errors present in using each of the circuits in Fig. 27-18 to determine R by the ammeter-voltmeter method.

11. An experimenter has two voltmeters: one has a range of 0–1000 v and the other 0–500 v. The 0–1000-v instrument is labeled as "20,000 ohms per volt." Similar information is not given about the other one. Without knowing anything further about the instruments, can the experimenter use the two in combination to measure correctly an unknown voltage that he knows is approximately 1200 volts?

Problems

Charge on proton or electron = 1.60 ×
10⁻¹⁹ coul
Mass of proton = 1.67 × 10⁻²⁷ kg

Charge on proton or electron $= 1.60 \times 10^{-19}$ coul

Mass of proton $= 1.67 \times 10^{-27}$ kg

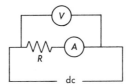 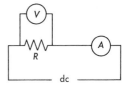

Fig. 27-18.

A

1. Calculate the magnetic moment of a square loop of wire 0.40 m on a side carrying a 6.0-amp current.

2. What is the energy (*a*) in joules and (*b*) in Mev (million electron volts) of protons accelerated in a cyclotron where the proton gains 100,000 electron volts each time it crosses the gap? Assume that the proton crosses the gap 50 times before emerging from the cyclotron.

3. What force is exerted on a 1.0-coul charge moving at 1.0 m/sec perpendicular to a magnetic field of 1.0 weber/m²? How likely is it that this set of conditions would occur in an experiment?

4. Calculate the frequency of a cyclotron designed to accelerate protons in a 1.5-weber/m² magnetic field.

5. A voltmeter has a resistance of 300 Ω, and its maximum reading is 1.50 v. What resistance must be connected in series with it to make its maximum reading 150 v?

B

6. The coil of a galvanometer has a resistance of 50 Ω. A current of 20 × 10⁻⁵ amp will produce a full-scale reading of 5.0 units. (*a*) What shunt resistance will convert the instrument into an ammeter with a range of 10.0 amp? (*b*) How much resistance connected in series will convert the galvanometer into a voltmeter of 100-v range?

7. What resistance must be connected in series with the galvanometer described in problem 6 in order to convert it into a voltmeter reading 15 v when the scale reading is 5.0 divisions?

8. The coil resistance of a galvanometer is 100 Ω, and the scale reading is 10 divisions when the current in the coil is 1.00 milliamp.

What is the resistance of the shunt that will convert the galvanometer into an ammeter reading 5.0 amp when the scale reading is 10 divisions?

9. A voltmeter *A* of resistance 10,000 Ω and one of resistance 5000 Ω are connected in series across a 100-v generator. What is the reading of *A*?

C

10. The coil of a galvanometer has an area of 5.0 × 10⁻⁴ m² and the number of turns of wire is 150. (*a*) What is its magnetic moment when the current through it is 1.00 milliamp? (*b*) What torque does it experience in a field of flux density 0.15 weber/m² when the plane of the coil is parallel to the field?

11. A voltmeter with a resistance of 200 Ω is connected to a dry cell of emf 1.50 v and internal resistance 0.050 Ω. What is the percentage error in assuming that the reading of the voltmeter will be the true emf of the cell?

12. Plot the percentage difference between the voltmeter reading and the true emf versus voltmeter resistance when the terminal voltage of a 1.5-v cell with internal resistance of 0.050 Ω is measured with voltmeters of resistances 5.0 × 10² Ω, 5.0 × 10³ Ω, 5.0 × 10⁴ Ω, 5.0 × 10⁵ Ω, and 5.0 × 10⁶ Ω.

13. One hundred turns of wire are wound onto the edge of a square board 0.20 m on each side. The current in the wire is 2.40 amp. (*a*) What torque acts on the coil when it is placed parallel to a magnetic field of induction 150 × 10⁻⁶ weber/m²? (*b*) What torque acts on the coil when it is at an angle of 60°

with the field? (*c*) What is the magnetic moment of the coil?

14. (*a*) What speed will a proton acquire in passing between two points that differ in potential by 2000 v? (*b*) What is the radius of curvature of the path on which the proton is constrained to move when traveling at right angles to a magnetic field of induction $B = 0.20$ weber/m^2.

15. Electrons move in an undeflected beam in crossed electric and magnetic fields, that is, in a region in which the electric field, the magnetic field, and the direction of the beam are at right angles to each other. If the velocity of the electrons is 2.0×10^6 m/sec and the magnetic induction is 5.0×10^{-2} weber/m^2, what is the electric field strength?

16. If an electron has a cyclotron frequency of 4.0×10^8/sec in a magnetic field, what is the magnetic induction?

17. A cyclotron uses an oscillator of frequency 1.0×10^7 cycles/sec. What magnetic induction is needed to accelerate (*a*) protons, (*b*) deuterons, (*c*) alpha particles? See Table 46-1, page 753.

For Further Reading

Cahill, Laurence J., Jr., "The Magnetosphere," *Scientific American,* March, 1965, p. 58.

CHAPTER TWENTY-EIGHT

Sources of Magnetic Fields

In the last two chapters you learned that magnetic fields accompany moving charges. The magnetic lines of flux surrounding a current-carrying straight conductor are circles concentric with the conductor. Magnetic lines close upon themselves. The right-hand rule predicts their direction. A permanent magnet sets up a magnetic field in its neighborhood, the external lines running from the north magnetic pole to the south. Currents are responsible for the magnetic field of a permanent magnet, too, but they are currents associated with the atomic magnets or dipoles. Magnetizing a bar of iron consists of aligning these elementary magnets in the direction of an external magnetic field.

Unvarying magnetic fields exert deflecting forces upon moving charged particles regardless of whether the particles are conduction electrons drifting along a wire, free electrons in a beam, or cosmic-ray protons. The right-hand motor rule predicts the direction of the force on moving charges. We saw that a small current-carrying straight conductor—or a small electron beam—could be used to measure the magnetic induction B: one turns the conductor in the field until the force on it is a maximum; B is defined as this maximum force per unit current per unit length. The direction of the magnetic field is the direction in which the electric current in the conductor must point to experience *no* force. Alternatively, B is the maximum force per unit charge per unit velocity acting upon charged particles in a beam. The number of magnetic lines per unit area perpendicular to the field, drawn to represent the magnetic field at a point, equals the magnitude of B at that point. B is therefore called the magnetic flux density as well as the magnetic induction.

In this chapter we shall discuss the sources of the magnetic field. First, we shall consider how to determine the magnetic induction from the current that sets up the magnetic field. Then we shall discuss magnetic materials from which electromagnets and permanent magnets are made and see how their presence changes the magnetic field of a current in a conductor. Naturally, we shall have to talk about the atomic basis of magnetism in doing so.

The Magnetic Fields of Currents in Conductors

Shortly after Oersted discovered that an electric current is accompanied by a magnetic field, the famous French scientist André Ampère derived equations for the magnetic induction of the field set up by currents in conductors of various shapes in a vacuum or (approximately the same) in air.

Figure 28-1 represents a wire carrying an electric current I. Consider the current in a short section of wire Δl. Ampère assumed that the tiny bit of current in Δl makes a contribution ΔB to the magnetic induction at the point A, where

$$\Delta B = k\frac{I\Delta l \sin \theta}{r^2} \qquad \text{(Element of current)}$$

The angle θ is the smallest angle between the short section of wire and a line connecting Δl and A. In the mks system of units,

$$k = \frac{\mu_0}{4\pi}$$

where μ_0, called the **permeability of a vacuum,** has the value $4\pi \times 10^{-7}$ kg-m/coul². This equation resembles that for the electric field strength of an isolated point charge in a vacuum

$$E = \frac{q}{4\pi\epsilon_0 r^2}$$

To find the magnetic induction at a point, we must use Ampère's equation to find the contribution ΔB of each portion of current-carrying conductor in the neighborhood and then add all such contributions. The resulting equations can be tested because, unlike Ampère's equation for an element of current, they apply to experimentally realizable situations. As we often do in physics, we judge the correctness of Ampère's assumption by the correctness of these equations derived from it.

Let us illustrate the use of Ampère's law by calculating the magnetic induction at the center of a circular loop of wire of radius r with a current I (Fig. 28-2). The magnetic fields of the currents in the straight portions of the wire by which the current enters and leaves cancel each other rather completely at the center. We have to concern ourselves only with the magnetic fields of the many tiny elements of length Δl in the circular loop. We have labeled three of these in the figure, but actually there are very many more in the entire circumference of the loop. Each contributes to the magnetic induction at the center an amount

$$\Delta B = k\frac{I\Delta l \sin \theta}{r^2}$$

Our problem is to add up all of these contributions to B:

$$B = (\Delta B)_1 + (\Delta B)_2 + (\Delta B)_3 + \cdots$$

It is rather easy to do this for the circular loop. First of all, $\theta = 90°$ for all of the tiny elements of current because the distance from each to the center is a radius of the circle and a radius is always perpendicular to an arc, however small the arc. Hence, $\sin \theta = \sin 90° = 1$. Also, r is the same for each element.

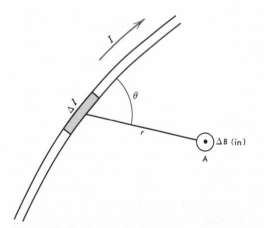

Fig. 28-1. *The magnetic induction due to a small section of a wire carrying an electric current.*

Since k, I, $\sin \theta$, and r are the same for each element,

$$B = \frac{kI}{r^2}[(\Delta l)_1 + (\Delta l)_2 + (\Delta l)_3 + \cdots]$$

But the sum $[(\Delta l)_1 + (\Delta l)_2 + (\Delta l)_3 + \cdots]$ is nothing more than the total distance around the circular loop or $2\pi r$. Thus

$$B = \frac{kI(2\pi r)}{r^2} = \frac{\mu_0 I(2\pi r)}{4\pi r^2}$$

$$B = \frac{\mu_0 I}{2r} \qquad \text{(Circular loop)}$$

Example. The radius of the circular loop in Fig. 28-2 is 0.50 m and the current is 10 amp. What are (a) the magnitude of the magnetic induction at the center of the loop and (b) the direction?

$$\text{(a)} \quad B = \frac{(4\pi \times 10^{-7}\ \text{kg-m/coul}^2)(10\ \text{amp})}{2(0.50\ \text{m})}$$

$$= 4\pi \times 10^{-6}[(\text{kg-m/coul}^2)(\text{amp})]/\text{m}$$

The combination of units $1[(\text{kg-m/coul}^2)(\text{amp})]/\text{m}$ equals 1 newton/amp-m. To show this we first substitute 1 amp-sec for 1 coul:

$$B = 4\pi \times 10^{-6}[(\text{kg-m/amp}^2\text{-sec}^2)(\text{amp})]/\text{m}$$
$$= 4\pi \times 10^{-6}\ (\text{kg-m/sec}^2)/\text{amp-m}$$
$$= 4\pi \times 10^{-6}\ \text{newton/amp-m}$$

(b) By the right-hand rule, the magnetic induction is directed *into* the page.

If, instead of a single circular loop, we have a thin, tightly wound circular *coil* of N turns, the magnetic induction at the center of the coil is simply N times that of a single turn or

$$B = \frac{\mu_0 NI}{2r} \qquad \text{(Thin coil)}$$

The physics of calculating the magnetic induction for *any* current distribution is exactly the same as in the procedure that we have used for a circular loop. The geometry is usually such as to make the calculation mathematically more complicated, however. We shall merely give the results of carrying out

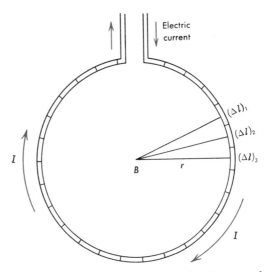

Fig. 28-2. *The magnetic induction at the center of the circular loop is given by* $B = \mu_0 I/2r$ *and is directed into the page.*

this procedure for two other common arrangements—a long straight wire and a solenoid.

The magnetic induction at a distance r from a long, straight wire is given by

$$B = \frac{\mu_0 I}{2\pi r} \qquad \text{(Straight conductor)}$$

Example. A power wire carries an electric current of 100 amp in a south-to-north direction. (a) What is the magnetic induction, due to the current, at a point 1 m below the wire? (b) What is the direction of the field at this point?

$$\text{(a)} \quad B = \frac{(4\pi \times 10^{-7}\ \text{kg-m/coul}^2)(100\ \text{amp})}{(2\pi)(1\ \text{m})}$$

$$= 2.0 \times 10^{-5}\ \text{newton/amp-m}$$
$$= 2.0 \times 10^{-5}\ \text{weber/m}^2$$

(b) By the right-hand rule, the magnetic field is found to be directed westward.

A solenoid, as you will recall, as a long, tightly wound, helical coil. The magnetic induction at any interior point, not too near either end (Fig. 28-3), in a long solenoid is

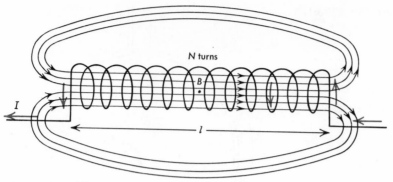

Fig. 28-3. *The magnetic field of a current-carrying solenoid. The magnetic induction in the interior of the long coil is* $B = \mu_0 NI/l$.

found by Ampère's equation to be

$$B = \mu_0 \frac{NI}{l} \quad \text{(Interior of solenoid)}$$

where l is the length of the coil and N the number of turns.

You can determine the direction of the magnetic field in the solenoid by applying the right-hand rule. Pretend that you grasp the right end of the coil with your right hand so that your thumb points in the direction of the current as indicated by the arrow. Your fingers will point out of the coil. Its right end is a north pole.

Example. A solenoid is 1.0 m long and has 1000 turns. If the current is 20 amp, what is the magnetic induction in the coil?

$$\frac{N}{l} = \frac{1000 \text{ turns}}{1.0 \text{ m}} = 10^3 \text{ turns/m}$$

$$B = (4\pi \times 10^{-7} \text{ kg-m/coul}^2)$$
$$(20 \text{ amp})(10^3 \text{ turns/m})$$
$$= 8\pi \times 10^{-3} \text{ newton/amp-m}$$
$$= 8\pi \times 10^{-3} \text{ weber/m}^2$$

The Forces Between Parallel Conductors

In Chapter 26, you learned that two long conductors, parallel to each other, exert forces on each other—attractive forces if the currents

are in the same direction, repulsive forces if the currents are oppositely directed. Now we are in a position to calculate the magnitude of these forces.

Let the current in conductor *a* (Fig. 28-4) be I_a and that in conductor *b*, I_b. Let the conductors be separated by a distance *r*. We shall calculate the force *F* exerted by *a* on a length l of I_b. According to our basic equation for the force on a current-carrying straight conductor we have

$$F = BI_b l \tag{1}$$

where *B* is the magnetic induction set up at conductor *b* by the current in *a*. *B* is perpendicular to the current I_b. For a straight conductor,

$$B = \frac{\mu_0 I_a}{2\pi r} \tag{2}$$

Hence, equation (1) becomes

$$F = \frac{\mu_0 I_a I_b l}{2\pi r} \tag{3}$$

Example. In Fig. 28-4, I_a and I_b are both 10 amp, *r* is 0.050 m, and l is 1.0 m. What is the attractive force between the two conductors?

$$F = \left[\frac{4\pi \times 10^{-7} \text{ kg-m}}{\text{coul}^2} \right]$$
$$\left[\frac{(10 \text{ amp})(10 \text{ amp})(1.0 \text{ m})}{2\pi(0.050 \text{ m})} \right]$$

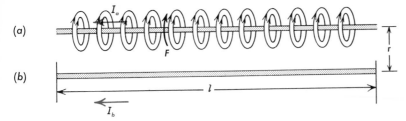

(a)

(b)

Fig. 28-4. *Calculating the attractive force between parallel currents in the same direction. Only the magnetic field due to I_a is shown. I_b has a similar magnetic field surrounding it.*

$$= 4.0 \times 10^{-4}\ (\text{kg-m/amp}^2\text{-sec}^2)(\text{amp}^2)$$
$$= 4.0 \times 10^{-4}\ \text{kg-m/sec}^2 = 4.0 \times 10^{-4}\ \text{newton}$$
$$= 4.1 \times 10^{-2}\ \text{gwt}$$

Forces between parallel conductors can be measured by a *current balance*. Knowing the force, one can determine currents from mechanical measurements with great precision. A current balance designed for use in the student laboratory is shown in Fig. 28-5. Research-grade current balances are used by laboratories such as the National Bureau of Standards for precise calibration of moving-coil electrical meters.

The Ampere

When we introduced the coulomb as the unit of charge in Chapter 21 and the ampere as the unit of current in Chapter 23, we promised to give more fundamental definitions of both of these units. Now we can do this.

Suppose that the *same* current is sent through two long parallel wires, separated in vacuum by a distance of one meter, and that the current is adjusted until the force exerted on one meter of either conductor is just 2×10^{-7} newton, a number arbitrarily chosen for reasons of convenience. Then by definition the current in either conductor is *one ampere*. The ampere is thus based fundamentally on the action of a magnetic field on a current-carrying conductor.

Having defined the ampere in this way, we can then logically define the coulomb as the charge transferred through any cross section of a conductor in one second by a current of one ampere. Definitions of the volt as the joule per coulomb, and of the other electrical quantities then follow logically.

What can we now say about the quantities μ_0, which appeared for the first time in Ampère's equation, and ϵ_0, which appeared in Coulomb's law of electrostatics?

The numerical value $4\pi \times 10^{-7}$ kg-m/coul^2 chosen for μ_0 is *arbitrary*, chosen for reasons of computational convenience. Having made that choice, we are led to the above definitions of the ampere and the coulomb. The numerical value of 8.85×10^{-12} coul^2/newton-m^2 for ϵ_0 is *not* arbitrary, but is experimentally determined on the basis of the definition of the ampere and the coulomb. How is this done? You recall that the capacitance C of a parallel-plate vacuum capacitor is

$$C = \frac{\epsilon_0 A}{d}$$

where A is the area of a plate and d is the separation between the plates. The capacitance is q/V, the charge on either plate divided by the potential difference. Hence

$$\epsilon_0 = \frac{d}{A} C = \frac{d}{A} \frac{q}{V}$$

By measuring d, A, q, and V, we can evaluate ϵ_0.

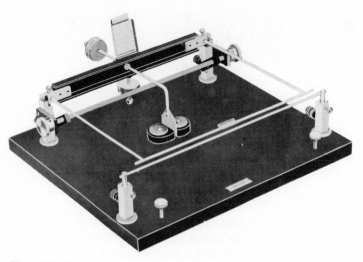

Fig. 28-5. *A student-model current balance. (Courtesy Sargent-Welch Scientific Company.)*

Magnetic Materials

So far, we have discussed magnetic fields set up by conduction currents; the fields were in free space or, what is almost the same, in air. Now we must consider the effect of bringing magnetic materials into the field. Magnetic materials in general change the value of B—the density of magnetic lines—that would otherwise exist at a point in the field. For most materials—zinc, aluminum, rubber, and glass, for example—the value of B is increased slightly; for some—iron, nickel, cobalt—B is greatly increased; and for a few—antimony, bismuth, copper—it is decreased slightly.

You will recall that in our earlier discussion of dielectrics (Chapter 22) we found that surface charges appeared on insulators in an electric field because the electric dipoles became aligned with the electric field. The effect of the surface charges was to *weaken* the field that would otherwise exist. We called the bound surface charge per unit area (taken perpendicular to the field) the polarization P. The parallel-plate capacitor with slabs of the dielectric material inserted between its plates

furnished us with a useful experimental device for studying dielectrics.

In magnetic materials, the elementary *magnetic* dipoles take the place of the electric dipoles in insulators. The atoms of most of the elements act like tiny magnets; they have a magnetic moment (Chapter 27) and experience a torque when placed in a magnetic field. Much evidence has been accumulated which convinces physicists that the magnetic moments are mainly associated with the spin of the electrons in the atoms, although the orbital motion of the electrons and the spin of the nuclei also contribute. The magnetic moments of most atoms are small because the electrons in an atom pair off—one electron with its spin axis pointing up and one with its spin axis down—thus canceling each other's magnetic fields. Usually no more than one electron spin is left uncanceled; such atoms are weakly magnetic. In an applied magnetic field, the magnetic dipoles align themselves with the field and, adding their magnetic fields to the external field, slightly increase the external density of magnetic lines. Atoms of this kind are called *paramagnetic*.

Atoms of iron, cobalt, nickel, and other

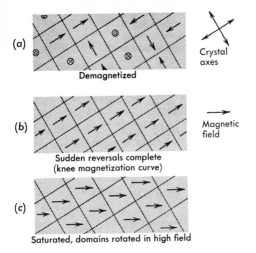

(a) **Demagnetized**

Crystal axes

(b) **Sudden reversals complete (knee magnetization curve)**

Magnetic field

(c) **Saturated, domains rotated in high field**

Fig. 28-6. Magnetic domains. (a) In unmagnetized iron, the dipoles in different domains point in different directions. (b) Partly magnetized iron. (c) Saturated iron. The dipoles line up parallel to the field. (Courtesy of Dr. R. M. Bozorth.)

elements near iron in the periodic table (Appendix I) are in a special category of magnetic materials. Each iron atom has *four* uncanceled electron spins and has a large magnetic moment. Moreover, groups of iron atoms, called domains, act cooperatively in a "unit voting" arrangement. In a magnetic field, the atoms in the domain—held together by what are called exchange forces—swing into line with the field as a unit. When one atom swings over, all the others in the domain follow it, greatly strengthening the magnetic field external to the magnet (Fig. 28-6). Materials like iron, cobalt, nickel, and their alloys are called *ferromagnetic*. Below a certain temperature—called the *Curie* temperature—these materials are magnetized even in the absence of a magnetizing field; above the Curie temperature, which is 770°C for iron, they lose this magnetization.

Since the magnetizing of iron consists in lining up billions and billions of spinning electrons, you might expect the change in the

angular momentum of the total sample to be observable upon magnetization. This effect—the Einstein–de Haas effect—has been observed.

A few materials—such as antimony, bismuth, and copper, mentioned above—are *diamagnetic.* Their magnetic moments due to electron spin are zero, and the effect of their orbital electron motion in a magnetic field is to oppose the applied field and reduce the external density of magnetic lines.

Ampère gave us a way of regarding magnetized materials that enables us to see the similarity between the fields caused by the elementary atomic magnets that we have just discussed and the fields caused by conduction currents. Consider a thin slice of a uniformly magnetized bar magnet (Fig. 28-7). The magnetic dipoles are replaced by *Amperian currents* whose total contribution to the external magnetic field is equivalent to that of the aligned iron atoms in the slice. Notice that adjacent Amperian currents are oppositely directed in the interior of the slice and cancel each other there. At the edge, there are uncanceled portions of Amperian currents. These uncanceled portions are equivalent to a single large Amperian current around the edge of the slice. *The effect is just the same as if the slice of magnetic material were a current-carrying circular conductor.* An entire cylindrical bar magnet can be con-

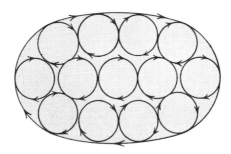

Fig. 28-7. Amperian currents in a slice of a bar magnet are equivalent to a large Amperian current around the edge.

sidered to be made up of stacked slices of this kind, each slice with its Amperian current around the edge. The total external magnetic field of the bar magnet is the sum of the contributions of these Amperian currents.

Consider now a long solenoid with an iron core. The magnetic induction that exists when a conduction current is sent through the turns of the solenoid arises from two sources: (1) the conduction currents themselves and (2) the Amperian currents that represent the aligned magnetic dipoles of the iron. The contribution to the magnetic induction of the Amperian currents adds to that of the conduction currents, giving us a flux density that is often thousands of times what it would be if the iron core were removed.

We define the *relative permeability* κ_m of the iron core as the ratio of the magnetic induction B when iron fills the core to the magnetic induction B_0 when the core of the solenoid is empty space or air.

$$\kappa_m = \frac{B}{B_0}$$

The relative permeability—measured by comparing the flux density in a solenoid before and after the material is inserted—is analogous to the dielectric constant for insulators, the ratio of the capacitance of a parallel-plate capacitor with the dielectric

Table 28-1. Relative Permeabilities[a] at a Magnetic Flux Density of 0.002 weber/m²

Aluminum	1
Copper	1
Lead	1
Nickel	100
Tin	1
Zinc	1
Magnetic iron	200
Permalloy (78.5% nickel, 21.5% iron)	8000
Mumetal (75% nickel, 2% chromium, 5% copper, 18% iron)	20,000

[a] From the *American Institute of Physics Handbook.*

material to that without it. We shall discuss in the next chapter how the flux density is measured by electromagnetic induction. The relative permeabilities of paramagnetic materials are slightly greater than 1, those of ferromagnetic materials in the hundreds or thousands, and those of diamagnetic materials a little less than 1. The relative permeabilities of a number of materials at a flux density of 0.002 weber/m² are given in Table 28-1.

Magnetization Curves

The relative permeability of ferromagnetic materials such as iron is not constant, but decreases as the magnetic materials approach saturation. In order to discuss this kind of behavior of magnetic materials, we first define the magnetizing field intensity *H*. The magnetic induction B_0 in the interior of a solenoid whose core is empty space, we saw earlier in the chapter, is

$$B_0 = \mu_0 \frac{NI}{l}$$

We define H_0 in empty space as

$$H_0 = \frac{B_0}{\mu_0} \qquad \text{(Empty space)}$$

Inside a magnetized iron core in a short coil, *H* is less than B/μ_0 by a term involving the strength of the magnetic poles at the end. However, if the coil and core are long ones, the poles at the ends do not affect the value of *H* in the interior. Alternatively, a toroidal (doughnut-shaped) coil and core can be used in which the poles do not appear because the core is continuous. *For solenoid or toroid cores in which the poles can be neglected, $H = H_0$, and*

$$H = \frac{B_0}{\mu_0} = \frac{NI}{l} \qquad \text{(Long coil or toroid)}$$

H is expressed in ampere-turns per meter. The ratio of the magnetic induction *B* in a sample of magnetic material to the magne-

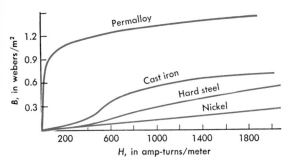

tizing field strength *H* is defined as the permeability μ of the material.

$$\mu = \frac{B}{H} = \kappa_m \mu_0$$

Plotting the magnetic induction *B* in the material against the magnetizing field strength *H*, we obtain the *magnetization curve* for the material (Fig. 28-8). At high magnetizing field intensities, the domains are almost all aligned in the direction of the magnetic field, and the material is said to be saturated. Figure 28-9 is a magnetization curve for

a wrought-iron specimen. The permeability is plotted on the same graph. We see that μ rises to a maximum at a magnetizing field intensity of about 200 amp-turns/m and then decreases as the iron saturates.

Example. The current in a long solenoid of 300 turns and length 1.0 m is 1.0 amp. The core of the solenoid is wrought iron of the kind whose magnetization curve is given in Fig. 28-9. The cross-sectional area of the core is 5.0×10^{-4} m². (*a*) What is *B* in the core? (*b*) How many magnetic lines pass through the core? (*c*) What is the permeability of the iron at this magnetizing field intensity?

$$(a) \quad H = \frac{NI}{l} = \frac{(300 \text{ turns})(1.0 \text{ amp})}{1.0 \text{ m}}$$

$$= 300 \text{ amp-turns/m}$$

At a magnetizing field intensity of 300 amp-turns/m, the magnetic induction *B*, according to the magnetization curve of Fig. 28-9, is approximately 1.3 weber/m².

(*b*) The magnetic induction is also the density of magnetic lines. Hence the number of magnetic lines ϕ (Greek *phi*) through the core is

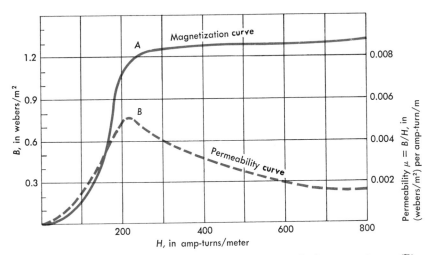

Fig. 28-9. (A) *Magnetization curve for one wrought-iron specimen.* (B) *Its permeability B/H for different magnetizing fields.*

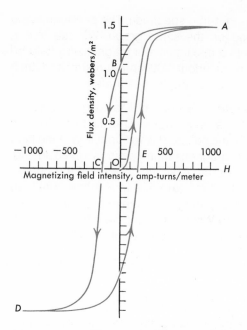

Fig. 28-10. Hysteresis. *The magnetization of the iron lags behind the magnetizing field.*

$$\phi = (1.3 \text{ weber/m}^2)(5.0 \times 10^{-4} \text{ m}^2)$$
$$= 6.5 \times 10^{-4} \text{ weber}$$

(c) $\quad \mu = \dfrac{B}{H} = \dfrac{1.3 \text{ weber/m}^2}{300 \text{ amp-turns/m}}$

$$= 4.3 \times 10^{-3} \text{ (weber/m}^2)/(\text{amp-turn/m})$$

When we gradually increase the current through the coil of a long solenoid having an iron core, the core gradually becomes magnetized as indicated by the curve OA (Fig. 28-10). When we then gradually decrease the current to zero, the flux density follows a *different* curve AB because the domains do not return to their original orientations when the field is reduced. When the magnetizing field is zero, the steel retains a flux density OB; it is a permanent magnet. By reversing the current and increasing it, we can gradually reduce the magnetic flux to zero, and then remagnetize the core in the opposite direction along BCD. Finally, by again reversing the current, we can increase the flux density

along DEA. (OB is called the remanence of the material and OC its coercive force.) The lagging of the flux density behind the magnetizing field intensity is called *magnetic hysteresis.*

When there is an alternating current in the windings of an electromagnet, the continual oscillations of the magnetic particles as they go through the hysteresis cycle generate heat and waste energy. In advanced textbooks it is shown that the area enclosed by the hysteresis loop ACDEA is proportional to the energy dissipated in one cycle of the magnetizing field intensity. The cores of an electromagnet used for alternating currents should have narrow hysteresis curves of small area to minimize this waste.

Uses of Magnetic Materials

Great effort has been exerted to devise new alloys that have magnetic properties best suited to special uses. Permanent magnets are made of hard steel or of other alloys. Their permeabilities are low, but they have high remanences: they retain their magnetism after the magnetic field is removed. The nickel-iron alloy *permalloy* can be magnetized to saturation by very weak fields (Fig. 28-8). It has many applications in telephony. *Ferrites*—of which the ancient lodestone is an example—are metallic compounds of iron, oxygen, and small amounts of other elements such as nickel, copper, manganese, cobalt, and aluminum. Ferrites combine high magnetization with high electrical resistance. Ferrite materials can be quickly magnetized and demagnetized with relatively small heating. They are used in computers, high-frequency transformers, and other devices.

An electromagnet consists of one or more coils wound on a core made of a material that has high permeability. We saw in the last chapter that large electromagnets are used in accelerators, such as cyclotrons, to bend the ions into circular paths. Electromagnets have many technological applications for two reasons: First, they can be strongly magnetized and hence exert great forces on their "armatures." Second, the strength of their magnetization can be readily controlled. One very important application of electromagnets is in *relays*. In Fig. 28-11 a weak current through the windings magnetizes the iron core so that it pulls the armature down, closing one electrical circuit and opening another. Thousands of relays like this are employed in automatic telephone systems, in computers, and the like.

We have already seen that the armatures and field magnets of electric motors are made of iron. The iron increases the density of magnetic lines and channels them through the current-carrying coils of the armature. As a result, the magnetic induction B in the vicinity of the armature windings is thousands of times greater than it would otherwise be. The force on the windings and the torque on the armature are correspondingly greater than they would be if the windings were surrounded by air.

In the next chapter, we shall see that the properties of magnetic materials play an important role in the design of electrical generators and transformers.

Fig. 28-11. *A relay. When a weak current is sent through the coil, the armature of the relay moves down, opening one electrical circuit and closing another. (Courtesy of Potter and Brumfield.)*

Summary

The magnetic induction at the center of a narrow coil of N turns, each of radius r, is given by $B = \mu_0 NI/2r$. The magnetic induction B at a distance r caused by a current I in a long straight wire is given by $B = \mu_0 I/2\pi r$. The magnetic induction at any point inside a solenoid of N turns and length l is given by $B = \mu_0 NI/l$.

Two long parallel conductors exert a force per unit length on each other given by $F/l = \mu_0 I_1 I_2/2\pi r$.

One ampere is the current in each of two long parallel conductors one meter apart such that either conductor exerts on the other a force of 2×10^{-7} newton per meter.

The magnetic induction in the iron core of a long solenoid is set up (1) by the conduction currents in the windings and (2) by the Amperian currents that represent the magnetization of the iron. The relative permeability of a magnetic material is the ratio of the magnetic induction within the solenoid when the core is made of the material to the magnetic induction when the core is of air.

The magnetic particles producing the strong magnetization of iron and nickel are spinning electrons. The relative permeability of a ferromagnetic material is much greater than unity. That of a paramagnetic material is slightly greater than unity, and that of a diamagnetic material is slightly less.

The magnetizing field intensity of a long solenoid is $H = NI/l$.

The magnetic permeability μ of a substance is the ratio of the flux density B inside a long core to the magnetizing field intensity H. $\mu = B/H = \kappa_m \mu_0$ where κ_m is called the relative permeability.

Questions

1. How can we test the correctness of Ampère's equation?

2. What are some of the similarities and differences between the process of magnetization of a ferromagnetic material and that of the polarization of a dielectric?

3. Distinguish between magnetic induction (B) and magnetizing field intensity (H). What units does each have?

4. How are the values of μ_0 and ϵ_0 determined?

5. What similarities did Ampère find between a slice of magnetized material and a current-carrying conductor? What is an Amperian current?

6. What is a current balance? How could it be used to determine whether the current in a circuit is one ampere?

7. What does the area within a hysteresis loop represent? What is the desirable shape of such a loop (a) in a transformer, (b) in the "memory core" of a computer?

8. How do these materials differ in their magnetic properties: ferromagnetic, paramagnetic, diamagnetic? How are the differences accounted for?

9. Prove that these units are equivalent: kilogram-meters per square coulomb; newtons per square ampere; webers per ampere-turn-meter.

10. What approximations are involved in computing B due to the current in a "long straight wire"?

Problems

A

1. A long wire carries a current of 1.00 amp. What is the magnetic induction at a point 0.15 m from the wire?

2. What is the magnetic induction at the center of a compact circular coil of 100 turns and radius 0.10 m carrying a current of 0.250 amp?

3. A solenoid has 1500 turns and is 0.50 m long. What current in the coil will produce a magnetic induction equal to that of the earth, $B = 48 \times 10^{-6}$ weber/m^2?

4. What is the magnetic induction inside a solenoid 1.00 m long having 250 turns if the current is 2.00 amp?

5. What is the permeability of a piece of steel which, in a magnetizing field intensity of 350 amp-turns/m, has a magnetic density of 1.2 weber/m^2?

6. What is the relative permeability of the piece of steel in problem 5?

B

7. On a wooden rod 100 cm long and 1.00 cm^2 in cross-sectional area, 5000 turns of wire are wound, evenly spaced along the coil. The current in the wire is 0.50 amp. (a) What is the magnetic induction? (b) What total flux through the coil does the current produce?

8. When a long steel rod is inserted in the solenoid described in problem 4, the flux density is 1.4 webers per square meter. What is the relative permeability of the steel?

9. What is the magnetic induction at the center of an air-core toroid of mean diameter 0.10 m and 500 turns? The current is 1.0 amp.

10. What is the magnetic induction at a point midway between two long parallel straight wires, 10 cm apart, carrying currents in opposite directions, each current being 2.0 amp?

11. Two long straight wires A and B are 20 cm apart, and each carries a current of 100 amp. Find (a) the magnetic induction at any point on B due to the current in A, (b) the force on a section of B 0.20 m long caused by this field.

C

12. Each of two wires, A and B, 2.00 m apart, carries a current of 50 amp. What is the

magnetic induction (*a*) at a point 50 cm from *A* and 150 cm from *B*, (*b*) at a point 200 cm from each wire? The two currents are directed oppositely.

13. A toroid coil 1.0 m in mean circumference, has a cross-sectional area of 25.0 cm² and 500 turns of wire. The current is 1.5 amp. If the toroidal iron core has a relative permeability of 320, find (*a*) the magnetizing field intensity, (*b*) the flux density, and (*c*) the total flux.

14. (*a*) Show how curve *B* in Fig. 28-9, page 453, is obtained from curve *A*. (*b*) From the appropriate curve in Fig. 28-8, sketch the permeability curve of nickel.

15. Calculate the energy loss in one cycle of the hysteresis loop in Fig. 28-10, page 454.

For Further Reading

Berge, Glenn L. and George A. Seielstad, "The Magnetic Field of the Galaxy," *Scientific American,* June, 1965, p. 46.

Bitter, Francis, *Magnets, The Education of a Physicist,* Doubleday and Company, Garden City, 1959.

Bozorth, Richard M., "Magnetic Materials," *Scientific American,* January, 1955, pp. 68–73.

Furth, H. P., M. A. Levine, and R. W. Waniek, "Strong Magnetic Fields," *Scientific American,* February, 1958, p. 28.

Hogan, C. L., "Ferrites," *Scientific American,* June, 1960, p. 92.

Kolm, H. H. and A. J. Freeman, "Intense Magnetic Fields," *Scientific American,* April, 1965, p. 66.

CHAPTER TWENTY-NINE

Induced Electromotive Force

After Oersted discovered that an electric current produces a magnetic field, physicists asked the question, "If electric currents produce magnetic fields, should not magnetic fields produce electric currents?" Among those who considered this matter was Michael Faraday. It took him several years to find the answer. His discovery of *electromagnetic induction* in 1831 is one of the great triumphs of science.

Induced Electromagnetic Force Due to Relative Motion

Electromagnetic induction can be illustrated by a few simple demonstrations. Connect a coil of insulated wire to a galvanometer (Fig. 29-1). Move a bar magnet toward the coil, and the galvanometer needle is deflected, revealing the existence of a current. The current indicates that there is an electromotive force in the circuit. When the motion ceases, the electromotive force and the current cease also. Move the magnet away from the coil, and the galvanometer needle will be deflected in the opposite direction, showing that the electromotive force and the current are reversed. Move the magnet slowly toward the

coil, and the deflection of the galvanometer needle will be small. Move it rapidly, and a much larger deflection results. Next, replace the magnet by a second coil connected to a battery (Fig. 29-2). When you move this coil toward the first one, or away from it, the galvanometer needle deflects as it did when the magnet was shifted. As a third experiment, keep the second coil at a fixed position, and open the circuit so as to interrupt the current. Then the galvanometer needle is deflected as though the coil were removed. When you close the circuit again, the deflection will be in the opposite direction.

Consider the field of the magnet used in the first experiment. When the north pole is at a considerable distance from the coil, few magnetic lines thread through the coil. When the pole is moved closer, the *magnetic flux,* or number of lines threading through the coil, increases; when the magnet is moved farther away, the number diminishes. When the magnet moves rapidly, the number of lines threading through the coil changes rapidly. Similarly, moving the current-carrying coil (Fig. 29-2) or establishing a current in it causes magnetic lines to be thrust into the other coil. In both cases emf's are induced.

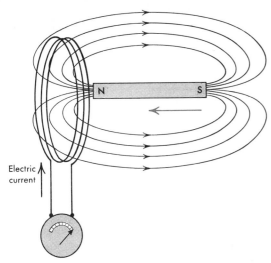

Electric current

Fig. 29-1. Induced electromotive force. Changing the magnetic flux through the coil induces an electromotive force in the coil, establishing a current in the circuit.

An electromotive force is induced in a circuit whenever the flux of magnetic lines threading through that circuit is changed; if the circuit is closed, there is an induced current.

What is the direction of the induced current? When a magnet pole is moved toward one face of a coil, the current induced in the coil produces a magnetic field. This field always *opposes the change of magnetic flux that is occurring.* For example, move the north pole of a magnet closer to one face of a coil. The induced electric current will be counterclockwise, as viewed from the direction in which the magnet is approaching, and will oppose the increase of flux through the coil (Fig. 29-3). Remove the bar magnet, and the induced electric current in the coil will be clockwise, opposing the decrease of flux. This rule is expressed by Lenz's law, as follows:

Whenever a current is induced, its magnetic field opposes the change of flux.

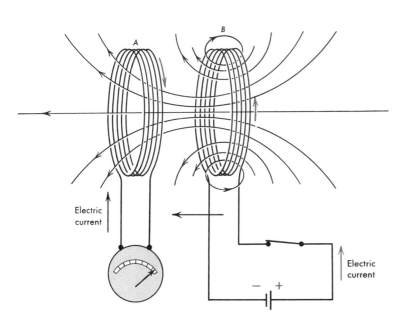

Electric current

Electric current

− +

Fig. 29-2. Induced electromotive force. Moving the coil B nearer to A increases the flux through A and induces an electromotive force in it.

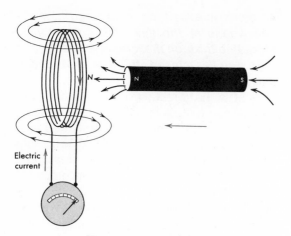

Fig. 29-3. *The magnetic field of the induced current opposes the change of flux.*

coil having *N* turns when a change of magnetic lines $\Delta\phi$ (delta phi) occurs in a time Δt is found experimentally to be

$$\mathscr{E} = -N\frac{\Delta\phi}{\Delta t}$$

When the magnetic flux is given in webers and the time in seconds, the induced electromotive force is in volts. One volt equals one weber-turn per second. The negative sign indicates that the induced emf is in such a direction as to establish a current that *opposes* the change of magnetic flux.

Example. By moving a bar magnet toward a coil having 1000 turns, you increase the magnetic flux threading through the coil from 0 to 1.2×10^{-5} weber in 0.20 sec. (*a*) What average emf is induced? (*b*) What average current is induced if the resistance of the coil is 15 Ω?

$$\frac{\Delta\phi}{\Delta t} = \frac{1.2 \times 10^{-5} \text{ weber}}{0.20 \text{ sec}}$$

$$= 6.0 \times 10^{-5} \text{ weber/sec, rate}$$
$$\text{of change of magnetic flux}$$

(*a*) $\mathscr{E} = -(1000 \text{ turns})(6.0 \times 10^{-5} \text{ weber/sec})$
$$= -0.060 \text{ v}$$

Lenz's law might have been predicted from the principle of the conservation of energy. When you move a magnet toward a coil and thus induce a current in its windings, the induced current heats the wire. In order to supply the energy to do this, you must *do work* in overcoming an *opposing force.* If the force did not oppose the motion, you would create energy. Thus the magnetic field of the induced current must oppose the change.

Lenz's law and the right-hand rule can be used to determine the direction of an induced electric current. When the north pole of a magnet is moved closer to a coil, the induced electric current is accompanied by a field which opposes the increase of flux through the coil. A magnetic north pole is produced on the nearer face. To cause this north pole, magnetic lines must emerge from this face of the coil as shown in Fig. 29-4. Now grasp the coil with your right hand, so that your fingers point in the direction of the induced magnetic field. Your thumb will point in the direction of the electric current, that is, counterclockwise.

Changing the number of magnetic lines that thread through a circuit induces a current in it *if the circuit is closed,* but the change *always* induces an electromotive force. The average electromotive force $\mathscr{E}$ induced in a

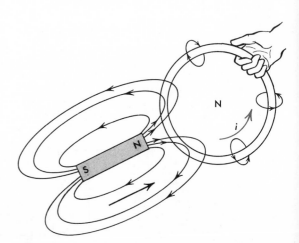

Fig. 29-4. *The direction of the induced electric current may be found by using Lenz's law and the right-hand rule.*

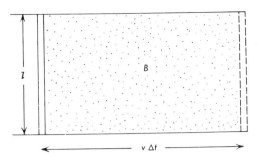

Fig. 29-5. *A conductor cuts magnetic lines and an emf is induced.*

(b) $\quad I = \dfrac{V}{R} = \dfrac{0.060 \text{ v}}{15 \ \Omega} = 0.0040 \text{ amp}$

Electromotive Forces Induced by "Cutting" Magnetic Lines

Electromagnetic induction may be viewed from a different standpoint. When a magnet is moved toward a coil, thrusting magnetic lines into it, we may think of the magnetic lines as being "cut" by wires of the coil. Thus, we say that an electromotive force of one volt is induced when a single conductor *cuts* one weber per second.

In Fig. 29-5, a conductor—for example, the metal wing of an airplane moving in the earth's magnetic field—moves across a magnetic field, cutting magnetic lines. In a time Δt, the conductor of length l, moving with a velocity v, sweeps across an area $(v\ \Delta t)l$. The magnetic flux density—the number of lines per unit area—perpendicular to the conductor is B.

$$B = \frac{\Delta \phi}{\Delta A}$$

The number of magnetic lines cut in time Δt is

$$\Delta \phi = B\ \Delta A = B(v\ \Delta t)l.$$

Therefore

$$\mathscr{E} = -\frac{\Delta \phi}{\Delta t} = -Bvl$$

An emf is also induced when a circuit is deformed in a magnetic field, thus cutting lines. Figure 29-6 represents a circuit, one side of which is movable. If *AB* slides to the right, the rate of decrease of magnetic lines threading through the circuit must be equal to the rate at which the conductor *AB* cuts the lines. Again, $\mathscr{E} = Blv$. Since current is induced during the motion, work must be done against an opposing force. According to Lenz's law, the induced current sets up a magnetic field that opposes the change of flux—in this example, the decrease of magnetic lines through the circuit. Use of the right-hand rule proves that in the figure the induced electric current in the sliding conductor must be directed toward you.

Let us summarize what we have learned about induced electromotive force. An electromotive force is induced in a conductor in any of these situations:

1. There is motion with respect to a stationary conductor of a permanent magnet or a coil carrying a current or when a current grows or declines in a coil near the conductor: that is, when the magnetic flux through a stationary conductor changes.
2. The conductor moves across a magnetic field.
3. The area enclosed by a circuit is changed

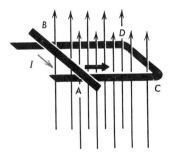

Fig. 29-6. *Cutting magnetic lines induces an emf. As* AB *slides to the right, the induced electric current along* AB *is directed toward you and out of the page.*

in a magnetic field, or a conductor is deformed — bent, twisted, crumpled — in a magnetic field.

In all of these cases, magnetic lines are "cut." The rate of cutting of magnetic lines equals the induced electromotive force in the single conductor; if N conductors are in series, the induced emf's in them are combined algebraically. At low velocities — appreciably less than the speed of light — the induced emf in a conductor moving across a magnetic field is Blv. Near the speed of light, relativity theory must be used to calculate the induced emf.

Measuring Magnetic Flux Density

In Chapter 26, we discussed a way to determine B, the magnetic induction or the magnetic flux density at a point in a magnetic field, by measuring the *force* per unit current per unit length exerted on a current-carrying conductor at that point. Now we shall discuss how to determine B by measuring the *charge* that circulates in a circuit when magnetic lines are cut by the circuit and an electromotive force is induced in it.

In Fig. 29-7, the plane of a small coil is initially at right angles to a magnetic field. If the magnetic flux density at that point is B, the number of magnetic lines passing through the coil is $B \cdot A$, where A is the cross-sectional area of the coil. Suppose the coil is

turned 90° in a time t to the position shown by dotted lines. All of the lines initially passing through it are cut. The average induced emf in the circuit is

$$\mathscr{E} = N\frac{\Delta\phi}{\Delta t} = N\frac{B \cdot A}{t}$$

where N is the number of turns on the coil. If the resistance of the coil and the galvanometer G is R, the average current induced is

$$I = \frac{\mathscr{E}}{R} = \frac{N}{R}\frac{(B \cdot A)}{t}$$

But the charge q transferred by a current I in time t is

$$q = I \cdot t$$

and therefore

$$q = \frac{N}{R}(B \cdot A)$$

Hence

$$B = \left(\frac{R}{NA}\right)q$$

We can measure the charge circulated when the coil turns 90° by a properly calibrated ballistic galvanometer, an electrical meter whose coil has a larger moment of inertia and whose spring is much weaker than in current-measuring galvanometers. When a charge is suddenly circulated through the ballistic galvanometer, the needle swings to a maximum deflection, whose magnitude is proportional to the change. Knowing the charge and the resistance, the cross-sectional area, and the number of turns, we can determine B. A device of this kind, consisting of a rotatable exploring coil and a ballistic galvanometer, calibrated to read webers per square meter, is called a *fluxmeter*.

Some Applications of Electromagnetic Induction

A laboratory *induction coil* or *spark coil* generates high voltages, drawing its energy from

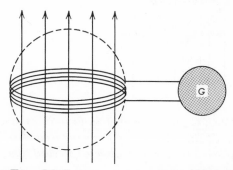

Fig. 29-7. *Measuring magnetic flux density.*

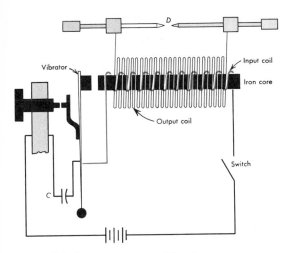

Fig. 29-8. *Spark coil. The battery current magnetizes the core so that it attracts the armature of the vibrator and opens the circuit periodically. The varying flux induces high emf's in the secondary coil.*

a battery or other direct-current source (Fig. 29-8). The induction coil consists of an input coil of a few turns of heavy wire, wound on an iron core, and an output coil of many thousand turns. When the switch in the battery circuit is closed, the iron core is magnetized, and a large number of magnetic lines passes through both the input and the output coil. The vibrator, like that of an electric bell, is at-

tracted by the iron core, opening the circuit. Opening the circuit suddenly demagnetizes the core, the magnetic lines are withdrawn from the output coil, and a large emf is induced in the output coil. The resulting large electric field strength at the gap *D* causes a gas discharge in the air—a spark. The vibrator opens and closes the circuit automatically, and sparks are produced at the output terminals *D*. The capacitor *C* acts like a reservoir into which charge flows when the vibrator contact is opened. Thus the capacitor prevents sparking at the contact points, which would vaporize the metal there.

In the engine of an automobile the explosive mixture of air and gasoline is ignited by means of an electric spark produced by an *ignition coil*. The ignition coil differs from an induction coil in that the contacts are closed and opened by a rotating wheel driven by the engine. In the simplified diagram of Fig. 29-9, the wheel *W* has projections which push in turn against the pivoted arm and thus open the primary circuit periodically. Each time that the circuit is opened, the core of the ignition coil is suddenly demagnetized just as in the spark coil, and a large emf is generated in the output windings. A single ignition coil produces the sparks in all the cylinders of an engine. In the distributor of a 6-cylinder car, a revolving arm passes close to six contacts in

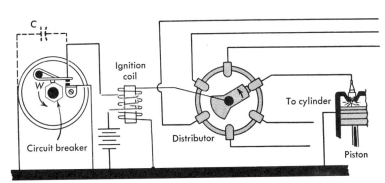

Fig. 29-9. *Ignition system of an automobile. The rotating wheel* **W** *opens the circuit periodically, inducing high voltage in the coil. The distributor connects the coil to the different spark plugs in sequence.*

turn, each being connected to the spark plug in a separate cylinder. This revolving arm is driven on the same shaft and at the same speed as the circuit breaker, so that sparks are produced in the various cylinders in sequence. When a car travels at full speed, the ignition produces more than 12,000 sparks per minute.

The *telephone* reproduces sounds at a distant point by means of varying electric currents. The telephone was invented in 1875 by Alexander Graham Bell and, independently, by Elisha Gray, who registered his patent application a few hours after Bell. A simple circuit, illustrating the principles of the telephone, is diagramed in Fig. 29-10. In the microphone, a flexible diaphragm pushes against one side of a box filled with carbon granules. When you speak into the microphone, the diaphragm vibrates, varying the pressure of the carbon granules. This changes their resistance and causes the current in the circuit to fluctuate. The changing current through the electromagnet coil of the receiver varies its magnetization. The flexible diaphragm in the receiver is forced to vibrate and to emit sound. In Chapter 31 we shall consider more sophisticated types of microphones and speakers. In practical telephone systems, the variations of line voltage are amplified so that messages can be sent over long distances. Modern telephones are carefully designed acoustically and electronically to transmit messages and reproduce sounds with as high a fidelity as is consistent with economical use of telephone cables.

Electromagnetic flowmeters, such as those used to measure the rate of flow of blood, are based on Faraday's law of electromagnetic induction. When a conducting fluid, such as blood, flows across magnetic lines, an emf is generated in the fluid. The emf is at a maximum in a direction perpendicular to both the direction of flow and the direction of the magnetic induction and is proportional to the fluid's velocity, the distance between the sensing electrodes, and the magnetic induction. A blood flowmeter includes an electromagnet to establish a known magnetic induction and two electrodes to detect the induced emf, all sealed in a thin plastic probe that can fit around the blood vessel. The probe is inserted surgically into the patient or the experimental animal, the electromagnet is energized by connecting leads, and the induced emf is measured with an amplifier and voltmeter. If the flowmeter has been properly calibrated, the voltmeter reading can be used to compute the velocity of flow. The *rate* of flow of blood is then the product of the velocity and the cross-sectional area of the blood vessel.

The *betatron,* used to give electrons energies of millions of electron volts in nuclear experiments, is based on electromagnetic induction. By changing the magnetic flux

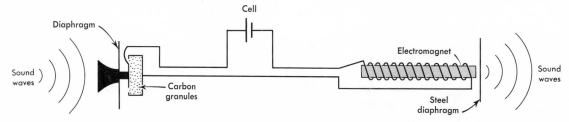

Fig. 29-10. Simple telephone circuit. Sound waves vary the pressure of the carbon grains in the microphone, thus varying their resistance. In consequence, the strength of the electromagnet varies so that the steel diaphragm vibrates.

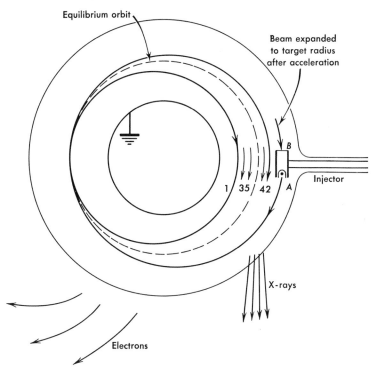

Fig. 29-11. *The betatron. Electrons, emitted at* A, *are accelerated by the increasing magnetic field normal to the orbit. The electrons move on paths 1, 2, 3, etc., and finally reach an equilibrium orbit where their energy is increased to 100,000,000 ev. They are then made to strike the target* B *and produce extremely penetrating X-rays. (Adapted with permission from an article by D. W. Kerst in G. A. Baitsell, ed.,* Science in Progress, *Series V,* Yale University Press, 1947.)

through a coil, you can generate an emf in the coil. An emf exists in a closed loop, through which the magnetic flux is changing, *even when no metallic conductor is present.* In the betatron, the metal conductor is replaced by a highly evacuated hollow porcelain doughnut-shaped ring. This is mounted between the jaws of a huge electromagnet like that of the cyclotron. Electrons are set free at *A* (Fig. 29-11). When the current suddenly increases in the magnet coils, magnetic lines are thrust through the doughnut-shaped tube, thus generating a large electromotive force and establishing an electric field within the

tube. While the magnet current is increasing, each electron, acted upon by this electric field, is driven around the ring hundreds of thousands of times and many hundreds of miles, gaining millions of electron volts of energy. The theory of design of the betatron reveals that the electrons will move in a stable circular orbit, gaining energy with each revolution, if the changing magnetic field is so arranged that the average magnetic induction in the magnetic field set up by the electromagnet is always *twice* the magnetic induction at the orbit. Careful design has made it possible to realize this condition.

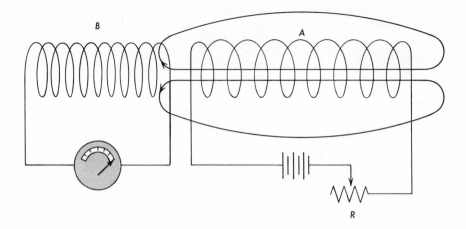

Fig. 29-12. *Mutual induction of two coils.*

At the end of the cycle, the electrons strike the target *B*, producing exceedingly energetic, penetrating X-rays, which can produce nuclear reactions.

Mutual Induction

When two coils are located near each other (Fig. 29-12), a change of current in coil *A*—made by changing the variable resistance of the rheostat *R*, for example—will vary the magnetic flux through *B*. Similarly, the battery and the meter can be interchanged, and a change of current in *B* will induce an emf in *A*. Thus varying the current in either coil will induce an emf in the other. We say that the combination has a *mutual inductance*. The mutual inductance *M* is defined as the electromotive force in one coil per unit rate of change of current in the other.

$$M = -\frac{\mathscr{E}_B}{\Delta I_A / \Delta t}$$

M for a given pair of coils depends upon their separation, the number of turns in each, the characteristics of any magnetic materials that link them, and so on. Except in the simplest cases, the mutual inductance is measured experimentally instead of being calculated.

The unit of mutual inductance, the henry,* is defined as follows:

One henry is the mutual inductance of two coils, when a rate of change of current of one ampere per second in one coil induces an emf of one volt in the other one.

Self-induction

You have learned that an emf can be induced by varying the number of magnetic lines threading through a circuit, and that the induced current always opposes the change that is occurring, no matter what causes the change of magnetic flux. It may be due to the motion of a magnet or to the change of current in a nearby electrical circuit. The change of magnetic flux may also be due to a change of current *in the coil itself.* We call this effect *self-induction.*

Suppose that several hundred feet of wire,

* Joseph Henry (1799–1878), the American physicist. Both he and Faraday were sons of poor parents. Both went to work when very young. Each of them independently discovered electromagnetic induction. Each refused to make money by applying his science to engineering problems. Both became directors of scientific institutions: Faraday of the Royal Institution in London, and Henry of the Smithsonian Institution in Washington.

in a single large loop, are connected in series with an incandescent lamp, a 120-v direct-current source, and a switch. When the switch is closed the current in the circuit will increase, in a few millionths of a second, to a steady value determined by Ohm's law. Now let this wire be wound onto an iron rod to form a coil. When the switch is again closed, the current will increase to the same final value as before, but the time required will be several hundredths of a second. In the coil there are hundreds of turns of wire, side by side. The current in each turn is accompanied by magnetic lines that thread through the other turns. An increase of current in any loop varies the flux through all the others, and the change of flux of magnetic lines generates an emf. This induced emf opposes the change of current.

Self-induction opposes not only the increase of current in a coil but the decrease also. When you open the circuit, the current will not stop instantly. The forward induced emf will cause a spark at the switch.

In order to demonstrate self-induction, connect a large electromagnet and an incandescent lamp, in parallel with each other, through a variable resistor to a direct-current source (Fig. 29-13). The resistance of the electromagnet is relatively low, that of the lamp relatively high. When you close the switch, at first the increasing current through the coils of the electromagnet increases the flux, thus generating an opposing emf. Self-induction impedes the current through the coil. Most of the charge flows through the lamp, which glows brightly. After the current becomes constant, most of the charge flows through the coil, and the lamp becomes dim. When you open the switch, the flux through the coil will decrease rapidly. The induced emf will make the lamp glow brightly for an instant. Hold the switch at right angles to the magnetic field of the electromagnet. When you open the switch, the magnetic field will deflect the ions in the arc sidewise and extinguish the arc so

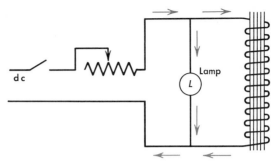

Fig. 29-13. *Self-induction. When the switch is first closed, self-induction opposes current through the coil; hence the lamp glows brightly. Afterward, most of the current flows through the coil. When you open the switch, the demagnetization of the coil induces an emf so that the lamp glows for an instant.*

quickly that a large emf will be generated. The induced current may even burn out the lamp.

Just as the inertia of a baseball opposes any change of its *velocity,* so the self-inductance of a circuit opposes change of *electric current.* Sometimes we call self-inductance *electrical inertia.* It is instructive to compare inertia in mechanics and self-induction, or electrical inertia, from the energy standpoint. When a pitcher throws a baseball, his hand experiences an opposing reaction force. He does work against this force in *augmenting* the kinetic energy of the ball. When the ball is caught by a player, his glove experiences a reaction force, and this force does work in pushing his glove backward, thus *decreasing* the energy of the ball. When the switch of an electromagnet is first closed, the increasing current builds up a magnetic field in the region surrounding the magnet, and a counter emf is generated, opposing the growth of current. This counter emf arises because the energy of the magnetic field in the space around the magnet is increasing. When the switch is opened, the counter emf, opposing the decrease of current, arises because the

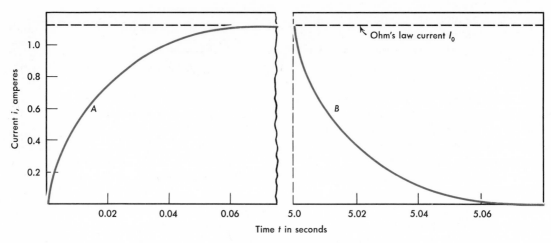

Fig. 29-14. (A) *Growth of current in coil.* (B) *Decay of current after switch was opened. Where did the energy to maintain the current come from?*

energy of the magnetic field is diminishing. The energy stored in the magnetic field reappears in the current.

The emf induced in a coil depends upon the rate of change of the current and upon the *self-inductance* of the coil. The unit of self-inductance is the same as that of mutual inductance. *A coil has a self inductance L of one henry if a counter emf of one volt is induced when the current changes at a rate of one ampere per second.*

$$\mathscr{E} = -L\frac{\Delta I}{\Delta t}$$

where the negative sign reminds us that the induced emf opposes the applied emf.

One henry equals one volt per ampere per second. The self-inductance of a coil depends upon the number of turns in the coil, the presence of magnetic materials, and so on.

Example. The current in an electromagnet decreased from 8 amp to 3 amp in 0.010 sec. The average induced emf was 1000 v. What was the self-inductance of the coil?

$$1000 \text{ v} = -L\frac{-5.0 \text{ amp}}{0.0100 \text{ sec}} = L\frac{500 \text{ amp}}{\text{sec}}$$

$$L = 2\frac{\text{v}}{\text{amp/sec}} = 2 \text{ henries}$$

Figure 29-14 shows the increase in current through the electromagnet when the switch is closed, and also the decrease when the switch is opened. The *decrease* or *decay* of current in a circuit (Fig. 29-14*B*) is called an exponential decay because mathematically the current *i* in the coil at any time *t* can be related to the resistance *R* of the circuit and its self-inductance *L* by the equation

$$i = I_0 e^{-(R/L)t}$$

where I_0 is the initial current in the circuit and *e* is a numerical constant, called the base of natural logarithms, which is approximately equal to 2.72. When a current *grows* in a circuit (Fig. 29-14*A*), the current *i* at any time is given by

$$i = I_0(1 - e^{-(R/L)t})$$

We shall encounter exponential growth or decay equations again. See Appendix E for a mathematical review of their properties.

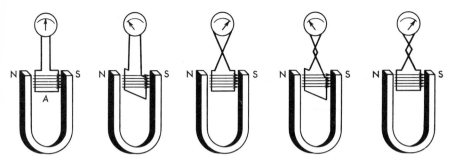

Fig. 29-15. *Generating an emf. Revolving the loop 360° cuts the magnetic lines four times.*

Generators

When Faraday discovered that an electromotive force can be produced by varying the number of magnetic lines threading through a coil, considerable interest was aroused. At a public lecture the prime minister of Great Britain, Mr. Gladstone, asked a question which we often hear today. "Mr. Faraday," he said, "this discovery is very interesting, but of what use is it?" Faraday replied dryly, "Perhaps it will give rise to a great industry on which you can levy taxes." Faraday's prediction came true half a century later when practical *generators* began to supply electrical energy to homes and factories, and the electrical industry began its rapid growth.

A generator consists of one or more coils that can be rotated in a magnetic field so as to produce an electromotive force. Connect a rectangular loop of wire to a galvanometer, and suspend the loop between the poles of a U-magnet (Fig. 29-15). Rotate the loop 90° so that the side *A* faces the north pole of the magnet. This operation thrusts magnetic lines through the loop and the galvanometer needle deflects, say, to the left. Turn the loop through another 90°, removing all the magnetic lines, and the galvanometer deflects to the right. A third rotation of 90° brings the face *A* next to the south pole of the magnet and causes a deflection to the left. Finally, bring the loop back to its initial position and the needle moves to the right. During the complete rotation, the magnetic lines were thrust through the loop twice and were twice removed from it. The loop cut the magnetic lines *four times.* By rotating the loop at constant speed, you could generate an alternating current that changed direction twice each rotation.

The loop just described could not be rotated continuously because of the twisting of the connecting wires. In Chapter 27 we saw that this difficulty could be avoided in motors by the use of "slip rings." Figure 29-16 shows slip rings mounted on the axle on which the loop of the simple generator revolves. Brushes pushing against the revolving rings connect the loop to the external part of the circuit. When the circuit is open and no charge flows, little effort is required to overcome friction in turning the loop. If the circuit is closed, the induced current produces a magnetic

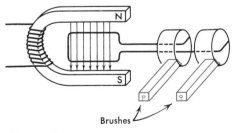

Brushes

Fig. 29-16. *Slip rings and brushes. They connect the coil to the external part of the circuit.*

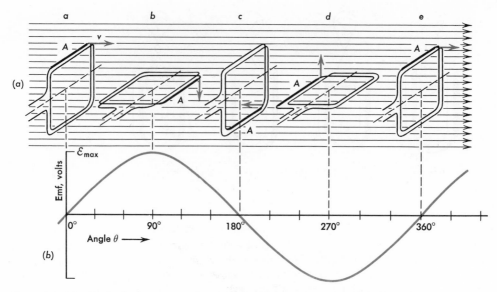

Fig. 29-17. (a) *Coil rotating in a magnetic field. During one revolution, each turn cuts* 4BA *magnetic lines.* (b) *The induced emf at any instant is given by* $\mathscr{E} = \mathscr{E}_{max} \sin \theta$.

field which opposes the motion (Lenz's law). More work must be done, and mechanical energy is transformed into electrical energy.

Figure 29-17a represents a rotating coil having N turns, each enclosing an area A. When the coil is at a, at right angles to a magnetic field of flux density B, the flux through each of its turns is $B \cdot A$. When the coil has rotated 90° to a position b, the flux through each turn is zero. While the coil rotates 90°, each turn cuts $(B \cdot A)$ magnetic lines. During each revolution, each turn cuts $4(B \cdot A)$ magnetic lines. If the coil makes n revolutions per unit time, lines are cut at a rate

$$N\frac{\Delta\phi}{\Delta t} = 4NBAn$$

Thus the average induced emf in volts is

$$\mathscr{E}_{av} = -4NBAn$$

when B is expressed in webers per square meter, A in square meters, and n in cycles per second. (Strictly speaking, this gives the *average* induced emf during one-half cycle.

During one complete cycle, the average emf is zero, for the current reverses twice during each revolution.*)

When the coil of a simple generator rotates at constant speed, the electromotive force varies as shown in Fig. 29-17b. Such an electromotive force is called an *alternating emf* and gives rise to an *alternating current* in a resistor. The emf is zero when the coil is at right angles to the field, and rises to a maximum $\mathscr{E}_{max}$ when it has turned 90°. Then the emf decreases to zero at 180°, reaches a negative maximum at 270°, and goes back to zero at 360°. This curve is repeated for each complete revolution of the coil. We call it a "sine" curve because its equation is $\mathscr{E} = \mathscr{E}_{max} \sin \theta$. $\mathscr{E}_{max}$ is the maximum value of the emf, and θ the angle through which the coil has turned from a position where the induced emf was zero. (This equation may also be written $\mathscr{E} = \mathscr{E}_{max} \sin 2\pi t/T$ in which T is the time for one

* The effective emf (Chapter 30) is given by

$$\mathscr{E}_{eff} = 0.707\mathscr{E}_{max} = 1.11\mathscr{E}_{av}.$$

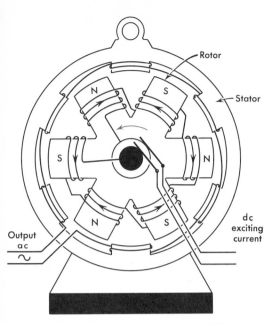

Fig. 29-18. An a-c generator. The rotor thrusts magnetic lines into the coils of the stator so as to induce an alternating emf.

revolution and *t* the time required for the coil to turn through an angle *θ*.)

Figure 29-18 represents a more practical, six-pole, *alternating-current (a-c) generator.* This revolving rotor thrusts magnetic lines in and out of the coils of the "stator" so as to generate an alternating emf in them. The stator has 3 pairs of poles; therefore, the rotor must revolve 60/3 = 20 times per second to generate a 60-cycle/sec alternating current.

Figure 29-19 shows a simple *direct-current (d-c) generator.* Like the d-c motors of Chapter 27, the generator has a *commutator,* which reverses the connections to the brushes twice for each revolution. If the ends of the coil were connected to two slip rings, as in the a-c generator, the output voltage would vary as represented by the dotted line in Fig. 29-20. However, the commutator reverses the connections at *A, B, C,* and *D* so that the output voltage at the brushes varies as shown by the heavy lines. The current is

not constant, but it never reverses in direction. Practical direct-current generators, like d-c motors, have "drum"-type armatures. Numerous coils are wound in slots in a cylindrical iron core. The commutator has many segments and the brushes make contact with the various coils in turn. Each coil makes contact with the brushes at the time when the induced emf in that coil is a maximum; hence the voltage across the brushes is nearly constant.

The Counter emf of a Motor

When a direct-current motor (Chapter 27) is running, the windings of its armature cut the magnetic lines of its field and generate an emf. This emf always opposes the current in the armature windings; hence we call it a *counter* (or *back*) emf. To demonstrate this effect, connect a small motor, in series with an incandescent electric lamp, to the lighting system. First, prevent the armature from rotating. The lamp will glow brightly. Release the armature. As it speeds up, the light will become more dim. This is because the armature cuts magnetic lines more rapidly at the higher speed and the counter emf is larger.

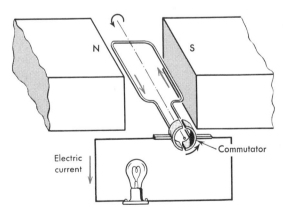

Fig. 29-19. Simple d-c generator. The commutator reverses the connections to the external circuit twice for each revolution. The current in the external circuit never reverses, but pulsates.

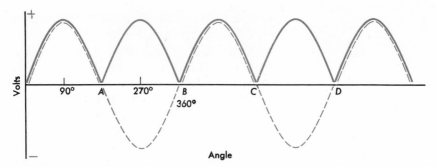

Fig. 29-20. "*Rectified*" *alternating current. The commutator reverses coil connections at* A, B, C, *and* D. *The current pulsates, but it never reverses.*

When a motor runs freely, at maximum speed, the counter emf is large and the current is small. As the load on the motor is increased, the armature slows down, and the counter emf becomes smaller. Thus, the input current (and power) increase with load.

Let the resistance of the armature be R, the potential difference across the armature V, and the counter emf $\mathscr{E}$. Then, by Ohm's law, the current I through the armature is given by

$$I = \frac{V - \mathscr{E}}{R}$$

Example. The armature coils of the starting motor of an automobile have a resistance of $0.050\ \Omega$. The applied potential difference is 6.0 v, and the counter emf is 3.00 v. (*a*) What is the current through the armature? (*b*) If the armature did not turn, what would the current be?

(*a*) $I = \dfrac{V - \mathscr{E}}{R} = \dfrac{6.00\ \text{v} - 3.00\ \text{v}}{0.050\ \Omega} = 60\ \text{amp}$

(*b*) $I = \dfrac{V}{R} = \dfrac{6.00\ \text{v}}{0.050\ \Omega} = 120\ \text{amp}$

For a given load, the armature of a motor turns at such a speed that the sum of the counter emf and the IR potential drop due to resistance are equal to the applied potential difference across the coils of the armature ($\mathscr{E} + IR = V$). The counter emf $\mathscr{E}$ is produced by the varying magnetic flux through the coils. If the strength of the magnetic field is increased, the armature will not have to turn so fast to generate this counter emf. Thus, strange as it may seem, *increasing* the field strength of a motor will slow down the motor. Conversely, the speed may be increased by diminishing the current through the coils so as to weaken the field.

In some motors the armature coils and the field-magnet coils are connected in series (Fig. 29-21*a*). When the armature turns slowly, the counter emf is small, and the current is large. This current energizes both the field windings and the armature windings so that the torque is very great. Motors of this type are used for machines in which the load varies greatly. The torque of a series motor at low speed is much greater than at high speed.

In "shunt" motors (Fig. 29-21*b*) the field-magnet coils and the armature coils are connected in parallel so that the full voltage is applied across the field magnets and their strength is constant. Hence the torque varies less with speed than does that of the series motor. Electric fans usually are driven by shunt-wound motors.

In the "compound" motor the field has two sets of windings, one connected in series with the armature coils, the other in parallel. The torque does not vary greatly with speed.

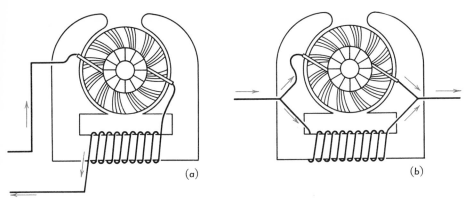

Fig. 29-21.(a) *Series-wound motor. Both the field current and armature current increase with load, as does the torque.* (b) *Shunt-wound motor. The field current does not increase as the armature turns more slowly under heavy load.*

Eddy Currents

We have been discussing induced electromotive force in rings, coils, and — in the betatron — around a loop in space through which the magnetic flux was changing. Electromagnetic induction also occurs whenever the magnetic lines threading through a solid conductor change. Support an aluminum ring, like a pendulum, around one pole of an electromagnet connected to a battery (Fig. 29-22). Close the switch, and magnetic lines will be thrust through the ring. The magnetic field of the current induced in the ring opposes this change; hence the ring is pushed outward. Soon the induced current dies down and the pendulum swings back to its initial position. Open the switch so that the coil is demagnetized. This is equivalent to withdrawing a magnet pole from the ring. The induced current opposes the change, and the ring is pulled toward the electromagnet. Push the ring quickly to and fro while the coil is magnetized. You will experience an opposing force.

Repeat these experiments after replacing the ring by a solid copper disk suspended in front of the iron core. The disk will behave as the ring did. *Currents are induced in a solid*

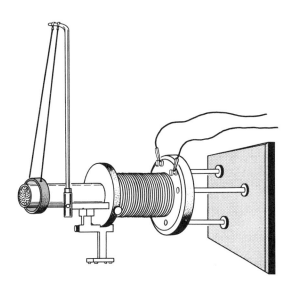

Fig. 29-22. Eddy currents. Sudden magnetization of the core induces a current in the ring, which is repelled. A solid copper disk will be repelled because of eddy currents.

conductor by a varying magnetic field. The currents of electrons generated in solid bodies by varying magnetic flux are called *eddy* currents.

Try to hold the aluminum ring over the pole of the electromagnet (Fig. 29-22) when the electromagnet is connected to an alternating-current source. Eddy currents surging to and fro will quickly heat the metal so that it will become too hot to hold. Eddy currents can be useful in producing heat. Induction heaters used in industry utilize eddy currents. However, in most electrical machinery, it is desirable to prevent eddy currents. The cores of transformers (Chapter 30) and of field magnets and the armatures of generators and motors are built up of varnished sheets of iron in layers parallel to the magnetic field. The insulation between the adjacent "laminated" layers minimizes eddy currents. Ferrites (Chapter 28) are magnetic materials that combine high electrical resistance with high magnetic permeability, making them very suitable for use when eddy currents must be avoided.

Summary

Whenever the number of magnetic lines threading through a coil or a circuit is changed, an electromotive force is induced, establishing a current in a closed circuit. The induced current is in such a direction as to cause a magnetic field opposing the change. The induced emf is proportional to the rate of change of magnetic lines through the circuit. One volt is induced in a single loop of wire when the rate of change of magnetic flux is one weber per second.

The mutual inductance of two circuits is the emf induced in one circuit per unit rate of change of current in the other circuit. Self-induction is the production of an emf in a circuit by variation of the current in that circuit. The induced emf always opposes the change of current. One henry equals one volt per ampere-per-second.

A generator consists of one or more coils rotating in the field of an electromagnet so as to generate an electromotive force.

When the armature of a motor rotates, a counter emf is induced which opposes the current.

Questions

1. Describe several different methods for inducing an emf in a circuit.

2. In what sense is a motor a generator?

3. Why are precautions against electrical shock needed in starting or stopping a large motor?

4. When are (*a*) slip rings, (*b*) commutators used in a generator?

5. State Lenz's law and describe an experiment by which it may be proved. What is the mechanical analogy to Lenz's law?

6. A loop of wire hangs on the wall of a room. When the south pole of a magnet is moved toward it, is the induced electric current counterclockwise or clockwise?

7. Are (*a*) an emf and (*b*) a current always induced when the number of magnetic lines passing through a coil is changed?

8. When the motor of an electric refrigerator is starting, the lights of a house may momentarily dim. Explain.

9. Why is it harder to turn the armature of a generator when a resistor is connected across its terminals than when it is on open circuit?

10. If a large coil were connected to a sensitive galvanometer, could it be moved without circulating a charge through the galvanometer?

11. Describe two ways in which the magnetic induction of a magnetic field can be measured.

12. When a coil turns 360° in a magnetic field, how many times are the magnetic lines cut? Derive the equation for the average induced emf.

13. Plot current *i* versus time for the circuit in Fig. 29-23 (*a*) when the circuit is completed

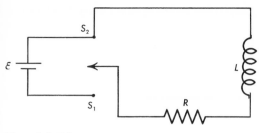

Fig. 29-23.

(by throwing the switch to S_1) and (b) when it is broken (by throwing the switch to S_2). Let $L = 20$ millihenries, $R = 10\ \Omega$, and $\mathcal{E} = 6$ v.

14. A copper disk vibrates like a pendulum near one pole of an electromagnet. Describe what happens (a) when the coil is magnetized, (b) when it is not magnetized.

Problems

A

1. The number of magnetic lines threading through a single loop of wire changes from 3.0×10^{-3} to 5.0×10^{-3} weber in 0.10 sec. What is the average induced emf?

2. An induction coil has 1.0×10^4 turns. The number of lines through it changes from 5.0×10^{-3} weber to 0 in 0.010 sec. What is the average induced emf?

3. A rectangular coil, 10.0 cm by 12.0 cm, has 100 turns of wire. The coil is mounted on an axle at right angles to a magnetic field of 0.15 weber/m². At what speed must the coil rotate in order to generate an average emf of 0.600 v during each half-revolution?

4. The armature coil of a simple generator has 1200 turns; a magnetic flux of 1.5×10^{-3} weber threads through the coil when it is at right angles to the field. What average emf is induced during each half-revolution if the coil makes 10 rev/sec?

5. A wire is fastened around the edges of a door, forming a 1.0-m by 2.0-m rectangle. The door is rotated 90° in 3.0 sec, turning from an east-west to a north-south plane. The strength of the horizontal component of the earth's field is 20×10^{-6} weber/m². (a) How many magnetic lines pass through the rectangle in the first position? (b) What average emf is induced in the wire by the rotation of the door?

B

6. When the generator of a car is connected to zero load (open circuit), the input power is 100 watts to overcome bearing friction. What is the input power when the generator supplies a current of 20 amp at an emf of 12.6 v? (Assume that, except for the frictional loss, the generator is 100% efficient.)

7. One thousand turns of wire are wound around the edge of a square board 40 cm on each side. The strength of the horizontal component of the earth's field is 20×10^{-6} weber/m². If the board is rotated about a vertical axis from a north-south plane to an east-west plane in 0.20 sec, what are (a) the average induced emf and (b) the average current if the resistance of the coil is 2.0 Ω?

8. A motor operating at full speed generates a counter emf of 110 v when connected to a 120-v line. If the motor current is 5.00 amp when running at full speed, what would be the current if the motor were prevented from turning?

9. The resistance of a small motor is 10 Ω. When the motor is connected to a 120-v source and runs at normal speed, the current is 0.40 amp. What is the counter emf then?

10. A shunt-wound motor has a field resistance of 200 Ω and an armature resistance of 0.16 Ω. When the applied voltage is 120 v, the current is 20 amp and the output power is 2.75 hp. Compute the efficiency of the motor.

11. An electromagnet has a self-inductance of 5.0 henries. What is the induced emf when the current is reduced from 10.0 amp to 0 in (a) 1.0 sec, (b) 0.010 sec?

12. The strength of the vertical component of the earth's magnetic field at a certain place is 56×10^{-6} weber/m². A boy's hoop having a cross-sectional area of 0.80 m² lies on a horizontal surface. If it is raised to a vertical posi-

tion in an east-west plane in 0.20 sec, what average emf is induced in the hoop?

13. A wire 20 m long is attached to an airplane at right angles to the line of motion. The horizontal velocity of the plane is 600 km/hr, and the strength of the vertical component of the earth's field is 50×10^{-6} weber/m². (a) At what rate does the wire "cut" the earth's magnetic lines? (b) What is the potential difference between the ends of the wire?

C

14. A circular loop of wire 40 cm² in cross section rotates on an axis at right angles to a uniform magnetic field of strength 0.15 weber/m². At what rate must it rotate in order to generate an average emf of 30 millivolts (during each half-revolution)?

15. The self-inductance of an electromagnet is 3.0 henries. When a switch is opened, the current through the magnet coils decreases from 10 amp to 0 in 0.20 sec. What average induced emf opposes the decrease of current?

16. The voltage across the terminals of the armature of a generator on open circuit is 120 v. When the current is 10 amp, the voltage is 115 v. What is the resistance of the armature coils?

17. A generator armature has 200 turns and moves across a pole with an area of 1000 cm², near which the flux density is 1.5 webers/m². The coil passes the pole in 0.40 sec. Determine the average emf induced in the coil.

18. What voltage is induced in a horizontal wire 2.0 m long placed in the east-west magnetic meridian, when it is moving horizontally north at the rate of 10 m/sec? The earth's field

has a vertical intensity of 60×10^{-6} weber/m², and the angle of dip is 60°.

19. A d-c series motor running at normal speed generates a counter emf of 110 v when connected to a 120-v line. The current is 0.60 amp. (a) What is the resistance of the coils? (b) What would the current be if the armature did not revolve? (c) What fraction of the input energy is wasted as heat?

20. A coil of wire has a resistance of 1.0 Ω and a self-inductance of 0.010 henry. If the coil is connected to a voltage source of 6.0 v, what is the current in the coil (a) 0.001 sec, (b) 0.01 sec, (c) 0.1 sec, after the circuit is closed?

21. Estimate the voltage induced between the ends of an iron bar 1.0 m long 0.25 sec after it has been dropped from rest from a height of 10 ft at your geomagnetic latitude. The bar falls in an east-west plane and remains horizontal as it falls.

For Further Reading

Babcock, H. W., "The Magnetism of the Sun," *Scientific American,* February, 1960, p. 52.

Cox, Alan et al., "Reversals of the Earth's Magnetic Field," *Scientific American,* February, 1967, p. 44.

Elsasser, W. M., "The Earth as a Dynamo," *Scientific American,* May, 1958, p. 44.

Kondo, Herbert, "Michael Faraday," *Scientific American,* October, 1953, p. 90.

Sharlin, Harold I., "From Faraday to the Dynamo," *Scientific American,* May, 1961, p. 107.

Wilson, Mitchell, "Joseph Henry," *Scientific American,* July, 1954, p. 72.

CHAPTER THIRTY

Alternating Currents

About 99% of the electrical energy that is generated in the United States is distributed by alternating current. Alternating currents also occur in telephones, in radio and television receivers, and in almost every other electrical communications device. Despite these facts, most of our attention has been given to direct-current theory. Alternating-current theory requires for its understanding more advanced mathematics than we have been using and cannot be mastered in a brief first-year course. We shall present only a few of its more elementary aspects, stating the mathematical results without formal proof.

Alternating Currents and Voltages

If you move a magnet pole in and out of a coil with simple harmonic motion, the cutting of magnetic lines will induce an *alternating emf.* Electrons in the wire will not drift in one direction as in direct current. Each electron will oscillate about an average position.

In the last chapter, you learned that the electromotive force in a loop of wire rotating at constant speed in a magnetic field changes sinusoidally with time. Suppose that a simple generator of this kind is connected to a piece of wire that acts like a *pure resis-*

tance, that is, its self-inductance and its capacitance are very small (Fig. 30-1*a*). The voltage across the wire and the current in it vary as shown in Fig. 30-1*b*. Along the *x*-axis we have plotted time; the time is also proportional to the angle through which the generator coil has turned. Notice: (1) the voltage and current are in phase, reaching their maxima, their zero values, and their minima at the same times; (2) the maximum voltage is 170 v, and the maximum current is 2.0 amp; and (3) the period of the current and voltage is $\frac{1}{60}$ sec, and their frequency is 60 cycles/sec. (The American standard is 60 cycles/sec; some countries use a frequency of 50 cycles/sec.)

We can write equations for the current *i* and voltage *v* at any time *t*:

$$i = I \sin \frac{2\pi}{T} t = I \sin 2\pi \, ft$$

$$v = V \sin \frac{2\pi}{T} t = V \sin 2\pi \, ft$$

where I is the maximum value or amplitude of the current, V the maximum value or amplitude of the voltage, T the period of rotation of the generator, and f the frequency of the generator.

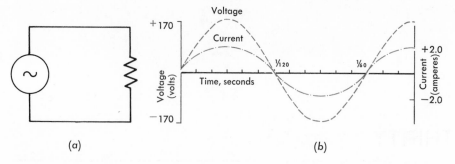

Fig. 30-1. *Alternating current and voltage.* (a) *A simple a-c generator is connected to a resistor.* (b) *The current in the resistor and the voltage across it are in phase.*

When the current and voltage are in phase—this is usually true in house lighting circuits—the instantaneous current and voltage are related by Ohm's law:

$$i = \frac{v}{R} \text{ and } I = \frac{V}{R} \quad \text{(Resistance only)}$$

where R is the resistance of the circuit.

The effective current I_{eff} and the effective voltage V_{eff} are convenient quantities to use in describing alternating-current circuits because, as we shall note later, they are related to the power delivered to the circuit. These are the quantities measured by alternating-current ammeters and voltmeters. We shall define them for sinusoidal currents and voltages as follows:

$$I_{\text{eff}} = 0.707 \ I$$
$$V_{\text{eff}} = 0.707 \ V$$

In order to compare time-varying alternating currents and voltages with steady direct currents and direct voltages, we define an effective ampere and an effective volt:

One effective (alternating-current) ampere is that alternating current which heats a conductor at the same rate as a direct-current ampere.

One effective volt is the alternating electromotive force in a circuit such that the ef-

fective current is one ampere when the resistance of the circuit is one ohm.

Impedance

The alternating current is not always in phase with the alternating voltage across a circuit. Figure 30-2 represents a circuit in which are included a resistor *A*—a lamp—and an inductor *B* having an iron core.

Suppose that when the system is connected to a 120-v direct-current source the current is 0.50 amp. By Ohm's law the resistance is found to be 240 Ω. Now connect the system to a 120-v alternating-current source of 120 v effective voltage. The current will be smaller than before, and the lamp will glow less brightly.

As was pointed out in Chapter 29, the decrease of current occurs because the flow of electricity is opposed not only by the resistance but also by the counter emf induced in the coil by variations of its own magnetic field.

The ratio of the effective potential difference between the terminals of a circuit to the effective current is the *impedance* (im pee′ dance) of the circuit. The impedance is expressed in ohms.

One ohm is the impedance of a circuit in which there is a current of one effective am-

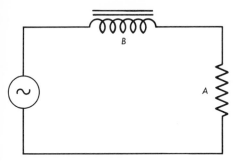

Fig. 30-2. Circuit including a resistor A and an inductor B. The inductor opposes change of current.

pere when the alternating potential dif-ference between the terminals of the con-ductor is one effective volt.

The effective current I_{eff} in a circuit of impedance Z is given by

$$I_{eff} = \frac{V_{eff}}{Z}$$

Example. If the effective current in the circuit of Fig. 30-2 is 0.4 amp and the effective alternating potential difference across the terminals is 120 v, what is the impedance of the circuit?

$$Z = \frac{V_{eff}}{I_{eff}} = \frac{120 \text{ v}}{0.4 \text{ amp}} = 300 \text{ }\Omega$$

Inductive Reactance

The impedance of a circuit depends upon its resistance and also its *reactance.* In Chapter 29 you found that the counter emf induced in a coil depends upon its self-inductance and also upon the rate at which the current changes. A coil has a self-inductance of 1 henry, that is, 1 v/(amp/sec), if 1 v is induced by a current changing at a rate of 1 amp/sec. The *inductive reactance* X_L of a circuit de-pends upon its self-inductance and the frequency f of the alternating current. The in-ductive reactance of an inductor having a

self-inductance L is given by

$$X_L = 2\pi f L$$

Example. What is the inductive reactance of a coil having a self-inductance of 0.50 henry if the frequency of the alternating current is 60 cycles/sec?

$X_L = 2\pi \times 60$ cycles/sec
$\qquad \times 0.50$ v/(amp/sec) $= 60\pi \text{ }\Omega$

Suppose that you connect an elec-tromagnet having a removable core in series with a lamp to an alternating-current source. By removing the core, you decrease the self-inductance and therefore the reactance of the circuit so that the current increases and the lamp glows more brightly.

To compute the impedance of a circuit, we do not add its resistance to its inductive reac-tance. Instead, we apply the parallelogram rule. In Fig. 30-3, the vector OA represents the resistance of a circuit and OB its inductive reactance. The diagonal vector OC repre-sents the impedance of the circuit. The mag-nitude Z of the impedance is given by

$$Z = \sqrt{R^2 + X_L^2} \qquad (1)$$

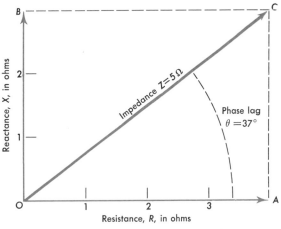

Fig. 30-3. Impedance of a circuit found by the parallelogram law.

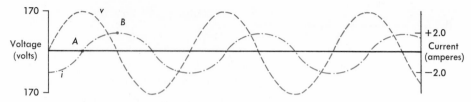

Fig. 30-4. *The current and total voltage in an inductive-resistive circuit. The current lags the voltage.*

Example. What is the inductance of the circuit in Fig. 30-2 if the impedance is 300 Ω, the resistance is 240 Ω, and the frequency 60 cycles/sec?

$$(300 \ \Omega)^2 = (240 \ \Omega)^2 + X_L{}^2$$
$$X_L = 180 \ \Omega$$
$$180 \ \Omega = 2\pi \times 60 \text{ cycles/sec} \times L$$
$$L = 1.5/\pi \text{ henry}$$

You may wonder why the resistance vector and the inductive reactance vector are at right angles. In Fig. 30-4, the *total* voltage *v* at any instant and the instantaneous current *i* are plotted against time. The potential difference IX_L due to inductive reactance is greatest at *A* when the current is changing most rapidly. The potential difference *IR* due to resistance is greatest at *B* when the current is a maximum. The phase difference between IX_L and *IR* is thus 90° with *IR* lagging IX_L. The total alternating voltage is the sum of these two alternating voltages differing in phase by 90°. The phase difference between *R* and X_L may be said also to be 90°. We use the parallelogram method to combine voltages in alternating current circuits and to combine resistance and reactance because of phase differences that are present.

Phase Lag, Power, and Power Factor

The angle θ (Fig. 30-3) is called the *phase lag angle* or the *lag*. When the total reactance of a circuit is zero, its phase lag is zero and its impedance is equal to its resistance. As the

inductive reactance increases, so does the lag.

Figure 30-5*a* represents the alternating current and voltage in a circuit that has zero reactance. The two are exactly in step. In Fig. 30-5*b* the inductive reactance opposes the increase or decrease of current so that the current lags behind the voltage. The angle of lag is 30°, or one-twelfth of a cycle.

If the current and the voltage in an alternating-current circuit are in phase, the power is computed by the formula

$$P = V_{\text{eff}} \times I_{\text{eff}} \qquad (2)$$

If they are not in phase, the maximum voltage and current occur at different times in each cycle. The power is not given by the product of the voltage and the current, but by

$$P = V_{\text{eff}} \times I_{\text{eff}} \times \cos \theta \qquad (3)$$

The ratio of the "true" power given by equation (3) to the "apparent" power $V_{\text{eff}} I_{\text{eff}}$ is called the *power factor*.

$$\text{Power factor} = \frac{\text{True power}}{\text{Apparent power}}$$

$$= \cos \theta = \frac{R}{Z}$$

The Transformer

Alternating current is convenient because transformers readily change its voltage. A transformer is simple, has no mechanically moving parts, and may be very efficient.

Figure 30-6 shows a simple transformer.

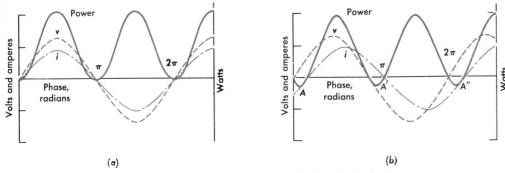

Fig. 30-5. *Power of a system.* (a) *Phase lag zero.* (b) *Lag 30°. During part of the cycle, the power is negative. The average power is less than that in* (a).

Separate coils are wound on an iron core. Suppose that the coil **A** is connected to a battery. When the circuit is first closed, the iron will be magnetized and the magnetic lines will follow the iron path and thread through the other set of coils **B** and **C**. The changing of the magnetic flux will induce an emf in each coil which will persist only while the field is changing. Open the circuit. As the magnetic field dies down, an opposite emf will be induced. Now connect the coil **A** to the house lighting system which provides an effective alternating voltage of, say, 120 v. The current through the coil will reverse 120 times per second (60 cycles/sec), and magnetic lines will repeatedly be thrust through the other coils and withdrawn. If the three coils have the same number of windings, the effective

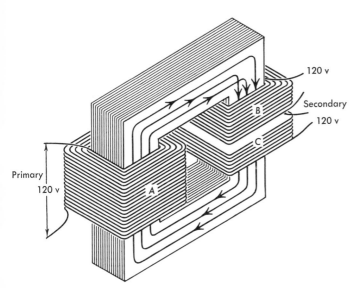

Fig. 30-6. *A transformer. Alternating current in coil* A *varies the flux through* B *and* C *so as to induce alternating emf's.*

emf induced in *B* and that in *C* will each be about equal to that in *A*. You can connect the two coils *B* and *C* in series so that the output emf will be twice that of the input circuit.

A transformer in which the output coils (secondary winding) have more turns than the input coils (primary winding) is called a *step-up* transformer. One in which the output coils have fewer turns is called a *step-down* transformer. Suppose that the primary coil of a transformer has N_1 turns and the secondary has N_2 turns. According to the principle of electromagnetic induction of the last chapter, the electromotive forces in the two coils are

$$\mathscr{E}_1 = -N_1 \left(\frac{\Delta\phi}{\Delta t} \right)_1$$

$$\mathscr{E}_2 = -N_2 \left(\frac{\Delta\phi}{\Delta t} \right)_2$$

If the iron core is well designed, almost all of the magnetic lines set up by the primary will thread through the secondary. Thus $(\Delta\phi/\Delta t)_1$ will equal $(\Delta\phi/\Delta t)_2$. Dividing one of the above equations by the other, we have

$$\frac{\mathscr{E}_1}{\mathscr{E}_2} = \frac{N_1}{N_2}$$

A doorbell transformer having one-tenth as many turns in the output as in the input lowers the effective voltage from 120 v to 12 v. In general, *the ratio of the number of turns on the two coils approximately equals the ratio of the electromotive forces.*

A transformer cannot create energy; hence the output power cannot be greater than the input power. If the voltage is stepped up by a transformer, the current must diminish in the same ratio:

$$\frac{(I_1)_{\text{eff}}}{(I_2)_{\text{eff}}} = \frac{N_2}{N_1}$$

The output power of a transformer operating at full load may be 98% of the input power.

Example. The effective input current to a transformer is 1 amp at 2000 v. What is the effective output current at 100 v if the efficiency is 100%?

$$2000 \text{ v} \times 1.0 \text{ amp} = 2000 \text{ watts}$$
$$= 100 \text{ v} \times (I_2)_{\text{eff}}$$
$$(I_2)_{\text{eff}} = 20 \text{ amp}$$

A transformer automatically controls the energy that it takes from the source. Suppose the secondary circuit has nothing connected to it, that is, is open. The primary circuit has a large self-inductance because many magnetic lines are cut each second; the inductive reactance and impedance of the primary are large. The current in the primary is therefore quite small; the transformer wastes little power on open circuit. Now close the secondary circuit by connecting a lamp to the secondary coil. There is an induced current in the secondary, and this current, by Lenz's law, sets up magnetic lines in the core that oppose the magnetic field changes near the primary. Fewer magnetic lines are cut each second by the primary. The reactance of the primary is reduced, and energy is taken from the source by the transformer.

A transformer, silently transmitting energy at a rate equal to that of 50,000 horses, is one of the most important devices in the electrical industry. In a typical electrical generating system (Fig. 30-7), the electrical energy is generated at 2400 effective volts. Then, it is stepped up to 120,000 v by huge transformers at the power station. Then the energy is transmitted to central points in the city, where transformers reduce the emf to 2400 v. Smaller step-down transformers, mounted in iron boxes on poles near the residences, lower the emf to 120 v. From these, the energy is delivered to the householders. Smaller transformers operate electric bells, and the filaments of vacuum tubes in radio and television receivers, and the like.

Energy is usually transmitted at high potential difference because the required current is small, and hence little energy is wasted in heating the wires. Suppose that energy is

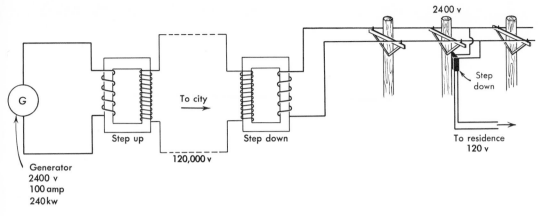

Fig. 30-7. *Electrical transmitting system. Transformers raise the voltage at the generator, lower it at the city, and again at the residences.*

transmitted along a wire of 2 Ω resistance at a rate of 2000 watts. If the potential difference is 100 v, the required current will be 20 amp, and energy will be wasted in heating the wire at a rate of (20 amp)^2 × 2 Ω = 800 watts. The efficiency is only 60%. If the energy is transmitted at 10,000 v, the current will be 0.20 amp. The loss will be (0.2 amp)^2 × 2 Ω = 0.08 watt, and the efficiency will be only slightly less than 100%.

Capacitive Reactance

We have seen that when the reactance of a circuit increases, the current required for a certain power must increase since $P = V_{eff} I_{eff} \cos \theta$. This increase in current wastes energy due to heating. For example, if the motors in an air-conditioning system in a theater cause a phase lag of 30°, the power lost in electrical heating is increased by about 30%. The phase lag of a system can be reduced by connecting capacitors in the circuit.

Suppose that we connect an incandescent lamp at *A* and a capacitor at *B* in series with a d-c source (Fig. 30-8). When the circuit is first closed, there will be a momentary flow of charge until the capacitor is charged, but afterward the current will cease. If the system is

connected to an a-c source, the capacitor will be charged and discharged continuously, electrons will surge to and fro through the filament, and the lamp will glow. If the plates of the capacitor are made larger or are put closer together, its capacitance will be increased, more charges will flow in the circuit for each alternation, and the current will be greater. A capacitor's opposition to current varies *inversely* as its capacitance. The alternating current in such a circuit depends not only upon the capacitance *C* of the circuit, but also upon the frequency, *f*. In general, the reactance of a capacitor varies inversely as the frequency. We define the *capacitive reac-*

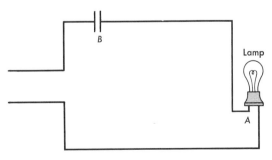

Fig. 30-8. *Circuit containing resistance and capacitance.*

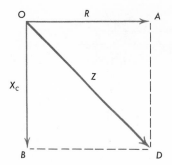

Fig. 30-9. *Impedance of a circuit containing resistance and capacitance.*

tance of a circuit X_C as

$$X_C = \frac{1}{2\pi f C}$$

in which C is the capacitance of the circuit and f the frequency of the alternations.

To compute the impedance of a circuit containing a resistor and a capacitor, we again apply the parallelogram rule. In Fig. 30-9, the vector OA represents the resistance of a circuit and OB its capacitive reactance. The diagonal vector OD represents the impedance of the circuit. The magnitude Z of the impedance is given by

$$Z = \sqrt{R^2 + X_C{}^2}$$

The resistance vector and the capacitive-reactance vector are at right angles, and the resistance vector leads. In charging a capacitor, the current is a maximum when the charge on the capacitor and its potential difference are zero. When the capacitor has its maximum charge, the current is zero and the potential difference is a maximum. In general, the potential difference IR leads the potential difference IX_C by 90°.

A Circuit Including a Resistor, an Inductor, and a Capacitor

The electrical circuit, Fig. 30-10, contains a resistor, an inductor, and a capacitor con-

nected to an a-c generator. The inductor and the capacitor always oppose each other in their effects on the current. The total reactance (Fig. 30-11) of the circuit is equal to the inductive reactance X_L minus the capacitive reactance X_C.

The impedance of the system is given by

$$Z = \sqrt{R^2 + (X_L - X_C)^2} \qquad (4)$$

In order that the total impedance of a circuit may be small, R should be small and X_L should be nearly equal to X_C.

Example. The resistance of an a-c circuit is 20 Ω, its inductive reactance is 40 Ω, and its capacitive reactance 30 Ω. What are (a) the impedance and (b) the phase lag of the circuit?

(a) $\quad Z = \sqrt{(20\ \Omega)^2 + (40\ \Omega - 30\ \Omega)^2} = 22.4\ \Omega$

(b) $\quad \tan \theta = \dfrac{10\ \Omega}{20\ \Omega} = 0.5$

$$\theta = 26°\ 34'$$

In a submarine telegraph cable the inner wire is surrounded by insulating material covered by a metal sheath—forming a long "capacitor." The capacitance of a long cable is so great that many seconds are required for a current to grow to full value. For this reason, only a few words per minute could be transmitted across the Atlantic over the first cable used.

We have been concerned with resistors, in-

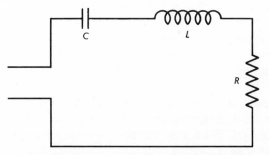

Fig. 30-10. *Resistor, inductor, capacitor.*

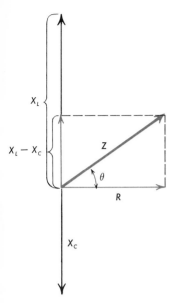

Fig. 30-11. *The impedance of a circuit containing resistance, inductance, and capacitance.*

ductors, and capacitors in **series** a-c circuits. Parallel connections are also possible. The theory of parallel a-c circuits is a somewhat complicated extension of the basic approach we have taken here and will not be discussed in this book.

Electrical Series Resonance

You can readily make a pendulum vibrate if you push it, time after time, with just the right frequency. This is an example of **mechanical resonance.** Similarly, the alternating current in a circuit having reactance depends upon the frequency of the alternations. According to equation (4), the alternating current in such a circuit is a maximum when the inductive reactance X_L is equal to the capacitive reactance X_C. This is the condition for **series electrical resonance.** Thus, at series resonance,

$$X_L = X_C$$

$$2\pi f L = \frac{1}{2\pi f C}$$

The period *T* of the oscillations is given by

$$T = \frac{1}{f} = 2\pi\sqrt{\frac{L}{1/C}} = 2\pi\sqrt{LC}$$

This resembles the equation for a mechanical oscillator

$$T = 2\pi\sqrt{\frac{m}{F/s}}$$

Notice that the mass *m* of the oscillator corresponds to the self-inductance *L* (electrical inertia) of the circuit. What corresponds to the spring-constant of the spring?

Alternating-Current Ammeter, Voltmeter, and Wattmeter

An ordinary d-c ammeter or voltmeter will not operate in an a-c circuit. The needle merely quivers about its zero point. The coil has so much inertia that it will not vibrate 60 times per second. Figure 30-12 shows the essential parts of one kind of a-c **galvanometer.** The pivoted coil **P** is connected in series with the two field coils **F** and **F'**. When the electric current through the coils is directed as indicated by the arrows, the magnetic field due

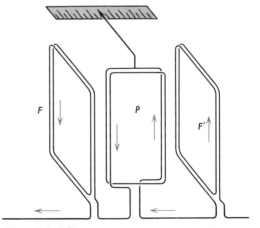

Fig. 30-12. *An a-c galvanometer. Reversing the current does not reverse the deflection of the needle.*

to the currents in *F* and *F'* is from right to left across *P*. The current-carrying conductors in the left side of *P* will experience a deflecting force *into* the page and those on the right will be deflected *out* of the page. The needle attached to *P* will be deflected to the *right*. Suppose that the current is reversed so that the current in the pivoted coil is reversed. Meanwhile, the magnetic field of the two field coils will also reverse so that the coil will be deflected *in the same direction as before.* Such an instrument will measure either direct current or alternating current.

An alternating-current galvanometer, like a direct-current galvanometer, can be converted into an *ammeter* by providing a shunt of appropriate resistance. It can be converted into a *voltmeter* by connecting a resistor of suitable resistance in series with it.

The structure of a *wattmeter* (Fig. 30-13) closely resembles that of the instruments just described. The chief modification is that the field coils are connected, as in an ammeter, directly in the main circuit; but the pivoted coil, in series with a resistor, is connected, like a voltmeter, *in parallel* with the "load." The field strength due to the field coils is proportional to the current, and the current in the pivoted coil is proportional to the voltage.

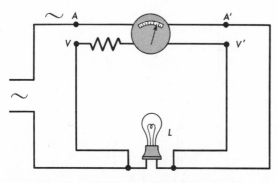

Fig. 30-13. Wattmeter connected in a circuit. The main current is in the field coils. The pivoted coil is connected, like a voltmeter, in parallel with the load L.

The torque acting on a coil is proportional to the product of the current and the voltage, that is, to the power. The alternating-current ammeter and the voltmeter can be used together to measure the true power of an alternating current only if the voltage and the current are in phase and the lag is zero. A wattmeter indicates the true power, *regardless of the lag.*

Alternating-Current Motors

The most widely used type of alternating-current motor is the *induction* motor. Before presenting its theory, we shall describe the "two-phase" alternating-current generator (Fig. 30-14*a*). It has two independent pairs of magnet coils, *a-b* and *c-d*, connected to two pairs of wires. The rotor or moving system is permanently magnetized by current from a battery or other direct-current source. The revolving rotor thrusts magnetic lines first through the vertical coils, then, a quarter cycle later, through the horizontal coils. The two output currents differ in phase by one-quarter cycle or 90°.

A simple induction motor is represented in Fig. 30-14*b*. Its vertical coils *a'-b'* are connected to one pair of wires from the two-phase generator, and the horizontal coils *c'-d'* to the other. As the rotor of the generator revolves, first the vertical coils of the motor are energized, then the horizontal coils. That is, first *a'* will be a north pole, then *c'*, then *b'*, then *d'*, and so on. In other words, a *rotating magnetic field* is produced by the fixed coils (Fig. 30-14*c*). Suppose, now, that a metallic cylinder is mounted on an axle between the poles. The rotating magnetic field will induce eddy currents in this cylinder, and will force it to rotate. In practical induction motors the currents are induced in copper wires or bars wound in the slots of a drum armature. However, these wires are not connected to any source of electrical energy as they are in an ordinary motor. Instead they are short-circuited so that the current generated by the ro-

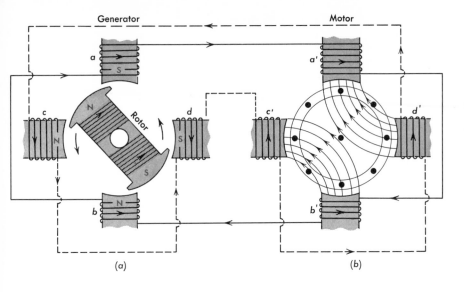

(a)   (b)

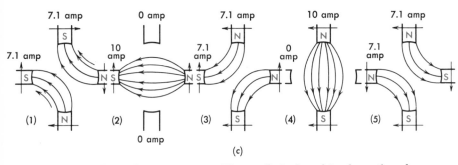

(c)

Fig. 30-14. (a) *Two-phase generator. The emf's induced in the coils* c-d *lag 90° behind those in* a-b. (b) *Induction motor. The two currents cause a rotating field.* (c) *The rotating field at one-eighth cycle intervals.*

tating magnetic field may circulate through them. This type of motor has no commutator, slip rings, or brushes.

(Most electrical energy is supplied by electrical power companies in "three-phase," with the phases differing by 120°, but the principle as here stated still holds.)

Some small induction motors are operated by ordinary two-wire, "single-phase" sources. The field system of a very simple motor of this type is diagramed in Fig. 30-15. Notice that the poles *c* and *d* are encircled by copper rings in which the alternating magnetic field

induces currents. These currents retard the magnetization of *c* and *d*. First *a* is a north pole, then *c*, then *b*, then *d*, then *a* again. Thus a rotating field is generated exactly as in the two-phase motor.

Suppose that the drum armature in Fig. 30-14*b* turned at the same speed as the rotor of the generator and the rotating field. Then the wires of the armature would not cut the lines of the revolving field and no current would be induced. Thus the drum would not be magnetized, the rotating field would exert no torque on the armature, and the motor

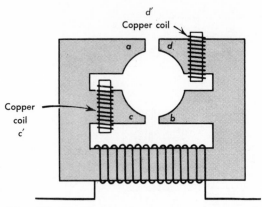

Fig. 30-15. *Single-phase induction motor. Induced currents in the rings make the magnetization of* c *and* d *lag behind that of* a *and* b. *Thus a rotating field is generated.*

could do no work. Hence the armature must rotate more slowly than the field. When the load on a motor is small, its armature rotates at high speed so that the armature windings cut few magnetic lines. As the load is progressively increased, the "slip" of the armature increases so that the induced current in the armature and the input power increase.

If the armature coils of an induction motor were not short-circuited but were connected through slip rings to a battery or other direct-current source, the rotor would be driven at exactly the same speed as the rotating field and would serve as a *synchronous motor*. Small synchronous motors having permanent magnet armatures drive electric clocks. The alternating current at the power station operates a "master" electric clock, and the frequency of the alternations is regulated so that it keeps correct time. Then all the electric clocks of the city, if once adjusted, will keep in step with the master clock and tell correct time. Another application is in controlling the stop-go traffic lights of cities. If all the traffic lights on an avenue are once synchronized so as to change color at the same instant, they will continue to do so as long as there is no failure of the supply of electrical energy.

Summary

In alternating current, the electric charges do not progress, but oscillate. An effective ampere heats a conductor at the same rate as would a direct-current ampere.

The impedance of a conductor measures its opposition to alternating current. One ohm is the impedance of a conductor through which a current of one effective ampere is produced by a potential difference of one effective volt.

The inductive reactance X_L of an inductor is given by $X_L = 2\pi f L$.

The capacitive reactance X_C of a capacitor is given by $X_C = \dfrac{1}{2\pi f C}$.

The total reactance of a circuit, including an inductor and a capacitor in series, is given by $X = X_L - X_C$, and the impedance by $Z = \sqrt{R^2 + X^2}$.

When a series electric circuit is in resonance, the inductive reactance and the capacitive reactance are equal, so that $f = 1/(2\pi\sqrt{LC})$. Thus the period $T = 2\pi\sqrt{LC}$.

Because of inductive reactance, the alternating current in an inductive circuit achieves its maximum value after the emf does. The lag θ is expressed in degrees. Power $= E_{eff} \times I_{eff} \times \cos\theta$.

$$\text{Power factor} = \frac{\text{True power}}{\text{Apparent power}} = \cos\theta = \frac{R}{Z}$$

A transformer consists of a primary coil and a secondary coil wound on an iron core. The ratio of the primary emf to the secondary emf approximately equals that of the number of turns of the respective coils. When the voltage is stepped up, the current is diminished, the power remaining approximately constant.

Questions

1. Discuss electrical resonance and compare it with mechanical resonance, explaining what terms are analogous.

2. Give some common examples of the use

of alternating and of direct current. What are the advantages of each?

3. Describe a step-up transformer. Can it be used with direct current? What is a typical efficiency?

4. Why may the alternating current in a circuit be smaller than the direct current, even though the effective emf and the d-c emf are the same in each instance?

5. How does doubling the frequency affect the reactance of (a) an inductor, (b) a capacitor?

6. Name devices that (a) conduct direct current, but not alternating current, (b) conduct alternating current, but not direct current.

7. Using an alternating-current ammeter, voltmeter, and wattmeter, how could you determine the power factor of a circuit?

8. Discuss (a) the reactance of a capacitor and (b) the reactance of an inductor, when connected to a d-c voltage supply.

9. The electric motors of the air conditioning system of a department store cause a power factor of 0.76. How can this power factor be increased?

10. How can lights in an a-c circuit be dimmed without varying a resistance, assuming constant voltage?

11. Estimate the maximum instantaneous voltage obtainable from (a) a "60-cycle 120-volt" line, (b) a "50-cycle 240-volt" line.

12. Show by sketches the phase relation between the a-c voltage and current for circuits that are (a) resistive, (b) inductively reactive, or (c) capacitively reactive.

13. Why do we not notice flicker of lamps that are operated on 60-cycle/sec voltage? What is the flicker frequency?

14. Estimate the upper frequency limit for a resonant circuit containing conventional inductors and capacitors.

15. What is the difference between a kilovolt-ampere and a kilowatt?

16. How could you verify that an a-c frequency is exactly 60 cycles/sec?

17. What happens when the switch of the circuit in Fig. 30-16 is closed?

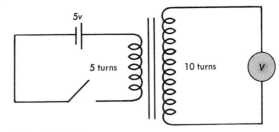

Fig. 30-16.

18. Is a low or a high power factor to be desired in an a-c circuit? Discuss.

Problems

A

1. When a certain coil is connected to a 120-v d-c source, the current is 0.30 amp. When connected to a 120-effective-volt a-c source, the alternating current is 0.24 effective amp. Find (a) the resistance, (b) the impedance, (c) the inductive reactance of the coil.

2. What is the self-inductance of the coil in problem 1 if the frequency is (a) 60 cycles/sec, (b) 6.0×10^4 cycles/sec?

3. In problem 1, find (a) the phase lag and (b) the power in each case.

4. What is the capacitive reactance of a capacitor of 6.0 microfarads if the frequency is (a) 60 cycles/sec, (b) 6000 cycles/sec?

5. What is the reactance of a coil whose impedance is 283 Ω and whose resistance is 200 Ω?

6. The input coil of a transformer has 1500 turns; the output coil has 100 turns. The effective output current is 2.0 amp at 120 v. (a) What are the effective input current and voltage if the efficiency is 100%? (b) What is the input power?

B

7. The voltage of a power transmission line is 20×10^4 effective volts, and the current is 100 effective amperes. What are (a) the power input to the system and (b) the power loss (I^2R) in heating each kilometer of wire? The resistance of the line is 0.30 Ω/km.

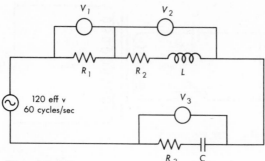

Fig. 30-17.

8. There are 400 turns on the input coil of a toy transformer and 40 turns on the output coil. Suppose that a youthful experimenter connected a 120-v alternating-current source to the output coil. What effective voltage would there be across the terminals of the input coil? What would be some of the probable consequences?

9. A capacitor of capacitance $5.00/\pi$ microfarads is connected to a 60-cycle, 120-effective-volt source. What is the capacitive reactance of the capacitor?

10. Find the impedance at a frequency of 60 cycles/sec of (a) a pure resistance of 20 Ω, (b) a coil of inductance $1/(2\pi)$ henry and zero resistance, (c) a capacitor of capacitance $50/\pi$ microfarads.

11. A coil of 0.1-henry inductance and 15-Ω resistance is connected across a 240-v, 60-cycle line. Find (a) the reactance of the coil, (b) its impedance, (c) the effective current in it, (d) the angle by which this current will lag the applied potential difference.

12. Make graphs plotting (a) X_L versus f for $L = 1/(2\pi)$ henries for frequencies of 10^4, 10^5, 10^6, 10^7 cycles/sec; (b) X_C versus f for $C = 1/(2\pi)$ farads for the same frequencies.

13. In a 60-cycle a-c circuit, a voltmeter reads 115 effective volts and an ammeter reads 1.00 effective ampere. If the phase lag is 21.5°, what is the power?

C

14. A capacitor having a capacitance of 5.0 microfarads is connected to an a-c source

of frequency 60 cycles/sec. (a) What is the capacitive reactance of the capacitor? (b) What inductance would give a numerically equal reactance at that frequency?

15. The alternating current in a circuit is 10.0 effective amperes when the voltage across its terminals is 120 effective volts. A wattmeter indicates that the power is 1100 watts. What are (a) the power factor, (b) the phase lag?

16. What is the capacitance of a capacitor whose capacitive reactance is numerically equal to that of a 0.40-henry coil at 60 cycles/sec?

17. What is the resonance frequency of a circuit that includes a coil of inductance 1.0×10^{-3} henry and a capacitor of capacitance 1.0 microfarad?

18. The resistance of a circuit is 60 Ω and its inductive reactance is 80 Ω. (a) What is its impedance? (b) What will be the impedance of the combination if a capacitor of capacitive reactance 80 Ω is connected in the circuit?

19. A series circuit includes a resistance of 4.0 Ω, a capacitive reactance of 3.0 Ω, and a variable inductance coil. What is the impedance when the inductive reactance is (a) 2.0 Ω, (b) 3.0 Ω, (c) 4.0 Ω, (d) 5.0 Ω? Make a reactance-impedance graph.

20. In Fig. 30-17, $R_1 = 10$ Ω, $R_2 = 20$ Ω, $R_3 = 30$ Ω, $L = 1/(4\pi)$ henry, and $C = 1/(6000\pi)$ farad. (a) What are the readings of the three voltmeters? (b) Why is their algebraic sum not equal to 120 effective volts?

21. (a) What is the ratio of the impedance of a 0.20-henry choke coil at 6.0×10^6 cycles/sec to that at 60 cycles/sec? (b) What is the ratio of the impedance of a 2.0×10^{-8} farad capacitor at 6.0×10^6 cycles/sec to that at 60 cycles/sec?

For Further Reading

Galambos, Robert, *Nerves and Muscles,* Doubleday and Company, Garden City, 1962.

CHAPTER THIRTY-ONE

An Introduction to Tube Electronics

Electronics is the study of the motion of electrons and the application of such knowledge to practical problems in communications and control. Electronics is firmly based on the fundamental principles of physics. Its applications, however, are an important branch of engineering.

In *communications,* the problem of electronics is to convey information—an intelligible series of signals—from one location to another. The information may be a telephone message carried over wires from one city to another as described in Chapter 29. In radiotelephone communication, electromagnetic waves, generated by the sending station, travel through space and are detected by the receiving station; a cable connecting the two stations is not needed. Information conveyed electronically need not be in the form of words; it may consist of an electronic report by a measuring instrument to a central control station by the methods of *telemetry.* As the space probe Pioneer IV, for example, traveled through interplanetary space on its orbit around the sun, it reported by radio to earth the magnitude of the magnetic fields through which it was passing. The information was in the form of coded signals impressed by the fluxmeter in the satellite upon the radio beam. Music and the television picture are also information. Radio and television broadcasting techniques and the equipment for sending and receiving these signals are an important part of electronics.

Electronics is also concerned with *controls.* The thickness of the tin plating on a moving steel sheet is electronically monitored in a factory that makes cans for the food industry. If the plating becomes too thin or too thick, the electronic monitor senses this and sends a correcting signal back up the production line, meanwhile cutting out and discarding the defective strip. Electronic devices control the flow rates and compositions of oil refinery products; some refineries are almost completely automatic. Radar, used for the control of air and sea traffic in crowded lanes and in unfavorable weather conditions, is an application of electronics. Many electronic safety devices are used in industry—for example, the photoelectric safety lock that prevents a press from operating until the workman's hands are out of danger.

Electronic devices are extremely useful to scientists, especially to physicists, but increasingly to biological scientists, chemists, and medical researchers. Amplifiers, particle counters used in nuclear physics, devices for measuring very large and very small currents and voltages, oscilloscopes for studying rapid changes in voltages, pH-meters, electrocardiographs, instruments for measuring residual pressures in highly evacuated containers—all of these and many more are based on electronics.

The field of electronics is so large that we cannot give you more than a brief introduction to it. We shall first discuss some of the basic elements in electronic devices. Following this, several of the more fundamental electronic circuits will be discussed. We shall stress the applications of the fundamental principles of physics. In this chapter we shall discuss electronic circuits in which vacuum tubes or gas-filled tubes are used, leaving solid-state devices, such as transistors, to a later chapter.

Some Common Elements in Tube Electronics

Electronic devices are available in a great variety of forms, designed to solve many special problems in the sciences and engineering. There are some elements common to all, however. Let us examine these.

Evacuated or gas-filled space. The electrons whose motions we are interested in usually move in a highly evacuated container (pressures usually less than 10^{-5} cm-Hg) made of glass or metal. In a vacuum, the electrons are free of the restraints of the atoms that bind them in conductors. They are also free to move without colliding with gas molecules. The result is that they are extremely responsive to forces applied to them and can quickly execute commands to speed up or slow down. This rapid response accounts for their ability to generate very-high-frequency radio waves, to count nuclear particles rapidly, and to operate controls with very little loss of time.

In some devices, however, mercury vapor,

argon, or some other gas is introduced into the operating space of the electrons. Larger currents can be drawn from the device, for reasons that we shall look into later, but the response of the device becomes much more sluggish.

A source of electrons. Given a suitable enclosure we must obtain a supply of electrons. There are three primary ways to do this: (1) by the thermionic effect (Chapter 25) in which electrons are "boiled off" a heated cathode (Fig. 31-1*a*); (2) by the photoelectric effect (Chapter 39) in which electrons are emitted by a suitably prepared metal plate (Fig. 31-1*b*) when light of a sufficiently high energy strikes it—the photocell is an example; and (3) by field emission (Fig. 31-1*c*) in which electrons are pulled out of a pointed rod by a large electric field at its surface. The first two methods are the most common. Thermionic cathodes are generally heated indirectly by heater filaments connected to an a-c source; this reduces hum or a-c interference.

Electric and magnetic fields. Fields are used to control the motion of the electrons: to make them speed up or slow down, to form narrow beams of them, to make the electrons move on circular paths, to send them pouring through an opening in a steady stream or in a stream that increases and decreases in density in a periodic fashion, or to group them into clusters. Electric fields, you will recall, exert on electrons forces that are oppositely directed to the electric field strength and have a magnitude $F = eE$. Magnetic fields exert sideways deflecting forces on moving electrons; the direction of the force is given by the right-hand motor rule, and the magnitude by $F = Bev \sin \theta$. We treat the electron current as one in the opposite direction to the conventional electric current.

Electrodes and coils supply the fields. In addition to cathodes, electronic tubes contain electrodes to supply the electric fields used in controlling the electrons' motion and to

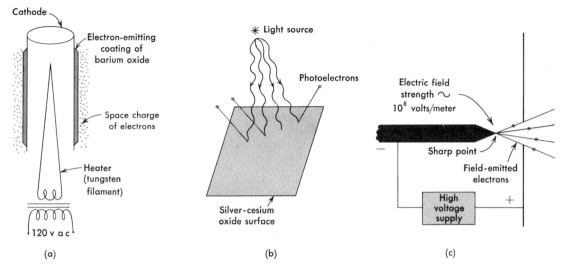

Fig. 31-1. *Electron emission:* (a) *thermionic emission,* (b) *photoelectric emission,* (c) *field emission.*

take the electrons out of the tubes (Fig. 31-2). The plate of a vacuum diode (Chapter 25) is a simple example: the electric field between the plate and the cathode regulates the flow of electrons in the tube, and the metal plate serves to collect the electrons and conduct them into the external circuit. Electrodes come in a variety of shapes: metal disks with holes in them, spirals of wire, cylinders, parallel plates, rectangular boxes, and screens, for example. All are connected to the outside of the tube by wires that pass through vacuum-tight seals in the glass or metal walls. By means of these external leads, the desired voltages can be applied to the electrodes. Magnetic fields are usually supplied by permanent magnets or electromagnet coils mounted on the outside of the tube.

Inputs and outputs. Electrical energy is supplied to vacuum tubes in a variety of forms: the feeble alternating currents set up in a radio antenna when electromagnetic waves cross it, the photoelectric current produced in a phototube when light strikes it, the electrical

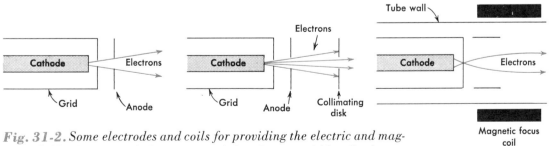

Fig. 31-2. *Some electrodes and coils for providing the electric and magnetic fields that control the motion of the electrons. (After Bachman,* Experimental Electronics, *John Wiley and Sons, 1948.)*

currents in the human brain or heart, the sharp voltage change—or pulse—produced when an energetic particle strikes a Geiger counter. An electronic device responds to the input that it is designed to handle by delivering output energy to a load. One device may amplify the feeble radio signal by adding energy from its own source (Fig. 31-3a). A phototube may receive an input of energy from the headlight beams of a car and send a current through a relay, setting into operation a motor that opens the garage door (Fig. 31-3b) in one type of garage door opener. Another device may receive the pulse from a Geiger counter and record the fact that the counter has been struck by an energetic particle.

Signals are the "language" of electronic devices. By a signal we mean a voltage varying with time in a characteristic way (Fig. 31-4). An alternating voltage varying sinusoidally with time is one kind of signal (Fig. 31-4a). More complex signals can be analyzed by Fourier analysis (Chapter 20) into the sum of many sine-wave signals. Pulses are more abrupt changes of voltage (Fig. 31-4e). An electronic device may be designed to amplify a signal, increasing the size of the voltage change and inevitably dis-

torting it somewhat in the process. Or the device may be designed to change the shape of the signal completely—producing square waves, for example, when the input consists of sine waves. As a signal passes through an electronic system, many different circuits affect it, changing its size, its shape, and so on. Complicated electronic circuit diagrams become comprehensible when one understands that they represent a connected series of operations on an electrical signal passing through the system.

Power Supplies

Most electronic equipment contains a power supply with a rectifier for converting the commercially available alternating voltages into the higher direct voltages that are applied to the electrodes of vacuum tubes. How is this done?

The simplest vacuum tube is the diode or two-element tube discussed in Chapter 25. The diode consists of a cathode, or electron-emitting electrode, and a plate, or electron-collecting electrode. Usually the cathode is a surface so treated that it emits electrons copiously when heated by an enclosed glowing heater filament. When the plate is at a positive

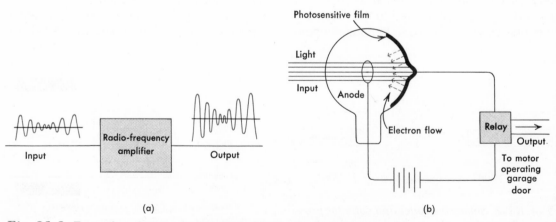

(a)   (b)

Fig. 31-3. Examples of inputs and outputs: (a) *a radio-frequency amplifier,* (b) *a phototube control for a garage door.*

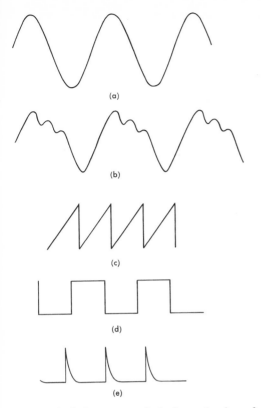

Fig. 31-4. Some signals (voltage is plotted vertically, time horizontally); (a) sine wave, (b) complex harmonic, (c) sawtooth, (d) square wave, (e) pulses.

potential with respect to the cathode, electrons from the negative space charge that surrounds the cathode are attracted to the plate. When the plate is at the same potential as, or at lower potential than, the cathode, electrons are not attracted.

Figure 31-5 shows how a diode can be used in a simple half-wave rectifier power supply. A step-up iron-core transformer provides an alternating voltage between the plate and the cathode of the diode. (The cathode is heated by a separate step-down winding on the power transformer.) When, during the alternating voltage cycle, the plate is more positive than the cathode, electrons

flow from the space charge to the plate, strike the plate, flow through the resistor *R*, and return to the cathode where they are re-emitted. During the negative-voltage half of the cycle, the current in *R* is zero. Thus the *iR* voltage across *R*—the output voltage—is a half-wave pulsating direct voltage, as shown at the right of Fig. 31-5.

The pulsating voltage produced by a half-wave rectifier is undesirable for many uses in electronics. The full-wave filtered rectifier, which makes use of the cooperative action of *two* diodes within one tube, produces a steady direct voltage (Fig. 31-6). The double diode contains two plates and one cathode. When plate P_1 is positive during the alternating voltage cycle, electrons flow from the space charge to P_1, through the upper half of the step-up winding on the transformer to the center tap *T*, through the resistor *R* at the right of the diagram, and back to the cathode. When plate P_2 is positive, the electrons flow through the *bottom* half of the step-up winding to *T*, but their path thereafter is the same as when P_1 is positive. *Electrons flow through R in the same direction during both halves of the cycle.*

What does the filter circuit do? The kind shown in Fig. 31-6 consists of two capacitors C_1 and C_2 and an inductor, or choke, *L*. All three serve as inertial elements and act against any decrease in the electron flow. The capacitors store charge when the electron flow is at its maximum and then give up electrons when the current diminishes. The inductor builds up its magnetic lines when the current is greatest. When the current diminishes, magnetic lines are cut self-inductively by the coil, and the electromotive force is such as to maintain the current. Another way of looking at the action of the filter circuit is to recall that, for high-frequency alternating current, the impedance of capacitors is small and the impedance of inductors is large. The inductor blocks from the output the high-frequency components of the pulsating cur-

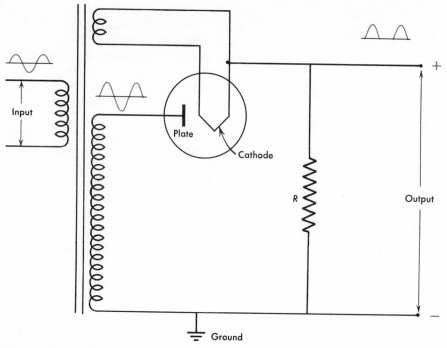

Fig. 31-5. *A half-wave rectifier. The diode conducts, that is, electrons are attracted to the plate, only when the plate is more positive than the cathode. The input is an alternating voltage and the output is a half-wave pulsating direct voltage.*

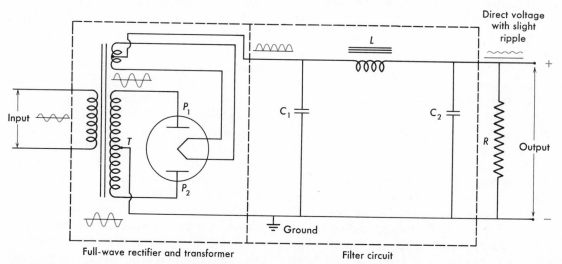

Fig. 31-6. *A power supply containing a full-wave rectifier and a filter circuit. The rectifier converts an alternating voltage into a pulsating full-wave direct voltage, and the filter circuit smooths out the fluctuations in the voltage, producing a steady, direct voltage.*

rents and the capacitors shunt them to ground. Thus only the low-frequency currents—direct currents with a little ripple in them—get through the filter.

Power supplies that are required to deliver more than a few milliamperes of current use gas-filled diodes instead of vacuum diodes. In a diode filled with mercury vapor, the mercury atoms become ionized and tend to cancel the effect of the electron space charge in limiting the tube current. Larger current outputs are thus possible.

Amplifiers

Much of the usefulness of electronic circuits depends upon their ability to amplify small alternating voltages or pulses. Amplification first became possible when the three-element vacuum tube, or *triode*, was developed by Lee de Forest in 1907.

Triodes contain a cathode and a plate separated by a *grid* (Fig. 31-7a), a spiral wire electrode. The nearness of the grid to the space charge makes changes in its potential with respect to the cathode extremely effective in controlling the current of electrons to the plate. Making the grid more negative decreases the plate current; making it less negative increases the plate current. Thus the grid acts like a valve in a water pipe.

To appreciate the extent of grid control, let us consider the results of measuring (1) the plate current as the plate voltage (the plate-to-cathode potential difference) is varied at several constant grid voltages and (2) the plate current as the grid voltage (the grid-to-cathode potential difference) is varied at several constant plate voltages (Fig. 31-7b). The first of these yields the plate characteristic $(V_p\text{-}I_p)$ curve (Fig. 31-8a) and the second the grid characteristic $(V_g\text{-}I_p)$ curve (Fig. 31-8b). An increase of grid voltage from -12 v to -8 v causes the plate current to change from about 1.5 milliamp to 4.5 milliamp (Fig. 31-8b) at a constant plate voltage of 120 v. To increase the plate current from 1.5 milliamp to 4.5 milliamp at a constant grid voltage of -8 v requires that the plate voltage be increased from 80 v to 120 v (Fig. 31-8a). Thus a change of only 4 v on the grid produces the same change in plate current as a change of 40 v on the plate. *The grid is 10 times as effective in controlling the plate current as the plate voltage.*

We define the amplification factor μ of the triode as the ratio of the change in plate voltage to the change in grid voltage to produce

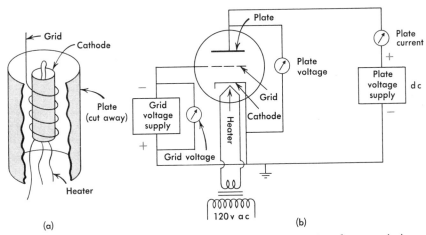

(a)

(b)

Fig. 31-7. (a) *A triode.* (b) *A circuit for determining its characteristic curves.*

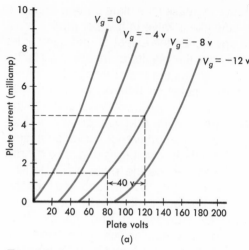

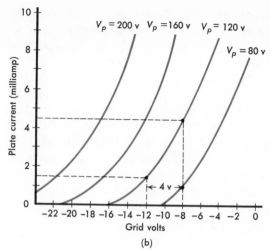

Fig. 31-8. (a) *A plate voltage-plate current curve for a certain triode.* (b) *A grid voltage-plate current curve for the same triode.*

the same change in plate current. In our example, the amplification factor is 10.

$$\mu = \frac{\Delta V_p}{\Delta V_g} \qquad (I_p \text{ constant})$$

What we are really interested in, however, is how the triode amplifies a small input signal into a larger output signal. Basically this is what happens: a small alternating voltage on the grid causes relatively large variations in the plate current. These changes in the plate current cause larger iR voltage changes across the load resistor R_L. The large voltage changes constitute the amplified signal.

In the amplifier circuit of Fig. 31-9*a*, the plate voltage supply is fixed at 200 v. The load resistor R_L is 20,000 Ω. The *grid bias voltage* is the grid-cathode voltage chosen to prevent the grid from becoming positively charged and so drawing electrons from the space charge; in the circuit shown, the grid bias is −8 v. The input voltage is an alternating sine-wave voltage whose amplitude is 4 v; thus the grid voltage swings up to −4 v and down to −12 v, always operating about its bias voltage of −8 v. What is the output signal?

To answer this question, we first draw the *load line* on the V_p-I_p curves of Fig. 31-8*a* and so obtain Fig. 31-9*b*. The load line is the graph of all corresponding values of plate voltage and plate current for a given load resistor and plate voltage supply, in our case 20,000 Ω and 200 v. One point on the load line is the plate voltage of 200 v and plate current of zero, because when electrons are not flowing to the plate the iR_L voltage drop is zero and the plate voltage equals the plate supply voltage. Another point on the load line is the plate voltage of zero and the plate current of 10 milliamp, because when the tube is short-circuited the supply voltage of 200 v is the potential drop across R_L and this potential drop, according to Ohm's law, produces a current of 10 milliamp in a 20,000-Ω resistor. We can connect these two points (200 v, 0 milliamp and 0 v, 10 milliamp) by a straight line—the load line for our circuit.

Where the load line intersects the V_p-I_p curve for a grid voltage of −8 v—our bias voltage—we have the operating point of the amplifier: a plate current of about 4.3 milliamp and a plate voltage of about 120 v. The

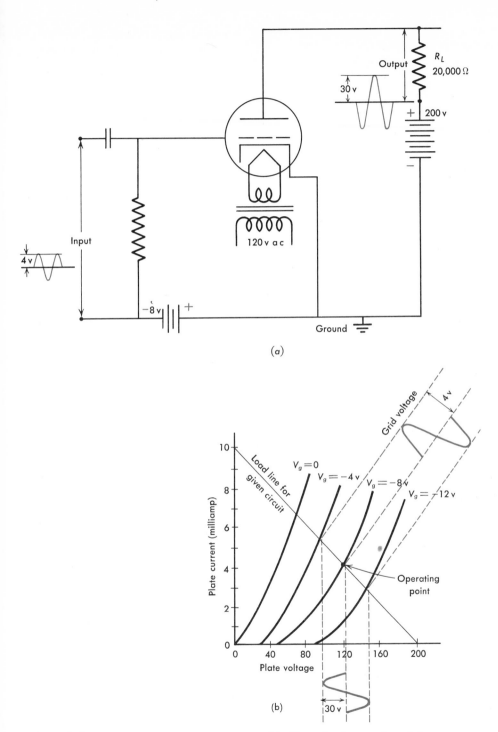

Fig. 31-9. (a) *A triode amplifier circuit.* (b) *How the input signal is amplified.*

plate current and voltage will change from their operating point values when the grid voltage changes from −8 v. When the input signal reaches the amplifier, the grid voltage varies, rising to −4 v and falling to −12 v sinusoidally. If you follow the dotted lines in Fig. 31-9b from the grid voltage changes at the top of the graph to the plate voltage changes at the bottom, you will see that the plate voltage variations are much larger than those of the grid. The amplitude of the plate voltage change is about 30 v compared with the grid voltage change of 4 v. Thus the input signal was amplified about 7.5 times to give the output signal. We say that the *gain* is 7.5. The energy to do this comes from the plate voltage power supply. Note that the plate voltage changes are 180° out of phase with the grid voltage changes. (Why?)

The output signal from the amplifier stage shown in Fig. 31-9 can be fed into another stage for a further gain in amplification. The total gain produced by several amplifier stages is the *product* of their separate gains. However, one cannot usefully amplify a signal indefinitely because of distortions produced in the amplified signal. Five-element tubes—pentodes—that contain three grids are used in place of triodes in amplifiers when distortion must be held to a minimum. One of these grids plays the same role as the grid of a triode; the others improve the frequency response of the tube so that it amplifies all frequencies approximately the same. In addition, the pentode has a higher amplification factor than a triode.

Oscillators

A device in which electric charges can surge to and fro repeatedly with a certain frequency is called an *electrical oscillator.* Circuits containing an inductor in parallel with a capacitor—called *LC* circuits or tank circuits—are often a part of an electronic oscillator. In Fig. 31-10 the two plates of a capacitor of capacitance *C* are electrically connected through the coil of self-inductance *L*. Suppose that electrons are suddenly added to one plate of the capacitor. They will repel one another and will flow along the connecting wire. The coil, having self-inductance or electrical inertia, will impede the growth of current at first, and afterward will oppose its decrease so that electrons will accumulate on the other plate, charging it negatively. Then a reverse current will carry the charges back again. This surging to and fro occurs repeatedly, but the alternating current gradually decreases in amplitude (is "damped") because of the resistance of the circuit.

The period of the oscillation depends upon two quantities: the capacitance and the self-inductance of the oscillator. Suppose that the two plates of the capacitor are moved closer together so that it has greater capacitance. The electrons will repel one another less strongly than before, the current will be smaller, and a greater time will be required for each oscillation. Thus the period of oscillation will be increased. If we replace the coil by one having more turns and therefore greater self-inductance, the coil will offer greater opposition to change of current so that more time will be required for the charges to flow from plate to plate. The period of oscillation may be increased by increasing either the capacitance or the self-inductance of the system. The period *T* of an oscillator of capacitance *C* and self-inductance *L* is given by

$$T = 2\pi \sqrt{LC}$$

Example. The capacitance of an oscillator is 9.0×10^{-11} farad, and its self-inductance is 0.40 henry. What are the period and the frequency of the system?

$$T = 2\pi \sqrt{0.40 \text{ henry} \times (9 \times 10^{-11} \text{ farad})}$$

$$= 2\pi \sqrt{\frac{0.40 \text{ v}}{\text{amp/sec}} \times \frac{9.0 \times 10^{-11} \text{ amp-sec}}{\text{v}}}$$

$$= 12\pi \times 10^{-6} \text{ sec}$$

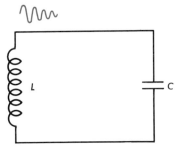

Fig. 31-10. An LC *circuit. Electrons can surge to and fro through the coil whose self-inductance is* L. *The period of oscillation is given by* $T = 2\pi\sqrt{LC}$. *The oscillations are damped as indicated by the wave form at the top of the figure. This is due to the resistance in* L *and* C.

$$f = \frac{1}{T} = 27{,}000 \,\frac{1}{\text{sec}}$$

$$= 27 \text{ kilocycles/sec} = 27 \text{ kilohertz}^*$$
$$= 27 \text{ kH}$$

Electronic circuits include capacitors of constant or of variable capacitance. "Fixed" capacitors are often manufactured by placing sheets of insulating paper between layers of tinfoil and connecting alternate metal sheets together to form sets of "plates" which may be oppositely charged. The capacitance of such a capacitor is relatively great because the total area of the plates is large and because they are close together. Variable capacitors are employed to tune a radio receiver so that it is in resonance with a transmitting station. One set of plates, rigidly mounted, is connected to one terminal (Fig. 31-11). The other set, connected to the other terminal of the *LC* circuit, is mounted on an axle and may be rotated so as to be completely within the first set. When the rotator is in this position, the areas of the plates opposed to each other are greatest, and the capacitance is maximum. The capacitance is varied by turning a knob attached to the axle. Thus the *LC* circuit is

---

* The *hertz* (H)—named for Heinrich Hertz (Chapter 32) is a unit of frequency equivalent to 1 cycle/sec.

tuned so as to have the same frequency as the incoming radio waves, that is, to be in resonance with them. The current in the coil of the *LC* circuit is then large.

In a complete electronic oscillator circuit (Fig. 31-12) used to generate high-frequency alternating currents, both the plate circuit and the grid circuit are in oscillation. Each contains an *LC* circuit or its equivalent. The oscillating grid voltage is amplified and applied to the plate circuit in phase with the oscillating plate current so as to maintain the oscillations in the plate circuit. To overcome the losses of energy in the grid circuit due to the resistance of the inductance coil, energy is *fed back* from the plate circuit to the grid circuit. Thus the grid "paces" the oscillations of the plate, and energy fed back from the plate to the grid maintains the grid oscillations.

There are various ways by which *feedback* from the plate to the grid occurs: by magnetic-flux coupling of the two circuits (transformer action), by a special feedback network, or by the capacitance between the plate and the grid within the tube itself. We shall illustrate the last of these. In Fig. 31-12, the L_1C_1 circuit and the L_2C_2 circuit are tuned to approximately the same frequency. Small random fluctuations in voltage soon set the whole oscillator into oscillation when it is turned on.

Suppose the grid voltage is increasing at

Fig. 31-11. Variable capacitor. In the position shown, it has minimum capacitance.

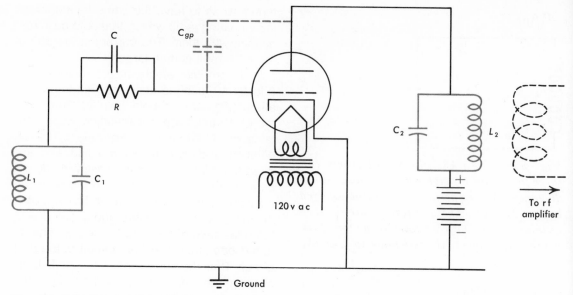

Fig. 31-12. *An oscillator. Energy is fed back from the plate circuit to the grid circuit by way of the plate-grid capacitance C_{gp} to maintain oscillations in the grid L_1C_1 circuit. The oscillating grid voltage is amplified and maintains the oscillations in the plate L_2C_2 circuit.*

one point in the grid oscillation. This will increase the plate current through L_2, the voltage across L_2 will increase, and the plate voltage will drop. The plate and the grid together within the tube act like the plates of a tiny capacitor C_{gp} so a slightly lower (less positive) voltage on the plate induces a slightly higher (less negative) voltage on the grid. This increase in voltage gives the grid L_1C_1 circuit a "push" in the right phase to counteract the damping of its oscillations. Thus the grid oscillations and the plate oscillations sustain each other. The energy for doing this, of course, comes from the plate voltage supply. The frequency of oscillations is slightly less than the natural frequency of the *LC* circuits. The resistor *R* helps to set the bias on the grid, and the capacitor *C* provides a low-impedance path around *R* for the alternating currents.

Ordinarily one does not attempt to draw much power from an oscillator to avoid changing its frequency. Hence, tiny high-frequency oscillations are produced by an oscillator and then amplified. The secondary of a transformer coupling the L_2 coil of the oscillator to the input of an amplifier is shown (dotted) at the right of Fig. 31-12. The transformer that couples the oscillator to the amplifier is an iron-core transformer when the frequencies of the oscillations are less than about 20,000 cycles/sec—*audio frequencies* (af). At *radio frequencies* (rf), frequencies above 100,000 cycles per second, air-core transformers are used because the changes in the magnetization of iron are too slow to follow the high-frequency oscillations.

At ultra-high frequencies—those used in radar exceed 10^9 H, a billion cycles per second, for example—techniques different from those described here must be used for generating the high-frequency currents. Capacitive impedances between electrodes become so low that the tubes we have been dealing with are essentially shorted out. Also, the time required for an electron to move from the space charge to the plate becomes unmanageably long at these very high frequencies. Special tubes—magnetrons, klystrons, and traveling-wave tubes—have been designed to generate microwave alternating currents. These tubes and their related circuits are described in articles referred to at the end of this chapter.

Modulation and Demodulation

As we shall see in the next chapter, electromagnetic waves are emitted when electric charges oscillate at high frequencies. The alternating currents on the antenna of a radio transmitter must have frequencies in the radio-frequency range if the broadcast radio waves are to be emitted at sufficiently large intensities so that they can be detected by radio receivers at distances of hundreds, thousands, or—in the case of space-probe transmitters—millions of miles. However, you will recall that audible sound waves have

frequencies between about 20 cycles/sec and 20,000 cycles/sec as do the audio-frequency alternating currents that are the electrical counterparts of the sound waves. So the electronic-circuit problem of transmitting sound by radio is how to add low-frequency audio signals to a high-frequency "carrier," generated by the oscillator of the transmitter, and how to recover the audio signals from the carrier in the receiver. The process of adding a message signal to the carrier is called *modulation* and that of recovering the message signal is called *demodulation* or *detection*.

Suppose that your radio receiver is tuned to a station that has a frequency of 700 kilocycles/sec. When no message is being sent, "carrier" signals of constant amplitude cause a constant current through your receiver (Fig. 31-13a,b) and no sounds emerge from the loudspeaker. Suppose that someone holds a vibrating tuning fork of frequency 1000 cycles/sec near the transmitter microphone, which—like the mouthpiece of a telephone (Chapter 29)—sets up audio-frequency cur-

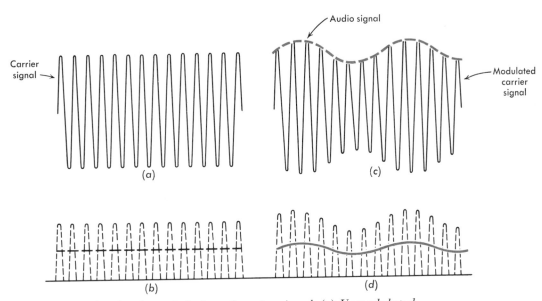

Fig. 31-13. Amplitude modulation of carrier signal. (a) Unmodulated carrier, (b) and rectified. (c) Modulated, (d) and rectified.

rents. Then the carrier signals are *amplitude modulated* so that their amplitude varies with a frequency of 1000 times per second, although their frequency is still 700 kilocycles/sec. When the modulated radio waves from the transmitter fall on the antenna of the receiver, the detector circuit of the receiver demodulates or detects the signal by removing the high-frequency oscillations due to the carrier signal. The current through the loudspeaker varies and reproduces the sound of the fork (Fig. 31-13c,d). In frequency modulation (FM), the amplitude of the modulated carrier is constant, but its frequency varies at a rate controlled by the audio signal. Let us see how each of these steps is carried out.

The first step in the process of modulation is the production of an electrical audio signal in the microphone of the transmitter. Many radio transmitters have *crystal microphones* instead of the carbon microphones used in telephones. A thin sheet of crystal, often Ro-

chelle salt, has metal films on its opposite faces. Increase of pressure distorts the crystal and develops a potential difference between the films (the piezoelectric effect). When someone speaks into a crystal microphone, the varying pressure caused by his voice waves generates a varying voltage which causes the current through the transformer to vary in the same way that a current through a carbon microphone varies. (Another example of the use of crystals in circuits is the crystal-type phonograph pickup, in which the needle varies the pressure on a crystal so as to generate voltages which, when amplified, make the phonograph loudspeaker emit sounds.)

The audio signal generated in the microphone is amplified in an audio amplifier and added to the alternating carrier voltage in the modulator circuit. In *plate* modulation (Fig. 31-14), the audio signal adds or subtracts from the plate voltage. The carrier signal from the high-frequency oscillator is applied, after amplification, to the grid of the modulator

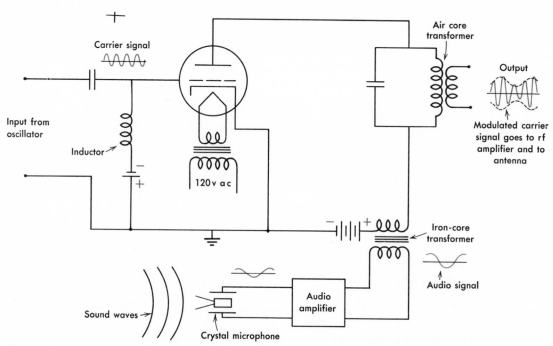

Fig. 31-14. Plate modulation.

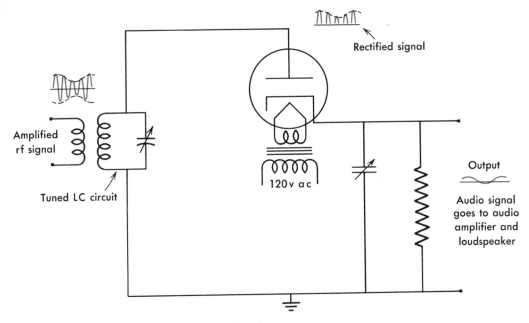

Fig. 31-15. A simple diode detector circuit.

tube. When the audio voltage is small, the plate voltage is relatively small, and the carrier signal on the grid produces a comparatively low-amplitude carrier-frequency signal in the plate circuit. When the audio voltage is large, the plate voltage is large and the carrier-frequency signal in the plate circuit has a large amplitude. As the audio signal rises and falls in amplitude, therefore, the amplitude of the carrier signal rises and falls similarly, as in Fig. 31-13c. This modulated high-frequency signal is then further amplified and fed to the antenna of the transmitter, from which radio waves are broadcast.

The process of demodulation in the radio receiver is essentially one of rectification of the high-frequency carrier signal that is set up in the antenna when radio waves sweep across it. A diode accomplishes this very well (Fig. 31-15).

A loudspeaker should reproduce sounds of all frequencies equally well so as to avoid distortion. In the *electrodynamic speaker* (Fig. 31-16), a small coil of wire is fastened to

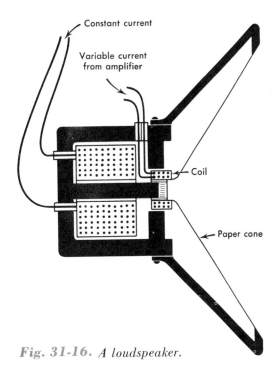

Fig. 31-16. A loudspeaker.

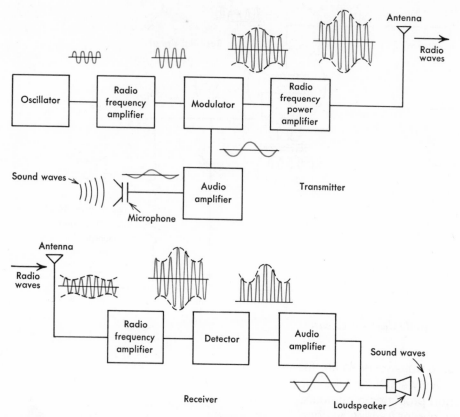

Fig. 31-17. A block diagram of a simple radio transmitter and receiver.

a cone made of paper so flabby that it will not vibrate with any definite "preferred" frequency. This coil is suspended in the field of a strong electromagnet—or, in some speakers, of a small permanent magnet. The varying electric currents from the audio amplifier force the coil and cone to vibrate so as to reproduce the sound.

The steps in radio transmission and reception are summarized in a block diagram in Fig. 31-17. The basic principles we have described apply to commercial transmitters and receivers. However, the circuits must be more complicated than those described here in order to provide high-fidelity radio transmission without interference from sources of radio "noise."

Sweep Circuits

Often we want the voltage between two electrodes to increase proportionately with time to a certain potential difference, to drop rapidly to zero, and to repeat this cycle at a desired frequency. The circuit that generates such signals is called a *sweep circuit.* The signals are used in cathode-ray oscilloscopes as a base for timing very small intervals. Your television receiver contains sweep circuits that cause the electron beam within the picture tube to move back and forth across the screen, scanning the screen and producing the television picture.

To understand the operation of a sweep circuit, let us consider a simple one that includes a capacitor that is charged by a battery or other voltage supply. The charging current passes through a resistor (Fig. 31-18a). The time to charge the capacitor to a given fraction of its final voltage depends upon both the capacitance C and the resistance R. The growth and decay of voltage across the plates of a capacitor follow exponential laws in a way similar to the growth and decay of currents in self-inductive circuits (Chapter 29 and Appendix E). When a capacitor of capacitance C in series with resistance R is charged by being connected to a battery of voltage V_0, the voltage v at time t is given by

$$v = V_0(1 - e^{-(t/RC)})$$

where e, as before, is the base of natural logarithms and equals approximately 2.72. When a charged capacitor is *discharged* through resistance R, the voltage across the plates v decreases with time as follows:

$$v = V_0 e^{-(t/RC)}$$

With switch S_2 open (Fig. 31-18a), suppose that switch S_1 is closed, connecting the battery to the capacitor through the resistor. The capacitor plates charge, eventually reaching the battery potential difference of 400 v; the charging current then stops. The time constant of the circuit, the time for the capacitor voltage to reach a certain fraction — $(1 - 1/e)$ or 63% — of its final voltage, always equals the product RC of the capacitance and the resistance. If C is 10^{-8} farad and R is $10^6 \Omega$, the RC time constant is 0.01 sec; it takes this long for the voltage across the capacitor to reach 63% of 400 v or 252 v.

The charging curve (Fig. 31-18b) is roughly a straight line at the beginning of the charging process. If we use only the beginning of the charging curve, we will have a voltage that increases in direct proportion to the time. For example, analysis shows that it takes about 1/10 of the time constant — 0.001 sec — for the voltage to rise to 1/10 of 400 v or

40 v. The 40-v increase will be quite linear with time. How can we do this? If some mechanism can be arranged to close switch S_2 0.001 sec after S_1 has been closed and every 0.001 sec thereafter, the capacitor will be discharged and returned to zero potential difference every 0.001 sec. The voltage across the capacitor, shown as a dashed curve in Fig. 31-18b, will be a sweep voltage, rising proportionately with time to 40 v, dropping rapidly to zero, and then repeating this with a frequency of 1000/sec.

A triode filled with argon gas or mercury vapor, called a *thyratron,* acts as an electronic switch (Fig. 31-19) to replace switch S_2 of Fig. 31-18. When the plate voltage of a thyratron rises to a critical value for a given grid voltage, the tube conducts a large current, because the positive ions created by stripping electrons from the gas atoms move toward the filament and lessen the effect of the space charge there. When the plate voltage is lowered to another critical value, the thyratron stops conducting. The conducting thyratron effectively short-circuits the capacitor, and the non-conducting thyratron allows the capacitor to charge. The voltage across the capacitor drops when the thyratron conducts and simultaneously the plate voltage of the thyratron decreases until the thyratron stops conducting. Then the charging of the capacitor starts again. The voltage across the capacitor rises linearly from the voltage at which the thyratron stops conducting — this is not zero, however — to that at which it "fires." By adjusting the time constant RC of the circuit, one can vary the frequency of the sawtooth sweep voltage; by adjusting the grid voltage of the thyratron, one can vary the amplitude of the sweep voltage.

The Cathode-Ray Oscilloscope

The shapes of signals — whether they are sinusoidal voltages or pulses — can be displayed on the screen of a cathode-ray os-

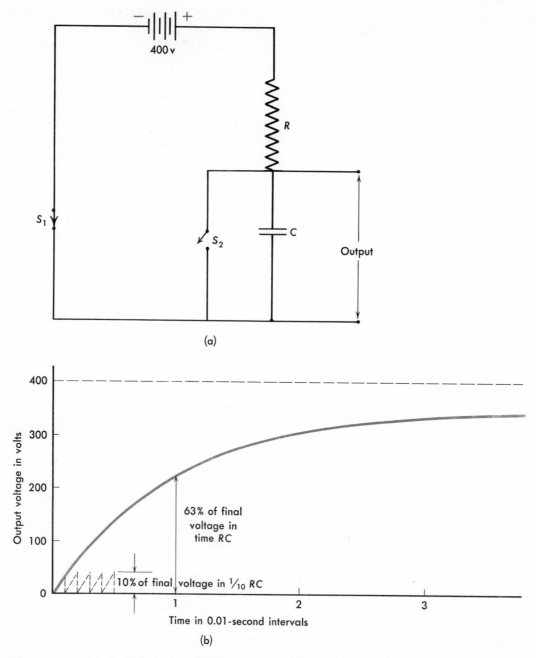

Fig. 31-18. (a) *An* RC *circuit.* (b) *The charging curve (solid line) and the sweep voltage (dashed line).*

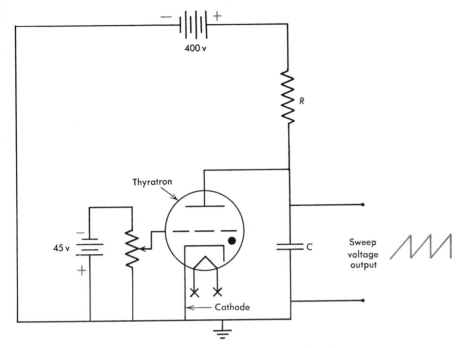

Fig. 31-19. A thyratron-controlled sweep voltage circuit.

cilloscope. Oscilloscopes are useful also for measuring voltages, frequencies, and short time intervals.

A typical electrode arrangement in an oscilloscope is shown in Fig. 31-20. The tube itself is a highly evacuated glass envelope, mounted in a metal cabinet that also contains the necessary power supplies, amplifiers, and sweep-voltage circuits. Electrons are emitted by the indirectly heated cathode, which is at a high negative voltage with respect to ground. The electrons are repelled by the cathode and attracted by the accelerating electrode. The electric field between the focusing and accelerating electrodes serves as an electron lens and prevents the electron beam from spreading; it serves to focus the beam. In the absence of voltages on the horizontal and vertical deflecting plates, the electron beam strikes the center of the fluorescent screen where it causes the emission of light—the spot that one sees when the beam

is not being deflected. One can make the spot brighter by making the intensity-control grid less negative and can also focus the beam; controls for doing this are provided on the front panel of the oscilloscope.

To display the wave form of a signal, the signal voltage is usually applied between the vertical deflecting plates—sometimes it is first amplified within the oscilloscope. A sweep voltage of controllable frequency and amplitude is applied to the horizontal deflecting plates. Thus the electrons in the beam are subjected to the electric fields of two capacitors—a vertically deflecting field and a horizontally deflecting one. The sweep voltage causes the beam to move horizontally at a uniform rate. At the same time the signal voltage causes the beam to move up and down in a way characteristic of the wave form of the signal.

The best analogy is provided by a boy who walks steadily beside a blackboard, touching

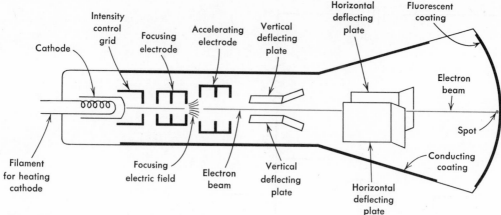

Labels in the figure:
Cathode · Intensity control grid · Focusing electrode · Accelerating electrode · Vertical deflecting plate · Horizontal deflecting plate · Fluorescent coating · Electron beam · Spot · Conducting coating · Horizontal deflecting plate · Vertical deflecting plate · Electron beam · Focusing electric field · Filament for heating cathode

Fig. 31-20. *A cathode-ray tube.*

a piece of chalk to the board, while he moves his hand up and down in simple harmonic motion. The chalk will trace a sine curve just as the electron beam would trace a sine wave form if the signal on the vertical deflecting plates were a sinusoidal voltage.

Feedback

Earlier in this chapter, we saw that energy from the plate circuit was fed back to the grid circuit of an oscillator to maintain the electrical oscillations. Feedback has many other uses, however, in electronics and in the automatic control of devices. Amplifiers, for example, can be made more stable if suitable feedback is provided.

Suppose that a fraction K of the output voltage V_o of an amplifier is fed back to the input side of the amplifier and there combined with the incoming input signal V_s. The relationships are shown in the block diagram of Fig. 31-21. The dashed line shows the feedback path by which a *correction* or *error* signal of voltage KV_o is added to V_s giving an effective input voltage $V_i = V_s + KV_o$.

Let us determine that the *overall* gain V_o/V_s of the amplifier system (including feedback) will be. We have

$$\frac{V_o}{V_s} = \frac{V_o}{V_i - KV_o}$$

In the *absence* of feedback, the gain is g where

$$g = \frac{V_o}{V_i}$$

or

$$V_o = gV_i$$

Therefore,

$$\frac{V_o}{V_s} = \frac{gV_i}{V_i - KgV_i}$$

and

$$\frac{V_o}{V_s} = \frac{g}{1 - Kg}$$

Let us next consider several possibilities. First suppose that K is negative (negative feedback). The feedback signal KV_o is 180° out of phase with the input signal V_s. The overall gain of the system will be less than g (because $1 - Kg$ will be greater than unity). In return for this decreased amplification, however, we have gained increased stability of the output. Should the output V_o increase because of some variation in the performance of the amplifier, the feedback would be greater in the negative sense and would act to reduce V_o. Should V_o decrease, feedback would operate to increase it again. Thus the output would be made independent of fluctuations within the amplifier itself. This is a characteristic of *negative* feedback.

What would happen if the feedback signal were positive (*K* positive)? Now the overall gain of the system is greater than *g* (since $1 - Kg$ is less than unity), and we have gained in amplification. However, the system is susceptible to instability because the error signal *reinforces* fluctuations in the output V_o.

An interesting special case occurs when *K* is positive and the product of *K* and *g* equals unity. For that condition, the system will oscillate. This may be desirable if the system is intended to serve as an oscillator, but it is undesirable in an amplifier. The "motorboating" or "popping" noises made by audio amplifiers at certain frequencies can be traced to this cause.

Summary

Electronics is the study of the motion of electrons and the application of such knowledge to practical problems of communication and control. Some of the common elements in tube electronics include: an envelope containing an evacuated or gas-filled space, a source of electrons, electric and magnetic fields and electrodes and coils for establishing them, inputs and outputs of energy, and signals whose shapes and sizes are changed as they progress through the electronic system.

Power supplies typically include a transformer, a rectifier diode, and a filter circuit. Full-wave rectification requires a double diode.

Triodes amplify signals; a small alternating voltage on the grid produces relatively large variations in the plate current and hence in the plate voltage.

The natural period of an *LC* circuit is given

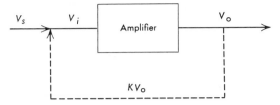

Fig. 31-21. *Block diagram for an amplifier with a feedback loop. The dashed line shows the feedback path.*

by $T = 2\pi\sqrt{LC}$. An electronic oscillator contains *LC* circuits or their equivalent in the grid circuit and in the plate circuit. Energy fed back from the plate circuit sustains the oscillations in the grid circuit.

Modulation is the process of adding an audio signal to the high-frequency carrier signal required for radio broadcasting. Demodulation or detection in a receiver is the process of recovering the audio signal from the modulated carrier.

The time base of a sweep circuit can be provided by the charging of a capacitor connected to a battery through a resistor. The time for the capacitor to charge to 63% of the battery voltage is *RC*, the time constant of the circuit. A thyratron, suitably biased and connected in parallel with the capacitor, acts like a switch in the generation of a sweep voltage—one that rises proportionately with the time.

In a cathode-ray oscilloscope, a suitably accelerated and focused electron beam produces a spot on a fluorescent screen. A sweep voltage on the horizontally deflecting plates causes the beam to swing back and forth, and the signal to be studied is placed on the vertically deflecting plates, producing the characteristic wave form of the signal.

Questions

1. Discuss the force on a beam of electrons in (*a*) a transverse uniform electric field, (*b*) a parallel uniform magnetic field.

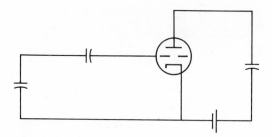

Fig. 31-22.

2. How does (a) an inductor, (b) a capacitor affect a full-wave pulsating rectified voltage?

3. Upon what quantities does the frequency of a resonant circuit depend? Compare these with the quantities that determine the frequency of a mechanical oscillator consisting of a body suspended by a spring.

4. How do electronic devices transmit information?

5. What is the source of the sound energy emitted by a loudspeaker?

6. Explain the action of a simple receiver circuit.

7. Explain the structure and operation of a power supply.

8. How do modulation and demodulation differ?

9. Make a block diagram for a cathode-ray oscilloscope. What are some uses of the CRO?

10. Do you increase or decrease the capacitance of a radio circuit in tuning from 600 to 1000 kH?

11. How are resonance and simple harmonic motion related?

12. How does feedback occur in an electronic oscillator? Compare this with the feedback system represented by a child on a swing who is being pushed by a friend.

13. Show by a sketch the load line for the amplifier circuit of Fig. 31-9 if the plate resistor is 40,000 Ω.

14. Design a circuit containing a thyratron, motor, and various electronic components to use the radio beam from a car for opening and closing a garage door.

Problems

A

1. What is the period of an oscillator having a frequency of 318 kH?

2. What is the capacitance of a capacitor that may be combined with a self-inductance of 0.080 henry to form an oscillator having a frequency of 1.59×10^5 hertz?

3. Calculate the amplification factor of a triode when the grid input change in voltage is 5.0 v and the plate output change is 50 v to produce the same change in plate current.

B

4. What is the natural frequency of a tank circuit that includes a 0.050-microfarad capacitor and a coil of inductance 50 microhenries?

5. A public-address system has two stages of amplification, each of which increases the intensity of the signal tenfold. If the input signal is 0.20 v, what is the output signal? What is the gain in decibels where the decibel of power level is defined in the same way as the decibel of sound level?

6. What is the voltage delivered by a plate supply to a triode amplifier if the plate-to-cathode voltage is 200 v and the 10,000-Ω load resistor carries 8 milliamp?

C

7. What is the time constant in milliseconds of a sweep circuit that contains a resistance of 10^6 Ω and a capacitor of 0.10 microfarad?

8. Calculate the frequency produced by a 0.020-henry inductor and a 0.001-farad capacitor. Would this frequency lie in the radio-frequency range?

9. Given a 0.15-millihenry inductor, what must be the maximum and minimum values of

a capacitor connected in parallel with it to tune stations from 540 to 1600 kH?

10. (a) Prove that the product RC has the dimensions of time. (b) Show that the voltage across the capacitor in a sweep circuit rises to 63% of its maximum value in a time RC.

11. Complete the oscillator circuit in Fig. 31-22 by adding the missing components.

12. What is the velocity of an electron that is accelerated by a potential of 4.52 v?

13. (a) Draw the load line in Fig. 31-9 for a plate-voltage supply of 120 v and a load resistor of 30,000 Ω. (b) If the grid bias is —4.0 v and the amplitude of the input signal is 4.0 v, what is the amplitude of the output signal?

14. A capacitor of capacitance 5.0×10^{-8} farad is connected to a 200-v battery for several minutes and is then discharged through a 10-megohm resistor. What is the charge on the capacitor 0.25 sec after the capacitor is connected to the resistor?

For Further Reading

The Editors of *Scientific American, Automatic Control,* Simon and Schuster, New York, 1955.

Fink, Donald G. and David M. Lutyens, *The Physics of Television,* Doubleday and Company, Garden City, 1960.

Ginzton, E. L., "The Klystron," *Scientific American,* March, 1954, p. 84.

Pierce, John R., *Electrons and Waves,* Doubleday and Company, Garden City, 1964.

Pierce, John R., *Waves and Messages,* Doubleday and Company, Garden City, 1967.

Light

CHAPTER THIRTY-TWO

Electromagnetic Waves

In preceding chapters we discussed electric fields in the neighborhood of static electric charges and magnetic fields near moving charges. A field, we found, depended upon its source, and changes in the source produced changes in the field. For example, the electric field between the plates of a parallel-plate vacuum capacitor vanishes when the plates are grounded, and the magnetic field within a solenoid increases as the current in the solenoid increases.

Now we shall see that electric fields and magnetic fields under certain conditions "cooperate" to form *electromagnetic waves* which, once emitted by a source, are thereafter independent of the source. The complete theory of electromagnetic waves summarizes most of classical electricity and magnetism—the subject of the last eleven chapters—and points the way to the development of modern wave mechanics, which we shall discuss later. We saw in Chapter 10 how basic the constancy of the velocity of electromagnetic waves in vacuum is in the special theory of relativity. In technology, the theory of electromagnetic waves is extremely important; a large part of the electrical power and communications industries is based upon it.

Historically, the development of our understanding of electromagnetic waves occurred rather rapidly from the time that their existence was predicted theoretically until they were detected experimentally. In 1863, Clerk Maxwell, a British mathematical physicist, published a paper developing the idea that the electromagnetic disturbances produced by high-frequency alternating currents travel with the same speed as light and are of the same nature as light.

Maxwell was one of the great synthesizers of physics. At the mathematical level of this book we can only hint at the grandeur and the wide applicability of his theory. He took the experimental results of Faraday and others and organized them into a consistent mathematical theory of electric and magnetic fields. Maxwell found that the idea of electromagnetic waves was a logical consequence of his theory, and he published this result, even though the electromagnetic nature of light was experimentally unproven—although suspected—and no one had detected the existence of other electromagnetic waves. Like Newton, Maxwell "stood on the shoulders of giants," and—also like Newton—he saw further than they.

Investigators set about discovering means of producing and detecting the electromagnetic waves predicted by Maxwell. In 1888, Heinrich Hertz, a German physicist, devised the first transmitter and detected its signals at distances of several yards. He found that the waves were transverse and that their wavelengths were a few meters. The gap between two hitherto separate fields of physics—electricity-magnetism and light—had been bridged.

Electromagnetic Waves in Vacuum

Faraday made the discovery that a *changing* magnetic field is accompanied by an electric field. When the magnetic flux threading through a copper ring increases or decreases, an electromotive force appears in the ring, the size of the electromotive force depending upon the rate of change of magnetic flux. The electric field acts on the conduction electrons in the copper and causes them to drift around the ring in an induced current. The electromotive force and the electric field accompany the changing magnetic field even if the copper ring is not there; if you imagine drawing a circle where the ring was formerly placed, the electromotive force around the circle still equals the rate of change of magnetic flux within the circle. We saw that in a betatron the electric field set up within the hollow doughnut-shaped vacuum chamber by the changing magnetic field of the electromagnet accelerated electrons to high energies. Whether or not matter is present, *a changing magnetic field is accompanied by an electric field.*

Is a changing *electric* field accompanied by a magnetic field? An alternative question, since we expect magnetic fields to originate in currents, is: Does a changing electric field constitute a current in any sense? Maxwell found that the answer to both of these questions was yes. He named the "current" that

accompanied the changing electric field a *displacement** current.

To understand in what sense a changing electric field can be considered a current, consider a vacuum parallel-plate capacitor connected to a battery through a switch (Fig. 32-1). When the switch is closed, a conduction current in the wires leads to an increase in the charge on the plates of the capacitor. The conduction current is largest just after the closing of the switch, dropping to zero when the capacitor has been charged to the battery voltage. At some instant during this buildup of charge, the magnitude of the charge on either plate is q, and the electric field strength E between the plates (Chapter 22) is

$$E = \frac{q}{\epsilon_0 A}$$

where A is the area of the plates and ϵ_0 is the permittivity of free space. Rearranging this equation, we have

$$q = \epsilon_0 A E$$

In a small time interval Δt, an amount of charge Δq will be added to the plates causing an increase ΔE in the electric field strength:

$$\Delta q = \epsilon_0 A \, \Delta E$$

and the rate of addition of charge is

$$\frac{\Delta q}{\Delta t} = \epsilon_0 A \frac{\Delta E}{\Delta t}$$

But $\Delta q / \Delta t$ is the conduction current i_c at that instant. We see that it equals $\epsilon_0 A (\Delta E / \Delta t)$, and hence the latter quantity has the dimensions of a current. If A is in square meters, ϵ_0 in square coulombs per newton-square meter, and $\Delta E / \Delta t$ in newtons per coulomb-second

---

* In advanced treatments of electric fields, the electric displacement is defined—in the mks rationalized system of units—as $\epsilon_0 E$. We shall not make explicit use of the electric displacement, preferring to emphasize the role of E, the electric field strength.

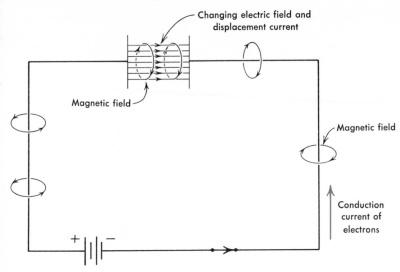

Changing electric field and
displacement current

Magnetic field

Magnetic field

Conduction
current of
electrons

+ −

Fig. 32-1. The conduction current in the wires and the displacement current between the plates are both accompanied by magnetic fields.

we see that $\epsilon_0 A(\Delta E/\Delta t)$ has the units coulombs per second or amperes.

We define the *displacement current i_D* as

$$i_D = \epsilon_0 A \frac{\Delta E}{\Delta t}$$

and see that it is proportional to the rate of change of electric lines between the plates. Thus the conduction current in the circuit, the charge transferred per unit time, ends at the plates of the capacitor and is replaced in the space between the plates by the displacement current.

Next let us note that *the displacement current is accompanied by a magnetic field whose magnitude depends upon the rate of change of the electric field.* Magnetic lines loop around the wires of the circuit, accompanying the conduction current, and magnetic lines also loop around the electric lines between the plates, accompanying the displacement current. The magnetic fields set up by displacement currents are the same in every way as those set up by the motion of charge; they are present only in non-static sit-

uations, however—that is, only when electric fields are changing. The principles of physics we have discussed can be used to calculate the magnitude of magnetic fields set up by displacement currents. The mathematical approach requires calculus, however, and we shall not carry out the calculation here.

Before going on, we should point out that tiny displacement currents exist in the wires of alternating-current circuits at power frequencies because the electric fields are changing in the wires. Similarly, tiny conduction currents exist between the plates of practical capacitors because of a small amount of electrical leakage.

We see that there is great symmetry in the processes of electricity and magnetism: *Changing magnetic fields are accompanied by electric fields, and changing electric fields are accompanied by magnetic fields.* Maxwell was struck by this symmetry in the logical structure he had created and wondered whether the two processes could be combined to give a single process that would be self-sustaining. If the magnetic field were changing at a *non-uniform* rate—sinusoidally, for example—the electric field induced would not be constant, but would similarly change with time. This changing electric field would induce a changing magnetic field which would induce a changing electric field and so on. Maxwell's ensuing calculations led him to electromagnetic waves.

Electromagnetic waves have their origins in oscillating charges—for example, those oscillating at high frequency on a radio antenna. Let us not worry about the origin of the waves for the moment, however, and consider the wave once formed. A graphical representation of an electromagnetic wave in space is shown in Fig. 32-2.

Remember that we are talking about electric and magnetic fields; the electromagnetic wave is a wavelike change in field strengths with an accompanying propagation of energy in the direction of the wave. These are not waves in the sense that material particles are moved, as water molecules are moved when a ripple passes over the surface of a pond.

Figure 32-2 shows the magnitudes and directions of *E* and *B* at any instant. However, we must imagine the whole graph to be traveling to the right. An observer who was at rest with respect to the wave and was equipped with instruments that could respond to these rapidly changing electric and magnetic fields would observe, as the wave passed, that the electric field strength at his location would first be directed upward and then downward, its magnitude varying sinusoidally. At right angles to the electric field, he would detect a magnetic field whose magnetic induction would be directed first out and then in, in a sinusoidal variation. The electric field variations and the magnetic field variations are in phase with each other; they sustain each other because—to repeat—sinusoidally varying magnetic fields produce sinusoidally varying electric fields, and sinusoidally varying electric fields produce sinusoidally varying magnetic fields. The electric field is always perpendicular to the magnetic field, and both fields are perpendicular to the direction of propagation. The wave is transverse.

The fields that constitute an electromagnetic wave have the properties of any other electric and magnetic fields that change with time. When an electromagnetic wave strikes a conductor, the electric field component causes alternating conduction currents. When an electromagnetic wave sweeps across an antenna, magnetic lines of the magnetic field component are cut and an alternating induced current of the frequency

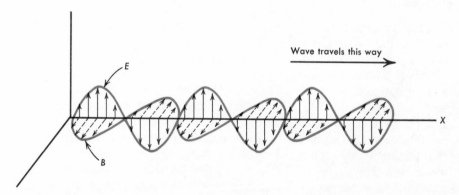

Fig. 32-2. *A graphical representation of the electric and magnetic fields in an electromagnetic wave.*

of the wave is induced in the antenna. We have seen that energy can be stored in electric fields and magnetic fields. Similarly, energy is carried by electromagnetic waves.

How Electromagnetic Waves Originate

Consider a *short* transmission line—two parallel conductors mounted on insulators—connected to a high-frequency oscillator (Fig. 32-3). The currents on the transmission line and the electric and magnetic fields near the conductors vary in phase with the oscillator

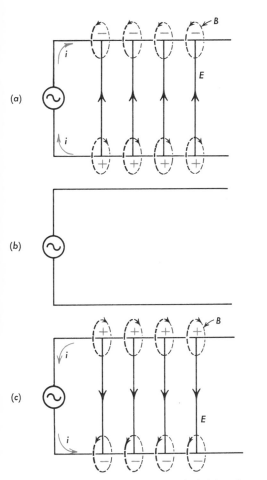

Fig. 32-3. The currents and fields along a short transmission line connected to an oscillator.

as the oscillator goes through its oscillations. When the lower terminal of the oscillator is positively charged, the entire lower conductor of the transmission line is positively charged; at the same time the upper conductor is negatively charged. The currents and the fields are as shown in Fig. 32-3a. In Fig. 32-3b, the oscillator voltage is zero and the currents and fields are zero. In Fig. 32-3c, the oscillator voltage has reversed, and the currents and fields have reversed. Thus during this half-cycle of the oscillator, the currents and fields on the *short* transmission line have kept in phase with the oscillator.

Contrast Figure 32-3 with Fig. 32-4, which shows the same oscillator connected to a *long* transmission line. Experimentally we find that the signals sent out by the oscillator do not travel down the line with an infinitely great velocity. Instead, the speed with which electromagnetic signals travel on a transmission line is almost the speed of light in free space—about 3×10^8 m/sec—a large velocity, to be sure, but not an infinitely large one. What this means is that currents and fields in parts of the transmission line that are distant from the oscillator do not change in phase with the currents and fields near the oscillator. The signals sent out by the oscillator, carrying "instructions" as to changes in currents and fields, do not reach the remote sections of the line instantaneously; a retardation is involved, and we expect the very distant parts of the line to lag behind the parts near the oscillator by times equal to the distances divided by the speed of light.

Figure 32-4 illustrates what happens. In Fig. 32-4a, the oscillator has just been turned on. The signal reaches the transmission line near the oscillator; the currents and fields are similar to those shown in Fig. 32-3a. The signal propagates along the line reaching point A (Fig. 32-4b) after a quarter cycle, point B (Fig. 32-4c) after a half cycle, and so on. Meanwhile, the oscillator continues to send out new signals. Thus we see in Fig. 32-4c that the

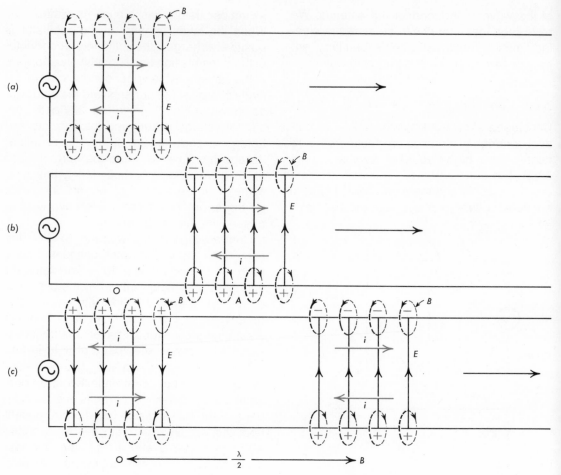

Fig. 32-4. *The propagation of an electromagnetic wave along a long transmission line. Currents and fields at distant parts of the line lag behind the oscillator.*

variations in currents and fields in the more distant sections of the transmission line are propagated down the line as a *wave*. The wave is in fact an electromagnetic wave; it is confined to the region between the conductors, however, instead of being free to spread outward through space. The distance from *O* to *B* is half the wavelength λ of the electromagnetic wave.

Electromagnetic waves on a transmission line are similar to elastic waves on a stretched string (Chapter 19). If one end of a

stretched string is vibrated, elastic waves are propagated along it, causing a variation of displacement and velocity of the particles of the string just as the electromagnetic wave causes a variation in the currents and the fields along the transmission line. At the end of the stretched string, the elastic waves are reflected, setting up standing waves along the string. Similarly, the reflection of electromagnetic waves at the end of a transmission line causes standing electromagnetic waves—with nodes and antinodes of current and voltage—to be set up on the transmission line.

Now we are ready to discuss how electromagnetic waves are emitted by an antenna. An antenna of length λ/2 (Fig. 32-5) is connected to a high-frequency oscillator. It can be thought of as the end of a transmission line that has been spread apart. Such an antenna constitutes an oscillating dipole. An electric dipole, as we saw in Chapter 22, consists of a positive charge and an equal negative charge separated by a certain distance. An oscillating dipole is one whose positive and negative charges exchange positions periodically. The high-frequency alternating current on the antenna causes the antenna to behave like an oscillating dipole whose oscillations occur at the frequency of the oscillator to which it is connected.

Notice that the electric lines near the antenna originate on positive charges and end on negative ones. The field is not confined to a small space, as in the earlier example of the

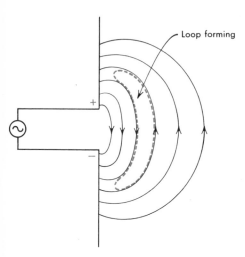

Fig. 32-5. The radiation field of an antenna. The electric lines near the antenna begin and end on charges, but at large distances form closed loops. The magnetic lines, not shown in the diagram, form closed loops perpendicular to the electric lines and center on the antenna.

transmission line, but is spread out. It is this spreading out of the field that allows electromagnetic waves to be broadcast. The magnetic lines near the antenna are not shown; they are circular lines perpendicular to the electric lines and centered on the antenna. Both the electric and the magnetic fields in the region near the antenna build up and collapse in phase with the oscillator. In this region, energy is fed into the fields and withdrawn; no energy is radiated.

Far from the antenna, the electric lines form closed loops. The reason for this is the retardation caused by the finite velocity of the signal. The signal from the antenna travels outward with the speed of light. The fields in this region far from the antenna lag behind the oscillations of charge on the antenna. The result is that, as the fields build up and collapse, a portion of the fields is detached. The electric lines no longer begin and end on the antenna but instead form closed loops that spread out into space. These loops are the electric-field component of the electromagnetic waves, which carry energy outward with the speed of light. Only the electric component of the waves is shown in Fig. 32-5. The magnetic lines form closed loops at right angles to the electric loops. Far from their source, the electromagnetic waves take the shape shown in Fig. 32-2.

As shown in Fig. 32-5, the electric lines are densest along the perpendicular to the antenna and have zero density along the line of the antenna itself. This means that the energy radiated per unit area per unit time is greatest at right angles to the antenna and zero along the line of the antenna, as one can readily verify experimentally. The waves emitted by the antenna are *polarized;* this means that the electric component is always parallel to the antenna and the magnetic component is always perpendicular to it. Mathematical treatment of the radiation of waves shows that the total energy per unit time radiated into space by the antenna depends upon the *fourth power* of the frequency of the oscillator; that is, if the frequency of the oscillator is doubled, the total power radiated becomes sixteen times as great. For this reason, high-frequency oscillators are a necessary part of a radio transmitter in order that sufficient energy will be radiated.

The Electromagnetic Spectrum

Electromagnetic waves travel through free space with the same speed, but may have different wavelengths and frequencies. So far, we have mentioned radio waves and light, two of the many electromagnetic waves that have been observed.

Table 32-1 summarizes a few of the facts about the electromagnetic waves that comprise the *electromagnetic spectrum.* This table should impress you with both the diversity and the unity of our knowledge of electromagnetic waves. The frequency limits *f* are given in cycles per second, and the wavelength limits λ in meters. A few of the sources and detectors used in each region are listed, and some examples of the scientific and technological uses of the radiations in each region are also given. Notice first that in frequency the observed waves run all the way from about 3×10^{21} cycles/sec (gamma rays) to 10^4 cycles/sec (very-low-frequency radio waves) and in wavelength from 10^{-13} m to 3×10^4 m. Different wavelength regions of the spectrum are given different names: gamma rays, X-rays, ultraviolet rays, light, infrared rays, microwaves, and the radio regions (ultrahigh frequency, very high frequency, high frequency, medium frequency, low frequency, and very low frequency). However, the extent of each region is arbitrary and mostly a matter of definition; the physical properties of the waves change gradually with wavelength so that one region merges into another.

The sources of the waves and the ways in which the waves are detected differ greatly from one part of the spectrum to another.

Gamma rays, as we shall see later, are emitted by the nuclei of atoms and by high-energy atomic particles as they slow down; the ultimate sources of gamma rays are dimensionally very tiny (approximately 10^{-16} m in diameter). Radio waves, as we saw in both this chapter and the preceding one, are emitted by the large antennas of radio transmitters, which consist of electronic equipment that, in the case of commercial broadcasting, fills entire rooms. Detectors of electromagnetic waves of different wavelengths also differ greatly, as we shall see. The human eye, the detector with which we are most familiar, responds to only a tiny portion of the electromagnetic spectrum, that lying between about 4×10^{-7} m and 7.7×10^{-7} m in wavelength. The eye is not sensitive to the other waves, and special detectors have to be used for gamma rays, ultraviolet rays, and radio waves, for example.

Each region of the electromagnetic spectrum has special interest for the scientist because of the particular way in which the electromagnetic waves in that region originate and in which they interact with matter, including living cells, as they pass through it. These different properties will occupy much of our attention for the remainder of this book. Engineers are extremely interested in the properties of the different regions of the electromagnetic spectrum also, but for a different reason. Engineers apply knowledge of these properties to solving various problems in communications, testing of materials, transmission of power, navigation, and medical technology, for example. A few of the scientific and the engineering uses of each region of the electromagnetic spectrum are given in Table 32-1.

We have been discussing differences among electromagnetic waves. In closing this section, we remind you that the most impressive fact about these waves is their *similarity:* all travel in free space with the same speed, all obey Maxwell's theory, all are transverse

Table 32-1. The Electromagnetic Spectrum and Typical Sources, Detectors, and Uses of Electromagnetic Radiations

Boundary markers (placed at the divisions between the bands, left to right):

f(Hz): 3×10^{21}, 3×10^{19}, 3×10^{16}, 7.5×10^{14}, 3.9×10^{14}, 3×10^{11}, 3×10^{9}, 3×10^{8}, 3×10^{7}, 3×10^{6}, 3×10^{5}, 3×10^{4}, 3×10^{4}

λ(m): 10^{-13}, 10^{-11}, 10^{-8}, 4×10^{-7}, 7.7×10^{-7}, 10^{-3}, 10^{-1}, 1, 10, 10^{2}, 10^{3}, 10^{4}, 10^{4}

	γ-rays	X-rays	UV	Light	Infrared	Microwaves	UHF	VHF	HF	MF	LF	VLF
Sources	Nuclei during energy transitions. High energy particles during deceleration.	Atoms during electron transitions. High-energy electrons during deceleration.	Atoms during electron transitions.	Atoms during electron transitions.	Molecules during rotational transitions. Objects above 0°K. Atoms during some electron transitions.	Electronic apparatus. Electrons absorb and emit these radiations in paramagnetic resonance.	Electronic apparatus. Interstellar hydrogen emits 21-cm radiation.	Electronic apparatus.	Electronic apparatus. Nuclei absorb and emit these radiations in nuclear magnetic resonance.	Electronic apparatus.	Electronic apparatus.	Electronic apparatus.
Detectors	Photographic emulsions. Geiger and scintillation counters. Ionization chambers. Semiconductor detectors.	Photographic emulsions. Geiger and scintillation counters. Ionization chambers. Fluorescent materials. Semiconductor detectors.	Photographic emulsions. Geiger counters. Photocells. Fluorescent materials.	The eye. Photographic emulsions. Photocells.	Photocells. Photographic emulsions (near infrared). Thermopiles (far infrared). Bolometers.	Electronic apparatus.	Electronic apparatus.	Electronic apparatus.	Electronic apparatus.	Electronic apparatus.	Electronic apparatus.	Electronic apparatus.
Scientific Uses	Nuclear structure. Elementary particle processes.	Atomic structure. Crystal structure. Interaction of radiation and matter, including living cells.	Atomic structure. Interaction of radiation with living cells. Photoelectricity. Gas discharges. Fluorescence.	Atomic structure. Vision.	Atomic structure. Molecular structure. Black-body radiation.	Electron paramagnetic resonance.	Radio astronomy.		Nuclear magnetic resonance.			
Technical Uses	Radiation therapy. Inspection of materials.	Medical examination. Therapy. Inspection of materials.	Germicide. Vitamin source. Chemical analysis.	Vision. Photography. Chemical analysis.	Remote sensing. Heat lamps. Photography. Chemical analysis.	Radar.	Television.	Television. FM broadcasting. Mobile communication.	International fixed radio communication. Amateur broadcasting.	Radio broadcasting.	Mobile radio communication. Radio navigation.	Mobile radio. Fixed radio communication.

waves having an electric-field component and a magnetic-field component, all originate in the high-frequency oscillations of charged particles, and so on. The electromagnetic spectrum is one of the great generalizations of physics.

Flux, Intensity, and Irradiance

Flux of radiant energy is defined as the energy passing through a given surface per unit time. Thus, a station that broadcasts radio waves so that the energy carried outward per unit time in all directions is 500 watts at a frequency of 1 megahertz emits a total flux of 500 watts at that frequency.

The *intensity* of radiation is the flux per unit solid angle. Consider drawing on the surface of a sphere an area equal to the square of the radius of the circle (Fig. 32-6). The area subtends a *unit solid angle,* called one steradian, at the center of the sphere. Thus, an entire sphere subtends a total solid angle of 4π steradians because the area of a sphere contains r^2 4π times. In general, the solid angle

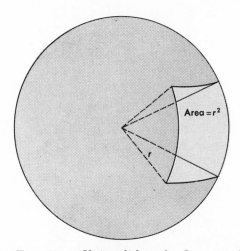

Fig. 32-6. Unit solid angle. One steradian is the solid angle subtended at the center of a sphere of radius r *by an area* r^2 *on the surface of the sphere. The total solid angle surrounding a point is thus 4π steradians.*

subtended by any surface area traced on a sphere is the area divided by the radius squared. Defining intensity as the flux per unit solid angle, we see that the intensity of the radio station referred to is 500 watts/4π steradians or about 40 watts/steradian — *if* the energy is radiated equally in all directions. This is not a realistic assumption in this case, as we saw earlier in this chapter, and the intensity thus calculated is really an average intensity.

The *irradiance* of a surface is the flux per unit area striking the surface. Suppose that a flux of 80 microwatts from a distant radio transmitter strikes a surface whose area is 2 square meters. The irradiance of the surface is 40 microwatts per square meter.

In free space, electromagnetic waves travel in straight lines. The hunter relies on this fact when he aims his rifle, and the carpenter when he sights along the edge of a board. We represent this graphically by drawing arrows called rays whose directions show the way in which the waves are traveling. We can demonstrate the rectilinear propagation of light by admitting sunlight to a darkened room through a small hole and observing the parallel beam, made visible by dust particles in the air.

Electromagnetic waves passing through narrow openings are deviated or diffracted; however, they are refracted in going from one medium into another in which their speed is different, and — according to the general theory of relativity — they are bent slightly in passing near massive bodies in space.

We use the fact that electromagnetic radiation travels in straight lines in free space in determining how the irradiance is related to the distance from the source of the waves. Imagine a very small "point" source of radiation *S* at the center of three transparent spheres of radii 1 ft, 2 ft, and 3 ft (Fig. 32-7). Suppose that the radiation from the source of intensity *I*, passing through the opening *A*, passes through *B* and *C*. The area of *B* is four

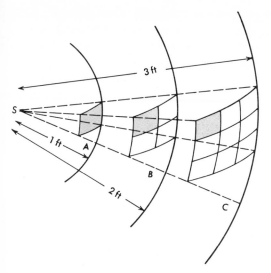

Fig. 32-7. Irradiance and distance. B, *twice as far from* S *as* A, *has four times as great an area. Therefore, unit area of* B *receives one-fourth as much flux as* A.

times that of *A*, hence unit area of *B* receives one-fourth as much flux as an equal area of *A*. Similarly, unit area of *C* receives one-ninth as much. *The irradiance of a surface perpendicular to the rays varies inversely as the square of its distance from a point source.**

We note that for electromagnetic waves from a point source at the center of a sphere,

Irradiance of the sphere

$$= \frac{\text{Flux}}{\text{Area}} = \frac{4\pi \times \text{Intensity}}{4\pi \times r^2}$$

$$= \frac{\text{Intensity}}{\text{Square of distance}} = \frac{I}{r^2}$$

This is the inverse-square law of irradiance due to a point source. It does not hold for sources whose sizes are appreciably large

---

* We have assumed that the electromagnetic waves strike the surface normally at an angle of 90° with the surface. In general, if the angle between the incident rays and the normal to the surface is θ,

$$\text{Irradiance} = \frac{\text{Intensity}}{r^2} \cos \theta$$

and whose intensities vary with direction, but it is useful in situations where the source is small compared to the distance to the point at which the irradiance is being computed. As an example, we can calculate the irradiance due to the radio station referred to above, treating the source as a point. At a distance of 1000 m, the irradiance is the intensity of 40 watts/steradian divided by $(1000 \text{ m})^2$ or 40 microwatts/m^2. The steradian—a ratio of an area to a distance squared, both of which are expressed in the same units—can be dropped from the units of the irradiance.

Light

The electromagnetic waves of most immediate interest to us are *light* waves, those electromagnetic waves that produce the sensation of sight in the eye and the brain. We shall emphasize light waves in our discussion of the properties of electromagnetic waves, but call your attention from time to time to the ways in which other electromagnetic radiations resemble and differ from light waves in their behavior.

Luminous (light) flux, intensity, and irradiance have special units. This is because light was the first electromagnetic radiation studied and because it was not known until comparatively recently in the history of physics that the eye responds only to a small portion of the electromagnetic spectrum. Later we shall talk about the relative response of the eye to electromagnetic waves of different wavelengths.

Scientists at one time used a wax candle of carefully prescribed structure as a standard of luminous intensity. In 1919, several incandescent electric lamp bulbs were selected and their luminous intensities were determined when operated at prescribed voltages. These lamps, carefully protected like the standard kilogram and meter, are occasionally used for testing other light sources which are the secondary standards. These practical standards of luminous intensity were unsatisfactory in that the lamps might be damaged, or their light-giving properties might vary with age. This difficulty was avoided by a new definition authorized in 1940:

The international standard candle, the unit of luminous intensity, is one-sixtieth of the luminous intensity of a "black-body" source, one square centimeter in cross-sectional area, at the melting point of platinum (1755°C). The source must be viewed at right angles to the area. You will recall from Chapter 14 that a small opening in a cavity whose walls are uniformly heated is a black-body source and is a maximum emitter of radiation at a given temperature. Lamps do not emit light equally in all directions. The intensity of a lamp may be 100 candles horizontally and 0 candles downward.

Having defined the unit of luminous intensity as the international standard candle, we are ready to define the unit of luminous flux—the *lumen.* It is a convenient unit in lighting engineering. Imagine that a standard 1-candle "point" source is at the center of a hollow sphere of radius 1 m having a window 1 m² in area (Fig. 32-8). Then the light flows through the opening at a rate of 1 lumen.

The surface of the sphere has an area 4π times that of the window. Let the luminous intensity of the source be I. Then the total luminous flux, transmitted through the sphere, is $4\pi I$ lumens.

Light obeys the inverse-square law of irradiance for point sources. However, the luminous flux per unit area is generally called the *illumination* or the *illuminance.* Illumination is expressed in lumens/ft²; 1 lumen/ft² is also called 1 *lux.*

Illumination is also often expressed in meter-candles. *One meter-candle is the illumination of a surface, every part of which is one meter from a point source whose intensity is one candle.* The name meter-candle is badly chosen, for illumination decreases with increasing distance. Candle per meter-squared would be a better term. *One meter-candle is one candle per meter-squared.*

We see in Fig. 32-8 that the illumination at the window is one meter-candle and is also one lumen per square meter. Thus one lumen per square meter produces an illumination of one meter-candle (one candle per meter-squared).

Example. A 60-candle lamp is at the center of a hemispherical dome, 2.0 m in radius. (*a*) What is the illumination at the surface of the dome? (*b*) What is the luminous flux through a window in the dome of area 0.20 m²?

(*a*)   Illumination $= \dfrac{60 \text{ candles}}{(2.0 \text{ m})^2}$

$= 15$ candles/m²
$= 15$ m-candles
$= 15$ lumens/m²

(*b*)   Luminous flux $= 15$ lumens/m² $\times 0.20$ m²
$= 3.0$ lumens

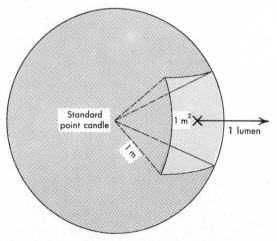

Fig. 32-8. Luminous flux. The standard point candle is 1 m from the window of unit area. The luminous flux through the window is 1 lumen.

Photometry

The *photometer* is used to compare the luminous intensities of two light sources. The

Fig. 32-9. Light meter. Light incident on the upper window striking a photovoltaic cell causes a current proportional to the illumination of the cell.

student-type Jolly photometer consists essentially of two paraffin blocks, separated by a sheet of aluminum foil. The pair of blocks, mounted between the two sources, is shifted to and fro until their edges are equally bright. Then the two sources illuminate the blocks equally. Let the known luminous intensity of the standard source A be I_s and that of the test lamp I. Let the respective distances be s_s and s.

$$\text{Illumination} = \frac{I_s}{s_s{}^2} = \frac{I}{s^2}$$

The light meter (Fig. 32-9) consists of a photovoltaic cell, in series with a galvanometer. A typical photovoltaic cell consists of an iron disk covered with a thin layer of selenium. The selenium layer is coated with a semitransparent film of silver. When light strikes the selenium layer, it causes electrons to move from the iron to the selenium, where they are collected by the silver coating. Thus the cell acts like a battery. The current through the galvanometer is proportional to the illumination. Photographers use

modified light meters to determine proper times of exposure. When the light meter is pointed in the same direction as the camera, the scale reading tells the proper time of exposure for a given lens-aperture or stop-opening.

Luminous Efficiency

The efficiency of a device is the ratio of its *useful* energy output to the *total* energy input. The *luminous efficiency* of a lamp is the ratio of its light output to the input power. Luminous efficiencies are expressed in average (in all directions) candles per watt or in lumens per watt. Table 32-2 gives luminous efficiencies and efficiencies of several light sources. Note that a paraffin candle has an efficiency of $\frac{1}{300}$ of 1 %. It wastes 99.997 % of the input energy. This wasted energy is mostly infrared radiation to which the eye is not sensitive. A 60-watt tungsten lamp is more efficient. It wastes about 98% of the energy. The efficiency of a fluorescent mercury lamp is about 6.5%.

Table 32-3 shows desirable illuminations for different activities.

Table 32-2. Luminous Efficiencies and Efficiencies of Different Sources

Light Source	Average Candles Per Watt	Lumens Per Watt	Efficiency, %
Paraffin candle	0.017	0.21	0.0035
Kerosene lamp	0.02	0.25	0.04
Open carbon electric lamp	0.96	12	2.0
Carbon-filament lamp (40 watts)	0.23	2.9	0.45
Gas-filled tungsten lamp (60 watts)	0.98	12.4	2.0
Gas-filled tungsten lamp (100 watts)	1.14	14.3	2.3
Mercury-vapor lamp (fluorescent)	3.2	42	6.5
Firefly	48	600	
Maximum theoretical efficiency, yellow light	51	650	100

Table 32-3. Illuminations

	m-candles
Noon sunlight	50,000–100,000
Bright moonlight	0.25
Needed for night baseball and football	300–500
Needed for street lighting	0.50–2.5
Desirable values for reading	300–500
Desirable values for work requiring close inspection	500–1000
Hospital operating rooms	1000

The Speed of Electromagnetic Radiation

You may remember that Galileo studied the laws of the pendulum and used it as a timer to measure small time intervals. He also tried to measure the speed of light. Galileo stationed two men, *A* and *B*, with lanterns covered with screens on hilltops less than a mile apart. The first observer, *A*, removed the screen of his lantern. As soon as possible after receiving the signal, *B* uncovered his lantern in turn, thus sending a signal to *A*, who tried to estimate the time required for the signals to make the complete journey. After repeated efforts, Galileo concluded that the speed was too great to measure by the means at his disposal.

Galileo constructed a telescope in 1601 and discovered the four principal moons of the planet Jupiter. Sixty-five years later, a Danish astronomer named Roemer (Rerm′ er) carefully studied the times at which the innermost of these moons disappears behind the planet Jupiter and is eclipsed. Roemer wanted to know whether the moon moved steadily enough so that it would serve as an accurate clock for navigators to use on sea voyages (to measure longitudes). He observed that during half the year, the Jovian "clock" ran too fast, and during the remainder, it ran too slow. He found that Jupiter's innermost moon has a period of about 42.5 hr. Roemer concluded that light is not infinitely fast, but has a finite speed.

Suppose that the time elapsing between successive eclipses of the moon of Jupiter is noted when the earth is at *A* (Fig. 32-10) and a timetable or schedule of eclipses is prepared for the coming year. Three months later the earth is at *B* and the eclipse occurs about 8 min behind schedule. After another three months the earth has moved to *C* and the eclipse is 16 min ≅ 1000 sec behind schedule. The light from Jupiter has to travel farther when the earth is at *C* than when it was at *A*. The maximum difference in distance equals

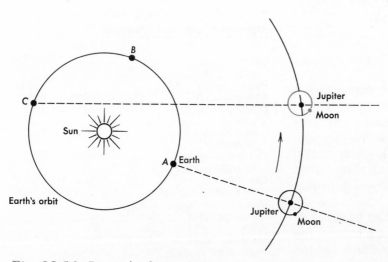

Fig. 32-10. Roemer's observations.

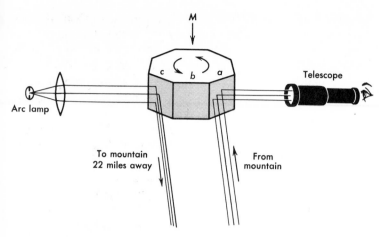

Fig. 32-11. Michelson's rotating mirror method. Light from the arc was reflected at c, *then to distant fixed mirrors, returned, and was reflected by* a. *If the mirror rotated precisely one-eighth of a revolution while the light journeyed to the distant mirrors and back, the observer saw the light as though the mirror had not rotated.*

the diameter of the earth's orbit, which we now know is about 186,000,000 mi. This distance divided by the time of travel, 1000 sec, gives the speed of light as 186,000 mi/sec. Thus a beam of light could travel around the earth about 7.5 times in 1 sec. Little wonder that Galileo's experiment failed, although it was correct in principle.

In 1850, nearly two centuries after Roemer's observations, Foucault (Foo ko'), the French physicist, measured the speed of light by a revolving mirror method which was improved by Newcomb, Michelson, and others. In Michelson's improved method (1925), an eight-sided mirror *M* (Fig. 32-11) mounted on a vertical axle was so adjusted that light from an intense source—an electric arc—was reflected from one face *c*, traveled to a mirror about 22 mi away, was reflected by it, and returned to another face *a*, whence it was reflected to the telescope. Thus an observer, looking through the telescope, saw the lamp like a star. When *M* was rotated uniformly, a series of flashes of light was sent to the distant fixed mirror. While a given flash was

making the 44-mi journey, the rotating mirror turned through some angle so that in general the returning light was not reflected into the telescope. Suppose that the speed of rotation were carefully adjusted so that the mirror turned exactly *one-eighth of a revolution* while a flash of light made the to-and-fro journey. Thus the face *b* moved to *a* and was in exactly the right position to reflect the light into the telescope. Each returning flash entered the telescope, and the observer saw the lamp as though the mirror were at rest.

The most difficult task was to measure the distance to the fixed mirror. To determine a distance of 22 mi uphill and downhill with an error of a few inches was an unequaled achievement in surveying. The United States Coast and Geodetic Survey did the work, hoping that later the revolving-mirror method might be employed to measure distances with great accuracy, using Michelson's value of the speed of light. This value was 299,790 km/sec, or 186,280 mi/sec. The probable error of the determination was about 1 part in 75,000 or $\frac{1}{750}$%.

Example. Suppose that, in a determination of the speed of light, the returning light was observed in the telescope when an eight-sided mirror turned at a speed of 500 rev/sec. Find the speed of light if the distance to the fixed mirror was 23.29 mi.

The time for the light to travel to and fro was

$$\frac{1}{8} \times \frac{1}{500} \text{ sec} = \frac{1}{4000} \text{ sec}$$

$$\text{Speed of light} = \frac{2 \times 23.29 \text{ mi}}{(1/4000) \text{ sec}} = \frac{186,300 \text{ mi}}{\text{sec}}$$

In free space, electromagnetic waves of all frequencies travel at the same speed. If red light traveled faster than blue light, when Jupiter's moon emerges from the shadow after an eclipse, the red light from it would reach the earth first. Thus, the moon, after the eclipse—contrary to observation—would at first appear to be red, and its color would gradually change as light of other hues reached the observer. Experiments in which radar signals from earth have been reflected from the moon (Fig. 32-12) and from the planet Venus added to the evidence that these signals, too, travel in space with the speed of light.

Foucault in his pioneer experiments proved that light travels slower in water than in air. Later, more precise experiments showed that blue light travels in water about 1% more slowly than red light. The ratio v_v of the speed of light in a vacuum to v, the speed in a medium, is called the *index of refraction n* of

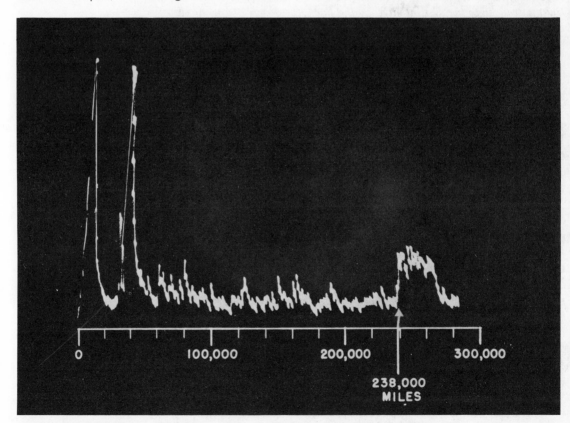

Fig. 32-12. Radar to the moon. The oscilloscope trace shows the emitted signal and the signal reflected from the moon 238,000 mi away. (U.S. Army Signal Corps Photo.)

the medium for that color (wavelength) of light.

$$\text{Index of refraction} = \frac{\text{Speed in vacuum}}{\text{Speed in medium}}$$

$$n = \frac{v_v}{v}$$

We shall consider the consequences of these changes in speed in the next chapter.

Summary

A changing magnetic field is accompanied by an electric field. A changing electric field is accompanied by a magnetic field. An electromagnetic wave consists of an electric-field component and a magnetic-field component. The two fields are at right angles to each other and to the direction of propagation of the wave, and they are in phase with one another.

Electromagnetic waves can be propagated down a long transmission line, a situation in which they are confined to the transmission line, but are not free to travel through space. The fields along the transmission line at large distances from the oscillator lag behind the oscillator; it is in this distant region that the electromagnetic waves are formed.

If the end of a transmission line is opened into a half-wave antenna, electromagnetic waves will be emitted and will travel through space. Retardation of the fields far from the antenna leads to the formation of the waves.

Electromagnetic waves of different wavelengths and frequencies form the electromagnetic spectrum. Although the properties of the waves in different parts of the spectrum are different, all the waves obey the requirements of Maxwell's theory.

Radiant flux is the energy crossing a given surface per unit time. Radiant intensity is the flux per unit solid angle. Irradiance is the flux per unit area striking a surface. The irradiance of a surface obeys the inverse-square law if the source can be treated as a point:

$$\text{Irradiance} = \frac{\text{Intensity}}{(\text{Distance})^2} \cos \theta$$

Light consists of electromagnetic waves that can cause the sensation of sight.

The international standard candle is defined as $\frac{1}{60}$ of the luminous intensity of a "black-body" source, 1 cm^2 in cross-sectional area, at the temperature of melting platinum (1755°C).

The illumination received from a small "point" source varies inversely as the square of the distance from the source. The meter-candle is the illumination of a surface which is 1 m from a standard international candle.

The lumen is the $1/4\pi$ part of the luminous flux in all directions from 1 international "point" candle.

The luminous efficiency of a source equals the ratio of the light output to the total energy input. It is expressed in average candles per watt and in lumens per watt.

Roemer observed that the speed of light is finite by measuring the time between successive eclipses of one of Jupiter's moons, and from this value predicting the time of successive eclipses for a year. When the earth was most distant from Jupiter the eclipses occurred later than the scheduled time. He assumed that the difference was the time required for light to travel across the earth's orbit.

In Michelson's method, a beam of light reflected from one face of a revolving mirror traveled to a second fixed mirror and thence returned to the first one. The speed of revolution was so adjusted that the returning beam was reflected into a telescope. The time of travel of the light was measured by noting the speed of revolution of the first mirror.

The speed of electromagnetic waves of all frequencies in free space is approximately 3×10^8 m/sec. Their speeds in all substances are less than that in a vacuum, and the speeds are different for electromagnetic waves of different frequencies.

Questions

1. In what frequency band did Hertz experiment? What considerations led him to make that choice?

2. Are television waves (a) polarized, (b) reflected? How could you find out by experiment?

3. How could you show that radio waves travel with the speed of light?

4. How could the velocity of light be measured in a laboratory? Describe the method and the equipment needed.

5. Are "60 cycles a-c" a part of the electromagnetic spectrum?

6. What quantities determine the illumination at a point on a screen?

7. Discuss whether the inverse-square law applies to the light from a large diffusing globe.

8. Discuss the effect of a hemispherical reflector on (a) the candlepower and (b) the luminous flux from a lamp behind which the reflector is placed.

9. Calculate the product $\mu_0\epsilon_0$ and determine its units. What significance do you attach to the result?

10. If in Michelson's method the distance to the distant fixed mirror were increased, would the rotating mirror have to turn faster or slower? What is the advantage of using an eight-sided mirror instead of a four-sided one?

11. Why is "meter-candle" a misnomer for a unit of illumination?

12. Discuss the origin of electromagnetic waves. How are they emitted by an antenna?

13. From the information in Table 32-1, estimate typical sizes of (a) an atomic nucleus, (b) the antenna of a commercial radio station, (c) the antenna of a television receiver, (d) an atom, (e) a molecule.

Problems

Speed of light ≅ 186,000 mi/sec
$$\cong 3.0 \times 10^8 \; m/sec$$

A

1. Find the distance in miles from the earth to the star Antares, the light of which requires 520 years to reach the earth.

2. When a book is held at a distance of 0.50 m from a lamp, the illumination is 100 m-candles. What is the illumination at a point 0.25 m from the lamp?

3. If a photographic print can be made in 16 sec when held 0.60 m from a lamp, what is the correct exposure when it is held 1.00 m away?

4. A book is adequately illuminated when it is 2.0 m from a 100-candle source. What is the illumination when it is moved 2.0 m farther away?

5. What is the luminous intensity of a source which, when placed at a distance of 0.50 m from a point, produces at that point illumination equal to that from a 60-candle source at a distance of 3.00 m?

6. How far from a 40-candle source must a book be held so that the illumination may equal that afforded by the full moon (0.25 m-candle)?

7. If the illumination of a card is 400 m-candles, what will it be when the card is moved to a distance twice as far from the source?

8. What is the intensity of a source that provides an illumination of 40 m-candles at a distance of 2.0 m?

B

9. Two lamps, A (10-candle intensity) and B (90-candle intensity), are mounted 80 cm apart in a photometer. The illumination of the two faces of the photometer head located on the line through the lamps is equal. Compute the distance of the photometer head from A.

10. A book 0.20 m² in area is placed at a distance of 1.0 m from a 100-candle incandescent lamp which, it is assumed, sends out light equally in all directions. How many lumens are received by the book (a) if the

light is incident normally, (*b*) if the angle of incidence is 60°?

11. When a photographic print is held at a distance of 0.30 m from a small incandescent bulb, an exposure of 3.0 sec is required. What exposure is required if the distance is 0.50 m?

12. In a determination of a distance, using Michelson's apparatus for measuring the speed of light, suppose that the eight-sided mirror had to be turned at the rate of 1200 rev/sec. Find the distance between this revolving mirror and the distant reflector.

13. A standard candle and a 4.0-candle source are 1.0 m apart. How far from the standard candle must a screen be placed to be equally illuminated by each source? (There are two answers. Show why, using a diagram.)

14. A 100-candle source is at the center of a sphere of radius 100 cm. How many lumens pass through an opening in the sphere of area 80 cm²?

15. Two 100-candle lamps are 160 cm apart and 100 cm above a table. What is the illumination at a point on the table (*a*) vertically under one lamp, (*b*) equidistant from each lamp?

16. A 100-watt lamp has a luminous efficiency of 1.00 candle/watt. How far above a table should this lamp be located to afford an illumination of 250 m-candles?

C

17. The sun supplies radiant energy to the earth at a rate of 2.0 cal/cm²-min. Its distance from the earth is 93×10^6 mi. At what rate does the brightest star, Sirius, supply energy per year per unit area to the earth, assuming that it emits energy at the same rate as the sun and that its distance is 8.7 light-years?

18. A 100-candle lamp is 1.00 m from a screen. A piece of smoked glass is interposed, and to secure the same illumination as before, the lamp must be brought to a distance of 60 cm from the screen. Find the percentage of the incident light absorbed by the smoked glass.

19. If the illumination of a horizontal surface at the equator at noon, when the sun is vertically overhead, is 100,000 m-candles, find the illumination of such a surface on the same day (*a*) at 30°N latitude, (*b*) at 60°N latitude, (*c*) at the North Pole. (Assume equal light absorption by the atmosphere in each case.)

20. An eclipse of one of Jupiter's satellites is found to occur when the earth is most distant from the planet. This is 1000 sec later than the time predicted from observations of the period made 6 months earlier. Estimate how much the eclipse is behind time, (*a*) 3 months, (*b*) 2 months, (*c*) 1 month after the earlier observations.

21. A 30-kw lamp has a luminous intensity of 100,000 candles in a certain direction. At a distance of 75 cm, it produces an illumination equal to that of noon sunlight. The distance of the sun is 150×10^6 km. (*a*) What is the luminous intensity of the sun? (*b*) What is the brightness of the sun in candles/cm², assuming it to be a disk of 700,000-km radius?

22. Suppose that in Fig. 32-4, the right-hand end of the transmission line is short-circuited by means of a heavy wire. The voltage supplied by the oscillator has a frequency of 300 megahertz, and the length of the line is 9.0 m. The speed of electromagnetic waves along the line is 3.0×10^8 m/sec. (*a*) Is the short-circuited end of the line a voltage node or a voltage antinode? (*b*) Is that end a current node or a current antinode? (*c*) Where is the wavelength of the waves? (*d*) Where are voltage nodes to be found along the line?

For Further Reading

Battan, Louis J., *Radar Observes the Weather*, Doubleday and Company, Garden City, 1962.

Henry, G. E., "Radiation Pressure," *Scientific American,* June, 1957, p. 99.

Jaffe, Bernard, *Michelson and the Speed of Light,* Doubleday and Company, Garden City, 1960.

Morrison, Philip and Emily, "Heinrich Hertz," *Scientific American,* December, 1957, p. 98.

Newman, J. R., "James Clerk Maxwell," *Scientific American,* June, 1955, p. 58.

Page, Robert Morris, *The Origin of Radar,* Doubleday and Company, Garden City, 1962.

CHAPTER THIRTY-THREE

Reflection and Refraction

When electromagnetic waves, such as light, strike the boundary between two media, waves are reflected into the first medium and may also, under certain conditions, pass across the boundary into the second. For example, if a narrow beam of light strikes nearly perpendicularly the upper surface of water in a glass, some light is reflected into the air (the first medium) and some enters the water (the second medium).

In this chapter, we shall consider what happens to light waves when they pass from one medium to another. Note, however, that the principles we shall discuss apply in some measure to all electromagnetic waves. Maxwell's mathematical theory of electromagnetic waves, referred to in the last chapter, allows one to explain mathematically the experimental facts of this chapter by a consideration of what happens to the electric and magnetic fields when electromagnetic waves strike the boundary between two dielectrics or between a dielectric and a conductor. The reflection of light waves from a bathroom mirror is explained by the same principles that explain the reflection of short radio waves from the conducting layers of the earth's ionosphere.

In reflection, the direction of travel of the waves changes. Change of direction can also occur in refraction when light passes from one medium into another. As we shall see, the formation of *images* accompanies these changes of direction. Thus this chapter introduces us to the principles of geometrical optics which underlie the design of many everyday aids to vision—lenses, plane and curved mirrors, telescopes, microscopes, and so on. We shall lay the groundwork for later discussion of these instruments by first discussing the behavior of light waves when they strike plane boundaries between media. Although we shall stress the bending of *rays*, this emphasis is only temporary. Light is a wave, and we shall soon return to a discussion of its wave properties.

Reflection

In order to demonstrate reflection, let sunlight enter a darkened room through a small hole in a piece of cardboard. You can trace the narrow beam of light by means of illuminated dust particles. Let the light strike a plane mirror, and adjust the mirror so that the reflected beam retraces its path and goes out at

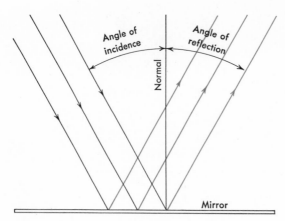

Fig. 33-1. *Light reflected from a plane mirror. The angles of incidence and of reflection are equal.*

the same hole through which it entered. If, next, the light strikes the mirror at an acute angle, the reflected beam and the incident one will make equal angles with the perpendicular and lie in the same plane with it (Fig. 33-1).

You will recall that we represent the directions in which light travels by lines with "barbs" or arrow points, called *rays.*

The angle of incidence is the angle between an incident ray and the normal to the surface at the point where the light is incident. The angle of reflection is the angle between the reflected ray and the normal.

To investigate the formation of images by a plane mirror, support a pane of unsilvered glass in a vertical position and place a wooden block in front of it. You will see a dim reflected *image* or *optical counterpart* of the block back of the mirror. A second block, placed behind the mirror, may be made to coincide with the reflected image. The sizes of the object and its image are exactly equal. Furthermore, the distance from any point on the object to the mirror is equal to the distance of the image of that point from it.

You can locate such an image by the *parallax* method. To see what this means,

sight past your two forefingers at a distant object. Move your head to the right and the *nearer* finger will seem to move oppositely. Repeat, after shifting your two fingers closer together. The amount of parallax will decrease. Finally, hold your fingers at the same distance from your eyes. When you move your head, the two fingers will stay together. Parallax will be zero.

Now mount a pencil behind the glass near the image of a corner of the block referred to above. Move your head sidewise and test for parallax. Shift the pencil until the parallax is zero. Now the pencil and the image coincide. The distance from the pencil to the mirror equals the distance from the corner of the block to the mirror.

In Fig. 33-2, light from the point *A* is reflected by a mirror and seems to come from the point *A'* behind the mirror. Mount a pin at *A* and you will see its image at *A'*. Place two other pins at *B* and *C* so that *A'*, *B*, and *C* are on the same straight line. Draw lines *AB* and *BC*. Draw a line through *B* normal to the mirror. Measurement will show that the angle of reflection *NBC* is equal to the angle of incidence *NBA*. More precise experiments show that *the angle of reflection is always equal to the angle of incidence. The incident ray, the reflected ray, and the normal to the mirror all lie in the same plane.* These laws of reflection have been known since ancient times.

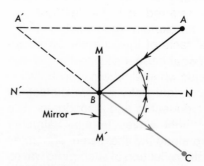

Fig. 33-2. *Locating an image formed by a plane mirror.*

Huygens' Principle

Christian Huygens (1629–1695), a great Dutch physicist and a contemporary of Newton, helped to develop the wave theory of light. He showed that the progress of a wave can be predicted by assuming that *every point in a wave front acts like a new source of waves.* Suppose that, in Fig. 33-3a, *AB* represents the position of a water wave from a distant source—a plane wave. Assume that at zero time circular wavelets start from the points marked by crosses and travel with the same speed in all directions. In a time *t*, each wavelet will travel a distance *vt* where *v* is the speed of the waves. The waves will destructively interfere with one another except along the "envelope" *A′B′*—that is, a line tangent to each wave. This line represents the position of the wave front after a time *t*.

Huygens' principle explains reflection. In Fig. 33-3b, *AB* represents the incident wave. If there were no change of medium, the wave would advance a distance $s = vt$ to *A′B′* in a time *t*. As each part of the wave reaches the reflecting surface, a wavelet starts outward. The reflected wave front is along the envelope *A″B′* of the wavelets.

Sound waves can be used to illustrate Huygens' principle. Figure 33-4 reproduces a photograph of a sound wave passing through slits. Notice how the separate wavelets merge to form the transmitted wave. The wavelets reflected by the barriers between the slits also merge.

Specular and Diffuse Reflection

Aesop tells a fable about a dog who saw his image in a pool and dropped the bone which he was carrying in order to grasp at its image. The dog did not perceive the surface of the water but only the image behind it. Often, in homes, mirrors mounted against the walls deceive the visitor, who thinks that he is looking into another room, so that an effect of spaciousness is achieved. Mirrors have im-

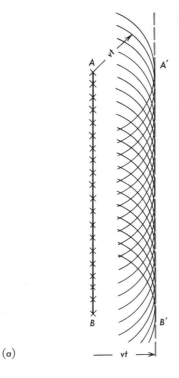

(a)

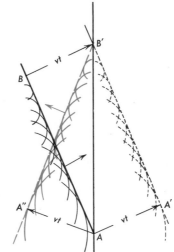

(b)

Fig. 33-3. Huygens' principle. (a) *Assume that each point on the wave* AB *is a source of waves. The envelope* A′B′ *shows the position of the wave after a time* t. (b) AB *is the incident wave,* A′B′ *shows the position of the wave if it had not been reflected, and* A″B′ *shows that of the reflected wave.*

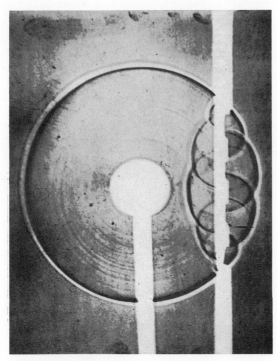

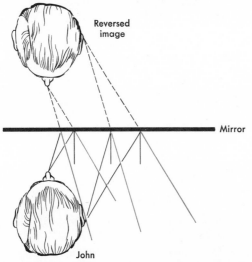

Fig. 33-5. *Reversed image. When John looks north, his image looks south. Therefore, his image is left-handed (if he is right-handed).*

perfections in their surfaces, so that they can be seen. A perfect reflector would be invisible. Reflection by a plane mirror, as in Fig. 33-1, is called *specular* (Latin *speculum*, mirror) *reflection.*

When you look into a mirror mounted on the north wall of a room and point east with your right hand, your image points apparently east with the left hand (Fig. 33-5). Rays drawn from each point on your body to the mirror are reflected according to the principle of reflection. You or a person standing beside you see your image reversed left and right. This reversal occurs because when you look north, your image looks south.

When a beam of sunlight strikes a good mirror, almost all the light is reflected in a single, parallel beam; little of it is scattered, and the mirror is nearly invisible. If the light falls on a piece of paper, diffuse reflection occurs and the paper is visible everywhere in the room. The paper reflects light diffusely in all directions because its surface has a multitude of microscopic hills and valleys (Fig. 33-6). *Objects viewed by reflected light are visible because they reflect light diffusely.*

Some Examples of Reflection

Suppose that in the experiment in the darkened room referred to above, you first adjust the mirror so that the reflected ray is incident normally and retraces its path, and the angles of incidence and reflection are zero. Tilt the mirror through an angle $\theta/2$. Now the angles of incidence and of reflection are both $\theta/2$ and the reflected beam is rotated through an angle θ. *Tilting the mirror through an angle $\theta/2$ rotates the reflected beam through an angle θ.*

This rule is applied in the *sextant,* which navigators use to measure the elevation above the horizon of the sun or a star. In Fig. 33-7, let light from the horizon, traveling along *AD*, pass through the unsilvered part of the mirror *C* and enter the telescope. Let another

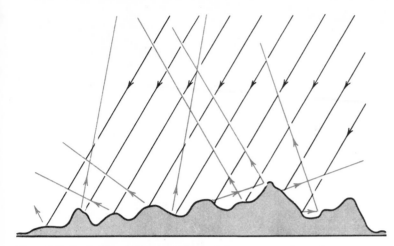

Fig. 33-6. *Light diffusely reflected from a rough surface.*

parallel beam from the horizon traveling along *A'B*, be reflected by the pivoted mirror at *B* and then by the silvered part of the mirror *C*. The twice-reflected beam will travel to the telescope so that the two images of the horizon coincide. Suppose that light from a certain star travels along *A"B*. Let the angle between *A"B* and *A'B* be θ. If we rotate the pivoted mirror through an angle $\theta/2$, the light from the star will enter the telescope. Then the star will seem to be on the horizon. The piv-

oted arm has an index which indicates the value of θ on a scale. The angle between Polaris—the North Star—and the horizon, for example, is (except for a correction of 1° or less) the latitude of the observer.

When light is reflected from two plane mirrors, several images may be seen. In Fig. 33-8 the two mirrors are at right angles to each other. The eye sees the object *O* itself and an image I_1. It also sees an image I_2 due to light once reflected from the second mirror.

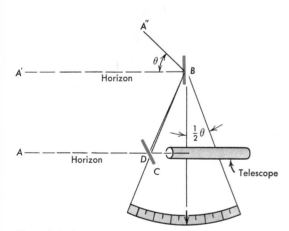

Fig. 33-7. *The sextant. Rotating the pivoted mirror* B *through an angle* $\theta/2$ *rotates the reflected beam through an angle* θ.

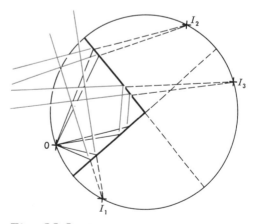

Fig. 33-8. *Two plane mirrors, mounted at right angles to each other, form three images of an object.*

Finally, it sees a third image I_3 due to light reflected by *both* mirrors. The object and the three images lie on a circle centered at the intersection of the two mirrors.

Refraction

When light passes from one medium into another, the beam usually changes direction at the boundary between the two media. This bending we call *refraction.* To demonstrate refraction, dissolve a little dye in the water of an aquarium and flood the air space with smoke. Let a narrow beam of light strike the water as in Fig. 33-9. Part of the beam is reflected specularly, and part enters the water. As the beam enters the liquid it is bent toward the normal. At the bottom of the tank the light is reflected by a horizontal mirror and travels back toward the surface. There, part is again reflected and part refracted into the air. But the beam as it enters the air is bent *away from* the normal. The angle between an incident ray and the normal to the surface is the *angle of incidence;* that between the refracted ray and the normal is the *angle of refraction.* Refraction occurs except when a narrow beam strikes the plane boundary between two different media perpendicularly or — as we shall see — at an angle of incidence greater than a certain critical angle.

We can determine the relationship between the angle of incidence of a beam and the

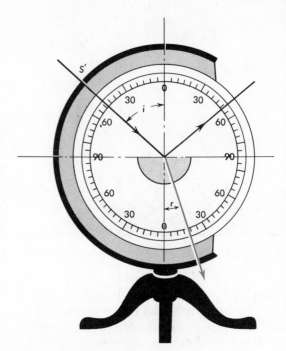

Fig. 33-10. *Apparatus to measure angles of incidence and refraction.*

angle of refraction by means of the device shown in Fig. 33-10. A narrow beam of light passing through the slit strikes the semicircular disk. Part of the light is reflected, and part enters the glass and is refracted toward the normal. The incident and the refracted rays lie in the same plane. On leaving the disk the beam is not deviated because it strikes the circular boundary perpendicularly. The angles of incidence i and of refraction r can be read from the scale, which is graduated in degrees. The position of the slit is then changed so as to vary the angle of incidence and the corresponding angle of refraction. Experiments such as this have established Snell's law:

The ratio of the sine of the angle of incidence to the sine of the angle of refraction in a substance is a constant characteristic of the substance and of the wavelength of the light.

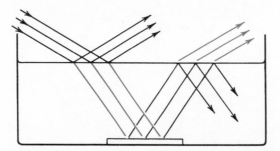

Fig. 33-9. *Refraction at a water surface. Rays entering the water are refracted toward the normal.*

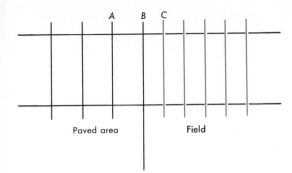

Fig. 33-11. *The marching men do not change direction when they enter the rocky field and change speed.*

Snell's Law and Wave Speed

In Chapter 32 we defined the index of refraction of a substance as the ratio of the speed of light in a vacuum to the speed in the substance. Now we shall prove that this index is also equal to the ratio of the sine of the angle of incidence in a vacuum θ_v to the sine of the angle of refraction in the substance θ. Thus

$$n = \frac{v_v}{v} = \frac{\sin \theta_v}{\sin \theta}$$

In general, if light goes from a medium of which the index of refraction is n_1 to a medium of which the index is n_2, then

$$n_1 \sin \theta_1 = n_2 \sin \theta_2$$

where θ_1 and θ_2 are the angles of incidence and refraction, respectively, at the boundary between the two media.

As an analogy, suppose that a column of marchers enters a rocky field (Fig. 33-11), leaving the paved area at right angles to its boundary. Assume that on the paved area each man has a speed of 3 ft/sec, but in the field his speed is only 2 ft/sec. While one rank of marchers goes from A to B, each man in the preceding rank will move a shorter distance from B to C, and the line of march will not change. Next, suppose that the men cross the boundary obliquely as in Fig. 33-12 and that each man travels at right angles to the rank. Then while the man at A marches to B, the one at A' will move only two-thirds as far, arriving at B" instead of B' in the same time. Hence the line of march deviates to the right. The slower the speed on the rocky ground, the greater will be the deviation of the line of march.

Now assume that Fig. 33-11 represents a train of water waves approaching the shore of a lake and that at B the waves pass over a ledge and the water becomes more shallow.

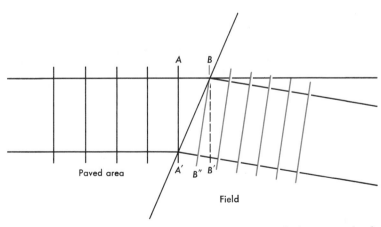

Fig. 33-12. *The line of march deviates toward the normal when the ranks enter the field.*

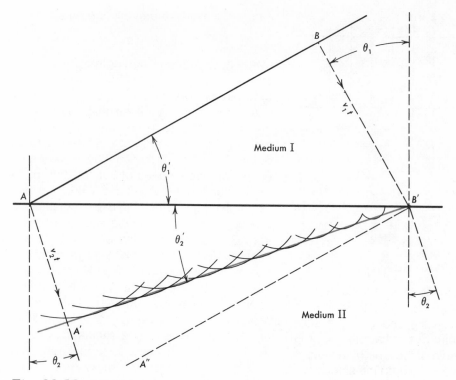

Fig. 33-13. *Refraction of wave:* $\sin\theta_1/\sin\theta_2 = v_1/v_2 = n_2/n_1$.

Because of this shallowing, the speed is found experimentally to diminish, and the waves will crowd together. In this illustration the rays or lines of advance of the waves are not deviated (or refracted). However, when the waves cross the ledge obliquely (Fig. 33-12), the end *B'* will be retarded and the waves will be deviated downward toward the normal to the boundary. The greater the change of speed, the greater will be the change of direction, or refraction.

Now we are ready to derive Snell's law for light. Assume that a narrow beam of light—its rays almost parallel to each other—traveling in a medium I with speed v_1 is incident on a plane surface bounding a material II in which the speed is v_2 (Fig. 33-13). Let *AB* represent a wave front. While the part of the wave at *B* advances to *B'*, *A*, traveling in medium II, advances to *A'*. The ratio *BB'/AA'* equals v_1/v_2,

the ratio of the speeds in the two media:

$$\frac{BB'}{AA'} = \frac{v_1}{v_2}$$

Divide numerator and denominator by *AB'*.

$$\frac{BB'/AB'}{AA'/AB'} = \frac{v_1}{v_2} = \frac{\sin\theta_1'}{\sin\theta_2'}$$

But θ_1' equals the angle θ_1 between the normal and the incident ray, and θ_2' equals the corresponding angle θ_2 in medium II. Hence

$$\frac{\sin\theta_1}{\sin\theta_2} = \frac{v_1}{v_2} \qquad (1)$$

The ratio v_1/v_2 is a constant, so that Snell's law is explained.

The relationship between Snell's law and the indices of refraction of the media is derived as follows:

Rewrite equation (1) thus:

$$\frac{\sin \theta_1}{\sin \theta_2} = \frac{1/v_2}{1/v_1}$$

Multiply numerator and denominator by v_v, the velocity of light in a vacuum. Then

$$\frac{\sin \theta_1}{\sin \theta_2} = \frac{v_v/v_2}{v_v/v_1} = \frac{n_2}{n_1}$$

or

$$n_1 \sin \theta_1 = n_2 \sin \theta_2$$

in which n_1 and n_2 are the indices of refraction of the two media (Table 33-1).

Suppose that light travels through several plane parallel surfaces that separate media I, II, III, etc., whose indices of refraction are n_1, n_2, n_3, etc. (Fig. 33-14). Then

$$n_1 \sin \theta_1 = n_2 \sin \theta_2$$
$$= n_3 \sin \theta_3 = n_4 \sin \theta_4 = \cdots$$

The index of refraction of the glass in a prism of vertex angle A may be determined by measuring in a *spectrometer* (Chapter 36) the total deviation D of the incident beam when the angles i of incidence at the first surface separating air and glass and of refraction r at the second surface separating glass and air are equal (Fig. 33-15). For this condition, the deviation D is a minimum and

$$n = \frac{\sin[(D + A)/2]}{\sin[A/2]}$$

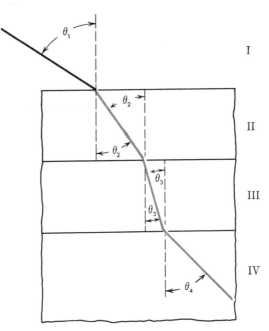

Fig. 33-14. *Refraction at several parallel surfaces:* $n_1 \sin \theta_1 = n_2 \sin \theta_2 = n_3 \sin \theta_3 = \cdots$.

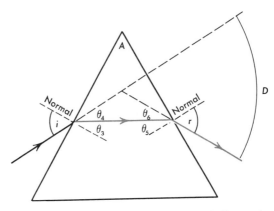

Fig. 33-15. *If* i *and* r *are equal,* D *is the minimum angle through which the prism deviates the light.*

Table 33-1. Refractive Indices

(Yellow light from a sodium source, $\lambda \cong 5.89 \times 10^{-7}$ m)

Air (standard conditions)	1.00029
Alcohol, ethyl	1.36
Canada balsam	1.53
Carbon disulfide	1.64
Diamond	2.47
Glass, crown	1.51–1.52
Glass, flint	1.57–1.89
Oil, cottonseed	1.47
olive	1.48
Rutile (diamond substitute)	2.62
Water	1.33

As we shall see in a later chapter, the index of refraction is different, not only for different substances, but for different wavelengths.

Example. If a ray is incident at an angle of incidence i on one face of a regular prism whose apex angle is A, what is the deviation angle D in terms of A, i, and the indices of refraction of the medium surrounding the prism and of the glass?

In Fig. 33-15, suppose that i does *not* equal r. We can determine D by tracing a ray through the prism. When the ray enters the prism, we have

$$n_1 \sin i = n_2 \sin \theta_3 \qquad (2)$$

and when it leaves, we have

$$n_2 \sin \theta_5 = n_1 \sin r \qquad (3)$$

Now θ_3 and θ_5 can be shown by simple plane geometry to be related to A:

$$A + (90° - \theta_4 - \theta_3 + \theta_4)$$
$$+ (90° - \theta_5 - \theta_6 + \theta_6) = 180° \quad \text{(Why?)}$$

Hence,

$$A = \theta_3 + \theta_5 \qquad (4)$$

We also have

$$\theta_3 + \theta_4 = i \qquad (5)$$

$$\theta_5 + \theta_6 = r \qquad (6)$$

and, adding,

$$\theta_3 + \theta_4 + \theta_5 + \theta_6 = i + r \qquad (7)$$

But

$$D = \theta_4 + \theta_6 \quad \text{(Why?)} \qquad (8)$$

Hence, substituting equations (4) and (8) into (7), we obtain

$$A + D = i + r$$

or

$$D = i + r - A \qquad (9)$$

From equations (2) (3) and (4) we can calculate r. Substituting values of i, r, and A into

equation (9) we obtain D, the angle through which the ray is deviated by the prism.

Although calculus is required for the proof and we shall not prove it here, D has its *smallest* value when $i = r$. We see that *in this special case*

$$r = \frac{D + A}{2}$$

and

$$\theta_5 = \frac{A}{2}$$

So, inserting these values into equation (3), we obtain

$$n_2 \sin\left(\frac{A}{2}\right) = n_1 \sin\left(\frac{D + A}{2}\right)$$

which, for $n_1 = 1$ (when air surrounds the prism) and $n_2 = n$ (the index of refraction of the glass), becomes

$$n = \sin\left(\frac{D + A}{2}\right) \Big/ \sin\left(\frac{A}{2}\right)$$

Total Reflection and the Critical Angle

Suppose that light travels from a medium I into a medium II in which the speed is greater. Then the refracted light is deviated away from the normal. In Fig. 33-16, light rays diverge from a lamp in the bottom of an aquarium. The ray *OF* merits close attention, for the refracted ray is parallel to the water surface. *The angle of refraction θ_2 for this particular ray is 90°.* Light incident at any point G, beyond F, is not transmitted. It is totally reflected. The angle of incidence θ_1 for which the angle of refraction θ_2 is 90° is called the *critical angle.*

The critical angle θ_c is the angle of incidence for which the angle of refraction is 90°.

Let θ_2 be the angle (90°) in the medium of higher speed and θ_c the critical angle.

$$n_2 \sin 90° = n_1 \sin \theta_c$$

$$\sin \theta_c = \frac{n_2}{n_1}$$

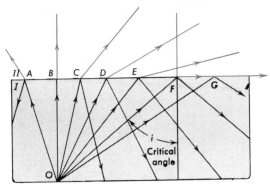

Fig. 33-16. Critical angle, the angle of incidence when the angle of refraction is 90°.

For light going from water into air, $n_1 = 1.33$, $n_2 = 1.00$; therefore $\sin \theta_c = 1/1.33$; $\theta_c \cong 48.6°$.

Notice that light is totally reflected at a boundary only when it is incident in the optically denser medium in which it travels more slowly. To demonstrate this fact, hold an electric lamp above a tumbler of water. Little light is reflected at the surface since most of it enters the water and is refracted. If the lamp is below the level of the surface (Fig. 33-17), light will be totally reflected as from a well-silvered mirror.

The reflecting prism (Fig. 33-18) offers an important application of total reflection. Light incident on the face AC is totally reflected,

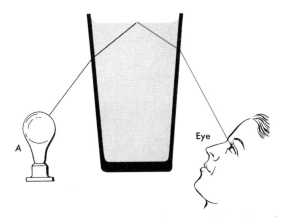

Fig. 33-17. Light from A is totally reflected.

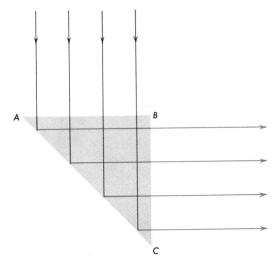

Fig. 33-18. Totally reflecting prism.

and so the prism acts like a mirror. Such prisms are often used in field glasses and in the periscopes of submarines because there is no metallic film which may become corroded.

Few fishermen realize that to a fish the world above water seems quite different from that which we see. In Fig. 33-19, let rays O and O' pass near the smooth, unruffled surface of a pond. At sunrise the light incident at a is refracted downward toward the normal, making an angle of refraction $\theta_c = 48.6°$. Between sunrise and sunset the fish sees the world above the water through a "window" whose diameter is the distance between a and a'. Except for what comes through this window, he sees the bottom of the lake reflected in the surface. Consider also the appearance of a fisherman as viewed by the fish (Fig. 33-20). Though the fisherman stands by the water's edge, his image is seen at $A'B'$.

Atmospheric Refraction

When the sun is near the horizon, the light coming from less dense to more dense air is refracted downward as it passes through the

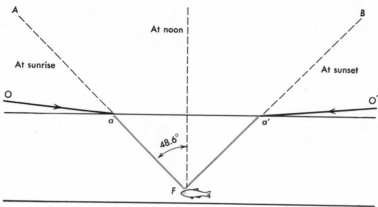

Fig. 33-19. *At sunrise the fish sees the sun high in the sky.*

atmosphere. Thus the sun seems to be higher in the sky than it really is. The sun is completely visible when, geometrically considered, it is slightly below the horizon. Because of this refraction, the time from sunrise to sunset is lengthened by about four minutes. The deviation of the light rays is usually rather small. However, just at sunset, the *difference* in atmospheric refraction of the light from bottom to top of the sun is enough to make the sun appear flattened or oval.

When the air near the earth's surface is colder than that higher up, the deviation is so pronounced that observers can see images of objects beyond the horizon (Fig. 33-21). This phenomenon, called *looming,* is often observed at sea or over great plains. In 1906, Peary, the discoverer of the North Pole, reported the discovery of land which he named Crocker Island. Later expeditions failed to find it; apparently Peary saw in the skies an image of some more distant region. Another explorer reported seeing an island which was 125 miles away from the observation point.

On still, summer days the layer of air above a concrete highway becomes strongly heated

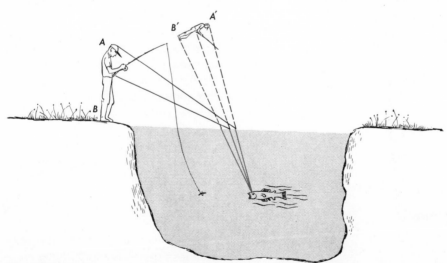

Fig. 33-20. *To the fish, the fisherman seems up in the air.*

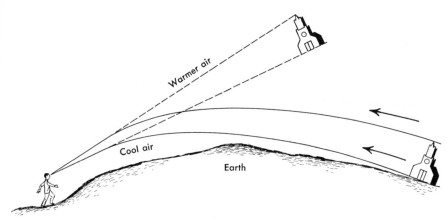

Fig. 33-21. *Looming (exaggerated).*

and less dense than the air farther up, and a mirage can be seen. Light from the sky is refracted upward near the roadway so that the surface seems to be covered with a reflecting layer. Under favorable conditions images of approaching vehicles may be seen reflected in this manner (Fig. 33-22). Often just as an automobile reaches the summit of a hill the image of the sky can be seen reflected at the roadway.

Atmospheric refraction is very troublesome for astronomers because it causes "twinkling" of the stars. The changing densities of air at different levels in the atmosphere cause the images of stars formed by telescopes to dance back and forth on the photographic plate. A blurred image results in spite of the greatest care taken by the astronomer to refine his instrument. Observatories are usually located on high mountains to avoid as much of this atmospheric turbulence as possible.

The best possible observatory is a well-stabilized satellite, orbiting around the earth far outside the earth's atmosphere, or an observatory on the moon. Astronomers are now using a satellite-mounted telescope that takes photographs and sends them back to earth.

Summary

A ray is a line indicating the direction of travel of waves.

The angle of incidence of light at a surface is the angle between an incident ray and the normal to the surface; the angle of reflection is the angle between the reflected ray and the normal. The angles of incidence and reflection are equal. The image of an object formed by a plane mirror is equal in size to the object. The object distance and image distance from the mirror are also equal.

Fig. 33-22. *Mirage. The layer of less dense air near the roadway reflects the light like a mirror.*

A perfect reflector would be invisible. Non-luminous objects are seen solely by diffused, reflected light.

When light passes from one medium to another, refraction occurs. The ratio v_v/v of the speed of light in a vacuum to the speed in a medium is the index of refraction n of that medium. By Snell's law, $n = \sin \theta_v/\sin \theta$.

When a beam of light passes through a series of media bounded by plane parallel surfaces, $n_1 \sin \theta_1 = n_2 \sin \theta_2 = n_3 \sin \theta_3 = \cdots$.

The critical angle is the angle of incidence of light for which the angle of refraction is 90°.

Looming and mirage are caused by atmospheric refraction.

Questions

1. Illustrate Huygens' principle by applying it in the spreading of ripples when a stone is dropped into water.

2. When light goes from one medium to another, do the velocity, frequency, and wavelength all change? Discuss.

3. Why are spots of sunlight under a tree shaped like ellipses instead of being irregular in shape like the openings among the leaves through which the sunlight passes?

4. Which has a greater critical angle when surrounded by air: glass, $n = 1.53$; or diamond, $n = 2.5$?

5. Using a drawing, prove that to see his complete image, a man 6 ft tall must use a mirror at least 3 ft high.

6. If a man of such height that his eyes are 5 ft above the floor is able to view his shoes in a small vertical mirror, how far above the floor must the top of the mirror be?

7. Explain why the image of a right-handed man seen in a plane mirror appears to be left-handed.

8. Locate four images of a candle placed between two plane mirrors making an angle of 60° with each other.

9. Plot the angle of incidence versus the

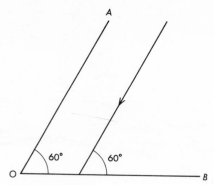

Fig. 33-23.

angle of refraction for light passing at various angles of incidence from vacuum into (a) glass, (b) water. Can any portion of the curves be used to determine the index of refraction of glass and water?

10. What happens to the ray in Fig. 33-23 if the plane mirrors *AO* and *OB* are at an angle of 60°?

11. If a cubical room had mirrors on all surfaces, how many rooms would appear to touch the original room?

Problems

A

1. To see an object most distinctly, a person with normal eyes holds it at a distance of about 25 cm. At what distance should he hold a mirror to see his image distinctly?

2. If a man walks toward a plane mirror with a speed of 2.0 m/sec, with what speed does he approach his image?

3. When light goes from a certain substance into air, the critical angle is 45°. What is the index of refraction of the substance?

4. Light from a lamp beneath the surface of a pool of water is incident on the surface at angles of (a) 20°, (b) 35°. What is the angle of refraction of each ray?

5. What is the critical angle in air of (a) flint glass, (b) rutile, (c) ethyl alcohol?

6. Light is incident on the surface of a pond

at an angle of 45°, traveling from air into water. What is the angle of refraction?

7. The speed of light in free space is 300,000 km/sec. What is its speed in water?

B

8. (a) Compute the speed of light in glass whose index of refraction is 1.60. (b) What is the wavelength of the light in the glass if the wavelength in air is 6.50×10^{-5} cm?

9. Two parallel mirrors face each other. A candle is placed between them at a point 0.10 m from mirror A and 0.20 m from mirror B. How far behind B is the second reflected image of the candle?

10. If a plane mirror A is 0.20 m to the left of a candle, how far to the right is a parallel mirror B if the second reflected image of the candle is 0.80 m behind mirror B?

11. To a fish in an aquarium at the equator, the morning sun on a day in March seems to be 30° from the zenith. What is the actual time of day as indicated by a sundial?

C

12. A nearly parallel beam from a flashlight enters the water at the edge of a pool completely filled with water. The angle of incidence is 30°. If the water is 2.0 m deep, at what distance from the vertical side wall will the beam of light strike the floor of the pool?

13. Two plane mirrors are placed at an angle of 60° with each other. A beam of sunlight is reflected first by one mirror and then by the other. What is the angle between the beam incident on the first mirror and that reflected by the second one?

14. Compute the lateral displacement of a ray of light that enters a rectangular glass plate 10 cm thick if the angle of incidence is

60°. The index of refraction of the glass is 1.50.

15. A thick glass plate is supported in a horizontal plane in a vessel of water. A ray of light is incident on the upper water surface at an angle of 55°. At what angle will it be incident (a) on the upper surface of the glass plate, (b) on the lower surface? Index of refraction of water, 1.33; of glass, 1.50.

16. What is the critical angle between olive oil and water if the index of refraction of water is 1.33 and that of olive oil is 1.48?

17. Plot a curve of n (index of refraction) for heavy flint glass versus λ (wavelength) and from it interpolate to determine the index of refraction at a wavelength of 0.502×10^{-6} m.

λ	n
0.656×10^{-6} m	1.644
0.589×10^{-6} m	1.650
0.502×10^{-6} m	?
0.486×10^{-6} m	1.664
0.434×10^{-6} m	1.675

18. Calculate the angle of minimum deviation of yellow light ($\lambda = 5.89 \times 10^{-7}$ m) by a 60° equilateral prism made of glass whose index of refraction at that wavelength is 1.52.

19. The range of vision of a scuba diver in water 60 ft deep is limited to rays that come within a certain cone. What is the diameter of the cone at the surface?

For Further Reading

Minnaert, M., *The Nature of Light and Colour in the Open Air,* Dover, New York, 1954.
O'Connell, D. J. K., "The Green Flash," *Scientific American,* January, 1960, p. 112.

CHAPTER THIRTY-FOUR

Lenses and Curved Mirrors

In this chapter, we shall apply to lenses and curved mirrors the principles of reflection and refraction at plane boundaries between media which we discussed in the last chapter. Lenses, we shall see, behave as if they were made of many small prisms placed side by side; curved mirrors similarly act like an array of tiny plane mirrors suitably oriented to form a curved surface. Snell's law applies to the former, and the law of reflection to the latter. Most optical devices such as cameras, telescopes, and microscopes have combinations of lenses and, less frequently, of curved mirrors.

Lenses

A lens is a piece of transparent material having two polished surfaces, at least one of which is curved (Fig. 34-1). Lenses are made of various materials, chosen for their transparency to the radiation involved, for their indices of refraction, and for their ability to take and keep the required shape. For example, glass lenses are suitable for light, and quartz lenses for ultraviolet rays (to which glass is opaque). The *principal axis* of a lens is the straight line through the centers of its two polished surfaces. Lenses are divided into two classes: *converging* (*positive*) and *diverging* (*negative*).

A convex glass lens in air converges light because (1) the light travels more slowly in glass than in air and (2) the light going through the relatively thick center of the lens is retarded more than light going through the relatively thin edges. In Fig. 34-2, plane light waves from a distant source are incident on a lens. While the outer parts of the wave travel from A to A' and C to C', being refracted as they pass through the curved surface of the lens at other than normal incidence, the central part travels the distance B-B' in the glass. The outer parts of the wave get ahead, and the wave converges at the point F, called a *principal focus* of the lens. (Notice that the incident rays are parallel to the axis of the lens.) *A principal focus of a lens is a point where rays parallel to the principal axis intersect after being refracted by the lens.* The *focal length* of a lens is the distance from its center to a principal focus.

An alternative way of thinking about the lens is to consider it as made up of small prisms stacked one above the other as shown by dotted sections in the lens of Fig. 34-2.

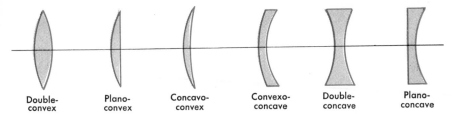

Fig. 34-1. Types of simple lenses.

| Double-convex | Plano-convex | Concavo-convex | Convexo-concave | Double-concave | Plano-concave |

Each prism produces a deviation of the ray passing through it, in accord with Snell's law, and the combined effect of all of these refractions is the focusing action.

The Focal Length of a Thin Lens

The focal length f of a thin lens (in air) whose two faces have radii of curvature r_1 and r_2 is given by

$$\frac{1}{f} = (n-1)\left(\frac{1}{r_1} + \frac{1}{r_2}\right)$$

in which n is the relative index of refraction of the material of which the lens is composed. The sign of the radius of a surface is taken as positive when convergence is produced at that surface. Thus for a double convex lens both r's are positive, and the focal length is positive. This equation—sometimes called the lens-maker's equation—is not exact but holds approximately for thin lenses, such as spectacle lenses, and for rays that remain within a few degrees of the axis.

To understand how the focal length of a curved surface depends upon the radius of the surface and upon the index of refraction of the material, consider (Fig. 34-3) a block of transparent material, one face of which is part of a sphere of radius r. The ray AB, refracted at B as it goes from the block into air, crosses the axis of the figure at F, which therefore is a principal focus of the system.

Let the index of refraction of the block be n. From Snell's law

$$\sin \phi = n \sin \theta$$

If the angles θ and ϕ are small and are

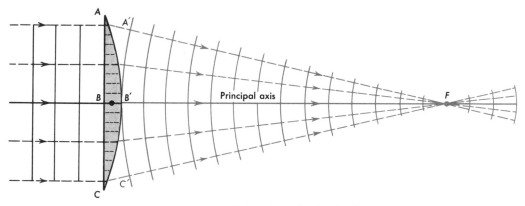

Fig. 34-2. Converging lens. Rays parallel to the principal axis converge at the principal focus F. The distance from the center of the lens to the principal focus is the focal length of the lens.

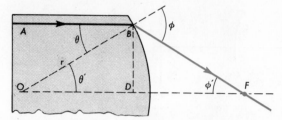

Fig. 34-3. *Refraction at a spherical surface.*

expressed in radians,

$$\sin \theta \cong \theta \quad \text{and} \quad \sin \phi \cong \phi$$

Then:

$$\phi = n\theta = n\theta'$$

$\phi = \theta' + \phi'$ (Exterior and interior angles of a triangle)

$$n\theta' = \theta' + \phi'$$

$$(n-1)\theta' = \phi'$$

$$\theta' \cong \frac{BD}{r} \quad \text{and} \quad \phi' \cong \frac{BD}{f}$$

$$\frac{(n-1)}{r} \cong \frac{1}{f}$$

This is an equation for the focal length of a plano-convex lens, the convex side of which has a radius of curvature r. For a thin lens having two convex surfaces of radii r_1 and r_2, a similar derivation yields the lens-maker's

equation,

$$\frac{1}{f} = (n-1)\left(\frac{1}{r_1} + \frac{1}{r_2}\right)$$

Figure 34-4 shows how rays parallel to the principal axis (dashed lines) converge at the principal focus. It also shows another set of rays (solid lines), not parallel to the principal axis. They converge elsewhere in the *principal plane* of the lens, a plane perpendicular to the principal axis and passing through the principal focus.

A lens of short focal length is said to be *stronger* than one of greater focal length. The *strength* of a lens is the reciprocal of its focal length. When the focal length is in meters, the strength is in *diopters. One diopter is the strength of a lens whose focal length is one meter.* The strength of a lens of focal length 0.5 meter is 2 diopters.

Let f be the focal length of two thin lenses, in contact, and let f_1 and f_2 be their individual focal lengths. It can be proved from the lens maker's equation that

$$\frac{1}{f} = \frac{1}{f_1} + \frac{1}{f_2}$$

For example, two lenses of strength +0.50 diopter and +1.00 diopter, in contact, have a strength of +1.50 diopters, and the focal length of the combination is 0.667 m.

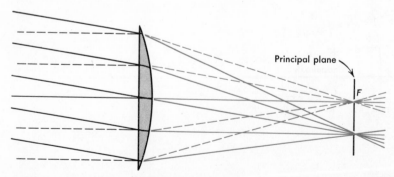

Fig. 34-4. *Principal plane. Rays parallel to the principal axis converge at the principal focus* F. *Other parallel rays converge in the principal plane.*

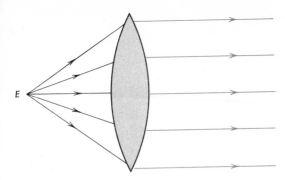

Fig. 34-5. Light from a source at the focus of the lens is converged so as to become parallel to the principal axis. The image of E is formed at a very great distance—at infinity.

Images Formed by Convex Lenses

Suppose that a light source is moved from a great distance along the principal axis toward a converging lens. Then the image will move from the principal focus farther away from the lens.

When the light source reaches the principal focus E (Fig. 34-5), the rays leaving the lens are parallel to the principal axis, and the image of the light source is formed at a very great distance (infinity). When the source is located between the principal focus and the lens, the rays do not converge beyond the lens; they diverge as if they originated at a point behind it (Fig. 34-6).

When light converges after passing through a lens, it forms a *real image.* Energy of the electromagnetic waves is concentrated at the real image. A thermometer placed at the image would be heated, and photographic paper would be blackened. The image can be seen on a screen placed in the image plane and is inverted. When rays diverge after passing through a lens, the image is *virtual,* is erect, and cannot be projected on a screen. *A real image is one formed by a converging beam and can be projected on a screen. A virtual image cannot be so projected.*

In order to locate the image of an extended object, draw at least two of the following lines from some point on the object (Fig. 34-7). (For convenience, we may imagine replacing the real lens by an imaginary lens that has the same focal length and a diameter large enough to construct these lines.) One of them, *AB,* parallel to the principal axis, is refracted so as to pass through the principal focus F_1. A second ray, *AC,* through the center of the lens, is not deviated. The third ray, *AD,* through the other principal focus, becomes parallel to the principal axis. All three rays meet at *A',* which is the image of *A.* (Any other ray *AG* will be refracted so as to pass through *A'.*) By locating the images of other points in the object by this same method, we can construct the entire image.

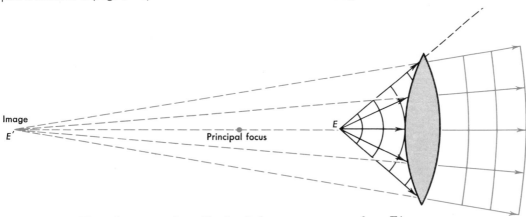

Fig. 34-6. When the source is at E, the light seems to come from E'.

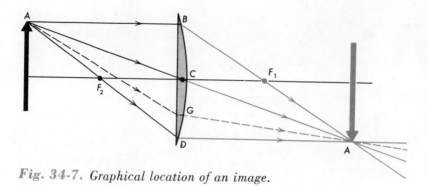

Fig. 34-7. Graphical location of an image.

A virtual image is located by the same method (Fig. 34-8).

The image of any object can be located by the graphical method, which we have discussed, but often it is more convenient to do so by computation. The relation between the distance *p* of the object from the lens, the distance *q* of the image from the lens, and the focal length *f* is given by

$$\frac{1}{\text{Object distance}} + \frac{1}{\text{Image distance}}$$
$$= \frac{1}{\text{Focal length}}$$

$$\frac{1}{p} + \frac{1}{q} = \frac{1}{f}$$

The focal length *f* is considered positive for a converging lens and negative for a diverging lens. The object distance is taken as positive (+) for any real object (light diverging when it strikes the lens). Most objects in examples discussed here are real. The image distance is considered positive (+) when the image is real and negative (−) when it is virtual.

To derive the image equation, consider the similar triangles *ACB* and *GCH* in Fig. 34-9.

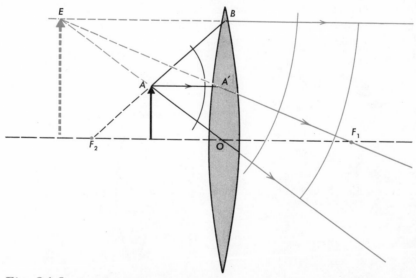

Fig. 34-8. Graphical location of a virtual image.

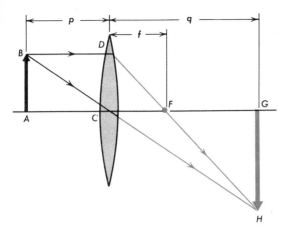

Fig. 34-9. *Derivation of the image equation.*

We see that

$$\frac{GH}{AB} = \frac{CG}{CA} = \frac{q}{p}$$

The triangles *CFD* and *GFH* are also similar; therefore

$$\frac{GH}{CD} = \frac{GF}{CF} = \frac{q-f}{f}$$

AB = CD; hence the left-hand terms of the two equations are equal, and the right-hand terms are also equal:

$$\frac{q}{p} = \frac{q-f}{f} = \frac{q}{f} - 1$$

We divide each term by *q* and obtain the image equation

$$\frac{1}{p} + \frac{1}{q} = \frac{1}{f}$$

Example. (a) A lamp is 36 cm from a converging lens of focal length 12 cm. What is the image distance, and what kind of image does the lens form?

$$\frac{1}{36 \text{ cm}} + \frac{1}{q} = \frac{1}{12 \text{ cm}}$$

$$q = +18 \text{ cm} \qquad \text{(Real image)}$$

(b) If the lamp were 8.0 cm from the lens, where would the image be?

$$\frac{1}{8.0 \text{ cm}} + \frac{1}{q} = \frac{1}{12 \text{ cm}}$$

$$q = -24 \text{ cm} \qquad \text{(Virtual image)}$$

Linear Magnification

The size of the image formed by a lens may be larger than, smaller than, or equal to, the size of the object. *The linear magnification m is the ratio of image length to object length:*

$$\text{Linear magnification} = \frac{\text{Image length}}{\text{Object length}}$$

In Fig. 34-9, by similar triangles,

$$m = \frac{GH}{AB} = \frac{q}{p}$$

This equation holds for all kinds of lenses and mirrors.

The linear magnification produced by a lens equals the ratio of the image distance to the object distance.

Images Formed by Diverging Lenses

When a beam of light passes through a diverging (negative) lens, the central part of the wave front gets ahead because it traverses a smaller thickness of glass than the outer part. In Fig. 34-10, the incident beam, parallel to the principal axis, is diverged, and the light leaving the lens seems to come from the point *F*, which is a principal focus of the lens. The focus at *F* is virtual, for the image cannot be projected on a screen.

In order to locate the image of an extended object, employ the same method as you used for convex lenses. Draw at least two rays from any point of the object such as point *A* (Fig. 34-11). One of these, *AB*, parallel to the principal axis, is deviated upward so that the light travels along *BD*, away from the focus F_2. The second ray, *AC*, through the center of the lens, is not deviated. (A third ray directed toward F_1 is deviated by the lens so as to become parallel to the principal axis.) To an eye beyond the lens, the top of the arrow seems to be at *A'*, which is the virtual image of *A*.

The image equation is used for concave lenses, but the focal length is considered as negative.

Concave lenses invariably give virtual images of real objects. These images are smaller than the objects and closer to the lens.

Example. A small lamp is 4.0 cm from a concave lens of focal length $f = -12$ cm. What is (*a*) the image distance, (*b*) the linear magnification?

(*a*) $$\frac{1}{4.0 \text{ cm}} + \frac{1}{q} = \frac{1}{-12 \text{ cm}}$$

$$q = -3.0 \text{ cm} \qquad \text{(Virtual image)}$$

(*b*) $$m = \frac{q}{p} = \frac{-3.0 \text{ cm}}{4.0 \text{ cm}} = -0.75$$

(We ignore the negative sign and consider the magnification as positive.)

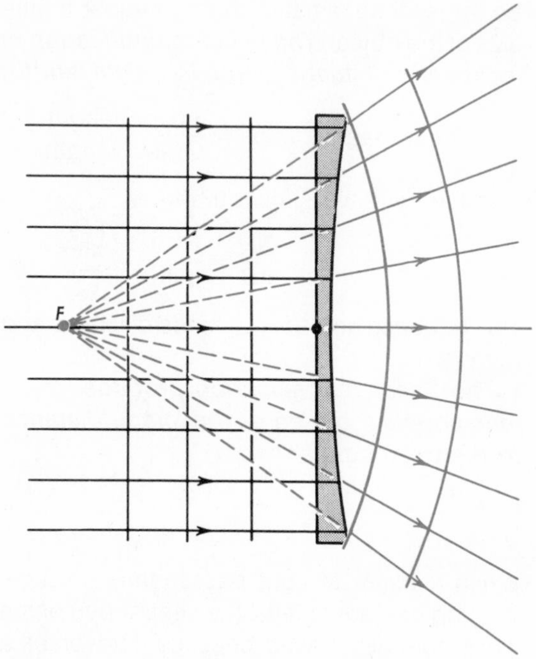

Spherical Aberration

The graphical methods which have been described and the associated formulas are satisfactory for thin lenses such as those of spectacles, but they fail for thick lenses like the one represented in Fig. 34-12 (or for thin lenses when incident rays making large

Fig. 34-10. *Diverging lens. Incident rays parallel to its principal axis are diverged. They seem to intersect at the principal focus* F.

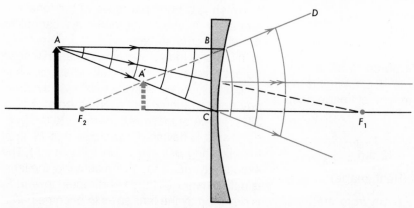

Fig. 34-11. *Image formation by a concave lens.*

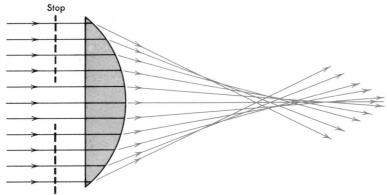

Fig. 34-12. *Spherical aberration. The thick lens does not have a point focus. Use of the stop decreases the spherical aberration.*

angles with the principal axis are considered). Parallel light, incident on a thick lens, for example, is not brought to a single focus, and therefore sharp images cannot be produced. This defect we call *spherical aberration.*

The image formed by a thick lens may be made sharper and better defined by interposing an opaque diaphragm or "stop" so as to cut off the outer parts of the beam, but this reduces the amount of light transmitted by the lens, and the image is less bright. In microscopes and telescopes, combinations of lenses reduce the aberration. In a later chapter we shall consider chromatic aberration, the failure of a lens to bring light of all wavelengths to a point focus.

*Curved Mirrors**

Some optical instruments use curved mirrors to collect light and produce images.

Mirrors with curved surfaces are used like lenses to converge or diverge light. Familiar examples of curved mirrors are the reflectors of automobile headlights and the rearview mirrors sometimes mounted on the fenders of trucks.

* The material on curved mirrors closely parallels that on lenses. Much of the following may be omitted without loss of unity.

A *spherical mirror* is one of which the surface is part of a sphere, and its *radius of curvature* (*R*, Fig. 34-13) equals the radius of the sphere. The *principal axis* of the mirror is a line drawn through the center of curvature C and the vertex or center of the mirror.

Image Formation by a Converging Mirror

If you hold a concave mirror so that it converges sunlight onto a sheet of paper and move the paper to and fro, you can find a position where the image of the sun is sharply

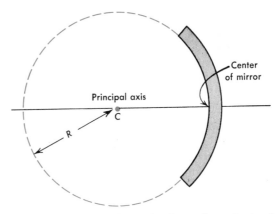

Fig. 34-13. *A spherical mirror. Its principal axis passes through its center and its center of curvature* C.

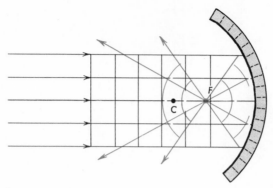

Fig. 34-14. *Light incident parallel to the principal axis is converged at the principal focus.*

focused. Repeat, using a small, distant source such as an arc lamp. In both examples, the image is formed at the principal focus of the mirror (Fig. 34-14). *The principal focus of a mirror is that point where rays incident parallel to the principal axis intersect after being reflected.* Note that we can think of a spherical mirror as made up of many small plane mirrors (dotted sections in Fig. 34-14), each of which reflects light in accord with the principle that the angles of incidence and reflection are equal.

In Fig. 34-15, *AB* represents a ray of light incident on a spherical mirror and reflected so as to pass through the principal focus *F*. The focal length *FG* is approximately equal to one-half of the radius *GO* of the mirror. The proof is left as an exercise for you at the end of the chapter.

To locate the image of the point *A* (Fig. 34-16), draw at least two of the following rays: *AB*, parallel to the principal axis, passes through the principal focus *F* after being reflected. *AK*, drawn through the principal focus, after being reflected is parallel to the principal axis. *AL*, drawn through the center of curvature of the mirror, is incident normally at *L* and retraces its path after being reflected. *AG*, drawn to the center of the mirror, is reflected so that the angle of incidence *AGO* equals the angle of reflection *OGA'*. The point *A'*, where the reflected rays intersect, is the image of *A*.

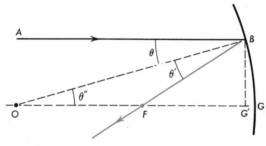

Fig. 34-15. *Focal length* GF *is approximately one-half of the radius of curvature* GO.

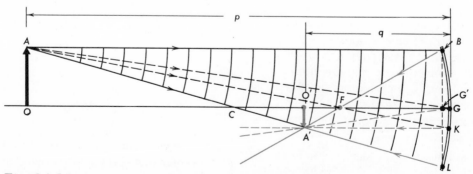

Fig. 34-16. *Locating the image formed by a curved mirror.*

The image formed by a curved mirror can be located by the graphical method which we have described, but often it is more convenient to employ the image equation,

$$\frac{1}{p} + \frac{1}{q} = \frac{1}{f}$$

where p is the distance of the object from the mirror, q is the image distance, and f the focal length. This equation is the same as that used for lenses, and the same rules for signs apply. Consider p as positive for any real object; q is positive for real images and negative for virtual ones. f is taken as positive for converging (concave) mirrors and as negative for diverging (convex) mirrors. The equation holds for shallow mirrors and for rays that remain within a few degrees of the axis.

To prove the image equation for mirrors, note that in Fig. 34-16

$$\frac{O'A'}{OA} = \frac{q}{p} \qquad (1)$$

The triangles BFG and $A'FO'$ are similar, and

$$OA = BG$$

$$\frac{O'A'}{OA} = \frac{O'A'}{BG} = \frac{O'F}{FG} \qquad (2)$$

For a very shallow mirror, $FG = FG'$ approximately. Therefore, from equations (1) and (2),

$$\frac{q}{p} = \frac{O'F}{FG'} \cong \frac{q-f}{f}$$

$$\frac{1}{p} \cong \frac{1}{f} - \frac{1}{q}$$

$$\frac{1}{p} + \frac{1}{q} = \frac{1}{f}$$

As for lenses, the linear magnification m is given by

$$m = \frac{O'A'}{OA} = \frac{q}{p}$$

Example. A man holds a spherical concave shaving mirror of radius of curvature 60 cm

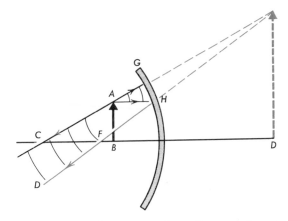

Fig. 34-17. A concave mirror forms a virtual image of an object located at a distance less than the focal length.

and focal length 30 cm at a distance of 15 cm from his nose. Locate the image and compute the magnification.

$$\frac{1}{15\ cm} + \frac{1}{q} = \frac{1}{30\ cm}$$

$$q = -30\ cm \qquad \text{(Virtual image)}$$

$$m = \frac{30\ cm}{15\ cm} = 2.0$$

When the object distance is less than the focal length of a concave mirror, the image seems to be formed behind the mirror (Fig. 34-17). The two reflected rays diverge and must be produced backward to find the point of intersection. The reflected light does not pass through the image, hence the image is virtual.

Convex Mirrors Diverge Light

When a beam of light parallel to the principal axis strikes a convex mirror, the reflected beam diverges from the principal focus, which is behind the mirror. The image of an object is located, as in concave mirrors, by drawing such rays as AB (Fig. 34-18), parallel to the principal axis, and AG, normal to the

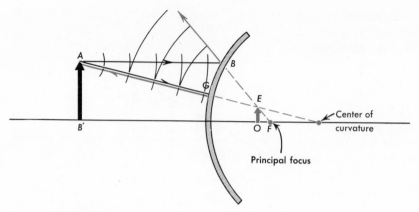

Fig. 34-18. Convex mirror forms a virtual image.

mirror. The two reflected rays diverge from a point *E*, which is the image of *A*. The images of real objects formed by convex mirrors, like those formed by concave lenses, cannot be projected on a screen. They are virtual images.

Sometimes a convex mirror is mounted on the left front fender of a truck to afford a view of the roadway behind. Because of the convexity, the image is reduced in size. The driver can see a wide field in the mirror; in a plane mirror the view would be narrowly restricted. The great disadvantage of such convex rearview mirrors is that they distort distances. Cars approaching from the rear seem to be farther away than they really are and, consequently, seem to be moving more slowly.

Spherical Aberration

The image equation and the graphical method for locating images are valid for shallow mirrors, but they fail when applied to deep mirrors—or to shallow mirrors when incident rays making large angles with the principal axis are considered—just as they do for thick lenses. In Fig. 34-19, the rays incident near the edge of the deep mirror do not converge near the principal focus. By restricting the incident rays to those near the axis—by

using a stop—we can reduce spherical aberration, but at the cost of reducing the amount of light that forms the image. To avoid spherical aberration without decreasing the brightness of the image, a parabolic mirror is used (Fig. 34-20). All rays parallel to the principal axis of such a mirror are brought to the same focus, and all rays originating at the focus leave the mirror in a parallel beam. The most familiar example is the reflector of an automobile headlight. The filament of the lamp bulb is mounted near the principal focus of the parabolic mirror, and the reflected beam is nearly parallel. Intense electric arcs are used for large searchlights. The parabolic

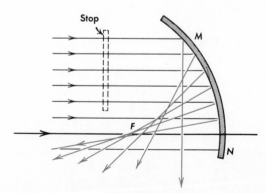

Fig. 34-19. Spherical aberration. A deep spherical mirror does not converge a parallel incident beam to a point focus.

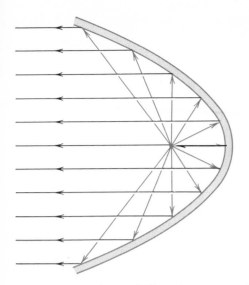

Fig. 34-20. A parabolic mirror as used in a searchlight.

mirrors are so well designed that the illumination of a distant object may be millions of times greater than it would be if the arc were used without the mirror.

Summary

A lens is a piece of transparent substance bounded by curved surfaces. A converging lens is thicker at the center than at the edge.

The principal axis of a lens is a line through the centers of curvature of its two faces. Its principal focus is a point where a beam of light, incident parallel to the principal axis, converges. The focal length of a thin lens is the distance from a principal focus to the lens.

A real image can be projected on a screen; a virtual image cannot be so projected.

The image of the head of an arrow may be located by drawing two rays from it, one parallel to the principal axis, the other passing through the center of the lens (or mirror). The second ray is not deviated by the lens (or mirror), but the first one is bent so as to pass through the principal focus. These rays intersect at the image of the point.

When the object distance is greater than the focal length of a converging lens (or mirror), the image is real. When it is less than the focal length, the image is virtual.

The image of an object produced by a diverging lens (or mirror) is always virtual.

Linear magnification is defined as the ratio of image length to object length:

$$m = \frac{\text{Image length}}{\text{Object length}} = \frac{\text{Image distance}}{\text{Object distance}}$$

The image formula is

$$\frac{1}{p} + \frac{1}{q} = \frac{1}{f}$$

The object distance is positive. The image distance is positive for real images and negative for virtual ones. The focal length is positive for converging lenses or mirrors and negative for diverging ones.

The focal length of a thin lens may be computed by the equation $1/f = (n - 1)(1/r_1 + 1/r_2)$, in which r_1 and r_2 are the radii of the two curved surfaces.

Spherical aberration is the failure of a thick lens or deep mirror to bring parallel rays to a point focus.

The radius of curvature of a spherical mirror is the radius of the sphere of which the mirror surface is a part. The focal length of a spherical mirror is one-half of its radius. The laws of image formation for lenses and mirrors are closely analogous.

Questions

1. Discuss the effect of a converging lens on plane light waves.

2. What is the physical difference between a real and a virtual image?

3. How may spherical aberration in (a) a lens and (b) a mirror be decreased?

4. Under what conditions, if ever, is the image formed by a converging lens (a)

smaller than the object and inverted, (b) smaller than the object and erect?

5. In what direction will the image move when an object moves from a distant point toward (a) a converging lens, (b) a diverging lens?

6. By graphical construction, locate the images of arrows placed at the following distances from a converging lens to focal length 5 cm: (a) 15 cm, (b) 6 cm, (c) 4 cm, (d) 0.5 cm.

7. Prove that the focal length of a concave mirror is approximately equal to half the radius of curvature (Fig. 34-15).

8. What approximations are made in deriving the lens and mirror image equations? Discuss the conditions under which the approximations would cease to be valid. Try to set some quantitative limits.

9. What is desired optically in a rearview mirror in a car? What dimensions are needed for an inside mirror?

10. Prove that a plane mirror gives a virtual image.

11. Can a real image be obtained with a diverging lens?

12. How can a virtual image serve as an object?

13. If you have a number of thin lenses, how can you determine which are converging and which are diverging? If two are in contact, how can the resultant focal length be calculated?

14. Does a hollow glass sphere containing air and placed under water cause a parallel beam of light in water to converge or diverge?

15. A small arrow on the principal axis of a converging lens is shifted from a point where p is slightly greater than f to one where it is slightly less. What happens to the image of the arrow?

16. Discuss the lenslike qualities of (a) a glass of water, (b) a drop of dew on a leaf, and (c) the sphere at the tip of a glass rod that has been melted in a flame.

Problems

(*Whenever possible, construct a ray diagram to check your answer.*)

A

1. A window 20 cm by 25 cm is 80 cm from a convex lens of focal length 40 cm. What is the area of the image?

2. Calculate the position of the image of an arrow 5 cm high located 5.0 cm from a concave mirror of radius 25 cm.

3. Where is the image formed and how large is it when an arrow 1.0 cm high is located 50 cm from (a) a converging lens whose focal length is 25 cm, (b) a diverging lens of focal length −25 cm?

4. The image formed on a screen by a converging lens of focal length 40 cm is 40 cm from the lens. Where is the object?

5. A small lamp is located 6.0 cm from a concave lens of focal length −6.0 cm. Where is the image formed? Make a graphical construction.

6. A converging lens of focal length 20 cm forms a real image at a distance of 30 cm. What is the object distance? What is the power of the lens in diopters?

7. A virtual image is formed 30 cm from a lens when the object distance is 15 cm. What is the focal length of the lens?

8. (a) The focal length of a lens is 8.0 in. How far from it must an object be located in order that the real image may be as large as the object? (b) How far away must the object be from the lens if the real image is to be twice as large as the object?

9. A lens, located 12 cm from an object, forms a virtual image one-half as large as the object. What is the power of the lens in diopters?

10. An arrow 3.0 cm high is located on the principal axis of a converging mirror of focal length 5.0 cm. By a graphical method, locate the image of the arrow when it is (a) 10.0 cm, (b) 7.5 cm, (c) 0.50 cm from the mirror.

11. A real image is formed 50 cm from a concave mirror when the object distance is 25 cm. What is the radius of curvature of the mirror?

B

12. Assuming the sun's diameter to be 1 million miles and its distance from the earth to be 100 million miles, find the size of the sun's real image formed by a convex lens of 20-m focal length.

13. An object is 5.0 cm from a concave mirror, and the virtual image is apparently 10 cm behind the mirror. Find the focal length of the mirror.

14. You can determine the approximate focal length of a thin converging lens by measuring the distance from the lens to the image of a fairly distant object. What percentage error would occur if you applied this method to a lens of focal length 20 cm, using an object distance of (a) infinity, (b) 60 cm, (c) 240 cm?

15. A lamp and a screen are 4.0 m apart. What is the focal length and strength in diopters of a lens that will form an image of the lamp twice as long as the lamp itself?

16. A lamp and a screen are 125 cm apart. At what distances from the screen may a convex lens of focal length 20 cm be located to form an image of the lamp on the screen? Find the linear magnification for each position.

17. A shaving mirror 15 cm from a man's face produces a virtual image with twofold magnification. Find the focal length of the mirror.

18. A dentist holds a concave mirror of radius of curvature 4.0 cm at a distance of 1.0 cm from a filling in a tooth. (a) Where is the image formed? (b) What is the magnification?

19. A lens maker wishes to make a plano-convex spectacle lens of strength 1.5 diopters. The index of refraction of the glass is 1.50. What must be the radius of curvature of the glass?

C

20. An electric lamp is 1.0 m from a plane mirror suspended against the east wall of a hall 3.0 m wide. Light emitted by the lamp is reflected by the mirror to a lens placed beside the lamp. What is the focal length of the lens if a real image of the lamp is formed on the west wall?

21. A converging lens *A* of focal length 15 cm is 16.5 cm from another converging lens *B* of focal length 2.0 cm. Find the distance from the lens *A* to the image of a distant star produced by the lens combination. The light passes through *A* and then through *B*.

22. Solve problem 21, assuming that the distance between the two lenses is 10.0 cm. (Assume that the object distance for the second lens is negative.)

23. Two lenses, each of focal length 8.0 cm, are mounted 70 cm apart, and an arrow 1.0 cm high is located 4.0 cm from the first lens. Where is the second image formed, and how high is it?

24. The radius of curvature of a thin-walled spherical glass container is 100 cm. What is the power in diopters of the lens formed by placing in this container (a) water, (b) carbon disulfide ($n = 1.64$)?

25. Light from a star passes through a converging lens of focal length 15 cm and then is incident on a converging mirror of focal length 30 cm. How far from the mirror is the final image when the distance between the lens and the mirror is (a) 45 cm, (b) 50 cm, (c) 25 cm, (d) 15 cm?

26. If the mirror of the Mt. Palomar telescope has a focal length of 18 m, what is the diameter of the sun's image as formed by the mirror alone? The sun has a diameter of 1.4×10^9 m and is 1.5×10^{11} m from the earth.

27. (a) What is the radius of curvature of the curved surface of a plano-convex lens of focal length 80 cm ($n = 1.53$)? (b) What would the focal length of this lens be when submerged (1) in water, (2) in carbon disulfide ($n = 1.64$)?

28. Parallel light is incident on a spherical surface of radius of curvature 20 cm polished on one side of a large block of optical glass. How far from the surface is the image formed within the glass?

29. Make a graph of the velocity of the image versus the object distance when a man walks toward a large concave mirror of radius r with a velocity v_0 relative to the mirror.

30. Make a graph of the image distance versus the object distance, from zero to infinity, for (a) a converging lens, (b) a diverging lens.

31. A convex air lens, made of two thin watch glasses, each of radius of curvature 20 cm, with air sealed between them, is immersed in water. (a) What is the focal length of the lens-water combination, disregarding the slight refraction in the glass? (b) If an underwater object is 40 cm from the lens, where will its image be formed?

32. A beam of light, converging toward a point 30 cm in back of a lens, passes through the lens and forms a real image 10 cm behind the lens. What is the focal length of the lens?

33. An object is 200 cm to the left of a converging lens (focal length, 100 cm). A plane mirror is 50 cm to the right of the converging lens. Locate the final image after light from the object has passed through the lens and been reflected by the mirror.

34. A spherical mirror is placed between points *A* and *B*. The concave side of the mirror faces toward *A*, which is 1.40 cm from *B*. The image of *A* is inverted at 4.0 cm from the mirror. When the mirror is reversed at the same position (concave side toward *B*), the image of *B* is erect at 6.0 cm from the mirror. Find the radius of the mirror.

For Further Reading

Harrison, George R., *Atoms in Action,* Chapters 4–10, William Morrow and Company, New York, 1949.

CHAPTER THIRTY-FIVE

Some Optical Instruments

The first artificial aids to vision were mirrors, which were used in the days of the Pharaohs in Egypt. Spectacles were invented 4500 years later, in the thirteenth century. The great scientific awakening in the time of Galileo gave us the microscope and the telescope; the photographic camera is a more recent invention.

In this chapter we shall apply the principles of Chapters 33 and 34 to a few common optical instruments. We cannot do more than suggest the wide range of useful optical instruments that are available to workers in the physical sciences, the life sciences, and engineering. In addition to optical instruments described here, physicists and astronomers use spectrographs of many different kinds (Chapter 36); chemists use spectrographs, refractometers, and colorimeters; physicians use opthalmoscopes (for viewing the retina of the eye) and gastroscopes (for viewing the lining of the stomach); and engineers use transits and Schlieren apparatus (for viewing flow patterns in wind tunnels) — to give a few examples.

The Camera

The camera (Latin *camera,* a room) is an opaque enclosure with an aperture at one end and a light-sensitive film at the other. In the *pinhole* camera (Fig. 35-1*a*) the aperture is small. Light from each point of the object to be photographed is incident on a small circle on the film. Thus the hole sorts out the light from different points in the object and directs it to the proper place in the image. However, these circles of light overlap; hence the image is not sharply defined. If one reduces the size of the aperture so that the circles of light do not overlap so much, the image is more sharply defined, but less light enters the camera and the exposure time is greatly increased. The lens of the ordinary camera (Fig. 35-1*b*) focuses the light, so that the aperture can be much larger than that of the pinhole camera without destroying the sharpness of the image.

When you take a picture with a camera, the time of exposure depends upon the diameter of the aperture or "stop opening" and also upon the focal length of the lens. In Fig. 35-2*a*, for example, the diameter of the stop is 1 cm, and the focal length of the lens is 10 cm. *A* is the image of a distant object. In Fig. 35-2*b* the focal length of the lens is twice as great; hence the area of the image is quadrupled. The diameter of the stop opening is also doubled, so that four times as much light reaches the image. Each square centimeter of

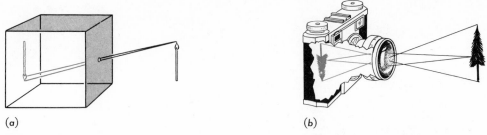

Fig. 35-1. (a) *The pinhole camera.* (b) *A simple lens camera.*

the image receives the same amount of light as before; hence the exposure time is unchanged. The lens systems are said to have equal *f*-values.

The *f*-value of a lens is usually specified as the ratio of the focal length to the diameter of the stop opening. Thus *f*/4.5 means that the stop diameter is 1/4.5 of the focal length. An *f*/9 stop setting would require four times as long an exposure.

A small stop opening is advantageous in that careful focusing is not required. A simple camera, maximum speed *f*/11, requires no focusing. Sharp-focus high-

speed cameras require highly corrected lenses to avoid spherical aberration that would otherwise be present at their relatively large stop openings (Fig. 35-3).

Usually the lens mounting is extensible so that the distance from the lens to the film may be varied until the image is sharply focused. To aid in this adjustment, in studio cameras the film may be replaced temporarily by a ground-glass screen on which the image is viewed while the focusing is accomplished. In other cameras a scale indicates the proper lens setting for an object at a known distance.

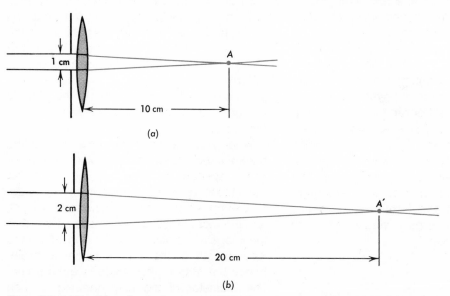

Fig. 35-2. *Doubling both the diameter of the aperture of a camera and its focal length leaves the illumination of the image unchanged.*

Fig. 35-3. A sharp-focus high-speed camera. Each of the lenses is compound, correcting for spherical and chromatic aberration.

The *shutter* of the camera is a movable diaphragm which covers the lens, except when the exposure is being made.

The light-sensitive coating of the photographic film consists in part of tiny crystals of a silver salt (usually silver bromide) embedded in gelatin. Light incident on one of these crystals frees electrons that wander through the crystal until they are trapped by a "sensitivity speck" where they attract and neutralize a small number of free silver ions, forming silver atoms. The pattern of affected crystals throughout the emulsion forms what is called the *latent image.* To develop the image, the emulsion is immersed in a developer solution which acts upon all of the silver bromide crystals that have been affected by light, reducing the silver ions in them to metallic silver, but not affecting the unexposed crystals. A second immersion in a fixing solution dissolves away the silver bromide that was not acted upon by the light. The bromine gas escapes, and the silver remains as a black deposit. We call the developed film a *negative* because the brighter parts of the view appear as darker parts of the image. To make a positive print, the negative, laid on a piece of sensitized paper, is placed in a beam of light. The more dense parts of the negative absorb more light than the less dense parts, hence the silver salts will be more affected under the lighter parts of the negative than under the darker regions. The exposed sensitized paper is developed and fixed in the same manner as the film. The brighter parts of the scene photographed appear as the brighter parts of the image.

The Human Eye

The human eye (Fig. 35-4) resembles a photographic camera. The eyelids correspond to the shutter. The iris diaphragm, corresponding to the stop, is an opaque screen with an aperture or pupil which widens in dim light and narrows when the illumination is intense. Instead of a photographic film, the eye

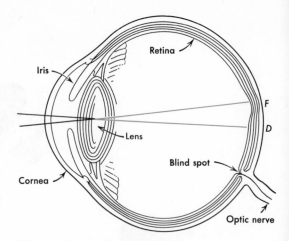

Fig. 35-4. The human eye.

has a retina in which there are thousands of nerve endings that send electric signals to the brain, where the image is interpreted.

Your eyes are more sensitive to light when you are in a dark room than when the room is strongly illuminated. When you go from a sunny street into a darkened motion-picture theater, at first you cannot perceive surrounding objects clearly. After a short time, your vision improves. The improvement in vision comes in part from the dilation of the pupils of the eyes but mostly from the accumulation in the retinas of a dye, the *visual purple,* which increases the sensitivity of your eyes. Strong illumination bleaches this dye. In a dimly lighted room, the rate of bleaching decreases, the dye accumulates, and the sensitivity of the eye gradually rises. Because of this control of sensitivity, you can read large print either by moonlight or by sunlight, an illumination almost half a million times greater.

Everyone who drives a car at night has experienced discomfort when a stab of light from an approaching automobile partially blinds him. The light falling on the retina quickly bleaches out most of the pigment. Afterward, several minutes are required for the pigment to accumulate again.

The *blind spot,* a small region of the retina

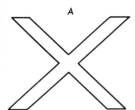

A   B   C

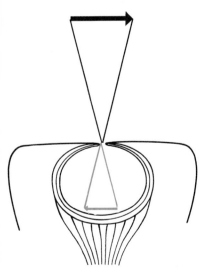

Fig. 35-5. *To locate the blind spot.*

where the optic nerve enters the eyeball, is insensitive to light. To demonstrate the existence of the blind spot, hold this book at arm's length. Close your left eye and look steadily at the *X* in Fig. 35-5. You will see the circle at *B* and the square at *C*. Move the page closer. At a certain distance, *C* will disappear because its image is formed on the blind spot. Move the page still closer, and at another position, *B* will disappear and *C* will be visible.

Accommodation is the change of focus of the eye for viewing near and far objects. The eye of the chambered nautilus is a pinhole camera, so it cannot get out of focus (Fig. 35-6). The eye of the fish is focused, like a camera, by shifting the lens to and fro so as to vary the distance between lens and retina.

Fig. 35-6. *The eye of the nautilus, a pinhole camera.*

The human eye is accommodated by changing the curvature of the crystalline lens. When the normal eye is relaxed, its lens system focuses images of distant objects on the retina. When a person examines nearby objects, the muscles *L* (Fig. 35-7) contract toward *S* in order to loosen the fibers or ligaments attached to the lens. Then the forward face of the lens bulges. A child 10 years old can see objects distinctly at a distance of 12 cm. For a man of 40, the distance is about twice as great. Most people past 60 are farsighted. Because of hardening of the lenses, they require spectacles with converging lenses to see near objects clearly (Fig. 35-8*a*). The eye lenses of people who are nearsighted are too strong; hence images of distant objects are focused in front of the retinas. These people need diverging spectacle lenses (Fig. 35-8*b*).

The cornea surface of the normal eye is part of the surface of a sphere. In astigmatism, the cornea, like the side of an egg, is more curved in one plane than in another at right angles to it. Glass lenses can also be astigmatic. In Fig. 35-9, the glass lens is more curved about the vertical axis than about the horizontal axis. It forms the image of a distant point source of light, not as a point, but as a vertical line at *A'B'* and as a horizontal line at *C'D'*. Astigmatic eyes behave in a similar way. Astigmatism is corrected by superimposing a cylindrical shape upon the spherical shape of spectacle lenses.

To test your eyes for astigmatism, look at Fig. 35-10. If your eyes are normal, you will see all the lines equally clearly. If you have astigmatism, some of them will seem clearer

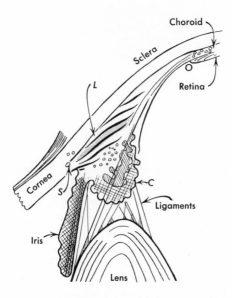

Fig. 35-7. *Accommodation mechanism of the human eye. For near vision, the muscles* L *contract so as to loosen the ligaments. The lens bulges.* (*Adapted with permission from O. Stuhlman,* Introduction to Biophysics, *John Wiley and Sons, 1943.*)

than others. Rotate the book about the line of vision. At different angles, different sets of lines will be most distinct. You can get the same effect by varying the distance of the page from your eyes.

"Seeing is believing," but it is comparatively easy to fool the eye. When the moon is high in the heavens it seems smaller than when it is near the horizon, though in fact the angle which it subtends at the eye is about the same. When the moon is high, it seems small compared with the clouds, or the vastness of space, but when it is near the horizon it subtends a greater angle than distant objects on earth which we know to be large.

The line *a* in Fig. 35-11 seems to be shorter than *b*, though they are equally long. The two long lines *c* and *d* are parallel, but they seem to approach each other. The three men in *e* are equally tall.

The Magnifying Glass

We judge the size of an object partly by the angle which it subtends at the eye. When we say that the moon looks "as large as a cartwheel," we mean that the moon subtends at the eye as large an angle as a cartwheel would at a certain distance (about 400 ft). Move this book toward you so that each letter seems larger. When the page is at a certain distance, your eye lenses will reach the limit of their accommodation by bulging as much as possible. You cannot see the letters clearly at any closer distance. We call this smallest object distance *the distance of most distinct vision.* It varies for different people and increases greatly as one grows older. In solving problems, we often use 25 cm as an average value.

A convex lens, held close to your eye, enables you to see an object clearly when it is closer than the distance of most distinct vision *Q* because you see its virtual enlarged image at a distance equal to or greater than

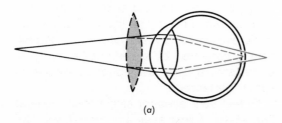

(a)

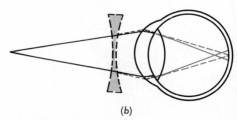

(b)

Fig. 35-8. (a) *A convex (positive) lens corrects farsightedness.* (b) *A concave (negative) lens corrects nearsightedness.*

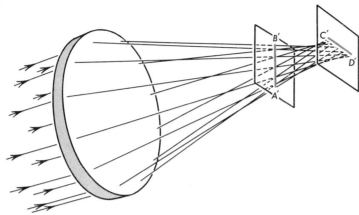

Fig. 35-9. *Astigmatism. The lens is more curved about the vertical axis than about the horizontal axis. (Reprinted with permission from J. A. Eldridge,* College Physics, *John Wiley and Sons, 1947.)*

Q. Suppose that in examining an arrow, you hold it at the distance of most distinct vision *OA* (Fig. 35-12). Then you hold a magnifying glass close to your eye and move the object to *B* at a distance from the lens such that the virtual image is at the distance *Q*.

The image subtends a larger angle at the eye than the object did when it was at a dis-

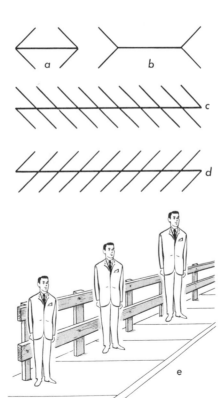

Fig. 35-11. Deceiving the eye. The horizontal lines in a *and* b *are of equal length. In* c *and* d *they are parallel. The men in* e *are of equal height.*

Fig. 35-10. Test chart for astigmatism.

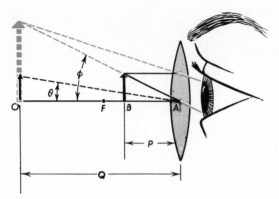

Fig. 35-12. *Angular magnification produced by a magnifying glass.*

tance Q. The **angular magnification M** is given by

$$M = \frac{\phi}{\theta}$$

Since the angles are small,

$$M \cong \frac{\tan \phi}{\tan \theta} = \frac{Q}{p}$$

But

$$\frac{1}{p} = \frac{1}{f} - \frac{1}{-Q} = \frac{Q+f}{Qf}$$

Therefore, multiplying by Q, we obtain

$$M = \frac{Q}{f} + 1$$

Example. A jeweler examines a diamond through a magnifying glass of focal length 2.5 cm. What is the angular magnification if the distance of most distinct vision is 25 cm?

$$M = \frac{25 \text{ cm}}{2.5 \text{ cm}} + 1 = 11$$

The Compound Microscope

The compound microscope has two lenses (or lens systems), which produce two magnifications. The object is placed just beyond the principal focus of the **objective** lens L_1 (Fig. 35-13), which produces a *real*, enlarged

image I_1. The linear magnification and the angular magnification caused by this lens are equal and are given by

$$M_1 = m_1 = \frac{q_1}{p_1}$$

This real image is at a distance p_2 from the eyepiece. The eyepiece forms a second, virtual image I_2 at the distance of most distinct vision* and produces an angular magnification

$$M_2 = \frac{Q}{f} + 1$$

The total magnification is the product $M_1 \times M_2$. The arrangement of lenses in a modern microscope is shown in Fig. 35-14. Note that the complete optical system shown includes an extremely important part—the eye of the observer.

Example. The objective lens of a compound microscope has a focal length of 2.0 cm. That

* The skilled observer, to avoid eyestrain, focuses his microscope so that the final image is at infinity. The angular magnification produced by the eyepiece can then be shown to be equal to just Q/f.

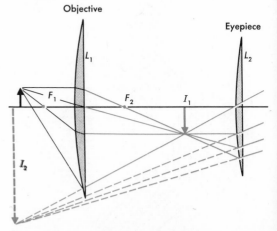

Fig. 35-13. *Simple compound microscope. The objective lens forms an enlarged real image I_1. The eyepiece forms a second, enlarged virtual image I_2. (Angles between rays and the principal axis are exaggerated.)*

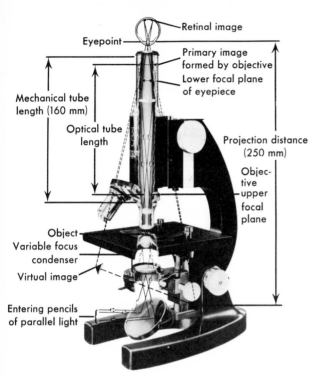

Labels on figure:
- Retinal image
- Eyepoint
- Primary image formed by objective
- Lower focal plane of eyepiece
- Mechanical tube length (160 mm)
- Optical tube length
- Projection distance (250 mm)
- Objective upper focal plane
- Object
- Variable focus condenser
- Virtual image
- Entering pencils of parallel light

Fig. 35-14. Modern microscope.

of the eyepiece is 5.0 cm. The first image distance is 20 cm and the virtual image distance is −25 cm. Find the total magnification.

Let *p* be the object distance of the real image from the objective lens.

$$\frac{1}{p} + \frac{1}{20 \text{ cm}} = \frac{1}{2.0 \text{ cm}}$$

$$p = 2.22 \text{ cm}$$

$$M_1 = \frac{20 \text{ cm}}{2.22 \text{ cm}} = 9.0$$

$$M_2 = \frac{25 \text{ cm}}{5.0 \text{ cm}} + 1 = 6.0$$

Total magnification = $M_1 \times M_2 = 9.0 \times 6.0 =$ 54.

The Refracting Telescope

The refracting telescope, like the compound microscope, has two lens systems, the objec-

tive and the eyepiece. The important difference between the two instruments is that the object viewed through the telescope is not close to the lens, but is at a relatively great distance. In Fig. 35-15, the beam from the top of a distant flagpole is converged by the objective lens to form a real image at *G*. Beams from other points of the flagpole are converged to form real images of those points elsewhere in the focal plane of the lens. The entire real image of the flagpole is viewed through an eyepiece as in the microscope. The final virtual image is not located at the distance of most distinct vision, but *at the same distance as the object.* Since the object distance is usually very great, the light rays emerging from the eyepiece are practically parallel. This means that *the real image must be located at the principal focus not only of the objective lens but also of the eyepiece.* The rays entering the eye of

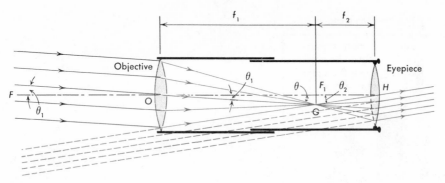

Fig. 35-15.Refracting telescope. The objective lens forms a real image of the distant flagpole. The eyepiece forms a virtual image far away. (Angles between rays and the principal axis are exaggerated.)

the observer are parallel, like those entering the objective lens, and the final convergence to an image on the retina is accomplished by the lens of the eye. The distance between the two lenses is f_1 plus f_2.

When a distant object is viewed through a refracting telescope, the total angular magnification M is given by

$$M = \frac{\text{Focal length of objective}}{\text{Focal length of eyepiece}} = \frac{f_1}{f_2}$$

To prove this relation, consider Fig. 35-16. The ray *KO* is directed from the top of a distant flagpole; the other ray *FO*, from its base. The real image of the flagpole is viewed through the eyepiece. θ_1, the angle subtended at the objective by the flagpole, is equal to θ, the angle subtended by the real image. θ_2 is the angle subtended by the

image at the eye of the observer. The angular magnification is the ratio of the two angles, that is

$$M = \frac{\theta_2}{\theta_1} = \frac{\theta_2}{\theta}$$

If the angles are small, they are approximately equal to their respective tangents, that is (expressed in radians), $\theta \cong \tan\theta$, and $\theta_2 \cong \tan\theta_2$. Therefore

$$M = \frac{\tan\theta_2}{\tan\theta} = \frac{GF_1/F_1H}{GF_1/OF_1} = \frac{f_1}{f_2}$$

The refracting telescope forms an inverted, reversed image. It can be modified by inserting a third lens between the real image and the eyepiece. This lens forms a second, real and erect, image. A better way is to use two 90° prisms to invert and reverse the

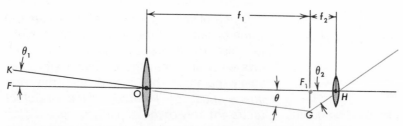

Fig. 35-16. Magnification produced by a refracting telescope.

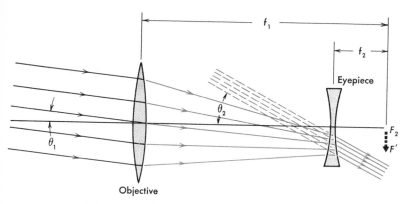

Fig. 35-17. *The opera glass.*

The Opera Glass

You read in Chapter 11 about Galileo's devising a telescope. It was what we now call an *opera glass,* having a diverging lens as an eyepiece. The eyepiece of an opera glass is located between the objective lens and its focus. In Fig. 35-17, the objective lens converges the light from the tip of a flagpole toward F' in the focal plane of the objective. The eyepiece deviates the beam and makes it parallel. The angle θ_2 through which the observer sees the final image is greater than the angle θ_1 which the distant object subtends at the objective lens. The total angular magnification of the opera glass is given by

$$M = \frac{\theta_2}{\theta_1}$$

If the focal length of the objective of an opera glass is 10 cm and that of the eyepiece is −3 cm, the total distance between the lenses is 7 cm.

The Reflecting Telescope

The largest astronomical telescopes, as well as most of the smaller ones made by ama-

teurs, have parabolic mirrors as objectives instead of lenses. The essential parts of a reflecting telescope are shown in Fig. 35-18. A beam of light from a star, incident on the parabolic objective mirror, after reflection, converges toward the principal focus F of the mirror. The star image is usually photographed on a photographic plate located at F, or, if desired, the beam may be reflected by a mirror M_2, forming a real image which may be viewed through an eyepiece.

At the beginning of the twentieth century, astronomers were principally concerned with studying the positions and velocities of the nearer stars of the Milky Way, which is a galaxy of a billion or more stars, including our own sun. During the last fifty years, the scope of investigation has widened so that other star clusters or nebulae are studied, some of them at distances greater than several billion light-years. The largest of these telescopes is in use at Mt. Palomar, California. Its objective mirror, nearly 17 ft in diameter (as wide as a small lecture room), weighs 18 tons. The new telescope extends our knowledge of the universe to galaxies that are several billion light-years away (Fig. 35-19).

The electromagnetic radiations received from the stars tell us not only their positions, but also their sizes, temperatures, and velocities. The methods of securing some of this information will be described in Chapter 36.

Binoculars consist of two refracting telescopes, each containing two 90° prisms, mounted side by side.

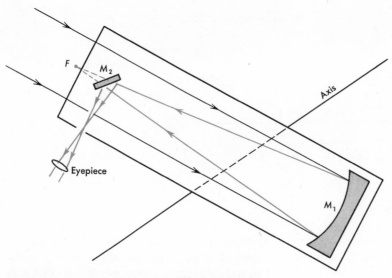

Fig. 35-18. A reflecting telescope.

Fig. 35-19. Nebulae at the limit of the Mt. Palomar telescope—approximately 2 billion light years. The brightest galaxy in the cluster indicated by the marker may be two galaxies in collision. (Courtesy Mt. Wilson-Palomar Observatories.)

The Radio Telescope

The reflecting telescope just described collects and focuses light waves from distant celestial objects. Other electromagnetic radiations are also of interest to astronomers. By means of radio telescopes, they can study sources of radio waves in the universe, obtaining information that supplements and extends that gathered by optical telescopes. One band of radio waves of great interest to astronomers is that whose wavelength is approximately 21 cm. These waves are emitted by clouds of neutral hydrogen gas.

Radio telescopes must be large in order to collect and concentrate the feeble radio signals reaching the earth from space and to resolve, or separate, radio sources whose angular separations are small. Radio telescopes, using parabolic reflectors, commonly have diameters up to 300 ft. The parabolic reflector (Fig. 35-20) concentrates the energy of the radio waves on an antenna. The signal received by the antenna is amplified by a high-selectivity amplifier, to discriminate against unwanted "noise" from terrestrial radio sources, and recorded as the telescope scans the sky.

Fig. 35-20. Twin radio telescopes. The parabolic array of conductors focuses the radio waves at the antenna mounted at the apex of the V-shaped structure in front of each disk. The signal is then sent to the receiver. Each telescope can be rotated about a vertical axis and also tilted about a horizontal one. (Courtesy Radio Observatories, California Institute of Technology.)

Astronomers have traced the shape of our galaxy by means of the 21-cm radio waves from clouds of hydrogen that lie along the arms of the galaxy. They have shown that our galaxy, like many others, is a rotating spiral (Fig. 35-21).

Summary

A camera is an opaque enclosure with an aperture that forms on a sensitive film an image of the object to be photographed.

Important parts of the eye are the cornea, the iris, the lens, the retina, and the optic nerve. When nearby objects are viewed, the lens of the eye bulges so that its converging power increases. In nearsightedness, the lens causes too great convergence and the image of a distant object is formed in front of the retina; diverging spectacle lenses are required. In farsightedness, the eye lens causes too little convergence and converging spectacle lenses are required. The distance of most distinct vision is the smallest distance from the eye at which an object can be held to produce a sharp image on the retina.

The simple magnifying glass is a lens which enables one to bring an object closer than the distance of distinct vision for the unaided eye.

In the compound microscope a real image of the object is produced by the objective,

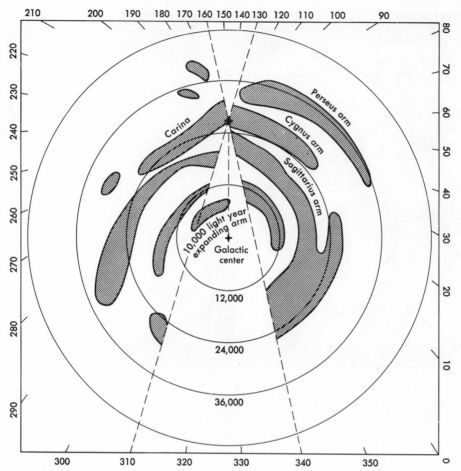

Fig. 35-21. Spiral arms of our galaxy charted in 21-cm hydrogen radiation. Clouds of neutral hydrogen gas are more dense in the galaxy arms than elsewhere. By analyzing the signal strength and the red shift of radio signals from the clouds, radio astronomers can locate them and hence the arms of the galaxy. (Courtesy of Scientific American.)

and this image is further magnified by an eyepiece.

In the refracting telescope, a real image is produced near the focus of the objective. Usually the eyepiece is so located that the second, virtual image is produced at the same distance as the object. The magnifying power when a distant object is viewed equals the ratio of the focal length of the objective to that of the eyepiece.

An opera glass has a converging objective lens and a diverging eyepiece.

Questions

1. Discuss the concepts of angular and linear magnification as applied to (*a*) a refracting telescope, (*b*) a compound microscope.

2. Estimate the approximate *f*-value of a typical pinhole camera.

3. State ways in which the eye and the camera are similar and ways in which they differ.

4. The images on the retina are inverted. Why doesn't the world seem upside down?

5. Explain why a swimmer cannot see objects clearly when he is submerged. Does a scuba diver with a diving mask have equal difficulty?

6. What simple tests could you make with a person's spectacles to tell whether he is farsighted, nearsighted, or astigmatic?

7. Why do older people wear bifocal lenses? What are trifocals?

8. Does a telescope that has an objective lens 5 cm in diameter form an image of the sun that is larger than, smaller than, or equal in size to the image formed by a lens of equal strength but 90 cm in diameter? What is the difference in the two images?

9. Design a simple telescope that could be used by an amateur in viewing earth satellites.

10. In listing f-values of a camera lens, these numbers are often used: 2, 2.8, 4, 5.6, 8. Can you show how f/2.8 and f/5.6 are related?

11. Which of these at arm's length has an angular size most nearly equal to that of the full moon: the bottom of a cola bottle, a nickel, or the end of a pencil?

12. On a clear night you photograph the full moon satisfactorily with an exposure of 1/50 sec. Estimate the exposure time for photographing a scene on the earth by the light of the moon on the same night at f/11.

Problems

A

1. In a camera the distance from lens to film is 6.0 cm. In order to form the sharpest image, the object photographed must be 250 cm from the lens. What is the focal length of the lens?

2. On a motion-picture film, the image of the nose of an actor is 1.0 mm long. The distance from the film to the center of the lens system of the projector is 10 cm. What is the size of the image formed on a screen that is 40 m from the lens?

3. A jeweler uses a simple magnifying glass of focal length 4.5 cm to examine a gem. (a) At what distance from the lens must the object be held in order to produce a virtual image at a distance of 25 cm? (b) What is the magnification?

4. In a refracting telescope the focal lengths of the objective and the eyepiece are 100 cm and 1.0 cm, respectively. (a) At what distance apart should the lenses be to view the moon? (b) What is the magnification?

5. In order to secure an f-value of f/6, how large must the aperture be (a) in a camera of focal length 5.0 cm, (b) in a camera of focal length 30 cm?

B

6. A nearsighted person has a distance of most distinct vision of 12 cm. What should be the focal length and strength in diopters of his spectacle lenses so that his distance of most distinct vision shall be 25 cm?

7. A nearsighted person cannot see clearly an object which is more than 3.0 m distant. What is the focal length of the lenses required in order that he may see very distant objects? (Parallel light rays after passing through the spectacle lens must diverge from a point 3.0 m distant, i.e., $p = \infty$, $q = -3.0$ m.)

8. A farsighted person must hold a book at a distance of 60 cm from the eyes. What is the focal length of the required lenses so that he may hold it at a distance of 25 cm? (The virtual image distance is −60 cm.)

9. A farsighted man cannot see clearly with his left eye objects that are closer than 40 cm, and with his right eye objects that are closer than 35 cm. What strength in diopters must the lenses of his spectacles have so that each eye will form a clear image of an object 25 cm away?

10. A nearsighted person cannot see

clearly objects that are more than 25 cm from the left eye and 30 cm from the right eye. What strength must his spectacle lenses be so that he can see very distant objects clearly?

11. If the objective lens of an opera glass (Galilean telescope) has a focal length of 20 cm and the eyepiece a focal length of −6.0 cm, (a) at what distance apart must the two be mounted in order that the system may form at an infinite distance an image of a star, (b) how far must the eyepiece be shifted in order to form an image at infinity of an object 300 cm away?

C

12. The focal lengths of the lenses of a refracting telescope are 75 cm and 1.25 cm. What should be the distance between them so that the eyepiece may form a real image of the sun on a screen 30 cm away from the eyepiece?

13. The objective of a low-power compound microscope has a focal length of 2.0 cm. The focal length of the eyepiece is 6.0 cm. (a) If the total magnification is 50, what is the distance between objective and eyepiece? (The virtual image is formed at the distance of distinct vision, 25 cm.) (b) How far must the eyepiece be moved to produce a real image on a photographic plate located 8.0 cm from the eyepiece?

14. In a compound microscope, the focal lengths of the objective and the eyepiece are, respectively, 0.80 cm and 1.20 cm. The real image is formed by the objective at a point 19 cm from it, and the distance of the virtual image from the eyepiece is −25 cm. Find (a) the distance of the object from the objective lens, (b) the two magnifications, (c) the total magnification.

15. The distance of the moon from earth is 2.4×10^5 mi. (a) At what distance would it have to be for its mountains to seem as large as they do when seen through the Lick Obser-

vatory telescope? The focal lengths of the objective and the eyepiece are, respectively, 1500 cm and 3.0 cm. (b) The real image produced by the objective lens is 13.0 cm in diameter. What is the diameter of the moon?

16. A human eye forms an image of a star on the retina; the image distance is 2.0 cm. (a) What is the power in diopters of the eye lens system? (b) How much must the power of the lens increase so that it may form on the retina an image of a pin held at a distance of 30 cm? (c) If the eye is farsighted and cannot form images of objects closer than 50 cm, what is the power of the lens required in order that the distance may be reduced to 30 cm?

17. If light from an object 30 cm to the left of the convex lens (Fig. 35-22) passes through the two lenses, calculate where the final image is located.

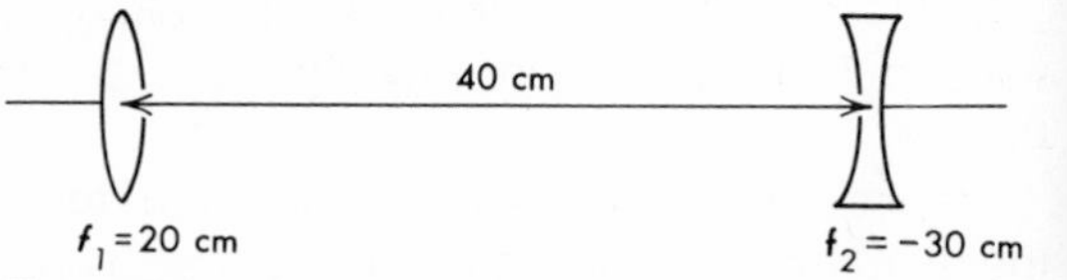

Fig. 35-22.

For Further Reading

Bok, Bart J., "The Arms of the Galaxy," *Scientific American,* December, 1959, p. 92.
Bondi, Hermann, *The Universe at Large,* Doubleday and Company, Garden City, 1960.
Kraus, J. D., "The Radio Sky," *Scientific American,* July, 1956, p. 32.
Milne, L. J. and M. J., "Electrical Events in Vision," *Scientific American,* December, 1956, p. 113.
Newhall, Beaumont, *Latent Image,* Doubleday and Company, Garden City, 1967.
Wald, George, "Eye and Camera," *Scientific American,* August, 1950, p. 32.

CHAPTER THIRTY-SIX

Dispersion, Spectra, and Color

In free space, electromagnetic radiations of all observed frequencies have been found to have the same speed. In matter, however, electromagnetic waves travel at different speeds that depend on their frequencies. When a mixture of electromagnetic waves of different frequencies travels through a medium, some travel at greater speeds than others. Under certain conditions, which we shall describe in this chapter, the radiation mixture of many frequencies is *dispersed* or separated by its passage through the medium into its component frequencies. (Dispersion can occur in other ways, as the next chapter will show, and it can occur for non-electromagnetic waves—sound waves, for example.)

The result of dispersion is a *spectrum* in which radiations of different frequencies occupy different positions on a detector such as a photographic plate. The electromagnetic spectrum, as we saw in Chapter 32, comprises *all* electromagnetic radiations. Usually, we deal at any one time with only a portion of the electromagnetic spectrum—for example, light, the microwave region, the infrared—adapting our experimental techniques to that portion. Light, of course, has

special importance to us because our eyes are sensitive to it, and the differing response of the eye-brain combination to light waves of different frequencies in the visible spectrum we call *color*.

Newton and the Spectrum

You may remember that when Newton was a 22-year-old student at Cambridge University, an epidemic forced him to take a long vacation. During those two years, he made a study of gravitation, invented the calculus, and investigated the spectrum of sunlight (Fig. 36-1a). In his own words: "In a very dark chamber, at a round hole about one-third part of an inch broad, made in the Shut of a window, I placed a Glass Prism."

A narrow beam of sunlight entered through the hole, passed through a glass prism and was spread out fanwise, forming a band of rainbow hues (Fig. 36-1b). This image, in which radiations of different hues and frequencies occupied different positions, was the visible spectrum. The effect of the prism upon the light of different frequencies was to disperse them. Next Newton placed another, reversed, prism beyond the first one, and brought the beams together again, producing white light. Thus, by two simple but painstaking experiments, he demonstrated that white light can be dispersed into rainbow hues and that radiations of different colors can be combined to produce white light. Also, he proved that the index of refraction of a glass differs for different spectral colors.

The Spectroscope

When Newton refracted sunlight by means of a prism, he formed an *impure* spectrum in which light of different colors (and frequencies) overlapped. The *spectroscope* (Fig. 36-2) forms a relatively *pure* spectrum. Light from the slit is made parallel by the first lens. If the light entering the slit were of one frequency only (monochromatic), all of it would be refracted through the same angle by the prism. The refracted beam would be focused by the lens of the telescope so as to form a single narrow image of the slit. When white light enters the slit, the lights of different hues (and wavelengths) are refracted through different angles, and light of each small range of wavelengths occupies a different position along *RB*. The spectrum may be viewed through the eyepiece (of a spectroscope) or it may be photographed (in a spectrograph). Both the telescope and table holding the prism can be rotated about the center of the table. If the angular position of the telescope can be read on a precisely divided circle, the instrument is called a spectrometer.

(a)

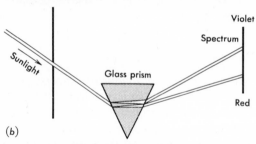

(b)

Fig. 36-1. (a) *Newton produces a spectrum.* (*Courtesy of Bausch and Lomb.*) (b) *Using a prism to disperse a beam of light.*

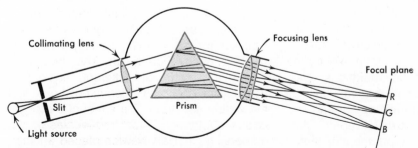

Fig. 36-2. *A spectroscope analyzes light striking the slit.*

The index of refraction of a given material for light of different frequencies can be measured by a spectrometer. If we measure the angle of mimimum deviation D produced by a prism of apex angle A for light of a given frequency, the index of refraction n can be calculated by the equation (Chapter 33)

$$n = \sin\left(\frac{A + D}{2}\right) \Big/ \sin\left(\frac{A}{2}\right)$$

Suppose that we measure the indices of refraction of a glass prism for light of different frequencies. A typical dispersion curve for the visible region is shown in Fig. 36-3. Notice that the index of refraction *decreases* as the wavelength in vacuum *increases*. This means that the index increases as the frequency increases since $f = v/\lambda$. Other materials yield similar dispersion curves for light. Since the index of refraction is the ratio of the speed of light in vacuum to the speed of light in the particular medium, we see that light waves of greater frequencies travel in material media with the smaller speeds. Red light is found

experimentally to have a smaller frequency than blue light, and the index of refraction of glass is greater for blue light than for red.

We have stressed frequency rather than wavelength in this discussion. Recall that the wavelength changes in proportion to the speed of waves, but that the frequency—for a source and an observer both at rest—is constant. When red light passes from air into water, its speed decreases, and its wavelength decreases, but its frequency remains constant. Later in this chapter we shall discuss Doppler effects that can cause the frequency of electromagnetic waves to change.

How Do Electromagnetic Waves
Travel Through Substances?

We know that electromagnetic waves are traveling electric and magnetic fields and that atoms consist of charged particles—electrons usually bound to nuclei. In addition, certain materials—the electrical con-

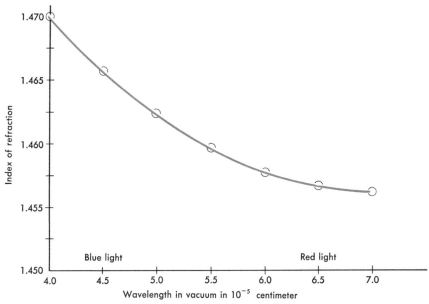

Fig. 36-3. A dispersion curve in the visible region. The material is fused quartz glass. (Data from the American Institute of Physics Handbook.*)*

ductors—have free electrons that are not tightly bound to nuclei, but that can wander from one atom to another. We might expect that there would be an interaction between the fields in an electromagnetic wave and the charged particles in the material that it strikes or through which it is traveling. This interaction actually occurs. Detailed mathematical treatment yields results that agree well with measurements of the reflection, transmission, and absorption properties of materials. We shall give here some of the qualitative results of this theory.

Electromagnetic waves penetrate metallic conductors only slightly and, instead, are reflected—as we have seen. The reason for this is that the high-frequency fields in the wave set up high-frequency currents of free electrons in conductors. The electromagnetic wave is damped out within a thickness of a few wavelengths in the metal by energy transfer to the free electrons and eventually into heat. However, the high-frequency alternating currents thus set up at the surface act themselves as electromagnetic radiators—small broadcasting antennas—and the emitted electromagnetic wavelets join to form the reflected wave. As we might expect, the reflected wave in mirror-like reflection is polarized with the electric field component parallel to the plane of the reflecting surface because that is the plane in which the small secondary radiators are predominantly vibrating.

Electromagnetic waves in fact travel through some non-conducting materials: window glass and air are transparent to light, for example; quartz, to a considerable portion of the ultraviolet spectrum; and lithium fluoride, to the infrared. Basically, the effect of an electromagnetic wave on a non-conductor is to excite vibrations of the bound charges in the atoms or the molecules. An electron bound to a nucleus constitutes a small electric dipole that can be made to oscillate at the frequency of the electric field component of the electromagnetic wave. As the incident electromagnetic wave progresses through the medium, it exerts a periodic force on the electric dipoles and excites them into oscillation as tiny secondary radiators. Then high-frequency oscillations result in the emission of electromagnetic wavelets that combine with the incident wave in the direction in which the wave is traveling. However, the excited oscillations differ in phase from the incident wave. For light waves traveling through glass, the oscillations lag behind the wave and this difference in phase results in a slowing down of the wave. In general, the index of refraction depends in a complicated way on the structure of the electric dipoles, their spatial arrangement, and on the frequency of the incident wave.

Since energy is radiated in all directions by the oscillating dipoles, energy is usually taken away from the electromagnetic wave as it travels through non-conductors—the medium acts as an absorber. When the frequency of the electromagnetic wave happens to correspond to a natural frequency of the electric dipoles, the wave and the dipoles are in resonance, and the energy transfer from the wave to the medium is very large; absorption of energy by the medium is then large and the transmission is small. We shall return to this point in a later chapter.

Chromatic Aberration and Achromatic Lenses

Chromatic aberration and its correction in optical instruments furnish good examples of dispersion. You may recall that Galileo, who died the same year that Newton was born, constructed crude telescopes and discovered the moons of Jupiter. Newton constructed telescopes with larger lenses than those of Galileo. These telescopes formed blurred images because the lenses deviated the light of different hues by different amounts. In Fig. 36-4, white light from a dis-

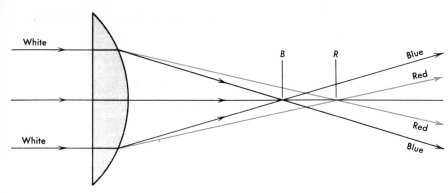

Fig. 36-4. *Chromatic aberration. The lens deviates blue light more than red light.*

tant source is incident on a convex lens. The blue light converges at *B*, forming an image there. The red light, less deviated than the blue, is focused at *R*, and radiation of other hues is focused at intermediate points. If a white card is placed at *R*, the red central image will be surrounded by bands of color, the outer edge being blue. If it is placed at *B*, the center will be blue and the outer edge red. This failure of a lens to converge light of different hues to the same point is called *chromatic aberration.*

After considerable effort, Newton abandoned the attempt to construct large refracting telescopes and invented the reflecting telescope. It is free from chromatic aberration because the law that the angles of incidence and reflection are equal holds for light of all frequencies.

Later, other workers discovered how to reduce chromatic aberration by combining lenses made of *different kinds of glass.* Assume that a convex lens of crown glass (common glass) (Fig. 36-5) refracts the blue light of ray *AB* inward, toward the principal axis, through an angle of, say, 16°, and the red light through 15°. The average deviation is 15.5°, and the dispersion or angular separation of the blue and red rays is 1°. The diverging lens made of flint glass, containing a large proportion of lead, refracts the ray of blue light outward 10°, the red light 9°. The average deviation is −9.5°, and the dispersion is 1° outward, away from the axis. The two lenses together produce an average refraction of 15.5° − 9.5° or 6° inward, and the dispersion is 0°, so that almost all the light converges at the same point. Usually the two

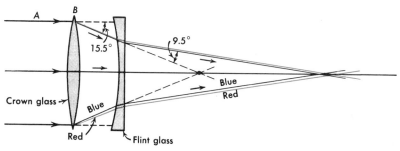

Fig. 36-5. *Achromatic lens combination.*

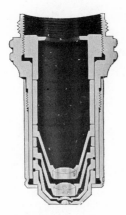

Fig. 36-6. Achromatic objective lens system of a modern microscope. It also corrects for spherical aberration and other defects. (Courtesy Bausch and Lomb.)

lenses are placed in contact with each other, and the combination is termed achromatic (Greek, *without color*). In spectacle lenses no attempt is made to eliminate chromatic aberration, but in many optical devices expensive combinations of lenses minimize the dispersion and also the spherical aberration (Fig. 36-6).

Some Spectra

Let us now consider the experimental results of dispersing radiations from various sources by means of suitable spectroscopes and of measuring the wavelengths and the intensities of these dispersed radiations. We shall limit ourselves to the experimental results. The theoretical interpretation of these spectra takes us into a consideration of the structures of atoms and molecules and will be discussed in later chapters.

The spectrum of a hot solid. Figure 36-7 shows how the intensities of the spectrum of a "black body" (maximum emitter) depend upon wavelengths at a given temperature, and how the entire spectral curve depends upon temperature. The source is a small cavity in a hot, glowing solid. The spectroscope of Fig. 36-2 will not serve to measure the in-

tensity of the radiation at various wavelengths in the spectrum of the black body because the glass prism and lenses are not transparent to infrared, and photographic plates are sensitive only in the near infrared. Instead, an infrared spectrometer is used in which the dispersion is produced by a lithium fluoride or rock salt prism. The dispersed radiation is converged by concave mirrors to a focus at which a thermocouple measures the radiation intensity at each wavelength studied.

When the black body source is at 2000°K, nearly all the radiation is in the infrared region. At 4000°K, about 10% is in the visible part of the spectrum.

In Chapter 14, you learned that the total radiation from a black body is proportional to the fourth power of its Kelvin temperature. Thus in Fig. 36-7 the total area under the 4000° Kelvin curve is $2^4 = 16$ times that under the 2000° curve. We can determine the tem-

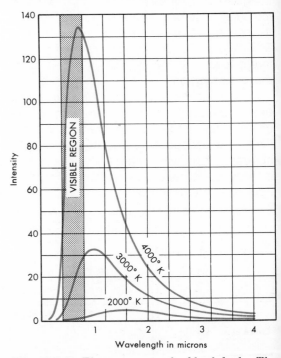

Fig. 36-7. The spectrum of a black body. The visible part of the spectrum lies between 0.4 and 0.8 μ (1 μ = 10^{-6} m).

perature of a furnace or a star by measuring the total radiation that is incident on the junction of a thermocouple in a *radiation pyrometer.*

Notice that the spectrum is a *continuous* one, radiation of one wavelength blending into radiation of a different wavelength with no gaps in the spectrum. The part of the continuous spectrum that lies in the visible region comprises the rainbow hues — violet, indigo, blue, green, yellow, orange, and red. Notice further that with rising temperature the position of the maximum of the curve shifts to the left, toward the region of shorter wavelengths. The wavelength λ at the maximum of the spectral curve when the black body is at a Kelvin temperature T is given by

$$\lambda T = k$$

in which $k = 0.289$ cm-K°. The light from an incandescent lamp filament or any other hot, glowing solid forms a continuous spectrum also. The temperature of a furnace or of a star may be estimated not only by measuring the rate at which total radiation comes from it, but also by observing the wavelength at which the maximum intensity of the spectrum occurs

and by applying the above equation. Appropriate corrections must be made because the assumption that the source radiates like a black body is in general not justified.

Bright-line spectra. The spectrum of a luminous gas or vapor consists of bright lines, with dark spaces between them. Dip a piece of asbestos into a solution of sodium chloride and insert it into the flame of a Bunsen burner placed before the slit of a spectroscope. The salt will vaporize, and the sodium atoms will emit yellow light. Two of the most prominent lines in the visible spectrum of sodium are two yellow lines, very close together in wavelength, but separated in wavelength from each other and from other radiations in the sodium spectrum. Each yellow line is a separate image, formed by light of a slightly different wavelength, of the slit of the spectroscope. Strontium and potassium chlorides, treated in this manner, give spectra consisting of other lines. Other metals vaporized in an electric arc or in an electric spark discharge yield even more complicated spectra (Fig. 36-8). Gases are studied by enclosing them in glass or quartz tubes through which electrical discharges are sent.

Fig. 36-8. Spectra of metallic elements: mercury, iron, barium, calcium (top to bottom). (Courtesy Bausch and Lomb.)

Spectroscopic chemical analysis is most useful in rapidly detecting and measuring traces of impurities and in estimating the concentrations of critical elements. Samples of steel are taken from open hearth furnaces before the steel is poured so that the composition can be adjusted before pouring. A pneumatic tube carries samples to the laboratory. Specks of the metal are placed in the crater of a rod of very pure carbon. An electric arc between this rod and another one vaporizes the steel. Light from the hot vapor passes through the slit of a spectrograph, is analyzed by the spectrograph, and is photographed. By examining the film, the spectroscopist quickly identifies the spectral lines of such elements as manganese, silicon, copper, nickel, molybdenum, chromium, and tin. He can determine their concentrations from the relative blackness of the lines. Rapid and reliable procedures of this kind reduce costs greatly.

Absorption spectra—dark band and dark line. Place an incandescent lamp in front of the slit of a spectroscope and, looking through the eyepiece, observed the continuous spectrum of the light emitted by the filament. Then interpose a glass bottle containing a dilute solution of a black dye. The various parts of the spectrum will be dimmed about equally. Replace the bottle with one containing blood in water. A dark band will be seen in the yellow and green part of the spectrum. Bubble ordinary manufactured illuminating gas through the solution, and the carbon monoxide in the gas will combine with the blood, producing a chemical change. Now you will see two dark bands instead of one. The spectroscope is often used in this manner to detect chemical transformations. Colored glass, gem stones, and many other solids produce the same type of band spectra as colored liquids.

Gases and vapors can absorb light of certain wavelengths. Suppose that you mount two or three lighted Bunsen burners in front of the slit of a spectroscope. Dip pieces of asbestos in sodium chloride and place them in the flames of the burners. The vaporized sodium will cause the flames to be yellow. Through the spectroscope you will see a yellow line (actually two lines close together). Now mount a lighted incandescent lamp beyond the burners so that its light will pass through the three yellow flames. Through the spectroscope you will see a continuous spectrum. However, there will be a dark line in exactly the position formerly occupied by the yellow line. The relatively cool sodium vapor of the flames absorbs some of the white light from the incandescent lamp and reradiates this light in all directions. The reradiated light is lost from the beam of white light, and the dark line results. As you would expect, *the frequency of the absorbed light is exactly the same as the light emitted by the excited sodium atoms.* This is a quantum effect whose nature we shall examine in a later chapter.

The spectrum of light from the sun tells us a great deal about the condition of its surface. The spectrum of sunlight is not continuous, but is crossed by hundreds of dark vertical lines called the *Fraunhofer* lines. It is a dark-line spectrum. This fact proves that the light must have passed through vapors or gases which absorbed radiations of certain frequencies. Two of these dark lines correspond exactly with the yellow lines of sodium; hence we reason that sodium exists in the sun's atmosphere. Other lines reveal the existence of calcium, magnesium, iron, copper, and more than fifty other elements. Helium was discovered by the detection of its absorption lines in the sun's spectrum.

The spectra of some stars have dark lines like the sun's. These stars are believed to have hot cores surrounded by gases and vapors. Others are known to be gaseous, for their spectra show bright lines only. Other stars have continuous spectra and are hot bodies without dense atmospheres. A few stars reveal themselves only by the infrared rays or radio waves that they emit.

The Doctor Effect for Electromagnetic Radiation

In Chapter 20 it was pointed out that, when an observer moves toward a source of sound, the frequency f of the sound that he hears is increased, and that, if the source moves toward the observer, the wavelength λ is decreased—the Doppler effect. Electromagnetic waves behave in a similar fashion. The frequency of the light emitted by a hydrogen atom moving at high velocity toward the observer, for example, is greater than that emitted by the atom at rest. Astronomers use the Doppler effect to determine the velocity v with which the earth and a star approach or move away from each other. When this velocity is small compared with c, the velocity of light, the change of wavelength $\Delta\lambda$ or of frequency Δf is given approximately by

$$\pm\frac{\Delta f}{f} \cong \mp\frac{\Delta\lambda}{\lambda} \cong \pm\frac{v}{c}$$

($+v$ means velocity of approach and $-v$ velocity of separation.)

Figure 36-9 shows the Doppler shift due to the motion of the rings of Saturn. Figure 36-10 shows the shift toward lower frequencies for stars in nebulae that are many millions of light-years away. Many astronomers interpret this fact to mean that the universe is expanding.

At the beginning of the twentieth century, astronomers knew the distances of about one hundred of the nearer stars. During the last seventy years, the telescope, the spectroscope, and other devices have been improved so that we can study stars that are millions of light-years away. Copernicus and Galileo showed that the earth is merely one of several tiny planets circling around the sun. We now know that the sun itself is a mediocre member of what we call the Milky Way, which is made up of a billion stars (or suns) dispersed through a disk shaped like a watch. The disk is many thousand light-years wide and perhaps one-tenth as thick. If you look toward the Milky Way, your line of sight is in the plane of the disk. You see myriads of stars, most of them too distant to be distinguished separately. If you look away from the Milky Way, you see relatively few stars, because you are looking through the narrow part of the disk.

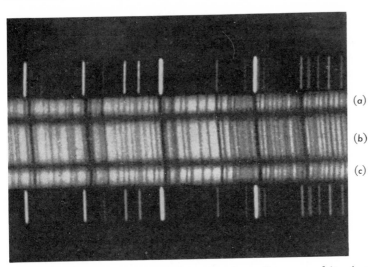

(a)

(b)

(c)

Fig. 36-9. Spectrum of light from the planet Saturn and its rings. Spectral lines at (a) *are displaced to the left because the edge of the ring is moving toward the observer. At* (c), *the lines are displaced to the right, indicating opposite motion.*

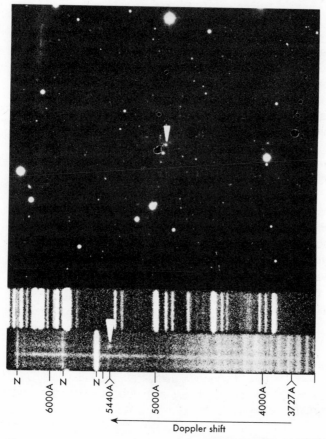

N 　6000A 　N 　N 　5440A 　5000A 　4000A 　3727A

← Doppler shift

Fig. 36-10. The Doppler shift for galaxy 3C295 (marked by pointer in top picture), the most distant galaxy yet measured with the 200-inch Mount Palomar telescope. The velocity of recession was computed as 110,000 km/sec from the red shift of the 3727-A oxygen spectral line to an observed wavelength of 5440-A (marked by pointer in lower spectrum). The three vertical lines marked (N) are emission spectra from the night sky: 6300-A oxygen, 5893-A sodium, and 5577-A oxygen. The upper spectrum is a comparison spectrum of helium. (Photograph from the Mount Wilson and Palomar Observatories.)

Stars appear as points of light even through the most powerful telescope. Some cloudlike objects called *galaxies* prove to be collections of stars like our Milky Way. The nearest of these, in Andromeda, is visible to the naked eye and is about two million light-years distant. Large telescopes, such as the Mt. Palomar telescope, reveal millions of galaxies. Some, like that in Andromeda, are gigantic spiral pinwheels revolving in space. Our own Milky Way is a galaxy with spiral arms (Chapter 35).

The Luminosity of Radiation

We have discussed many of the physical properties of light that would exist even if no living beings could see. In this section and the next, we shall comment briefly upon light as it causes the sensation of seeing. How sensitive is the human eye to electromagnetic radiation? You may recall that the ear cannot hear air vibrations of frequency less than about 20 per second or greater than about 20,000 per second. Thus the range of hearing covers about ten octaves. The eye is much more restricted. Earlier in this book you learned that the eye is totally insensitive to all except a tiny portion of the electromagnetic spectrum—that with wavelengths between about 3800 angstroms and about 7700 angstroms (one angstrom, abbreviated A, equals 10^{-10} meter)—a range of about one octave. Within the visible spectrum, the response of the eye varies greatly from one wavelength band to another.

Vision is the sensation produced within the eye-brain combination when electromagnetic waves in the visible region strike the nerve endings in the retina. There are two kinds of nerve endings, the *cones* and the *rods*. The cones, which enable us to discriminate colors, are most numerous in the "yellow spot" (*fovea centralis*). This is a tiny region the size of a pinhead, where light is focused in direct vision. Cones are lacking in the regions far removed from the yellow spot. You cannot distinguish the color of an object that is near the boundary of the field of vision. Look at a red pencil and a white paper, side by side, and shift them gradually sidewise until they vanish from view. Just before they disappear, you will fail to distinguish their colors. In dim light, we cannot distinguish colors, for the rods only, but not the cones, are sensitive. When the illumination is less than about 0.25 m-candle (bright moonlight) we see by rod vision only. According to an old saying, "Where all the lights be out, all the cats be gray."

The rods are less dense in the yellow spot than outside it; hence in dim light we see more distinctly by direct vision. Look directly at the second star from the end of the handle

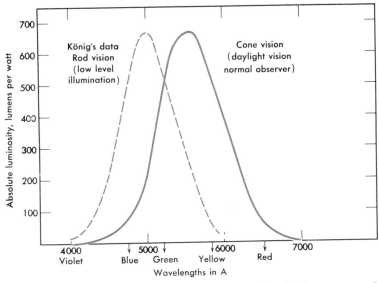

Fig. 36-11. Wavelength-luminosity curves for the human eye. In dim light (rod vision), the sensitivity of the eye is greatest at 5100 A; in bright light (cone vision), at 5800 A. (Adapted by permission from O. Stuhlman, Introduction to Biophysics, John Wiley and Sons, 1943.)

of the Big Dipper and you may see a very faint companion star. Look a little to one side of it, and you will perceive it more clearly. The sensitiveness of the rods is greatly decreased by a deficiency of vitamin A (formed by the body from carotene) in the diet. It is claimed that about one automobile driver in twenty-five is partially blind at night because of vitamin A deficiency. Some airplane pilots and truck drivers supplement their diets by vitamin A preparations.

Figure 36-11 shows how relative luminosity for an average observer depends on wavelength. Notice that for strong illumination (cone vision) the maximum efficiency is at about 5800 A (yellow light). The luminous efficiency of light of this wavelength is about 685 lumens per watt, more than 50 times that of a 100-watt incandescent lamp! In dim light, the eye is most sensitive at about 5100 A.

Color

In sound, pitch corresponds to frequency, loudness to intensity, and quality to wave form or blend of frequencies. In light, *hue* corresponds to pitch, *illumination* to loudness, and *color* to quality. The eye, unlike the ear, cannot resolve the vibrations that it receives. For example, green light and red light, blended together, are not perceived as two colors, but as a single color, yellow.

The colored lights or hues of the spectrum can be blended in varying proportions to give cerise, magenta, pink, lavender, heliotrope, and hundreds of other colored lights that are not in the spectrum. White light can be produced not only by mixing all the spectral hues, but also by blending two properly chosen hues. Two colored lights that blended give white are called *complementary* colors. Examples of complementary colors are: yellow and blue, red and bluish green. To demonstrate complementary colors, stare steadily at a very strongly illuminated red object for a minute or more, then quickly shift your gaze to a brightly illuminated white screen. After a few seconds, you will see a blue-green image on the white background. When you stare intently at the red object, the nerves in the retinas of your eyes that are sensitive to red light become fatigued or partially blinded. When you look at the white surface, these nerve endings do not respond fully. The other nerve endings, sensitive to the complementary color, respond more strongly; hence the after-image shows the complementary color.

The three hues red, green, and blue—so-called *primary* colors—may be combined to yield not only the other spectral colors but also any color such as magenta, cerise, or lavender. The theory of color vision, developed by Helmholtz in the nineteenth century, held that the color-sensitive nerve endings responded to just these three colors; some nerve endings responded to "red" waves, some to "green," and some to "blue." The total response of the eye to a mixture of waves was the sum of the responses of the specialized nerve endings to these colors. Modern studies, which we shall refer to shortly, have shown that the physiology of color vision of everyday objects is much more complicated than this simplified theory of colored lights.

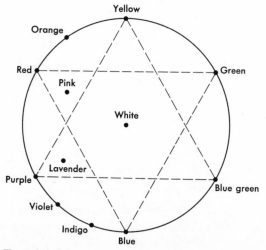

Fig. 36-12. Newton's color star.

Certain facts as to the mixing of colored lights may be made clear, however, by a study of Sir Isaac Newton's color star, on which the hues of the rainbow and purple are arranged in their proper order (Fig. 36-12).

1. White appears at the center of the circle because light of all the rainbow hues can be mixed to produce this sensation.
2. The eye cannot distinguish between the whiteness of sunlight and that produced by the mixing, in certain proportions, of light of any two complementary colors whose names are printed at opposite points on the circle. Examples are red-bluish green and yellow-blue.
3. The tints, such as pink and lavender, are mixtures of white light and colored light. In the circle they are located between the center and the boundary.
4. All known colors can be produced not only by mixing lights of the spectral colors but also by mixing lights of the primary colors, blue, green, and red.
5. A colored light causing nearly the same sensation as a spectral colored light may be produced by mixing lights of the two colors that are adjacent to it on the star. For example, red and green light produce yellow; blue-green and yellow produce green.
6. When placed side by side, colored objects blend well when their colors are adjacent on the circle. Strong contrasts are produced by colors that are widely separated. Keep these facts in mind in choosing clothing.

Most of the colors that we see are produced not by the mixing of pure colored lights but by *selective absorption.* When sunlight passes through red glass, most of the yellow, green, and blue light is absorbed and most of the red light is transmitted. Disperse a beam of sunlight by a prism so as to form a spectrum on a screen. Interpose a plate of red glass in the beam and the yellow, blue, and green light will mostly be blotted out. Similarly green glass will absorb most of the red, yellow, and blue light. The two plates together will absorb nearly all the light. A piece of white paper reflects all the colors of sunlight about equally. In the red part of the spectrum the paper seems red; in the blue part it seems blue. Red paper reflects red light and absorbs the green and blue; hence it shows its natural color in the red part of the spectrum. In the other regions it appears almost black. When you buy a necktie or a scarf, remember that the color of a piece of cloth depends upon the illumination. The light from an ordinary tungsten lamp is more yellow than daylight; hence the appearance of the cloth may alter when you take it out-of-doors. "Daylight" tungsten lamps have slightly blue bulbs. The glass absorbs part of the yellow and red light so that transmitted radiations have nearly the same relative intensities as sunlight.

Studies by Edwin Land have shown that the process by which we perceive the color of everyday objects—a collection of minerals, a colored advertisement in a magazine, a basket of fruit, for example—depends upon the response of the eye to a long-wavelength image and a short-wavelength image of the scene. Red and green combined would not, according to the Helmholtz theory, produce colors such as yellow, white, brown, and blue. Yet a simple experiment shows that under certain conditions a red image (long wavelength) and a green image (shorter wavelength) superimposed on a screen do just this.

Imagine that a basket containing an apple, an orange, a banana, and some walnuts is photographed first through a green filter on ordinary black-and-white film and then, in a separate exposure, through a red filter on black-and-white film. The two negatives are developed, and black-and-white positive lantern slides are made from the negatives. There is no color evident in these lantern slides, only variations in black and white. If the lantern slides are placed in two projection lanterns—the "green" slide in one with a green filter over the objective lens and the "red" slide in one with a red filter over the objective—and the two images superimposed upon a screen, the combined image has the full range of colors. The apple is red and yellow, the orange is greenish orange, the banana speckled yellow, and the walnuts brown. However, if the "red" projector is turned off, the image appears green, and if the "green" projector is turned off, the image appears red.

Land has experimented with different colors in this way and has found that any two narrow wavelength bands, if separated by a suitably large difference in wavelength, produce color in the combined image. For example, two yellow images—one photographed at 5790 A and one at 5990 A—produce a surprisingly wide range of colors. Land has interpreted these and other related results as meaning that color vision is the response of the eye to two "records" of a scene—one made in long-wavelength light and one in shorter-wavelength light, each record varying randomly in intensity from point to point over a natural image.

Summary

Dispersion is the separation of a mixture of waves into its component waves of different wavelengths. A pure spectrum is one in which radiations of different wavelengths do not overlap.

Chromatic aberration is the failure of a lens to converge at a single point light of different colors. Chromatic aberration is avoided by using a combination of lenses made of glasses having different dispersive powers.

The light from an incandescent solid is dispersed by a prism into a continuous spectrum. That from a luminous gas forms a bright-line spectrum. A gas or vapor strongly absorbs light of the same frequencies that the gas or vapor would emit. A liquid or solid absorbs light of a wide range of frequencies; hence the transmitted light forms a dark-band spectrum. Fraunhofer lines are dark lines in the sun's spectrum caused by absorption in the sun's atmosphere.

The velocity of a star with respect to the earth can be determined by the Doppler equation, $\pm v/c \cong \mp \Delta\lambda/\lambda \cong \pm \Delta f/f$.

Lights of three primary hues, indigo, green, red, or violet, yellow, blue, can be mixed in varying proportions so as to produce all other colors. Two colored lights which, combined, cause the sensation of white are said to be complementary.

Questions

1. Explain the production of the rainbow.
2. Discuss color mixing and the colors of objects seen by reflected light. What is a negative after image?
3. Sketch the spectral distribution curve of a black body at 5000°K, using the data in Fig. 36-7 and the two black-body radiation laws we have studied.
4. Radiation from a star forms a bright-line spectrum. What is the physical state of the surface of the star?
5. What is the color of a spectral line that has a wavelength twice that of 3889 A?
6. Why can you see through glass but not through a sheet of copper?
7. Cite evidence that all colors of light travel with the same velocity in free space.
8. If you were given an iron electrode and a mixture of chemicals, how could you find out (a) whether or not iron is present in the mixture, (b) how much is present?
9. Devise an arrangement of mirrors by which the light of the continuous visible spectrum could be concentrated at one part of a screen. What would be the color of the image thus produced?
10. Make a table classifying the apparent colors of "blue," "green," "yellow," and "red" cards as viewed under sodium (yellow), neon (red), and mercury (blue) light.

11. An artist uses red, green, violet, white, and black paints, placing dots close together. How can he produce the following colors: (a) yellow, (b) purple, (c) pink, (d) brown?

12. Estimate the velocity at which you would have to approach a traffic light so that, according to the Doppler effect, it would turn from red to green.

13. At what velocity would a galaxy be moving away from us so that a spectral line of standard wavelength 4000 A would be just outside the visible region?

14. Under what conditions is (a) wavelength, (b) frequency, of light invariant?

15. An earth satellite emits radio waves. From observations on the Doppler shift of their frequency, how could (a) the velocity, (b) the orbit radius of the satellite be determined?

16. Redraw the spectrum (Fig. 36-13) with correct spacing and positions of the lines (a) if the source moves toward us at a high speed, (b) if the source moves away at a high speed.

Fig. 36-13.

17. Under what conditions is a spectral "blue shift" possible? What would be the percentage blue shift that one might expect to encounter in our galaxy?

Problems

$1 A = 10^{-8} cm$

A

1. What is the frequency of the ultraviolet beryllium line at 2651 A?

2. What is the wavelength of maximum intensity for a black body at 6000°K?

3. Calculate the frequencies of these argon lines (wavelength in air): 8115 A, 4348 A.

4. At what temperature (°C) is 4000 A the wavelength of maximum intensity for black-body radiation?

5. What are the wavelengths in air (a) in angstroms and (b) in meters for these frequencies: 2.0×10^{14} H, 3.0×10^{14} H, 8.0×10^{14} H?

B

6. A star moves toward the earth with a speed of 8250 km/sec. What is the wavelength of a hydrogen spectral line emitted by the star if the standard wavelength is 6563 A?

7. Because of the earth's motion away from a star, light of frequency 56.7×10^{13} H seems to have a frequency of 56.5×10^{13} H. What is the speed of the earth away from the star?

8. A distant galaxy emits yellow light of wavelength 5890 A which, observed by our spectroscopes, seems to have a wavelength of 6520 A. (a) What are the actual frequency of the light and the apparent frequency? (b) What is the relative velocity of the earth with respect to the star?

9. A certain double "star" consists of two stars of equal mass which revolve around their common center of gravity. One star is dark, the other incandescent. When the bright star moves away from the earth, the Balmer α line in its hydrogen spectrum has a wavelength of 6565 A. What is the speed of the star relative to the earth if the standard wavelength of this hydrogen line is 6563 A?

10. A spectral line whose standard wavelength is 4000 A is found to have a wavelength of 4030 A when emitted by a star. At what speed is the star moving away from the earth?

C

11. In problem 9, what is the period of the bright star if the two are equal in mass and their centers are 160×10^7 km apart?

12. The radius of curvature of a plano-convex lens is 40 cm. The index of refraction of the crown glass for red light is 1.511; for blue light it is 1.528. (a) What are the focal lengths of the lens for light of the two colors? (b) If the lens is 6.0 cm in diameter, what is the diameter of the red image of a star formed on

a photographic plate when the blue image is sharply focused on it? Let λ of blue light be 4860 A.

13. A galaxy moving away from the earth with 50% of the speed of light would shift a violet 3934-A line in what direction? What would be the wavelength of the line as received on earth?

14. The spectral lines at the edges of the sun show a relative Doppler shift corresponding to a velocity of 4.0 km/sec. The radius of the sun is 7.0×10^5 km. What is the period of rotation of the sun?

For Further Reading

Hewish, Antony, "Pulsars," *Scientific American,* October, 1968, p. 25.

Land, Edwin, "Experiments in Color Vision," *Scientific American,* May, 1959, p. 84.

Minnaert, M., *The Nature of Light and Colour in the Open Air,* Dover Publications, New York, 1954.

Murray, B. C. and J. A. Westphal, "Infrared Astronomy," *Scientific American,* August, 1965, p. 20.

Newton, Sir Isaac, *Optiks,* Dover Publications, New York, 1952.

CHAPTER THIRTY-SEVEN

Interference and Diffraction

In Chapters 19 and 20, we discussed interference phenomena as characteristic of all kinds of wave motion. Interference, we saw, is the coming together of waves to produce a resultant vibration whose amplitude may be greater or less than that produced by any of the waves alone. Waves on a vibrating stretched string, for example, can interfere at certain points to produce maximum displacement of the string—called antinodes or loops—and at other points to produce minimum displacement—nodes. Ripples of water can interfere, and so can sound waves in an organ pipe. The succession of maximum and minimum amplitudes observed in these phenomena is called an *interference pattern.*

We also noted the conditions under which interference patterns could be observed. When waves of the same frequency travel different distances by different paths from the same source before reaching the point at which they combine, constructive interference results when crest meets crest, that is, when the difference in path is 0, 1, 2, 3, . . . wavelengths. Destructive interference occurs when crest meets trough, that is, when the difference in path is 1, 3, 5, 7, . . . half-wavelengths.

Do electromagnetic waves interfere? We shall consider in this chapter evidence for believing that they do. Indeed, interference and the related phenomenon of diffraction provide some of the best evidence we have for the wave nature of electromagnetic radiations. Light will serve as our chief example, but all electromagnetic radiations show interference and diffraction effects.

Diffraction

In the preceding chapters we assumed that light waves travel in straight lines except when they are reflected or refracted. However, you will recall that waves of all kinds can change direction in other circumstances, namely, *when passing through a narrow aperture or near an obstacle.* This bending or spreading of waves when they pass near an obstacle is called *diffraction.*

Sound waves afford many examples of diffraction. When music from a band enters an open window, the sound does not travel forward in a well-defined beam, but is diffracted sidewise. If you try to shut off the noise by holding a book between your ear and the

(a)

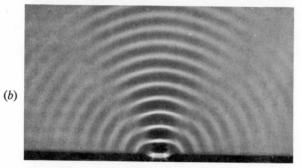

(b)

Fig. 37-1. *Diffraction at slits.* (a) *Short wavelength,* (b) *longer wavelengths.*

source, the waves pass around the edges of the book readily.

The amount of diffraction that an opening produces depends upon the *width* of the aperture and upon the *wavelength* of the waves. The diffraction is more marked the *greater* the wavelength and the *smaller* the aperture. We can demonstrate these facts by letting ripples on the surface of a pool of water pass through a slit. In Fig. 37-1a, the aperture is greater than the wavelength, and the waves spread out slightly at the edges. In Fig. 37-1b, the opening is smaller than the wavelength, and the diffraction is very great. *In all kinds of wave motion, diffraction effects are considerable when the aperture is small compared with the wavelength.*

The megaphone offers an excellent example of diffraction control in sound. A cheerleader's mouth is small compared to the length of the sound waves; hence the sound

of his voice is diffracted through a wide angle. A megaphone increases the effective width of his mouth and concentrates the beam of sound in the desired direction.

The wavelengths of light are only a few hundred-thousandths of an inch; hence ordinary apertures cause little diffraction. A hunter can aim his rifle by peeping through the V-shaped opening of the sight mounted on the gun barrel. In order to demonstrate the diffraction of light, look between two fingers of one hand at a small, intense light source such as the filament of an unfrosted, incandescent lamp. (View the U-shaped filament edgewise.) Gradually move your fingers toward each other until the opening between them is completely closed. Just before the light is cut off, the source will seem to broaden out into a band. This is because the light spreads out sidewise in passing through the narrow opening.

Diffraction is really one aspect of interference. Applying Huygens' principle to diffraction at an opening helps us understand why this is so. A stone, dropped into a quiet pool, generates waves that spread out in everwidening circles. Suppose that many stones are simultaneously dropped into the water along *af* (Fig. 37-2). After a time each wave will have traveled a distance *aa'* and all the waves will reinforce each other along *a'f'*. In other regions the crests and troughs will cancel each other and the disturbance will be small. When the waves reach the barrier *a"f"*, those passing through the wide aperture progress straight forward except at the edges where they deviate into the "shadow" of the barrier. They deviate into the shadow at the edge nearer *f"*, for example, because the wavelets that would otherwise destructively interfere with this portion of the advancing wave front have been cut off by the edge of the aperture. The waves passing through the narrow slit are greatly diffracted because only a few wavelets get through, as though a single stone were dropped at the slit. Once

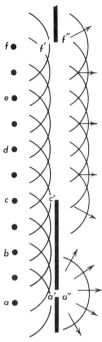

Fig. 37-2. Diffraction at wide and narrow openings. Many wavelets, originating from points a-f on a wave front pass through the wide opening c'f' with slight diffraction. The narrow opening a″ lets through few wavelets and therefore causes great diffraction.

again, we see that *each point on a wave front may be considered as a source of independent wavelets which interfere with each other to produce a resultant disturbance.*

Light as Waves or Particles

The problem of the nature of light has puzzled scientists for several centuries. Newton favored the hypothesis that light is corpuscular. He believed that particles of light, once emitted by a source, were subject to the laws of mechanics. For example, they could be reflected by a plane mirror just as tennis balls are reflected when they strike the tennis court. In refraction, according to this view, light cor-

puscles traveling from air into water would be gravitationally attracted by the water as they approached its surface and would *speed up,* moving with a *greater* speed in water than in air. Christian Huygens, Newton's contemporary, ably argued that light is a form of wave motion. Waves can be reflected, as we have seen, and Huygens' wave theory gives the correct law of reflection. It also explains refraction, but with the significant difference from Newton's theory that the wave speed in water must be *less* than in air in order to explain the bending of the beam toward the normal. Both theories explained reflection of light, but, in explaining refraction, they differed in their predictions of what the speed of light should be in the denser medium.

As in all such situations in physics, where two theories are logically consistent but give different predictions, experiment must decide. The crucial experiment was not performed until the middle of the nineteenth century by Foucault, who measured the speed of light in water, but the result was conclusive: light travels more slowly in water than in air. However, several decades before this, Thomas Young (1773–1829) had discovered convincing evidence for the wave behavior of light by his experiments on the interference of light; Fresnel and Fraunhofer had also demonstrated that light has a wave nature by their successful explanations of diffraction.

The wave theory of light seemed by the end of the nineteenth century to have defeated the corpuscular theory, but this is not the end of the story. In later chapters, we shall see the particle or photon of light restored to good standing in physics as a complement of the light wave in a development that opened the age of modern physics. It has turned out that Newton and Huygens were both partly right. A lamp sends out corpuscles called photons, but the concentration of these photons is governed by waves. Let us now, however, consider the work of Young, Fraunhofer, and Fresnel.

Young's Experiment

In order to repeat Young's experiment, cut one narrow slit with a sharp blade in an opaque piece of developed photographic film and two parallel slits about 0.5 mm apart in another opaque film. Mount the two films parallel to each other and about 20 cm apart. Place a piece of red glass in front of the single slit and focus a beam of sunlight on it (Fig. 37-3a). The light passing through this aperture is diffracted and illuminates the two slits very close together. Light from the two openings falls on a screen, producing interference. The point D is equidistant from each slit; hence the two beams arrive at D in phase so that crest meets crest and trough meets trough, causing maximum illumination. F is one wavelength farther from one slit than the other. The two beams are again in phase at F, producing maximum illumination. E is one half wavelength farther from one slit than from the other. The waves are opposite in phase; crest meets trough, and the illumination is a minimum. Figure 37-3b is a photograph of interference bands produced by two apertures.

Replace the red glass in front of the first slit by blue glass. The average wavelength of the transmitted light will be shorter and the interference bands will be closer together. Remove the glass, so that the double slit is illuminated by white light. The dark bands for some wavelengths will coincide with the bright bands for others; hence the pattern will be colored.

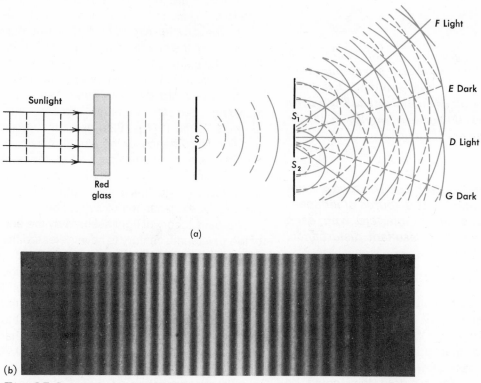

Fig. 37-3. *Interference of light.* (a) F *is one wavelength farther from* S$_2$ *than* S$_1$; *hence maximum illumination results.* (b) *Interference bands due to two parallel slits.* (*Reprinted with permission from F. Jenkins and H. White,* Fundamentals of Optics, *McGraw-Hill Book Company, 1950.*)

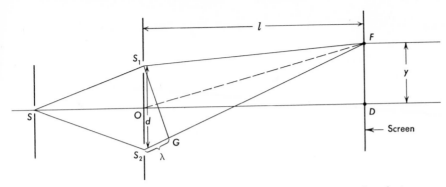

Fig. 37-4. *Calculating the position of the bright interference bands in Young's experiment.*

What is the separation between the bright interference bands in the two-slit experiment? The answer depends upon the wavelength, the separation of the slits, S_1 and S_2, and the distance from the slits to the screen on which the interference pattern is formed. Suppose the slits S_1 and S_2 (Fig. 37-4) are separated by a distance d and the perpendicular distance from the slits to the screen is l. At D, which is equidistant from S_1 and S_2, the light waves arrive in phase, and a central bright band is produced. Consider the point F which is farther from S_2 than S_1 by a distance equal to λ, the wavelength of the light. Light waves from S_1 and S_2 travel different distances to F, but since the difference in path $S_2 G$ is one wavelength, the waves are in phase at F and reinforce each other, producing a bright band there. (A similar bright band exists at an equal distance below the center line in Fig. 37-4.) Ordinarily, the distance l is large compared to y, the distance from the central bright band to the one at F. Hence, the triangles OFD and $S_1 S_2 G$ are similar and

$$\frac{\lambda}{d} = \frac{y}{l}$$

or

$$y = \frac{l\lambda}{d}$$

In general, bright bands will be found at dis-

tances y above and below the central band such that

$$y = \frac{n\lambda l}{d}$$

where n is 1, 2, 3, ... There will be dark bands at

$$y = \frac{m\left(\frac{\lambda}{2}\right)l}{d}$$

where $m = 1, 3, 5, \ldots$ (Why?) In practice, because of the diffraction pattern of the individual slits, the bright bands are observable only for fairly small values of n; the bright bands "wash out" to the right and left of the central bright band as you will notice in Fig. 37-3b.

The presence of dark bands in the interference pattern does not mean that the energy of the electromagnetic waves is destroyed at those points. The interference pattern represents a redistribution of energy. The energy that would ordinarily reach the dark-band portion of the screen is redirected to the bright bands. There the amplitude of the electromagnetic vibration — the magnitude of the electric field vector — is greater than it would be under uniform illumination, and the radiant energy per unit area there is correspondingly greater.

Coherence

We started with light from a single source at S in Fig. 37-3 and divided this light by slits S_1 and S_2 into two beams which traveled different distances to reach the screen. We could *not* have produced an interference pattern by putting *different* sources of light behind S_1 and S_2. The reason has to do with the *coherence* of the light beams reaching a given point on the screen. Coherent sources emit waves that always have the same phase relations between them.

Two kinds of coherence are important in preserving the phase relationship: *frequency coherence* and *spatial coherence.* Frequency coherence is the extent to which the source emits waves of a narrow range of constant frequencies or wavelengths. For example, the pure tone of a tuning fork would be highly frequency-coherent; the noise produced when a chain saw cuts down a tree would have very little frequency coherence because the sound waves would be a jumble of many sounds of different frequencies. Spatial coherence refers to the regularity of the waves in space. Spherical sound waves originating from a small point source are spatially coherent; sound waves that have been scattered by reflection from a rocky hillside are not. Waves such as those emerging from slit S in Fig. 37-3a are spatially coherent; light waves that have been scattered by a rough piece of cardboard are not.

Coherence is of fundamental importance in interference experiments. Two radar transmitters of the same frequency could be electronically "locked" together so that their radar beams always had the same phase relation and were coherent. A radar interference pattern could then be observed in a two-source experiment. Light waves, however, are emitted by most sources in a random process by atoms. Atomic sources of light waves are usually incoherent because the phase relations between the emitted light waves change randomly. Hence, two slits must be used to

divide the light from a *single* source so that as the random changes in phase occur in the source, they will occur at both of the slits in the same way and the changes in phase of the two beams at a point on the screen will also occur in unison. An exception to the statement that atomic sources of light are incoherent has been found in what are called *lasers,* devices for producing coherent beams of light. We shall discuss them in Chapter 44.

Diffraction by a Single Slit

Consider light that passes through a *single* slit. Can it produce an interference pattern? Experimentally, we find that it can. The arrangement for Fraunhofer diffraction is shown in Fig. 37-5a. Light from a slit S is rendered parallel by lens L_1, passes through the rectangular diffracting slit s, is collected by lens L_2 and focused on the screen. The interference pattern on the screen consists of a broad central bright band O accompanied by fainter secondary bright bands separated by dark bands—as at M.

Let us consider how the interference pattern is produced and how one can calculate the angular separation θ between the central bright band and the first dark band below it. Suppose that the slit s of total width a is divided into small openings of equal width Δa; two of these, Δa_1 at the top and Δa_2 at the middle of the slit, are shown in Fig. 37-5b. The intensity of the light waves reaching point M will be a minimum if the waves going through all of the pairs of little openings Δa arrive at M 180° out of phase. In the experimental arrangement shown, the optical paths—in glass and air—of all of the waves reaching M are the same except for the portions between the plane of the slit and the lens L_1. For example, the plane wave fronts from Δa_1 have to go a greater distance than the wave fronts from Δa_2. Suppose that the optical path for waves from Δa_1 is greater than for waves from

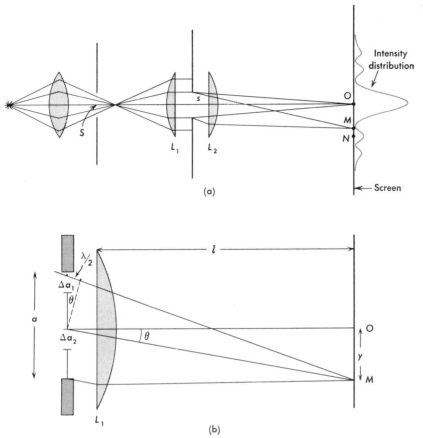

(a)

←—— Screen

(b)

Fig. 37-5. Fraunhofer diffraction by a single slit of width a. *Light passes through the entire slit. Rays drawn to point* M *show the direction of waves reaching* M *from the top, middle, and bottom portions of the slit.*

Δa_2 by just *one-half wavelength* $\lambda/2$. Then the waves reaching *M* from Δa_1 and Δa_2 will be out of phase. But if this is so, then waves reaching *M* from all of the small openings Δa will cancel because we can pair off all such openings just as we did Δa_1 and Δa_2, and the geometrical arrangement of these pairs will correspond to that between Δa_1 and Δa_2. Thus there will be total destructive interference at *M* and a dark band will result there.

The angular separation θ between *O* and the first dark band *M* (or between the plane of the slit and the dotted line in Fig. 37-5*b*) is given by

$$\sin \theta = \frac{\lambda/2}{a/2}$$

$$\sin \theta = \frac{\lambda}{a}$$

For small angles,

$$\sin \theta = \frac{y}{l}$$

and

$$y = l\frac{\lambda}{a}$$

By dividing the slit into thirds, each of which is subdivided into small openings Δa, we can show that the wave contributions of the pairs Δa in two of the thirds will cancel each other at N, leaving the waves passing through the remaining third to produce a bright band (of lesser intensity than the central bright band) at N in Fig. 37-5a.

Diffraction by a single slit can also be observed without the lenses L_1 and L_2. Fresnel (Fre nell') diffraction, as this phenomenon is called, was historically the first to be observed. The interpretation of Fresnel diffraction is more difficult than that of Fraunhofer diffraction, because the wave fronts in Fresnel diffraction are divergent, rather than plane, and the simplifications we made in the foregoing discussion are not allowable.

The Diffraction Grating

If, instead of using a single slit to produce a diffraction pattern, we use many parallel narrow slits, we have a *diffraction grating*. Strictly speaking, we should call it a "diffraction-and-interference" grating, because both diffraction and interference must be considered to understand how the grating disperses light. The diffraction grating is widely used to measure wavelengths of light. One type of grating consists essentially of an opaque screen which has many equidistant apertures or "slits" (Fig. 37-6). Let a parallel beam of light of wavelength λ be incident normally on this screen. Most of the light goes straight through and produces a central bright image. Fig. 37-6 shows that part of the radiation passing through the slits which is diffracted through a certain angle θ, so chosen that the distance from the upper slit to the lens is exactly one wavelength greater than the distance of the next slit. Then each wave from slit D will reach the lens at the same instant as the preceding wave that left

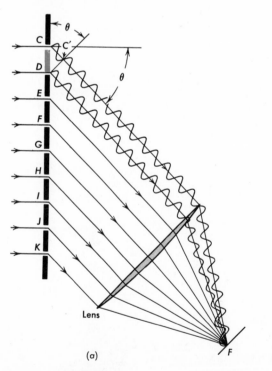

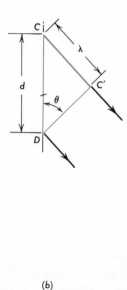

(a)   (b)

Fig. 37-6. *Diffraction grating. Light reaching* F *from all the apertures is in phase if the difference in path for adjacent waves is one wavelength.*

C, and the two will reinforce each other at the focus *F*. Similarly, waves from the third slit will reinforce those from the second one. Moreover, waves from *all* the slits will reinforce one another, and a bright image will be formed at the focus *F*. If a person's eye were substituted for the lens shown, he would seem to see the source of light in the direction from which the diffracted beam is coming. If the angle of diffraction is even slightly greater or less than θ, the wavelets from the slits will destructively interfere.

If monochromatic light (of a single wavelength) is incident on the grating, a very narrow line at *F* will be brightly illuminated, but the parts of the screen nearby will be dark. If the light is of several wavelengths, maxima will occur for different values of θ, and several bright lines will appear side by side. If white light is used, a continuous spectrum will be produced. Thus the grating, like a prism, can produce a spectrum. Moreover, a diffraction grating, unlike a prism, can be used to measure the wavelengths of the light. You can easily show by examining Fig. 37-6*b* that the relationship between the wavelength λ, the distance *d* between adjacent slits, and angle of diffraction θ is given by

$$\frac{n\lambda}{d} = \sin \theta$$

n is called the *order* of the interference band, and *n*λ is the difference in path of light rays from adjacent slits. For instance, in the second order, the difference in path is 2λ.

Example. The distance between adjacent slits of a grating is 0.000200 cm. The angle of diffraction θ is 16°. (*a*) What is the wavelength of the light? (First-order interference.) (*b*) What is θ for the second order?

(*a*) $\sin \theta = \dfrac{n\lambda}{d}$

$0.276 = \dfrac{1 \times \lambda}{0.000200 \text{ cm}}$

$\lambda = 0.0000552 \text{ cm} = 5520 \text{ A}$

(*b*) $\sin \theta = \dfrac{2 \times 0.0000552 \text{ cm}}{0.000200 \text{ cm}} = 0.552$

$\theta = 33° \ 30'$

In order that the angle of diffraction θ shall be large for the first-order spectrum, it is necessary that *d*, the distance between adjacent apertures, be nearly as small as the wavelength of the light. Gratings ordinarily have 400 to 12,000 apertures or "lines" per centimeter. *Transmission* gratings are made by ruling parallel lines on a plate of glass by means of a diamond point that is moved back and forth by a ruling engine. The transparent regions between adjacent lines are the apertures. *Reflection* gratings have fine lines ruled on polished metal mirrors. An inexpensive "replica" grating is made by coating a grating with a thin solution of collodion which dries and afterward is peeled off and mounted on a plate of glass. The parallel ridges, formed where the liquid collodion solution flowed into the grooves of the grating, are opaque. The regions between the ridges are the apertures.

Table 37-1. Wavelengths of Some Visible Spectral Lines (Stated to Four Significant Figures)

(1 angstrom = 10^{-8} cm; 1 micron = 10,000 angstroms = 10^{-4} cm = 10^{-6} m)

	Centimeters	Angstroms	Microns
Red Fraunhofer A line	0.00007594	7594	0.7594
Orange krypton-86 line (Primary International Standard)	0.00006058	6058	0.6058
Red cadmium line	0.00006438	6438	0.6438
Yellow sodium D lines			
D$_1$	0.00005896	5896	0.5896
D$_2$	0.00005890	5890	0.5890
Yellow mercury lines	0.00005791	5791	0.5791
	0.00005770	5770	0.5770
Green mercury line	0.00005461	5461	0.5461
Fraunhofer K line in extreme violet	0.00003934	3934	0.3934
Violet helium line	0.00003889	3889	0.3889

Resolution of Images by a Lens

When light from a star goes through a lens, the light is diffracted, forming an interference

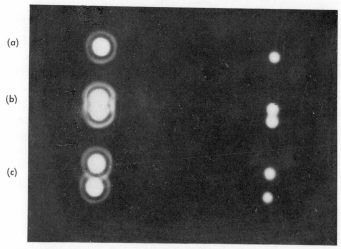

(a)

(b)

(c)

Fig. 37-7. Resolution of images of point sources of light. Images on the left were formed by a lens of small diameter; images on the right by a lens of larger diameter. (a) One source. (b) Two sources quite near each other; the images are just resolved. (c) Two sources somewhat further apart; the images are completely resolved. (Reprinted with permission from F. Jenkins and H. White, Fundamentals of Optics, McGraw-Hill Book Company, 1950.)

pattern such as we see when light passes through a slit. The image of the star covers a greater area than it would if diffraction did not occur. If two objects are sufficiently close together, their image diffraction patterns, formed by a lens, overlap. When the objects are farther apart, the images are seen separately, and they are said to be "resolved" by the lens. Figure 37-7 shows the resolution of images of adjacent point sources of light by a lens of small diameter and then by a lens of somewhat greater diameter. Note that the separate images are not points, but diffraction patterns. Because of the circular geometry, the diffraction patterns are not bright and dark lines, but bright and dark rings. Many stars that seem to be single, because of the merging of their diffraction patterns when viewed with the naked eye, are found to be double when observed through a large telescope. The giant 200-in. reflecting telescope at Mt. Palomar, California, can separate astro-

nomical details—of the surface of the moon, for example—that would be indistinguishable if viewed through a smaller telescope.

In Fig. 37-8, *F* is the focus of the lens and *a* is the width of the aperture. Were it not for diffraction, all the light from one star, incident parallel to the principal axis of the lens, would converge at *F*, forming a point image. Because of diffraction, the image is not a point, but a circular patch of light, the "circle of confusion," surrounded by faint rings. Let us think of the lens opening as equivalent to a single slit. The lens forms a diffraction pattern (solid curve) with the light from the star. The angular separation θ between the central bright area and the adjacent dark area, we found earlier in this chapter, is given by

$$\sin \theta = \frac{\lambda}{a}$$

This equation holds for a rectangular slit. For a circular aperture, the equation is found by

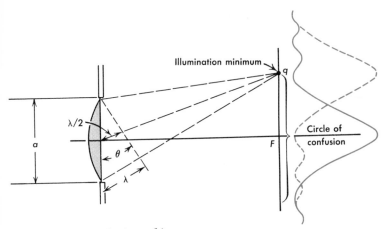

Fig. 37-8. *Resolution of images.*

diffraction theory to be

$$\sin \theta = \frac{1.22\lambda}{a} \qquad (1)$$

Now suppose that the source of light is a double star. The diffraction pattern formed by the lens from light from the second star (dashed intensity curve) will overlap the diffraction pattern due to the first star (solid curve). If the central bright ring in the pattern of the second star falls on the first dark ring in the pattern of the first star, the two images will not be resolved. Thus two point sources of light can barely be resolved by a lens if they subtend at the lens an angle equal to θ in equation (1). The angle θ affords a measure of the limit of resolution of the lens, and objects separated by angles less than θ at the lens are not resolved.

Example. (a) What is the resolution of a man's eye when the diameter of the opening in the iris is 0.30 cm?
  Assume that

$$\lambda = 0.000060 \text{ cm}$$

$$\sin \theta = \frac{1.22\lambda}{a} = \frac{1.22 \times 60 \times 10^{-6} \text{ cm}}{3.0 \times 10^{-1} \text{ cm}}$$

$$= 2.44 \times 10^{-4}$$

$$\theta \cong 1'$$

(b) At what distance could the man's eyes resolve the wires of a window screen if the distance between successive wires were 0.24 cm?
  Adjacent wires must subtend at the man's eyes an angle θ such that

$$\sin \theta = \frac{0.24 \text{ cm}}{s} = 2.44 \times 10^{-4}$$

$$s = 981 \text{ cm}$$

The biologist uses a microscope to study the details of the structures of cells, so he needs an instrument which has high resolution. To this end, the aperture of the objective lens should be large compared with the focal length. Blue light is less diffracted at the edges of the lens than red light of greater wavelength; hence the diffraction patterns are smaller, and better resolution is obtained with blue light.

A decrease in wavelength may be secured by using a microscope having an "immersion" objective lens. The space between the specimen and the lower face of the lens system is filled with a liquid having about the same index of refraction as the glass. The light travels more slowly in the liquid than in the air, its wavelength is correspondingly decreased, and the resolution of the microscope is increased about 50%.

Interference in Thin Films

Strange to relate, the colors of the peacock's feathers or soap bubbles, and the iridescence of the pearl are not produced by pigments or other coloring matter, but by thin, nearly transparent surface films. Part of the incident light is reflected at the first surface and part at the second one. The eye receives two reflected beams which have traveled different distances and may therefore either reinforce or oppose each other. After a rain, pools of oil, spread out into thin films on a highway, display gorgeous colors. Suppose that blue light of a certain wavelength, incident on the film, is reflected partly at the upper surface of the film and partly at the lower one. The light from the second surface is retarded by its extra journey, and the two reflected wave trains may either reinforce or oppose each other. If they are in phase, the surface seems blue; if they are opposite, it seems black; hence we see alternate bands of blue and black. If the surface is illuminated with red light, more widely separated bands of red are seen. With white light, several colors are seen. The radiations of different wavelengths are selectively reflected at different regions.

Figure 37-9 represents, on an exaggerated scale, the reflection of monochromatic light from two surfaces of a wedge-shaped oil film on a glass plate. The speed of light in oil is greater than in glass. When one looks down on the wedge, bright interference bands, parallel to the thin edge of the wedge, appear at A, C, E, G, I, and K, etc., with relatively dark bands between them. At A, the waves reflected from the upper and lower surfaces are in phase, for the distance between them is zero. At the next bright interference band C, the difference in path is one wavelength of the light. Since the path down and back again is one wavelength, the thickness of the film at C must be one-half wavelength. Similarly, the film thickness at E is one-half wavelength greater than that at C. The thickness increases one-half wavelength for each interference band as we move to the right. Note that the wavelength is that of the light *in the medium in which the light is traveling* — in this example, oil.

The light that is not reflected must be transmitted. Remember that an interference pattern represents a redistribution of energy, not a destruction of energy. At B, the thickness of the film is $\frac{1}{4}\lambda$, hence the difference in path between the two rays is $\frac{1}{2}\lambda$. They oppose each other in phase, so that no light is reflected.

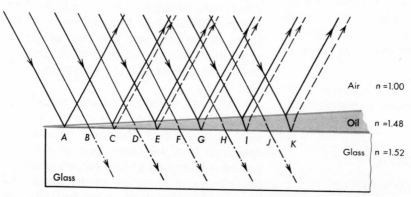

Fig. 37-9. Reflection from oil-on-glass film. Maximum reflection occurs at A, C, E, G, I, K, *where the thickness of the film is* 0, 1, 2, 3, 4, 5, 6 *half-wavelengths of the reflected light. At* B, D, F, H, J, *the light is transmitted.*

Fig. 37-10. Non-reflecting glass. The left, coated, third of the surface reflects about one-fourth as much light as the right third.

The light incident at *B* is transmitted. Similarly, maximum transmission occurs at the points *D*, *F*, *H*, and *J*, where the difference in path is 1, 3, 5, etc., half wavelengths.

Let us look more closely at what happens during reflection. When light is reflected at *C* in Fig. 37-9 at the air-oil surface and also at the oil-glass surface, in each case the light travels from one medium to another in which the speed is *less*. A phase change of 180° occurs at *both* the air-oil surface and the oil-glass surface (Fig. 19-15, page 310), but the *difference* in phase introduced by reflection is zero. Consider, however, light reflected from the two surfaces of a soap film in air. At the first air-water surface, transmitted light travels from a medium of higher speed to one of lower speed, and the change of phase of reflected light is 180°. At the second water-air surface, the transmitted light passes from the water, in which its speed is slower, into air, in which it is higher; the change of phase of the reflected light is zero. *Reflection at the two surfaces thus causes a phase difference of 180° between the two beams.* Thus, when the difference of path is very small—less than one-half wavelength—little light is reflected. In the previous case, no phase difference was introduced; hence a very thin film of oil on glass caused *maximum*, rather than minimum reflection.

The thickness of a film may be so adjusted that light beams reflected from the two surfaces will interfere either constructively so as to reinforce each other or destructively so as to give minimum reflection. Lenses of many cameras, microscopes, and the like are coated with non-reflecting films (Fig. 37-10). Suppose that in the lens system of a motion picture projector there are 5 lenses with 10 polished faces. Non-reflecting coatings on the lens surfaces will reduce the loss of light by reflection from 30% to about 4%.

Another use of interference is in the detection of "hills and valleys" on surfaces. The plate to be tested is laid on another very flat test plate, and a thin piece of paper is inserted between them at one end so that one surface is slightly tilted with respect to the other. If the adjacent surfaces are quite flat, the interference bands will be parallel to one another (Fig. 37-11a). If either surface has

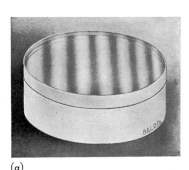

(a)

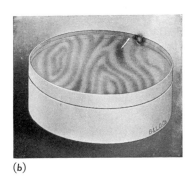

(b)

Fig. 37-11. Interference bands. Produced by (a) optically flat surfaces, (b) window glass. (Courtesy of Bausch and Lomb.)

hills and valleys, the bands will be irregular (Fig. 37-11*b*). Surfaces that have no valleys deeper than a few millionths of a centimeter are said to be "optically flat." (If a person's thumb is held against one of the glass plates for a few seconds, it will heat the glass sufficiently that its expansion will distort the fringes.)

Interferometers

Devices that employ light waves as "foot rules" are called *interferometers.* In Chapter 10 we saw that an interferometer invented by A. A. Michelson was used in the famous Michelson-Morley experiment that demonstrated that the speed of light in free space was the same in all frames of reference. Now we are able to discuss the optical principles involved. A Michelson interferometer is shown viewed from above in Fig. 37-12. M_1 and M_2 are mirrors adjusted to be accurately at right angles to each other. M_1 can be moved back and forth by a finely divided screw. P_1 is a glass plate that is lightly silvered—half-silvered—on its right-hand face so that it both reflects and transmits at that face. P_2 is a similar glass plate placed in the beam to equalize the optical paths. Light from a sodium source (a nearly monochromatic source) passes through a ground-glass screen that effectively broadens the source. At the half-silvered surface of P_1, the beam splits, part being transmitted toward the mirror M_2 and part being reflected to M_1. Reflected by the mirrors M_2 and M_1, the two beams recombine at O and are sent to the eye of the observer. Since different portions of the beams have traveled different distances, there are phase differences among them and

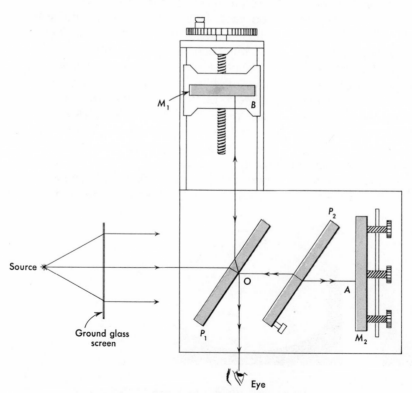

Fig. 37-12. *A Michelson interferometer, viewed from above.*

a series of circular bright and dark bands will be seen by the observer.

Suppose that in Fig. 37-12 you move M_1 backward by means of the screw. Then all the interference bands move, one band shift occurring for every half wavelength that the plate is moved. If you know the wavelength of the light, you can measure the displacement of M_1 by counting the number of interference bands that move past any point. Tiny displacements can be readily measured. If thin films are introduced into one of the beams, the optical thickness of the films can be measured. Physicists, using interferometers, have determined the distance between the two fine marks on the standard meter at Paris. This distance is 1,650,763.73 wavelengths of orange light from a krypton-86 source.

In his famous experiment, Michelson assumed that the earth's motion around the sun causes an ether wind of about 18 mi/sec (about $\frac{1}{100}$% of the velocity of light). He mounted his interferometer on a rotatable platform and turned it so that OA (Fig. 37-12) was parallel to the assumed ether drift and OB at right angles to it. He set the cross hair of a telescope on a certain bright band. Then he rotated the apparatus 90° so that OB was parallel to the wind. Michelson confidently expected this change to displace the bands slightly. He was disappointed. The expected change did not occur. Many additional experiments of great precision by Michelson and his colleagues failed to detect the effect he sought.

Michelson's failure to detect an ether wind aroused great interest among scientists. Some explained it by assuming that the earth dragged part of the ether along with it through space so that we would not be able to detect an ether wind by an experiment on the earth. Others assumed that Michelson's apparatus contracted in the direction of the ether flow by a few millionths of one per cent so as to compensate for the ether wind.

As we have seen, Albert Einstein attacked the problem more boldly. He felt that it was foolish to believe in an ether if Michelson's experiment and others failed to detect it. In his special theory of relativity, Einstein assumed that the speed of light is the same to all observers moving relative to each other with constant velocity. This is equivalent to saying that the speed of light is independent of the motion of the source and the observer and that it is not necessary to assume the existence of an "ether" as a medium for the transmission of light. We have already considered some of the results of this theory (Chapter 10), and we shall return to further applications of it in a later chapter.

Summary

When waves pass near an obstacle, they are diffracted and some of the wave energy travels into the shadow of the obstacle. Waves passing through an aperture are diffracted, the amount depending on their wavelength and the size of the opening. Diffraction is great when the aperture is smaller than the wavelength.

When the images of two objects do not overlap, the two are said to be resolved. The ability of a lens to resolve two point sources depends upon the size of the aperture and upon the wavelength of the light.

When waves which have traveled different distances arrive at a point they interfere, and the resultant effect may be greater or less than that produced by either train of waves alone.

The diffraction grating is a series of equidistant apertures in a screen. When light waves pass through it and are focused by a lens, they produce maximum illumination (a) along the direction of the incident beam and (b) along other directions at an angle θ to the incident beam such that $\sin \theta = n\lambda/d$, λ being the wavelength, d the distance between adjacent slits, and n the order of the interference.

Interference of light may be produced by two partially reflecting surfaces. If the light in the beams from the surfaces is in phase, maximum illumination results. When white light is reflected by the surfaces, rainbow colors appear.

Questions

1. Explain how the colored bands seen on soap bubbles are produced.

2. Derive the working equation for the diffraction grating.

3. How could a Michelson interferometer be used to measure the thickness of a soap film?

4. Explain why the two headlights of a distant automobile are seen as one. Why does a telescope enable the observer to see them separately?

5. What do we mean when we state that a lens "resolves" the images of two stars?

6. How would you expect the eyes of an underwater swimmer without a mask to perform in regard to (a) resolution, (b) distance of distinct vision?

7. Discuss Michelson's attempt to detect the earth's motion through the "ether." If you cannot detect a physical phenomenon, should you assume that it exists?

8. How does a diffraction grating utilize (a) diffraction, and (b) interference?

9. Is it possible to have interference without diffraction? If so, give an example.

10. What would be the appearance of a street lamp viewed through a wire screen?

11. Do (a) interference, (b) diffraction occur for electromagnetic waves at television frequencies? Give examples.

12. How could color photographic film be used to separate overlapping orders in a diffraction grating?

13. Measure the limit of resolution of your eye by backing away from a television picture tube until the horizontal lines are no longer visible.

14. Cite experiments that prove that light consists of waves. Cite others that can be explained by either a wave theory or a particle theory of light.

15. What was the significance of Foucault's crucial experiment in the wave-particle debate about the nature of light?

16. Discuss the significance of coherence for experiments on diffraction and interference.

17. Can evidence of life on the earth be seen with the unaided eye from an altitude of 200 km?

18. How does the diffraction pattern of monochromatic light passing through two slits (Young's experiment) change when the slits are brought closer together? Make sketches to illustrate the changes.

19. How does the diffraction pattern of monochromatic light passing through a single slit (Fraunhofer diffraction) change when the slit is narrowed? Illustrate the changes.

Problems

A

1. An observer looks through a fine handkerchief at a distant vertical slit, illuminated with red light ($\lambda = 6.5 \times 10^{-5}$ cm) and sees a bright diffracted image on each side of the image of the slit. The sine of the angle between the central image and the first diffracted image is 0.0020. (a) What is the distance between adjacent apertures in the cloth? (b) How many apertures are there per centimeter?

2. Two glass plates, separated by a thin wedge of air and illuminated from above, produce a system of interference fringes. How far must the upper plate be raised in order that 10 fringes may pass a certain point on the glass? The wavelength of the light is 6.23×10^{-5} cm or 6230 A.

3. In a diffraction grating the distance between adjacent apertures is 0.00080 cm.

What is the sine of the angle at which light of wavelength 6.0×10^{-5} cm can be diffracted and form an image if the difference of path for beams from adjacent slits is one wavelength?

4. A flat plate of glass is placed on another horizontal plate, but not parallel to it. The system is illuminated by a vertical beam of light of wavelength 4.0×10^{-5} cm. The distance from the first bright interference band to the eleventh one is 1.5 cm. How much farther apart are the plates at one end of the space than at the other?

5. Monochromatic light of wavelength 5500 A is incident normally on a grating having 4000 lines/cm. What is the angle of deviation, θ, for the first order, second order, third order?

6. How many lines per centimeter does a grating have if a first-order image of a 6560-A hydrogen line is formed at an angle θ of $10°10'$?

B

7. The two slits used in an interference experiment like that of Young were 0.020 cm apart. The screen was 100 cm from the slits. What was the distance from the midpoint of the central maximum band to the midpoint of the next band? ($\lambda = 6000$ A.)

8. What is the width of the slit in a Fraunhofer diffraction experiment if red light of wavelength 6500 A has its first minimum at an angle of 5° from the central maximum?

9. A film of olive oil lies on a glass plate. (a) What is the wavelength of red light (λ_{air}, 6500 A) in the oil film? (b) What minimum thickness of the film, in angstroms, will reflect red light (λ_{air}, 6500 A)? (c) What minimum thickness will reflect blue light (λ_{air}, 4000 A)? The index of refraction of olive oil is 1.47 at 6500 A and 1.49 at 4000 A.

10. A film of olive oil of variable thickness floating on water is illuminated by yellow sodium light incident normally. What is the thickness of the film at the eighth bright band from the point of zero thickness?

11. What is the distance from the central image to the first diffracted image when sodium yellow light passes through two narrow slits 0.080 cm apart and falls on a screen 150 cm away?

12. What is the distance between two lamps, 500 m away, that can barely be resolved by an opera glass of aperture 4.00 cm? Assume that λ equals 6.0×10^{-5} cm.

13. If your eyes can just resolve two light sources when the angle between them is 1 minute of arc, at what distance can you just barely resolve the two headlights of a car that are 165 cm apart?

C

14. If the opening in the pupil of a man's eye has a diameter of 0.30 cm, what is the angle subtended by the "circle of confusion"? Assume that λ is 5500 A.

15. At what maximum distance in meters could the eye in problem 14 examine a human hair of diameter 0.0050 cm?

16. A beam of monochromatic light is incident normally on a diffraction grating having 4000 lines/cm. The sines of the angles of diffraction for the different orders are 0.252, 0.501, and 0.755. (a) What is the average deviation from the mean of the values of $(\sin \theta)/n$? (b) Compute the average value of the wavelength of the light in angstroms.

17. What is the angular separation of two stars, whose images are barely resolved by a refracting telescope, if the objective has a diameter of 10 in. and a focal length of 120 in.? ($\lambda = 4000$ A.)

18. A student holds his handkerchief in a vertical plane and looks through it at the vertical filament of an incandescent lamp 3.00 m away. He observes that the two first-order diffracted images are located 0.5 cm away from the central image. (a) What is the average distance between the threads of the cloth? ($\lambda = 6000$ A.) (b) What is the distance between the third-order images?

19. Light of wavelength 6000 A from a star

passes normally through a slit 0.200 mm wide and is incident on a screen that is 20.0 cm distant. (*a*) What is the width of the central diffraction band? (*b*) What angle does it subtend at the slit?

20. What is the distance between adjacent apertures of a grating that deviates the red standard cadmium spectral line (first-order) through an angle of 0.523 radian?

21. What is the size of the smallest terrestrial object that could be seen from an altitude of 100 mi by a normal unaided eye?

22. What is the thickness of a film ($n = 1.38$) that causes a shift of 6.0 fringes of yellow sodium light when placed in the arm of a Michelson interferometer?

For Further Reading

Boys, C. V., *Soap Bubbles,* Doubleday and Company, Garden City, 1959.

Kock, Winston E., *Lasers and Holography,* Doubleday and Company, Garden City, 1969.

Kock, Winston E., *Sound Waves and Light Waves,* Doubleday and Company, Garden City, 1965.

Ruechardt, Eduard, *Light, Visible and Invisible,* University of Michigan Press, Ann Arbor, 1958.

Scientific American, Issue on Light, September, 1968.

CHAPTER THIRTY-EIGHT

Polarized Light

Diffraction and interference experiments yield evidence favoring the wave theory of light, but they give no hint as to whether the waves are transverse, like those in a stretched rope, or compressional, like sound waves. We find the answer to this question in the phenomenon of *polarization.* We have already mentioned the polarization of the electromagnetic waves emitted by an oscillating dipole (Chapter 32). Now we shall discuss ways in which light waves can be polarized.

What are Polarized Waves?

Make one end of a long horizontal rope vibrate in a vertical plane so that waves travel along it. Each particle of the rope oscillates up and down on a straight line. All these lines lie in the same vertical plane, and the waves are *plane polarized.* Make the rope vibrate up and down and sideways, at all possible angles. Each particle will also move in a random manner: the waves are *unpolarized.*

If you move the end of the rope in a circle, each particle of the rope will also move in a circle and the waves will be *circularly* polarized. If each particle moves in an ellipse, they are *elliptically* polarized.

Polarized waves are transverse waves in which each particle vibrates continually on the same line or path.

Pass the rope through a grating with its slots vertical (Fig. 38-1a). Vibrate the end up and down and sideways in random directions as before, causing unpolarized waves. The slot will prevent the sideways motion, but it will permit the vertical components of the vibrations to pass through. Waves polarized in a vertical plane will pass through the opening.

Place a second vertical grating beyond the first one so that their slots are parallel. Vertical vibrations will pass through both slots freely. Turn the second grating to a horizontal position and the waves will no longer be transmitted (Fig. 38-1b). Perform a similar experiment, replacing the cord by a long spiral spring, but set up compressional waves in it. In this experiment the waves will pass freely through both gratings whether the slots are parallel or at right angles to each other. An arrangement of slots can cut off transverse waves. It can*not* cut off those that are compressional (or longitudinal).

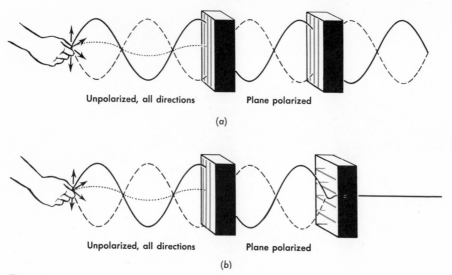

Unpolarized, all directions   Plane polarized

(a)

Unpolarized, all directions   Plane polarized

(b)

Fig. 38-1. *Plane polarized waves in a rope.* (a) *Those passing through the first grating can pass through the second one.* (b) *The second grating blocks the waves.*

Light Waves are Transverse

The following experiment supports the belief that light waves are transverse. Tourmaline, a cyclosilicate, is a semiprecious gem that can be polished to form flat crystals. It has interesting optical properties. Look through two tourmaline crystals, having their longer sides parallel (Fig. 38-2a), at a lamp. Then light is transmitted. Rotate the second crystal through 90° about the light beam as an axis (Fig. 38-2b). The light transmitted by the first crystal will not pass through the second one. The combination will be opaque.

To interpret this experiment, we must assume that light waves from the atoms in the source are a mixture of electromagnetic waves with the electric field vectors pointing in random directions. Thus, the light incident on the first crystal is unpolarized; its vibrations are directed up and down and sideways in random directions. The tourmaline crystal has the remarkable property of absorbing all the sideways components of the vibrations and transmitting only plane-polarized light,

that is, light whose electric field vector is in the vertical direction. This plane-polarized light passes through the second crystal when it is vertical, but is absorbed when the crystal

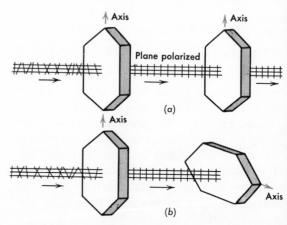

Fig. 38-2. *Tourmaline crystals. Light polarized by one crystal is* (a) *transmitted and* (b) *absorbed by the other.*

Fig. 38-3. *Two crossed polarizers are opaque.*

is rotated 90°. If the waves were compressional, like sound, they would pass through in both cases.

Light may also be polarized by a sheet of Polaroid linear polarizer. It consists of a plastic film of polyvinyl alcohol which is first stretched and then saturated with iodine. The iodine atoms line up in long strings parallel to the direction of stretching. Two sheet polarizers whose axes are parallel transmit light. When they are crossed, with axes at right angles, they are opaque (Fig. 38-3).

Double Refraction

When light is incident obliquely on a block of glass, part of the light is refracted in a single beam. When incident on a crystal of Iceland spar (calcium carbonate), the light is *doubly*

refracted so that the beam entering the crystal divides into two beams (Fig. 38-4). If you look through the crystal at an illuminated pinhole, you will see two images. Examine the two beams through a polarizer, and rotate the polarizer about them. When the axis of the polarizer is vertical, the beam O is extinguished. When it is horizontal, the beam E disappears and only O is seen. This proves that the light in each beam is plane polarized and that in the two beams the planes of vibration are at right angles to each other. Place a block of Iceland spar on a screen in which there is an illuminated pinhole. When you rotate the crystal, the ordinary beam forms an image of the pinhole which remains fixed in position as though it were viewed through a plate of glass. The image formed by the extraordinary beam revolves in a circle.

The fact that the beams are refracted at different angles indicates that the light in the two travels at different speeds, for the Iceland spar evidently has different refractive indices for the two beams. An analogy will aid in explaining why the speeds are unequal. You can cut a piece of wood more easily parallel to the grain than across the grain. It can be compressed more easily in one direction than the other. If it were transparent, like glass, we should expect light waves vibrating parallel to the grain to travel at a speed different from that of the waves vibrating at right angles to it.

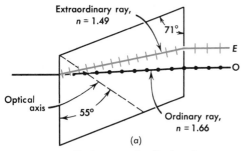
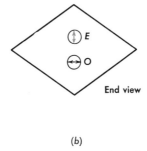

Fig. 38-4. *Refraction by Iceland spar crystal. The two beams travel at different speeds (unless incident along the optical axis). The extraordinary beam vibrates in the plane of the page.*

A piece of Iceland spar—or of other optically active materials, such as quartz—resembles a piece of wood in that it has different elastic properties in different directions. It is not surprising, therefore, that light waves vibrating in one direction may have a different speed from those vibrating at right angles to that direction. The index of refraction of Iceland spar is 1.486 for the extraordinary ray and 1.658 for the ordinary ray. There is one direction, called the *optical axis,* along which the two components travel at equal speeds and therefore are not doubly refracted.

The Nicol prism, based on the double refraction of Iceland spar, is sometimes employed for producing and detecting polarized light. It is constructed by sawing a crystal of Iceland spar into two wedge-shaped pieces which are glued together with a transparent cement called *Canada balsam* (Fig. 38-5). Ordinary unpolarized light, incident on the crystal, is doubly refracted, and the two beams strike the film at different angles. The index of refraction of Canada balsam is 1.530 and is intermediate between the indices of Iceland spar for the ordinary ray and the extraordinary ray. The lower beam (extraordinary ray), vibrating in the plane of the paper, is *incident* at an angle smaller than the critical angle of the crystal-balsam surface and is mostly transmitted. The upper (ordinary) beam, vibrating at right angles to the plane of the paper, is incident at an angle *greater* than the critical angle of the surface. Hence it is totally reflected. The transmitted beam is completely polarized.

Polarization by Reflection

Polarized light can be produced by reflection. Illuminate a glass plate or a water surface, and, using a sheet polarizer, examine the light reflected at an oblique angle. Rotating the polarizer will cause the intensity of the transmitted beam to vary; hence the light must be partly polarized. Change the angle of reflection until, by adjusting the polarizer, the reflected beam can be cut off entirely. In this condition, all of the reflected light is plane polarized (Fig. 38-6) in a plane parallel to the surface of the reflector. The refracted beam is partly polarized in a plane at right angles to that of the reflected light.

When all of the reflected light is plane polarized, the tangent of the angle of incidence is equal to the index of refraction of the medium (Brewster's law). For water, the polarizing angle of incidence is 53°. For common (crown) glass, it is 57°.

Applications of Polarized Light

Chemists use polarized light in a saccharimeter to measure concentrations of sugar solutions. Sugar, like quinine, quartz, and other substances, rotates the plane of vibration of light. Cross two polarizers so that they cut off the light from a sodium lamp. Interpose between them a flask containing a solution of ordinary cane sugar, and the illumination will be restored. The sugar molecules have rotated the plane of vibration of the light through some angle, for instance 30°. The second polarizer may be turned through this angle, again cutting off the light. Suppose that the flask is replaced by a similar one containing a less-concentrated solution which rotates the plane of vibration through only 15°.

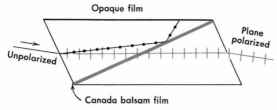

Fig. 38-5. Diagram of a Nicol prism. The extraordinary ray, incident at an angle less than the critical angle, is mostly transmitted. The ordinary ray, incident at an angle greater than the critical angle, is totally reflected.

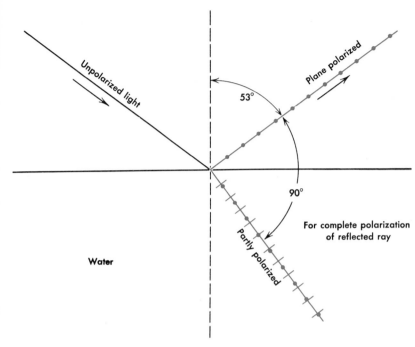

Fig. 38-6. Light reflected from water at a certain angle (53°) is plane polarized.

Then the concentration of sugar is half as great as before. The angle of rotation θ can be calculated by the formula

$$\theta = \alpha l c$$

in which α is a constant characteristic of the kind of sugar, l is the length of the path of light in the solution, and c is the concentration of the sugar in solution.

Fishermen and lifeguards find polarizing glasses helpful in absorbing the glaring light polarized by reflection from the surface of the water of a lake or ocean. Automobile drivers use them to absorb sunlight reflected from pavements.

The Polarization of Scattered Light

A beam of sunlight admitted to a darkened room reveals small motes or dust particles which scatter the light sidewise. Tobacco smoke, blown into the air, increases the number of particles and therefore the scattering.

The color of the light that is scattered with greatest intensity depends upon the sizes of the particles. Very small particles scatter blue light, of shorter wavelength, more than red light, of longer wavelength. Lord Rayleigh was able to show that the intensity of light scattered by small particles depends inversely upon the *fourth* power of the wavelength. The smoke from a cigarette ordinarily is blue because the particles are so small that they scatter blue light more than the other parts of the spectrum. The sky is blue because the air molecules and other particles scatter blue light much more than red. Red sunsets are seen because the clouds in the west through which the sunlight passes scatter the blue light sideways and transmit only red light. The eruption of the volcano

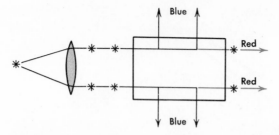

Fig. 38-7. Scattering of light by small particles suspended in water. The blue light is scattered sidewise, and the red light is transmitted. The scattered light is plane polarized even when the incident and transmitted light are unpolarized.

Krakatau in 1883 threw dust into the atmosphere that gave rise to gorgeous sunsets all over the world for many months.

Particles that are large compared with the wavelength scatter light of all colors about equally. Cigarette smoke that has been inhaled appears white because the particles have grown larger by deposition of water vapor. The whiteness of the lily and of seafoam and of a well-known brand of soap is due to the same cause. Cumulus clouds are white because large water droplets scatter light of all colors equally.

To demonstrate the dependence of scattering on particle size, project a beam of light, simulating sunlight, through a vessel containing a water solution of photographer's hypo (Fig. 38-7). Add a little concentrated sulfuric acid, and stir. After a minute or so, particles, precipitated in the water, will scatter blue light sidewise. The patch of transmitted light on the screen will be red, like the setting sun. Later, the precipitate will become so dense that all the light will be shut off, blotting out the "sun."

Light scattered by small particles is plane polarized. One can easily demonstrate this by rotating a sheet polarizer in the scattered beam (Fig. 38-7). To understand this, recall that electromagnetic waves are scattered by small particles because the waves induce oscillations in the bound electrons. A vibrating electron acts like an electric dipole, radiating at greatest intensity along a line perpendicular to its line of vibration. Thus the light scattered at right angles to the incident beam will have its electric-field vector at right angles to *both* the direction of the incident light and to a line at right angles to that direction.

A Crystal Restores Illumination

A piece of doubly refracting material such as mica, inserted between crossed polarizers, restores the illumination. In Fig. 38-8, a vertical beam of light, polarized in a northeast-southwest plane, is doubly refracted by the

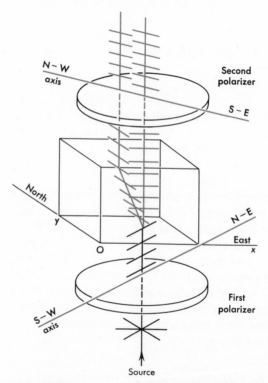

Fig. 38-8. A crystal between two crossed polarizers doubly refracts the incident beam into beams vibrating in the planes Oy and Ox. The second polarizer transmits part of each beam.

crystal. The vibrations in the transmitted beams are parallel to *Ox* (in an east-west plane) and to *Oy* (in a north-south plane). Light in the two beams travels at slightly different speeds in the crystal; hence the vibrations that emerge from the crystal and are transmitted by the second polarizer may be in phase or out of phase with each other. Whether they are in phase or out of phase depends upon the thickness of the crystal and the difference in the indices of refraction of the crystal for the ordinary and extraordinary beams. If they are in phase, the resulting illumination will be a maximum; if they are out of phase, it will be zero. Thus, the crystal may or may not restore the illumination. The relative speeds of the two beams depend upon frequency so that if light of certain frequencies is extinguished, light of other frequencies is transmitted. The absorption or transmission of the light depends upon both its frequency (or wavelength in vacuum) and the thickness of the crystal. Therefore a crystal of varying thickness interposed between polarizers exhibits different colors in different parts of its surface.

Practical application of the restoration of illumination by a crystal is made in the testing of glass. If a glass dish, for example, has been properly annealed, it will have small internal strains and will have the same optical properties in all directions. It will not restore the illumination when placed between crossed polarizers. If the glass is insufficiently annealed, it will be strained, will cause double refraction like a crystal, and therefore will restore the illumination. Engineers study stress patterns in structures by constructing plastic models and subjecting them to loads when placed between crossed polarizers.

Summary

A plane-polarized wave is a transverse wave train in which all the vibrations are in a single plane.

In ordinary light the electric-field vectors at any point are in random directions.

Plane-polarized light may be produced by passing ordinary light through a tourmaline crystal or a sheet polarizer, or by reflection from a surface.

If light from a sheet polarizer is incident on a second polarizer, the beam can be cut off by setting the second one with its axis at right angles to the first. This fact proves that a polarizing filter transmits light vibrating in one plane only, and that light waves are transverse.

A beam of light incident on the surface of a crystal may be doubly refracted into two beams that travel at different speeds in the crystal.

A sheet of doubly refracting substance placed between crossed polarizers may restore the illumination.

Questions

1. Can sound waves be polarized? Explain.

2. Explain the blueness of the sky, the redness of the sunset, and the whiteness of some clouds.

3. What is the common property of a tourmaline crystal, a sheet polarizer, and a Nicol prism that enables them to polarize light?

4. Cite evidence indicating that light waves are transverse.

5. Describe several means by which polarized light can be produced and detected.

6. Are (a) ripples, (b) deep-water waves polarized?

7. Can a Nicol prism transmit more than 50% of the incident light? Why?

8. Is the light from the sky polarized? How could you prove it?

9. How could you prove that TV waves are polarized?

10. How could you determine the direction of the optical axis of a doubly refracting crystal?

11. When light is doubly refracted into an ordinary ray and an extraordinary ray, which ray obeys Snell's Law?

12. What color is seen when two polarizing sheets are crossed? Why?

13. Two polarizing sheets, placed 10 cm apart, transmit no light to an observer behind them. If a third polarizer is placed between them, light is then transmitted to the observer. For what angle between the axis of the first polarizer and the axis of the new polarizer would you expect the maximum amount of light to be transmitted to the observer?

14. When a crystal is interposed between crossed polarizers and the second polarizer is rotated 90°, the transmitted light changes in hue to the complementary color. Explain.

Problems

A

1. What is the index of refraction of a medium for light for which the polarizing angle of incidence is 50°?

2. What is the polarizing angle for sodium-yellow light incident on (a) diamond, (b) ethyl alcohol? (See Table 33-1.)

3. What is the polarizing angle for sodium-yellow light incident on (a) carbon disulfide, (b) crown glass? Assume that n for crown glass is 1.515.

4. Calculate the critical angle for total reflection of the ordinary ray at the Canada-balsam layer in a Nicol prism. ($\lambda = 5890$ A.)

B

5. For yellow light of wavelength 5890 A, the index of refraction of the ordinary ray in Iceland spar is 1.658. For the extraordinary ray, it is 1.486. What are the critical angles for the two rays if incident on a crystal-air surface?

6. Calculate the wavelength of 5890-A sodium light for the ordinary ray in calcite.

7. A 10-cm tube contains dextrose with a concentration of 0.15 gm/cm^3. Through what angle will this solution rotate polarized light of wavelength 5890 A? (For dextrose, $\alpha = 5.25°$/cm-unit concentration.)

8. A sugar solution in a tube 20 cm long contains sucrose. If polarized light of wavelength 5890 A is rotated 13.3°, what is the concentration of the sugar solution? (For sucrose, $\alpha = 6.64°$/cm-unit concentration.)

9. What is the length of a tube containing a 0.10-gm/cm^3 solution of sucrose if polarized light of wavelength 5890 A is rotated 3.32°? (For sucrose, $\alpha = 6.64°$/cm-unit concentration.)

C

10. The indices of refraction of quartz for yellow light of wavelength 5890 A are: for the ordinary ray, 1.544; for the extraordinary ray, 1.553. What are the two wavelengths of the yellow light in quartz?

11. The index of refraction of the ordinary ray in quartz for light of wavelength 5890 A is 1.544 and that for the extraordinary ray is 1.553. What thickness of quartz will retard the extraordinary ray one-half wavelength more than the ordinary ray so as to introduce a phase difference of 180°?

12. What are the indices of refraction for the ordinary and extraordinary rays in Iceland spar if red cadmium light (6438 A) has wavelengths of 3890 A and 4330 A in it?

13. What minimum thickness of calcite will produce destructive interference for sodium-yellow light if the plate is placed between two crossed Nicol prisms?

14. What thickness of mica ($n = 1.605$ and 1.612) is needed to form a quarter-wave plate (one that introduces a phase difference of 90° between the ordinary and extraordinary rays) for light at 5890 A?

15. Plot the polarizing angle of incidence i versus the index of refraction n (Brewster's law):

i	n
52.4°	1.30
54.5°	1.40
56.3°	1.50
58.0°	1.60
59.5°	1.70

16. Plot the polarizing angle versus wavelength for light falling on a quartz surface, using the following data:

λ	n
6560 A	1.456
5890	1.458
5090	1.462
3610	1.475

17. A 60° prism is made from Iceland spar in such a way that the optical axis is parallel to the refracting edge. Determine the separation of the ordinary and extraordinary rays when they emerge from the prism if the ordinary ray undergoes minimum deviation.

For Further Reading

LaMer, V. K., and M. Kerker, "Light Scattered by Particles," *Scientific American,* February, 1953, p. 69.

Shurcliff, W. A. and Stanley S. Ballard, *Polarized Light,* D. Van Nostrand, Princeton, 1964.

Wood, Elizabeth A., *Crystals and Light,* D. Van Nostrand, Princeton, 1964.

Atomic
and
Nuclear
Physics

CHAPTER THIRTY-NINE

The Photon

In the short period of forty years, roughly from 1890 to 1930, physics underwent a profound change. This change is still going on in contemporary physics. Better apparatus and sharper experimental techniques allowed physicists to penetrate new realms of nature. Many discoveries were made about the atom, the molecule, and radiation. Revolutionary theories emerged to unify this new knowledge. The philosophical outlook of most physicists changed also. They were less ready to accept a mechanical universe—one somewhat resembling a clock whose "gears" were expected to operate in accord with Newtonian mechanics. They began to question familiar concepts and adopted a wary attitude toward what was clearly a mysterious universe. Familiar concepts—space, time, position, causality—broke down when one tried to apply them in the realms of nature that the physicists were then entering. As for the universe, it was indeed a mysterious one. The only prediction one could make about it with any confidence was that it held stunning surprises for the scientist. Things changed rapidly in physics during those years; the period is sometimes called the Second Scientific Revolution, the first being that launched by Galileo and Newton.

The remainder of this book will deal with some of the changes that occurred during the transition to contemporary physics. We shall introduce you to new experimental knowledge and to new theories about the atom and its constituent particles, about radiation, and about physical systems containing these things. We cannot treat the whole of modern physics in a few chapters, of course. We shall concentrate on things that are basic—particularly the philosophy with which physicists approach their problems these days. Many details and whole areas important in contemporary physics will be left to your later reading.

Many of these ideas will at first seem strange to you or at least inconsistent with what you have already learned. We shall consider, for example, the problem of specifying the position and velocity of an electron. We find that we cannot say with precision *where* an electron is in an atom without necessarily becoming quite vague about its *velocity*. This means that we cannot construct a time table for electrons in an atom as we can for planets in the solar system. Sharply defined orbits for electrons are ruled out. We are reduced to making statements like this: "It is rather probable that the electron of the unexcited hydrogen atom will be found 0.5 angstrom from the nucleus, but there is a slight probability that it could be found within *any* tiny volume that you care to specify." Clearly this is a different situation from that in astronomy, where the astronomer knows with great precision where the planets are at any instant and how fast they are moving. But the physicist—and you as a student of physics—must live with this indefiniteness as the price of knowledge of the atomic world.

The ideas of physics you have already encountered generally apply in modern physics, but many of them must be reinterpreted. We shall still use the principles of conservation of energy and momentum, for example, and the concepts of velocity, mass, position, and time. However, our literal everyday interpretation of these concepts often fails us in the atomic world. Be on your guard against assuming that principles derived from observations on falling bodies and colliding billiard balls, for example, can be applied without change to electrons and other atomic particles. Many of the principles we shall consider are familiar ones, but are extended or reinterpreted to apply to the atomic world. Other principles will be wholly new. Sometimes they will seem to be contradictory to what "common sense" leads us to believe. Remember, however, that "common sense" is largely based on what we observe with our senses in the world of everyday objects around us. Beyond the range of our senses—accessible only to instruments of greater range and sensitivity—is the atomic world of the very small and the world of vast astronomical space. We must be prepared for surprises when we investigate these worlds.

Black-body Radiation

To begin our study of modern physics we return to an old question: "Is light composed of waves or corpuscles?" The experiments and theories of Young, Fresnel, Foucault, and others on interference and diffraction early in the nineteenth century led physicists to decide in favor of the wave theory. But surprises were in store for them, beginning with a revolutionary new interpretation of the process of radiation by a black body. Let us review some of the facts about black-body radiation.

In Chapter 14 and again in Chapter 36 you learned that black-body radiation is the electromagnetic radiation that exists inside an enclosure or cavity whose walls are at some uniform temperature above 0°K. In order to study the properties of such radiation, we make a tiny hole into the cavity so that a small amount of radiation—not enough to upset appreciably the thermal equilibrium of the interior—can reach the outside. The cavity with the tiny hole constitutes a *black-body source* of radiation. Such a source has a maximum emissivity ($e = 1$) and a maximum absorptivity ($a = 1$), according to Kirchhoff's law. The total radiant energy per unit time emitted by it is directly proportional to the fourth power of the Kelvin temperature of the cavity (Stefan-Boltzmann law). Black-body radiation forms a continuous spectrum. The wavelength at which the radiation has maximum intensity shifts toward shorter wavelengths as the temperature of the black-body source increases—the radiation gets "bluer" as the temperature rises. The way in which the energy is distributed among the wavelengths of a black-body spectrum is shown by the graph of spectral intensity (the intensity associated with a tiny band of wavelengths) versus wavelength (Fig. 36-7).

As we have repeatedly emphasized, the physicist wants to "explain" knowledge by relating it to basic and general concepts. The spectral energy distribution and the other properties of a black body were facts requiring explanation by nineteenth-century physicists. Black-body properties had to be explained in terms of the basic concepts of heat and thermodynamics, electricity and magnetism, and optics. (You remember that in the second half of the nineteenth century, optics began to be considered a part of electricity and magnetism.) For a while, all went well. Treating the radiation in a cavity as the working substance in a Carnot engine, Boltzmann was able to show from the principles of thermodynamics that the total energy emitted per unit time depended upon T^4. Also using thermodynamics, Wien showed that the wavelength of maximum emission depended inversely upon the Kelvin temperature ($\lambda_{max\,I}\,T = k$). But classical thermodynamics was *not* able to account satisfactorily for the shape of the spectral distribution curve. One equation, derived by Rayleigh and Jeans, fitted the experimental curve approximately at large wavelengths, but failed at small wavelengths (Fig. 39-1). Another, due to Wien, fitted the curve quite well everywhere *except* at large wavelengths. Classical assumptions seemed unable to cope adequately with black-body radiation.

Planck's Theory

The classical approach to black-body radiation had assumed that the electromagnetic radiation in a cavity was emitted and absorbed in a continuous process by the countless tiny micro-oscillators that constituted the walls of the cavity. An oscillator dribbled energy away in the form of an electromagnetic wave. At some later time, it absorbed a wave, storing up energy like a tank being filled with water. Max Planck, in a paper published in 1901, questioned this assumption. He asked whether it was not possible for a tiny oscillator to emit energy in discrete amounts—in packages of radiant energy, or *quanta.* The quanta were assumed to have energy *proportional to the frequency of the radiation.* The proportionality constant, which he designated as *h* and which soon acquired the name of *Planck's constant,* was valid for quanta of all frequencies. Planck's constant has the value $h = 6.625 \times 10^{-34}$ joule-sec.

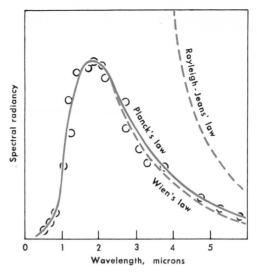

Fig. 39-1. *The spectral-energy distribution curve for a black body at 1600°K. At small wavelengths, Wien's equation fits the experimental points (circles), but the Rayleigh-Jeans equation does not. At large wavelengths (not shown), the Rayleigh-Jeans equation fits better than the Wien equation. Planck's equation fits well over the entire spectrum.* (O. H. Blackwood, T. H. Osgood, A. E. Ruark, et al., An Outline of Atomic Physics, *Third Edition, John Wiley and Sons, 1955*).

Here is Planck's picture of the black-body radiation process. Assume (for simplicity) that a single oscillator in the cavity walls can have one of two energies, E_A or E_B, which determines the amplitudes of its oscillations. It can exist upon one of two *energy levels* (Fig. 39-2), of which level *B* is of greater energy than level *A*. To move from the energy state represented by *A* to that represented by *B*, the oscillator must absorb additional energy represented by the energy difference $E_B - E_A$. To return from *B* to *A*, the oscillator must give up this much energy. Unlike a baseball falling from a shelf to the floor and giving up potential energy in a continuous process, the oscillator in Planck's theory gives up energy in a

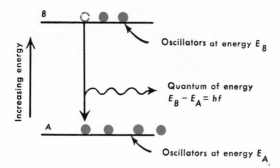

Fig. 39-2. A two-level energy level diagram, illustrating Planck's hypothesis.

single sharp burst by emitting a *quantum* or bundle of electromagnetic energy equal to $E_B - E_A$. Indeed one must imagine that the oscillator exists *only* on energy level *A or* on level *B*, but not at an energy in between. An energy of 0.263 times $E_B - E_A$, for example, is not an allowable energy. The theory says nothing about what goes on during the transition from *A* to *B* or *B* to *A*, only that the transition can occur. Planck, in effect, asked his colleagues to allow him to make this drastic assumption and to see where it would lead. The assumption was completely unlike the assumption that one encounters in classical—that is, pre-quantum—physics.

The quantum of energy emitted has a frequency *f* that can be calculated from the energy difference $E_B - E_A$, Planck assumed, by the equation*

$$E_B - E_A = hf$$

Planck proceeded next to calculate the spectral energy distribution of black-body radiation. He had to consider the huge number of oscillators in a cavity as the oscillators made energy "jumps" downward from their higher energy levels to lower ones. (Planck did not claim that the quanta, once emitted, traveled through space as particles

---

* The frequency is often represented by the Greek letter ν in this equation. We shall continue to use *f* for frequency.

of light; that development, as we shall see, occurred later.) He computed for a given temperature how many oscillators were likely to be on a given level. Energy changes of the micro-oscillators could only be whole-number multiples of *hf*.

The situation is somewhat similar to the distribution of velocities among the molecules of a gas—the Maxwell-Boltzmann distribution that we considered in Chapter 15. There we had a continuous distribution of velocities from zero velocity to very large velocities. A certain molecule *A* might make a favorable collision with another molecule *B* and have its kinetic energy and velocity increased. Upon colliding with a third molecule *C*, molecule *A* might give up energy to *C* and drop back to a lower velocity. Similarly, one of the atomic oscillators could receive energy from the cavity and be boosted to a higher energy, dropping back when it radiated. The difference—and it is an essential one—is that in the classical Maxwell-Boltzmann situation a molecule could receive or give up energy in *any* amount, depending upon the dynamics of the collision. In the Planck theory, the only changes of energy allowed were in units of *hf*.

The results of Planck's theoretical calculation are shown as the solid curve in Fig. 39-1. Notice that where the Rayleigh-Jeans and Wien curves, based on classical assumptions, failed to fit the experimental points, Planck's curve fits them very well. Moreover, the Planck theory accounted for the total energy emission and the shift of the curve as the temperature changed. By giving up the idea that physical processes always had to occur continuously—in arbitrarily small steps—and by introducing in its place the idea of the "quantum jump," Planck had not only explained black-body radiation, but had set modern physics upon a new course of development.

Example. Within a narrow band of wavelengths whose average wavelength is

Table 39-1.

Analogy	BB Shot	22-Short Bullet	38-Caliber Bullet	2-Pound Shell	20-Pound Shell
Region of spectrum	Infrared	Visible	Ultraviolet	X-ray	Gamma ray
Wavelength of quantum	12,340 A	$\frac{1}{2}$(12,340 A)	$\frac{1}{6}$(12,340 A)	10^{-4}(12,340 A)	10^{-6}(12,340 A)
Energy of quantum	1 ev	2 ev	6 ev	10,000 ev	1,000,000 ev
Principal effects	Heat	Photographic Sight Heat	Fluorescent Chemical Photographic	Ionization Fluorescent Photographic	Ionization Fluorescent Photographic

6000 A, a black-body source emits energy at the rate of 10^{-10} watt. (*a*) What is the energy of a quantum at that wavelength? (*b*) At what rate are quanta emitted at that wavelength?

(*a*) For a quantum of wavelength λ and frequency *f*

$$E = hf = h\frac{c}{\lambda}$$

$$E = 6.6 \times 10^{-34} \text{ joule-sec} \frac{3.0 \times 10^8 \text{ m/sec}}{6.0 \times 10^{-7} \text{ m}}$$

$$= 3.3 \times 10^{-19} \text{ joule}$$
(the energy of a single quantum)

This tiny amount of energy would lift a mosquito vertically a few million millionths of a centimeter.

(*b*) $\dfrac{10^{-10} \text{ joule/sec}}{3.3 \times 10^{-19} \text{ joule/quantum}}$

$$= 3 \times 10^8 \text{ quanta/sec}$$

The energy* of an infrared quantum of wavelength 12,340 A is 1 ev, that of an orange-light quantum of wavelength 6150 A is 2 ev, and that of an X-ray quantum of wavelength 1.23 A is 10,000 ev. An X-ray quantum having a very short wavelength has thousands of times as much energy as a quantum of light. If we think of a quantum of red light as a

---
* It is often convenient to express the energy of a quantum of known wavelength in electron volts. Since

$$E = hf = h\frac{c}{\lambda}$$

$$E = 12,340 \text{ A-electron-volts } \frac{1}{\lambda}$$

speeding 22-short rifle bullet, then an X-ray quantum would be a 2-pound shell (Fig. 39-3 and Table 39-1).

The Photoelectric Effect

Physicists eventually discovered that one must assume not only that radiant energy is emitted as quanta, but that quanta are propagated through space and are absorbed as "packets" of energy. The discovery and explanation of *photoelectricity* led to this further understanding of the nature of quanta. Photoelectric emission is the process by which electromagnetic radiation liberates electrons from surfaces.

The photoelectric effect was first noticed by Hertz in 1887 during the course of his experiments with the first simple radio transmitter. He observed that a spark jumped more readily between the terminals of his high-voltage source when they were irradiated by ultraviolet rays. It was soon shown that negatively charged particles were emitted by some metals under irradiation. The charge-to-mass ratio of the particles was that of electrons, and physicists soon agreed that the particles were indeed electrons.

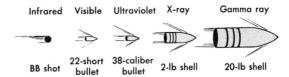

Infrared	Visible	Ultraviolet	X-ray	Gamma ray
BB shot	22-short bullet	38-caliber bullet	2-lb shell	20-lb shell

Fig. 39-3. Projectiles illustrate the relative energies of quanta.

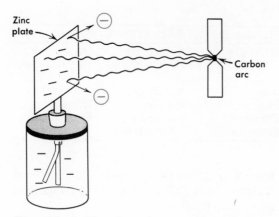

Fig. 39-4. The photoelectric effect. Quanta, striking the negatively charged zinc plate, eject electrons.

To demonstrate the photoelectric effect by an experiment due to Hallwachs, attach a zinc plate, cleaned with sandpaper, to the knob of an electroscope and charge it negatively. Illuminate the plate with light and ultraviolet rays from a naked carbon arc (Fig. 39-4). The radiations will liberate excess electrons from the zinc, so as gradually to discharge the electroscope. The plate will not be discharged if it is charged positively. If the electroscope is initially uncharged, it acquires a positive charge when it is irradiated.

Some Experimental Results Concerning Photoelectricity

A number of facts about the photoelectric effect were established by painstaking experiments by Millikan and other physicists. In these investigations, different metallic surfaces were irradiated with electromagnetic radiations of different wavelengths, and the emission of photoelectrons was studied. Photoelectrons can be collected by a positively charged collector or anode (Fig. 39-5). The photoelectric current is measured by an ammeter A—generally an exceedingly sensitive ammeter, called an electrometer, that can

measure currents as small as 10^{-12} amp. The chief results of such experiments are these:

1. At a given frequency, the photoelectric current is accurately proportional to the illumination of the photosensitive surface over a very wide range of illuminations. Moreover, *the maximum kinetic energy with which the photoelectrons leave the surface is entirely independent of the illumination.*
2. Photoelectrons are emitted by most metals *provided that the frequency of the radiation exceeds a certain critical threshold frequency.* The threshold frequency for the metals sodium and potassium lies in the visible region of the spectrum.
3. Any time delay between the turning on of the source and the appearance of the photoelectric current is extremely short, less than a few billionths of a second.
4. The kinetic energy of the photoelectrons emitted ranges from zero to a maximum. *If the maximum kinetic energy is plotted against the frequency of the radiations, a straight-line graph results.*

Observation 4 was obtained by measuring the *stopping potential* of photoelectrons for light of known wavelengths. The stopping potential is the *negative* collector-to-cathode voltage that just barely stops all photoelec-

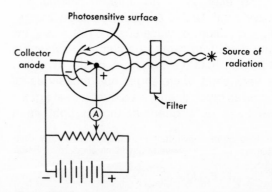

Fig. 39-5. An apparatus for studying the photoelectric effect.

trons emitted by those radiations. As one reduces the normally positive voltage between the anode and the photosensitive surface, approaching zero voltage, the photoelectric current decreases. If the voltage is then made increasingly negative, a voltage is reached at which the current is just barely detectable by the electrometer. This voltage is the stopping potential. Figure 39-6 shows a graph of the stopping potential versus frequency for a given photosurface. Since the most energetic photoelectrons are just stopped by this negative voltage, the kinetic energy with which they leave the photosurface just equals the work they do in moving against the opposing electric field. In equation form:

$$\tfrac{1}{2}mv_{max}^{2} = V_{s}e$$

where *m* and *e* are the mass and charge of the electron, v_{max} is the maximum velocity with which photoelectrons are emitted, and V_{s} is the stopping potential for that frequency.

Einstein's Interpretation of the Photoelectric Effect

How are we to interpret these results? In particular, do they require the wave theory of light or the quantum theory for their correct interpretation? In 1905, even before Millikan's precise data were available, Albert Einstein showed that the quantum theory gave the only satisfactory interpretation of the photoelectric effect. Einstein assumed that the quanta of a given frequency (Notice that we are still talking about frequency—a wave property!) traveled from the source to the photosurface and interacted in some fashion with the metal, liberating electrons. The photoelectrons could escape from the metal, if sufficient energy were given to them by the quanta. The energy given up by a single quantum—according to Planck's hypotheses—was *hf*. Hence, if we let ω_0 represent the energy lost by a photoelectron in escaping through the

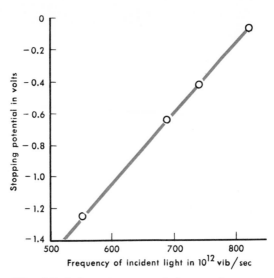

Fig. 39-6. Stopping potential versus frequency of incident light for a photosensitive surface, uncorrected for contact potential. (*Courtesy of A. Capecelatro and M. Mainardi.*)

surface of the metal—a quantity usually called the **work function**—the maximum kinetic energy with which the photoelectron emerges is

$$\tfrac{1}{2}mv_{max}^{2} = hf - \omega_0$$

Why then are photoelectrons with smaller velocities observed? Some photoelectrons are liberated beneath the surface of the metal—at a depth of a few atomic layers, for example. They lose greater amounts of energy than ω_0 in leaving the metal, emerging finally with less than the maximum kinetic energy given by the equation above.

Example. Quanta of wavelength 6000 A strike the surface of a metal whose work function is 1.0 ev. What is the greatest kinetic energy that a photoelectron can have?

$$E = h\frac{c}{\lambda} = (6.6 \times 10^{-34} \text{ joule-sec})$$

$$\left(\frac{3.0 \times 10^{8} \text{ m/sec}}{6.0 \times 10^{-7} \text{ m}}\right)$$

$$= 3.3 \times 10^{-19} \text{ joule} \quad \text{(the energy of quanta at 6000 A)}$$

$$\tfrac{1}{2}mv_{\max}^2 = E - \omega_0$$

$$= 3.3 \times 10^{-19} \text{ joule}$$

$$- (1.0 \text{ ev})(1.6 \times 10^{-19} \text{ joule/ev})$$

$$= 3.3 \times 10^{-19} \text{ joule}$$

$$- 1.6 \times 10^{-19} \text{ joule}$$

$$= 1.7 \times 10^{-19} \text{ joule} \quad \text{(the maximum kinetic energy of the photoelectrons)}$$

Let us see now whether Einstein's interpretation satisfies the experimental facts given above.

1. Since a quantum gives up its energy in one sudden event, for every quantum that disappears *in the photoelectric effect* one photoelectron appears. To be sure, not every quantum liberates a photoelectron; that is, the efficiency of the photoelectric process is much less than one photoelectron per incident quantum. However, we would expect the number of photoelectrons that are collected per unit time to be proportional to the rate at which quanta fall on the metal, that is, to the illumination at a given frequency. This we have seen to be experimentally correct. Moreover, since the maximum kinetic energy of the photoelectrons depends only upon the frequency of the light and the work function and *not* upon the rate at which quanta strike the surface, we must again conclude that a photoelectron absorbs only *one* quantum — and not several — before being liberated.

2. According to Einstein's equation, no photoelectrons will be emitted unless hf is greater than ω_0. (Why?) This tells us that we must expect a threshold frequency for the photoelectric effect, below which no photoelectrons will be emitted. Since ω_0 depends upon the nature and condition of the metal, we would expect the threshold frequency to be lower for some metals than others.

3. According to the *classical* wave theory of light, energy is delivered to a metal surface by electromagnetic waves in a continuous process, the metal "soaking up" energy. One can calculate how long it would take an electron to absorb enough energy by this non-quantum process to escape as a photoelectron; it turns out to be not less than about a minute. Delays as long as this are not observed in the photoelectric effect. We must conclude that radiant energy is absorbed as quanta, and the photoelectric current is established in the very brief time that it takes the quanta to travel from the source to the metal and for the photoelectrons to reach the collector.

4. One can readily see that Einstein's equation $\tfrac{1}{2}mv_{\max}^2 = hf - \omega_0$ fits the straight-line experimental plot of stopping potential V_s versus frequency (Fig. 39-6). Since

$$V_s e = \tfrac{1}{2}mv_{\max}^2$$

Einstein's equation becomes

$$V_s e = hf - \omega_0$$

or

$$V_s = \frac{h}{e} f - \frac{\omega_0}{e}$$

Here h/e, the ratio of Planck's constant to the charge on the electron, is the *slope* of the line in Fig. 39-6, and ω_0/e, the ratio of the work function to the electronic charge, is the intercept of the line on the vertical axis. We know e from Millikan's oil droplet experiment. To compare theory with experiment, we can obtain h/e from this experimental graph. We find that it yields values of h in good agreement with those obtained from the black-body spectral curve. The values of ω_0 obtained from the photoelectric curve are consistent with values of the work function obtained from studies of the thermionic effect—a related process, as we shall see in Chapter 44. Some experimental values of the work function for different metals are given in Table 39-2.

Table 39-2. Photoelectric Work Functions (in Electron Volts)

Aluminum	4.20
Cadmium	4.10
Calcium	2.71
Palladium	4.97
Potassium	2.24
Silver	4.46
Sodium	2.28
Strontium	2.74

These explanations of the photoelectric effect convinced physicists that radiant energy was not only emitted—in black-body radiation—as quanta, but that quanta traveled through space and interacted with matter as packets of energy. We shall now call the quanta *photons* and go on in the following chapters to consider other situations in which photons are produced, are absorbed, and interact with matter.

Applications of Photoelectricity

There are many technological applications of the photoelectric effect. *Photocells* are evacuated tubes, similar to that shown in Fig. 39-5, in which the negative electrode is coated with potassium, cesium, or some combination of these elements in a light-sensitive layer. Electrons ejected from this layer by photons are collected by the collector anode; the photoelectric current is proportional to the illumination. Photocells are used as control devices. Whenever the illumination of the photocell is changed, the photoelectric current changes, the voltage across a resistor in series with the cell changes, and an electric signal is sent to an amplifier. Thus the photocell can be used as a burglar alarm, as a safety device on industrial machinery, and as a counter of objects moving on a conveyor belt.

Another application is in *sound motion-picture films.* When the motion pictures are taken, the film is drawn at constant speed past a variable aperture illuminated by a beam of light. The width of the opening and hence the blackening of the film of the sound track is controlled by the sounds incident on the microphone. When the film is projected, a beam of light passes through the sound track into a photoelectric cell. The variable light absorption causes pulsations of current in the photoelectric cell which, after being amplified, actuate the loudspeaker and reproduce the sounds.

In one form of the *television camera* (Fig. 39-7a), an optical image of the object that is being televised is formed on a mica screen (Fig. 39-7c) at the end of the tube. Its front surface is covered with a multitude of specks of a cesium-silver photosensitive compound and its back surface with a metallic film. The cesium specks on the front surface and the metallic film form a capacitor. Any change of charge on a speck will induce an opposite charge on the film. Each cesium speck acts like a tiny photoelectric cell from which photons liberate electrons. Suppose that the entire mica screen is illuminated. Then each speck gets discharged by the light as soon as the speck is charged by the electron stream which scans the screen of the camera. Each such discharge of electrons from a speck induces a negative charge on the metal film. Thus thousands of electrical pulses are sent to the grid of tube *T* per second, and the outgoing signal is modulated. Suppose that the entire screen is dark. Then the screen remains strongly charged, and incoming electrons are repelled to the collector ring. Since the charge on the specks remains constant, the outgoing television waves are unmodulated, and the entire screen of the receiver picture tube (Fig. 39-7b) is dark. When an image is focused on the screen of the cam-

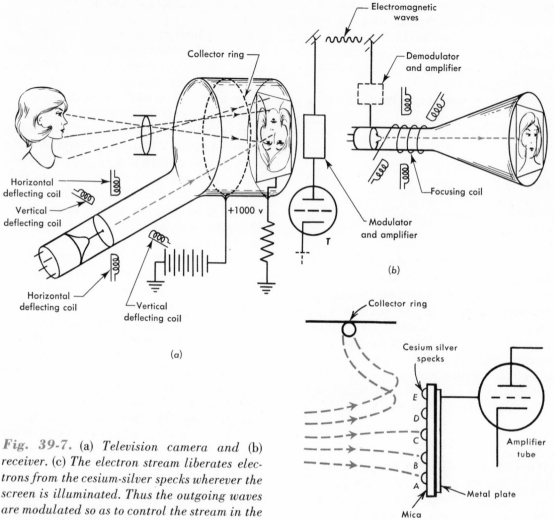

Fig. 39-7. (a) *Television camera and* (b) *receiver.* (c) *The electron stream liberates electrons from the cesium-silver specks wherever the screen is illuminated. Thus the outgoing waves are modulated so as to control the stream in the receiver tube.*

era, impulses will open the "valve" of the intensity-controlling grid of the receiver tube whenever the electron stream of the camera touches a bright part of the image. The electron "paint brush" of the picture tube of the television receiver reproduces the picture that the television camera sees.

Summary

New areas of knowledge were explored by physicists during the transition period from

about 1890 to 1930 that saw physics move from classical physics to contemporary physics. New interpretations were given to old concepts during this Second Scientific Revolution. The nature of radiation was one of the most important problems requiring solution. Theoretical and experimental investigations of black-body radiation and photoelectricity led to the concept of the photon.

Planck proposed that the energy of a micro-oscillator in the wall of a black-body cavity could change only by a tiny discrete amount or some integral multiple of this. The energy was added or removed—that is, the oscillators moved from one energy level to another—by the absorption or emission of packets of energy—or quanta—whose energy $E = hf$. Using the idea that radiant energy came in packets of this kind, Planck was able to account quantitatively to a high degree of accuracy for the shape of the spectral energy curve for a black-body source as well as for the other black-body properties.

Studies of the black-body radiation failed to reveal whether radiant energy traveled through space in the form of quanta. Experimental investigation of the photoelectric effect, in which electrons are emitted by surfaces when illuminated, and theoretical interpretation by Einstein of the results indicated that quanta indeed travel through space. The maximum kinetic energy of photoelectrons emitted by a surface whose work function is ω_0 when illuminated by radiation of frequency f is

$$\tfrac{1}{2}mv_{\max}{}^2 = hf - \omega_0$$

The stopping potential is the negative voltage V_s between the cathode and the collector in a phototube such that the most energetic photoelectrons just barely fail to reach the collector.

$$V_s e = \tfrac{1}{2}mv_{\max}{}^2$$

Questions

1. Are different principles needed to describe events in the world of atoms and molecules than in the "everyday" world?

2. How is the change of color of steel, when heated in a furnace, related to Wien's law of the wavelength of maximum intensity?

3. What is meant by (a) the "Second Scientific Revolution," (b) "classical physics," (c) "contemporary physics?"

4. Distinguish between Einstein's and Planck's contributions to our understanding of the quantum.

5. Why do you suppose nineteenth-century physicists first resorted to classical thermodynamics to account for black-body radiation?

6. What is an energy-level diagram? How is it used to explain black-body radiation?

7. Draw a five-level energy diagram for a micro-oscillator and also a curve showing how the oscillator's energy might change with time as it emits quanta.

8. Calculate the conversion factor relating (a) the Mev to the joule, (b) the Mev to the kilogram, and (c) the Mev to the angstrom.

9. Is the wave theory of light more fundamental than the corpuscular theory? Discuss.

10. Which would you expect to be more highly photosensitive: iron, a red dress, aluminum, nitrogen gas, copper?

11. Describe an experiment to measure the work function of a material.

12. Differentiate between the thermionic and the photoelectric effects.

Problems

A

1. Calculate h/e in the mks system.

2. The temperature of a black-body source is initially 500°K. If the temperature is doubled, what happens to the rate at which energy is emitted?

B

3. What is the energy of a quantum of radiation with the following wavelength: (a) a radio wave of 1.2×10^9 A; (b) a light wave of 5.5×10^3 A; (c) an X-ray of 0.80 A?

4. What is the wavelength of a 10-Mev gamma ray? (1 Mev $= 1.6 \times 10^{-13}$ joule.)

5. Calculate the energy of a 0.020-A gamma-ray quantum.

6. Calculate the voltage required to stop a

photoelectron moving with a velocity of 8.0 × 10^5 m/sec. ($e/m = 1.76 × 10^{11}$ coul/kg.)

7. Calculate the energy associated with (a) a radio wave of frequency 10^3 kH, (b) a light wave of frequency 10^{14} H.

c

8. At what rate are quanta being emitted within a narrow band of wavelengths at 7000 A by a black-body source that radiates energy within that band at the rate of 10^{-5} watt?

9. If a photosensitive surface has a work function of 2.0 ev, what will be the maximum kinetic energy in electron volts of the photoelectrons emitted if the surface is illuminated by radiation of frequency $1.5 × 10^{15}$ H?

10. What is the wavelength of radiation that will give photoelectrons a kinetic energy of $2.5 × 10^{-19}$ joule if the surface has a work function of 4.0 ev?

11. Red light of wavelength 6800 A is barely able to liberate electrons from metallic sodium. What is the work function of sodium in electron volts?

12. If ultraviolet light of wavelength 3200 A strikes a sodium-coated surface (work function 2.0 ev), with what maximum energy in electron volts will the electrons escape?

13. According to Richardson's equation, the thermionic electron current per unit area of a metal surface whose Kelvin temperature is T is given by

$$i = 120 \text{ amp/(cm}^2\text{-}K^{°2})T^2 e^{-\omega_0/kT}$$

where ω_0 is the same work function as that obtained from the photoelectric effect and k is Boltzmann's constant, $1.38 × 10^{-23}$ joule/°K. (a) Calculate the value of i for palladium (see Table 39-2) at a temperature of 1500°K. (b) Explain why you would expect the work function to be of importance in the thermionic effect as well as in the photoelectric effect.

14. Wien's equation for R_λ, the rate of emission of radiant energy by a black body per unit area per unit wavelength interval, is given by

$$R_\lambda = c_1\lambda^{-5}e^{-c_2/\lambda T}$$

Planck's equation for R_λ is

$$R_\lambda = c_1\lambda^{-5}\frac{1}{e^{c_2/\lambda T} - 1}$$

In both of these equations, λ is the wavelength, T the Kelvin temperature, c_1 a constant whose value is $3.74 × 10^{-16}$ watt-m^2, and c_2 a constant equal to $1.44 × 10^{-2}$ m-K°. Plot the curve for R_λ versus λ at 3000°K from $0.5 × 10^{-10}$ m to $4.0 × 10^{-10}$ m (a) by Wien's equation and (b) by Planck's equation. Since Planck's equation is in good agreement with experiment, compare the two curves to see where Wien's equation fails.

15. Show that Wien's equation for black-body radiation reduces to Planck's equation (see preceding problem) except in the longer-wavelength region.

For Further Reading

Blackwood, O. H., T. H. Osgood, A. E. Ruark, et al., *An Outline of Atomic Physics,* third edition, John Wiley and Sons, New York, 1955.

Bronowski, J., *The Common Sense of Science,* Harvard University Press, Cambridge, 1953.

Gamow, George, *Thirty Years That Shook Physics,* Doubleday and Company, Garden City, 1966.

Hoffman, Banesh, *The Strange Story of the Quantum,* Harper and Brothers, New York, 1947.

Oldenberg, Otto, *Introduction to Atomic Physics,* McGraw-Hill Book Company, New York, 1954.

CHAPTER FORTY

The Photon and the Atom

The photon was an essential part of Planck's interpretation of the continuous spectrum of a black body. Is the photon equally essential in understanding bright-line spectra—an even more intriguing kind of spectrum? By the sharpness and separateness of the wavelengths present, bright-line spectra hinted at photons. Physicists of the early twentieth century sought to use the new idea of the photon to understand the simplest bright-line spectrum, that of atomic hydrogen.

The hydrogen spectrum contains three principal *series* of lines—one in the ultraviolet, one in the visible, and one in the infrared. Figure 40-1 shows the visible, or Balmer, series of spectral lines of hydrogen. Note that the lines seem to show a regularity of wavelength, the lines converging toward a limit at the shorter wavelength end of the spectrogram. The intensity is greatest for the first line in the series, H_α or Balmer alpha, and becomes less for the shorter wavelengths. Johann Balmer and other physicists of the late nineteenth century were impressed by regularities of this kind within each series. They were able to write formulae containing experimentally derived constants with which one could calculate the wavelengths of the

lines in each series. For example, Balmer found that the wavelengths of the series of hydrogen lines in the visible part of the spectrum obeyed the relation

$$\lambda = 3645.6 \text{ A } \frac{n^2}{n^2 - 4}$$

where n was equal to 3, 4, 5, etc. When n is 3, λ equals 6562.08 A, in good agreement with the measured wavelength of 6562.10 A for the Balmer α line of hydrogen. However, Balmer was unable to explain why these regularities existed.

Assumptions of the Bohr Theory of the Hydrogen Atom

In 1913, Niels Bohr,* a Danish physicist, announced a theory which related the hydrogen spectrum to the structure of the hydrogen

---

* Niels Bohr (1885–1962), perhaps more than any other physicist of his generation, can be said to have laid the groundwork for the era of quantum physics. In addition to his own contributions—he received the Nobel Prize in physics in 1922 for his theory of the atom—he had a profound influence on other physicists, directing their attention to significant problems, asking penetrating questions, and inspiring them to greater accomplishments.

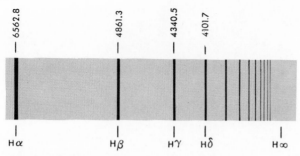

Fig. 40-1. *The visible spectrum of atomic hydrogen. Wavelengths are in angstroms (Courtesy of Dr. Gerhard Herzberg.)*

atom. Bohr's new theory accounted for spectral wavelengths so satisfactorily that physicists realized that they were on the threshold of great new understanding of both radiation and matter.

Bohr's theory starts from basic principles of classical mechanics and electricity and magnetism but goes on to make startlingly new assumptions. First of all, Bohr assumes that the hydrogen atom consists of a tiny positively charged nucleus—a proton—with a single electron rotating about it. This model of the atom was essentially the Rutherford model (Chapter 21) that had been developed several years earlier. Electrostatic attraction between the proton and the electron holds the atom together. For simplicity, Bohr assumes that the nucleus is stationary and that the electron travels in a circular orbit. At this point, he introduces several revolutionary ideas:

1. More than one orbit is possible for the electron, according to the theory, each possible orbit having a different radius. However, when the electron is in a given orbit, *the energy of the atom is constant.* The electron is assumed *not to lose energy* even though it has a central acceleration. According to classical theory, the accelerated electron should emit electromagnetic waves continuously, spiraling into the nucleus as its energy becomes

depleted. Bohr assumes that this does not happen.

2. When an electron makes a "quantum jump" from one orbit to another nearer the nucleus, the atom emits a photon. The energy of the photon is given by Planck's equation

$$E_B - E_A = hf$$

where $E_B - E_A$ is the change in the energy of the atom. The theory says nothing about the manner in which the electron goes from one orbit to another.

3. Only a limited number of orbits are possible. They are picked to satisfy a new *quantizing condition:* the angular momentum $I\omega$ of the electron in any orbit has to be an integral number times Planck's constant divided by 2π. In equation form,

$$I_n \omega_n = n\frac{h}{2\pi}$$

where I_n is the moment of inertia of the electron and ω_n its angular velocity when it is in the *n*th orbit and $n = 1, 2, 3, \ldots$. *n* is called the quantum number of the orbit. *No other orbits are possible.*

Calculating the Total Energy of the Electron

The total energy E_n of the hydrogen atom when the electron is in the *n*th orbit of radius r_n (Fig. 40-2) is the sum of the potential en-

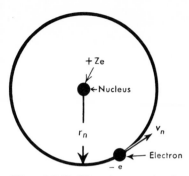

Fig. 40-2. *The electron in the nth Bohr orbit.*

ergy $(E_P)_n$ and the kinetic energy $(E_K)_n$. Let us treat the nucleus and the electron as point charges. The electron has a charge $-e$ and the proton a charge $+Ze$, where Z, the atomic number, is 1 for hydrogen, and e is the charge on the electron. We recall that the potential energy $(E_P)_n$ for two charges in this situation (Chapter 22) is

$$(E_P)_n = \frac{-k_0 Ze^2}{r_n}$$

where k_0 in the mks system of units approximately equals 9×10^9 newton-m²/coul². The negative sign means that work must be done *on* the atom to pull the electron away from the nucleus. The kinetic energy $(E_K)_n$, associated with the motion of the electron as it moves around the nucleus, is:

$$(E_K)_n = \tfrac{1}{2}mv_n^2$$

where m is the mass of the electron and v_n its linear velocity. Thus the total energy is

$$E_n = (E_K)_n + (E_P)_n$$

$$E_n = \tfrac{1}{2}mv_n^2 - \frac{k_0 Ze^2}{r_n} \qquad (1)$$

We should like to obtain an equation that will allow us to calculate the change in energy of the atom—and hence the energy of the photon emitted—when the electron jumps from one orbit to another. Our mathematical strategy will be to express v_n and r_n in terms of physical constants and of the quantum

number of the orbit and to substitute these values for v_n and r_n in equation (1), finally obtaining an equation for E_n in terms of n. (Check each of the following algebraic steps yourself!)

The centripetal force mv_n^2/r_n acting on the electron is supplied by the Coulomb attraction between the proton and the electron:*

$$\frac{mv_n^2}{r_n} = \frac{k_0 Ze^2}{r_n^2} \qquad (2)$$

and this reduces to

$$v_n^2 = \frac{k_0 Ze^2}{mr_n} \qquad (3)$$

The angular momentum of the electron must equal $nh/2\pi$, according to assumption number 3 in the last section:

$$I_n \omega_n = n \frac{h}{2\pi}$$

But the moment of inertia of a mass point is $I_n = mr_n^2$ and the angular velocity is $\omega_n = v_n/r_n$. Thus

$$I_n \omega_n = (mr_n^2)\left(\frac{v_n}{r_n}\right) = \frac{nh}{2\pi}$$

$$r_n = \frac{nh}{2\pi v_n m} \qquad (4)$$

Solving equations (3) and (4) simultaneously, we obtain

$$r_n = \frac{n^2 h^2}{4\pi^2 k_0 Ze^2 m} \qquad (5)$$

and

$$v_n = \frac{2\pi k_0 Ze^2}{nh} \qquad (6)$$

Substituting these values of r_n and v_n into equation (1) we obtain

$$E_n = \frac{1}{2}\left(\frac{4\pi^2 k_0^2 Z^2 e^4 m}{n^2 h^2}\right) - \frac{4\pi^2 k_0^2 Z^2 e^4 m}{n^2 h^2} \qquad (7)$$

---

* We assume here that the proton is at rest. It would be at rest only if its mass were infinitely greater than that of the electron. Actually, the proton and the electron revolve about their common center of mass—like the earth and the moon—and a more exact calculation would take this into effect.

$$E_n = -\left(\frac{2\pi^2 k_0^2 Z^2 e^4 m}{h^2}\right)\left(\frac{1}{n^2}\right) \qquad (8)$$

This is the equation we were seeking. Notice that all of the quantities within the first pair of parentheses are constants. Therefore E_n is inversely proportional to the square of the quantum number n. Notice also that the total energy is negative because the (negative) potential energy in any orbit is twice as great (equation 7) as the (positive) kinetic energy in that orbit. (The situation is similar to that of an earth satellite or planet whose orbit is a closed one and whose total energy is negative.) Thus the electron is in a "potential well." Energy must be given *to* the atom for the electron to jump from an orbit near the nucleus to one further out. Energy must be given up *by* the atom when the electron falls to a lower orbit.

Inserting the values of the constants into equation (8), we obtain

$$E_n = -\frac{\begin{array}{c}2\pi^2(9 \times 10^9 \text{ newton-m}^2/\text{coul}^2)^2 \\ \times (1.60 \times 10^{-19} \text{ coul})^4 \\ \times (9.11 \times 10^{-31} \text{ kg})\end{array}}{(6.62 \times 10^{-34} \text{ joule-sec})^2}\frac{1}{n^2}$$

$$E_n = -\frac{\begin{array}{c}(2\pi^2)(9 \times 10^9)^2(1.60 \times 10^{-19})^4 \\ \times (9.11 \times 10^{-31}) \text{ joule}\end{array}}{(6.62 \times 10^{-34})^2}\frac{1}{n^2}$$

(where we have used the relation 1 kg-m²/sec² = 1 joule, after canceling units). Or

$$E_n = -2.18 \times 10^{-18} \text{ joule}\,\frac{1}{n^2} = -13.60 \text{ ev}\,\frac{1}{n^2}$$

Figure 40-3 is an energy-level diagram—a picture of an "energy well"—for the electron of the hydrogen atom according to the Bohr theory. When the electron is in the innermost orbit ($n = 1$)—at the bottom of the "well"—its energy E_1 is −13.60 ev. This is the ground state or unexcited state of the hydrogen atom. We must put in 13.60 ev of energy to pull the electron out of the well from that level. On the second Bohr orbit, $E_2 = [1/(2)^2][-13.60 \text{ ev}] = -3.40$ ev, and we must supply 3.40 ev

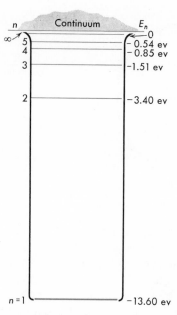

Fig. 40-3. *The energy well of the electron in the hydrogen atom.*

to free the electron, and so on. Above the value $E_n = 0$ (as n approaches infinity) lies a continuum of energy states. These are energy levels so close together that the electron, in moving from one to the other, has the "classical" motion of a free electron.

Wavelengths of the Hydrogen Lines

Suppose that the atom has been excited—by a collision with another atom, for example—so that the electron is in the third ($n = 3$) orbit and that it makes a quantum jump to the second ($n = 2$) orbit on its way back to the innermost one ($n = 1$). The electron gives up energy equal to 3.40 ev − 1.51 ev or 1.89 ev. A photon of energy 1.89 ev is emitted. The wavelength associated with the photon is found from Planck's condition $E = hf$ to be approximately 6530 A. Similarly, calculating the wavelength of the photon emitted when the electron jumps from the fourth ($n = 4$) to the second ($n = 2$) orbit, we obtain about 4850 A. In fact, let us assume that *all* of the spectral lines of the Balmer series consist of photons emitted when the hydrogen electron jumps from more distant orbits to the second ($n = 2$) orbit. We find that the wavelengths calculated from the Bohr theory are in agreement—to an accuracy of about 1%—with the experimentally determined wavelengths (Fig. 40-1). Similarly the wavelengths of the ultraviolet series of spectral lines—the Lyman series—can be calculated by assuming that they are the result of electron transitions to the innermost ($n = 1$) orbit. The series in the infrared—the Paschen series—is produced by electron transitions to the third ($n = 3$) orbit. Figure 40-4 illustrates some of these transitions.

We can write a formula that summarizes these steps. If an electron moves from an outer orbit whose quantum number is n' to an inner one of quantum number n, the hydrogen atom emits a photon whose wavelength λ can be calculated by

$$h\frac{c}{\lambda} = \left(\frac{2\pi^2 k_0^2 Z^2 e^4 m}{h^2}\right)\left(\frac{1}{n^2} - \frac{1}{n'^2}\right)$$

$$\frac{1}{\lambda} = \frac{2\pi^2 k_0^2 Z^2 e^4 m}{h^3 c}\left(\frac{1}{n^2} - \frac{1}{n'^2}\right)$$

For the Balmer series, $n = 2$ and $n' = 3, 4, 5, \ldots$; for the Lyman series, $n = 1$ and $n' = 2, 3, \ldots$; and for the Paschen series, $n = 3$ and $n' = 4, 5, \ldots$. The quantity $1/\lambda$ is called the **wave number** of the photon. The quantity $2\pi^2 k_0^2 Z^2 e^4 m/h^3 c$, as you can readily show by inserting values of the constants, has the approximate value 1.10×10^7 1/meter; it is called the Rydberg constant.

Extension of the Bohr Theory

Bohr had achieved a startling success in explaining the spectral series of hydrogen. Would his theory succeed in explaining more complicated spectra—the spectra of heavier elements, of ionized atoms, or of molecules? In general, the answer to this question was no. The theory accounted satisfactorily for certain

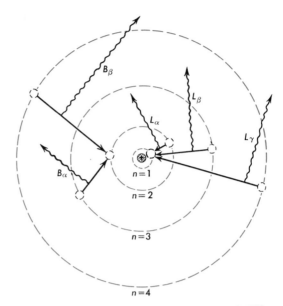

Fig. 40-4. Bohr hydrogen atom model. When the electron moves from an outer (higher-energy) to an inner (lower-energy) orbit, the atom emits a photon. Notice transitions producing some of the lines in the Balmer spectrum (B) and the Lyman spectrum (L) of hydrogen.

other special situations. For example, it could be used to calculate rather satisfactorily spectral wavelengths for *one-electron ions* that resembled the hydrogen atom in their structure. Among these were singly ionized helium (helium that has lost one of its two electrons), doubly ionized lithium (the normal lithium atom minus two electrons), and a few others. Since the energy of an energy level depends upon the square of the atomic number Z (equation 8 above), the energy levels of singly ionized helium should be $(2)^2$ or 4 times deeper in the energy well than those of hydrogen. Singly ionized helium should produce spectra similar to those of hydrogen except that the frequencies of the helium lines should be *four* times those of the corresponding hydrogen lines. Similarly, the frequencies of the lines of doubly ionized lithium ($Z = 3$) should be *nine* times those of the corresponding hydrogen lines. To the great pleasure of physicists, these predictions turned out to be in good agreement with experiment. There were also some successes of the Bohr theory in explaining the X-ray spectra of elements, as we shall see later in this chapter.

The Bohr theory, even when modified to permit elliptical orbits, failed to account for the spectra of heavier elements such as iron, vanadium, and uranium, which contain hundreds or thousands of lines, or even for the simpler spectra of elements such as sodium and potassium. To solve the general problem of atomic spectra, more sweeping changes had to be made in the physicist's approach to the atom. These changes, as we shall see, eventually took place and produced what we call *quantum mechanics.* In the Bohr theory, the description was semi-classical. The electron moved like a planet in orbit except for its mysterious and unexplained behavior during quantum jumps. In quantum mechanics, definite orbits would be ruled out. They would be replaced by a probability description of the electron's behavior as being more in keeping with our difficulty in specifying the electron's velocity and position at any time. The idea of stationary, quantized states, corresponding to Bohr's orbits of no radiation, survived; so did the photon. Since we are concerned in this chapter with the photon, we shall leave the question of atomic spectra for the moment and return to it later.

The Production of X-Rays

The black-body spectrum, the photoelectric effect, and the hydrogen spectrum were made more understandable by the idea of the photon—the packet of radiant energy. There is one characteristic that we have come to expect of a particle of any kind—its ability to collide with other particles and exchange energy and momentum with them in billiard-

ball fashion. Do photons show this kind of behavior? Evidence from the field of X-rays soon revealed to physicists that photons collide with electrons in just this way. To understand the experiments that disclosed this behavior, we must first consider how X-rays are produced and what some of their properties are.

Near the end of the nineteenth century, a few physicists devoted themselves to studies of the electrical conduction of gases. A controversy arose as to whether electrons were charged particles or radiation. Phillip Lenard of Germany proved that the electrons, which had been accelerated to high velocity in a high-vacuum discharge tube similar to a cathode-ray tube, could penetrate a thin sheet of metal foil, escape from the tube, and make a screen *fluoresce*, or emit light. As a result of these long-continued efforts, Wilhelm Roentgen (Rent'gen) in Germany made a great discovery in 1895. He thought that the fluorescent light from a discharge tube might be able to pass through paper. Accordingly, he enclosed the tube in a pasteboard box and found that his "hunch" was wrong. He could not see the tube even in a darkened room. However, he was delighted to observe that something *did* penetrate the pasteboard and caused a fluorescent screen to emit light. Roentgen was uncertain as to the nature of the radiations, and so he called them "X-rays" because *x* represents an unknown quantity in algebra. Sometimes years elapse between a discovery by the scientist and its adoption for practical purposes. X-rays are unique, for they were used by surgeons for examining broken bones only a few weeks after their discovery. For example, an injured hand may be held between the X-ray tube and a "fluoroscope" or glass plate coated with a material sensitive to the radiations. The bones absorb more of the X-rays than the flesh; hence their shadows are seen on the fluorescent screen.

X-rays generally are produced by the de-

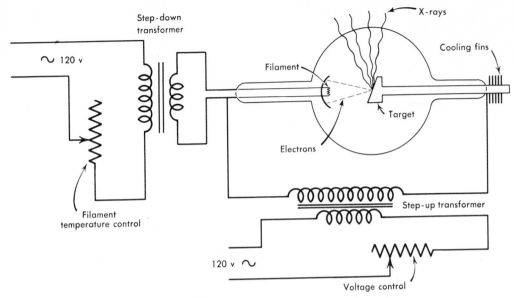

Labels in figure:
Step-down transformer
~ 120 v
Filament temperature control
Filament
Electrons
Target
X-rays
Cooling fins
Step-up transformer
120 v ~
Voltage control

Fig. 40-5. *A Coolidge X-ray tube.*

celeration of very energetic electrons when they strike a target. The Coolidge X-ray tube (Fig. 40-5) is highly evacuated. Electrons are emitted from a heated filament as in other vacuum tubes. A potential difference of several hundred thousand kilovolts accelerates the electrons as they move toward the target and they acquire velocities of thousands of miles per second. When they strike the metal target, most of their kinetic energy is converted into heat, but a fraction of a per cent is emitted as X-rays. A step-up transformer provides the voltage to operate the tube. The accelerating voltage is rectified either by a rectifier circuit, similar to those used in radio receivers, or by the X-ray tube itself. An X-ray tube of this kind can produce photons of energies reaching hundreds of thousands of electron volts. X-rays of greater energies — in the millions of electron volts — are produced by allowing the energetic electrons of a betatron beam (Chapter 29) to strike a target.

X-rays — like gamma rays, which we shall study later — are themselves invisible. They are detected by fluorescence (the emission of

light by certain materials such as zinc sulfide when struck by energetic radiations), by the blackening of a photographic film, and by ionization, which they produce in gases and which can be detected by Geiger counters or scintillation counters (Chapter 45).

X-rays are electromagnetic radiation like light, but of much shorter wavelength, their wavelengths lying between the generally accepted limits of 100 A and 0.1 A. X-rays can be reflected, polarized, and diffracted. They penetrate matter to greater depths than do radiations of longer wavelength. The absorption of X-rays increases with increasing thickness of the absorber, with the density of the material, with wavelength, and with increasing atomic number. The absorption per atom is proportional to $\lambda^3 Z^4$.

Suppose that an X-ray beam of initial intensity I_0 passes through an absorber of thickness t. The beam is exponentially absorbed (Appendix E), and the intensity of the emergent beam is given by the equation

$$I = I_0 e^{-\mu t}$$

Atomic and Nuclear Physics 647

Fig. 40-6. Model of a sodium chloride crystal. The atoms are at the corners of the cubes.

where μ is the linear absorption coefficient and e is the base of natural logarithms.

Measuring the Wavelength of X-Rays

Ordinary diffraction gratings cannot be conveniently employed to measure the wavelengths of X-rays. First, the X-rays would penetrate the grating. Second, the grating lines are too widely spaced. Fortunately, nature provides crystal gratings. Sodium chlo-

ride crystals, you will recall, are built up of parallel layers of atoms at a known distance d apart (Fig. 40-6). When X-rays strike such a crystal, the waves "scatter" when they strike the electrons surrounding each atom. In certain directions, the scattered waves reinforce each other. In other directions they cancel each other. A convenient way to think of this scattering process—in order to calculate the angles at which reinforcement occurs—is to assume that each of many atomic layers in the crystal reflects a little radiation, much as equally spaced parallel sheets of mosquito netting reflect sound. The reflected beam contains radiation reflected from many layers. These reflected beams interfere so that maximum or minimum reflection may occur. Figure 40-7 represents X-rays of wavelength λ incident on the crystal at an angle i and reflected, as from a mirror, at an equal angle from a particular set of planes. Consider radiation reflected from the first and second layers. The waves will be in phase and will reinforce each other, provided the difference in path is one wavelength. Moreover, if the waves reflected by the first two planes are in

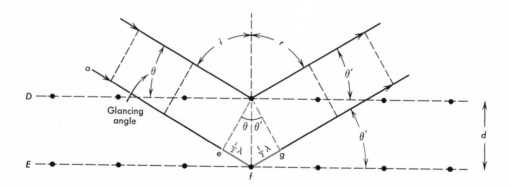

Fig. 40-7. Reflection (first-order) of X-rays by a set of planes in a crystal. The separation of the planes is d. *The path of one ray is* 1λ *longer than that of the other. Thus* $ef = \frac{1}{2}\lambda = d \sin \theta$.

phase, those reflected from hundreds of other parallel crystal planes will be also. The crystal will reflect the X-rays abundantly. If the path difference is even a trifle different from one wavelength, the many reflected waves will annul one another.

For maximum reflection,

$$\tfrac{1}{2}\lambda = d \sin \theta$$

To see why this is so, let i be the angle of incidence of the ray and $\theta = (90° - i)$ be the "glancing" angle between the incident ray and the plane. The difference of path, for radiation reflected from adjoining planes, is *efg*. Thus, if we let l be the distance *ef*,

$$l = \tfrac{1}{2}\lambda = d \sin \theta = d \cos i$$

Reinforcement will also occur if this difference of path is 1, 2, 3, 4, 5 wavelengths. In general, for reinforcement,

$$m\tfrac{1}{2}\lambda = d \cos i = d \sin \theta$$

$$\sin \theta = \frac{m\tfrac{1}{2}\lambda}{d}$$

$$m\lambda = 2d \sin \theta$$

This is called the Bragg equation. θ, the *glancing angle,* is *not* the angle of incidence. m, called the *order* of the interference, is the number of wavelengths, 1, 2, 3, etc., contained in the difference of path for X-rays reflected from adjoining planes in a particular set of planes.

Sodium chloride crystals were used when the diffraction of X-rays was first discovered because the distances between the layers in a set of crystal planes can be computed rather easily. If one diffracts a beam of X-rays with a sodium chloride crystal, he can use Bragg's equation to determine the wavelength of the X-rays. An X-ray spectrometer contains a crystal instead of a diffraction grating for dispersing X-rays of different wavelengths; the detector can be a photographic film, an ionization chamber, or a Geiger counter, as we shall see later.

Example. What is the grating space of a sodium chloride crystal? To obtain the distance between atoms for a sodium chloride crystal, we proceed as follows: The molecular weight of sodium chloride is 58.44 gm. Imagine a cubical sodium chloride crystal of this mass. It contains 6.02×10^{23} sodium atoms and an equal number of chlorine atoms. We shall assume that sodium atoms and chlorine atoms scatter X-rays in exactly the same way, although this is not quite true. The total number of **atoms** is $2 \times 6.02 \times 10^{23}$ **atoms.**

The density of sodium chloride is 2.163 gm/cm^3; therefore the volume of this crystal must be

$$\frac{58.44 \text{ gm}}{2.163 \text{ gm/cm}^3} = 27.05 \text{ cm}^3$$

Thus a cubical crystal of volume 27.05 cm³, 3.01 cm on each side, contains $2 \times 6.02 \times 10^{23}$ atoms. The number of atoms along one edge of the crystal is $(2 \times 6.02 \times 10^{23})^{1/3} = 1.064 \times 10^8$.

The distance between adjacent atoms, the grating space, is therefore

$$\frac{3.01 \text{ cm}}{1.064 \times 10^8} = 2.83 \times 10^{-8} \text{ cm} = 2.83 \text{ A}$$

However, it is possible for other layers of atoms—those cutting diagonally across the cube, for example—to reflect X-rays also. These other reflecting layers are separated, in general, by a different grating space, and a different Bragg angle holds for them for a given wavelength.

If a crystalline substance is pulverized and a powder specimen placed in a beam of X-rays of a narrow range of wavelengths, a *powder diffraction pattern* can be photographically recorded (Fig. 40-8) in a cylindrical camera with the specimen at the center, the beam traveling along a diameter, and the film along the wall of the cylinder. Some of the many tiny crystallites in the sample are always oriented so that their crystal planes are in a position for Bragg

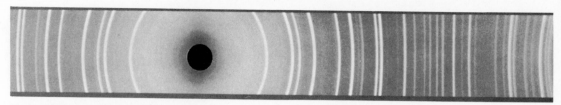

Fig. 40-8. An X-ray powder diffraction pattern of corundum (Al_2O_3). (Courtesy W. L. Kehl, Gulf Research and Development Company.)

reflection. As a result, conical beams of constructively interfering X-rays leave the specimen and are intercepted by the film. By measuring the diameters of the circular photographic images, it is possible to determine the grating spaces of the crystalline substance and to identify the substance or study its properties.

X-Ray Spectra

The spectrum emitted by the target of an X-ray tube resembles the spectrum of a black body in that it is continuous (Fig. 40-9). However, there is a sharp cut-off at the short wavelength (high-frequency, high-energy) end of the spectrum. Assuming that all of the

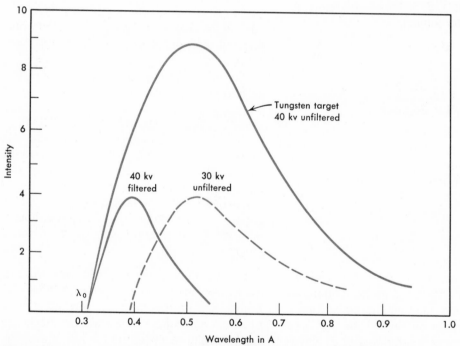

Fig. 40-9. X-ray continuous spectrum. For a tungsten-target tube operated at 40 kilovolts, λ_0 is 0.31 A. The wavelength of maximum intensity is 0.5 A. For "filtered" rays, the maximum is at 0.4 A. The filter absorbs most of the radiations of wavelength greater than 0.6 A. At 30 kilovolts, λ_0 is 0.39 A. (Reprinted with permission from O. Stuhlman, Introduction to Biophysics, John Wiley and Sons, 1943.)

kinetic energy of a decelerating electron in the X-ray tube goes into creating a single photon—a possible, but not very probable event—we can calculate what this short wavelength limit should be. Let the potential difference between the cathode and the target of the tube be V. In speeding through this potential difference, an electron acquires kinetic energy equal to Ve. The energy of the photon is hc/λ_0. Therefore,

$$Ve = \frac{hc}{\lambda_0}$$

or

$$\lambda_0 = \frac{hc}{Ve}$$

For a tube operated at 40,000 volts, λ_0 is approximately 0.31 A, in good agreement with the measured value (Fig. 40-9).

Superimposed on the continuous X-ray spectrum, there are bright lines like those in the optical spectrum of a gas (Fig. 40-10). The bright-line X-ray spectra of all the elements have the same general pattern, but the wavelength of the lines decreases regularly with increasing atomic number. The wavelengths of these characteristic bright lines do not depend upon whether the element is in a pure form or is a part of a chemical compound.

The Bohr atomic model helps to explain how atoms emit bright-line X-ray spectra. According to the Bohr model and the evidence from chemical reactions, in an atom of large atomic number, each inner orbit is occupied by a limited, definite number of electrons. The innermost, K orbit or shell ($n = 1$) has 2 electrons, the L shell ($n = 2$) has 8, etc. Suppose that a high-speed electron, striking the tungsten target of an X-ray tube, knocks one of the K electrons out of an atom. Then other electrons will fall in from outer shells, step by step, giving out photons of definite energies and wavelengths. Thus the characteristic bright-line series of spectral lines is produced in the X-ray region of the spectrum.

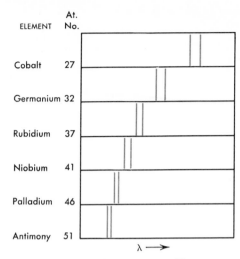

Fig. 40-10. *Characteristic K spectra of several elements. The wavelengths decrease with increasing atomic number.*

The radii of the orbits of a heavy atom are smaller than those of hydrogen, and the nucleus has a greater charge. For both reasons, the energy to ionize the K shell of an atom increases with atomic number. When one electron has been removed from the innermost orbit, the atom effectively has an atomic number of $(Z - 1)$. Since the ionization energy depends upon Z_{eff}^2 (equation 8 above), the ionization energy E_{ion} required to eject an electron from the K shell of an atom of atomic number Z is

$$E_{ion} = 13.6 \text{ ev } (Z - 1)^2$$

Thus to eject a K electron from tungsten, $Z = 74$, requires an energy input $E_{ion} = (13.6 \times 73^2) \text{ ev} = 72,800 \text{ ev}$. The wavelengths of the X-ray lines in the characteristic K spectra of an element of atomic number Z can be calculated approximately by a modified Bohr formula. For example, assuming that the first line in the K series—K_α—is emitted when an electron falls from the second orbit ($n = 2$) into the first ($n = 1$), one has:

$$\frac{1}{\lambda_{K\alpha}} = 13.60 \text{ ev } (Z - 1)^2 \left(\frac{1}{1^2} - \frac{1}{2^2} \right)$$

Another Look at $E = mc^2$

In Chapter 10 when we discussed the special theory of relativity, we promised to provide later a proof of the formula for the equivalence of mass m and energy E. We can now do this. A famous "thought experiment," devised by Einstein, uses the concept of the photon, some fundamental principles of mechanics, and the principles of special relativity to show that $E = mc^2$, where c is the speed of light in vacuum.

First let us consider the momentum $\mathcal{M}$ of a photon of energy E. Special relativity theory shows that

$$\mathcal{M} = \frac{E}{c}$$

This tells us that a stream of photons of energy E carries with it momentum $\mathcal{M}$, which is directly proportional to their energy. If the photons strike a body and are absorbed by it, they deliver momentum $\mathcal{M}$ to it. This relationship has been experimentally tested over a wide range of photon energies and found to hold true. For example, it explains the *radiation pressure* exerted by a beam of light—a stream of photons—on a moving vane mounted in an evacuated container.* The relationship also holds for the collision between a single photon and a material particle and explains the *Compton effect,* which we shall discuss in the next section.

Einstein's thought experiment asks us to consider an opaque, evacuated tube (Fig. 40-11) of mass m_t and length l which is mounted on frictionless bearings and is ini-

---

* You have undoubtedly seen toy radiometers in which a rotating set of vanes, one side of which is bright and one dark, is kept spinning within an illuminated glass container. This effect is *not* principally due to radiation pressure but rather is due to the effect of molecular bombardment of residual gas molecules upon the vanes. Indeed, you can show that the spinning vanes turn in the *opposite* direction to that expected if radiation pressure were dominant. The radiation pressure we have referred to above can be demonstrated experimentally only within highly evacuated containers.

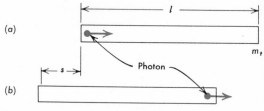

Fig. 40-11. *Deriving* $E = mc^2$.

tially at rest. Let us imagine that a lamp mounted on the left-hand end of the tube emits a single photon of energy E and sends it down the tube. The photon has a momentum

$$\mathcal{M}_{\text{ph}} = \frac{E}{c}$$

According to the law of conservation of linear momentum, the tube will recoil to the left with an equal and opposite momentum $\mathcal{M}_t$. If the recoil velocity of the tube is v we have

$$\mathcal{M}_t = m_t v$$

and

$$\mathcal{M}_{\text{ph}} + \mathcal{M}_t = 0$$

or

$$\frac{E}{c} + m_t v = 0$$

and

$$v = -\frac{E}{m_t c}$$

In a very short time—approximately l/c—the photon will strike the opposite end of the tube, be absorbed, and bring the tube to rest. However, during this time the tube will have moved a distance s given by

$$s = (v)\left(\frac{l}{c}\right) = \left(-\frac{E}{m_t c}\right)\left(\frac{l}{c}\right) = -\frac{El}{m_t c^2} \qquad (1)$$

What has happened as a result of the emission and absorption of the photon? The tube—an isolated system with no external forces acting on it—has moved a distance s.

The only way this can happen is for some redistribution of the mass of the tube-photon system to occur so that *the location of the center of mass of the system will be unchanged.* Put in specific terms, this means that if the mass m_t shifts to the left a distance s, some other mass within the system must shift to the right to compensate. This other mass, Einstein said, is that associated with the photon; let us call it m_{ph}. We then have the following condition that must be satisfied for zero motion of the center of mass:

$$m_t \cdot s + m_{ph} l = 0 \qquad (2)$$

Solving equations (1) and (2) simultaneously, we obtain

$$m_{ph} = -\frac{m_t s}{l} = \left(\frac{m_t}{l}\right) \frac{El}{m_t c^2} = \frac{E}{c^2}$$

or

$$E = m_{ph} c^2$$

This result tells us that radiation that has energy E must be understood to have a mass m_{ph} and to have the inertial properties that are associated with mass.

This simple proof was limited to mass-energy equivalence for the photon. Special relativity theory, however, shows that $E = mc^2$ holds for all physical systems.

The Compton Effect

Let us now return to the question we posed earlier in this chapter. Can photons collide with electrons, exchanging energy and momentum in the collision? In 1922, A. H. Compton showed by a series of experiments that "billiard-ball" collisions could indeed occur between photons and electrons. Applying the laws of conservation of mass-energy and of momentum, he was able to calculate the expected change in the wavelength associated with a photon and to show that the calculated results were in agreement with measurements of the wavelengths of photons before and

after they were scattered by collisions with electrons.

To understand Compton's procedure, we apply some familiar principles to these collisions, remembering at the same time that the electron will be moving so fast after a collision that its mass will not be its rest mass. Some of the results of special relativity theory (Chapter 10) must be used.

The energy of a photon is given by hf. Since the photon moves at the velocity of light c in the medium, we can think of this energy as the kinetic energy of the photon.

$$E = hf = \frac{hc}{\lambda}$$

As we saw in the last section the momentum of the photon $\mathcal{M}_{ph}$ is

$$\mathcal{M}_{ph} = \frac{E}{c}$$

or

$$\mathcal{M}_{ph} = \frac{hc}{\lambda c} = \frac{h}{\lambda}$$

The kinetic energy of an electron, according to special relativity theory, is

$$E_K = m_0 c^2 \left(\frac{1}{\sqrt{1 - v^2/c^2}} - 1 \right)$$

where m_0 is the rest mass of the electron and v is its velocity relative to the observer. The momentum of the electron is its relativistic mass m (not m_0!) times its velocity v.

$$\mathcal{M}_e = mv = \frac{m_0 v}{\sqrt{1 - v^2/c^2}}$$

In a collision (Fig. 40-12) with an electron that is initially at rest, the photon yields energy E_K to the electron which leaves the scene of the collision with energy E_K and momentum $\mathcal{M}_e$. The wavelength associated with the scattered photon is λ, a greater wavelength than the λ_0 associated with the incident photon. The equation for conservation of energy is

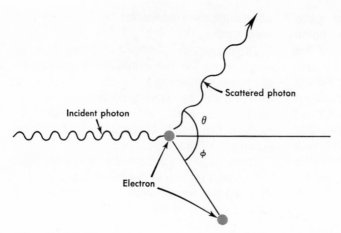

Fig. 40-12. *The Compton scattering of an X-ray photon by an electron.*

Energy of incident photon
= energy of scattered photon
+ kinetic energy of electron

or

$$\frac{hc}{\lambda_0} = \frac{hc}{\lambda} + m_0 c^2 \left(\frac{1}{\sqrt{1 - v^2/c^2}} - 1 \right)$$

For conservation of momentum, we have:

$$(\overrightarrow{\mathscr{M}}_{ph})_0 = (\overrightarrow{\mathscr{M}}_{ph}) + \overrightarrow{\mathscr{M}}_e$$

where the arrows remind us that momentum is a vector quantity.

We shall not go through the straightforward, but laborious, algebra required for a general solution of these equations for an arbitrary angle of scattering. If the angle between the direction of the incident photon and that of the scattered photon is θ, Compton found that the change of wavelength to be expected was

$$\lambda - \lambda_0 = \frac{h}{m_0 c} (1 - \cos \theta)$$

where $h/m_0 c$, called the *Compton wavelength,* has the value 0.0242 A. When θ is 90°, the change in wavelength is 0.0242 A *whatever the wavelength of the incident X-rays.* Figure 40-13 shows some experimental results when X-rays are scattered by the electrons in graphite. The measured

change in wavelength is 0.0236 A, in quite good agreement with the predicted change.

Example. What is the change in wavelength expected when photons are scattered at 180°? Use basic principles to obtain the answer.

We apply the principles of conservation of mass-energy and momentum.

Conservation of mass-energy:

$$\frac{hc}{\lambda_0} = \frac{hc}{\lambda_r} + mc^2 \left(\frac{1}{\sqrt{1 - v^2/c^2}} - 1 \right) \qquad (9)$$

Conservation of momentum:

$$\frac{h}{\lambda_0} = -\frac{h}{\lambda_r} + \frac{mv}{\sqrt{1 - v^2/c^2}} \qquad (10)$$

where λ_0 is the incident wavelength and λ_r the wavelength of photons scattered at 180°. Dividing c out of equation (9), transposing the term of h/λ_r, and squaring the resulting equation, we have

$$\frac{h^2}{\lambda_0{}^2} + \frac{h^2}{\lambda_r{}^2} - 2 \frac{h^2}{\lambda_0 \lambda_r} + 2mch \left(\frac{1}{\lambda_0} - \frac{1}{\lambda_r} \right) + m^2 c^2$$

$$= \frac{m^2 c^2}{(1 - v^2/c^2)} \qquad (11)$$

Transposing the term $-h/\lambda_r$ and squaring equation (10), we obtain

$$\frac{h^2}{\lambda_0^2} + \frac{h^2}{\lambda_r^2} + \frac{2h^2}{\lambda_0 \lambda_r} = \frac{m^2 v^2}{(1 - v^2/c^2)} \qquad (12)$$

Notice that

$$\frac{m^2 v^2}{(1 - v^2/c^2)} = \frac{m^2 c^2}{(1 - v^2/c^2)} - m^2 c^2$$

and equation (12) can be written

$$\frac{h^2}{\lambda_0^2} + \frac{h^2}{\lambda_r^2} + \frac{2h^2}{\lambda_0 \lambda_r} = \frac{m^2 c^2}{(1 - v^2/c^2)} - m^2 c^2 \qquad (13)$$

Subtracting equation (13) from (11), we obtain

$$-4\frac{h^2}{\lambda_0 \lambda_r} + 2mch\left(\frac{1}{\lambda_0} - \frac{1}{\lambda_r}\right) = 0$$

or

$$\lambda_r - \lambda_0 = 2\frac{h}{mc}$$

Inserting the value 0.0242 A for h/mc, we find that $\lambda_r - \lambda_0$ at 180° is 0.0484 A.

Light: Waves or Particles?

The evidence for photons is strong. Light does seem to behave like particles in many phenomena: black-body radiation, the photoelectric effect, the emission of spectral radiation by hydrogen, X-ray spectra, and Compton collisions. But so is the evidence strong for the wave nature of light in such phenomena as diffraction, interference, and polarization (Chapters 37 and 38). Physicists have been jokingly accused of believing in light waves on Mondays, Wednesdays, and Fridays and in photons on Tuesdays, Thursdays, and Saturdays. Is there any way to decide whether light is *either* waves or photons?

Physicists are generally agreed that the strong evidence for *both* waves and photons illustrates a fundamental fact about nature. When we question nature by means of one class of experiments about the properties of light—experiments on diffraction and interference, for example—nature replies in terms of waves. When our experiments probe other kinds of phenomena—black-body radiation

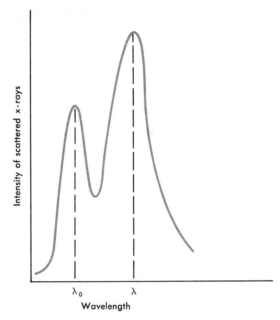

Fig. 40-13. *The Compton shift of wavelength due to scattering of X-rays by electrons in graphite. λ_0 was 0.708 A, and $\lambda - \lambda_0$ was 0.236 A for an angle of 90°.*

and the Compton effect, for example—nature replies in terms of particles. Both waves and particles are needed to interpret light: this is one aspect of what is called the *wave-particle duality* of nature.

Waves and particles complement each other in furnishing a complete explanation of light. A photon has energy E and momentum $\mathcal{M}$, and a wave has a wavelength λ and a frequency f. These quantities are not unrelated—as we see in studying the Compton effect. The energy $E = hf$, and the momentum $\mathcal{M} = h/\lambda$. We can describe natural phenomena in terms of either waves described by f and λ or particles described by E and $\mathcal{M}$.

Notice, however, that the wave explanations often seem connected with the question of the *intensity* of radiation and the particle explanations with the *energy* of radiation. We explain the bright and dark lines in a diffraction pattern by discussing the interference of waves. We explain the maximum energy of photoelectrons by referring to the energy of the incident photons. Einstein proposed to reconcile the wave theory of light and the particle theory by assuming that the amplitudes of the electromagnetic waves determined the numbers of photons at each point in a radiation field. A "ghost field" determined the probability of finding photons at a certain point. At the bright center of a single-slit diffraction pattern, the light waves interfere constructively and the amplitude of the electric and magnetic fields there is large. In Einstein's interpretation, this large amplitude indicates a large probability of finding photons there. In a dark fringe of the diffraction pattern, the amplitude is small, and the probability of finding photons in the dark fringe is also small. Thus the wave interference provides "traffic rules" for the photons, but the photons carry the energy of the radiation. Later, Dirac provided a more complete and elegant explanation of the role of photons in the electromagnetic field, using the principles of quantum mechanics.

Summary

Bohr provided a successful explanation of the bright-line spectrum of atomic hydrogen in terms of a new model of the hydrogen atom:

1. The electron travels around the nucleus on certain circular orbits, held in its orbit by electrostatic attraction between the nucleus and the electron.
2. Only those orbits are possible for which

$$I_n \omega_n = \frac{nh}{2\pi}$$

3. Except when making a transition from one orbit to another, the energy of the electron is constant.
4. If the change in energy of the electron during a "quantum jump" from one orbit to another is E, the energy of the photon emitted is $E = hf$.

The energy levels of the hydrogen atom have negative total energy associated with them. The energy is inversely proportional to the square of the quantum number n.

Both the continuous and the bright-line X-ray spectra provide evidence for the correctness of the photon hypothesis. The short-wavelength limit λ_0 of the continuous spectrum emitted by an X-ray tube with a voltage V across it is obtained from

$$Ve = h\frac{c}{\lambda_0}$$

The wavelength of the first line in the K series of X-ray lines of an element of atomic number Z is given approximately by

$$\frac{1}{\lambda_{K\alpha}} = 13.60 \text{ ev}(Z-1)^2\left(\frac{1}{1^2} - \frac{1}{2^2}\right)$$

Crystals act as satisfactory diffraction gratings for X-rays because the distance d between atomic layers is about the same size as X-ray wavelengths. Bragg reflection takes place at angles θ such that

$$m\lambda = 2d \sin \theta$$

Einstein's "thought experiment," dealing with a photon in a tube, provides a derivation of $E = mc^2$.

Compton was able to calculate the change of wavelength that occurs when a photon collides with an electron. His success furnished further support for the validity of the particle explanation of radiation.

Light exhibits both wave properties and particle properties. A complete understanding of light requires us to assume that the properties are complementary to each other.

Questions

1. Prove that the Compton wavelength equals 0.0242 A.
2. What fundamental quantities enter the equation for the Rydberg constant? Show that the Rydberg constant equals approximately 1.10×10^7/m.

3. Describe the Bohr atom, being very careful in stating Bohr's assumptions. What are "energy states"?

4. Why did Bohr find it necessary to assume that when the electron is in a given orbit in the hydrogen atom the energy is constant?

5. What does it mean to say the angular momentum is quantized in the Bohr model?

6. Why is the production of X-rays sometimes called the inverse photoelectric effect?

7. In the Bohr theory, how does the production of bright-line X-rays differ from that of bright-line spectra of light?

8. Do the principles of conservation of energy and momentum hold in atomic physics? Can the mechanics of billiard balls be applied to electrons and atomic particles? Cite some evidence.

9. Devise an experiment to demonstrate and measure the Compton effect.

10. Can the Bohr theory successfully be extended to atoms of greater atomic number than that of hydrogen? Discuss the successes and failures of such extensions.

11. Is it probable that any of the spectral lines of atomic hydrogen appear in the microwave region? Describe the transitions that would produce them.

12. We no longer think of the electron in the hydrogen atom as moving in sharply defined orbits. Why, then, do we use the Bohr atom model?

13. (a) How can a rock-salt crystal be used to measure the wavelengths of X-rays? (b) How can the distance between the planes of another kind of crystal then be determined?

14. What is the relationship between kinetic energy and momentum for a particle, according to classical mechanics? Compare this result with the relativistic one.

Problems

A

1. Calculate the lower limit of the wavelength of X-rays produced when an X-ray tube is operated at 6.0×10^4 v.

2. What minimum voltage would be required to produce (a) 5.0-A X-rays, (b) 0.05-A X-rays?

3. Calculate the frequency and wave number of radiations of these wavelengths: (a) 1.23 A, (b) 1216 A, (c) 6562 A.

B

4. Calculate the principal grating space for a potassium chloride (KCl) crystal (density = 1.984 gm/cm³).

5. Calculate (a) the wavelengths of the first three lines of the Balmer series in angstroms, and (b) the wavelengths of the first line of each of the Lyman and Paschen series in angstroms.

6. Make an energy-level diagram for atomic hydrogen. What are the energy and the wavelength of the photon emitted when the electron goes (a) from level 3 to 2, (b) from 4 to 2, (c) from ∞ to 2, (d) from 2 to 1, (e) from 3 to 1, (f) from ∞ to 1?

7. The potential difference between the cathode and the target of an X-ray tube is 1.50×10^4 v. (a) Find the energy in electron volts and the velocity of an electron when it strikes the target. (Neglect relativity effects.) (b) If all the energy were used in the emission of a single photon, what would be its frequency and wavelength?

8. Compute the momentum of a photon of the hydrogen line whose wavelength is 6563 A.

C

9. At what velocity is the mass of an electron twice its rest mass?

10. According to the Bohr theory, what energy is required to eject a K-electron from cobalt (atomic number 27)?

11. When X-rays of wavelength 0.80×10^{-8} cm are incident on a crystal at glancing angle $\theta = 30°$, the rays are copiously reflected at an equal angle. Find the distance between adjacent planes of the crystal.

12. A certain crystal reflects monochro-

matic X-rays strongly when the Bragg glancing angle (first-order) is 12°. What are the glancing angles for the second and third orders?

13. The distance between the adjacent reflecting planes of a certain crystal is 2.73 A. What are the Bragg glancing angles for X-rays of wavelength 1.23 A for the first, second, and third orders?

14. Show that a 300-Mev photon possesses about the same momentum as a 50-Mev proton. (Include relativity effects.)

15. Calculate the energy of a photon that has the same momentum as a 1-Mev electron. (Include relativity effects.)

16. Calculate according to the Bohr theory the wavelengths of the first three lines of the spectral series for *singly ionized* helium for which $n = 4$ and $n' = 5, 6, \ldots$ (Pickering series). Note that the helium atom has four times the mass of the hydrogen atom and twice its atomic number.

17. The spectral lines for singly ionized helium (see preceding problem) for $n = 4$ and $n' = 6, 8, 10$ should correspond approximately in wavelength with the first three lines of the hydrogen Balmer series. (a) Show why there should be such an approximate correspondence. (b) The lines do not correspond exactly. To explain this, investigate the assumption in the Bohr theory that the nucleus is at rest with respect to the observer.

18. Calculate (a) the gravitational attraction and (b) the electrical attraction between a proton and an electron separated by a distance of 0.5 A. Is gravity a major factor in the atomic world?

19. Make a graph showing the relative intensity of a beam of X-rays whose linear absorption coefficient in aluminum is 14 cm⁻¹ as the beam passes through 10 successive sheets of aluminum, each 0.50 mm thick.

For Further Reading

Bondi, Hermann, *Relativity and Common Sense,* Doubleday and Company, Garden City, 1964.

Bragg, Sir Lawrence, "X-Ray Crystallography," *Scientific American,* July, 1968, p. 58.

Friedman, Herbert, "X-Ray Astronomy," *Scientific American,* June, 1964, p. 36.

Gamow, George, *Thirty Years That Shook Physics,* Doubleday and Company, Garden City, 1966.

Giacconi, Riccardo, "X-Ray Stars," *Scientific American,* December, 1967, p. 36.

Holden, Alan and Phylis Singer, *Crystals and Crystal Growing,* Doubleday and Company, Garden City, 1960.

CHAPTER FORTY-ONE

Matter Waves

To explain the behavior of light and other electromagnetic radiations, we assume that they have both wave characteristics and particle characteristics. The wave behavior of light is shown most strongly by interference and diffraction. Its particle nature is shown by such phenomena as the photoelectric effect and Compton scattering. Other phenomena—reflection and refraction, for example—can be explained by *either* a wave theory or a particle theory. What finally emerged from the work of Planck, Einstein, Bohr, and other physicists was a *wave-particle* view of light. Photons of energy $E = hf$ and momentum $mc = h/\lambda$ are emitted and absorbed discretely by matter. Their distribution in space—the probability of finding a photon within a given small volume—is determined by the amplitude of waves that "accompany" the photons.

The wave properties and the particle properties of electromagnetic radiations are inseparably mixed in nature. Whether we choose to examine a phenomenon from the wave viewpoint or the particle viewpoint depends simply upon the kind of question we ask. In investigating the distribution of intensity in a diffraction pattern, we find that wave properties stand out (although photons carry the energy into the diffraction pattern). In explaining the maximum energy of a photoelectron, the particle properties of the incident photons predominate (although the diffraction of the accompanying waves at apertures determines the distribution of photons over the photoemissive surface).

These developments in the early twentieth century led physicists to consider the nature of particles of matter—electrons and entire atoms, for example—with renewed interest. Could it be that such entities, which certainly have particle properties, also have *wave properties?* In this chapter we shall investigate *matter waves* and find that the answer to our question is yes. The establishment of a wave-particle view of matter is one of the most important events in twentieth-century physics. It has led to the development of *quantum mechanics,* a powerful method of investigation and a philosophical approach to nature that must be mastered by anyone who wishes to understand the atomic world.

DeBroglie Waves

Thinking about the wave-particle nature of light, the French physicist Louis deBroglie (duh Brō-yee) began to speculate whether an electron might be accompanied by a wave, just as a photon is accompanied by one. As in many discoveries in physics, an analogy played an important role in deBroglie's thinking. If light were both wave-like and particle-like, he reasoned, might not matter be both particle-like and wave-like? DeBroglie was also convinced that nature was basically simple. Nature was not likely to have one set of rules for electrons and an entirely different set for photons. DeBroglie did not merely speculate, however, but proceeded to investigate mathematically the properties of matter waves, publishing his theory in 1922.

The wavelength λ of a matter wave is clearly one of the first quantities of interest to physicists. The wavelength determines the way in which waves interfere with each other and the way in which they undergo diffraction. What is the wavelength of a matter wave?

You will recall that the wavelength of a photon can be simply related to the photon's momentum (Chapter 40) by the following process. If we equate the relativistic mass energy ($E = mc^2$) of the photon to its energy, as given by Planck's equation ($E = hf$), we have

$$mc^2 = hf$$

or

$$mc^2 = h\frac{c}{\lambda}$$

and

$$\lambda = \frac{h}{mc} \qquad \text{(Photon)}$$

Substituting v, the velocity of a *material particle,* for c, the velocity of the photon, we might expect that the wavelength of the matter wave associated with the particle would be given by

$$\lambda = \frac{h}{mv} \qquad \text{(Particle)}$$

In fact, this is exactly the result that deBroglie developed by detailed analysis. The last

equation is called the deBroglie equation, and the wavelength given by it is called the *deBroglie wavelength.* Let us look more closely at the meaning of λ and v in the deBroglie equation. By doing so, we shall understand the essential steps that deBroglie had to take in going from speculation to a firm theory.

First of all, it is clear that a particle such as an electron cannot be simply a wave. A wave spreads as it travels from its source, diffusing its energy over an increasingly great wave front. Such unlimited diffuseness is at odds with the basic notion of a particle as an entity that is more or less localized in space. So we must direct our thoughts toward matter waves that somehow accompany particles, but do not replace them.

The next question is how fast the matter waves travel. Do they move with the velocity of the particle? One would think so, if they are to "accompany" the particle, but we must be careful. The basic equation relating the speed u with which the crest of a wave moves, the wavelength λ, and the frequency f was shown in Chapter 19 to be

$$u = f\lambda$$

Let us apply this equation to matter waves, substituting for f its equivalent E/h from Planck's equation and for λ its equivalent h/mv from deBroglie's equation. We then have

$$u = \left(\frac{E}{h}\right)\left(\frac{h}{mv}\right) = \frac{E}{mv}$$

But $E = mc^2$, by special relativity theory, and therefore*

$$u = \frac{mc^2}{mv} = \frac{c^2}{v}$$

According to the special theory of relativity (Chapter 10), the mass and the energy of a body become large without limit as the veloc-

---

* Note that E is the *total* energy of the particle and includes the energy equivalent of its rest mass.

ity of the body relative to the observer approaches the speed of light c. This means that although photons can and do move at the speed of light, the velocity of light cannot exceed c and the velocity of material particles can only approach that of light and cannot equal it. Thus v in the last equation is less than c, and u must therefore be greater than c. In other words, *matter waves seem to be required to move faster than light even though the particle that they accompany must have a velocity less than that of light.* This is a puzzling result. To explain it we must consider carefully the difference between simple harmonic progressive waves and wave packets.

Simple Harmonic Waves and Wave Packets

Simple harmonic progressive waves of a single frequency and constant amplitude are the simplest kind of wave motion we can imagine. Figure 41-1 shows a part of a train of such waves on an idealized and infinitely long rope, for example. The waves originate far away from the observer (at negative infinity), move monotonously past him, and vanish again from his observation (moving toward positive infinity). They are always the same. We saw in Chapter 31 that carrier radio waves are somewhat like this. Carrier waves must be modulated in amplitude or frequency before they can carry a message. But even radio carrier waves can be modulated by

turning them on and off at the transmitter—to produce a code message. Our idealized simple harmonic waves are not even interrupted. They flow on and on, carrying no message. They tell the observer *nothing.*

The special theory of relativity allows such waves to move at speeds greater than that of light. Special relativity requires only that *messages* or *energy* not be transmitted at speeds exceeding that of light. So our first difficulty about matter waves is overcome. They *can* travel at speeds greater than that of light because they do not carry a message or transmit energy. However, a more formidable difficulty remains. If waves are thought to be related to a particle, surely they must begin somewhere and end somewhere. Simple harmonic waves, as we have just seen, simply cannot be imagined to be localized in any sense.

Let us then take a restricted group of waves—a packet of waves—like the ones shown in Fig. 41-2. The packet moves from left to right. It clearly has a beginning and an end, so we can begin to think of it in connection with a particle. It has a wavelength λ. However, we have really not eliminated the infinitely long trains of simple harmonic waves. We have simply superimposed many such wave trains to produce our wave packet. *The localized wave packet can only be produced as a result of interference among progressive simple harmonic waves of different frequencies.* The simple harmonic waves cancel each other to the left and right

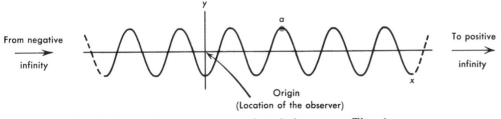

Fig. 41-1. *Simple harmonic waves on an infinitely long rope. The phase of a moves to the right with a phase velocity.*

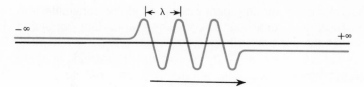

Fig. 41-2. A wave packet. The packet moves with a group velocity.

of the wave packet in Fig. 41-2. They interfere constructively only in the center of the figure, where they produce the effect we see as a traveling group of three waves. To make this clear, Fig. 41-3 shows the results of analyzing the wave packet of Fig. 41-2 into its component simple harmonic waves—a Fourier analysis. Our picture then of the wave packet of Fig. 41-2 is that it is a traveling interference pattern produced by progressive simple harmonic waves, like the ones in Fig. 41-1. The waves travel from left to right, canceling each other except where the wave packet is momentarily located.

What is the velocity of the wave packet? You might at first think that it would be the same as that of the progressive simple harmonic waves, but this is not so. A mathematical analysis, which we will not give here, shows that the velocity of the wave group or wave packet is *less* than that of the component simple harmonic waves, provided a certain condition, discussed below, is met. We call the velocity of the wave group or wave packet the *group velocity.* The velocity with which a crest of a simple harmonic wave moves is called the *phase velocity.* Thus the velocity (Fig. 41-1) of crest *a*—if we could mark it in some way!—is the phase velocity *u*, and the velocity (Fig. 41-2) of the wave packet is the group velocity *v.*

What is the mathematical condition for the velocity of the wave packet to be less than that of the simple harmonic waves? It is that the velocity of the simple harmonic waves *increases* with wavelength. In examining the dispersion of light in Chapter 36 we found

that in material media—such as glass and liquids—the index of refraction decreases as the wavelength increases. Since the speed of light is inversely proportional to the index of refraction, the speed of light in most dispersive media *increases** as the wavelength increases. Thus the group velocity of light in these media (the velocity with which the energy-carrying photons travel in a pulse of light emitted by an atom) is less than the phase velocity (the velocity of a crest of the simple harmonic waves forming the light wave packet). For example, in ordinary crown glass, the group velocity of light is about 2.4 % less than the phase velocity. In free space, on the other hand, light of all wavelengths travels with the same velocity. In free space, the group velocity (photons) equals the phase velocity (simple harmonic waves). Sound waves of all frequencies have the same velocity in the medium in which they are traveling. The group velocity of a pulse of sound waves equals the phase velocity.

That the velocity of a group of waves can be less than the velocity with which a given phase travels is demonstrated by waves on deep water. Motion picture films of a group of such waves show that individual waves form in the rear of the group, pass through it, appear in front of the group, and then fade away, traveling about twice as far in a given time as the group. Thus, the phase velocity of the

* There are exceptions to this at certain wavelengths throughout the electromagnetic spectrum where the frequency of the waves approaches the frequency of vibration of particles that are bound by elastic forces in the medium.

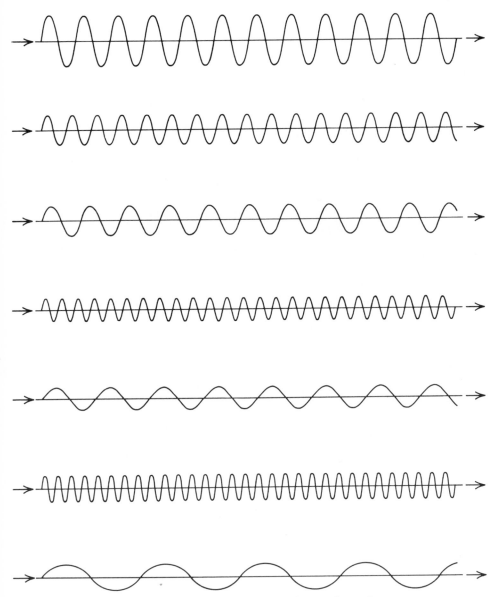

Fig. 41-3. The simple harmonic waves that interfere to form the wave packet in Fig. 41-2. Only a few of the component waves are shown.

progressive simple harmonic waves whose interference forms the group is about twice the group velocity. It is difficult to show this effect in static diagrams for a group such as that of Fig. 41-2, but the difference between group velocity and phase velocity can be seen in Fig. 41-4. You see there the groups formed when two simple harmonic progressive waves of slightly different wavelengths interfere. When we encountered this phenomenon before, we called the groups *beats,* and that is indeed what we have in Fig. 41-4.

Matter Waves in Free Space

Let us now summarize and apply to matter waves what we have said about wave groups and simple harmonic progressive waves. We shall list the important points of deBroglie's theory of matter waves and end this section with a brief description of matter waves in free space.

1. If material particles behave in the way that photons do, we might expect the particles to be associated with – not replaced by – waves. For particle waves, $\lambda = h/mv$ (deBroglie's equation). We have not yet said *what* is undergoing wave motion.

2. DeBroglie's equation and Planck's equation ($E = hf$) require that $u = c^2/v$ where u is the phase velocity of the matter waves, v is their group velocity, and c is the velocity of light in vacuum. The group of waves that

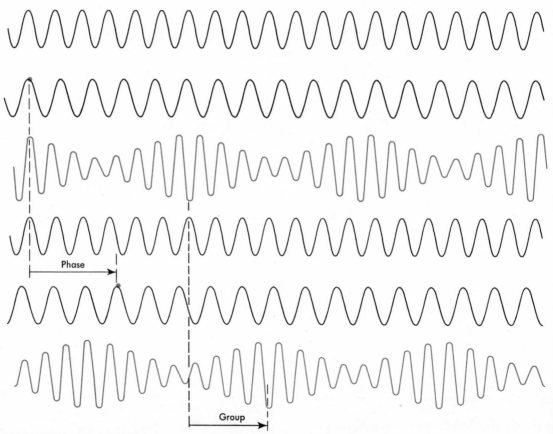

Fig. 41-4. *Phase velocity* u *versus group velocity* v.

travels with the particle (a) is localized in space and (b) has a velocity less than c because of the requirements of special relativity theory. But this requires u to be greater than c.

3. Simple harmonic progressive waves *can* have phase velocities u that are greater than c because they carry no energy and no message. Such waves interfere to form a group whose group velocity is less than c and is equal to v, the velocity of the particle. The group has a wavelength λ that we will ascribe to the particle and that we can hope to measure.

Our picture of matter waves in free space at this point in our discussion is one of a wave packet or wave group, created by the interference of simple harmonic progressive waves. The wave packet travels with a group velocity v, equal to that of the particle, and has a wavelength λ = h/mv. Because the simple harmonic progressive waves travel with different speeds in free space, the shape of the wave packet changes with time. We have localized the particle somewhere within the wave packet and restricted its velocity to values less than c, but we have accomplished this by bringing in many simple harmonic progressive waves. This will lead to a fundamental problem in our description of the motion of the particle, as we shall see shortly.

Example. What is the wavelength associated with (a) an electron that has been accelerated by passing through a voltage difference of 50 v and (b) a jet airliner of mass 110 tons (about 10^5 kg) flying at a velocity of 500 mi/hr (about 0.2 km/sec)?

(a) Neglecting the (slight) relativistic effect at this energy, we can obtain the velocity of the electron:

$$\tfrac{1}{2}mv^2 = Ve$$

$$\tfrac{1}{2}(9.1 \times 10^{-31} \text{ kg})v^2 = (50 \text{ v})(1.60 \times 10^{-19} \text{ coul})$$

$$v^2 = \frac{1.60 \times 10^{-17} \text{ (joule/coul)(coul)}}{9.1 \times 10^{-31} \text{ kg}}$$

$$v = 4.2 \times 10^6 \text{ m/sec}$$

Then

$$\lambda = \frac{h}{mv}$$

$$\lambda = \frac{6.6 \times 10^{-34} \text{ joule-sec}}{(9.1 \times 10^{-31} \text{ kg})(4.2 \times 10^6 \text{ m/sec})}$$

$$\lambda = 1.7 \times 10^{-10} \text{ m} = 1.7 \text{ A}$$

Evidently, to measure the wavelength of electrons, we would need a diffraction grating whose grating space is around 2 A.

(b)
$$\lambda = \frac{6.6 \times 10^{-34} \text{ joule-sec}}{(10^5 \text{ kg})(200 \text{ m/sec})}$$

$$= 3.3 \times 10^{-41} \text{ m} = 3.3 \times 10^{-31} \text{ A}$$

the wavelength of the jet airliner! The wavelengths of "everyday" objects are evidently extremely small, and the wave nature of matter becomes observable only on the atomic scale.

Experimental Observation of Matter Waves

C. J. Davisson and L. H. Germer, two physicists of the Bell Telephone Laboratories, were studying the reflection of electrons from metal surfaces at the time deBroglie's theory was first being discussed. Several of their metal samples had by chance been heat-treated to the point of becoming single crystals—solids in which the crystal regularities extend uninterruptedly through the entire sample (Chapter 12). Davisson and Germer were surprised to observe that electrons were not just scattered in random directions by the single-crystal samples. Instead, the reflected electrons were more concentrated in certain directions than others. They were able to account satisfactorily for this preferential scattering only by assuming that *the single crystal acted as a three-dimensional diffraction grating for the deBroglie electron waves.*

Figure 41-5 is a diagram of the apparatus of

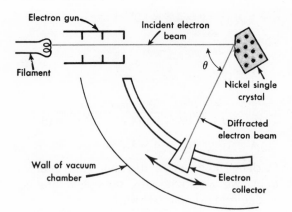

Fig. 41-5. *The apparatus of Davisson and Germer.*

Davisson and Germer. Electrons emitted by the hot filament were accelerated by passing through a voltage difference and formed into a beam by the slits of the electron gun. The electrons were incident perpendicularly upon the polished face of the nickel single crystal in which the atomic planes were arranged as shown. Scattered electrons were picked up by the electron collector. The output current of the collector was observed as the collector was rotated within the vacuum chamber through an angle θ with respect to the incident beam.

Davisson and Germer observed that if the crystal were randomly oriented, the collector current at a given accelerating voltage was generally weak. As the collector was rotated, however, the current went through a peak. Increasing the acceleration voltage, they repeated the measurements and found that at a voltage of about 54 v and an angle θ of about 50°, a sharp peak occurred in the collector current. They calculated the deBroglie wavelength of the electrons of that energy and found that the peak occurred just where they would expect to find it if the crystal were acting as a diffraction grating for electron waves. Later experiments, in which the beam was incident at an oblique angle, yielded

results that were quite analogous to those obtained in experiments on the Bragg scattering of X-rays (Chapter 40). The electron waves were reflected by atomic layers with strong constructive interference occurring at the Bragg angles θ given by the equation $m\lambda = 2d \sin \theta$ where λ was the wavelength of the electrons *in the crystal* and d was the grating space between atomic layers.

Electron waves can also be diffracted by thin metal foils. G. P. Thomson, a British physicist who shared the Nobel prize with Davisson and Germer for the independent discovery of matter waves, passed high-energy electrons through a thin gold foil, obtaining the beautiful electron diffraction pattern shown in Fig. 41-6. Each crystallite in the metal foil diffracted the electron beam so that constructive interference occurred at certain angles. Since the crystallites were randomly oriented with respect to the incident beam, the spots produced on the film by the diffracted electrons formed the circles that you see. Compare this figure with Fig. 40-8, which shows the X-ray diffraction pattern produced when X-rays of a narrow range of

Fig. 41-6. *Electron diffraction pattern obtained by G. P. Thomson by passing an electron beam through a gold foil. (O. H. Blackwood, T. H. Osgood, A. E. Ruark, et al.,* An Outline of Atomic Physics, *third edition, John Wiley and Sons, 1955.)*

wavelengths pass through a many-crystallite sample.

Even neutrons and atoms display experimentally observable wave characteristics. Neutrons from a nuclear reactor (Chapter 46) constitute a "white" source of neutrons having a wide range of energies and wavelengths. By allowing the neutrons to be diffracted by a crystal, experimenters can obtain diffracted neutron beams of a selected narrow band of wavelengths. The structure of solids can be studied by these monochromatic beams just as it is studied by X-ray beams — with complementary kinds of information. The diffraction of hydrogen atoms and helium atoms has also been observed. Their measured wavelengths agree well with those predicted by the wave theory.

All of these experimental results helped to establish firmly the idea that matter — like light — has a wave aspect.

The situation was summarized by Niels Bohr in the *principle of complementarity:* one can describe physical phenomena — for example, the diffraction of light or the diffraction of electrons — in terms of particles or in terms of waves. The particles are described by their momentum (mv) and their energy (E); the waves, by their wavelength (λ) and frequency (f). The two descriptions are connected by Planck's equation ($E = hf$) and by deBroglie's equation ($mv = h/\lambda$).

So far we have talked only about *free* particles. The evidence for the wave nature of particles held within atoms or of those moving within a crystal adds additional confirmation — as we shall see in later chapters.

The Uncertainty Principle

Our next step might well be to apply the idea of matter waves to electrons and other entities in many different situations. We could study electrons bound to atoms, electrons within a crystal, atoms of liquid helium, protons and neutrons in nuclei, and so on to see what new

insights into nature such investigations would give us. In fact, this is what happened in physics following the success of deBroglie's theory and the experimental confirmation of it. In the process, the theory itself underwent great changes. However, before we can deal with such further developments, we must understand an idea of great practical and philosophical importance — the *principle of uncertainty* or *indeterminacy*, first stated by Werner Heisenberg in 1927.

Earlier in this chapter we saw that we could limit the extent of an electron wave by replacing a simple harmonic progressive wave train of a single frequency by a traveling wave packet. Within a simple harmonic wave train, the electron could not be imagined to be localized at all. If we replace the waves of a single wavelength by a wave packet, we can localize the electron as narrowly as we please, but we do it at the cost of superimposing more and more simple harmonic waves (Fig. 41-7). You will recall that in Chapter 40 we defined the wave number as $1/\lambda$, the reciprocal of the wavelength. According to deBroglie's equation $mv = h(1/\lambda)$, the momentum of the particle is directly proportional to its wave number. If we are uncertain about the wavelength λ (because we have had to mix together waves of many different wavelengths to form the wave packet), we are correspondingly uncertain about the momentum of the particle. From deBroglie's equation we thus have the result $\Delta(mv) = h\,\Delta(1/\lambda)$, or *the uncertainty in the momentum equals Planck's constant times the uncertainty in the wave number.*

A fundamental theorem of mathematics tells us that the extent of the wave packet in space is inversely proportional to the range of wave numbers covered by the simple harmonic waves that are superimposed to form the wave packet. To localize the wave packet more and more, we have to add more and more waves of different wave numbers — representing different momenta (Fig. 41-7). Since

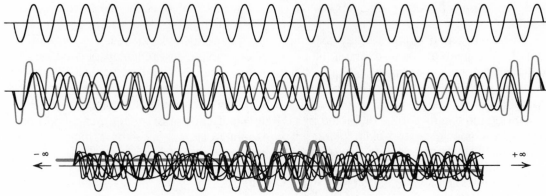

Fig. 41-7. Limiting the extent of a wave packet by superimposing waves of different wave numbers. (a) *An unlimited wave extending from* $-\infty$ *to* $+\infty$, *consisting of a simple harmonic wave of one frequency or wave number.* (b) *Two simple harmonic waves of slightly different wave numbers are superimposed, limiting the wave somewhat to a series of beats.* (c) *Many simple harmonic waves are superimposed (see Fig. 41-2) to produce a limited wave packet of three waves.*

the extent of the wave packet in space is a measure of the uncertainty in our knowledge of the position of the particle and since the range of wave numbers is a measure of the uncertainty in the wave number, we see that the uncertainty in position equals the reciprocal of the uncertainty in the wave number.

Summarizing, we have the following situation:

Uncertainty in momentum
$$= \text{(Uncertainty in wave number)}h$$
(From the deBroglie equation)

Uncertainty in position
$$= \frac{1}{\text{Uncertainty in wave number}}$$
(A characteristic of the wave group)

Hence we must conclude that

(Uncertainty in position)
$$\times \text{(Uncertainty in momentum)} = h$$

The principle of uncertainty: The uncertainty in position times the uncertainty in momentum approximately equals Planck's **constant.** The more certain we are concerning the location of a particle, the less certain we are concerning its momentum.

Locating Particles

Let us now consider the uncertainty principle from a different point of view. Suppose that we wish to determine the position and the momentum of an electron. To do this, we must use a "probe" of some kind—just we we might use a flashlight beam as a probe in a darkened room to locate an object by observing the light scattered to our eyes by the object. To locate the electron we could first try a beam of radio waves of relatively long wavelength. We would encounter the difficulty that the long radio waves would pass the tiny electron without being affected by it—or affecting it either. An analogy may be helpful: long-wavelength waves on a lake that pass a small pole driven into the lake bottom are not appreciably altered by diffraction at the pole, but just ride past it. We could not tell by examining the waves subsequently that they had passed it. Similarly we could not locate the electron by using long-wavelength radio waves because they would not be diffracted by the electron.

Let us then use electromagnetic waves of a smaller wavelength—X-rays or gamma rays. Just as ripples of short wavelength would be diffracted by the pole in the lake and would signal its presence, the short-wavelength X-rays would be diffracted by passing the electron. We can imagine studying the resulting diffraction pattern as we did in Chapter 37 and locating the electron within a small "ring of uncertainty." However, although we have considerably reduced the uncertainty in the position of the electron, *we have increased the uncertainty in its momentum.* In being scattered by the electron, a photon of the X-ray beam—notice how inseparable are the wave and particle aspects!—collides with the electron in a Compton collision, delivering momentum to it. In the act of locating the electron, we have changed its momentum. The shorter the wavelength we employ, the greater will be the kick given to the electron by the photon. If we backtrack, going to longer wavelengths, at which the photons have less momentum, we can reduce the uncertainty in the momentum of the struck electron, but the electron once again cannot be located without a large uncertainty.

Example. Show that the product of the uncertainty in the position of the electron and the uncertainty in its momentum is about as large as Planck's constant.

In Fig. 41-8, an X-ray photon is scattered into an "X-ray microscope" after colliding with the electron. The ability of the microscope to resolve images can be determined by an extension of the method used in Chapter 37 for determining the angular resolving power of a lens. The result is that the limit of resolving power is reached when two objects are closer to each other than a dis-

Fig. 41-8. *An X-ray microscope for locating an electron struck by an X-ray photon.*

tance Δx where

$$\Delta x = \frac{\lambda}{2 \sin \phi}$$

Here ϕ is half of the angle subtended by the lens at the electron, and λ is the wavelength of the X-rays. This means that we have an uncertainty Δx in our knowledge of the electron's position if we observe that the scattered X-rays have entered the microscope and formed a diffraction pattern at the focus of the objective lens.

Since the scattered photons of momentum h/λ could have entered the microscope *anywhere* within the cone whose apex angle is 2ϕ, the uncertainty in our knowledge of the *direction* of the scattered photon implies an uncertainty in our knowledge of its momentum equal to $(2h/\lambda) \sin \phi$. Since we can imagine that the electron was initially at rest, the law of conservation of momentum requires that an uncertainty of $(2h/\lambda) \sin \phi$ in the momentum of the scattered photon be accompanied by an equal uncertainty in the momentum of the struck electron.

Thus we have

Uncertainty in position of electron

$$= \Delta x = \frac{\lambda}{2 \sin \phi}$$

Uncertainty in momentum of electron

$$= \Delta mv = \frac{2h}{\lambda} \sin \phi$$

and

$$\Delta x \cdot \Delta mv \cong h$$

Another example of the uncertainty principle concerns the width of a diffraction pattern. When photons pass through a narrow slit, they are distributed into a diffraction pattern. The narrower the slit (that is, the smaller the uncertainty in the position of the entering photons), the greater the width of the diffraction pattern (that is, the greater the uncertainty in the momentum of the photons after they pass the slit).

Applications of the uncertainty principle to objects on the everyday scale of things—jet airliners, baseballs, etc.—will be left to the problems at the end of the chapter. Because of the small magnitude of Planck's constant, uncertainties in position and momentum of these large objects are negligibly small.

Philosophical Consequences of the Uncertainty Principle

These considerations led physicists to the recognition that the uncertainty principle expresses a fundamental truth about the atomic world. Our effort to obtain more and more detailed knowledge about one physical quantity leads to such a disturbance of the system that we cannot expect to know a related physical quantity with arbitrarily great precision. Newtonian mechanics, you remember, was based on the notion that, given precise information about the initial position and velocity of a body, we could hope to work out its motion thereafter by applying the laws of mechanics. The program of Newtonian mechanics will not work in the atomic world because we cannot with precision know corresponding initial values of the position and momentum of an atomic particle as we can those of a planet in the solar system.

Other sciences and philosophy have been profoundly affected by the realization that the uncertainty principle sets fundamental limits upon our ability to describe nature. In biophysics, biochemistry, and psychology, for example, efforts are being made to understand the basis of human thought. Much has been learned about the physical nature of the brain. Will it be possible to relate physical behavior—the electrical currents in the brain, for example—to the thought patterns and behavior patterns studied by the psychologist? Some scientists believe that the uncertainty principle may limit our knowledge in this area, as in others, because the act of measuring the physical quantities that describe the action of the brain may change the thought processes being studied. In philosophy, the uncertainty principle and other developments in modern physics have led to a reexamination of causality, according to which a precisely determined set of conditions will always produce exactly the same effects. As we have seen, we cannot

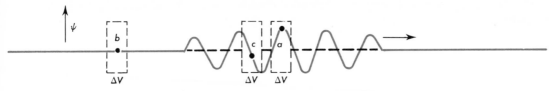

Fig. 41-9. The amplitude ψ of the wave function. Where ψ² is large—as at a—the probability of finding the particle within a small volume ΔV is large.

precisely determine a complete set of initial conditions in the atomic domain. Further discussion of these questions will be found in the references at the end of the chapter.

Waves of What?

Previously we studied displacement waves in strings and on water, compressional waves of sound, and wavelike variations of electric and magnetic fields in electromagnetic radiation. What kinds of periodic changes do matter waves describe?

Max Born made the suggestion in 1926 that the matter waves are a mathematical device to describe the *probability** of finding a single particle within a specified small volume or to compute the probable distribution of a large number of particles. The magnitude of the *wave function* that describes the matter waves associated with a particle is a

---

* You will recall from elementary mathematics that in a large number of tosses of a coin, one-half of them will turn up "heads." Although the concept of probability is not a simple one, this result is usually interpreted to mean that the probability of throwing a "head" or a "tail" is 1/2. The probability P of throwing a "head" is then the ratio of n_H, the number of "heads," to N, the total number of throws. Similarly, the probability that a "two" will come up in many throws of a die is 1/6. If a roulette wheel has 16 spaces with odd numbers and 16 with even numbers, the probability of finding the ball in a space with an even number is 1/2. Again the probability is the number of desired outcomes (16) divided by the number of possible outcomes (32). Most physical events and almost all sociological and economic ones cannot so easily be assigned probabilities because the outcomes are not independent of each other and are not "equally probable."

quantity ψ (Greek *psi*) such that the probability of finding the particle within a small volume ΔV is

$$\text{Probability} = \psi^2 \, \Delta V$$

Where the wave function instantaneously has a large value—as at *a* in Fig. 41-9—the probability of finding the particle within a small volume ΔV is large. The probability of finding the particle at *b* or at *c* at that instant is small.

The situation is similar to that for electromagnetic waves. We started by discussing electric and magnetic fields in terms of the forces exerted on particles. Then we considered electromagnetic waves in which the electric and magnetic field components could have an existence independent of the charges and currents that produced them. Finally we came to regard the magnitude of the electric field at a given point as describing the probability of finding a photon at that point. The electromagnetic wave was a mathematical way of stating concisely and in a way consistent with experiment the "behavior" of photons. Similarly, the *wave function,* which is the term we shall apply from now on to matter waves, describes the "behavior" of material particles. In the next chapter, we shall consider some of the simpler properties of wave functions and some of the information they can give us.

According to the uncertainty principle, it is impossible to determine today the exact position and momentum of even one electron. Hence we cannot speak with certainty about its position yesterday or tomorrow. We *can*, however, estimate *probable* positions and momenta by means of the electron's wave function. A life insurance company from statistical studies confidently predicts the *probable* life span of its clients — and manages to stay in business. So the physicist uses the laws of statistics and probability to predict the probable behavior of electrons, atoms, and even the universe and makes progress in understanding the nature of things.

Born has summarized the situation as follows:

"We now have a *new form* of the law of causality, which has the advantage of explaining the objective validity of statistical laws. It is as follows: if in a certain process the initial conditions are determined as accurately as the uncertainty relations permit, then the probabilities of all possible subsequent states are governed by exact laws."

The Electron Microscope

The optical microscope cannot reveal the structures of bodies much smaller than the wavelength of light, a few hundred-thousandths of an inch. The development of the electron microscope made it possible to study the structures of objects 1/50 as large as this.

The greater resolving power of the electron microscope arises from the fact that the wavelengths of electrons are so much smaller than those of light waves that diffraction effects are much less important. In a common type of electron microscope, the electrons are refracted by electric fields or magnetic fields instead of by lenses (Fig. 41-10). In practice, the resolution of images by the electron microscope is not limited by the wavelength of the electrons so much as by the stability requirements for the currents that establish the magnetic fields within the microscope. Specimens must be thin enough to transmit the electron beam. The electron microscope

reveals a wealth of detail about the structure of bacteria and viruses (Fig. 41-11).

Summary

DeBroglie proposed that particles, such as electrons, have associated with them a "matter wave" of wavelength $\lambda = h/mv$ where h is Planck's constant and mv the momentum of the particle. Germer and Davisson and Thomson subsequently observed the matter waves of electrons in the laboratory.

DeBroglie's equation and Planck's equation ($E = hf$) require that $u = c^2/v$ where u is the phase velocity of the matter waves, v is their group velocity, and c is the velocity of light in vacuum. The group of waves that travels with a particle is localized in space and has a velocity less than c because of the requirements of special relativity. This requires u to be greater than c.

The phase velocity is the velocity with which a given phase of a wave — a crest, for example — moves through space. The group velocity is the velocity with which the group of waves travels.

Bohr's principle of complementarity: one can describe physical phenomena in terms of particles or in terms of waves. The particles are described in terms of their momentum and their energy; the waves, by their wavelength and frequency. The two descriptions are connected by $E = hf$ and $mv = h/\lambda$.

Matter waves in free space are a wave packet or wave group, created by the interference of simple harmonic progressive waves. The wave packet travels with a group velocity v, equal to that of the particle.

The principle of uncertainty: the uncertainty in the position of an object times the uncertainty in its momentum approximately equals Planck's constant. The uncertainty principle reminds us that if we use short-wavelength radiation to locate a particle with precision, we disturb the momentum of the particle.

The matter waves are described by wave

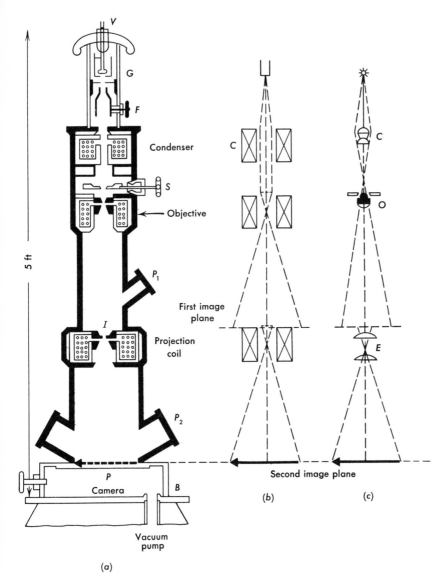

Fig. 41-10. The electron microscope. (a) *Arrangement of the focusing coils. The electrons, emitted by a hot filament* G, *are converged to the specimen at* S. *The magnetic lenses marked "Objective" and "Projection coil" form images of the specimen by focusing the transmitted electrons.* (b) *How the magnetic lenses focus electrons.* (c) *The corresponding parts of an optical microscope. (Reprinted with permission from O. Stuhlman,* Introduction to Biophysics, *John Wiley and Sons, 1943.)*

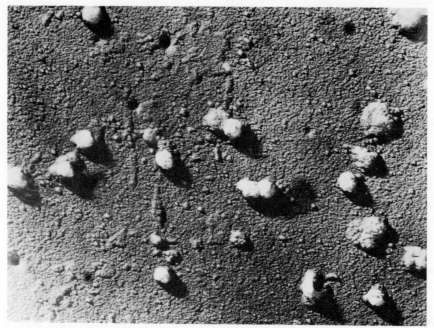

Fig. 41-11. *An electron micrograph. Influenza virus magnified 44,000 times. The specimen has been coated with uranium to give the effect of a shadow. (Courtesy of A. J. Tousimis.)*

functions. The amplitude ψ of the wave function at a given instant and at a given point is such that $\psi^2 \, \Delta V$ is the probability of finding the particle within a small volume ΔV at that point.

Questions

1. What is a deBroglie wave? Derive the equation for the deBroglie wavelength of a particle.

2. Discuss the wave-particle view of light and compare it with that for particles.

3. What was the process of reasoning that led deBroglie to think that electrons might have wave properties? Discuss the strengths and weaknesses of that process as a tool for scientists.

4. Why do we not observe the wavelengths of bodies the size of baseballs?

5. Can a car moving at 60 mi/hr be diffracted?

6. What is your wavelength when you run the 100-m dash in 10.0 sec? What would be the uncertainty of your hitting the tape?

7. How do group and phase velocity differ?

8. How can phase velocity exceed that of light in vacuum?

9. Why does an electron microscope have a higher usable magnification than an optical one?

10. Which has a shorter wavelength: a 10-ev photon or a 10-ev electron?

11. Can neutrons, protons, and electrons be diffracted? Describe experimental arrangements for observing their diffraction.

12. What effect has the uncertainty principle had on philosophy?

13. How do these examples illustrate the uncertainty principle: (*a*) trying to determine in visible light what an unexposed photographic plate looks like, (*b*) trying to get the exact temperature of a cup of coffee with a

thermometer that has been at room temperature?

14. What oscillates to give matter waves? If you tried to study the "medium" of matter waves, what do you think would happen?

Problems

(1 ev = 1.60 × 10⁻¹⁹ joule. Neglect relativistic effects.)

A

1. Calculate the mass of a 5000-A photon.
2. What wavelength is associated with a particle of mass 10^{-20} kg that is moving at a speed of 10^5 m/sec?
3. Calculate the wavelength of a 70-kg man who is walking with a velocity of 1.5 m/sec.
4. What is the wavelength of a proton moving with a velocity of 10^5 m/sec?

B

5. What is the velocity of a proton with a deBroglie wavelength of 0.15 A?
6. What is the wavelength of an electron that has been accelerated by 15 kv?
7. What is the energy of a proton that has a deBroglie wavelength of 1.5 A?
8. What are (a) the velocity and (b) the wavelength of a 5.0-Mev alpha particle?
9. What is the deBroglie wavelength of a 10-Mev deuteron?
10. Calculate the wavelength of a 2.0-ev electron.
11. Compare (a) the energy and (b) the momentum of a 1.0-A electron and a 1.0-A photon.

C

12. The velocity of a moving proton is measured with an uncertainty of 10^{-4} m/sec. With what uncertainty can we locate its position?
13. If the location of the center of gravity of a 100-kg cylinder has been measured with an uncertainty of 1000 A, what is the uncertainty in its velocity?
14. Calculate the uncertainty of the loca-

tion of the center of gravity (a) of a 10^5-kg airplane whose velocity of 0.25 km/sec is known with an uncertainty of 0.001 km/sec; (b) of a 0.15-kg ball with a velocity uncertainty of 1.0 mm/sec.

15. If we are certain that an electron is bound to an atom of diameter 3 A, what is the uncertainty in the momentum of the electron?
16. An alternative form of the uncertainty principle is

$$\Delta t \cdot \Delta E \cong h$$

where Δt is the time interval during which a system is studied and ΔE is the corresponding uncertainty in a measurement of energy of the system. If the time it takes for an excited atom to radiate (its radiation lifetime) is 5.0×10^{-9} sec, what uncertainty must be associated with the energy of the atom?

17. Using diffraction theory (Chapter 37), compare the resolution of an image formed by 10,000-volt electrons in an electron microscope with that of 5890-A light if the size of the aperture is the same for both.

For Further Reading

Born, Max, *The Restless Universe,* Dover Publications, New York, 1951.

Darrow, Karl K., "Quantum Theory," *Scientific American,* March, 1952, p. 47.

Dirac, P. A. M., "The Physicist's Picture of Nature," *Scientific American,* May, 1963, p. 45.

Gamow, George, *Mr. Tompkins Explores the Atom,* Macmillan Company, New York, 1944.

Gamow, George, *Mr. Tompkins in Wonderland,* Macmillan Company, New York, 1942.

Gamow, George, "The Principle of Uncertainty," *Scientific American,* January, 1958, p. 51.

Kac, Mark, "Probability," *Scientific American,* September, 1964, p. 92.

Schrödinger, E., "What is Matter," *Scientific American,* September, 1953, p. 52.

CHAPTER FORTY-TWO

The Methods of Quantum Mechanics

In regions where there are no fields of force, the wave function (matter wave) associated with an electron can be represented as a moving wave packet. If we know the magnitude of the wave function at a given point, we can compute by the methods of *quantum mechanics* how likely it is that the electron will be found within a tiny volume at that point.

Some of the most interesting situations in quantum physics, however, are those in which electrons and other atomic particles move in regions where there are fields. In these regions the potential energy of the particle changes. An electron making a transition between energy levels in a hydrogen atom, an alpha particle approaching an atomic nucleus, a conduction electron in a crystal lattice — all move under conditions of changing potential energy. How can these physical systems be described by the methods of quantum mechanics?

In planning an experiment or developing a theory of electron motion, the physicist must not forget the uncertainty principle. It says that one cannot specify exactly where the electron is without being completely ignorant concerning its momentum. If we know the position of the electron with an uncertainty Δx, we can expect the uncertainty in the momentum to be approximately $h/\Delta x$. Because of these basic uncertainties in our knowledge of the initial conditions, we describe the atomic world by statistical laws. We shall see *how* in more detail in this chapter.

We shall first discuss some simple examples — to give you an idea of the strategy and the objectives of quantum mechanics. Later we shall give the results of the theory in more complicated cases. The quantum theorist can obtain quantitative results of great generality from his elegant and powerful mathematical methods. The basic ideas of physics underlying quantum mechanics are simple, but they are stated in mathematical language requiring a greater knowledge of mathematics than we have assumed in this book.

We shall ask you in this chapter to accept many of the results of the theory in qualitative form and without mathematical proof because advanced mathematics is required for the derivations. We hope that you will decide later to acquire the necessary background in mathematics for a detailed study of quantum mechanics.

Quantum mechanics is one of the two major basic developments in physics in this century. The other is relativity theory. Why has quantum mechanics come to play such an important role in physics? (1) It is logically complete and elegant. (2) It contains built-in formal restrictions that prevent us from overlooking the limitations set by the uncertainty principle. (3) It is more powerful than the theories that preceded it. For example, although the Bohr theory correctly predicted the wavelengths of the spectral lines of hydrogen, quantum mechanics not only predicts those wavelengths, but also predicts the observed *intensities* of the lines.

Refraction of Waves

Before we discuss matter waves further, let us review some of the things we learned earlier about the refraction of waves. Figure 42-1a shows ripples on water that move obliquely across the boundary from deep water to shallower water. The ripples have a smaller phase velocity in shallow water than in deep water. They are refracted toward the normal to the boundary, their wavelength in shallow water being less than, and their frequency the same as that in deep water.

Light waves (Fig. 42-1b) are refracted to-ward the normal in moving from air into a flat slab of glass. In glass, their phase velocity is less than in air, their wavelength is less, and their frequency is unchanged.

Consider now the beam of electrons in Fig. 42-1c that leaves the electron gun, moves into the upper metal box at zero potential, and then travels through small openings from the upper box into the lower one, whose electrical potential is +100 v. As the electrons cross the gap between the two boxes, they are accelerated by an electric field (just as protons in a cyclotron are accelerated in crossing the gap between the "dees"). The electrons undergo an increase in the downward component of their velocity. When they enter the lower box, they have been refracted *toward the normal,* although their velocity has *increased.* If the potential of the lower box were greater than +100 v, evidently the electron beam would be bent more strongly toward the normal, indicating a greater index of refraction.

What about the deBroglie wavelength of the electrons? Since $\lambda = h/mv,$ since *v* has increased on crossing the gap, and since *m* can only increase with increase in velocity, we conclude that the wavelength of the electrons has *decreased.*

The beam of electrons can be shown to have an index of refraction which depends

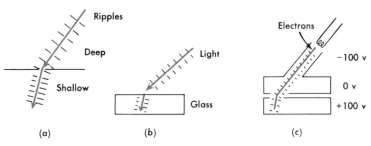

(a)   (b)   (c)

Fig. 42-1. Refraction. (a) *Ripples are refracted when they enter shallower water, in which their speed is less.* (b) *Light is refracted on going from air to glass, in which its speed is less.* (c) *Electrons are refracted in moving from a region of lower to one of higher electrical potential, in which their speed is* greater.

upon the electrical potential in which the electrons move, or more generally, upon the *potential energy* of the electrons. The deBroglie wavelength of each electron also depends upon the potential energy of the electron in a given region, as we can readily see from the following argument.

Let E be the total energy of the electron and E_P its potential energy. Then the kinetic energy E_K is

$$E_K = E - E_P$$

and, in nonrelativistic situations,

$$\tfrac{1}{2}mv^2 = E - E_P$$

or

$$v = \sqrt{2(E - E_P)/m}$$

Hence the deBroglie wavelength is

$$\lambda = \frac{h}{m\sqrt{2(E - E_P)/m}} = \frac{h}{\sqrt{2m(E - E_P)}}$$

If the total energy E of the electron is constant, λ varies with the potential energy E_P of the electron.

Knowing how the index of refraction and the wavelength of a matter wave change with potential energy in electric and magnetic fields, for example, one can work out the laws of "particle optics." He can treat the refraction, reflection, and diffraction of electron beams in a way that is similar to that followed in discussing the behavior of electromagnetic waves. Such an approach is followed, for example, in designing electron microscopes

(Chapter 41). The subject of electron optics, however, is a large one, and, except for some questions and problems at the end of this chapter, we shall not pursue it further here.

Matter Waves and the Quantization of the Hydrogen Atom

What would we expect to happen when an electron approaches a "bare" proton to form a hydrogen atom? In Fig. 42-2 we show the *refraction* of the electron wave toward the proton where the positive potential is great. In the "particle language" of the early Bohr theory of the hydrogen atom, we say that the electron is attracted into an orbit near the proton. In "wave language," we say that the waves are refracted by entering a region of higher electrical potential (lower potential energy) and lap around the proton. The refracted waves might be expected to close upon themselves and to interfere, setting up a standing wave pattern around the proton, provided that a certain relationship existed between the wavelength and the distance around the proton. This is in fact a fruitful prediction that is in quantitative agreement with the Bohr theory.

DeBroglie first noticed that the wavelengths of an electron matter wave fitted into the corresponding Bohr orbit *an integral number of times.* Using the results of Chapter 40, we calculate the radius r_1 of the first Bohr orbit and find it to be

$$r_1 = \frac{h}{2\pi v_1 m}$$

Now the circumference of the first Bohr orbit must therefore be

$$2\pi r_1 = \frac{h}{mv_1}$$

But the deBroglie wavelength of an electron whose velocity is v_1 is

$$\lambda = \frac{h}{mv_1} = 2\pi r_1$$

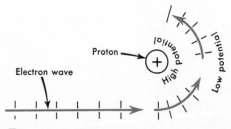

Fig. 42-2. The refraction of an electron as it approaches a proton.

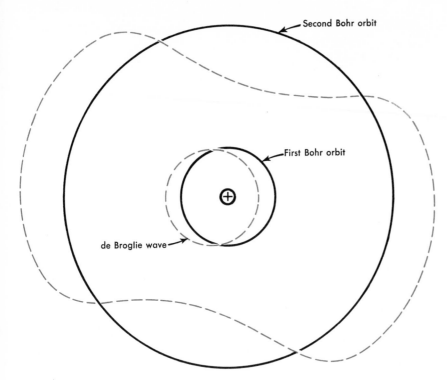

Fig. 42-3. *Quantization of the hydrogen atom. The deBroglie wavelength fits once into the first Bohr orbit, twice into the second, and so on.*

Thus we see that the ratio of the deBroglie wavelength (at the velocity of the electron on the first Bohr orbit) to the circumference of the first Bohr orbit is **unity. The matter wave fits into the circumference just once** (Fig. 42-3).

If we calculate the deBroglie wavelength and the orbit circumference when the electron is on the second Bohr orbit, we find that the wavelength is contained just *twice* in the circumference (Fig. 42-3). In general, the circumference of the nth Bohr orbit is exactly n times the deBroglie wavelength at the velocity of the electron upon that orbit. Just as the number of loops in the standing wave pattern on a vibrating string (Chapter 19) fits into the length of the string an integral number of times, the deBroglie wavelength fits into the circumference of the corresponding Bohr orbit an integral number of times. If a thin, highly elastic metal hoop is suspended and struck a sharp blow with a hammer, it will vibrate in a standing wave pattern that is somewhat similar to the one we visualize for the deBroglie waves.

This astonishing and pleasing result showed that quantization—one of the most successful features of the Bohr theory—could be understood in terms of the behavior of matter waves. The implications of this were worked out by Erwin Schrödinger, Werner Heisenberg, Max Born, Wolfgang Pauli, and others in the development of quantum mechanics. Before we continue our discussion of the hydrogen atom from the viewpoint of quantum mechanics, let us consider a particularly simple example of standing matter waves in a quantized system.

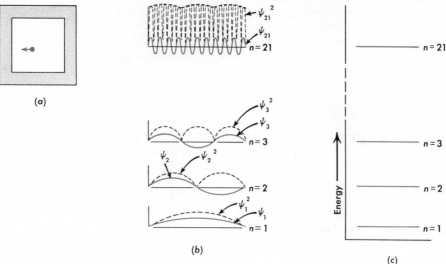

Fig. 42-4. A particle in an impenetrable box.

A Particle in a Box

A moving particle confined within an impenetrable enclosure—a "box"—represents one of the simplest systems we can consider. Where would we expect to find the particle (Fig. 42-4a) and what would its expected energy be at any instant? Classical mechanics would answer: "Anywhere, with equal probability, within the enclosure" and "Any energy within a continuous range of values, as determined by the initial situation." Quantum mechanics has quite different answers.

Since the particle has to be within the box, the value of the wave function ψ—the amplitude of the particle wave—beyond the walls of the box must be zero. You will understand why this is so if you remember that we have agreed to interpret ψ^2 as a quantity proportional to the *probability* of finding a particle within a tiny volume ΔV at the point for which ψ is given. If the particle *cannot* be outside the walls, the probability of finding it outside the walls is zero, and ψ^2 and ψ must be zero always and everywhere outside the box.

In this example, as in deBroglie's interpretation of the electron waves near the hy-

drogen nucleus, it is just as if the wave function represented a standing wave. In this case, it is like a standing wave on a vibrating stretched string that is clamped at the walls of the box. Various modes of vibration are possible provided they are consistent with the requirement that the ends of the string never vibrate. Remember that the "vibrating string" is merely an analogy. What we are really saying is that the wave function that describes the particle has to satisfy certain reasonable special requirements, including the requirement that its value beyond the walls be always and everywhere zero.

Energy Levels for a Particle in a Box

We saw in Chapter 19 that a standing wave system is characterized by nodes and antinodes. The distance between two successive nodes is $\lambda/2$, and this distance must fit into the length of the vibrating string an integral number of times. Applying that idea to our standing matter waves, letting λ_n be the wavelength of the matter wave that fits into the length l of the box just n times, we have

$$\frac{l}{\lambda_n/2} = n$$

or

$$\lambda_n = \frac{2l}{n}$$

A few of the many matter waves that meet this requirement are sketched in Fig. 42-4b. Matter waves of this deBroglie wavelength are associated with a particle of mass m whose velocity v_n is

$$v_n = \frac{h}{m\lambda_n} = \frac{nh}{2ml}$$

Therefore the kinetic energy of the particle when its wavelength is λ_n is

$$(E_K)_n = \tfrac{1}{2}mv_n{}^2 = \frac{n^2 h^2}{8ml^2}$$

Since the potential energy of the particle is constant and we may choose the zero of potential energy where we please, let us set the potential energy of the particle equal to zero. Therefore, the total energy of the particle is

$$E_n = \frac{n^2 h^2}{8ml^2} \qquad (1)$$

When $n = 1$, $E_1 = h^2/8ml^2$; when $n = 2$, $E_2 = 4h^2/8ml^2$; and so on. The energy can have no other values than those given by equation (1) because—to repeat—other energies would be associated with matter waves whose half-wavelengths would not fit into the length of the box an integral number of times. The system is *quantized*, and n is the quantum number. An energy-level diagram is given in Fig. 42-4c.

Where is the Particle?

Figure 42-4b shows the wave functions ψ (solid curve) and the squared wave functions ψ^2 (dashed curve) for the quantum states, or conditions, characterized by $n = 1$, 2, 3, . . . , 21, ... Although the figure shows

all of these wave functions, the particle is in only one of these states at one time.* Consider the state $n = 1$. ψ^2 has a large amplitude at the center of the box and zero amplitude at the walls. Our quantum mechanical prediction of the location of the particle would be that the particle is most likely to be found near the center of the box and least likely near the walls. If we should give the particle just the right amount of additional energy—enough to take it from energy $h^2/8ml^2$ to energy $h^2/2ml^2$—the particle would move from state $n = 1$ to $n = 2$. The probability of finding the particle near the center would now be zero and the probability would be greatest that the particle would be about $\tfrac{1}{4}l$ from either wall.

When the particle is given enough energy to enter one of the very high quantized states ($n = 21$ is shown, but you might also imagine $n = 501,328$), the standing matter wave has so many nodes and antinodes that the curve for ψ^2 is hardly distinguishable from a straight horizontal line. Since a straight horizontal line for ψ^2 would represent equal probability of finding the particle anywhere within the box, we see that *the quantum mechanical prediction merges into the classical one at high quantum numbers.* The principle that the quantum mechanical solution agrees with the classical solution at high quantum numbers is called the *correspondence principle* and was first stated by Niels Bohr.

A baseball rolling in a box does not show observable quantum mechanical effects, although its behavior cannot violate the laws of quantum mechanics. The explanation is that for everyday energies and for bodies of everyday size, the system is in such a high quantum state that the extremely small variations in ψ^2 cannot be detected by any experiments that we can perform—even if one could realize the conditions of no energy dissipation within the box.

* To be more exact, the particle will be only in that state for which a precise measurement of the energy yields the quantized energy corresponding to that state.

Example. A 0.10-kg baseball rolls in a 1-m cubical box with a velocity of 0.10 m/sec. In what quantum state is it? (Order of magnitude.)

$$E_n = \tfrac{1}{2}mv^2 = \tfrac{1}{2}0.10 \text{ kg}(0.10 \text{ m/sec})^2$$
$$= 5 \times 10^{-4} \text{ joule}$$

$$E_n = \frac{n^2h^2}{8ml^2}$$

$$5 \times 10^{-4} \text{ joule} = \frac{n^2(6.6 \times 10^{-34} \text{ joule-sec})^2}{8(0.10 \text{ kg})(1 \text{ m})^2}$$

$$n \cong 10^{31}$$

Our description of the particle in the box is consistent with the uncertainty principle. The uncertainty Δx in the position of the particle in the box is roughly equal to l. According to the uncertainty principle, the uncertainty in momentum is therefore

$$\Delta mv \cong \frac{h}{\Delta x} \cong \frac{h}{l}$$

or

$$\Delta v \cong \frac{h}{ml}$$

and the uncertainty in the kinetic energy is approximately (order of magnitude) h^2/ml^2. Since the kinetic energy does not become less than zero, we would expect the least value of kinetic energy that is consistent with an uncertainty of h^2/ml^2 to be around h^2/ml^2. This is approximately the quantized energy of the lowest energy state ($n = 1$).

A Particle in a "Leaky" Box

The situation of an electron in an atom or of an alpha particle that is emitted by a radioactive nucleus is quite different from that of the baseball in the box. The masses and the energies are much less, and the quantum-mechanical effects are very prominent, as we shall see. Even the quantum-mechanical behavior of small particles is different from that of the idealized "particle in a box." We

have assumed that the walls of the enclosure are impenetrable. In fact, small particles are not in a tight box—one that represents an energy well with vertical walls that go to infinity—but rather in a more or less "leaky" box (Fig. 42-5a). Given enough energy, the particle can escape. This possibility modifies the energy levels and the shape of the wave functions *even when the particle has an energy less than that required to escape from the box according to classical mechanics* (Fig. 42-5b).

Figure 42-5c indicates the kind of results one obtains from quantum mechanical calculations, in this case for the lowest state ($n = 1$). The ψ curve does not become zero at the walls of the box. Instead, it has a tail of non-zero values in the space beyond the box. This means that according to quantum mechanics there is a small (but finite) possibility of finding the particle *outside the box*. This is so even though its energy may be less than E_0, the energy required to escape from the well according to classical mechanics. *The particle can "tunnel through" the wall.*

This puzzling result seems at first to be completely at odds with common sense. We must remember, however, that we are attempting to describe the world of the atom. We must be led by the experimental evidence—the hints that nature gives us—and by logical deductions from our mathematical model for the situation. In fact, we shall see that the idea that a particle can "leak" out of an energy well is a very fruitful one in many situations, including radioactive decay, and leads to results that are in excellent agreement with experiment.

The Hydrogen Atom

We conclude this chapter by examining the way in which quantum mechanics describes the hydrogen atom. Like the Bohr theory, quantum mechanics predicts that the energy of the hydrogen atom is quantized. In fact, the

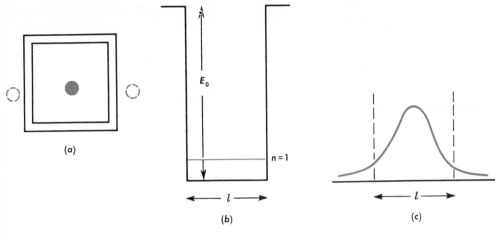

Fig. 42-5. (a) *A particle in a "leaky" box.* (b) *The corresponding energy well with the energy level for* n = 1. (c) *The wave function for* n = 1. *It does not become zero at the walls of the box.*

quantized energies—the energy levels—are the same in quantum mechanics as in the Bohr theory. However, instead of the Bohr picture of a classical particle traveling on a planetary orbit, quantum mechanics concerns the calculation of the wave function when the electron is in the potential-energy well of the hydrogen atom (Fig. 42-6).

Just as a stretched string has potential energy because of an elastic force, the electron-proton system has potential energy because of the inverse square law force of electrostatic attraction. We saw an analogy between the modes of vibration of a stretched string and the wave function of a particle in a box. However, we had to remember that the vibrations of the string represented displacements in space and the variations in the wave function represented a probability distribution. Similarly we could compare the wave function for the hydrogen atom to a vibrating system. Since a three-dimensional description is required to bring out the significant features, we would have to think of standing waves in a vibrating ***three-dimensional*** system. Two-dimensional standing waves can readily be observed. One can spread

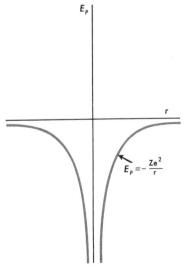

Fig. 42-6. *The potential energy of the hydrogen atom.* r *is the separation between the electron and the proton.*

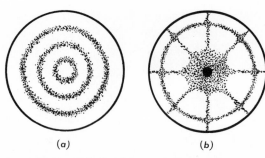

Fig. 42-7. Chladni figures. Nodes and loops of (a) *a vibrating drumhead clamped at the edge and* (b) *a vibrating cymbal plate clamped at the center.* (*From White's* Modern College Physics, *Fourth Edition, copyright 1962, D. Van Nostrand Co., Inc., Princeton, N.J.*)

cork dust on a metal disk and set the disk into vibration. The cork dust collects along nodal lines of zero vibration, forming characteristic patterns called *Chladni figures* (Fig. 42-7) that show the mode of vibration. Such two-dimensional standing waves exhibit nodal and antinodal lines just as a vibrating string shows nodal and antinodal points. They have some resemblance to the wave functions for a two-dimensional system. Three-dimensional vibrations are more difficult to observe, however. Since the analogy does not carry us very far, we shall not discuss it further. Instead, we shall discuss the kind of mathematical program one follows in calculating the wave function for the hydrogen atom.

Procedures in Quantum Mechanics

In solving problems earlier in this book, you found solutions to equations under a particular set of physical circumstances. Consider this very simple example: A freely falling body is released from rest from the top of a building 64 ft high. When will it strike the ground? The equation $s - v_0 t - \frac{1}{2} g t^2 = 0$ is set up from fundamental principles to represent the relation between s and t when the

motion is *uniformly accelerated.* (It does *not* hold for a simple harmonic oscillator.) The "answers" $t = 2$ sec and $t = -2$ sec are *mathematical solutions* of the equation. The values ($s = 64$ ft, $t = 2$ sec) and ($s = 64$ ft, $t = -2$ sec) reduce the equation to zero for the condition that v_0 is zero. We *reject* the solution $t = -2$ sec because it does not make physical sense, although mathematically it is unobjectionable.

Our problem in quantum mechanics is rather similar. The theorist finds solutions to a governing equation under the conditions set by the physical problem. He examines the mathematical solutions and accepts those that make physical sense. The governing equation that meets the requirements for physical and mathematical acceptability is called Schrödinger's wave equation. It is a differential equation involving ψ, the time t, the potential energy E_P, the space coordinate r, Planck's constant, and the mass of the electron. Schrödinger's equation requires a knowledge of advanced mathematical procedures to understand its statement and the methods for solving it, and we shall not go into these topics. The condition under which it is to be solved in dealing with the hydrogen atom is that the potential energy be given by $E_P = -Z e^2 / r$ where Z is the atomic number ($Z = 1$ for hydrogen) and e is the charge on the electron.

From the mathematical solutions, the wave function is formulated. *Not all of the mathematical solutions are physically acceptable.* We can use only those that are consistent with our ideas of how a respectable wave function should behave. For example, a wave function cannot become infinitely great anywhere because the total probability of locating a particle *somewhere* cannot exceed 1 (a probability of unity represents a certainty). Any mathematical solution to Schrödinger's equation that becomes infinite anywhere must be rejected. Likewise, any mathematical solution that cannot be summed over all space to give a total probability of 1 cannot be accepted for the same reason.

Quantum Numbers

What is particularly important and pleasing is that the mathematical conditions that are necessary to make the mathematical solutions to Schrödinger's equation "well-behaved" turn out to have physical significance. Certain quantities control the well-behavedness of the solutions. When these quantities—called quantum numbers—take on certain allowed values, the solutions in which they appear act in a well-behaved fashion. For other values, the solutions are not well-behaved and must be rejected. The *principal quantum number* *n* is associated with the energy levels and assumes the same values in quantum mechanics as it does in the Bohr theory: for *n* = 1, 2, 3, the mathematical solutions to the wave equation are well-behaved. The energy values corresponding to these well-behaved solutions are given—just as in the Bohr theory (Chapter 40)—by

$$E_n = -\left(\frac{2\pi^2 k_0{}^2 Z^2 e^4 m}{h^2}\right)\frac{1}{n^2}$$

The difference is that in the Bohr theory the integral values for *n* came from an arbitrary assumption about the values of angular momentum. In quantum mechanics, the integral values of *n* drop out of the mathematical calculations as conditions upon the physical interpretation of the solutions to Schrödinger's equation.

The principal quantum number *n* is not the only quantity that determines the nature of the solutions. A second quantum number—the *orbital quantum number* *l*—assumes integral values that are simply related to those of the principal quantum number:

$$l = 0, 1, 2, \ldots, n - 1$$

This means that for a given value of *n*, *l* may take on all positive integral values from and including zero to *n* − 1. For example, in the lowest energy state, *n* = 1 and *l* = 1 − 1 = 0. When *n* = 2, *l* = 0 or *l* = 1, and so on. We shall see in the next chapter that the orbital quantum number, as its name suggests, is related to the angular momentum of the atom in that state.

For each value of *n*, there is (except for *n* = 1) more than one possible value of the orbital quantum number *l* for which the solutions of Schrödinger's equation will be well-behaved. Similarly, for each value of *l*, a third quantum number, the *magnetic quantum number m,* is allowed to assume any integral value from −*l* to +*l*. For example, when *l* = 0, *m* = 0. When *l* = 1, *m* = −1, 0, or +1. When *l* = 2, *m* = −2, −1, 0, 1, 2. The magnetic quantum number will turn out to be related to the orientation of the atom in a magnetic field.

Later we shall find it necessary to introduce a fourth quantum number—*the spin quantum number.* For the purposes of this chapter, the three quantum numbers *n, l,* and *m* will suffice. Table 42-1 summarizes the various sets of values of *n, l,* and *m,* each set on a horizontal line representing an allowed solution to Schrödinger's equation. We have placed brackets around the different sets of quantum numbers that correspond to each value of *n.* Notice that there is *one* solution corre-

Table 42-1.

Principal Quantum Number *n*	Orbital Quantum Number *l*	Magnetic Quantum Number *m*
1	0	0
2	0	0
2	1	−1
2	1	0
2	1	1
3	0	0
3	1	−1
3	1	0
3	1	1
3	2	−2
3	2	−1
3	2	0
3	2	1
3	2	2

sponding to $n = 1$; *four* solutions corresponding to $n = 2$; *nine* solutions corresponding to $n = 3$. In general, there are n^2 different allowed solutions for each value of n. For each value of n, in the absence of an external magnetic or electric field, the n^2 solutions will all correspond to the same energy. When external fields are imposed, the energy levels break up into sublevels of slightly differing energies, as we shall see in the next chapter.

Hydrogen Wave Functions

We have been concerned with the wave function ψ that describes statistically the whereabouts of the electron in the hydrogen atom. The explicit forms of the hydrogen wave functions go beyond the mathematical scope of this book, but we can give some graphs of these mathematical expressions. Figure 42-8 gives the *radial probability density* versus distance, that is, the variation in ψ^2 with increasing distance from the nucleus—for several quantum states. Notice that the most probable location of the electron in the lowest energy state is at a distance r_1 from the nucleus roughly equal to the radius of the first Bohr orbit. Otherwise there is no suggestion of a planetary orbit in the quantum mechanical description of the hydrogen atom. Likewise, for the state $n = 2$, $l = 0$, there is a large probability that the electron will be found at a distance r_2 equal to the radius of the second Bohr orbit, but the electron may also be located at $r = r_1$ with a certain probability. For $n = 2$, $l = 1$, the radial probability density changes as shown in the third graph. Quantum states with $l = 0$ are called—largely for historical reasons—*s*-states and those with $l = 1$ are called *p*-states.

The "fuzziness" inherent in our quantum mechanical description of the atom is illustrated by the drawings of Fig. 42-9, which are

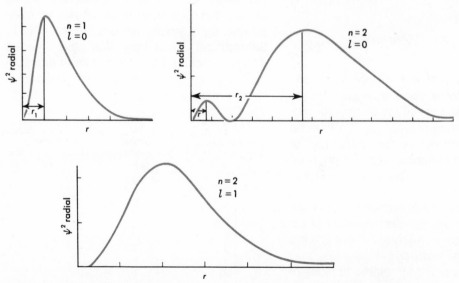

Fig. 42-8. *The radial probability density for the hydrogen atom for the states* $n = 1$, $l = 0$; $n = 2$, $l = 0$; *and* $n = 2$, $l = 1$. *The unit of* r *is* $r_1 = 0.529\,A$, *the radius of the first Bohr orbit. (Adapted from F. K. Richtmyer and E. H. Kennard,* Introduction to Modern Physics, *McGraw-Hill Book Company, 1942. Used by permission.)*

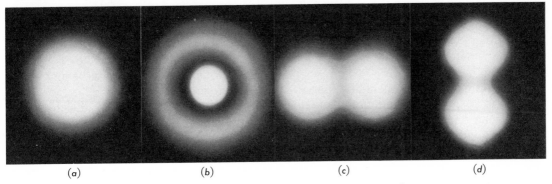

Fig. 42-9. *The charge distribution in the hydrogen atom for several states:* (a) *The 1-s state, when the electron is on the lowest energy level.* (b) *The state for which* n = 2, l = 0. (c) *The state for which* n = 2, l = 1, *and* m = −1 *or* +1. (d) *The state for which* n = 2, l = 1, *and* m = 0.

meant to suggest the matter distribution—the way in which the electron is "spread out"—in the hydrogen atom in various states. Since the electron has an electrical charge, this is also the charge distribution. Contrast this electron cloud with the sharpness of the orbits of the Bohr atom in Chapter 40 or as illustrated in the familiar "atom" symbol of the popular magazines ⊕. The change from the sharp Bohr orbits to probability distributions takes us over the line between classical-quantum physics and quantum mechanics. This fuzziness of the wave function solutions to our problems comes from what Max Born has called the "restlessness" of the universe of the atom:

"The completest knowledge of the laws of nature does not carry with it the power of prediction, nor of mastery over Nature. *If* the universe is a machine, its levers and wheels are too fine for our hands to manipulate. We can learn and guide its large-scale motions only. Beneath our veiled sight it quivers in eternal unrest."

Summary

Electron waves are refracted upon passing from one region to another of different potential energy E_P. The deBroglie wavelength of the electron is given by

$$\lambda = \frac{h}{\sqrt{2m(E - E_P)}}$$

DeBroglie showed that the wavelength of the electron fits into the circumference of the corresponding Bohr orbit an integral number of times. This related the wavelength of the electron to the quantization process.

The wave functions for a particle in a one-dimensional impenetrable "box" resemble the standing waves of a string with its ends clamped at the walls of the box. The energy of the particle is quantized: $E_n = n^2h^2/8ml^2$.

Bohr's correspondence principle: the quantum mechanical solution to a physical problem agrees with the classical one at high quantum numbers.

The wave functions for a particle in a "leaky" box do not become zero at the walls. This indicates that the particle can "tunnel through" the wall of the box even though its energy may not be sufficient to make this possible according to classical mechanics.

In general, to solve a problem such as that of the hydrogen atom according to the methods of quantum mechanics, one must: (1) insert the appropriate potential-energy function into Schrödinger's equation, (2) find mathematical solutions to the equation, and (3) accept for constructing the wave function those mathematical solutions that are "well-behaved."

Quantum numbers are quantities that appear in the solutions to Schrödinger's equation and can be assigned values that make the solutions well-behaved. The quantum numbers have physical significance: the principal quantum number n is related to the energy ($n = 1, 2, 3, 4, \ldots$); the orbital quantum number l to the angular momentum ($l = 0, 1, 2, \ldots, n - 1$); and the magnetic quantum number m to the orientations of the atom in a magnetic field ($m = -l$, $-l + 1, \ldots, 0, 1, 2, \ldots, l$).

The quantized energies of the hydrogen atom are

$$E_n = -\left(\frac{2\pi^2 k_0^2 Z^2 e^4 m}{h^2}\right)\frac{1}{n^2}$$

The wave functions of the hydrogen atom can be used to calculate the charge density of the electron "cloud" that surrounds the nucleus.

Questions

1. How can a beam of electrons be refracted? How does this differ from the refraction of light?

2. Describe some applications of electron optics in the cathode-ray oscilloscope or the television receiver.

3. If we know exactly where an electron is, what do we know about its momentum? Discuss.

4. Is the atom like the solar system? Explain.

5. In what sense is a baseball game quantized?

6. Interpret this statement quantum mechanically: When a kitten is stroked, either it feels nothing or its neck is broken.

7. Could a car "leak out" of a garage?

8. Why is quantum mechanics more useful than the Bohr theory in a study of the spectral lines of hydrogen?

9. Compare and contrast the pictures of the hydrogen atom in the Bohr theory and in quantum mechanics.

10. In the description of the hydrogen atom, where is the reference point for the potential energy of the electron?

11. For high quantum numbers, how do the classical and quantum mechanics predictions compare?

12. Why can the summed values of the wave function over all space not exceed 1? Why can the probability not be infinite in value at some point?

13. How are the wave functions changed when a particle is in a leaky box rather than in an impenetrable box?

14. What is quantum "tunneling"?

15. What do we predict about the position and energy of a particle in a closed box (a) from classical theory, and (b) from quantum mechanics?

Problems

A

1. What is the wavelength of a 200-ev free electron? Sketch and label the wave packet.

2. Calculate the kinetic energy of an electron of wavelength 8 A.

3. What is the wavelength of an electron that is (a) in the first Bohr orbit of hydrogen, (b) in the fourth orbit? Assume the radius of the first orbit is 0.5 A.

B

4. Write down all the possible values of the quantum numbers l and m for (a) $n = 4$, (b) $n = 5$.

5. What are the n values possible when (a) $l = m = 0$, (b) $l = m = 1$, and (c) $l = m = 2$.

C

6. Calculate the quantum state for a 0.20-kg croquet ball moving in a "one-dimen-

sional" 1.0-m enclosure with a velocity of 0.40 m/sec.

7. A small steel ball with a mass of 1 gm rolls in a "one-dimensional" 2-cm box. If it has a velocity of 3.0 cm/sec, what is its quantum state?

8. If electrons outside a hollow metal electrode have a kinetic energy one-fifth of that inside, what is the angle of refraction if the angle of incidence is 45°?

9. (a) Show that the equation for a progressive wave on a string (Chapter 19) can be rewritten to represent a progressive wave on a circular hoop as follows:

$$y = A \sin \left(\frac{2\pi t}{T} - \frac{2\pi a \theta}{\lambda} \right)$$

where A is the amplitude of the wave, T the period, a the radius of the circle, and θ the central angle. (b) For deBroglie waves in the Bohr atom, show that

$$y = A \sin \left(\frac{2\pi t}{T} - n\theta \right)$$

where n is the quantum number of the orbit.

10. Calculate the energies of the four lowest quantum levels for a particle of mass 10^{-34} kg in a tight box of length 10^{-7} m.

For Further Reading

Born, Max, *The Restless Universe,* Dover Publications, New York, 1951.
Gamow, George, *Thirty Years That Shook Physics,* Doubleday and Company, Garden City, 1966.

CHAPTER FORTY-THREE

Atoms and Quantum Numbers

In the last chapter, we saw how matter waves, which account for the behavior of free particles, can also help us to understand particles that are confined to a "box"—and a potential well. Free particles are described by traveling matter-wave packets; confined particles, by standing matter waves. Free particles can have any of a wide range—a continuum—of energies. Confined particles can have only certain allowed energies; the energy is *quantized*. The square of the amplitude of the matter wave or wave function at a point is proportional to the probability of finding a particle within a tiny volume at that point.

By using the principles of quantum mechanics, the physicist can calculate from the potential energy curve for the system the wave functions that meet certain reasonable mathematical and physical requirements. The probability description provided by the wave functions is the best knowledge we can hope to have about atomic systems. *Expected* or *"probable"* values of physical quantities—velocity, momentum, energy—take the place in quantum mechanics of the "certain" values of these quantities in classical physics.

In this chapter we shall go on with our discussion of what quantum mechanics tells us about the atom. We shall also look at experimental evidence for results of quantum mechanics that at first glance puzzle us. These strange results of the theoretical physicist may be logically correct, but are they a correct description of nature? We shall find that they are.

Angular Momentum in Quantum Mechanics

In classical mechanics we defined the angular momentum of a particle of mass m, rotating at angular velocity ω, on a circular orbit of radius r, as $I\omega$ or $mr^2\omega$ (Fig. 43-1). In a closed system, with no torque acting from the outside, the angular momentum is constant. It is represented by a vector at right angles to the plane of the orbit, pointing in the direction of advance of a right-handed screw that rotates in the same sense as the particle. The angular momentum vector in classical mechanics can assume any magnitude and direction in space. Its actual magnitude and direction are determined by the initial conditions and by the torques that act on the system.

In the Bohr theory, angular momentum played an important role. Electrons had angular momentum because they traveled around the nucleus on definite classical orbits, as planets travel around the sun. What was different about angular momentum in the Bohr theory was that it was *quantized*. Each energy state of the atom had a definite angular momentum associated with it: for $n = 1$, $I_1\omega_1 = h/2\pi$; for $n = 2$, $I_2\omega_2 = 2h/2\pi$, and so on. No other values of the angular momentum were possible. We saw in Chapter 40 how the energies were derived from this quantizing condition by applying the laws of classical mechanics and electricity and magnetism.

In quantum mechanics, the idea of a particle traveling on an orbit has to be given up. For one thing, according to the uncertainty principle, we cannot hope to verify any theoretical prediction we might make about the shape of the orbit. We would have to use such short-wavelength, high-energy photons to probe for the electron that we would knock it out of its orbit. We would destroy our system in the process of observing it. For a more technical reason, locating a particle on an orbit in quantum mechanics involves difficulties that come from representing the particle by a wave packet. The component simple harmonic waves of a wave packet travel at different speeds (Chapter 41). The wave packet changes its shape with time, spreading out and becoming progressively less localized. In one turn around the orbit, the wave packet would spread so much that it would cease to represent a particle in any reasonable way.

Angular Momentum and the Orbital and Magnetic Quantum Numbers

In quantum mechanics, instead of dealing with a particle moving on an orbit, we have wave functions—standing matter waves—for the different states of the atom. Although they represent probability distributions, they can

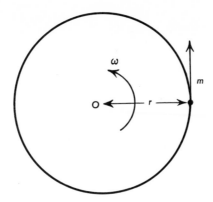

Fig. 43-1. The angular momentum of the mass point m *about the center* O *is* $mr^2\omega$. *The angular momentum vector, for the direction of rotation shown, points* out *of the page.*

be interpreted in terms of the angular momentum of the system in different states. We shall not go into the nature of the mathematical solutions to Schrödinger's equation that justifies such an interpretation.*

Let us first discuss a one-electron atom, such as hydrogen. The theory tells us that the orbital angular momentum associated with a given state of the atom is equal to l times $h/2\pi$, where l is the orbital quantum number for that state (Chapter 42) and h is Planck's constant. Not only is the magnitude of the orbital angular momentum of the atom quantized, but when there is an external magnetic field present, the *direction* of the orbital angular momentum is quantized also. When

---

* The *radial* variation (variation with distance from the nucleus) of the wave functions in different states shows that Bohr orbits can be associated with regions of high probability density (Fig. 42-8, page 686). The changes in sign ($+$ to $-$ or $-$ to $+$) of the *angular* part of the wave functions can be related to angular momentum. In an *s*-state ($l = 0$), the wave function does not change sign at all around a circle enclosing the nucleus: the angular momentum of an *s*-state is zero. In a *p*-state ($l = 1$), the wave function changes sign once, and the angular momentum associated with that state is $1(h/2\pi)$. In a state for which l is 2—a so-called *d*-state—the wave function changes sign twice, and the angular momentum is $2(h/2\pi)$. And so on.

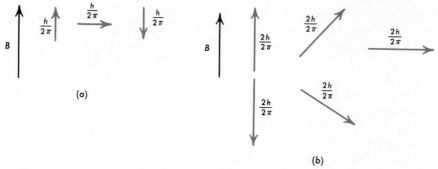

Fig. 43-2. *Allowed directions of the angular momentum vector with respect to a magnetic field of magnetic induction* B. *(a) For* $l = 1$, *three directions are allowed.* *(b) For* $l = 2$, *five directions are allowed.*

an atom is placed in a magnetic field, the arrow representing its total orbital angular momentum can point only in certain directions. The number of different possible directions depends upon the magnitude of the angular momentum. This is called *space quantization.* If the angular momentum is $1(h/2\pi)$, the angular momentum vector can point: in the direction of the field, in the opposite direction, or at right angles to the field (Fig. 43-2*a*). If the angular momentum is $2(h/2\pi)$, there are five possible directions (Fig. 43-2*b*). In fact the number of allowed directions equals the number of possible values of the magnetic quantum number *m* that we encountered in the last chapter.

Example. Compute the orbital angular momentum *L* of the hydrogen atom (*a*) according to the Bohr theory for the electron on the second orbit and (*b*) according to quantum mechanics for $n = 2$.

(*a*) For the second Bohr orbit:

$$L = I_2\omega_2 = \frac{2h}{2\pi} = \frac{6.6 \times 10^{-34} \text{ joule-sec}}{\pi}$$

$$= 2.1 \times 10^{-34} \text{ (kg-m}^2\text{)(rad/sec)}$$

(*b*) In the quantum-mechanical atom:

$$L = \frac{lh}{2\pi} \quad \text{where } l = 0, 1, \ldots n - 1$$

When $l = 0$, $L = 0$

When $l = 1$, $L = 1\frac{h}{2\pi}$

$$= 1.05 \times 10^{-34} \text{(kg-m}^2\text{)(rad/sec)}$$

An Analogy: the Gyroscope

We can relate what has just been stated about space quantization to a more familiar situation.* When we discussed the precession of an isolated gyroscope or a top in Chapter 9 we saw that the axis of spin of the gyroscope remains fixed in space if the gyroscope is well balanced and if frictional torques are negligible so that the external torque acting on it is zero. However, suppose that we apply an unbalanced torque. We can do this by loading one of the mounting rings of the gyroscope, for example, or by supporting the gyroscope at one end of its axis so that the weight of the gyroscope itself sets up an unbalanced torque (Fig. 43-3). When this is done, the gyroscope precesses around the direction of the gravitational field. The axis of the spinning gyroscope sweeps over the sur-

---

* In making use of analogies from classical physics, as we shall often do, let us remember that they are useful only as devices to help us remember the results of quantum mechanics. We cannot trust the analogies to hold even qualitatively beyond a certain point.

face of a cone. There are two motions: the spinning of the gyro wheel and the precessional turning of the axis under the effect of the gravitational torque.

However, there is no limitation on the *angle* between the spin axis and the gravitational field. The gyroscope can "lean" at any angle with the field from 0° to 180° although the rate of its precession will, of course, be different at different angles. What we are saying now is that the angular momentum of an atom is somewhat similar to this. In a magnetic field, the atom precesses like a tiny gyroscope around the direction of the field. The crucial difference between the gyroscope and the atom is that *the atom's angular momentum is allowed to have only certain magnitudes and to point only in certain allowed directions.*

The Conservation of Angular Momentum

When the forces on a rotating system are all forces directed toward the center—this is nearly true for the electrons bound to a nucleus—angular momentum is conserved. What happens when a photon is emitted by an atom? We would expect that the photon would carry off angular momentum equal to the change in the angular momentum of the atom in going from its initial to its final state. This turns out to be true: the expected angular momentum of a photon emitted in dipole radiation is $h/2\pi$ when the change in the angular momentum of the atom is $h/2\pi$. The sum of the angular momentum of the photon and that of the atom after it radiates equals the angular momentum of the atom before it radiates; angular momentum is conserved.

Angular Momentum and Energy

We learned in Chapter 42 that when $n = 2$, $l = 0$ or $l = 1$, and we have just associated different amounts of angular momentum with

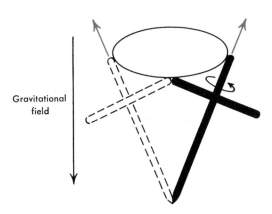

Fig. 43-3. *A precessing gyroscope.*

Gravitational field

the states $n = 2$, $l = 0$ and $n = 2$, $l = 1$. Does the atom have different energies for states that have the same principal quantum number n, but different orbital quantum numbers l? The answer is that the energy of a one-electron atom—such as hydrogen—is only slightly different* for different values of l *unless the atom is in an external magnetic field.* The presence of the magnetic field causes a splitting of the energy levels into several sublevels whose separation depends upon the magnitude of the magnetic induction. This effect, called the Zeeman effect, will be discussed later.

Spin

The total angular momentum of the solar system is the vector sum of (1) the angular momentum associated with the orbital motions of the planets, (2) the angular momentum associated with the rotation of the

---

* Even in the absence of a magnetic field, spectral lines of one-electron atoms show a *fine structure*—a separation into several lines of slightly different wavelengths—when examined with a spectroscope of high resolving power. The fine structure originates in a slight splitting of the energy levels because the electron in states of different angular momentum has slightly different energies. The effect is a relativistic one. It has been explained by a branch of physics called quantum electrodynamics.

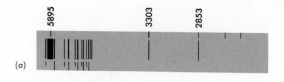

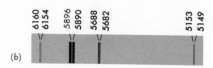

Fig. 43-4. *The bright-line spectrum of sodium, showing doublet spectral lines.* (a) *The spectrum from 6157 A to 2594 A.* (b) *Three doublets taken with larger dispersion. The 6160 A–6154 A doublet belongs to the* sharp *series, the 5896 A–5890 A doublet to the* principal *series, and the 5688 A–5682 A doublet to the* diffuse *series.* (*Herzberg,* Atomic Spectra and Atomic Structure, *Dover Publications, 1944, and O. H. Blackwood, T. H. Osgood, A. E. Ruark, et al.,* An Outline of Atomic Physics, *third edition, John Wiley and Sons, 1955.*)

planets about their axes, and (3) the angular momentum associated with the rotation of the sun about its axis. We can refer to (2) and (3) as *spin* angular momentum.

We have just seen that, in a one-electron atom, the wave functions for different states or conditions of the atom can be interpreted physically in terms of the quantized *orbital* angular momentum of the atom in those states. Is there anything in atomic theory that corresponds to *spin* angular momentum?

Before attempting to answer this question, let us recall that quantization was first introduced into our description of the hydrogen atom to explain the discreteness—separateness—and the regularity of the bright-line spectra of hydrogen. There is another kind of

regularity that appears when we consider the spectra of atoms of higher atomic number. These spectra contain *multiplet lines*—spectral lines that appear as doublets (pairs), triplets (groups of three), and so on. The brightest line in the sodium spectrum is a doublet line, a pair of bright lines with wavelengths of 5890 A and 5896 A. Further examination of the sodium spectrum shows that other lines are also doublets (Fig. 43-4). There is also a series of single lines. The spectrum of mercury contains both singlet lines and triplet lines. Multiplet lines point to a more complicated group of energy levels than we have in hydrogen, where the lines (except for the fine structure noted above) are singlet lines. This suggests that some new principle of quantization is operating.

The Spin of the Electron

S. A. Goudsmit and G. E. Uhlenbeck proposed in 1925 that electrons may possess *quantized angular momentum of spin.* The interaction of the spin motion with the orbital motion gives rise to multiple energy levels. To visualize the spin of the electron, let us think of the electron as a tiny ball of negative charge rotating on an axis at the same time as it revolves about the nucleus. A spinning charge acts like a tiny Amperian current and possesses a magnetic moment (Chapter 28). Thus an electron behaves like a tiny classical magnetic dipole. As it revolves on its orbit about the nucleus, the electron is in a magnetic field caused by the relative motion of the charged nucleus and other electrons about it.*

In this magnetic field, according to the theory, the electron aligns itself in one of two ways: with the spin vector *parallel* to the orbital angular momentum vector or with the spin vector *antiparallel* (pointing oppositely)

---

* The effect is the same as the apparent motion of the sun as observed from the revolving earth.

Fig. 43-5. The spinning electron. Its spin angular momentum is $+\frac{1}{2}$ (h/2π) when it points in the same direction as the orbital angular momentum 1(h/2π) and $-\frac{1}{2}$(h/2π) when it points oppositely to the orbital angular momentum. The picture is a classical one, drawn to make the relation of the vectors clear, not to represent the atom accurately.

to the orbital angular momentum vector (Fig. 43-5). We assign a spin angular momentum $\pm\frac{1}{2}(h/2\pi)$ to the spinning electron, using the plus sign when the spin vector and orbital vector point in the same direction and the negative sign when they are oppositely directed. Corresponding to the quantized spin angular momentum we have a spin quantum number **s** which can have one of two values: $+\frac{1}{2}$ or $-\frac{1}{2}$.

Combining Spin and Orbital Angular Momentum

The spin angular momentum and the orbital angular momentum are combined *vectorially*. Thus, when the electron possesses orbital angular momentum of 1($h/2\pi$) (a *p*-state), its total angular momentum is either $1(h/2\pi) + \frac{1}{2}(h/2\pi) = \frac{3}{2}(h/2\pi)$ or $1(h/2\pi) - \frac{1}{2}(h/2\pi) = \frac{1}{2}(h/2\pi)$. The energy of the electron is slightly greater when the total angular momentum is $\frac{3}{2}(h/2\pi)$ than when it is $\frac{1}{2}(h/2\pi)$ by the amount of energy required to flip the electron from the parallel to the antiparallel condition. Thus there are two slightly different energy levels corresponding to a given set of values of the principal quantum number and the orbital quantum number. We shall see in the next section how this doubling of levels leads to the doublet lines of the sodium spectrum.

Our visualization of the electron as a spin-

ning ball of negative charge is useful in first thinking about the electron spin, but it is much too sharply drawn to be verifiable. According to the uncertainty principle, we cannot probe an object as small as the electron to see whether it is spinning and at what rate. Nevertheless, quantum mechanics can generalize the idea of an electron spin and make good use of it in the more complete form of the mathematical theory—the combination of relativity theory and quantum mechanics developed by P. A. M. Dirac and others. The results are in good agreement with experiment. We shall see later that the idea of spin has been successfully applied to other fundamental particles as well. Protons and neutrons, for example, have spins. In many nuclei, these spins add to give a resultant spin angular momentum to the nucleus.

Many-Electron Atoms

Hydrogen is a one-electron atom; sodium has 11 electrons; uranium has 92. Although the complexity of the calculations increases greatly when the theory of quantum mechanics is extended from the light atoms to the heavier ones, the procedures remain very much the same. One approaches the problem in much the same way that the astronomer approaches the problem of calculating the orbits of the planets about the sun from observations of their positions. He first treats each planet separately, as if it were the only planet to be attracted by the sun, and calculates the elliptical orbit of each planet. The resulting orbits represent a "zero approximation" because the effects of the gravitational attractions among the planets have been left out. The astronomer then improves these zero approximations by a perturbation method. In applying this method, he treats the interplanetary attractions as slight perturbing influences on the zero-approximation orbits.

Similarly in quantum mechanics, the theorist first attempts to develop a one-electron wave function for each electron in the atom. Then he combines all of these one-electron wave functions into a single mathematical expression for ψ that will contain the coordinates of all of the electrons and represent the entire atom. In the first step, he assumes that the Coulomb attraction between the positively charged nucleus and each electron is dominant, reduced only by the screening effect of the average negative charge of electrons lying between that electron and the nucleus. In the second step, the theorist must introduce the Coulomb repulsion between electrons and the interaction between the spin and the orbital motion of each electron. The difficulties in making a good approximation to the wave function for the atom are great. We shall not go into the details here. There is one extremely important generalization, however, that we must point out—the Pauli exclusion principle.

The Pauli Exclusion Principle

Each of the one-electron wave functions that we referred to in the last paragraph is characterized by a set of four quantum numbers $n, l, m,$ and s. Recall that n (principal quantum number), l (orbital quantum number), and m (magnetic quantum number) came out of the theory of the hydrogen atom (Chapter 42) as conditions for the one-electron wave functions to be mathematically well-behaved. The

spin quantum number s was added in the present chapter to explain the multiplet energy levels.

We might ask ourselves whether we could have the same set of four quantum numbers appearing several times among the different one-electron wave functions that are put together to form the wave function for the many-electron atom. The answer—first suggested by Wolfgang Pauli in 1925 and abundantly confirmed by spectroscopic evidence —is no. *Pauli's exclusion principle states that no two electrons in the same atom can be described by wave functions characterized by the same set of quantum numbers $n, l, m,$ and s.* For example, if two electrons have the n, l, m quantum numbers 2, 0, 0, their s quantum numbers must be different: they will be respectively $+\frac{1}{2}$ and $-\frac{1}{2}$, corresponding to spin vectors oppositely directed. They are forbidden by this principle both to have the set 2, 0, 0, $\frac{1}{2}$. Likewise if they have the same $l, m,$ and s quantum numbers, their n quantum numbers must differ.

Pauli arrived at the exclusion principle by first calculating the energies of the energy levels that one would expect to have in a many-electron atom in which two electrons had the *same* set of quantum numbers. He then examined the spectra of such atoms to see whether spectral lines showed up with the correct energies to represent transitions between such energy levels. Such lines did *not* appear, indicating that the exclusion principle held true.

Although the verification of the exclusion principle rests upon careful comparisons with experiment, we can give a rough picture that will help you remember the exclusion principle. Suppose that two electrons *had* the same set of quantum numbers 2, 0, 0, $\frac{1}{2}$. According to our interpretation of wave functions, this would mean that we would expect on the average to find them *superimposed*— occupying the same position in space. Since the electrons are charged, the Coulomb repulsion force between them would be enormous and would blow them apart immediately. Hence, we do not expect them to have the same set of quantum numbers.

Electrons do not have identities. Fundamental particles cannot be described by such labels as "John Jones, Social Security Number 189-26-2252." A set of four quantum numbers specifies a wave function that locates an electron for us. That set cannot locate two electrons, because we have no way of telling them apart.

Building a Many-Electron Atom

By applying the procedures of quantum mechanics and the exclusion principle, we can imagine a process of building up a many-electron atom. We add electrons, one after the other, each with a different set of quantum numbers. Hydrogen (atomic number $Z = 1$) has its one electron in a state characterized by the quantum numbers 1, 0, 0, $-\frac{1}{2}$ when in the lowest-energy or "ground" state. Helium ($Z = 2$) adds a second electron with quantum numbers 1, 0, 0, $+\frac{1}{2}$. When this electron is added, we have exhausted the possibilities (Chapter 42) of additional *different* sets of quantum numbers which include $n = 1$. The $n = 1$ (or K) shell of the atom is filled by two electrons.

The next heaviest atom is lithium ($Z = 3$), which has its K shell filled and in addition has a single electron in the next-higher-energy shell — the L shell — with quantum numbers 2, 0, 0, $-\frac{1}{2}$. Then comes beryllium ($Z = 4$), which adds an electron with quantum numbers 2, 0, 0, $+\frac{1}{2}$. And so on until the L shell is filled with eight electrons.

Carefully examine Table 43-1, which summarizes the way in which atoms of the first eighteen elements, arranged according to increasing atomic number, are built up by the addition of electrons. The table is cumulative. For each element we have given the quantum numbers of the "last" electron added. The

atom in its ground state also contains all of the electrons whose quantum numbers are listed above that element in the table. The table can be extended to the heaviest elements by the procedure we have indicated. However, quantum mechanical peculiarities cause some of the heavier elements not to follow exactly the simple sequence shown in Table 43-1.

Table 43-1. Building the Light Atoms ($Z = 1$ to $Z = 18$)

| Element | Z | Shell | Quantum Numbers of "Last" Electron | | | |
			n	l	m	s
H	1	K	1	0	0	$-\frac{1}{2}$
He	2	K	1	0	0	$+\frac{1}{2}$
Li	3	L	2	0	0	$-\frac{1}{2}$
Be	4	L	2	0	0	$+\frac{1}{2}$
B	5	L	2	1	-1	$-\frac{1}{2}$
C	6	L	2	1	-1	$+\frac{1}{2}$
N	7	L	2	1	0	$-\frac{1}{2}$
O	8	L	2	1	0	$+\frac{1}{2}$
F	9	L	2	1	1	$-\frac{1}{2}$
Ne	10	L	2	1	1	$+\frac{1}{2}$
Na	11	M	3	0	0	$-\frac{1}{2}$
Mg	12	M	3	0	0	$+\frac{1}{2}$
Al	13	M	3	1	-1	$-\frac{1}{2}$
Si	14	M	3	1	-1	$+\frac{1}{2}$
P	15	M	3	1	0	$-\frac{1}{2}$
S	16	M	3	1	0	$+\frac{1}{2}$
Cl	17	M	3	1	1	$-\frac{1}{2}$
Ar	18	M	3	1	1	$+\frac{1}{2}$

The quantum numbers of electrons are shorthand or code designations of wave functions. If pictorial representation were possible, we would show the space distribution of electrons rather than give the quantum numbers. However, pictures are misleading — as we have already seen — because they suggest a degree of definiteness that is not justified.

The Spectrum of Sodium

Let us apply these ideas about the many-electron atom to the sodium atom in order to see whether we can understand how its spectrum is formed. As Table 43-1 indicates, sodium atoms have the *K* and *L* shells completely filled and also have one electron in the *M* shell. The *K* and *L* electrons of sodium ordinarily do not participate in the emission of the spectral lines of sodium, being more tightly bound to the nucleus than the *M* electron. In fact, sodium—like the other alkali metals (lithium, potassium, rubidium, cesium, and francium)— is to some extent a one-electron atom resembling hydrogen. However, the single "active" electron in an unfilled *M* shell is partially *shielded from the nuclear charge* by the "inert" electrons in the intervening *K* and *L* shells. The situation is *not* like that of singly ionized helium, doubly ionized lithium, etc., as we saw in Chapter 40. Moreover, the energy levels associated with different values of l, m, and s differ somewhat in energy, even though the principal quantum number n may be the same. We thus have the possibility (Fig. 43-6) that the "active" electron of sodium can be excited to a higher energy level in one of several series of energy levels. The *S* levels are those characterized by zero angular momentum: $l = 0$. For the *P* levels, $l = 1$; for *D* levels, $l = 2$; and for the *F* levels, $l = 3$. For example, the 3-*S* level is characterized by $n = 3$ and $l = 0$; this is the *ground state* or lowest level occupied by the "active" electron in sodium. A 5-*D* level is one with $n = 5$ and $l = 2$. When the electron is in a state for which the angular momentum is not zero, the angular momentum and the spin momentum interact, as we saw above, to produce a doubling of the energy levels. Thus the *P*, *D*, and *F* levels of sodium are all doublet levels.

Consider the transitions from the 3-*P* levels to the 3-*S* level. In terms of the Bohr atom, we would say that the electron jumps from the upper 3-*P* levels to the 3-*S* level, a transition characterized by a decrease of the energy and the angular momentum of the atom. When the electron starts from the upper 3-*P* level, the emitted photon of wavelength 5890 A

carries off the energy difference of approximately 2.10 ev and the angular momentum difference of $h/2\pi$. When the electron starts from the lower 3-*P* level, the wavelength of the emitted photon is 5896 A. These doublet spectral lines—5890 A and 5896 A— are the two most prominent lines of the sodium spectrum. Other transitions starting from one of the *P* levels and ending on the 3-*S* level produce the other doublets of the so-called *principal* series. Transitions starting from the 4-*S* or higher *S* levels and ending on the 3-*P* level produce the doublets of the *sharp* series, and so on.*

The quantum mechanical description of the emission of photons by the sodium atom is characteristically different from the description we have just given using the terminology of the Bohr atom. Using the language of quantum mechanics, we would say that the emission of the 5890-A photon accompanies the dying down of the matter wave that includes a one-electron wave function for the "active" electron characterized by quantum numbers 3, 1, 0, $+\frac{1}{2}$ and the growing up of the matter wave that puts the active electron in the 3, 0, 0, $+\frac{1}{2}$ state. One can think of the emission of the 5890-A photon as a kind of beat phenomenon. The frequency of the dying matter wave (3, 1, 0, $+\frac{1}{2}$) beats against the frequency of the growing matter wave (3, 0, 0, $+\frac{1}{2}$), and the emitted photon has the difference frequency or beat frequency that one would expect. Quantum mechanics is capable not only of calculating this frequency, but also of calculating the relative *probability* that the change from the excited state (3, 1, 0, $+\frac{1}{2}$) to the ground state (3, 0, 0, $+\frac{1}{2}$) will occur. Since this relative probability tells us how likely it is that, say, of one million atoms, half

---

* We shall not go into the reason why transitions between, say, 3-*D* levels and 3-*P* levels actually lead to three spectral lines rather than four, as one might think. For further discussion of this matter, consult an atomic physics book under the headings "selection rules" or "forbidden spectral lines."

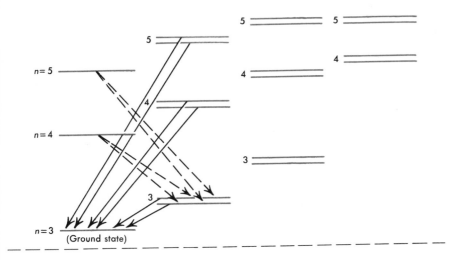

5 ⎯⎯⎯⎯ 5 ⎯⎯⎯⎯

5 ⎯⎯⎯

4 ⎯⎯⎯⎯

n = 5 ⎯⎯⎯ 4 ⎯⎯⎯⎯

4 ⎯⎯⎯

n = 4 ⎯⎯⎯ 3 ⎯⎯⎯⎯

3 ⎯⎯⎯

n = 3 ⎯⎯⎯
(Ground state)

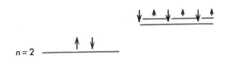

n = 2 ⎯⎯⎯

n = 1 ⎯⎯⎯

S levels	P levels	D levels	F levels
$l = 0$	$l = 1$	$l = 2$	$l = 3$

Fig. 43-6. An energy-level diagram for sodium. Electron transitions from the S levels to the P levels produce the sharp *series (– – –). Transitions from the P levels to the S levels produce the* principal *series (⎯⎯). Other series are also possible.*

of them will "de-excite" and emit photons in a certain time, the quantum mechanical procedures allow the theorist to calculate the relative *intensity* of the spectral lines observed. Here we have two important differences between the Bohr theory of hydrogen and the quantum mechanical theory: (1) instead of electron "jumps," we have a calculable process of the decay and growth of wave functions accompanied by the emission of photons to conserve energy and momentum, and (2) quantum mechanics, unlike the earlier Bohr theory, enables us to calculate how bright the spectral lines are relative to each other as well as what their wavelengths are.

Excitation and Ionization Energies

How do we know that the predictions of quantum mechanics about the energy levels of an atom are correct? First of all, data from spectroscopy—the wavelengths and the intensities of the spectral lines—can be checked against the energy levels. We can determine whether the differences between the energy levels correspond to the energies of the emitted photons and whether the predicted lifetimes of the excited states agree with the observed intensities. If all of the observed photon energies "fit," the physicist gains increased confidence in the correctness of the energy-level scheme.

In spectroscopic observations, many emission spectral lines are produced at one time. The spectroscopist places atoms of the substance in an electric arc or in an electric spark, bombards them with electrons, and causes them to collide violently with each other at the high temperatures of the discharge. Thus, some atoms are excited to each of the higher energy levels. As they return to the ground state—often by a series of transitions—the energy released appears in the emitted photons. The process of analysis is a non-selective one, somewhat like pushing a piano off the top of a building and analyzing

the sound produced in the crash to deduce the fundamental tones of the strings of the piano. An alternative procedure—one used more often by piano tuners—is to strike each key selectively, excite each string separately, and analyze each fundamental tone. The latter procedure is followed in atomic physics in measuring the *excitation* energies of atoms.

James Franck, G. Hertz, and other physicists, around 1914, developed a number of experiments to do this. Figure 43-7 shows the apparatus for one of these experiments. The three-element tube contains mercury vapor whose density is controlled by maintaining the temperature of the tube at about 250°C. Electrons emitted from the cathode are accelerated toward the grid by a positive cathode-grid voltage whose magnitude can be adjusted by the potentiometer P_1. The grid-plate voltage is a small *negative* voltage of about 0.5 v, set by potentiometer P_2. If there were no mercury atoms in the tube, the electrons would move without incident across the tube. Some would be collected by the grid; others, by the plate, provided the positive cathode-grid voltage is somewhat larger than the negative grid-plate voltage. The electrons acquire enough energy in going from cathode to grid to fight their way "uphill" against the retarding electric field set up by the negatively charged plate. The plate current is measured by the galvanometer G.

The presence of the mercury atoms changes the situation by providing the possibility of *inelastic resonance collisions* between the electrons and the mercury atoms. Consider first what happens at A in Fig. 43-7, where an electron that has just escaped from the cathode collides with a mercury atom. The electron has not had a chance to acquire much energy by moving through the accelerating field between cathode and grid. As a result, the electron cannot excite the mercury atom from its ground state to a higher excited state. The collision is *elastic*.

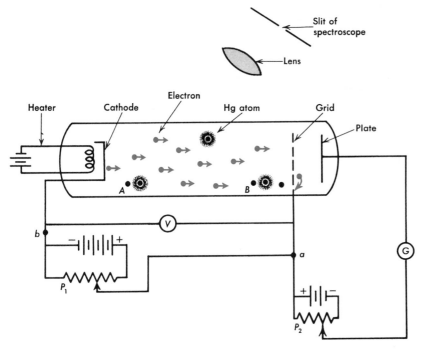

Fig. 43-7. Apparatus for measuring the excitation energy and ionization energy of mercury.

The electron recoils without loss of energy and soon is accelerating toward the grid again.

Suppose now that the electron makes no collisions until it strikes a mercury atom at *B* just before reaching the grid. If the cathode-grid voltage is greater than 4.86 v, the electron will have more than 4.86 ev of kinetic energy when it collides with the mercury atom. *An inelastic collision will occur.* The mercury atom* will be excited from the ground state (Fig. 43-8) to the lowest excited state, and the electron will lose almost all of its kinetic energy.

---

* Mercury has *two* "active" electrons in the outermost unfilled shell. Its energy-level diagram is accordingly more complicated than that for sodium, which has one active electron. We show in Fig. 43-8 only a portion of the mercury diagram—the ground state and the lowest excited state involved in the inelastic collisions we are considering.

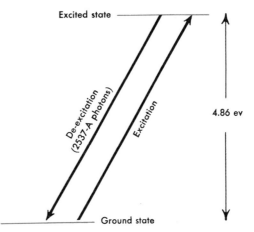

Fig. 43-8. A portion of the energy-level diagram for mercury showing the energy of the ground state and that of the lowest excited state.

How will we know when this happens? Here is where the negative retarding grid-plate voltage comes in. If the cathode-grid voltage becomes high enough so that a significant number of electrons begin to make inelastic collisions, they will not be able to get across from the grid to the plate because they will have lost the energy needed to coast "uphill" against the retarding voltage. Thus they will be driven back to the grid, and the plate current registered by the galvanometer will drop sharply. If we increase the cathode-grid voltage beyond that point until it is about twice 4.86 v, the electron can make an inelastic collision, acquire more energy in the field, and make a *second* inelastic collision. Hence, we can measure the excitation energy by plotting the plate current against the cathode-grid accelerating voltage (Fig. 43-9). The curve will start to dip each time the electron's energy (voltage times electronic charge) is an integral number times the excitation energy of the mercury atom. The voltage difference between two peaks, multiplied by the electronic charge, gives us the *excitation energy* of the lowest excited state of mercury—4.86 ev.

After the mercury atoms have been excited to a higher state, they de-excite—tumble back to the ground state—by emitting photons whose energy equals the difference between the energies of the two levels. We can examine the electromagnetic radiation emitted by the mercury by collecting some of it with a lens and sending it into a spectroscope (Fig. 43-7). We find that spectral lines appear, as predicted, with photon energies that are *less* than the energy of the bombarding electrons, but no spectral lines appear with photon energies that are *more* than the energy of the bombarding electrons. In fact, the 2537-A spectral line appears just as the first dip occurs in the plate current as the accelerating voltage is increased because it is at this critical accelerating voltage that the mercury atoms begin to be excited.

Measuring the Ionization Energy

To determine the *ionization energy*—the energy needed to pull one of the electrons in the unfilled shell away from the atom—we can modify the apparatus in Fig. 43-7 to record the onset of a current of positive mercury ions at the plate. Suppose that we disconnect the positive end of the retarding battery from the grid at *a* in Fig. 43-7 and connect it to the cathode at *b*. We now have the plate maintained at about 0.5 v *negative* with respect to the cathode so the plate will collect *no* electrons. However, it can attract positively charged mercury ions that result from ionizing mercury atoms and can give an electron to each of these that reaches it. Thus, if we raise the cathode-grid voltage toward 10 v or so, a voltage will be reached at which the electrons, accelerating from the cathode toward the grid, pass beyond the grid with enough energy to ionize mercury atoms in the space between the grid and plate. At this voltage—about 10.4 v—the galvanometer records the arrival of positive ions at the plate, and we know that this is the first ionization voltage of mercury. Then the energy required to remove the first electron from mercury is 10.4 ev, and we can check the theoretical prediction that the energy of the ground state of mercury is −10.4 ev.

By extension and refinement of these experiments, we can determine the energies of the various levels of atoms. Ionization measurements tell us how tightly bound are the electrons in the ground state. The study of inelastic resonance collisions tells us how far above the ground state are the various excited states. We check these direct and selective collision studies by means of the spectrographic data. The whole procedure is roughly comparable to determining the altitude of the top of a building with reference to sea level if the building were located in a valley that is below sea level. We could determine the altitude of the base of the building

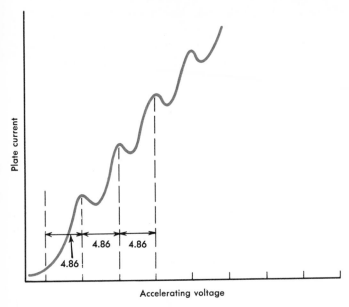

Fig. 43-9. Excitation-potential curve for mercury.

with reference to sea level (comparable to ionization energy), the height of the top of the building above its base (excitation energy), checking perhaps by separate measurements of the height of each story of the building (photon energies).

The Periodic Table of the Elements

Figure 43-10 is a graph of the ionization energies versus atomic numbers of the elements. These energies are those required to remove an electron from the outermost shell of a neutral atom. They are measured by a number of different methods, including the experiments on gases and vapors described in the last section. Notice the rise and fall of the ionization energy as one goes from hydrogen ($Z = 1$) to the higher atomic numbers. At the elements hydrogen and the alkali metals, lithium, sodium, potassium, rubidium, cesium, and francium, the ionization energy reaches minima. These elements, as we have seen, have a single electron in the outermost

shell. A core of filled shells lies between the average position of the outermost electron and the nucleus, screening the nuclear charge. The ionization energies tell us what we might expect. These atoms yield their outermost electron very readily to other atoms and will usually be found as ions with a net positive charge of 1. On the other hand, for the elements helium, neon, argon, krypton, xenon, and radon, the ionization energy reaches maxima. These elements contain completely filled outer shells. The high ionization energies confirm what we know of the chemical

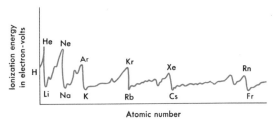

Fig. 43-10. The periodicity of the ionization energy of the elements.

properties of these elements. They are very non-reactive. In fact, we call them the inert gases or noble gases. It is difficult to ionize them, and they are liquefied or frozen only at very low temperatures.

Between the extremes of the alkali metals and the inert gases, for example, from sodium to argon, there is a progressive rise in the ionization energies of the other elements as the atomic number increases. These data suggest a *periodicity* of the properties of the chemical elements paralleling the periodicity that we noticed in the quantum numbers of the electrons. The first period extends from hydrogen to helium; the second, from lithium to neon; and so forth. Other physical properties show a similar periodicity (Fig. 43-11).

The periodicities observed in the chemical properties of the elements are even more striking. We approach here a common frontier of physics and chemistry. The story of how we acquired our knowledge of chemical compounds and how the study of these properties by chemists led to the periodic table of the elements is a fascinating one, but one that we shall not be able to go into here. We wish merely to point out that the developments in quantum mechanics that we have considered in physics and the developments in chemistry merged late in the 1920's to produce new understanding of the different kinds of chemical bonds in compounds. Thus a knowledge of the nature of the wave functions for atoms in different states is as indispensable these days for the chemist as for the physicist.

Atomic Magnetic Moments

In Chapter 27, we saw that the magnetic moment of a current loop is given by IA where I is the current and A the area enclosed by the loop. Let us calculate the magnetic moment μ associated with the orbital motion of an electron in the Bohr atom. For a circular orbit of radius r the current due to the motion of the electron of charge $-e$ is

$$I = \frac{q}{t} = \frac{-e}{2\pi r/v}$$

where v is the velocity of the electron. Thus the magnetic moment μ_0 associated with orbital motion is

$$\mu_0 = IA = \frac{-e}{2\pi r/v}(\pi r^2)$$

$$= -\tfrac{1}{2} erv$$

But the angular momentum L_0 of the electron is

$$L_0 = mvr$$

Therefore

$$\mu_0 = \frac{1}{2}\frac{e}{m}L_0$$

We see that *the magnetic moment associated with the orbital motion is proportional to the angular momentum*. Application of quantum mechanics to this problem leads to the same result.

There is also a magnetic moment μ_s associated with the *spin* of the electron. Advanced theory gives the result

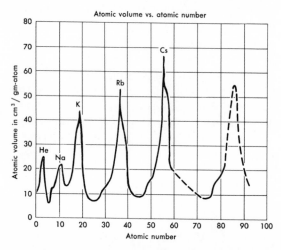

Fig. 43-11. The periodicity of the atomic volume of the elements. (After O. H. Blackwood, T. H. Osgood, A. E. Ruark, et al., An Outline of Atomic Physics, third edition, John Wiley and Sons, 1955.)

$$\mu_s = -\frac{e}{m} L_s$$

where L_s is the angular momentum of spin. The magnetic moment associated with *spin* is *twice* that associated with orbital motion.

To obtain the *total* magnetic moment for an atom, one must combine the orbital magnetic moments and spin magnetic moments for all of the electrons.* To do this by general procedures of quantum mechanics is quite complicated. One must know what the wave functions for the electrons are because these wave functions determine the angular momenta which, in turn, determine the individual magnetic moments. However, knowing that the electronic spins pair off because of the Pauli exclusion principle, we would expect the magnetic moments of the individual electrons to tend to cancel each other—spins up canceling spins down. This turns out to be correct. The resultant magnetic moment depends mainly on the "uncanceled" electronic spins.

Assuming that the total magnetic moment of an atom is μ, what would be the result of placing the atom in a magnetic field? We saw earlier in this chapter that, according to quantum mechanics (and the Bohr atom theory, also), the atom is *space quantized* in a magnetic field. The arrow representing its angular momentum is permitted to point only in certain allowed directions. For the silver atom, for example, there are two allowed directions: the angular momentum vector can point parallel to the magnetic field or antiparallel to it. Since we have just seen that the magnetic moment is proportional to the angular momentum, we would expect for the magnetic moment to point in just one of these two directions in a magnetic field. But how do we know this happens?

O. Stern and W. Gerlach performed an ingenious experiment in 1921 to test the con-

cept of space quantization. They allowed a flat beam of silver atoms to pass through a specially shaped magnetic field that would separate the silver atoms into two groups *if* their magnetic moments were oppositely directed. The apparatus (Fig. 43-12a) consisted of a vacuum chamber containing an oven from which a beam of silver atoms emerged through a horizontal slit, a pair of magnetic pole pieces shaped as shown in Fig. 43-12b, and a plate on which the silver atoms could be collected after passing through the magnetic field. Unlike the uniform magnetic fields which we have generally considered before, this magnetic field was *inhomogeneous.* The magnetic lines were more dense near the knife-edged pole piece than near the hollow pole piece. Whereas no net force acts on a magnetic dipole in a uniform magnetic field, a net force does act on a dipole in an inhomogeneous field. Dipoles with their magnetic moments pointing up are pulled upward and those with their magnetic

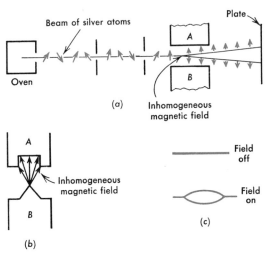

Fig. 43-12. *The Stern-Gerlach experiment.* (a) *The apparatus within a vacuum chamber.* (b) *The magnetic pole pieces that produce the inhomogeneous field.* (c) *The deposits on the plate when the magnetic field is turned off and when it is on.*

---

* Strictly speaking, one must also include the nuclear magnetic moment. This is much smaller than the electronic magnetic moment.

moments pointing down are pulled downward in this kind of field.

Stern and Gerlach found that the silver atoms behaved just as predicted. When the field was turned off, the flat beam came straight through and made a narrow line on the collecting plate (Fig. 43-12c). When the inhomogeneous field was turned on, the silver atoms, their magnetic moments initially pointing in random directions, entered the field and split into two beams. Those with magnetic moments pointing up moved upward, and those with magnetic moments pointing down moved downward. The plate showed two clearly separated lines of atoms (Fig. 43-12c). (If the atoms were not sharply space quantized, instead of two sharp lines, there would have been merely a smudged line when the magnetic field was applied.) This was a great triumph for the new physics. I. I. Rabi and his coworkers in the years following the Stern-Gerlach experiment refined the atomic-beam techniques and were able to measure atomic and nuclear magnetic moments with great precision.

Zeeman Effect

The Zeeman effect also gives us good evidence for space quantization. When a sodium

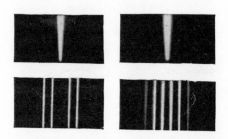

Fig. 43-13. *The Zeeman effect for the* D *lines of sodium.* (*F. A. Jenkins and H. E. White, Fundamentals of Optics, McGraw-Hill Book Company, 1950. Used by permission.*)

lamp is placed in a weak magnetic field, each of the doublet *D* lines breaks up into several spectral lines: the 5896-A line into four and the 5890-A line into six lines (Fig. 43-13). This indicates that the energy levels involved in the production of those lines must have been split into several energy levels clustering around the original level. Although the details of how this comes about are beyond the scope of this book, we can understand qualitatively why this should be so. Remember that in a magnetic field an atom behaves like a small gyroscope, precessing about the direction of the magnetic field with its gyroscopic spin axis taking one of the allowed angles set by the rules of space quantization. The potential energy of the atom-gyroscope is least when its spin axis is aligned with the magnetic field and greatest when it is antiparallel to the field. Corresponding to each angle made by the spin axis with the magnetic field, there are additional amounts of potential energy that must be added to the energy that the atom would possess in the absence of the field. Thus, the energy levels are split when a magnetic field is applied and, correspondingly, the spectral lines are split.

Summary

Quantized amounts of orbital angular momentum are connected with different l quantum numbers. For $l = 1$, the orbital angular momentum is $h/2\pi$; for $l = 2$, it is $2h/2\pi$; and so on.

The orbital angular momentum is space quantized. The angular momentum vector can point only in certain directions with respect to an external magnetic field. For $l = 1$, the vector can point parallel to the field, oppositely directed, or at right angles to it. The atom in a magnetic field behaves like a precessing top or gyroscope.

The electron has quantized *spin* angular momentum given by $\pm\frac{1}{2}(h/2\pi)$. The spin quantum number can have the values $\pm\frac{1}{2}$. The spin and orbital angular momenta combine vectorially. The energy of the atom when the spin vector and orbital angular momentum vector are parallel is slightly greater than when they are oppositely directed. This results in a splitting into two levels of the energy levels of "one-electron" atoms such as sodium. The spectral lines of these elements are doublets.

The Pauli exclusion principle: only one electron in an atom can have a given set of the four quantum numbers—n (principal), l (orbital), m (magnetic), and s (spin). We can "build" many-electron atoms by adding electrons, one after the other, filling the lowest energy level each time, giving each electron a different set of the allowable four quantum numbers.

The spectrum of sodium includes, among other series, the *principal* series of doublet lines (formed by transitions to the 3-S level from higher P levels) and the *sharp* series of doublets (transitions to the 3-P level from higher S levels).

The excitation energy is the energy required to boost an electron from the normal state to the lowest excited state. The ionization energy is the energy required to remove an electron entirely from the atom. Both can be measured by the method of inelastic collisions between electrons and atoms.

Many of the fundamental physical properties of atoms are periodic, going from a minimum value to a maximum periodically as one considers atoms of increasing atomic number. Detailed study of these periodic properties and of the chemical combinations of atoms leads to the periodic table of the elements.

An atom has a magnetic moment due to the combined magnetic moments of its electrons and its nucleus. The orbital magnetic moment μ_0 of a single electron is given by

$$\mu_0 = \frac{-eL_0}{2m}$$

where L_0 is the orbital angular momentum. The spin magnetic moment μ_s of an electron is

$$\mu_s = -\frac{e}{m} L_s$$

where L_s is the spin angular momentum.

Stern and Gerlach showed that the magnetic moment—as well as the angular momentum—of an atom is space quantized as quantum mechanics predicts. A beam of silver atoms which passed through an inhomogeneous magnetic field split into two components—one with spins up and one with spins down—as predicted.

Questions

1. To what extent does the sodium atom resemble the hydrogen atom?

2. What is meant by electron spin? Why was the concept of spin added to quantum theory?

3. In what respect is the following statement analogous to the Pauli exclusion principle: No two persons can have exactly the same latitude, longitude, and altitude at the same time?

4. What are the dimensions of angular momentum in the mks system?

5. How are (a) the sharp series, (b) the principal series, (c) the diffuse series of the sodium spectrum characterized?

6. Why are the P, D, and F levels of sodium double?

7. In quantum mechanics, what corresponds to the "certain" values of quantities in classical physics?

8. In quantum mechanics, why is it necessary to abandon the idea of definite orbits for the electrons in the atom?

9. In the d-state, how does the wave function change in going around a circle that encloses the nucleus?

10. Why was an inhomogeneous magnetic field, rather than a uniform field, used in the Stern-Gerlach experiment?

11. Does an atom have different energies for states with the same n, but different l's?

12. Discuss periodicity in the periodic table of the elements. Give some examples of periodic properties that were not mentioned in this chapter.

13. If one electron of an aluminum atom is removed, would you expect the resulting ion to behave as a magnesium atom in its spectral characteristics?

Problems

A

1. Which of the following electron pairs would not occur in the same atom, according to the Pauli exclusion principle and the rules for forming quantum numbers? The quantum numbers refer to n, l, m, s in that order. Explain.

(a) 1, 0, 0, $-\frac{1}{2}$; 1, 0, 0, $+\frac{1}{2}$
(b) 2, 1, 0, $-\frac{1}{2}$; 2, 1, 0, $-\frac{1}{2}$
(c) 2, 0, 0, $-\frac{1}{2}$; 2, 1, 0, $-\frac{1}{2}$
(d) 2, 1, 1, $+\frac{1}{2}$; 2, 1, 1, $-\frac{1}{2}$
(e) 2, 1, -1, $+\frac{1}{2}$; 2, 1, 1, $+\frac{1}{2}$

2. How are these levels usually identified: $n = 5, l = 2$; $n = 4, l = 3$; $n = 3, l = 1$; $n = 2, l = 0$; $n = 5, l = 0$?

B

3. List the number of electrons in each of the K, L, M, and N shells.

4. Compute the orbital angular momentum for the hydrogen atom according to the Bohr theory for the electron in the fourth Bohr orbit.

C

5. If the quantum numbers of the "last" electron for elements $Z = 19, 20, 21, 22, 23$ followed in regular order, what would be their values? (*Note:* actually there is irregularity, for overlapping begins with 19.)

6. Given these energy levels for sodium, calculate several possible spectral wavelengths (l must change by 1 in each transition).

$$n = 3, l = 0, -5.12 \text{ ev}$$
$$n = 4, l = 0, -1.94 \text{ ev}$$
$$n = 3, l = 1, -3.02 \text{ ev}$$

7. Sodium has a resonance potential of 2.10 v and an ionization potential of 5.14 v. (a) What two energy levels in the structure of sodium do these values establish? (b) Calculate the wavelength of the spectral line associated with the transition from the higher of these two levels to the lower.

8. The ionization potentials for hydrogen and the alkali metals (all having one electron outside closed inner shells) are as follows: hydrogen, 13.6 v; lithium, 5.39 v; sodium, 5.14 v; potassium, 4.34 v; rubidium, 4.18 v; and cesium, 3.89 v. Plot these values against (a) the atomic number of the atom, (b) the mass number of the atom, (c) the radius of the Bohr orbit corresponding to the ground state of the atom. Discuss any relationships that are suggested by the curves.

9. The numbers of atoms of the different elements in a typical living cell in a mammal have been estimated as follows: zinc, lithium, rubidium, copper, manganese, aluminum, iron, and bromine, 10^8–10^{10}; carbon and nitrogen, 10^{12}–10^{14}; sulphur, phosphorus, sodium, potassium, magnesium, chlorine, calcium, iron, and silicon, 10^{10}–10^{12}; hydrogen and oxygen, more than 10^{14}; all other elements, less than 10^8. According to these data, which elements are more likely to have a functional role (that is, to enter into the biochemical processes) in the living cell? From your knowledge of the biological processes involved comment on the role of each of the named elements and, if possible, relate this to the atomic structure of the atom.

For Further Reading

Bloch, Felix, "Nuclear Magnetism," *The American Scientist,* January, 1955, p. 48.

Bloom, Arnold, "Optical Pumping," *Scientific American,* October, 1960, p. 72.

Gamow, George, "The Exclusion Principle," *Scientific American,* July, 1959, p. 74.

Pollack, Gerald L., "Solid Noble Gases," *Scientific American,* October, 1966, p. 64.

Romer, Alfred, *The Restless Atom,* Doubleday and Company, Garden City, 1960.

CHAPTER FORTY-FOUR

Solids

Some of the greatest achievements of physics in the last forty years have occurred in the study of solids. Quantum mechanics has enabled us to have a greater understanding of the nature of physical phenomena in solids—how they conduct heat and electric charges and how they interact with electromagnetic radiation, for example. In Chapter 12 you learned that the forces between atoms—or in ionic solids, such as sodium chloride, between ions—result in a high degree of organization within crystals. The forces are short-range forces, acting strongly only at the small interatomic distances characteristic of solids. There are seven basically different crystal lattices. Each represents a kind of solution that nature reaches in stacking atoms of a particular size and type of atomic interaction. The ions of sodium and chlorine in solid sodium chloride organize themselves on a cubic lattice; the forces between these ions are mainly electrostatic in nature. In the metal copper—also possessing a cubic lattice—the atoms exert on each other forces of a van der Waals type—similar to those acting between the molecules of a real gas. Carbon in the form of graphite has a hexagonal crystal structure.

The carbon atoms exert covalent forces on each other—forces arising from the sharing of two electrons by two atoms. We also saw that a perfect crystal—one in which the crystal regularities extend without interruption throughout the crystal—does not exist in nature. Solids contain numerous imperfections—dislocations, lattice vacancies, and impurity atoms, for example. These imperfections explain many of the physical properties of the solid.

In our study of electricity and magnetism, we thought of electrical conduction in solids as being due to an "electron gas" that filled the spaces between the lattice sites in a crystal. When an electric field was applied to a copper wire by connecting the wire to a battery, the free electrons that constituted the electron gas drifted through the wire, forming an electron current. When a tungsten wire was heated to incandescence, some of the electrons "evaporated," escaping from the attraction of the metallic atoms; this was the thermionic effect. We also found that metallic surfaces reflected electromagnetic waves because the rapidly varying electric and magnetic fields associated with the incident wave caused surface currents in the metal that reradiated energy in the form of a reflected wave. Much of the classical theory of electron conduction in metals and of the interaction of electrons and electromagnetic waves was developed early in this century by the German physicist P. Drude and the Dutch physicist H. A. Lorentz.

In this chapter we shall consider how quantum mechanics gave physicists fresh insights into the nature of solids. Both theoretical and experimental advances led to a deeper understanding of solids. Many new experimental resources became available: purer materials than had existed before, so that the complicated effects introduced by impurities could be isolated and studied separately; apparatus for studying materials at very low temperatures; sources of radiation of controllable wavelength and intensity; and electronic measuring instruments of greater precision, sensitivity, and dependability.

Solid state physics is a large topic. We shall limit our discussion of it mainly to consideration of electrical conduction in solids. The emphasis will be on applying the principles of basic physics to these phenomena to the extent that the mathematical level of this book permits. We shall have occasion, however, to refer to some of the solid state devices that are based on these new developments, such as transistors. Let us first consider some of the experimental facts about electrical conduction in solids.

Resistivity

Ohm's law was the first quantitative fact known about conduction in metallic solids. We have seen that the electrical current I through a metallic conductor within a limited range of temperature is directly proportional to the potential difference V between the ends of the conductor. In symbols,

$$I = \frac{V}{R}$$

where R, the electrical resistance of the conductor, is essentially an experimentally determined proportionality "constant." In fact, the resistance itself depends upon the temperature and the structural characteristics of the metallic conductor—its external shape, its crystal structure, the mixture of atoms present in the solid, etc. Ohm's law can also be written as

$$E = j\rho$$

where E is the electric field strength, j the current density, and ρ the resistivity of the material (Chapter 23). In Chapter 24, we discussed several experimental methods of measuring the resistance of a conductor and the resistivity of the substance of which it is composed.

Since metals commonly consist of many tiny crystallites rather than a single crystal, the resistivity ρ of a polycrystalline conductor is the same in all directions. It is an average of the resistivities that would be measured in different directions in a single non-cubic crystal. ρ (Table 44-1) depends upon the temperature and the nature of the atoms present, but not upon the shape of the conductor.

Table 44-1. Some Data Associated with Electrical Conduction in Metals[a]

Metal	Electrical Resistivity at 0°C, Ω-m	Hall Coefficient at Room Temperature, m³/coul
Aluminum	2.8×10^{-8}	-0.30×10^{-10}
Beryllium	5.5×10^{-8}	$+2.44 \times 10^{-10}$
Copper	1.7×10^{-8}	-0.55×10^{-10}
Iron	10×10^{-8}	$+0.24 \times 10^{-10}$
Lead	19×10^{-8}	$+0.09 \times 10^{-10}$
Silver	1.5×10^{-8}	-0.84×10^{-10}
Sodium	4.3×10^{-8}	-2.50×10^{-10}
Tungsten	5.5×10^{-8}	$+1.18 \times 10^{-10}$

[a] Source: *Handbook of Physics,* McGraw-Hill Book Company, E. U. Condon and Hugh Odishaw, editors. Article by John Bardeen, pp. 4–74.

In the classical theory of electrical conduction in metals, we imagine that the electrons are acted upon by the electric field; that they accelerate until they collide with an atom, whereupon they are decelerated; and that in this fashion they drift slowly along the conductor in a direction opposite to that of the electric field strength. The magnitude of the force F on a single electron is

$$F = Ee = ma$$

where e is the charge on the electron, m its mass, and a its acceleration. Suppose that the electron starts from rest when the electric field is applied and accelerates for an average time τ before making a collision. τ is called the *relaxation time.* The velocity v of the electron is then

$$v = a\tau = \frac{Ee}{m}\tau$$

The current density j equals *nev* where n is the number of conduction electrons per unit volume. (Why?) Thus, on the average,

$$j = \frac{nEe^2}{m}\tau$$

Since $E = j\rho$, we finally have

$$\rho = \frac{m}{ne^2\tau}$$

This tells us what the resistivity—an *electrical property*—is in terms of various *mechanical* quantities and the electronic charge.

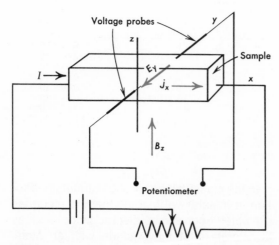

Fig. 44-1. An experiment for measuring the Hall coefficient.

The Hall Effect

To obtain some information about n, the number of charge carriers per unit volume of a given material, the physicist measures the *Hall coefficient* of the substance. A sample of the substance—a strip of bismuth, for example—is mounted between the poles of an electromagnet in a magnetic field whose induction is about 0.4 weber/m² directed at right angles to the axis of the sample (Fig. 44-1) in what we shall call the z-direction. A current of about 0.5 amp is sent through the sample in the x-direction shown in Fig. 44-1. When the current is established, *a potential difference appears* across *the sample in the y-direction.* This transverse potential difference is called the Hall voltage, and the electric field strength associated with it in the sample is called the Hall field strength.

Why does the Hall voltage appear? In Chapter 27 we learned that a beam of electrons in a magnetic field experiences a force that deflects the beam in a direction at right angles to both the beam and the magnetic field. Similarly, the slowly drifting electrons in a current in a metallic conductor that has been placed in a magnetic field are acted upon by the magnetic force and given a sidewise deflection. A charge builds up on the faces of the sample that are perpendicular to the y-axis until the resulting electric field strength is large enough to oppose further sidewise drift of the electrons. When this happens, a steady state is reached.

Let the magnetic induction be B_z, the Hall electric field strength E_y, and the current density j_x. Then the Hall coefficient R_H is defined as the Hall electric field strength per unit magnetic induction per unit current density:

$$R_H = \frac{E_y}{j_x B_z}$$

If the thickness of the sample in the y-direction is l, then $E_y = V_y/l$ where V_y is the Hall voltage measured with a potentiometer (Fig. 44-1). j_x can be computed from the current I in the sample and its cross-sectional area. B_z can be measured with a fluxmeter (Chapter 29). From these experimentally determined quantities R_H is computed (Table 44-1).

For metallic conductors for which the "electron-gas" model is valid, R_H is related to n, the number of charge carriers per unit volume. Let the drift velocity of the electrons in the x-direction be v_x. When the Hall electric force on the electron $E_y e$ equals the magnetic deflecting force $B_z e v_x$, we have

$$B_z e v_x = E_y e$$

or

$$E_y = B_z v_x$$

We also have

$$j_x = n e v_x$$

Therefore

$$E_y = \frac{B_z j_x}{ne}$$

But $E_y/B_z j_x$ is R_H, the Hall coefficient. Thus

$$R_H = \frac{1}{ne}$$

Example. For sodium at room temperature, the Hall coefficient is found by measurement to be -2.5×10^{-10} m³/coul, the minus sign indicating that the Hall voltage is directed as shown in Fig. 44-1 and that the current carriers are therefore negatively charged. The resistivity is 4.3×10^{-8} Ω-m. (a) What is the number of conduction electrons per unit volume? (b) What is τ, the average time between collisions (the relaxation time)?

(a)
$$R_H = \frac{1}{ne}$$

$$-2.5 \times 10^{-10} \text{ m}^3/\text{coul} = \frac{1}{n(-1.6 \times 10^{-19} \text{ coul})}$$

$$n = 2.5 \times 10^{28}/\text{m}^3 = 2.5 \times 10^{22}/\text{cm}^3$$

(b)
$$\rho = \frac{m}{ne^2 \tau}$$

$$\tau = \frac{m}{ne^2 \rho}$$

$$= \frac{9.1 \times 10^{-31} \text{ kg}}{2.5 \times 10^{28}/\text{m}^3 (1.6 \times 10^{-19} \text{ coul})^2}$$

$$(4.3 \times 10^{-8} \Omega\text{-m})$$

$$\tau = 3.4 \times 10^{-14} \text{ kg-m}^2/\text{coul}^2\text{-}\Omega$$

For 1 kg-m², write 1 joule-sec², and for 1 Ω, write 1 joule-sec/coul². (Why are these substitutions allowable?) Then 1 kg-m²/coul²-Ω equals 1 sec and $\tau = 3.4 \times 10^{-14}$ sec.

We see from the Hall effect measurements and the resistivity measurements in sodium that the charge carriers are negative (electrons), their density is about $2.5 \times 10^{28}/\text{m}^3$, and the average time between collisions is roughly 10^{-14} sec.

Note the experimental result that some metals (Table 44-1) have a positive Hall coefficient. This means that there are positive charge carriers present in the conductors and that they are greater in number than the negative conducting electrons present. According to the electron-gas theory, the electrons are solely responsible for conduction. What can these positive carriers be? The explanation takes us into the conduction-band theory of solids.

Electron Waves and Energy Levels

In a solid, conduction electrons experience forces exerted on them by the nuclei of atoms and by other electrons. The greater of the two effects is the Coulomb attraction of the nuclei, which causes the potential energy of the electrons to increase and decrease in a periodic way throughout the crystal. The electrons move through a *periodic field*.

To determine the effect of this on their motion, let us first recall that a free electron can be represented by a traveling wave packet (Fig. 44-2a) and that the kinetic energy of a free electron can take any one of a continuum of possible values (Fig. 44-2b). On the con-

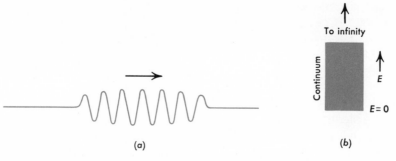

Fig. 44-2. *The free electron.* (a) *It is represented by a traveling wave packet.* (b) *Its energy levels form a continuum.*

trary, within a potential well—such as that set up by the proton in an isolated hydrogen atom—the wave function of the electron is a standing wave. The electron is allowed to have only certain discrete energies (Fig. 44-3).

The next step is to consider what happens to the wave function when the electron is near *two* attracting centers—such as an electron in the hydrogen molecular ion (H^{2+}), in which *both* protons attract it. Such a situation would be represented by a *double* potential well (Fig. 44-4). The results of solving Schrödinger's equation for this kind of situation are surprising from the classical viewpoint.

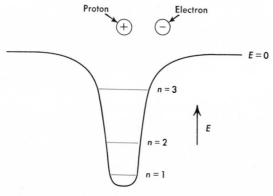

Fig. 44-3. *The electron in an isolated hydrogen atom. A few of the energy levels are shown.*

point, although completely understandable from that of quantum mechanics.

First of all, if the two potential wells are about one angstrom apart, there is a definite probability that an electron that had been placed in one potential well could be found in the other at some later time. In everyday terms, this is equivalent to saying that if a marble were placed in one of two water glasses, standing side by side, the marble would, unaided, move from the bottom of one glass to that of the other—without going over the top! Needless to say, everyday objects do not noticeably behave this way. Electrons, however, exhibit measurable wave properties and show strongly the effects of the uncertainty principle. Suppose the electron is initially in potential well *A*—that is, near proton *A*. Its momentum is limited (made less uncertain) by virtue of its being in the well. Therefore, according to the uncertainty principle, its position must be more uncertain and cannot be precisely restricted to the region within potential well *A* (Chapter 42). Detailed investigation shows that the wave function, which describes the probable location of the electron, spills over from *A* into *B*. Quantum mechanically, the electron tunnels through the classically forbidden region between *A* and *B* and occupies an energy level in *B* half of the time on the average and one in *A* half of the time.

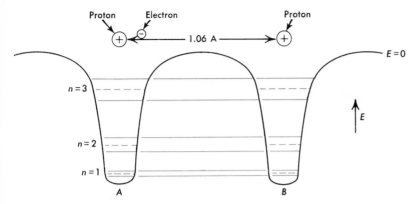

Fig. 44-4. The electron in the hydrogen molecule ion (H_2^+). The energy levels are doubled. The electron can tunnel horizontally from one level to the other.

The rules of quantum mechanics further show that each single energy level ($n = 1, 2$, etc.) in a single potential well (Fig. 44-3) splits into *two* energy levels in each of wells *A* and *B*. In Fig. 44-4, the dashed lines indicate the energy levels that would exist if only one potential well—one proton—were present. The solid lines show the actual double levels when two potential wells—two protons—are present. Incidentally, this "sharing" of the electron by two hydrogen nuclei is of basic importance in explaining the force between the nuclei—the covalent bond that is one of the chemical bonds by which atoms join together to form molecules.

A "One-Dimensional Crystal"

If we now turn to a "one-dimensional crystal"—a line of atoms regularly spaced (Fig. 44-5)—we approach the situation that exists in a solid. In fact, it is often useful to think of a crystal as being one huge molecule that possesses around 10^{22} atoms per cubic centimeter. The potential wells in Fig. 44-5 are shown periodically spaced with a separation representing the lattice spacing in a given direction in a crystal. The shape of the potential wells shown is not to be taken too seriously, but is meant to suggest the kind of field in which a given electron finds itself.

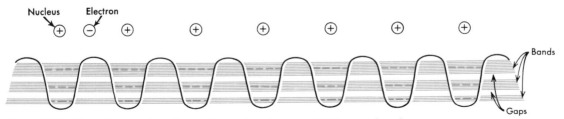

Fig. 44-5. The electron in a "one-dimensional crystal." Energy bands replace the double energy levels of Fig. 44-4 and the single energy levels of Fig. 44-3. An electron with an energy within one of the bands can pass from one potential well to another.

What does quantum mechanics have to say about this? The first important result is that what were formerly sharp energy levels in the isolated atom (Fig. 44-3) are now *bands* (Fig. 44-5), each consisting of a large number of energy levels very close together. The bands get wider as one moves to higher electron energies.

So long as an electron has an energy corresponding to that of one of the many energy levels within a band, the electron can travel from one potential well to another—tunneling through the classically forbidden regions between energy wells. We might say that these electrons have passports to travel. The spreading of the electron wave down the line of atoms is suggested in Fig. 44-5 by extending the solid lines representing energy levels through the regions separating potential wells. We have not tried to suggest the wave functions themselves, however.

Next we note that the bands of energies that have travel privileges associated with them are separated by bands containing no energy levels. These are *gaps* or *forbidden bands* that correspond to canceled "passports." If an electron should acquire one of these energies, its wavelength would be such that the wave would not spread down the line of atoms, that is, tunneling would not occur.

The electron waves at wavelengths corresponding to energies in forbidden bands are reflected back upon themselves at the "wall" of the potential well in such phase as to cancel the wave that would otherwise spread through the wall. At these wavelengths, strong electron-wave *reflection* occurs in the same way that reflection occurred in the Davisson-Germer or Thomson experiments.

Many Electrons in a Crystal

So far, we have considered only one electron in a crystal. However, in metallic sodium—to take an example—there are eleven electrons per neutral atom and thus approximately $11 \times$ 6×10^{23} electrons in each gram-atomic mass of sodium (23 gm). How do all of these electrons fit into the energy bands? Several considerations help us to decide this: (1) the Pauli exclusion principle restricts the number of electrons in one energy state to *two*; (2) the bands are formed around each energy level of an isolated atom with one energy level within a band for each atom in the crystal; and (3) in the ground state, the electrons fill the lowest-lying energy levels in the bands first. To illustrate: suppose that we imagine a crystal "empty" of electrons and that we slowly "pour in" enough electrons to make the crystal electrically neutral. The first electrons will go to the bottom energy levels in the lowest band. Each of these levels will accommodate not more than two electrons and the electrons must have opposite spins (Pauli exclusion principle). After these levels have been filled, the next higher levels will gain electrons until each level in the lowest band has its two electrons. Then the next higher band will begin to fill, and so on. For sodium, the last of the bands to contain electrons will be only half filled. This is so because the isolated sodium atom contains only one valence electron in the 3-S state, but it could contain two—with opposite spins. Therefore, the energy band in the solid corresponding to the 3-S energy level in the isolated sodium atom has room for twice as many electrons as are needed to make the metal electrically neutral. Above the last of the levels containing electrons are empty levels to which electrons can be excited.

In some solids the highest electron-containing band is not half filled as in sodium, but is completely filled. Diamond is an example. The difference between a partly filled band and a filled band has the greatest significance for us. It turns out—with certain important exceptions—to be the principal difference between conductors and insulators, as we shall see shortly. This highest normally occupied level of electrons in a given solid at

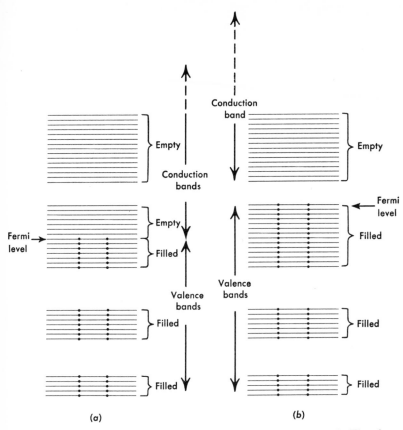

Conduction band

Empty

Conduction bands

Empty

Filled

Fermi level

Filled

Valence bands

Filled

Filled

(a)

Conduction band

Empty

Fermi level

Filled

Valence bands

Filled

Filled

(b)

Fig. 44-6. Energy bands in (a) *sodium and* (b) *diamond. The figure suggests the large number of levels within each band by showing a few of the levels.*

low temperatures is called the *Fermi level,* after Enrico Fermi, the great Italian physicist (1901–1954) who made both theoretical and experimental discoveries in many fields of physics. The energy distributions corresponding to the two situations we have described are shown schematically in Fig. 44-6.

Conduction

Let us see how the band theory explains conduction in a metal such as sodium. So far we have not said anything about the direction in which the electrons move through the crystal. It turns out in the theory that the electrons are paired in velocities when they occupy the lowest energy levels open to them. For every electron that moves in the x-direction in the ground state, there is one that moves in the $-x$-direction. Likewise, velocities in the y- and z-directions are paired. Thus, in the ground state, there is *no net current observable.* As much charge is flowing in one direction as in the opposite direction (Fig. 44-7a).

Now let us imagine that an electric field is applied to the metal by connecting the ends of the conductor to a battery (Fig. 44-7b). In the electric field, some of the electrons in the band levels just below the Fermi level pick up

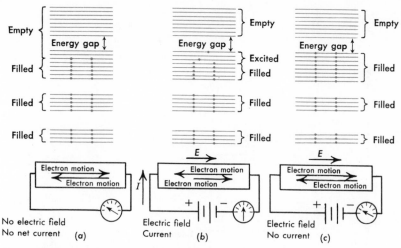

Fig. 44-7. *Conductors and insulators.* (a) *and* (b) *are conductors, and* (c) *is an insulator. The figure suggests the large number of levels within each band by showing a few of the levels.*

energy from the electric field and are excited to unoccupied levels lying above the Fermi level. In the excited state, the electrons are not paired in velocity. More are now moving in the direction opposite to the electric field strength *E* than in the direction of the electric field strength. There is a net electron current. As the electric field is increased, more electrons are excited into unoccupied levels, where they contribute to the conduction current. The current increases, as Ohm's law predicts.

Why do insulators not conduct? Figure 44-7c shows the energy bands of a material such as diamond. In the ground state, the Fermi level is at the top of a filled band— there are no unoccupied energy levels lying slightly above the last filled level. As a result, the electrons must either be given enough energy—about 7.2 ev in the case of diamond—to take them across the *energy gap* to the next unfilled band, or they will not be excited at all. *There are no energy levels for them to go to within the energy gap.* Diamond is therefore ordinarily an insulator. However, if a large enough electric field is

applied—about 10^{10} v/m—there *will* be a current. The establishment of a current in insulators at high field strengths is called *dielectric breakdown.*

Solids formed from atoms with only one valence electron outside the inner electron shells—such as sodium—are conductors because their highest occupied band is only half-filled. What about elements, such as zinc, which have two valence electrons and whose highest occupied band is filled in the ground state? These elements are metallic conductors because of a peculiarity in the atomic energy levels which causes the empty band partially to overlap the highest occupied band. There are thus energy levels to which the electrons can be excited without having to cross a large energy gap; the material conducts. In general, however, the good conductors have an odd number of electrons per atom, and the poor conductors an even number.

Resistance

If electrons can be so easily excited into conduction levels in a metal by applying a moderately large electric field, what limits the size of the conduction current? The limiting processes are all contained in what we call *resistance.* The picture of conduction that we have presented so far in discussing the band theory is too idealized. We must remember that crystals are *not* perfectly regular assemblages of atoms—single crystals of one kind of atom whose nuclei are at rest. Instead, the order of the crystal is disturbed by: (1) impurity atoms, (2) vacancies in the crystal lattice, (3) thermal vibrations of the lattice, (4) grain boundaries between crystallites, (5) dislocations, and (6) various distortions in the crystal structure caused by compression, tension, torsion, and so on. The electron waves are frequently scattered as a consequence of these irregularities—on the average of once each 10^{-14} sec, we saw above—and the current produced by the application of a given field is much reduced.

In general, whatever increases the amount of disorder in the solid increases the electrical resistance. *Resistance is due to disorder.* There is an important exception, however. The addition of tiny amounts of certain impurity atoms can cause what would ordinarily be an insulator to become a conductor. Such materials are called *semiconductors.* We shall discuss them next, finding in their behavior further evidence for the correctness of the band theory.

Semiconductors

Since germanium has four electrons in the outermost shells—an even number of electrons—the band theory predicts that the uppermost band containing electrons will be completely filled in the ground state and that germanium will be a poor conductor. This turns out to be experimentally true. *Pure* germanium has a resistivity at room temperature that is about a million times greater than that of copper. However, under certain conditions, germanium can become a conductor—by the

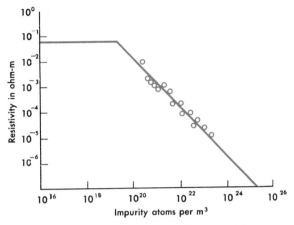

Fig. 44-8. *The dependence of the resistivity of germanium-antimony alloys on antimony concentration. (Adapted from E. U. Condon and H. Odishaw,* Handbook of Physics, *McGraw-Hill Book Company, 1958. Data from G. L. Pearson.)*

addition of tiny amounts of impurity atoms or by raising the temperature of the metal. If the first of these conditions exists, the germanium sample is called an *impurity* semiconductor; if the second, an *intrinsic* semiconductor.

Figure 44-8 shows how the resistivity of pure germanium decreases as tiny amounts of antimony are added. Note that the concentration of antimony atoms is extremely small, but effective in changing the resistivity. Germanium itself has a density of about 10^{30} atoms/m³, yet the effect on the resistivity of germanium of adding 10^{24} atoms/m³ of antimony—one antimony atom for each *million* germanium atoms—is quite marked.

A considerable part of the successful development of experimental solid state physics is due to the physicist's ability to produce very pure materials and to "dope" the germanium by adding tiny controllable amounts of "foreign" atoms. In the *zone refining* process for purifying germanium, for example, a bar of germanium is heated by induction heating coils so that a molten ring of germanium is formed near one end of the bar by the heating produced by eddy currents (Chapter 29). Impurities and germanium are both present in the molten zone. If the heating coils—and the molten zone—are then moved slowly along the length of the bar, the germanium freezes after the coil passes, but the impurities remain in the liquid phase. Thus the molten zone is made to travel along the bar, gathering up impurities as it goes. When the molten zone reaches the other end of the bar, the heater is turned off. The end of the germanium bar is then sawed off, taking most of the impurities with it and leaving the rest of the bar pure to a few parts in a billion. The foreign atoms are then added to the pure germanium in controlled amounts—usually when both the germanium and the additive atoms are molten, but occasionally by exposing the heated germanium to a vapor of the foreign atoms, which diffuse into the germanium crystal lattice.

N-Type Germanium

Germanium that has been doped in this way with antimony atoms is called N-type (negative-type) germanium because Hall effect measurements show that the carriers of charge in this semiconductor are negative. The band conduction theory has the following explanation of the increase in conductivity caused by the addition of antimony. The antimony atom has *five* electrons outside the last closed shell. Germanium has four. When antimony enters the crystal lattice of germanium, the antimony has *one more* electron than it needs to pair off its electrons with those of germanium and thus to form the covalent bonds that hold the crystal together. This surplus electron can be *donated* to the conduction band of electrons in the crystal.

Figure 44-9 shows how this is done. The presence of the antimony atoms disturbs the regularity of the periodic potential energy variation within the crystal. This results in the setting up of a *donor level*, characteristic of antimony, at an energy that lies within the energy gap and just below the empty conduction band. An electron that might momentarily occupy the donor level—thus being effectively attached to an antimony atom—is easily excited to the conduction band lying directly

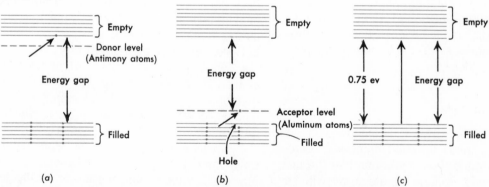

Fig. 44-9. *Energy bands of three kinds of semiconductors. (a) N-type germanium. The foreign antimony atoms* donate *electrons to the conduction bands of germanium. (b) P-type germanium. The foreign aluminum atoms* accept *electrons from the filled bands of germanium, creating "holes" that conduct. (c) Pure germanium is an intrinsic semiconductor. At high enough temperatures, electrons from the filled bands acquire enough thermal energy to jump the energy gap.*

above it. Thus the antimony atoms *inject electrons into the conduction band,* where they behave like the conduction electrons of any good conductor. Other impurity atoms having five valence electrons—phosphorus and arsenic, for example—have a similar effect upon the conductivity of germanium.

P-Type Germanium

Impurity atoms that have *three* electrons outside the last closed shell (such as aluminum and boron) also can increase the conductivity of germanium, but for a different reason. Figure 44-9*b* shows the energy-level diagram for germanium to which aluminum atoms have been added. Aluminum lacks the fourth electron needed to pair off electrons with germanium in forming covalent bonds. It will tend therefore to take up or accept electrons. The disturbance of the periodic field, caused by the presence of the aluminum atoms in the crystal lattice, leads to an *acceptor level,* characteristic of aluminum, that lies just above the last filled band of electrons in germanium. Electrons can be easily excited from the filled band into this acceptor level. This leaves an empty spot or "hole" in the hitherto "filled" levels and *results in conduction by the hole in the electron population of those levels.*

Let us examine this notion of a "hole" rather carefully. It will occur again in our discussion of nuclear physics. Basically, the vacancy caused when an electron moves into the acceptor level leads to a reshuffling of the other electrons in the almost filled band. The electrons, of course, are the entities that carry the charge and are responsible for conduction. Rather than talk about all of these electrons, however, it is more convenient to talk about the vacancy—the hole—and to treat it like a particle similar to the electron, but possessing a *positive* charge. Thus, saying that a certain number of holes move to the left through a semiconductor is equivalent to

saying that an equal number of electrons move to the right, so far as the final charge distribution is concerned. When an electron drops into a hole, the hole is wiped out. But if the electron leaves a vacancy at its previous position, obviously a hole appears there. The situation is somewhat similar to the motion of cars in a traffic jam (Fig. 44-10).

Hall effect measurements show that the current in germanium to which aluminum has been added is carried by positive charge carriers—the holes that we have been discussing. Such materials are therefore called *P*-type (positive-type) semiconductors.

Intrinsic Semiconductors

Finally, there are such things as *intrinsic* semiconductors, whose conductivity depends mainly on temperature. Pure germanium is an intrinsic semiconductor. The energy gap between the top of the filled band and the bottom of the empty conduction band in germanium is only about 0.75 ev. As the temperature rises above room temperature, more and more electrons acquire enough energy from interactions with the lattice vibrations of the crystal to pass the energy gap and move up into the empty conduction band. There they contribute to the current. Of course, every electron leaving the filled band creates a hole there, so both the hole current and the current due to electrons in the "empty" band contribute to the current in an intrinsic semiconductor.

Thermionic Emission

The band theory has led to explanations of many seemingly unrelated phenomena and to the development of many useful technological devices. Space will not permit us to give more than a few examples.

Thermionic emission of electrons from a hot filament can be interpreted as the passage of electrons in the conduction band over or

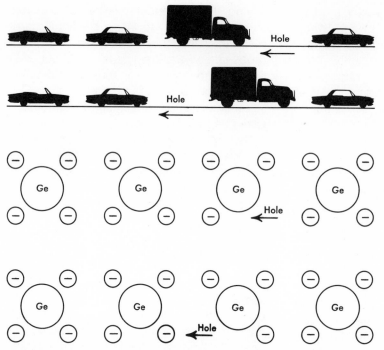

Fig. 44-10. *Cars moving ahead in a traffic jam cause the hole in the line of cars to move backward just as electrons moving into a hole cause the hole to move backward.*

through the potential-energy barrier that exists between the surface of the metal and the vacuum surrounding it. The potential-energy barrier shown to the right in the energy-level diagram of Fig. 44-11 is due to the attraction that the metal surface exerts on an escaping electron whose departure leaves the metal with a net positive charge; the force is sometimes called an image force. At 0°K, all of the electrons are in their lowest energy states, and the highest occupied level — the Fermi level — is well below the escape energy they need to leave the metal surface. As the temperature is raised, however, some electrons acquire energy from the increasing lattice vibrations and move to higher energy levels* in the conduction band. Eventually the

temperature becomes high enough that electrons have been raised to energy levels at which there is a significantly large probability that the electron wave will be transmitted over the barrier. Electrons escape into the vacuum, and a thermionic current is emitted. Quantum mechanics permits one to calculate how the thermionic emission current varies with temperature, but the details are beyond the scope of this book. Note the dashed potential-energy curve which shows the barrier that exists if an electric field is applied in such a way as to aid the electrons in their escape. The effect is to *lower* the potential-energy curve and to make it possible for tunneling through the barrier to occur with a consequent increase in thermionic current.

Similar quantum mechanical calculations lead to an explanation of the *photoelectric effect*. The photoelectric effect also repre-

---

* Although we did not stress this in discussing semiconductors, this temperature effect occurs in their conduction bands also.

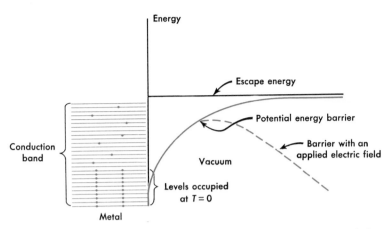

Energy

Escape energy

Potential energy barrier

Conduction band

Vacuum

Barrier with an applied electric field

Levels occupied at $T = 0$

Metal

Fig. 44-11. Thermionic emission over or through a potential-energy barrier.

sents the release of an electron from the conduction band; the energy input, as we saw in Chapter 39, is the absorption of a photon.

Junction Rectifier

In the early days of radio, the detector in the radio receiver consisted of a crystal (usually galena — lead sulfide) with a small metal wire or "cat's whisker" touching it. The operator of the receiver had to move the contact between the cat's whisker and the crystal over the surface of the crystal until a "sensitive" place was reached, whereupon the device detected the incoming radio signal. This device was really a crystal rectifier — a point-contact rectifier. The development of solid state physics has made it possible to construct more reliable crystal rectifiers for many purposes, using semiconducting materials. We shall discuss the *junction rectifier* here.

Figure 44-12 shows a *P-N* junction, a layer of *P*-type germanium in contact with a layer of *N*-type germanium. Such a junction has the property of rectification in that its resistance to current in one direction is low and its resistance to current in the opposite direction is high. In this respect it is similar to the vacuum

tube diode in which electrons flow readily from cathode to plate, but in which there is an extremely high resistance to flow from plate to cathode. To understand how the junction rectifier operates, consider first Fig. 44-12*a*. The *P*-type germanium contains predominantly holes (majority carriers) plus a few electrons (minority carriers) that are formed by the creation of electron-hole pairs described above. Likewise the *N*-type germanium contains mostly electrons (majority carriers) plus a few holes (minority carriers). In the absence of an external electric field, equilibrium is maintained by space charges of the donor atoms in the *N*-type germanium that have a net positive charge and those of the acceptor atoms in the *P*-type germanium that have a negative charge. These space charges set up an energy barrier or **barrier layer** between the *N*-type and *P*-type germanium that keeps the holes and electrons from mixing. The barrier layer is only 10^{-3} to 10^{-5} cm thick.

Now suppose the crystal is connected to a battery as shown in Fig. 44-12*b* so that the positive terminal is connected to the *P*-type germanium and the negative terminal to the *N*-type. The electric field is said to be in the *forward* direction. Electrons are attracted toward the *P*-type germanium and holes toward

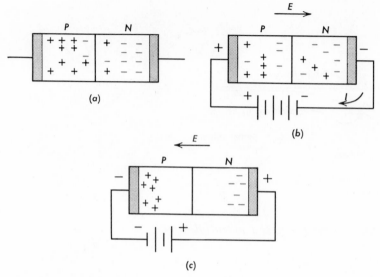

Fig. 44-12. *A junction rectifier.* (a) *No field.* (b) *Field in the forward direction. Conduction takes place.* (c) *Field in the reverse direction. No conduction.*

the *N*-type. The electrons are injected across the junction into the *P*-type germanium, where they recombine with holes. Similarly, holes move into the *N*-type germanium and recombine there with electrons. This drift of electrons to the left and holes to the right constitutes the observed current; it rises sharply with increasing voltage.

Finally, let the crystal be connected to the battery (Fig. 44-12c) so that the *P*-type germanium is connected to the negative terminal and the *N*-type germanium to the positive. Both holes and electrons are swept back from the junction. The current is very small, consisting only of the current of thermally generated holes and electrons that cross the junction and recombine.

Thus the direction from *P* to *N* is the *low-resistance, high-current, forward* direction of the electric field strength, and the direction from *N* to *P* is the *high-resistance, low-current, reverse* direction. Figure 44-13a shows a typical germanium junction rectifier cell, and Fig. 44-13b a typical characteristic curve.

The Transistor

If the junction rectifier is the solid state equivalent of a vacuum tube diode, the *transistor* is the equivalent of a vacuum tube triode. Like the triode, the transistor serves as an amplifier, controlling the output signal by the effect of the input signal upon the energy source. The vacuum triode does this by grid action upon the plate current; the transistor, by the effect of the emitter upon the collector. Unlike the vacuum triode, the transistor is extremely small (Fig. 44-14) — about the size of an aspirin tablet — and requires no power to operate a cathode, which in the triode requires at least several watts. The transistor can perform almost all of the functions of a vacuum tube and is increasingly used in amplifiers, oscillators, computer circuits, and so on.

A typical junction transistor consists of three layers of semiconducting material: a layer of *N*-type germanium, a very thin layer of *P*-type, and then another layer of *N*-type, all in

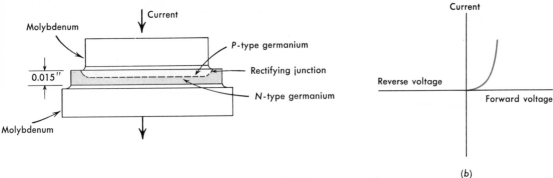

Fig. 44-13. (a) *A germanium junction rectifier.* (b) *Its current-voltage characteristic curve.*

contact at barrier layers and provided with external leads. For this *N-P-N* transistor—other kinds are possible—we thus have two *N-P* junctions within one device (Fig. 44-15). The emitter is biased negatively with respect to the base, and the collector positively with respect to the base. Thus the direction of low resistance is from the emitter to the base, and the direction of high resistance, from the collector to the base. Suppose now that the input signal becomes more negative. This will inject electrons across the emitter-base junction into the base. A few will combine with holes in the base or leave the base, but *most of the electrons—which are minority carriers in the P-type material—will move through the base and into the collector.* Of 1000 electrons that leave

Fig. 44-14. *Miniaturization of electronic components. From left to right are a tiny integrated circuit, a transistor in its protective case, and the base of a vacuum tube. (Courtesy Bell Telephone Laboratories.)*

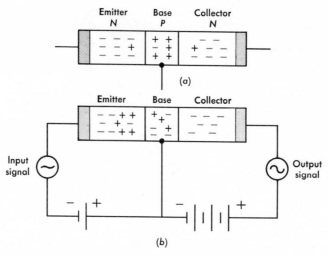

Emitter Base Collector
  N   P   N

(a)

Emitter Base Collector

Input signal

Output signal

(b)

Fig. 44-15. (a) *An* N-P-N *junction transistor.* (b) *A simple transistor amplifier.*

the emitter, about 980 reach the collector. In what sense, then, can we say that amplification has taken place? The answer is that the *resistance* through which the electrons pass in going from the base to the collector is several thousand times that between the emitter and the base. Thus essentially the same current going through a greater resistance yields a greater $V = IR$ voltage gain. The output signal is correspondingly amplified over the input signal.

Lasers

Turning from the subject of how solids conduct electricity, let us consider how they can be used in *lasers.* The word "laser" is an acronym for *light amplification by stimulated emission of radiation.* As the name suggests, lasers can be used to produce extremely intense beams of light. What is much more important for their future use in science and technology is that lasers furnish the first reliably coherent (Chapter 37) sources of radiation in the optical region of the spectrum.

The active element of the pulsed ruby laser (Fig. 44-16) is a cylindrical rod of pink ruby

(corundum with chromium added to it) about 4 cm long and 0.5 cm in diameter. The ends of the rod are optically polished to be quite flat and perpendicular to the axis of the cylinder. The ends are silvered—one of them heavily and the other lightly—to serve as plane mirrors. Wound around the ruby rod is a high-power gas-discharge tube that can be pulsed repeatedly by means of the power supply to emit intense bursts of light, each lasting for a few milliseconds. When the light from the gas-discharge tube strikes the ruby rod, some of the chromium atoms in the ruby are excited to a higher-energy state. When they return to the ground state they emit photons of red light, and the rod glows softly. This is the well-known phenomenon of fluorescence. Suppose now that the intensity of light from the discharge tube is gradually increased. At a certain threshold level of intensity, laser action begins: a thin pencil of red light emerges from the half-silvered face of the rod. The laser beam is plane parallel, highly coherent over the plane wave front, and restricted to a very narrow band of frequencies.

What causes laser action to take place?

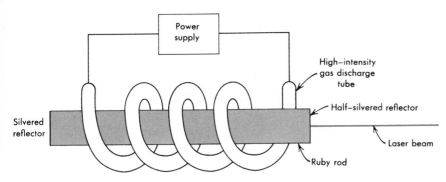

Fig. 44-16. *A simplified diagram of a pulsed ruby laser. The cooling mechanism is not shown.*

Four phenomena are involved: (1) optical pumping, (2) metastable levels, (3) inverted populations, and (4) stimulated emission. Figure 44-17 shows the pertinent energy levels of chromium atoms, which are responsible for laser action in the ruby laser. On the left are shown the transitions made by chromium atoms from the ground state to higher-energy bands in the crystal when photons from the gas-discharge tube strike the chromium atoms and are absorbed by them. This process of raising atoms to higher-energy states is similar to pumping water from a lake to a water tank on a hill and in fact is called *optical pumping.* The chromium atoms almost immediately lose some of their energy to the crystal lattice and drop back to a lower energy level—level 2 on the right side of Fig. 44-17—*without* emitting photons. There they

remain for an average time of about 3 milliseconds before dropping back to the ground state by emitting photons of wavelength 6934 A. Although 3 milliseconds seems very short, it is quite a long time for a radiation process: usually atoms lose their excitation energies in a few billionths of a second or less. Level 2 is called a *metastable* level or state: it has a relatively long lifetime. The metastable level serves to delay the atoms in their return to the ground state. The result of this delay is that atoms begin to pile up in the higher-energy state 2.

The growth in the number of atoms on metastable level 2 causes an *inversion* of the normal energy distribution of the population of chromium atoms in ruby. Ordinarily we expect most of the chromium atoms to be in the ground state, and only a few to be in excited

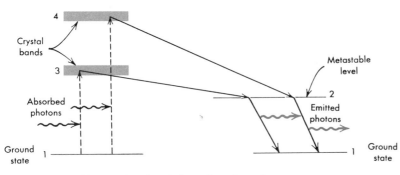

Fig. 44-17. *Energy levels of chromium in ruby.*

states. The result of optical pumping is to put *more* atoms on metastable level 2 than are in the ground state, thus turning the normal population distribution upside down.

Stimulated emission occurs when photons of the frequency that would ordinarily be emitted by an atom strike the atom. The atom is thereby induced to radiate, and two photons of that frequency result. Thus if photons of frequency 6934 A strike the atoms on metastable level 2, light of that wavelength will be amplified because each incident photon will stimulate the emission of other photons. This will continue until the population of atoms in the metastable state has been depleted.

Let us put these ideas together. When the short high-intensity burst of light of many different wavelengths from the gas discharge tube strikes the ruby rod, the chromium atoms absorb photons within several bands of wavelengths and are "pumped" up to levels 3 and 4 from which they return to metastable level 2. The population of level 2 increases until it exceeds that of the ground state. The right conditions then exist for the stimulated emission of light of wavelength 6934 A. A few photons of that light, moving in the direction of the axis of the rod, are reflected back and forth by the silvered ends and start an oscillation within the rod. As the photons move along, they stimulate the emission of other photons of the same frequency and a "snow-balling" effect occurs, the number of photons steadily increasing. Some of the photons escape through the half-silvered end of the rod and constitute the laser beam. The others are trapped in the rod, and their back-and-forth oscillations continue until the metastable level has been emptied. Then another burst of light from the gas-discharge tube starts the optical pumping again, and the process repeats itself.

The beam of a pulsed ruby laser typically has an average power density of 10^9 watts/cm^2 during its short burst. It can be focused to a spot 0.1 mm in diameter at which the power density reaches the enormous value of 10^{13} watts/cm^2. Even the most refractory material is vaporized by such beams, and they can be used to drill tiny holes in steel, diamond, and graphite. The wavelength band in the ruby laser beam is about 0.02 A wide at a wavelength of 6934 A. The spatial coherence of the radiation is so high that a two-slit diffraction pattern (Chapter 37) can be demonstrated by allowing the entire wave front of the laser beam to illuminate the two slits without first passing the light through a single slit as in Young's experiment. The laser beam diverges less than a few minutes of arc.

Many other kinds of lasers have been devised. Some operate continuously rather than in bursts like the pulsed ruby laser. Others use gases, glasses, or semiconductors as the active material.

Summary

Solid state physics has made great progress by applying quantum mechanics to the motion of electrons in crystals and by developing better experimental techniques such as the production of pure materials and materials with controlled amounts of known impurities.

The Hall effect is the appearance of a potential difference across a conductor in the *y*-direction when a magnetic field is applied in *z*-direction and there is a current in the *x*-direction. The Hall coefficient R_H is defined as $E_y/j_x B_z$. A negative Hall coefficient indicates that the conduction current is carried by negative charge carriers; a positive Hall coefficient indicates positive charge carriers.

By measuring the resistivity of a conductor and the Hall coefficient, we can estimate the number of charge carriers per unit volume and the relaxation time or average time between collisions of an electron.

The conduction band theory of solids states that there are bands of energy levels for the electrons within a solid. Motion from one potential well to another within the crystal is permitted by the laws of quantum mechanics to electrons upon one of those energy levels. Motion is denied for energies lying within the forbidden energy regions. In a crystal, the lowest energy bands are filled, two electrons of opposite spin occupying each level. The highest band occupied by electrons is half filled in conductors and completely filled in insulators.

Conduction results when energy levels are open to electrons that acquire tiny increments of energy from an electric field. Such electrons move away from their electron partners and contribute to the conduction current. Conductors then are substances which provide accessible empty levels to electrons, and insulators are those which do not.

Resistance is caused by disorder. Impurities, lattice vibrations, vacancies, grain boundaries, dislocations, and distortions of the crystal structure contribute to the scattering of the electron waves. Such scattering is resistance.

Materials such as germanium that are not good conductors in pure form become conductors (a) when certain impurities are added or (b) when the temperature becomes high enough. N-type germanium is germanium to which tiny amounts of 5-valency atoms, such as antimony, have been added. Such atoms set up *donor* energy levels that contribute electrons to the conduction band of germanium. P-type germanium contains 3-valency atoms, such as aluminum, which set up *acceptor* energy levels.

Thermionic emission of electrons can be interpreted quantum mechanically as the transmission of electron waves over or through the potential-energy barrier at the surface of a heated metal.

The junction rectifier and the junction transistor are two important technological applications of solid state physics. The boundary layer of the junction rectifier gives it a high reverse resistance and a low forward resistance—like the vacuum-tube diode. The emitter of the transistor controls the current to the collector—as the grid of a triode controls the plate current. The transistor and the triode can both produce voltage amplification.

Lasers are devices that produce intense beams of highly monochromatic light as a result of stimulated emission in certain solids and gases.

Questions

1. Explain the Hall effect in terms of the deflection of moving charges by a magnetic field.

2. What would be the characteristics of a perfect crystal? Discuss the possibilities of achieving a perfect crystal.

3. What is the classical theory of electron conduction in metals?

4. Compare and contrast insulator, semiconductor, conductor.

5. What causes electrical resistance?

6. What design considerations would influence you if you wanted to design (a) a resistor of very high resistance and (b) a conductor of very low resistance?

7. About how many conduction electrons per atom are there in sodium? How is this estimate supported by (a) experimental evidence, (b) atomic theory?

8. Test the rule that, in general, atoms with an odd number of electrons are good conductors and those with an even number of electrons are poor conductors by comparing atomic numbers and resistivities for the first twenty atoms in the periodic table (Appendix I). Look up the resistivities in a handbook of physics data.

9. Show that the current density $j = nev$ where n is the number of conduction electrons per unit volume, v is the drift velocity of electrons in a conductor, and e is the electron charge.

10. What is meant by (a) donor level, (b) acceptor level?

11. How does the band theory explain thermionic emission?

Problems

A

1. Calculate the number of atoms in a gram of potassium chloride.

2. How many atoms of antimony are there per million atoms of germanium if 0.10 mg of antimony is added to 0.3 kg of germanium?

3. List other elements with electron configurations similar to those of aluminum, germanium, and arsenic. See Appendix I.

B

4. Plot the following values of current versus voltage for a germanium rectifier:

I	V
14 ma	+1.0 v
7	+0.5
0	0
−1	−20
−2	−40

5. At what temperature does the mean kinetic energy of an electron equal 6.0 ev?

6. What is the voltage gain in a transistor if the emitter and collector currents are approximately equal? Assume that the emitter-circuit impedance is 100 Ω and the collector impedance is 100,000 Ω.

C

7. Show that the units for the Hall coefficient (R_H) are cubic meters per coulomb.

8. The Hall coefficient for copper is -0.55×10^{-10} m³/coul, and the resistivity is 1.7×10^{-8} Ω-m. Find (a) the number of conduction electrons per unit volume, (b) the relaxation time.

9. An electron is trapped in a vacancy in a crystal. Assuming that the electron is in a tight

box with a length of 4.0 A, determine the uncertainty in the velocity of the electron.

10. How many ping-pong balls, each 3.0 cm in diameter, could be stored in a room 5.0 m by 4.0 m by 3.0 m without crushing them?

11. Calculate for aluminum (a) the number of conduction electrons per unit volume, (b) the relaxation time. See Table 44-1, page 711.

12. A current of 150 amp is sent through a strip of copper 3.0 cm wide and 0.10 cm thick. If the magnetic flux density is 2.0 webers/m², what is the Hall potential difference?

For Further Reading

Buchhold, T. A., "Applications of Superconductivity," *Scientific American,* March, 1960, p. 74.

DeBenedetti, Sergio, "The Mössbauer Effect," *Scientific American,* April, 1960, p. 72.

Kock, Winston E., *Lasers and Holography,* Doubleday and Company, Garden City, 1969.

Kunzler, J. E. and Morris Tanenbaum, "Superconducting Magnets," *Scientific American,* June, 1962, p. 60.

Matthias, B. T., "Superconductivity," *Scientific American,* November, 1957, p. 92.

Pierce, John R., *Quantum Electronics,* Doubleday and Company, Garden City, 1966.

Schawlow, Arthur L., "Optical Masers," *Scientific American,* June, 1961, p. 52.

Shockley, W., "Transistor Physics," *The American Scientist,* January, 1954, p. 41.

Stewart, Alec T., *Perpetual Motion,* Doubleday and Company, Garden City, 1965.

CHAPTER FORTY-FIVE

Radioactivity

The history of our knowledge of the atomic nucleus begins with the discovery of radioactivity. Roentgen's discovery of X-rays in 1895 (Chapter 40) aroused wide interest and stimulated scientists to search for other penetrating radiations. Only a few months later, the French physical chemist, Henri Becquerel, "struck gold." As often is true, Becquerel was guided by a false hypothesis. He knew that sunlight makes some substances fluoresce like the glass walls of Roentgen's X-ray tube. He falsely guessed that the sunlight might stimulate such phosphorescent materials to emit penetrating rays. Accordingly, he exposed several minerals to sunlight until they glowed and placed them on photographic plates wrapped in paper. Only one of them, a uranium salt (uranium nitrate) blackened the plate. The winter sun did not shine for several days, and Becquerel left the uranium crystals on a photographic plate wrapped in paper. When he developed the plate, it was blackened even though the crystals had not been exposed to sunlight. Later experiments proved that anything containing uranium emitted penetrating rays.

A young Polish refugee, Marie Sklow-dovska, asked Becquerel's permission to investigate the new phenomena for her doctor's degree. She studied all the known elements and proved that only uranium and thorium emitted the radiations that discharged an electroscope. Then she tried many compounds and mixtures. Pitchblende, a uranium ore, was more active than uranium. Reasoning that the ore must contain a new element, she and the eminent professor, Pierre Curie, whom presently she married, decided to isolate the element responsible for the radiations. Hampered by poverty, they begged the use of an old dissection shed at the School of Chemistry and Physics of the City of Paris. They begged a ton of pitchblende from the government of Austria. They gave two years of joyful service to science and isolated two new elements. The first they called *polonium* for Madame Curie's native land; to the second they gave the name *radium.*

Radioactivity was found to be a spontaneous nuclear change causing radiations or particles to be emitted. Radioactive changes, occurring in the nuclei of atoms, are not affected by ordinarily used chemical processes. For example, radium emits its radiations just as rapidly in liquid air as in the hottest furnace. Chemical changes, having to do

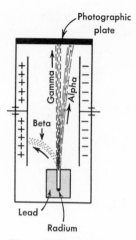

Fig. 45-1. *Influence of electric field on alpha particles (helium ions), beta particles (electrons), and gamma rays.*

with electrons in the outer shells of atoms, are very sensitive to temperature variations.

The Curies and the other early experimenters found that radioactive materials emit three kinds of radiations, not all of them being emitted at the same time. Imagine that you place pitchblende at the bottom of a narrow, deep hole in a block of lead and mount a photographic plate above the hole. Let the system be at low pressure. The radiations from the pitchblende will blacken the plate in a small spot just above the hole. Suppose that two oppositely charged vertical plates cause an electric field. The beam will divide into three parts, proving the existence of three kinds of rays. We call them *alpha particles, beta particles,* and *gamma rays.* The electric field between the plates forces the alpha particles in one direction and the beta particles oppositely (Fig. 45-1).

The *alpha particles* are helium ions, each carrying two positive charges ($2 \times 1.60 \times 10^{-19}$ coul). Each ion has a mass about four times that of a proton. They travel a few thousand miles per second and have energies as high as ten million electron-volts. A single sheet of writing paper will stop alpha par-

ticles. The *beta particles* are electrons and have widely variable speeds, as we shall see. Some of them travel nearly as fast as light. Beta particles, because of their high speeds, can pass through a millimeter or so of aluminum. The *gamma rays* are not deviated by the electric field. They are photons, like X-rays, but more energetic and of shorter wavelength. Gamma rays from radium can penetrate several millimeters of lead. Others are even more penetrating.

A Brief Chronology

The dates of some of the important developments in nuclear physics indicate the acceleration of the growth of knowledge in this field. Some of these developments we discussed in earlier chapters; others we shall discuss in this and the following chapter. The Curies announced the isolation of radium in 1898. Rutherford and Soddy in England in 1903, after careful investigations of the chemical properties of the radiating elements and studies of the radiations themselves, proposed that the radiations accompanied atomic disintegrations, resulting in changes in the atomic species. Later, they were able to capture enough of the emitted alpha particles to show that ordinary helium gas was formed when the alphas captured electrons. In 1905, Einstein announced his special theory of relativity with its development of the equivalence of mass and energy (Chapter 10) — a particularly important step in understanding the source of nuclear energy. By means of experiments on the scattering of alpha particles, Rutherford established the nuclear model of the atom (Chapter 21), showing that most of the mass and all of the positive charge of the atom were concentrated in its tiny nucleus. J. J. Thomson in England measured the masses of stable isotopes for the first time in 1913, using a method similar to the one used in measuring the mass of the electron (Chapter 27). Cosmic rays were discovered in 1914 by

Victor Hess. In 1919 Rutherford induced the first transmutation of an atom from one species to another by converting nitrogen nuclei into oxygen under bombardment by alpha particles. Better instruments were developed: the Geiger counter in 1928, the Cockcroft-Walton generator of energetic particles in 1930, and the Van de Graff generator (Chapter 21) in 1931.

In 1932, the pace of development quickened. The neutron, the positron, and deuterium were discovered in that year, and the first cyclotron was put into operation (Chapter 27). The coming of the nuclear machines had a profound effect upon physics. Not only did they make higher-energy particles available in greater numbers than could be derived from naturally radioactive substances, but they changed the working conditions of many physicists. Nuclear physics laboratories became large work centers, organized around accelerators costing many millions of dollars and employing hundreds of scientists, engineers, and technicians. In high-energy nuclear physics at least, the day of the individual experimenter, working in a tiny laboratory, was over. Teams of physicists shared the available operating time of the large accelerators and planned and carried out experiments together (Fig. 45-2).

Artificial radioactivity was discovered by Irene and Frederic Curie-Joliot, the daughter and son-in-law of Marie Curie, in 1934, making it possible to create radioactive isotopes that were not found in nature. In 1936, Neddermyer and Anderson in the United States discovered the meson in cosmic rays. Rabi made the first precise measurements of nuclear magnetism in 1937. In 1939, Hahn and Strassman in Germany made the discovery of nuclear fission. The betatron (Chapter 29) was invented by Kerst in 1941, the first chain-reacting pile put into operation in 1942 by Fermi, and the synchrotron principle for accelerating particles was announced by Vecksler in Russia and McMillan in the

United States. Purcell and Bloch discovered the principle of nuclear magnetic resonance in 1946. The heavy meson was discovered in cosmic radiation in 1947, and the first mesons were produced by an accelerator in 1948.

Since 1950, events in nuclear physics have crowded upon one another. Among the more spectacular of these were the overthrow of the principle of conservation of parity by Yang and Lee, the construction of huge synchrotron accelerators capable of accelerating particles to several hundred Gev (giga-electron volts $= 10^{12}$ ev) of energy, and the production of the antiparticles. Less spectacular developments have been equally important: the patient performance of many experiments to yield good data about the nuclear properties and the construction of increasingly successful theories based upon these data.

In this chapter, we will provide an introduction to some of the basic properties of the nucleus, concentrating on those that are revealed by radioactivity. In the next chapter, we shall review some of the things we know about the structure of the nucleus.

Experimental Methods

A stranger entering a nuclear physics laboratory (Fig. 45-2) is impressed by the apparent complexity of the equipment for detecting the particles being studied. There is usually a great variety of electronic black boxes intricately connected by cables and displaying flashing lights and moving oscilloscope traces.

Actually, however, a relatively small number of basic measuring techniques and devices are used. The energetic particles that have been emitted by radioactive nuclei or are the result of nuclear reactions are usually subjected to thicknesses of absorbing materials or to magnetic or electric fields that cause them to reveal their energies or momenta. The particles are then counted by a detector whose properties are matched to those of the

Fig. 45-2. *Victory at Batavia. The happy physicists at the National Accelerator Laboratory in Batavia, Illinois, crowd around the control panel as the accelerator reaches its design energy of 200 billion electron volts. (Courtesy National Accelerator Laboratory.)*

particle. The great variety of devices comes from the different requirements of sensitivity and counting speed imposed by the experiments. We shall next consider some of the basic devices and techniques.

Detectors

The *electroscope* (Chapter 21) is one of the simplest detectors of radiations. An energetic charged particle creates ions along its path by ripping electrons away from the atoms that it passes. Both positive and negative ions are produced in a gas by this means. If an electroscope is negatively charged, positive ions produced by the passage of an alpha particle or beta particle through it will be attracted to the gold leaf and discharge the electroscope. The rate at which the leaf collapses can be related to the activity of the radioactive source. The *dosimeter* (Chapter 21) is a sturdy, portable electroscope used to measure the amount of radiation to which laboratory workers are exposed. Electroscopes are not sensitive enough to detect individual par-

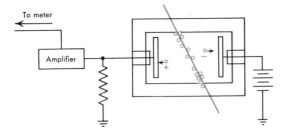

To meter

Amplifier

Fig. 45-3. *An ionization chamber.*

ticles; rather they measure flux of particles.

The *ionization chamber* consists of a pair of conductors insulated from each other and separated by a gas in which ions are produced by the passage of energetic charged particles (Fig. 45-3). The ions are collected very efficiently by an electric field between the conductors. The resulting steady current or pulse is amplified; currents are then displayed on a meter, and pulses are displayed on an oscilloscope or recorded on a scaler (see below). Individual particles or a flux of particles can be detected.

One of the most useful, rugged, and simple devices for counting energetic charged particles is the *Geiger counter* (Fig. 45-4). The glass (or metal) envelope contains a gas or a vapor at a pressure of a few centimeters-of-mercury. The central wire is at a positive potential of 800 to 2000 v above that of the cylindrical electrode. The large electric field strength near the wire is almost but *not quite*

sufficiently large to ionize the gas and thus start a discharge across the tube. A single energetic particle such as an alpha particle entering the tube will produce thousands of ions and electrons. Thus a single particle can upset the equilibrium and trigger an avalanche of electrons toward the central wire. The central wire gets charged negatively, and negative pulses pass through the capacitor to an amplifier and to a scaler where they are counted. This type of Geiger tube can count alpha particles and beta particles. It can also count X-ray and ultraviolet photons—with low sensitivity, however. If you hold a burning match before the window, photons striking the metal cylinder will knock out myriads of photoelectrons and cause a roar. Very weak light will cause a series of clicks.

The *scintillation counter* (Fig. 45-5) contains a crystal (NaI doped with thallium, for example), which fluoresces when struck by an incident alpha, beta, or gamma ray, and a very sensitive phototube, called an electron-multiplier tube, which measures the light emitted. If a gamma ray or an energetic charged particle passes through the crystal, light produced by fluorescence in the crystal reaches a photosensitive surface on the face of the electron-multiplier tube and knocks out electrons. These electrons are pulled from one electrode to another within the tube because the electrodes are made increasingly

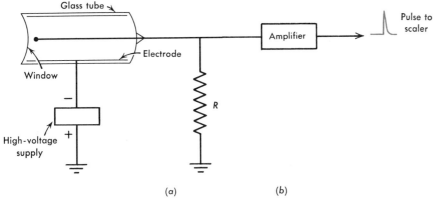

Fig. 45-4. *A Geiger counter.*

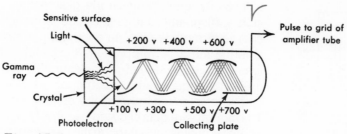

Fig. 45-5. A scintillation counter.

positive by a battery or power supply. When each electron strikes an electrode, it causes several electrons to be emitted by the process of secondary electron emission. These electrons join the others. The number of electrons gets greater and greater. The tube *multiplies* the number of electrons leaving the first photosensitive surface. When all the electrons reach the collector plate, there are enough of them to produce a negative pulse on the grid of the amplifier tube. The pulse is passed along to the scaler where the incident particle is counted. Scintillation counters are more sensitive to gamma rays than Geiger counters are. They are also very useful because the solid crystals "soak up" more of the energy of the incoming particle. The intensity of the light produced in the crystal is dependent upon the energy of the incident particle. Hence the size of the output pulse from the photomultiplier tube measures the energy of the particle.

Scalers

Scalers are used to cope with the rapid arrival of pulses from a Geiger counter, scintillation counter, or other detector. At counting rates less than about 60 counts per minute, the pulses can be counted by means of an electromechanical counter—a relay that operates a counter similar to that used by gatekeepers at football games. At higher counting rates, the electromechanical counter jams and misses pulses; the rate of arrival of pulses must be scaled down.

The simplest scaler is a scale-of-two electronic circuit whose output consists of one pulse for every *two* input pulses (Fig. 45-6). For example, an input counting rate of 120 counts/min would yield an output rate by the scaler of 60 counts/min, which the electromechanical recorder could record without jamming. (The operator knows to multiply the output rate by two to obtain the rate at which the detector detects particles.) If the output of this scale-of-two is fed into another scale-of-two, both together become a scale-of-four. Scales-of-ten, from which there is an output pulse for every ten input pulses, are also possible. By extension of this arrangement, one can arrange a scale-of-1000 or a scale-of-4096 very readily so that very high counting rates can be measured conveniently and accurately. The details of the electronic circuits for doing this can be found in books dealing with experimental nuclear physics.

Fig. 45-6. The action of a scaler.

Coincidence Circuits

Often the physicist wants to determine the rates at which *related* nuclear events occur. For example, he might want to determine how frequently a radioactive nucleus emits a beta particle that is immediately (within 10^{-10} sec, say) followed by a gamma ray. A *coincidence circuit* and two detectors are used for this. A Geiger or scintillation counter is placed on each side of the radioactive source. One counter is covered with a thick aluminum shield which beta rays cannot penetrate. The physicist then measures the *coincidence rate*—the number of times per second that a beta particle is detected by the unshielded counter at the same time that a gamma ray is detected by the shielded one.

A schematic diagram that illustrates the principle of the coincidence circuit is shown in Fig. 45-7. Amplifiers 1 and 2 are normally conducting a current. The voltage at point O is therefore relatively low. Suppose now detector 1 detects a gamma ray at the same time that detector 2 detects a beta particle. Negative pulses from the two detectors will be sent

to amplifiers 1 and 2 and will bias them so that the amplifiers will *cease to conduct.* The voltage at O will go up immediately and a positive output pulse will be sent to the scaler, which will record a coincidence. Suppose, on the other hand, that only detector 1 detects a particle, sending a negative pulse to amplifier 1. Amplifier 2 is still conducting, and the relatively small decrease in current through R, caused by cutting off amplifier 1, increases the potential at O only slightly. The resulting positive output pulse is not enough to trigger the scaler. *A recordable output pulse occurs only when both detectors detect particles at the same time.*

Cloud Chamber, Bubble Chamber, Spark Chamber, and Nuclear Emulsion Plates

These devices enable physicists to see or to photograph nuclear events. A simple lecture-room *cloud chamber*—operating on the same principle as large research cloud chambers—is shown in Fig. 45-8. A speck of

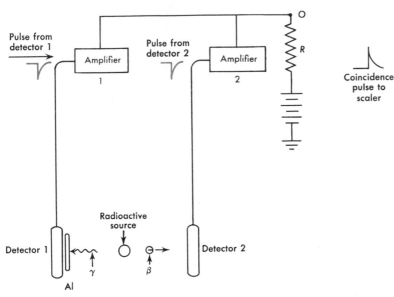

Fig. 45-7. *The principle of a coincidence circuit.*

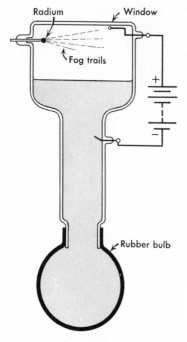

Fig. 45-8. *A lecture-room cloud chamber.*

event. Figure 45-9 shows cloud-chamber tracks produced when the space contains oxygen. Most of the tracks are fairly straight. This proves that each alpha particle plowing through thousands of atoms strikes no massive obstacle. The Y-shaped trail tells a different story. An alpha particle (helium ion) struck an oxygen nucleus whose mass is about four times that of the alpha particle. The more massive particle acquired the smaller sidewise velocity. By measuring the angles of deflection in such elastic collisions, we can determine the mass of the struck nucleus relative to that of the projectile.

The *bubble chamber* operates upon a principle similar to that of the cloud chamber, but contains a superheated liquid instead of a supersaturated gas. Liquid propane or liquid hydrogen is kept under pressure in the chamber at a temperature above its normal boiling point. When the pressure is suddenly

radioactive material, an alpha emitter, is mounted on the end of the rod. When each radioactive nucleus disintegrates, it hurls out an alpha particle. Suppose that you squeeze the rubber bulb so as to compress the air around the radium. When you release the bulb, the expanding air does work in pushing back the atmosphere. Having lost energy, the air gets chilled far below room temperature. Thus, the air becomes supersaturated with water vapor. In a supersaturated space, water vapor readily condenses on nuclei such as dust particles and ions. An alpha particle, plowing through the air, produces many thousands of ions. Water condenses on each of them so that you will see the path, marked out by the fog trail. The battery causes an electric field between the water surface and an invisible conducting film on the window. This field clears away the ions formed in the air space to prepare the cloud chamber for the next

Fig. 45-9. *Cloud chamber tracks show the paths of alpha particles. The Y-shaped track was produced when an alpha particle struck an oxygen nucleus. Which way did the alpha particle recoil?*

reduced—corresponding to the expansion in a cloud chamber—the liquid becomes superheated and vapor bubbles form upon locally heated spots formed by the passage of an energetic particle through the liquid. Thus the track of a high-energy particle is shown by a trail of bubbles and may readily be photographed (Fig. 45-10). The use of the bubble chamber is particularly advantageous in large accelerator laboratories of high-energy nuclear physics. The density of the liquid is such that high-energy particles are more likely to interact with the protons in the liquid. Also the products of the interaction will be slowed down and often stopped within the chamber itself so that the whole event can be seen.

The *spark chamber* (Fig. 45-11) combines some of the features of Geiger counters and some of those of bubble or cloud chambers. It makes the path of an energetic particle visible as a series of electrical sparks that are triggered by the ionization produced by the particle. The spark chamber contains a series of parallel plates separated by a few millimeters in an atmosphere of neon. Alternate plates are grounded, and the other plates can be raised to a voltage of several thousand volts relative to the grounded plates. Suppose an energetic cosmic ray particle passes through the spark chamber, leaving a trail of ions in the neon. As the particle passes through the counter, it fires the counter, causing it to send out a pulse to the high-voltage supply. The high-voltage pulse reaches the plates before the ions have disappeared. A spark jumps between each pair of plates, following the ionized trail of the energetic particle and making it possible to see and photograph its trail. Spark chambers are particularly useful in studying very-high-energy particles, because a large number of plates can be put into the path of the particles to slow them down.

Nuclear emulsion plates are used to photograph nuclear events by allowing energetic particles to pass through the fine-grain photographic emulsion (Chapter 35) and to affect the sensitive grains directly. The paths of the particles are shown after photographic development by lines of developed grains which can be seen with a microscope. By measuring the range of an unknown particle and by counting the number of developed grains per unit length of path, the experimenter can measure the mass of the particle if he knows the characteristics of the emulsion for a known particle.

Analyzing Magnets

Analyzing magnets should also be mentioned as one of the most useful tools of the nuclear physicist. Electromagnets are not only a part of high-energy accelerators, but are used to study the properties of the particles involved in nuclear events—the sign and magnitude of charge, the mass, and the energy. A charged particle in a magnetic field, you will recall from Chapter 27, is acted upon by a sidewise deflecting force given by $F = Bqv \sin \theta$. By measuring the curvature of the path of a particle in a known magnetic field (Fig. 45-10) and combining this information with other data, the physicist can often identify the particle and measure its properties.

Example. A proton enters a cloud chamber, crossing a uniform magnetic field of 1.0 weber/m^2 at right angles to the field. The radius of curvature of the proton's path, as measured from the photograph of the event, is 0.2 m. Calculate the velocity of the proton, assuming relativistic change of mass is negligible. The rest mass of the proton is 1.67×10^{-27} kg.

The centripetal force mv^2/r for the proton is supplied by the magnetic deflecting force $Bqv \sin \theta$.

$$Bqv \sin \theta = \frac{mv^2}{r}$$

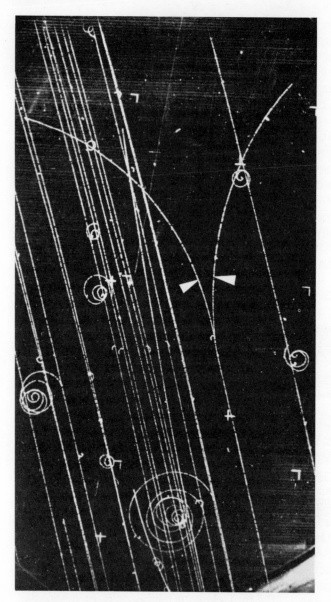

Fig. 45-10. *A bubble chamber photograph. The curved tracks indicated by arrows are those of a positron and electron that were created by the decay of a neutral pi meson. Note the tightening spiral tracks of electrons in the magnetic field. (Courtesy University of California Radiation Laboratory.)*

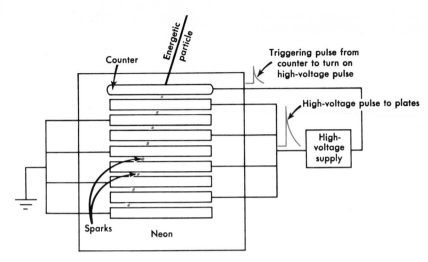

Fig. 45-11. A spark chamber.

$$(1.0 \text{ weber/m}^2)(1.60 \times 10^{-19} \text{ coul})(1)$$

$$= \frac{(1.67 \times 10^{-27} \text{ kg})v}{0.2 \text{ m}}$$

$$v = 2.0 \times 10^7 \text{ m/sec}$$

Radioactive Decay

When the nucleus of a radioactive atom disintegrates, it may emit an alpha particle or a beta particle. Gamma rays may precede or follow either kind of particle. When an alpha particle is emitted, the mass number *A* (the total number of protons and neutrons in the nucleus) decreases by 4 and the atomic number *Z* by 2, because the positively charged alpha particle carries off two electronic units of charge, leaving the positive nuclear charge *less* by two electronic units (conservation of energy). Emitting a beta particle does not alter the mass number; it *increases* the atomic number by one, because the negatively charged beta particle carries off one electronic unit of charge, leaving the positive nuclear charge **greater** by one electronic unit. Figure 45-12 shows the descendants of uranium-238. When a radium

atom ($A = 226$, $Z = 88$) hurls out an alpha particle and changes into the gas *radon*, its mass number decreases to 222 and the atomic number to 86. The disintegration continues, step by step, resulting in the final stable product, lead-206. Notice that the lead, bismuth, and polonium all have isotopes (of the same atomic numbers but different mass numbers). Other decay chains, similar to the uranium series, begins with thorium-232, neptunium-237, and actinium-235, respectively.

Beside each name the *half-life* is indicated This is the time required for half the atoms to be converted into the next element. The greater the half-life, the *less* probable is radioactive decay. For example, the half-life of uranium-238 is 4.51 billion years. This means that out of 2 million atoms of uranium-238, 1 million will remain 4.51 billion years from now. Radium is much less stable; its half-life is only 1622 years. The half-life of radon is less than four days. The *activity* of radioactive materials, the number of disintegrations per unit time, is measured in *curies.* One curie is the activity of a source which undergoes 3.7×10^{10} disintegrations per second.

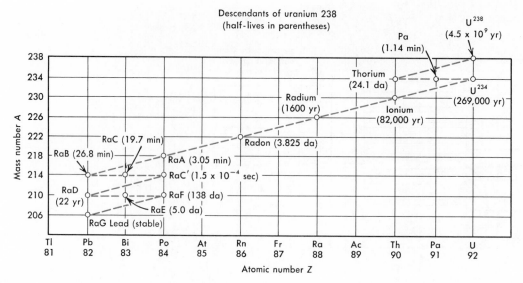

Descendants of uranium 238
(half-lives in parentheses)

Fig. 45-12. Descendants of uranium-238. Whenever an alpha particle is emitted, the mass number decreases by 4 and the atomic number by 2. When a beta particle is emitted, the atomic number increases by 1. (Beside each name, the "half-life" is indicated—the time in which half of the atoms will disintegrate.)

The number of radioactive nuclei ΔN of a given kind that undergo disintegration in a time Δt is proportional to the number of nuclei present and to Δt:

$$\Delta N = -\lambda N \, \Delta t$$

where λ is called the *disintegration constant* and where the negative sign means that the number of nuclei of that kind is decreasing due to radioactive decay. Let N_0 be the number of nuclei at time $t = 0$. (In our experiment, $t = 0$ would be the time when the radioactive material was separated chemically from the parent radioactive nuclei or when the beam of energetic particles producing the radioactive material in the laboratory was turned off.) Then, using calculus (Appendix E), we find* that

* The situation is not always this simple. If the radioactive material being studied is itself a decay product and has not been separated from the parent material, "new" nuclei are being created during the decay, and a different mathematical expression must be used.

$$N = N_0 e^{-\lambda t}$$

where e is the base of natural logarithms. Since the half-life T is the time for N to equal $N_0/2$, it turns out that $T = 0.693/\lambda$.

Example. If 8×10^{10} atoms of radon are separated from radium, how many disintegrations will occur in 11.46 days?

First, notice that 11.46 days is three times the half-life (3.82 days) of radon. We could solve the problem as follows: during the first 3.82 days, half of the radon atoms will decay, leaving 4×10^{10}; during the second 3.82 days, half of these will decay, leaving 2×10^{10} and during the final 3.82 days, half of these will decay, leaving 1×10^{10}. Thus 7×10^{10} disintegrations would occur in 11.46 days.

Using the formula for radioactive decay, we have:

$$N = 8 \times 10^{10} e^{-(0.693/3.82 \text{ days})(11.46 \text{ days})}$$
$$= 8 \times 10^{10} e^{-2.079}$$

Mathematical tables of exponential functions show that $e^{-2.079}$ equals 0.125. Thus $N = 1 \times 10^{10}$, and the number of disintegrations is 7×10^{10}.

We noted earlier that radioactive nuclei can be produced in the laboratory by bombarding stable nuclei with energetic particles from an accelerator. Figure 45-13 shows the decay scheme of sodium-24 (Na^{24}), a radioactive nucleus that results from bombarding natural sodium with deuterons. Na^{24} decays with a half-life of 14.9 hours, emitting beta particles whose maximum energy is 1.390 Mev and then in rapid succession gamma rays of energies 2.758 Mev and 1.380 Mev. The daughter nucleus is stable magnesium-24 (Mg^{24}).

Finally, we should mention that not only nuclei, but "single" particles can decay. As we shall see later, neutrons, mesons, and heavier particles decay into one or more lighter particles with a characteristic half-life and the release of energy.

Alpha Particles

The *alpha particle* was historically one of the earliest known to be emitted by radioactive nuclei. Alpha-particle emitters occur among the heaviest of the nuclei where, as we shall see in the next chapter, the energy that binds the nucleus together is somewhat less than for nuclei of smaller A. This means that these heavier nuclei are unstable and decay, emitting one or more alpha particles as they lose mass and energy and descend the mass scale toward the more stable nuclei.

Alphas are emitted with well-defined energies, indicating that the nucleus moved from one energy level to another. The tracks of alpha particles in a cloud chamber are quite straight — except for change of direction caused by nuclear collisions near the ends of the tracks — with a definite range associated with each energy. The alphas — like all energetic charged particles — lose energy by ionizing the atoms of the medium through which

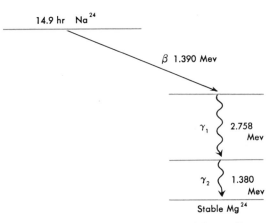

Fig. 45-13. *The decay scheme of the radioactive nucleus sodium-24.*

they pass. The range of a 5.3-Mev alpha particle in air at standard conditions is about 3.8 cm, and the number of ion pairs it creates is about 160,000.

Alpha-particle emission furnished one of the earliest indications that quantum mechanics could be applied to the nucleus as it had been to the electron clouds surrounding the nucleus. Gamow, Condon, and Gurney were able to show that the emission of an alpha particle could be explained by quantum mechanical tunneling through a potential-energy barrier. Figure 45-14 shows an idealized diagram of the potential-energy well of a nucleus such as radium with one of the energy levels indicated. Within the nucleus, an alpha particle, such as that on the level shown in Fig. 45-14, is bound by a specific nuclear attractive force that gives it a (negative) potential energy E_P. The alpha particle also has kinetic energy E_K due to its motion *within* the nucleus. Hence, its total energy is E, which equals E_K plus E_P. Outside the nucleus, the alpha experiences the Coulomb repulsion between the positive charge on the nucleus and the positive charge on the alpha. The combination of this repulsion and the nuclear attraction gives the potential barrier shown in Fig. 45-14.

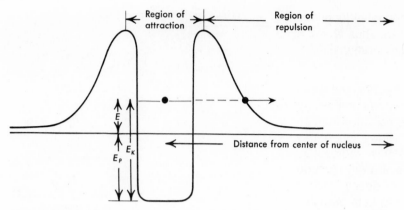

Fig. 45-14. *The potential-energy well of the nucleus of an alpha particle emitter.*

According to classical physics, an alpha particle could not escape from the nucleus because it lacked the energy to go over the top of the barrier. But quantum mechanics permits tunneling through the barrier, as we saw in the last chapter in the case of thermionic emission. If one solves Schrödinger's equation for the alpha particle in a potential well of this kind, the wave function for the alpha does not go to zero at the wall of the "crater," but extends through the barrier to the region outside the nucleus. This means that there is a definite probability that the alpha can tunnel through the barrier, appear on the outside with energy E, and speed away. Obviously the larger E is, the thinner the barrier, and the more probable the escape of the alpha. The calculations yield a definite relation between the half-life for alpha emission (related to the probability of escape) and the energy of the alpha; the greater the energy, the shorter the half-life. Experimental measurements amply verify this relationship.

More recent investigations make it clear that although there are energy levels for protons and neutrons in the nucleus, alpha particles as such do not exist as particles inside the nucleus. Instead, the alpha is apparently "made up" at the moment of emission.

Beta Particles

The cloud-chamber tracks of beta particles are much lighter and more straggling than those of alpha particles because the betas ionize less heavily and undergo frequent changes of direction in collisions as they traverse matter. The range of a 1.0-Mev beta particle in aluminum is about 2.2 mm.

Betas are emitted from the nucleus with a continuous spectrum of energies (Fig. 45-15), unlike the groups of alphas of uniform energies. This was puzzling at first to physicists because the emission of a beta particle corresponded to the transition of a nucleus from one definite energy level to another. What

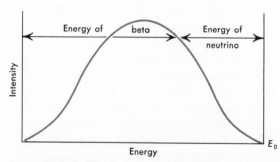

Fig. 45-15. *The continuous spectrum of beta particles.*

could have happened to the missing energy when a beta particle was emitted with energy less than E_0? What was further mystifying was where the beta particles came from in the first place. Electrons of the energies of beta particles have wavelengths many times the known nuclear diameter. In quantum mechanics this means that the probability is small that an electron would be found independently and stably existing inside the nucleus. Where then did the beta rays come from?

The first part of the explanation was suggested by Pauli, who proposed that the beta particle *shared* the energy of the transition with another particle of zero charge and very small mass, which he called the neutrino. Thus, if the energy of a beta transition was E_0 (Fig. 45-15), the nucleus emitted both a beta particle and a neutrino, whose energies added up to E_0. Since the neutrinos could carry off energies from zero to E_0, the corresponding betas had energies from E_0 to zero and a continuous spectrum of beta energies resulted. The neutrinos* were subsequently detected by detailed coincidence studies of the decay of heavier particles. Neutrinos are fantastically hard to detect; they have a mean free path in lead estimated at one astronomical unit (the mean distance from the earth to the sun)!

To explain where the beta particles and the neutrinos come from, we must introduce two more particles: the pi (π) meson or pion and the mu (μ) meson or muon. Both of these were first observed in cosmic rays, where their presence was detected by nuclear-emulsion and cloud-chamber studies. Later both were produced by high-energy accelerators by allowing energetic protons to hit nuclei. The pi meson has a mass about 270 times that of the electron; the mu meson, a mass of about 210 times the electron mass. (What would the wavelength of a mu meson be at a given

energy, compared with that of an electron at the same energy?) We shall see in the next chapter that the pi meson is thought to be involved in the attractive nuclear force that holds the nucleons together. Here we are interested in the fact that the pi meson has been observed to decay into a mu meson (plus a neutrino for conservation purposes) and the mu meson to decay into a *beta particle plus a pair of neutrinos.* It seems likely that the beta particle and neutrino suggested by Pauli are the same particles that are involved in the decay of the mu meson. The complete story has not yet been written and will have to await further experimental and theoretical investigation.

Positrons

So far, we have discussed the negative beta particle or negatron. There is also a positive beta particle — the *positron* — which is the *antiparticle* of the negatron. An antiparticle is one having the same mass and magnitude of charge as a particle, but a charge of opposite sign. The existence of the positron was first suspected as a result of a theoretical investigation by Dirac, who was developing a quantum mechanical theory of the electron which would agree with relativity theory. Dirac found it necessary in order to make his theory logically complete to assume that electrons could exist in states of *negative total energy.* If the theory was correct, there were thus in nature electrons in two kinds of situations: electrons with positive total energy (these are the ones we observe in atomic shells, in vacuum tubes, etc.) and electrons with negative total energy that we could not observe. However, we could observe vacancies among these electrons of negative energy, and this is where the positron came in.

In Dirac's theory, the highest of the negative energy states (Fig. 45-16a) is at -0.511 Mev. We know that the lowest of the positive energy

---

* It has been shown that there are two neutrinos: one kind which interacts with nucleons to produce mu mesons and another kind that produces beta particles.

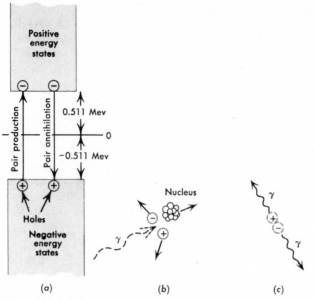

Fig. 45-16. *The positron.* (a) *Electron states of positive and negative energy.* (b) *The creation of a negatron-positron pair by a gamma ray.* (c) *The annihilation of a negatron-positron pair.*

states of ordinary electrons is +0.511 Mev because this is the energy equivalent of the rest mass of the electron. Suppose a gamma ray of 1.02-Mev energy should lift an electron out of the highest negative energy state and boost it into the lowest positive energy state, where it would become an ordinary electron. A "hole"—somewhat similar to the holes in semiconductors (Chapter 44)—would be left among the negative energy states. *The hole would behave like an electron of positive charge, and we could observe it.* The *creation* (Fig. 45-16b) of an electron-positron pair by the action of a gamma ray in the neighborhood of a nucleus could be reversed by *pair annihilation* (Fig. 45-16c). In the annihilation process an electron would fall into the vacancy in the negative-energy states and wipe out the positron—thus removing itself from observation—with the emission of two gamma rays which, in the simplest case, would speed off in opposite directions to conserve energy and momentum.

Positrons were discovered in cosmic radiation by C. D. Anderson, who observed (Fig. 45-17) a particle which was bent in the cloud-chamber magnetic field with a curvature in the reverse direction to that expected of an electron. Subsequently, the positron appeared in beta radiation from some radioactive nuclei produced in the laboratory. The positron is one of the decay products of a *positive* mu meson. Thus Dirac's theory was found to be correct, and physics had a new particle.

Gamma Rays

Gamma rays are high-energy electromagnetic radiation. There is no sharp dividing line between gamma rays and X-rays, and many of the things we have said about X-rays and other electromagnetic radiations apply to gamma rays—except as to their origins.

Gamma rays are emitted (1) by nuclei in an excited state or (2) by high-energy charged particles during their acceleration or deceleration. The emission of a gamma ray is a means by which an excited nucleus can get rid of energy if the available energy is not large enough for the emission of a material particle. Thus the emission of a "bright-line" spectrum of gamma rays often precedes or follows the emission of betas, alphas, and particles hurled out in a nuclear reaction. When high-energy particles — protons or electrons, for example — strike matter and are abruptly slowed down, they emit a continuous spectrum of gamma rays in a manner similar to the production of X-rays in an X-ray tube.

As they traverse matter, gamma rays are absorbed exponentially. The absorption law (Appendix E) is

$$I = I_0 e^{-\mu t}$$

where I is the intensity of the gamma-ray beam after it has passed through a thickness t of an absorber whose linear absorption coefficient is μ, and I_0 is the initial intensity of the beam.

Gamma rays give up their energy to other particles in three processes: (1) Compton scattering of electrons (Chapter 40), (2) the production of photoelectrons, and (3) (at gamma energies above 1.02 Mev) the production of positron-negatron pairs.

Summary

Radioactivity is a spontaneous nuclear transformation accompanied by the emission of radiations. It may be detected and measured by its effects in producing (a) ionization, (b) blackening of photographic films, and (c) fluorescence.

Alpha particles are doubly charged helium ions; beta particles are positive or negative electrons; and gamma rays are photons.

Energetic particles can be detected by the ionization that they produce. Ionization

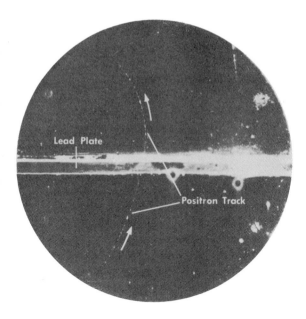

Fig. 45-17. The positron. The magnetic field points into the page, indicating that the positron is positively charged. Why does the track change curvature above the lead plate? (Courtesy California Institute of Technology.)

chambers, Geiger counters, and scintillation counters can be used to detect individual particles. Particle tracks can be made visible and photographed in cloud chambers, bubble chambers, spark chambers, and nuclear emulsion plates. Scalers reduce counting rates to proportionately lower ones. Coincidence circuits are used to determine the rate of related nuclear events. Magnetic fields can be used to deflect charged particles and to determine their charge and mass.

In radioactive disintegration, when an alpha particle is ejected, the mass number decreases 4 units and the atomic number decreases 2. If a negative beta particle is emitted, the mass number is not changed, and the atomic number increases by 1 unit.

The half-life of a radioactive element is the period required for one-half of the atoms to disintegrate. The law of radioactive decay is

$$N = N_0 e^{-\lambda t}$$

Alpha-particle decay was explained as tunneling of the alpha particle through the potential-energy barrier of the nucleus, thus showing that the laws of quantum mechanics apply to the nucleus.

Beta particles share the decay energy with neutrinos emitted at the same time, thus producing a continuous spectrum of beta-particle energies. The positron—the antiparticle of the electron or negatron—is a "hole" in the sea of negative energy states of electrons.

Gamma rays give up their energy by (1) Compton scattering, (2) photoelectron production, and (3) the production of electron-positron pairs (when the gamma-ray photon has an energy of not less than 1.02 Mev).

Questions

1. From what mineral ores are naturally radioactive materials derived?

2. What is pair production? Under what conditions does it occur?

3. During pair production, what changes occur in electric charge, momentum, and energy?

4. What is an alpha particle? What is an important source of alpha particles?

5. List some nuclear detectors and state a specific use of each.

6. What is the basic principle of the coincidence counter? Sketch a circuit to detect triple coincidences.

7. Sketch an experimental arrangement to detect *anticoincidences*—events in which one particle appears, but another is not present. Give an example of such an event.

8. What are some practical limits on the construction of bubble chambers?

9. Why does a Geiger counter detect alpha particles more efficiently than gamma rays?

10. Define half-life. If your half-life were 40 years, does this mean that you would live to be 80? Discuss.

11. The half-life of uranium equals approximately 5×10^9 years. Suppose that in a certain rock the number of lead atoms descended from uranium-238 is equal to the number of uranium-238 atoms. How old is the rock?

12. Historically, how were these discovered: cathode ray, beta ray, electron? Are they related?

13. Discuss the changes of mass number and atomic number when alpha particles, negative beta particles, positrons, and gamma rays are emitted by a nucleus.

Problems

A

1. If a laboratory had 100 mg of tritium (radioactive H^3) in 1975, how much would remain in 2000? (Half-life = 12.5 years.)

2. If the half-life of radium is 1600 years, what fraction of a sample of radium will have decayed after 4800 years?

3. A certain Geiger tube is certified by its maker as having a lifetime of 10^8 counts. If it

is used at an average counting rate of 1000 counts/min for a 40-hour week, when would one expect to replace it?

4. If a biologically hazardous radioactive material of half-life 10 days is spilled on a laboratory floor and the radiation level is 16 times that permitted, how long should the laboratory be closed if it is impossible to clean up the material?

5. The capacitance of the detecting element in an ionization chamber is 0.2 microfarad. The voltage sensitivity is 5 divisions/volt. How many ions, each singly charged, will be required to give a deflection of 1 division?

B

6. What is the energy of an alpha particle that has been accelerated through a potential difference of 60,000 v?

7. What is the ratio of m/m_0 for a beta ray moving at half the speed of light?

8. Calculate the disintegration constant for radium. What percentage of a sample of radium will remain after 16,000 years?

9. What must be the minimum wavelength of a gamma ray in order for it to produce an electron-positron pair?

10. An electron and a positron, each with negligible kinetic energy, combine to produce two gamma-photons. What is the energy in joules and in Mev of each photon?

C

11. The average energy required to form an ion-pair in air is 35 ev. A beta ray of energy 1 Mev produces about 25 ion-pairs per centimeter, but at 0.05 Mev, about 200 ion-pairs are produced. Calculate the approximate range in air of a 1-Mev beta ray. Assume the ion production is the average of the above values.

12. A proton (mass 1.67×10^{-27} kg) enters a cloud chamber perpendicular to a magnetic field of 2.0 webers/m². If the velocity of the proton is 4.0×10^7 m/sec, what is the radius of the path of the proton?

13. Estimate the time required for 90% of uranium-238 to decay below radon-222.

14. An alpha particle whose energy is 5 Mev has a range in standard air of 3.51 cm. About how many ion-pairs are produced during its absorption, assuming 35 ev per pair is required? What is the total charge of one kind produced by the alpha particle?

15. A Geiger counter is used to measure the decay of a radioactive element. The background counting rate is 30 counts per minute. With the sample in position, the measurements are:

Time of day (hours)	Counting rate (counts per minute)
0 (start)	130
6	80
8	70
20	40

Estimate the half-life of the element.

For Further Reading

Benedetti, Sergio, "The Mössbauer Effect," *Scientific American,* April, 1960, p. 72.

Knedler, John W., Jr., ed., *Masterworks of Science,* Doubleday and Company, Garden City, 1947. (Marie Curie on radioactivity.)

Lang, Daniel, *From Hiroshima to the Moon, Chronicles of Life in the Atomic Age,* Simon and Schuster, New York, 1959.

Marshak, Robert E., "Pions," *Scientific American,* January, 1957, p. 84.

Morrison, Philip, "The Overthrow of Parity," *Scientific American.* April, 1957, p. 45.

Penman, Sheldon, "The Muon," *Scientific American,* July, 1961, p. 46.

Romer, Alfred, *The Restless Atom,* Doubleday and Company, Garden City, 1960.

Wilson, Robert R. and Raphael Littauer, *Accelerators,* Doubleday and Company, Garden City, 1960.

CHAPTER FORTY-SIX

The Nucleus and the Nucleon

The study of radioactivity is only one approach to knowledge of the atomic nucleus. It was the only approach in the early days of nuclear physics. When high-energy accelerators and more sensitive detectors became available, however, physicists were able to probe the nucleus and cause it to undergo changes that revealed its properties. As even higher energies were achieved, it was possible to study the properties of the *nucleons*—the elementary particles that manifested themselves as protons or neutrons in the nucleus.

In this chapter, we shall review some of the things that we now know about the structure of the nucleus and about the elementary particles. Our discussion of these important subjects will have to be brief—partly because the detailed treatment requires the use of advanced mathematics and partly because this is an unfinished story. Much is not known, especially about the properties of the nucleon and the way these properties are related to those of the nucleus.

Nuclear Species

There are almost 1700 different nuclear species or *nuclides.* A nuclide is a nucleus characterized by a certain charge and mass. About 285 of these are stable or non-radioactive nuclides. About 40 or so are radioactive nuclides that are found in nature—such as radium. The rest are radioactive nuclides that have been created in the laboratory by bombarding stable nuclides with energetic particles. We shall next discuss some of the characteristics of all nuclides.

Charge

The nuclear charge is positive and equals the sum of the charges of the protons in the nucleus. Each nuclear proton has a charge equal in magnitude to that of the electron.* The number of protons in the nucleus we have already defined as the atomic number Z. The atomic number is characteristic of the *chemical species,* and the chemical properties of the elements change with atomic number. The arrangement of atoms by atomic number is, of course, the periodic table of the elements. Our knowledge of the nuclear charge thus comes from what we know of chemical

* Why does charge exist only in units of the electronic charge wherever we find it in the universe? Ever since Millikan's discovery, this constancy of the elementary charge has been one of the unsolved mysteries of nature.

reactions involving the atomic electrons, from X-ray spectra (Chapter 40), and from scattering experiments, of which those of Rutherford were the prototype, on the nucleus itself. Z ranges from 1 for hydrogen to over 100. Elements beyond uranium ($Z = 92$) are not found in nature and have been created in the laboratory by bombarding heavy nuclei.

Mass

Nuclear masses are determined by mass spectrographs, the first of these measurements having been performed by J. J. Thomson in England in 1913. The method is to subject beams of atoms of the species being studied to the effect of deflecting electric and magnetic fields so that the atoms are sorted according to their masses. In Chapter 27, we saw how a procedure of this kind was used to determine the ratio of charge to mass for the electron.

Figure 46-1 shows a mass spectrograph of the kind invented by Dempster and Bainbridge. Suppose that one wishes to determine the masses of the different isotopes of lithium. The spectrograph is evacuated by a vacuum pump. Some lithium chloride is placed on the heated filament at W, where it evaporates and ionizes. The positively charged lithium ions are repelled by the positively charged filament, pass through the slit S_1, and enter the velocity selector. Here they are in two fields: a magnetic field of induction B_1 (directed out of the plane of the paper) and an electric field E (directed from left to right). The electric force (directed to the right) opposes the magnetic force (directed to the left—as you can verify by applying the right-hand rule). The two forces will cancel each other if

$$Eq = B_1 q v_{cr} \qquad (1)$$

where q is the charge on the ion. *Only ions of this critical velocity v_{cr} will move straight through the slit S_2.* All of the others (such as those shown by curving dotted lines near N and M) will be deflected and strike the walls.

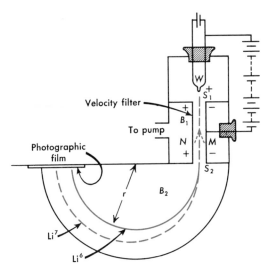

Fig. 46-1. *The mass spectrograph of Dempster and Bainbridge. (After O. H. Blackwood, T. H. Osgood, A. E. Ruark, et al.,* An Outline of Atomic Physics, *third edition, John Wiley and Sons, 1955.)*

The ions with velocity v_{cr}—but with different masses—then enter the space below S_2, where they are in a magnetic field of induction B_2. There is no electric field here. The magnetic deflecting force supplies the needed centripetal force to bend them around to the photographic plate on circular paths whose radii depend upon the masses of the ions—strictly speaking, upon the ratio of mass to charge. For an ion of mass m, the equation governing its motion here is

$$B_2 q v_{cr} = \frac{m v_{cr}^2}{r} \qquad (2)$$

The ions deposited on the photographic plate create a line spectrum from which the radii can be measured. Then, knowing B_1, B_2, E, and r, the experimenter can determine the ratio of m to q. Knowing q in terms of the electronic charge, he can determine m.

Figure 46-2 shows the mass spectra of some molecular ions, of the germanium isotopes, and of the tellurium isotopes. Table

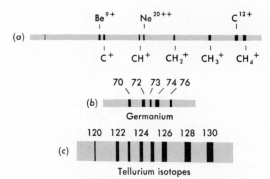

Be^{9+} Ne^{20++} C^{12+}

(a)

C^+ CH^+ $CH_2{}^+$ $CH_3{}^+$ $CH_4{}^+$

70 72 73 74 76

(b)

Germanium

120 122 124 126 128 130

(c)

Tellurium isotopes

Fig. 46-2. Mass spectra: (a) *after Bainbridge;* (b) *and* (c) *after Dempster.* (O. H. Blackwood, T. H. Osgood, A. E. Ruark, *et al.*, An Outline of Atomic Physics, *third edition, John Wiley and Sons, 1955.*).

46-1 contains the masses of some isotopes including two lithium isotopes. The nuclear masses are slightly less than the atomic masses, the contribution of the bound electron masses representing a small correction. The masses are expressed in atomic mass units. One atomic mass unit (amu) equals $\frac{1}{12}$ of the mass of the atom of C^{12}, the most abundant isotope of carbon.

It is customary to assign a *mass number A* to each isotope. *A* is the integer that is nearest to the atomic mass of the nuclide when the masses are expressed in atomic mass units. Thus the mass number of the lithium isotope of mass 6.01512 amu is 6. This is also equal to the sum of the number of nucleons—neutrons and protons—in the nucleus, lithium-6 containing three neutrons and three protons. The symbol for the lithium-6 nucleus is $_3Li^6$ with the mass number 6 written as a superscript to the right and the atomic number 3 written as a subscript to the left. Usually the latter is omitted since the chemical symbol indicates the atomic number.

The *binding energy per nucleon* is an important characteristic of a nuclide. *The total mass of a nucleus is less than the sum of the masses of the (free) neutrons and pro-* tons of which it is composed. The mass difference, converted into energy by Einstein's relation $E = mc^2$, is the binding energy. This is the energy which would have to be provided to the nucleus to pick it apart into its constituent nucleons against the forces that hold the nucleus together.

Example. Calculate the binding energy of the nucleus of deuterium ($_1H^2$) using the data of Table 46-1.

The deuteron is composed of one proton and one neutron. Their total mass, if free, would be 1.00782 amu + 1.00867 amu = 2.01649 amu. The mass of deuterium is 2.01410 amu. Thus 0.00239 amu of mass was converted into binding energy when the proton and neutron joined to become a deuteron. According to $E = mc^2$, 1 amu is equivalent to 931 Mev of energy. (We shall leave it to you in one of the problems at the end of the chapter to prove this.) Thus 0.00239 amu is equivalent to 2.22 Mev, the binding energy of the deuteron.

It is interesting to note that this energy is very close to the minimum energy that a gamma ray must have to split the deuteron into a proton and a neutron, as determined by photodisintegration experiments, in which deuterium is bombarded by gamma rays and splits into a proton and a neutron. The deuteron is unstable in states of high angular momentum: when set to spinning rapidly, the deuteron flies apart because the binding force is not large enough to supply the needed centripetal force.

The binding energy per particle is the binding energy divided by the number of nucleons or *A*, the mass number. Figure 46-3 is a graph of the binding energy per nucleon for the nuclides. Note that it is relatively small for small values of *A* (less than 20), becomes fairly constant at about 8.5 Mev/nucleon for medium nuclides (*A* from 40 to about 100), and then decreases for the heavier nuclides

Table 46-1. Masses of a Few Isotopes

(Atomic mass unit $= 1.660 \times 10^{-24}$ gm $= 931$ Mev)

Atomic No.		Atomic Mass	Mass No.	Atomic No.		Atomic Mass	Mass No.
	Electron	0.000548			[a]O^{15}	15.0030	15
	Neutron	1.00867		8	O^{16}	15.9949	16
1	H^1	1.00782	1		O^{17}	16.9991	17
	H^2 (deuterium)	2.01410	2		O^{18}	17.9991	18
	[a]H^3 (tritium)	3.01605	3	17	Cl^{35}	34.9688	35
2	He^3	3.01603	3		Cl^{37}	36.9658	37
	He^4	4.00260	4	47	Ag^{107}	106.9049	107
	[a]He^5	5.01229	5		Ag^{109}	108.9047	109
3	Li^6	6.01512	6	54	Xe^{134}	133.9051	134
	Li^7	7.01601	7	74	W^{184}	183.9478	184
6	C^{12}	12.00000	12		[a]U^{234}	234.0397	234
	C^{13}	13.0033	13	92	[a]U^{235}	235.0428	235
7	[a]N^{13}	13.0057	13		[a]U^{238}	238.0496	238
	N^{14}	14.0031	14				
	N^{15}	15.0001	15				
	[a]N^{16}	16.0061	16				

[a] Radioactive.

(*A* from 100 to 240). This means that nucleons in the light and heavy nuclides are less tightly bound than those in the medium range of mass. We would expect these light and heavy nuclides to be less stable than the others. We have already seen that the instability called alpha-particle emission occurs in the heaviest nuclei. We shall find a number of other instances to support this expectation.

The Size of the Nucleus

As Rutherford's early scattering experiments indicated, the nucleus is a small part of the atom. Atomic diameters are approximately 10^{-8} cm; those of nuclei, about 10^{-13} cm or 1 *fermi,* where the fermi named after Enrico Fermi, equals 10^{-13} cm.

Nuclear diameters can be measured in a

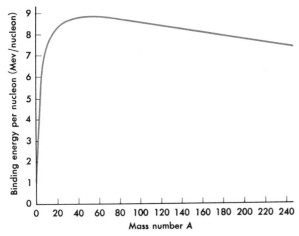

Fig. 46-3. *The binding energy per nucleon versus mass number.*

number of ways, for example, by the scattering of fast neutrons or by the diffraction of high-energy, short-wavelength electrons by the nucleus. Electron-scattering experiments by R. Hofstadter and others show that the nuclear radius R is proportional to the cube root of the mass number A of the nucleus:

$$R = (1.07 \times 10^{-13} \text{ cm})A^{1/3}$$

Here R is the distance from the center of the nucleus to the point at which the positive charge density has decreased to 50% of its maximum.

The electron-scattering experiments show that (1) the density of charge in the central region of a nucleus is quite constant and (2) the thickness of the surface layer of nuclear charge is approximately the same for all nuclei.

Nuclear Spin

The *spin angular momentum* of the nucleus has been measured for many nuclei. Like the electron, the proton and neutron each have a spin equal to $\frac{1}{2}(h/2\pi)$. When protons and neutrons combine to form nuclei, their spins combine to give the nucleus a spin angular momentum. Exceptions occur in nuclei where there is an even number of protons and an even number of neutrons. In these instances, the spins pair off to cancel each other, and the resultant spin of the nucleus is zero. When the nucleus has a spin, it also has a magnetic moment.

Nuclear Reactions

In 1919, Rutherford made the first observation of a nuclear reaction. He bombarded oxygen gas in a tube with alpha particles from a radioactive source and observed how far the oxygen atoms traveled after being struck. Then he replaced the oxygen with nitrogen and observed that the "range" was many times greater. Rutherford proved that the particles of great range were protons emitted by struck nitrogen nuclei which had been converted into stable oxygen nuclei of mass number 17. *For the first time in history, a scientist had transmuted one element into another* (Fig. 46-4).

Here is the reaction:

$$_2\text{He}^4 + {_7}\text{N}^{14} + E_1 = {_8}\text{O}^{17} + {_1}\text{H}^1 + E_2$$

Notice that the total nuclear mass number on the left equals that on the right, and the total atomic number on the left equals that on the right. E_1 is the kinetic energy of the alpha particle, and E_2 that of the system after the collision. E_1 is greater than E_2; hence this is an *endothermic* (or *endoergic*) reaction. It does not set free nuclear energy. (Other nuclei, bombarded with suitable projectiles, do liberate energy in *exothermic* reactions.) The equation is balanced as to charge and mass energy.

Notice also in Fig. 46-4 that the alpha particle does not retain its "identity" within the nucleus that it has penetrated. In chemical reactions, the atoms forming a nucleus combine to form a structure, yet remain discrete. In nuclear reactions, on the other hand, totally new nuclei are formed in which the nucleons show none of the structural detail of previous combinations.

Eleven years after Rutherford's discovery, Bothe and Becker in Germany bombarded beryllium with alpha particles. The beryllium emitted a radiation that could penetrate many centimeters of lead. It was much too penetrating to be X-rays from the beryllium. In 1932, Chadwick in England proved that the penetrating particles had about the same mass as the proton and that they were uncharged. The neutron had been discovered. In 1951, it was found that a *free* neutron undergoes beta decay into a proton, an electron, and a neutrino. The half-life of the neutron decay is about 15 minutes.

Neutrons exert no forces on nuclei or electrons except when they approach very closely

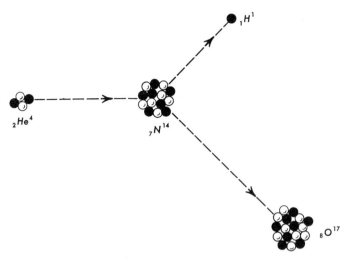

Fig. 46-4. Transmutation of nitrogen into oxygen. Neutrons are shown as white circles, protons as black ones.

(only short-range, specific nuclear forces act). Thus a fast-moving neutron can travel through billions of atoms without losing much energy. Both fast-moving and slow-moving neutrons can be captured by nuclei. Figure 46-5 shows a nuclear reaction in which a neutron is captured by a nitrogen nucleus and converts the nucleus into a nucleus of radioactive C^{14} which subsequently undergoes beta decay into ordinary nitrogen. The reaction is:

$$_0n^1 + _7N^{14} + E_1 = _6C^{14} + _1H^1 + E_2$$

The two reactions given illustrate the kind of experiment that is carried out to study the nuclear energy levels or to produce radioactive nuclei for radioactive tracers (see below). The bombarding particles include protons, neutrons, deuterons, alpha particles, electrons, gamma-ray photons, and—more recently—heavy ions, such as those of oxygen. The accelerators include Van de Graaff generators, cyclotrons, and betatrons. The energies range up to about 50 Mev in these experiments. Higher-energy particles are used to study the nature of the fundamental particles themselves.

In some experiments, the bombarding particle is absorbed by the nucleus it strikes, and one or more gamma rays are emitted. In others, one particle goes into the nucleus, and another one comes out. The emergent particle may be of the same kind as the bombarding particle, but with a different energy, or it may be a different kind of particle. By studying the gamma rays emitted by the excited target nucleus, the energy changes of the particles, and their distribution in direction as they scatter off the nucleus, the physicist can deduce what the energy levels are to which a nucleus can be excited. Measurements of the radioactivity—if any—of the struck nucleus provide additional information about the energy levels.

Cross Section

The cross section σ (sigma) of an event is a useful term that originated in bombardment experiments, but can be applied to other kinds of events. The cross section indicates the effective—not the geometric— size of the target that a nucleus presents to a bombarding particle. Suppose that a target foil

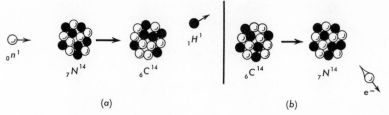

Fig. 46-5. (a) *A neutron (white circle), captured by a nitrogen nucleus, converts it into a radioactive C^{14} nucleus and knocks out a proton (black circle). (b) The radioactive C^{14} nucleus decays with a half-life of 5570 years into ordinary nitrogen by emitting a beta particle.*

that is to be bombarded in a cyclotron contains n_t nuclei per unit surface area so that the surface area per nucleus is $1/n_t$. (Note that this is *not* the same as the geometric cross-sectional area of the nucleus.) Let the number of energetic particles striking the foil per unit time per unit area be n_0 and let N be the number of "hits" per unit time per unit area in which a particle strikes a nucleus and produces the event being studied. The probability of a hit is N/n_0. We define the cross section of the event as that probability times the area $1/n_t$ or

$$\sigma = \frac{N}{n_0}\frac{1}{n_t} = \frac{N}{n_0 n_t}$$

The cross section is measured in *barns* (a barn is 10^{-24} cm²).

Example. When 5-Mev deuterons bombard a target 0.1 cm thick, 1 transmutation occurs for every 100,000 deuterons hitting the target. Assuming that there are about 6×10^{22} nuclei/cm³ in the target, what is the cross section of the reaction?

If there are 6×10^{22} nuclei/cm³, the number per unit surface area is this times the thickness, 0.1 cm:

$$n_t = 6 \times 10^{22}(1/\text{cm}^3)(0.1\text{ cm}) = 6 \times 10^{21}(1/\text{cm}^2)$$

$$\frac{N}{n_0} = \frac{1}{100,000}$$

$$\sigma = \frac{1}{100,000}\frac{1}{6 \times 10^{21}\ (1/\text{cm}^2)}$$

$$= 1.6 \times 10^{-27}\text{ cm}^2$$
$$= 1.6 \times 10^{-3}\text{ barn}$$

Radioactive Tracers

Radioactive atoms obey the same laws of chemistry as do non-radioactive ones. This makes possible the use of radioactive "tagged" atoms, which are mixed with other atoms of that species and enter into chemical or biochemical processes in the normal way. The presence of the tracer atoms is indicated by the emission of particles which can be detected by a counter.

Anthropologists find out the age of objects associated with ancient man by means of carbon-14 isotopes, whose nuclei emit beta particles with a half-life of 5570 years. Suppose that a door post made from the trunk of a tree is found in a buried village that flourished centuries ago. Carbon-14 is produced in the atmosphere when nitrogen in the air is bombarded by cosmic-ray neutrons. During its lifetime, the tree breathed in carbon dioxide and captured the carbon. To find out how long ago the tree lived, we can put some of the carbon from it into a Geiger counter and measure the rate of emission of beta particles. If the rate is one-half that of carbon from a contemporary tree, then the ancient tree grew about 5570 years ago, and that is the age of the village.

An important use of artificial radioactivity is in the investigation of physiological processes. In the study of goiter, radioactive iodine-131, fed to a patient, is carried through the body by the blood stream and is collected in the thyroid gland. I^{131} has a half-life of 8.0 days. The amount of radioactive iodine in the gland is estimated by holding a Geiger tube against the throat of the patient and determining the number of counts per minute. The rate at which the iodine accumulates depends upon the type of disease.

The processes of digestion are studied by giving a white rat some food that contains radioactive C^{14}. Afterward, the animal is killed, and an organ of its body is removed and placed near a Geiger counter. The counter's rate of clicking measures the amount of the activated carbon that has been deposited in that organ.

Fission

Great quantities of energy are imprisoned in the nuclei of atoms. The binding-energy curve of Fig. 46-3 indicates that some of this energy may be liberated by the *fusion* of hydrogen to form helium or by the *fission* of heavy atoms such as uranium and thorium. When light elements fuse, the nucleons become more tightly bound, and energy is released. Likewise, nucleons become more tightly bound when heavy elements split into lighter ones, again releasing energy. Formerly the liberation of nuclear energy in paying quantities seemed nearly hopeless. Alpha particles were expensive and millions of them were required to break up even one atom in an exothermic reaction. As Einstein remarked, "It is like shooting birds, blindfolded, in a country where there are not many birds."

The discovery of the neutron in 1932 led physicists to study neutron reactions. During a seven-year period, Hahn and Strassman in Germany, Fermi in Italy, and others bombarded uranium with neutrons. They hoped to produce new elements of atomic number greater than 92. They got evidence of new atoms, but found them difficult to identify. Finally, in 1939, Hahn realized what was happening. Instead of producing *heavier* atoms, the neutrons were splitting (fissioning) uranium atoms into unequal parts. *Neutrons can cause heavy atoms to undergo fission.* One of Hahn's co-workers, Miss Lise Meitner, driven from Germany by Hitler, carried the news to scientists in Denmark. They forwarded it to Neils Bohr. He announced it at a meeting of atomic physicists at Washington and broke up the convention. Everyone rushed home to experiment! Physicists thought that the fission process might be self-sustaining. It might be a *chain reaction.* That is, an atom, split by one neutron, might supply neutrons to split neighboring atoms. It was eventually shown that this did occur.

Basically, fission of certain heavy nuclides — isotopes of uranium and thorium — occurs because the delicate equilibrium of these nuclei is upset when they absorb a neutron. In any nucleus, the Coulomb forces among protons tend to blow the nucleus apart. The Coulomb forces are opposed by specific attractive forces among nucleons. The attractive forces succeed in holding the nucleus together against disintegration as the atomic number Z increases until Z reaches approximately 60. Here the first signs of instability appear: some of the nuclides are alpha-particle emitters. From $Z = 83$ on up, *all* of the chemical species have one or more isotopes that are alpha emitters. Finally for $Z = 90$ (thorium) and $Z = 92$ (uranium), the instability becomes so great that fission results if the nucleus is given a large enough "stimulus."

When a nucleus of U^{235} absorbs a *slow* neutron (with an energy of about $\frac{1}{40}$ ev), the nucleus acquires an excitation energy — due to binding the neutron — of about 6.8 Mev. It turns out that an excitation energy of this

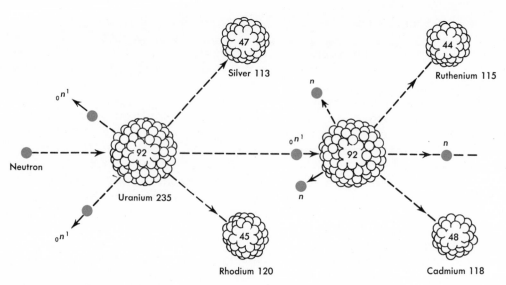

Fig. 46-6. *Fission of U^{235}. A slow neutron splits a U-235 nucleus, forming (unstable) isotopes of silver and rhodium. Three neutrons are also liberated. One of them fissions another U-235 atom, forming ruthenium and cadmium. Usually the fission fragments do not have nearly equal masses as shown above. The most probable fission yields fragments of mass numbers 97 and 137 roughly.*

amount sets the nucleus into an oscillation in which it is unstable against splitting. As a mercury droplet splits when it is made to vibrate with a greater and greater amplitude, the U^{235} nucleus splits when it receives this excitation energy. When U^{235} fissions, it not only forms two radioactive fission fragments (Fig. 46-6), but it also releases two or three neutrons. The fission fragments have considerable kinetic energy—the heat of the fission. About 200 Mev of energy are released in the fission of each U^{235} nucleus.

A U^{238} nucleus requires a greater amount of excitation before it will split. *Fast* neutrons with over 2 Mev of energy must be absorbed to cause fission in U^{238}. A neutron with less energy than this, captured by a U^{238} nucleus, excites it so that it emits an electron and is converted into the element **neptunium,** of atomic number 93—Np^{239}. This nucleus, in turn, emits a second electron and becomes

plutonium, atomic number 94. Pu^{239} can be made to fission by absorbing a slow neutron and serves as a substitute for U^{235} in the nuclear reactor.

Nuclear Reactor

A chain reaction can be sustained by providing very special conditions so that, of the two or three neutrons released in the fission of a uranium nucleus, at least *one* will cause another nucleus to fission. These conditions exist in a *nuclear reactor.* In order to understand how a reactor operates, you must know how atoms of low atomic mass slow down neutrons. Suppose that U^{235} is embedded in a hydrogen-containing material, such as paraffin or water. Each neutron that strikes a hydrogen nucleus of nearly equal mass head on will stop like a billiard ball hitting another squarely. The neutron will lose nearly *all* of its kinetic energy. In a slanting impact, the neutron will lose a considerable fraction of its energy. A substance (of low atomic weight) that slows down neutrons is called a *moderator*. Graphite is also used as a moderator in the nuclear reactor. It is cheap, can be highly purified, and can stand high temperatures. In Fig. 46-7 the aluminum tubes initially contain uranium-238 and -235. Some of the neutrons transform U^{238} into neptunium, which becomes plutonium. Other neutrons bumping into the graphite nuclei get slowed down so that they become slow or "thermal" neutrons, split U^{235} nuclei to liberate more neutrons, and thus keep the "fire burning." The boron steel or cadmium rods absorb neutrons. Inserting these rods farther into the reactor

decreases the activity and dampens the "fire."

Coal must be supplied to a furnace, and ashes must be removed. Similarly, in the nuclear reactor, U^{235} gets used up and plutonium accumulates. Also, the residues of the split atoms, which are now highly radioactive, increase in number. Many of them absorb neutrons and slow the reaction. From time to time, the aluminum tubes are removed. Untouched by human hands, the contents are purified. Many of the by-product radioactive isotopes are useful in science and industry. Also, if liquid sodium is circulated through tubes that run through the reactor, the thermal energy can be removed from the reactor and supplied by a heat exchanger to the working

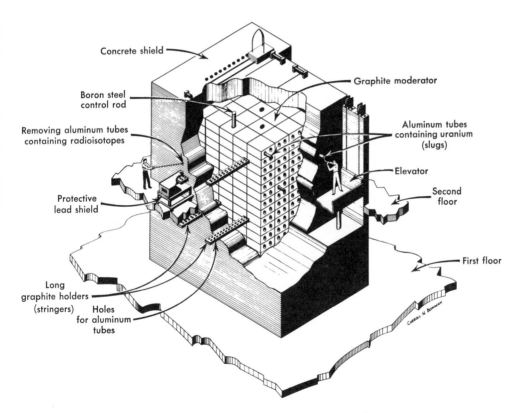

Fig. 46-7. A nuclear reactor or pile. A controlled chain reaction goes on in the uranium slugs embedded in the graphite moderator. (Courtesy of Atomic Energy Commission.)

substance of a turbine. The turbine will generate electric power for commercial use or, in ships, for propulsion.

The nuclear reactor pours out many particles, including neutrons and gamma rays. Usually the neutrons are absorbed by water and the gamma rays by concrete walls.

Not all reactors use graphite as a moderator. Many use heavy water. In such reactors, the fuel elements (aluminum-clad uranium rods) are placed in the proper geometric pattern in a tank filled with the liquid moderator.

Since nuclear reactors increasingly are located near large centers of population, nuclear engineers must give careful attention to reactor safety. A number of means are used to make it extremely unlikely that a runaway reactor would release highly radioactive materials. For one thing, reactors are designed so that they have "negative temperature coefficients" to the greatest extent possible. If the temperature of the reactor should rise suddenly—because the coolant is cut off, for example—the efficiency of the fission process is reduced, and the power level of the reactor drops. Also, safety devices can cause the control rods to drop all the way into the reactor in the event of a power surge, effectively shutting down the reactor. Finally, reactors are surrounded by gas-tight spherical or cylindrical structures that would contain any radioactive materials that might be released, thus preventing their escape into the atmosphere.

Cosmic Rays

In 1910, the Swiss physicist Gockel made a balloon ascension which resulted in a remarkable discovery. Gockel wished to study the variation with altitude of the intensity of the gamma rays from radioactive matter in the earth. He used a gold-leaf electroscope in a lead-covered box, so that its discharge rate would measure the ionization produced in the confined air by the penetrating gamma rays. It was supposed that, as the balloon rose higher, more of the radiations from the earth would be absorbed by the intervening air and that at an elevation of about a mile the charge of the electroscope would remain constant. The actual condition was astonishingly different. Even when the balloon had risen 3 miles, the discharge rate of the electroscope was at least as great as that at the earth's surface. Then a German physicist, Hess, found that, at greater elevations, the discharge rate of the electroscope actually increased. These results led to the opinion that the electroscope was discharged by *cosmic rays,* which come from the depths of space. Millikan constructed ingenious electroscopes weighing but a few ounces. Each instrument was self-recording and included a barometer and clock. These devices, lifted by small balloons, were carried to elevations of 10 miles or more. When they descended and were returned by the persons finding them, the records told the story of the discharge rate at each altitude (indicated by the recording barometers), and also the intensity of the rays. It was proved that these radiations are 6 to 100 times more penetrating than the hardest gamma rays. Later, cosmic rays were found occasionally to have energies greater than 10^{18} ev.

Measurements by instruments carried by sounding rockets or satellites indicate that nearly all primary cosmic-ray particles entering the earth's atmosphere are charged particles, protons or heavier nuclei, and are not electromagnetic radiation. The principal justification for this view is that the radiations, at sea level, are more intense at the magnetic poles than at the magnetic equator. As we saw in Chapter 27, charged particles that travel toward the magnetic poles are not deviated, for their motion is directed parallel to the earth's magnetic field. When they travel toward the earth in the plane of the equator, they are pulled sidewise by the earth's magnetic field so that the slower particles fail to

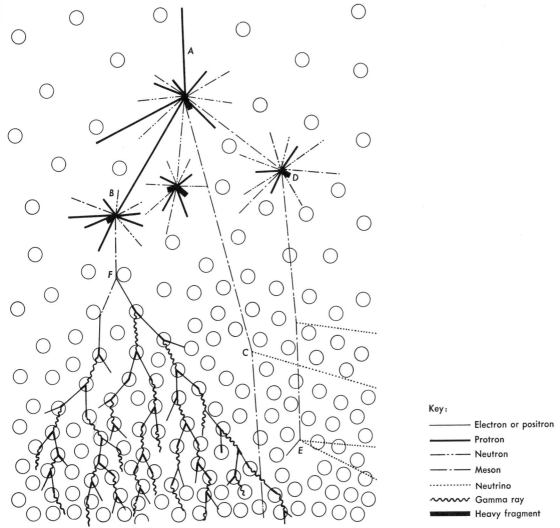

Key:

——————— Electron or positron

▬▬▬▬▬▬ Protron

—·—·—·— Neutron

—·—·—· Meson

··········· Neutrino

∿∿∿∿∿ Gamma ray

▬▬▬▬▬▬ Heavy fragment

Fig. 46-8. Secondary cosmic rays are produced when the primaries interact with nuclei in the earth's atmosphere. (Courtesy of Scientific American.)

reach the earth. Thus, the so-called latitude effect explains the difference between the cosmic-ray intensity at the poles and at the equator and justifies the view that the cosmic rays are charged particles.

When cosmic rays reach our atmosphere, they strike atomic nuclei and knock out several kinds of particles. Figure 46-8 tells a fascinating story. A primary particle *A* strikes a nucleus and forms a "star," hurling out mesons, protons, neutrons, and large fragments of the original nucleus. One of the protons shatters another nucleus at *B*. One of the positive π mesons *C* decays into a lighter μ meson by emitting a neutrino. At *D*, a negative π meson causes a nucleus to disintegrate. A

μ meson decays into an electron and two neutrinos at *E*. At *F*, a meson initiates a "shower" in which an ejected electron is accelerated near a nucleus, producing a gamma ray, which in turn produces an electron-positron pair. These produce gamma rays, and so on. The secondary cosmic rays are those that we observe on the surface of the earth.

Nuclear Forces

The nature of the forces between nucleons is not completely understood although much progress in theory and experiment has been made. It is clear that the attractive force between protons and neutrons is one of the major kinds of force in the universe. Physicists find it convenient to classify the various forces as follows in order of increasing strength:

1. *Gravitational interactions* are those which are described by the law of universal gravitation. Gravitation is the weakest of the interactions.
2. *Weak interactions* occur in beta decay and in the decay of mesons. They are stronger than gravitation by a factor of 10^{25}.
3. *Electromagnetic interactions* are those between charged particles and electric and magnetic fields. The Coulomb repulsion between two protons is an example of this interaction. It is stronger than the gravitational attraction between the protons by a factor of 10^{37}.
4. *Strong interactions* are the forces between nucleons. They are 10^{39} times as strong as the gravitational forces.

The gravitational forces that explain the stability of our solar system are important whenever one of the attracting bodies is of "astronomical" size, but ordinarily play a comparatively small role in the world of the atom and the nucleus. The reason, of course, is that the masses of the interacting bodies—two nucleons, for example—are so small that even at separations of only a few fermis, the gravitational force is insignificant compared to the other forces listed above. An exception apparently occurs in astrophysics: when a star has used up its nuclear fuels and the internal pressures due to high temperatures, radiation, and charge density diminish, it begins to contract because of gravitational attraction. In some circumstances this contraction continues, theorists believe, until the nucleons are crowded so closely together that *gravitational collapse* occurs. The gravitational force dominates all other forces, the nucleons are crushed out of existence, and the star disappears from observation.

The weak interactions allow transitions from one energy state of a nucleus to another energy state *of different charge*—and hence to a different nuclide. They govern beta decay and account for decay schemes of the kind shown in Fig. 45-13 on p. 743. These nuclear transitions differ from the atomic transitions studied earlier in that, when an atom moves from one energy state to another, the total charge does not change. In contrast, the emission of positive or negative beta particles causes the nuclear charge to change.

Electromagnetic interactions, of course, are dominant in the atomic world. We saw how the Coulomb attraction between the electrons and the nucleus of an atom accounts—through the rules of quantum mechanics—for the stability of the atom and for its spectrum and chemical properties. The electromagnetic interaction, as we saw when we studied electrostatics, even reaches out into the everyday world. If we assemble enough free electrons or positive ions in ordinary matter, the effects can be observed in the laboratory with electroscopes and other instruments. Within the nucleus, however, the Coulomb force has a lesser role. Its effect does not become dominant until the nuclear charge becomes quite large, as it does in the heavy nuclei. Then the instabilities of alpha emission and nuclear fission appear, due largely to Coulomb repulsion between protons.

The strong interaction force has been shown by nuclear-scattering experiments to be a short-range force, effective only at distances of a few fermis from the center of the nucleus. Only particles approaching closer than this are affected by the force, as we saw earlier in the discussion of Rutherford scattering. For this reason, the strong interaction cannot be observed on an external scale—with a "nucleoscope" similar to an electroscope, for example. *Pairs* of particles in the nucleus interact strongly—as one can show from the fact that the binding energy per particle is roughly constant. At distances of less than 1 fermi, the strong interaction force becomes one of *repulsion* and accounts for the observed stability of the nucleus against collapse.

The strong interaction seems to be due to the exchange of π mesons between pairs of nucleons. A proton emits a positive pi (π^+) meson and is transformed into a neutron. The π^+ is absorbed by a neighboring neutron which becomes a proton. The act of exchanging the meson binds the two nucleons together. Similarly, a neutron can emit a π^- and become a proton. Another proton absorbs the π^- and becomes a neutron. Finally, two protons can exchange a neutral meson (π^0) and bind each other together by so doing. (This π^0 is necessary because the strong interaction is *not* dependent upon the charge of the nucleon. The strong interaction between two protons—not the Coulomb force—is as large as that between a neutron and a proton or between two neutrons.) The π^+, π^-, and π^0 have all been observed experimentally outside the nucleus.

The attraction between two nucleons within the nucleus is approximately the same as the Coulomb attraction between two particles of the mass of the nucleon and of opposite charges $+3e$ and $-3e$.[*] If we insert a charge of $3e$ and a mass equal to that of the proton into the formula for the first Bohr orbit (Chapter 40) and into that for the energy of the atom in the lowest energy state (Chapter 40) we can obtain estimates for the size of the nucleus and for nuclear energies, respectively.

Example. Under the assumption given above, obtain "ball-park" estimates of the nuclear size and of nuclear energies.

The radius of the first Bohr orbit is

$$r = \frac{h^2}{4\pi^2 k_0 e^2 m}$$

Inserting the nucleonic mass and a charge of $3e$, we have:

$$r = \frac{(6.62 \times 10^{-34} \text{ joule-sec})^2}{(4\pi^2)(9 \times 10^9 \text{ newton-m}^2/\text{coul}^2) \times (3 \times 1.60 \times 10^{-19} \text{ coul})^2 (1.66 \times 10^{-27} \text{ kg})}$$

$$\cong 7 \times 10^{-15} \text{ m} = 7 \times 10^{-13} \text{ cm}$$
$$= 7 \text{ fermis, approximate radius of the nucleus.}$$

The energy of the lowest-energy Bohr state is:

$$E = -\frac{2\pi^2 k_0^2 e^4 m}{h^2}$$

Inserting the nucleonic mass and a charge of $3e$ we obtain:

$$E = -\frac{(2\pi^2)(9 \times 10^9 \text{ newton-m}^2/\text{coul}^2)^2 \times (3 \times 1.60 \times 10^{-19} \text{ coul})^4 (1.66 \times 10^{-27} \text{ kg})}{(6.62 \times 10^{-34} \text{ joule-sec})^2}$$

$$\cong -2 \times 10^{-13} \text{ joule} = -2 \times 10^6 \text{ ev} = -2 \text{ Mev,}$$

estimate of energy of nucleons in the nucleus. Thus the radius of the nucleus is about 10^{-4} times the radius of the atom, and nuclear energies are about 10^5 times atomic energies.

* See V. F. Weisskopf, "Three Steps in the Structure of Matter," *Physics Today*, August, 1970. We are indebted to Professor Weisskopf's articles on the "three spectroscopies" for much of the approach taken here.

The nucleons in the nucleus exist on different energy levels, as we noted previously in discussing alpha emission, and the rules of quantum mechanics apply to the nucleus. Various theoretical models have been proposed to account for the results of scattering experiments, investigations of radioactive decay and fission, and other studies of nuclear properties. None of these models has proven successful in explaining all of the known facts. For nuclei that are only slightly excited, a *shell model* of the nucleus has been successful in explaining the lower energy levels. The nuclear shell model resembles the shell model of the atom. Just as there are closed electronic shells for certain numbers of electrons — 2, 10, 18, . . . — there are closed nuclear shells at certain (different) numbers of nucleons of one kind — 2, 8, 20, 50, ... The atoms at these closed-shell electron numbers are relatively inert chemically; these are the inert gases — helium, argon, and so on. Similarly, the nuclides at the closed-shell nucleon numbers are remarkably stable against decay and are somewhat more abundant in nature. Another nuclear model is the *liquid-drop* model. As we noted above, a heavy nucleus can be likened to an oscillating liquid drop as it undergoes fission.

The Nucleon

The properties of the nucleus are those of an assembly of nucleons — the protons and neutrons of which the nucleus is composed. Just as the spectra and chemical behavior of molecules can be explained by the properties of the constituent atoms and the quantum-mechanical principles of their combinations into molecules, so we hope in time to be able to explain nuclear properties — the energy levels, radioactivity, fission, and so on — in terms of the properties of the nucleons. What can be said about the nucleons?

High-energy accelerators, capable initially of supplying particles of energies greater than three or four hundred Mev and soon reaching tens of Bev, came into use in the 1950's and were used to bombard nuclei. The results were astonishing. Instead of producing nuclear reactions similar to those observed at lower energies, the experiments yielded a bewildering variety of apparently *new* particles, unlike the familiar electron, proton, and neutron, or even the mesons observed in cosmic rays. The new particles had different masses — sometimes larger, sometimes smaller — than the proton. Some had unit positive charge, some were negatively charged, and some were neutral. All were unstable, decaying with relatively short half-lives by the emission of π mesons or heavier particles, positrons or negatrons and neutrinos, into protons or neutrons. The complexity was so great that physicists referred to it as the "nuclear zoo" and wondered uneasily what had happened to the concept of a fundamental particle.

Table 46-2 lists the known subatomic particles and some of their properties. For completeness, we have included such familiar particles as the electron, photon, neutron, and proton. The "new" particles in the table are the kappa (K), sigma (Σ), eta (η), xi (Ξ), lambda (Λ), delta (Δ), and omega (Ω) particles.

The antiparticles of many particles have been observed. The *antiproton* (negative proton) was produced and detected experimentally in 1956. It bears the same relation to the proton that the positron bears to the negatron. To create a positron-negatron pair requires 1.02 Mev of energy. Since the proton is about 1840 times more massive than the electron, an energy not less than 1840 times 1.02 Mev or about 2 billion ev is required to create an antiproton. However, since all of the energy of the bombarding particle is not delivered to the bombarded nucleus, it turns out that the bombarding protons must have an energy of about 6 Mev. High-energy accelerators called proton synchrotrons (Fig. 46-9) are easily capable of these proton energies.

Table 46-2. Subatomic Particles[a]

	Particle	Antiparticle	Spin	Mass Equivalent, Mev		Some Decay Modes	Mean Life (1/λ), Seconds
Photon		γ	1	0		stable	
Leptons	ν_e, ν_m	$\bar\nu_e, \bar\nu_m$	$\frac{1}{2}$	0		stable	
	e^-	e^+	$\frac{1}{2}$	0.511		stable	
	μ^-	μ^+	$\frac{1}{2}$	106		$\mu^\pm \rightarrow e^\pm + \nu + \bar\nu$	2.2×10^{-6}
Mesons	π^0		0	135		$\pi^0 \rightarrow 2\gamma$	0.7×10^{-16}
	π^+	π^-	0	140		$\pi^+ \rightarrow \mu^+ + \nu$ $\pi^- \rightarrow \mu^- + \bar\nu$	1.8×10^{-8}
	K^+	K^-	0	494		$K^\pm \rightarrow \mu^\pm + \begin{cases} \nu \\ \nu \end{cases}$	0.8×10^{-8}
	K^0	$\bar K^0$	0	498	equivalent to	$K_1^0 \rightarrow \pi^+ + \pi^-$	0.7×10^{-10}
						$K_2^0 \rightarrow \pi^\pm + e^\mp + \begin{cases} \bar\nu \\ \nu \end{cases}$	4.0×10^{-8}
	η^0		0	548		$\eta^0 \rightarrow \pi^+ + \pi^- + \pi^0$	$<10^{-16}$
Baryons — Nucleons	p^+	p^-	$\frac{1}{2}$	938		stable	
	n^0	$\bar n^0$	$\frac{1}{2}$	939		$n^0 \rightarrow p^+ + e^- + \bar\nu$	0.7×10^3
Baryons — Hyperons	Λ^0	$\bar\Lambda^0$	$\frac{1}{2}$	1115		$\Lambda^0 \rightarrow \pi^- + p^+$ $\Lambda^0 \rightarrow \pi^0 + n^0$	1.7×10^{-10}
	Σ^+	Σ^-	$\frac{1}{2}$	1190		$\Sigma^+ \rightarrow \pi^+ + n^0$ $\Sigma^+ \rightarrow \pi^0 + p^+$	0.8×10^{-10}
	Σ^0	$\bar\Sigma^0$	$\frac{1}{2}$			$\Sigma^0 \rightarrow \Lambda^0 + \gamma$	$<10^{-16}$
	Σ^-	Σ^+	$\frac{1}{2}$			$\Sigma^- \rightarrow \pi^- + n^0$	1.6×10^{-10}
	Δ	$\bar\Delta$	$\frac{3}{2}$	1238		$\Delta \rightarrow \pi^\pm + p^\mp$	$<10^{-16}$
	Ξ^0	$\bar\Xi^0$	$\frac{1}{2}$	1310		$\Xi^0 \rightarrow \Lambda^0 + \pi^0$	3.0×10^{-10}
	Ξ^-	Ξ^+	$\frac{1}{2}$	1320		$\Xi^- \rightarrow \Lambda^0 + \pi^-$	1.7×10^{-10}
	Ω^-	Ω^+	$\frac{3}{2}$	1676		$\Omega^- \rightarrow \Lambda^0 + K^-$	1.3×10^{-10}

[a] The Δ (charge -1, 0, $+1$, $+2$) is not listed as a particle in some tables.

Their energies now reach several million million (10^{12}) ev. Shortly after the antiproton was produced, the *antineutron* was observed.

Astronomers have speculated as to whether there are worlds out in space made of *antimatter* containing positrons (antielectrons), antiprotons, and antineutrons. These worlds could exist undisturbed as long as ordinary matter were far away. If antimatter and matter were to collide, they would annihilate each other as the positron and negatron do.

Leptons (Table 46-2) are the light particles: electrons (e^+ and e^-), neutrinos (ν) and antineutrinos ($\bar\nu$), and the muons (μ^+ and μ^-). *Mesons* are particles of intermediate mass: the pi mesons (π^+, π^-, and π^0), the kappa particles (K^+, K^-, K^0, and $\bar K^0$), and the eta particles (η). There are two types of neutrinos, one kind associated with electrons and the other with muons. The *baryons* are the heavy particles: the protons (p^+ and p^-) and neu-trons (n and n), the lambda particles (Λ^0 and $\bar\Lambda^0$), the sigma particles (Σ^+, Σ^-, and Σ^0), the xi particles (Ξ^- and Ξ^0), the delta particles (Δ), the omega particles (Ω), and the antiparticles of the sigma and xi particles.

Although the explanation is far from complete, great progress has been made in accounting for the new particles. It seems clear that some of them are *excited states* of the proton or neutron—or, to put it more concisely, excited states of the *nucleon,* since the proton and neutron are simply different states of that fundamental particle. For example, the Δ, Σ, Ξ, and Ω particles result when the nucleon receives enough energy from a bombarding particle to raise it to one of the higher energy states or spin states represented by those particles. Just as a sodium atom in an excited electronic state—just before emitting a 5896-A photon, for example—is still a sodium atom, a Λ particle seems really to be an excited nucleon.

(a)

Fig. 46-9. The Brookhaven Alternating Gradient Synchrotron. Magnets on a ring with an 843-ft diameter bend protons into a circular orbit and apply restoring forces when the protons leave the central orbit. The AGS provides the restoring force by reversal of the shape of the magnetic field many times around the ring. The focusing forces are so effective that the channel or aperture through which the protons travel is only 2.75×6 in. The working aperture is in a huge vacuum pipe to prevent the scattering of the protons by gas.

The protons are injected in a pulse into the huge ring after they have been accelerated to 50 Mev by a linear accelerator. As they travel around the ring they pass successively through the accelerating fields of twelve radio-frequency accelerating stations which are synchronized with the motion of the pulse. The particles gain about 100,000 ev per revolution, traveling almost at the speed of light, and emerge after reaching an energy of about 33 billion electron volts. The intensity with which the protons strike the target is around 3×10^{11} protons per pulse.

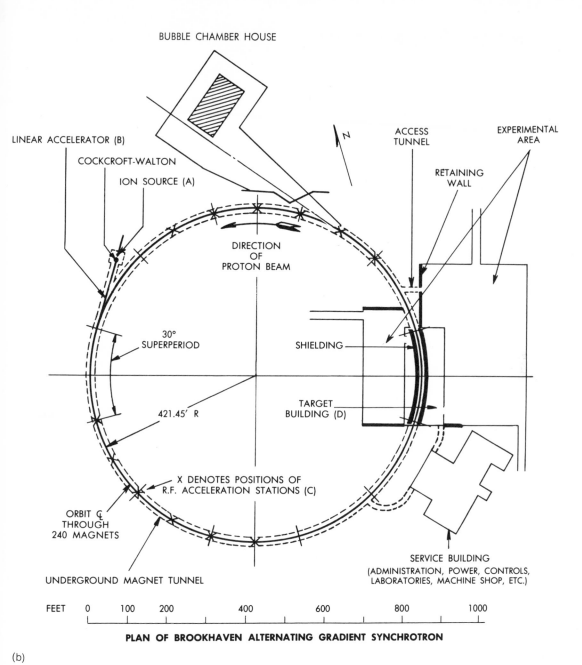

BUBBLE CHAMBER HOUSE

LINEAR ACCELERATOR (B)

COCKCROFT-WALTON

ION SOURCE (A)

N

ACCESS
TUNNEL

EXPERIMENTAL
AREA

RETAINING
WALL

DIRECTION
OF
PROTON BEAM

30°
SUPERPERIOD

SHIELDING

421.45′ R

TARGET
BUILDING (D)

X DENOTES POSITIONS OF
R.F. ACCELERATION STATIONS (C)

ORBIT ℄
THROUGH
240 MAGNETS

UNDERGROUND MAGNET TUNNEL

SERVICE BUILDING
(ADMINISTRATION, POWER, CONTROLS,
LABORATORIES, MACHINE SHOP, ETC.)

FEET   0   100   200   400   600   800   1000

PLAN OF BROOKHAVEN ALTERNATING GRADIENT SYNCHROTRON

(b)

Fig. 46-9. (a) *A view inside the tunnel of the AGS, showing where the linear accelerator, passing through the shielding wall (left rear), joins the main ring (right foreground). (b) The arrangement of the AGS. (Courtesy Brookhaven National Laboratory.)*

Others of the new particles—the K mesons, for example—apparently provide the mechanism for decay of the excited nucleon to its ground state—just as an atom loses excitation energy by the emission of a photon or the emission of a beta particle may de-excite a nucleus. A brief explanation of this kind cannot do justice to the wealth of experimental evidence that states the problem of the elementary particle or to the ingenuity with which it has been attacked. We must leave the details to your further reading.

Once again in physics, order has been perceived in disorder, and the complex is seen to be simple. But other challenges lie ahead as man pursues his search for knowledge of the world about him. The task of the physicist is never done. The frontiers of physics are always open.

Summary

There are about 1700 different nuclides or nuclear species: 285 are stable, about 40 are naturally radioactive, and the rest have been produced in the laboratory.

Among the properties that distinguish one nuclide from another are: charge, mass, binding energy per particle, size, and spin. Nuclear masses are measured by the mass spectrograph in which beams of the ions being studied are sorted according to mass by electric and magnetic fields.

Nuclear reactions can leave the resulting nucleus stable or radioactive. Reactions can be endothermic or exothermic. The cross section σ of a reaction is given by

$$\sigma = \frac{N}{n_0 n_t}$$

Fission is the splitting of a nucleus of uranium or thorium after it has absorbed a neutron. In a nuclear reactor, a chain reaction takes place, the splitting of one uranium nucleus yielding neutrons that cause the splitting of other nuclei.

Cosmic rays interact with nuclei in the atmosphere, producing secondary particles.

The various forces in nature can be classified as: strong interactions, electromagnetic interactions, weak interactions, and gravitational interactions.

The nuclear attractive force seems to be due to the exchange of pi mesons (π^+, π^-, or π^0) between pairs of nucleons.

Questions

1. Which elements are the most stable? Why?

2. List the stable subatomic particles.

3. How can the ages of these be determined: the earth, an Egyptian sarcophagus, a peat bed, an old Indian campsite? Discuss the accuracy of the determination.

4. What is the approximate diameter of (*a*) an atom, (*b*) a nucleus?

5. Discuss the stability of nuclei whose atomic numbers are greater than (*a*) lead, (*b*) uranium.

6. Why is an electron a poor bombarding particle in one sense and a good one in another?

7. Why is the law of conservation of energy expanded to the law of conservation of mass-energy in nuclear reactions?

8. Make a table of the basic types of accelerators and their characteristics.

9. Where do cosmic rays originate? What is a "cosmic-ray spectroscope"? What is a "cosmic-ray telescope"?

10. Examining the key to Fig. 46-8, explain why (*a*) the tracks of some particles are shown as long and others short, and (*b*) some tracks are thick and some thin.

11. What are the by-products when fission of U^{235} occurs? Why is the energy released in fission so great? What is its source?

12. What kinds of studies do radioactive tracers make possible (*a*) in medicine, (*b*) in chemistry, (*c*) in the biological sciences, and (*d*) in engineering?

13. How are heat and radioactive waste from reactors disposed of? Is this an environmental problem? Discuss.

14. Why does the sum of the masses of two neutrons and two protons not equal the mass of the alpha particle?

15. What role do π mesons play in nuclear theory?

16. Why was the discovery of plutonium so important? How can it be obtained?

17. The binding energy per particle is about constant for nuclides in the middle of the nuclidic chart (or periodic table). Why does this support the idea that nucleons interact in small groups within the nucleus?

18. Explain why the ratio of neutrons to protons increases as atomic number increases.

19. How do these differ: (a) lepton and meson, (b) lepton and baryon, (c) proton and neutron, (d) proton and antiproton?

20. What does it mean when we say a particle has a spin of $\frac{1}{2}$? $\frac{1}{2}$ of what?

Problems

A

1. If wood from an Egyptian tomb contains half as much radioactive carbon as that from a living tree, how old is the tomb?

2. Identify the chemical species of these isotopes: $_{48}X^{105}$, $_{26}X^{55}$, $_{80}X^{191}$.

3. Show that 1 atomic mass unit (amu) is equivalent to 931 Mev.

4. What is the binding energy per particle of N^{16}?

B

5. Calculate the energy of protons produced when 10.5-Mev alpha particles bombard N^{14}. Assume that the nuclear reaction is $_2He^4 + _7N^{14} \rightarrow _8O^{17} + _1H^1$.

6. Complete these equations, balancing charge and mass:

$$Be^9 + He^4 \rightarrow C^{12} + (\quad)$$
$$B^{10} + n^1 \rightarrow Li^7 + (\quad)$$

$$Li^7 + H^1 \rightarrow Be^7 + (\quad)$$
$$B^{11} + H^2 \rightarrow C^{12} + (\quad)$$

7. Complete the following equations, balancing charge and mass:

$$H^2 + B^{10} \rightarrow C^{11} + (\quad)$$
$$He^4 + _{13}Al^{27} \rightarrow _0n^1 + (\quad)$$
$$He^4 + (\quad) \rightarrow _8O^{17} + _1H^1$$

8. When a carbon-12 nucleus is bombarded by an alpha particle, a radioactive nuclide and a neutron result. The nuclide disintegrates with the ejection of a positron. What nuclide is the final product of this transmutation?

C

9. (a) How much energy would be liberated if 4.0313 gm of hydrogen were transformed into 4.0026 gm of helium? (b) How much gasoline, heat of combustion 11,000 cal/gm, would liberate this amount of heat in burning? (Speed of light $= 3.0 \times 10^{10}$ cm/sec.)

10. Assume that the fission of uranium-235 liberates 200 Mev/atom. (a) How much energy is liberated by the fission of a kilogram of the uranium? (b) The burning of how much gasoline, heat of combustion 11,000 cal/gm, would provide this amount of energy?

11. How many transmutations per million incident deuterons occur when a target 0.5 mm thick is bombarded by 2-Mev deuterons? Assume that the target contains 6×10^{22} nuclei/cm³ and that the cross section is 2 barns.

12. The specific gravity of the sun is 1.4; its diameter is 8.6×10^5 mi. The solar radiation constant at the earth is 0.135 joule/cm² per sec and the average distance to the sun is 93×10^6 mi. (a) At what rate does the sun lose radiation? (b) At what rate does the total mass change? (c) If this energy is derived from the formation of helium from hydrogen, at what rate is helium being produced in the sun? (Diameter of earth: 8.0×10^3 mi.)

13. In the theory of the synthesis of the heavier nuclei (beyond iron), neutron-capture processes in stars are thought to predominate at the prevailing temperatures of star interiors (10^8 to 10^9 °K). (*a*) Why would charged-particle interactions not be as effective? (*b*) Neutron cross sections with heavy elements are known to be relatively large at energies of 10–100 ev. Show that these neutron energies might be expected to occur at those temperatures according to simple kinetic theory.

14. Assuming that 5 of every 10^5 alpha particles from radon in a 50-millicurie radon-beryllium sample (used as a source of neutrons) produce a disintegration in the surrounding beryllium, how many neutrons are emitted per second?

15. Radioactive potassium-40 (K^{40}) decays to argon-40 (Ar^{40}) by capturing a K-shell electron. The half-life for the decay is 1.3×10^9 years. Archaeologists use the ratio of K^{40} to Ar^{40} in a sample, determined chemically and by the use of a mass spectrometer, as a means of dating the sample. (*a*) For which of the two isotopes could ordinary chemical analysis be used and for which would mass spectrometry be used? (*b*) If the K^{40}/Ar^{40} ratio of a sample is 4.0×10^{-4}, what is the age of the sample?

16. In the hot (10^8°K), dense (10^5 gm/cm^3) cores of red giant stars, carbon nuclei are synthesized from helium nuclei according to the equation

$$3 \, He^4 \rightarrow C^{12} + E$$

How much energy E is released in the above reaction?

17. Calculate the approximate energy of a proton which, upon collision with another proton, produces a proton-antiproton pair, 2 protons, and 4 Bev of energy.

18. What is the probability that a meteor that penetrates the atmosphere will fall in water? What is the chance that it will hit you? What is the "collision cross section" for the meteor's hitting you?

For Further Reading

Burbridge, G., and F. Hoyle, "Anti-Matter," *Scientific American,* April, 1958, p. 34.

Hahn, Otto, "The Discovery of Fission," *Scientific American,* February, 1958, p. 76.

Hughes, Donald J., *The Neutron Story,* Doubleday and Company, Garden City, 1959.

Korff, S. A., "The Origin and Implications of the Cosmic Radiation," *The American Scientist,* September, 1957, p. 281.

Rossi, Bruno, "High-Energy Cosmic Rays," *Scientific American,* November, 1959, p. 134.

Thorne, Kip S., "Gravitational Collapse," *Scientific American,* November, 1967, p. 88.

Weinberg, Alvin M., "Breeder Reactors," *Scientific American,* January, 1960, p. 82.

Weisskopf, V. F., "The Three Spectroscopies," *Scientific American,* May, 1968, p. 15.

Wilson, Robert R., and Raphael Littauer, *Accelerators, Machines of Nuclear Physics,* Doubleday and Company, Garden City, 1960.

Answers to Odd-Numbered Problems*

Chapter 1

1. 28 gm; 3. 9.46×10^{15} m; 5. 402 m; 7. 11.0 sec; 9. A; 11. (a) 42 liters, (b) 5.21 ft²; 13. 10^{17}; 15. 7.485 yd/sec; 17. 1.21 yd³; 19. A gains 1 min/day, B gains 4 min/day (sidereal clock), C mean solar clock, D loses 5 min/day.

Chapter 2

1. 1.648×10^5 kgwt; 3. 5.00×10^4 kgwt; 5. 1.91×10^4 lb; 7. 152 cm; 9. 1320 lb-wt; 11. 26.7×10^3 cm²; 13. 0.273; 15. 6.12×10^4 lb-wt/ft²; 17. (a) 267 lb/in², (b) 1067 lb; 19. 7.38 km; 21. B: 70.9 cm-Hg, C: 67.5 cm-Hg; 23. 104 gwt; 25. 8.2 cm³; 27. 23.3 kgwt; 29. 9.36×10^3 m³; 31. 40 cars; 33. 11.8×10^2 gwt (total); 35. 207 gwt down.

Chapter 3

1. 14.1 kgwt east; 3. 13.7 kgwt, equal in magnitude, opposite in direction; 5. 0.50 kgwt; 7. (a) 4000 kgwt, (b) 250 kgwt; 9. 60 kgwt; 11. 58 kgwt; 13. 346 kgwt; 15. 30.9 kgwt; 17. (a) 2.3 kgwt, (b) 1.7 kgwt; 19. (a) 120°, (b) 151°; 21. (a) 7.1 kgwt, (b) 13.7 kgwt; 23. 50 kgwt; 25. 4.35 kgwt.

* Only problems with numerical answers are included.

Chapter 4

1. (a) 16.7 m/sec²; (b) 10 m; 3. 1.50×10^3 m/sec²; 5. 33.3 m/sec²; 7. 24.5 m/sec; 9. (a) 24.5 m, (b) 19.5 m/sec down; 11. 0.6 m/sec²; 13. (a) 350 m/sec, (b) 1.71×10^{-3} sec, (c) 4.08×10^5 m/sec²; 15. 7.5 ft/sec; 17. (a) 3.87 sec, (b) 18.7 m, (c) 57.8 m; 19. (a) 20.4 m, (b) 5.25 sec, (c) 31.5 m/sec; 21. 491 m; 23. (a) 5.55 sec, (b) 11.1 m/sec; 25. (a) 72.7 sec, (b) 7.1×10^2 m/sec; 27. (a) 49.3 m/sec², (b) 210 sec; 29. 19.4 m.

Chapter 5

1. (a) 6.53 m/sec², (b) 6.04 m/sec²; 3. 2.94 m/sec²; 5. 7.0 kgwt or 68.6 newt; 7. 4.9 m/sec²; 9. 80 newt; 11. (a) 0.49 m/sec², (b) 97.8 newt; 13. (a) 1.33×10^7 cm/sec², (b) 4.0×10^8 dynes, 4.0×10^3 newt, 408 kgwt; 15. (a) 4.17 m/sec², (b) 3.17×10^3 kgwt; 17. 2.0 newt; 19. 20.2 m/sec² upward; 21. (a) 83.3 cm/sec², (b) 417 dynes, (c) 0.085; 23. (a) 600 m/sec², (b) 120 newt; 25. 11.5 m; 27. (a) 2.45 m/sec², (b) 1.35 m/sec², (c) 0.245 m/sec²; 29. (a) 367 newt, (b) 524 newt; 31. 39.2°; 33. 18.4 m; 35. (a) 3.0 kgwt, 0 m/sec², (b) 3.47 kgwt, 3.01 m/sec²; 37. 106 cm ($a = 212$ cm/sec²).

Chapter 6

1. (a) 4.0 m/sec, (b) 0 m/sec, (c) 2.83 m/sec, (d) 1.53 m/sec; 3. (a) 60°, (b) 0.87 m/sec; 5. 4.9 m; 7. 8.2°; 9. 4.43 m/sec; 11. 4.00×10^4 cm/sec^2; 13. (a) 49 newt, (b) 0.99 m/sec; 15. 500 cm/sec; 17. (a) 7.9×10^3 m/sec, (b) 1.41 hr; 19. (a) 2.0×10^3 newt, (b) 7.7° or 1.34 m; 21. 13.1 m; 23. 151 m.

Chapter 7

1. (a) 4.90×10^3 j, (b) 9.80×10^2 j; 3. 5.10 m/sec; 5. (a) 28.0 m, (b) 7.14×10^3 newt; 7. 0.49 newt; 9. 3.12×10^6 j; 11. 392 watts; 13. (a) 7.2×10^5 j; (b) 45.9 m; 15. (a) 6.40×10^3 j, (b) 2.56×10^5 newt; 17. (a) 0.33, 1.33 m/sec, (b) −1.0, 2.0 m/sec; 21. bullet 9.6 kg-m/sec, bear 1.20×10^3 kg-m/sec; 23. (a) 78.4 j, (b) 784 newt; 25. 26.6° east of north, 5.6 m/sec; 27. (a) 1.0×10^5 newt, (b) 3.0×10^4 newt; 29. 23.1 m/sec; 31. 45° from the original path of B; 33. (a) 92% efficient, (b) estimate 40 to 80; 35. $(m^2v^2)/(2\mu g[m + M]^2)$; 37. estimate 14 yrs.

Chapter 8

1. (a) 80 dynes, (b) 4.0×10^3 ergs/sec; 3. (a) 2.4×10^4 cm^3/min, (b) 4.0 m/sec; 5. 1.2×10^3 cm^3/min; 7. 10 m; 9. 470 cm-water; 11. 11.1×10^4 dynes/cm^2; 13. 82 cm-Hg; 15. (a) 1000, (b) 20 cm/sec; 17. (a) 396 cm/sec, (b) 4.48×10^3 cm/sec.

Chapter 9

1. 120 kgwt; 3. (a) 103 rev/sec, (b) 206π rad/sec; 5. (a) 0.105 rad/sec, (b) 0.84 cm/sec; 7. (a) 10.0 m-newt, (b) 1.60 rad/sec^2, (c) 0.80 rad; 9. 60 j; 11. (a) 49.0 m-newt, (b) 42.4 m-newt, (c) 24.5 m-newt, (d) 0 m-newt; 13. 20 kgwt; 15. 2.25 m; 17. (a) 0.16 kg-m^2, (b) 2.84×10^3 j, (c) 284 watts; 19. 2.19 rad/sec^2; 21. 20.0 rev/min or $2\pi/3$ rad/sec; 23. 2.57 m; 25. 176 kgwt; 27. (a) 0.687 cm, (b) 0.428 cm; 29. (a) 25 kgwt, (b) 17.65 kgwt, 45° from horizontal; 31. 15 kgwt; 33. 1.6π rad/sec^2; 35. (a) 3.57 m/sec, (b) 3.31 m/sec^2,

(c) 4.38 m/sec; 37. 5.6 m/sec; 39. (a) 11.2 cm/sec^2, (b) 1.98×10^5 dynes, (c) 5.6 cm/sec.

Chapter 10

1. negligible; 3. 0.41 c; 5. 1.005 sec; 7. 0.99995 c; 9. 0.113 c; 11. A 0.404 hr, B 0.402 hr.

Chapter 11

1. (a) 2.45 m/sec^2, (b) 2.72×10^{-3} m/sec^2, (c) 0 m/sec^2; 3. 3.13 m/sec; 5. 3.27×10^{-7} newt; 7. 3.67×10^{-47} newt; 9. 3.86×10^4 km; 11. 140 kgwt; 13. (a) 1.48 hr, (b) 7.77 km/sec, (c) 2.11×10^{13} j; 15. 268 newt; 17. 7.86×10^8 km; 19. 3.19 mi/sec; 21. approx. 6.0×10^{24} kg; 23. 4.3 yr; 27. 0.108 mass of earth; 29. (a) 2.73×10^{43} kg-m^2/sec, (b) 4.82×10^{41} kg-m^2/sec, (c) 11.4×10^{41} kg-m^2/sec, (d) 4% (e) approx. 100%.

Chapter 12

1. (a) 7.84×10^5 newt/m^2, (b) 2.00×10^{-2}; 3. (a) 2000 lb, (b) 3750 lb; 5. 294 dynes; 7. 1.44 cm; 9. 2.5 cm; 11. (a) 5.00×10^4 gwt/cm^2, (b) 1.00×10^{-3}, (c) 5.00×10^7 gwt/cm^2 or 4.9×10^{10} dynes/cm^2; 13. 0.162 cm; 15. 1.85×10^3 dynes; 17. 1.29×10^9 cm-dynes or 129 m-newt; 19. 100 dynes; 21. 1.04×10^{-2} ft or 0.125 in.; 23. 11.1 newt; 25. 8.0 m^3.

Chapter 13

1. 2.0066 cm^3; 3. 0.728 cm; 5. 1.04 m; 7. 3.72 cm^3; 9. 13.2×10^{-6}/C°; 11. 5.00×10^{-6}/C°; 13. 104 cm^3; 15. 9.29°C; 17. 12.0 cm; 21. 0.0497 cm.

Chapter 14

1. 1657 cal; 3. 3.33×10^{-2} cal/sec; 5. 101.7°C; 7. 21.6 cal/(sec-cm^2); 9. 12.4°C; 11. 15.15 kg; 13. 4.00×10^3 cal/gm; 15. (a) 2.78×10^8 cal, (b) 3.97×10^4 gm or 39.7 kg, (c) \$47.64; 17. 67%; 19. (a) 470°K, (b) 4 cal/sec, (c) 9.25×10^3 cal/sec; 21. (a) 5.6×10^8 cal, (b) 229 kg; 23. (a) 86.4 cal/sec, (b)

653.6 cal/sec, (c) 210 cal/sec; 25. (a) 5870°K, (b) 1.03×10^4 cal/(m²-min).

Chapter 15

1. 225 gm; 3. 9.0×10^4 cm³; 5. 8.29 km; 7. (a) 0.10 m³, (b) 0.80 m; 9. 493 m/sec; 11. (a) 295 cm-Hg, (b) 0.748; 13. 1.45 gwt/liter; 15. (a) 10.34 m, (b) 5.17 m; 17. (a) 10.5 cm, (b) 12.1 cm, (c) 15.3 cm, (d) 20.7 cm, (e) 28.0 cm, (f) 32.2 cm; 19. 1.59×10^9 km; 21. (a) 5.56 km/sec, (b) 10.7 km/sec.

Chapter 16

1. 2.24×10^5 cal; 3. 299 gm; 5. (a) 54.3%, (b) 73.3%; 7. (a) 9.41 gm/m³, (b) 54.4%; 9. 25.1 gm; 11. 0°C; 13. 101 gm; 15. 57.3°C; 17. 78.6 cal/gm; 19. 10.05°C; 21. 4.00 cm-Hg; 23. 51 cm-Hg; 25. liquid air 95.7 cal/gm, ice 92.8 cal/gm; 27. 71.36 cm-Hg; 29. $a =$ newt-m⁴, $b = $ m³; 31. 14.9 cm-Hg.

Chapter 17

1. (a) 2450 j, (b) 585 cal; 3. 3.44×10^{11} cal; 5. 68.1 km; 7. (a) 2.09 kw, 2.81 hp, (b) 209 w; 9. 102 m; 11. (a) 0.70, (b) 3.5; 13. 6.82×10^5 cal/min; 15. 34%; 17. 352 m/sec; 19. (a) 341 cal, (b) 136 cal, (c) 205 cal, (d) 3.0 cal/(mole-K°), (e) 0.75 cal/(mole-C°); 21. 3.5×10^6 cal; 23. 3.2×10^{10} joules; 25. (a) 8.8×10^{-3} j, (b) 334 j.

Chapter 18

1. (a) 3.43 sec, (b) 0.291 sec⁻¹; 3. (a) 1.0 sec, (b) 1.0 sec⁻¹, (c) 40 cm; 5. 3.17 sec; 7. 1.55 sec; 9. (a) 0.174 sec, (b) 0.201 sec; 11. (a) 0, (b) 2.51 m/sec, 2.17 m/sec, 1.26 m/sec; 13. (a) 10 cm, (b) 0 cm, (c) 20 cm, (d) 0 cm; 15. (a) 62.8 cm/sec, (b) 31.4 cm/sec; 17. 1.63 sec; 19. 6.53 j; 21. (a) 24 cm/sec, (b) 12 cm/sec, (c) 0, 600, 1200 dynes; 23. 16.5 cm; 25. 13.6 gm/cm³.

Chapter 19

1. 500 m; 3. 2.0 sec⁻¹; 5. 20 cm; 7. 4.0 m/sec; 9. 0.45 sec; 11. 0.798; 13. (a) 3.80, 0,

−3.80, 2.35 cm, (b) 3.80, −2.35, −2.35, 3.80, 0 cm; 17. (a) 1.27×10^3 newt, (b) 0.65×10^3 newt.

Chapter 20

1. (a) 0.425 m, (b) 1.70 m; 3. 1.27×10^3 m/sec; 5. 600 rev/min; 7. (a) 17.2 m, (b) 1.72 cm; 9. 4.56×10^3 m/sec; 11. (a) 86 sec⁻¹, (b) 430 sec⁻¹; 13. 494 sec⁻¹; 15. 32.2, 25.8, 21.5, 16.1 cm; 17. 467 sec⁻¹; 19. 1020 sec⁻¹; 21. 3.3°C; 23. 14.96 m/sec; 25. approx. 1600 m/sec; 27. 0.542 sec⁻¹.

Chapter 21

1. (a) 3.6×10^4 newt, (b) 0.9×10^4 newt, (c) 3.6×10^6 newt; 3. 4.44×10^{-3} μcoul; 5. (a) 1.56×10^3 newt, (b) 8.5 cm; 7. (a) 7.2×10^{-4} newt, (b) 7.06 m/sec²; 9. 5.75×10^{13} coul; 11. 1.14 m; 13. 13.9×10^{18} m/sec²; 15. (a) 9.22×10^{-8} newt, (b) 2.26×10^6 m/sec, (c) 7.2×10^{15} sec⁻¹.

Chapter 22

1. 2.67×10^3 newt/coul; 3. 3.0×10^3 j; 5. 12 v; 7. 12 μcoul; 9. 0.115 μf; 11. (a) 1.15×10^9 newt/coul, (b) 2.28×10^9 newt/coul, (c) 9.00×10^9 newt/coul; 13. (a) 2000 v/cm, 2.0×10^5 newt/coul, (b) 9.64×10^{12} m/sec²; 15. 4.52×10^5 lines; 17. (a) 1.33 v, (b) before, 3.00×10^6 j, after, 2.67×10^{-6} j; 19. $(4 k_0 Ze^2)/(mv^2)$; 21. (a) $(v_0 d)/(V_0[e/m])^{1/2}$, (b) $(v_0^2 + V_0^2[e/m]^2)^{1/2}$.

Chapter 23

1. 3.24×10^3 coul; 3. 4.0 amp; 5. 103°C; 7. 80 Ω; 9. 1.83×10^{-3}/C°; 11. (a) 16.0 Ω, (b) 14.0 Ω; 13. 0.149 kg/hr; 15. 0.819 gm; 17. (a) 4.0 faradays, (b) 8.0 faradays; 19. 74.4 m; 21. (a) 0.75 v, (b) 118.5 v.

Chapter 24

1. (a) 16 v, (b) 8.0 amp; 3. (a) 0.46 amp, (b) 2.77, 1.38, 1.85 v; 5. (a) 23.1 amp, (b) 4.62 v, (c) 115.4 v; 7. (a) 12 v, (b) 8.0 amp; 9. 10.4 v; 11. 19.78 v; 13. 24 v; 15. (a) 1.0 amp, (b) 0.50 amp; 17. (a) 0.50 amp, (b) 0.24 v, (c)

0.06 amp; 19. 4.0 Ω; 21. series aiding 1.53 amp, series opposing 0.51 amp, parallel 0.77 amp through 10 Ω; 23. 9/16 amp; 25. 26/15 amp; 27. top branch 0.77 amp to right, middle branch 2.15 amp to left, bottom branch 1.38 amp to right.

Chapter 25

1. 72 w, 450 w; 3. (a) 7.2 w, (b) 5.04 kw-hr, (c) 30¢; 5. 3.0×10^6 ev, 4.8×10^{-13} joule; 7. 28.7%; 9. 5.7 C°; 11. 698 sec; 13. (a) 1.32 kw, (b) 7.9¢; 15. (a) 0.30 amp, (b) 0.45 w, (c) 1.62×10^3 j; 17. (a) 3.67×10^6 kg, (b) 12 kg; 19. (a) 2.17×10^6 j, (b) 8.65 kg; 21. (a) 9.72×10^4 j, (b) 2.33×10^4 cal; 23. (a) 140°C or 400°C, (b) 0 to 250°C.

Chapter 26

1. 0.20 m; 3. (a) 5.2×10^{-5} weber/m^2, (b) 3.0×10^{-5} weber/m^2; 5. 0.50 newt; 7. (a) 4.0 newt, (b) west; 9. (a) 62°, (b) 5.3×10^{-5} weber/m^2; 11. 1.96 amp to repel; 13. $\mu(mg)/(Il)$, perpendicular to rails.

Chapter 27

1. 0.96 m^2-amp; 3. 1.0 newt; 5. 2.97×10^4 Ω; 7. approx. 7.5×10^4 Ω; 9. 66.7 v; 11. 2.5×10^{-2}%; 13. (a) 14.4×10^{-4} m-newt, (b) 12.5×10^{-4} m-newt, (c) 9.6 m^2-amp; 15. 1.0×10^5 newt/coul.

Chapter 28

1. 1.33×10^{-6} weber/m^2; 3. 1.27×10^{-2} amp; 5. 3.43×10^{-3} weber/(amp-turn-m); 7. (a) 3.14×10^{-3} weber/m^2, (b) 3.14×10^{-7} weber; 9. 2.00×10^{-3} weber/m^2; 11. (a) 1.00×10^{-4} weber/m^2, (b) 2.00×10^{-3} newt; 13. (a) 750 amp-turn/m, (b) 0.30 weber/m^2, (c) 7.5×10^{-4} weber; 15. approx. 1000 j/m^3.

Chapter 29

1. 2.0×10^{-2} v; 3. 0.83 rev/sec; 5. (a) 4.0×10^{-5} weber, (b) 1.33×10^{-5} v; 7. (a) 1.60×10^{-2} v, (b) 0.80×10^{-2} amp; 9. 116 v; 11. (a) 50 v, (b) 5.0×10^3 v; 13. (a) 0.167 weber/sec, (b) 0.167 v; 15. 150 v; 17. 75 v; 19. (a)

16.7 Ω, (b) 7.2 amp, (c) 8.3%; 21. 86 μv for $B_{hor} = 3.5 \times 10^{-5}$ weber/m^2.

Chapter 30

1. (a) 400 Ω, (b) 500 Ω, (c) 300 Ω; 3. (a) 0°, 36.9°, (b) 36 w, 23 w; 5. 200 Ω; 7. (a) 2.0×10^4 kw, (b) 3.0 kw; 9. 1.67×10^3 Ω; 11. (a) 37.7 Ω, (b) 40.6 Ω, (c) 5.91 amp, (d) 68.5°; 13. 107 w; 15. (a) 0.917, (b) 24°; 17. 5.03 kilohertz (kH); 19. (a) 4.12 Ω, (b) 4.00 Ω, (c) 4.12 Ω, (d) 4.47 Ω; 21. (a) 10^5, (b) 10^{-5}.

Chapter 31

1. 3.15 μsec; 3. 10; 5. (a) 20 v, (b) 20 db; 7. 100 millisec; 9. 5.8×10^{-4} μf, 6.6×10^{-5} μf; 13. (b) approx. 23 v.

Chapter 32

1. 3.05×10^{15} mi; 3. 44 sec; 5. 1.67 candles; 7. 100 m-candles; 9. 20 cm; 11. 8.33 sec; 13. 0.33 m toward 4.0-candle source, 1.00 m away from 4.0-candle source; 15. (a) 114.9 m-candles, (b) 95.3 m-candles; 17. 3.5×10^{-6} cal/cm^3; 19. (a) 8.66×10^4 m-candles, (b) 5.00×10^4 m-candles, (c) zero; 21. (a) 4.0×10^{27} candles, (b) 26×10^4 candles.

Chapter 33

1. 12.5 cm; 3. 1.41; 5. (a) 38°, (b) 23°, (c) 47°; 7. 2.25×10^5 km/sec; 9. 0.40 m; 11. 9:13 a.m.; 13. 60°; 15. (a) 38.0°, (b) 33.1°; 19. 136 ft.

Chapter 34

1. 20 cm × 25 cm; 3. (a) 50 cm, 1.0 cm high, (b) 16.7 cm, 0.33 cm high; 5. $q = -3.0$ cm; 7. 30 cm; 9. −8.3 diopters; 11. 33.3 cm; 13. 10 cm; 15. 0.88 m, 1.1 diopters; 17. 30 cm; 19. 0.33 m; 21. 10.5 cm; 23. 8.91 cm beyond second lens, 0.23 cm high; 25. (a) ∞, (b) 210 cm, (c) −15 cm, (d) 0 cm; 27. (a) 42.4 cm, (b) 282 cm, −632 cm; 31. (a) −40 cm, (b) −20 cm; 33. 50 cm to left of lens.

Chapter 35

1. 5.86 cm; 3. (a) 3.81 cm, (b) 6.5; 5. (a) 0.83 cm, (b) 5.0 cm; 7. −3.0 m; 9. 1.50 diopters,

1.14 diopters; 11. (a) 14 cm, (b) 1.43 cm; 13. (a) 26.07 cm, (b) 19.16 cm; 15. (a) 480 mi, (b) 2080 mi; 17. 60 cm to right of negative lens.

Chapter 36

1. 1.13×10^{15} H; 3. 3.70×10^{14} H, 6.90×10^{14} H; 5. (a) 15,000, 10,000, 3750 A, (b) 15×10^{-7}, 10×10^{-7}, 3.75×10^{-7} m; 7. 1.06×10^3 km/sec; 9. 91.4 km/sec; 11. 1.74 yr; 13. 5901 A.

Chapter 37

1. (a) 3.25×10^{-2} cm, (b) 30.7 per cm; 3. 7.5×10^{-2}; 5. 12.7°, 26.1°, 41.3°; 7. 0.30 cm; 9. (a) 4420 A, (b) 2210 A, (c) 1340 A; 11. 0.11 cm; 13. 5.68 km; 15. 20.5 cm; 17. 1.92×10^{-6} rad; 19. (a) 0.12 cm, (b) 20.7'; 21. 130 ft.

Chapter 38

1. 1.19; 3. 58.6°, 56.6°; 5. 37.1°, 42.3°; 7. 7.9°; 9. 5.0 cm; 11. 3.27×10^{-3} cm; 13. 1.71×10^{-4} cm; 17. 15°.

Chapter 39

1. 4.14×10^{-15} joule/amp; 3. (a) 10.3×10^{-6} ev, (b) 2.24 ev, (c) 15.4×10^3 ev; 5. 0.62 Mev; 7. (a) 6.62×10^{-28} joule, (b) 6.62×10^{-20} joule; 9. 4.21 ev; 11. 1.82 ev; 13. (a) 5.1×10^{-9} amp.

Chapter 40

1. 0.205 A; 3. (a) 2.45×10^{18} H, 8.12×10^9 m^{-1}, (b) 2.48×10^{15} H, 8.23×10^6 m^{-1}, (c) 4.60×10^{14} H, 1.53×10^6 m^{-1}; 5. (a) 6563 A, 4861 A, 4340 A, (b) 1215 A, 18,750 A; 7. (a) 1.5×10^4 ev, 7.25×10^7 m/sec, (b) $3.64 \times$ 10^{18} H, 0.825 A; 9. 0.866 c; 11. 0.80 A; 13. 13.0°, 26.7°, 42.5°; 15. 2.28×10^{-13} j; 17. (a) 6520, 4830, 4320 A.

Chapter 41

1. 4.4×10^{-36} kg; 3. 6.3×10^{-36} m; 5. 2.64×10^4 m/sec; 7. 5.8×10^{-21} j; 9. 6.40×10^{-15} m; 11. (a) 2.52×10^{-17} j, 1.98×10^{-15} j, (b) both 6.62×10^{-24} kg-m/sec; 13. 6.62×10^{-29} m/sec; 15. 2.21×10^{-24} joule-sec/m; 17. 2.09×10^{-5} ratio.

Chapter 42

1. 0.87 A; 3. (a) 3.14 A, (b) 12.6 A; 5. (a) $n = 1, 2, 3, \ldots$, (b) $n = 2, 3, 4, \ldots$, (c) $n = 3, 4, 5, \ldots$; 7. approx. 10^{27}.

Chapter 43

1. (b); 3. 2, 8, 18, 32; 5. 3, 2, −2, −$\frac{1}{2}$; 3, 2, −2, $\frac{1}{2}$; 3, 2, −1, −$\frac{1}{2}$; 3, 2, −1, $\frac{1}{2}$; 3, 2, 0, −$\frac{1}{2}$; 7. (a) the two lowest, (b) 5900 A.

Chapter 44

1. 1.61×10^{22} atoms; 3. Ga, Si, P; In, Sn, Sb; 5. 4.64×10^4 °K; 9. 0.182×10^7 m/sec; 11. (a) 2.08×10^{29}; (b) 6.13×10^{-15} sec.

Chapter 45

1. 25 mg; 3. 41.7 wks; 5. 2.5×10^{11} ions; 7. 1.16; 9. 0.012 A; 11. 255 cm; 13. greater than 10^{10} yr; 15. 6 hr.

Chapter 46

1. 5570 yr; 5. 9.3 Mev; 7. n^1, P^{30}, N^{14}; 9. (a) 2.58×10^{12} joules, (b) 5.6×10^4 kg; 11. 6000; 15. (a) K, (b) 14.7×10^9 yr; 17. approx. 6×10^3 Mev.

Appendices

Appendices

1. Atomic

(Constants are rounded to three significant figures. The symbol $\cong$ means "is approximately equal to.")

Speed of light in vacuum, $c = 3.00 \times 10^8$ m/sec = 186,000 mi/sec

λ of 1-ev photon = 12,300 A

Avogadro's number, $N_0 = 6.02 \times 10^{23}$ molecules/mole

Gas constant, $R = 8.31$ joules/mole-K°

Planck's constant, $h = 6.63 \times 10^{-27}$ erg-sec = 6.63×10^{-34} joule-sec = 4.14×10^{-15} ev-sec

Electronic charge, $e = 1.60 \times 10^{-19}$ coul

Rest mass of electron, $m_0 = 9.11 \times 10^{-31}$ kg

Atomic mass unit = 1.66×10^{-27} kg

2. Electricity and Magnetism

Permittivity constant, $1/(4\pi\epsilon_0) \cong 9 \times 10^9$ newton-m²/coul²

Permeability constant, $4\pi\mu_0 \cong 12.6 \times 10^{-7}$ kg-m/coul²

B. Numerical Equivalents

1. Length

1 in. $\cong$ 2.54 cm

1 ft $\cong$ 30.5 cm

1 mi = 5280 ft $\cong$ 1.61 km

1 cm $\cong$ 0.394 in.

1 m $\cong$ 1.094 yd $\cong$ 39.4 in.

1 km = 1000 m $\cong \frac{5}{8}$ mi

1 A = 10^{-8} cm

2. Volume

1 (liquid) qt $\cong$ 0.946 liter

1 liter $\cong$ 1.06 qt

3. Mass

1 pound mass (avoirdupois) $\cong$ 454 gm

1 kg = 1000 gm $\cong$ 2.20 pounds (mass)

4. Force

1 lb $\cong$ 454 gwt

1 gwt $\cong$ 980 dynes = 980 gm-cm/sec²

1 kgwt $\cong$ 9.8 newtons = 9.8 kg-m/sec²

5. Work, Energy, and Power

1 ft-lb $\cong \frac{1}{3}$ joule

1 Btu $\cong$ 778 ft-lb $\cong$ 252 cal

1 horsepower-hr $\cong$ 198,000 ft-lb

1 horsepower = 550 ft-lb/sec ≅ ¾ kw ≅ 0.71
  Btu/sec
1 erg = 1 dyne-cm
1 joule = 1 newton-m = 10^7 ergs ≅ 10,200
  gwt-cm ≅ ¾ ft-lb
1 cal ≅ 4.19 joules ≅ 42,700 gwt-cm
1 kw-hr = 3,600,000 joules ≅ 2,700,000 ft-lb
1 kw ≅ ⁴⁄₃ horsepower

6. Miscellaneous

1 ft^3 of water at 70°F weighs 62.4 lb ≅ 1000 oz
1 cm^3 of mercury at 0°C weighs 13.6 gwt
76 cm-Hg ≅ 1034 cm-water ≅ 30 in.-Hg ≅ 14.7
  $lb/in.^2$
$\pi ≅ 3.142 ≅ 3\frac{1}{7} = \frac{22}{7}$
$\pi^2 ≅ 9.87$

C. Formulae

Area of circle $= \pi r^2$
Surface of sphere $= 4\pi r^2$
Volume of sphere $= \frac{4}{3}\pi r^3$
$\sqrt{2} ≅ 1.41; \sqrt{3} ≅ 1.73$

Circumference of circle $= 2\pi r$
Volume of cylinder, prism, or parallelopiped =
  Area of base × Altitude measured perpendicular
  to base

D. Mathematical Review

1. Fractions

    a. Addition and subtraction.
      1. Make the denominators of all the fractions
         the same.
      2. Then add or subtract the numerators only.

Example:

$$\frac{1}{2} + \frac{2}{3} = \frac{1 \times 3}{2 \times 3} + \frac{2 \times 2}{3 \times 2} = \frac{3}{6} + \frac{4}{6} = \frac{7}{6} = 1\frac{1}{6}$$

    b. Multiplication.
    Multiply numerators and denominators res-
    pectively to get the numerator and denom-
    inator of the product.

Examples:

$$\frac{a}{b} \times \frac{c}{d} = \frac{ac}{bd}$$

$$\frac{2}{3} \times \frac{4}{5} = \frac{8}{15}$$

    c. Division.
    Invert the divisor and multiply.

Examples:

$$\frac{\frac{a}{b}}{\frac{c}{d}} = \frac{a}{b} \div \frac{c}{d} = \frac{a}{b} \times \frac{d}{c} = \frac{ad}{bc}$$

$$\frac{\frac{2}{3}}{\frac{4}{5}} = \frac{2}{3} \div \frac{4}{5} = \frac{2}{3} \times \frac{5}{4} = \frac{10}{12} = \frac{5}{6}$$

2. Exponents

    A positive exponent indicates the number of
times that a quantity must be taken as a factor.
Thus

$$3^4 = 3 \times 3 \times 3 \times 3.$$

    a. In multiplying, add the exponents to find the
exponent of the product.

Examples:

$$a^m \times a^n = a^{(m+n)}$$
$$10^2 \times 10^3 = 10^5$$

    b. In division, subtract the exponent of the di-
visor from that of the dividend to find the exponent
of the quotient.

Examples:

$$a^m \div a^n = a^{(m-n)}$$
$$10^5 \div 10^2 = 10^3$$
$$10^2 \div 10^2 = 10^0 = 1$$

    A number having a negative exponent equals
the reciprocal of that number with a positive exponent.

$$a^{-m} = \frac{1}{a^m}$$

$$10^{-2} = \frac{1}{10^2} = \frac{1}{100}; \quad 10^{-6} = \frac{1}{10^6} = \frac{1}{1,000,000}$$

    To simplify computations involving complicated
numbers, employ these rules.

Example:

$$\frac{0.00007 \times (1,000,000,000) \times (300,000)}{(500)^2} =$$

$$\frac{(7 \times 10^{-5})(1 \times 10^9)(3 \times 10^5)}{5 \times 5 \times 10^4} =$$

$$\frac{(7) \cdot (1) \cdot (3)}{(5) \cdot (5)}\, 10^{-5+9+5-4} = \frac{21}{25} \times 10^5 = 8.4 \times 10^4$$

$$= 84,000$$

3. Ratio and Proportion

The ratio of two numbers a and b is the result a/b of dividing one by the other. A proportion is the equality of two ratios; e.g., $a/b = c/d$ (sometimes written $a:b = c:d$). From this equation, $ad = bc$.

If two variable quantities are so related that their ratio is constant, they are said to be directly *proportional* to each other. Suppose that a uniform plank is cut into pieces of unequal length, l_1, l_2, l_3, etc. Then, since the area of cross section is constant the ratio of the volume V of each piece to its length l is constant.

$$\frac{V_1}{l_1} = \frac{V_2}{l_2} = \frac{V_3}{l_3} = k_1$$

or, in general,

$$\frac{V}{l} = k_1$$

$$V = k_1 l$$

In these equations, k_1 is called the *constant of proportionality,* and in this quantity it is an area.

For example, a piece of the plank 20 in. long has a volume 120 in.^{3}; then the constant of proportionality is $k_1 = 120$ in.^{3}/20 in. $= 6$ in.^{2}

Often one quantity is proportional to the square of another. For example, the areas A of different circles are proportional to the squares of their respective diameters d.

$$\frac{A_1}{d_1{}^2} = \frac{A_2}{d_2{}^2} = \frac{A_3}{d_3{}^2} = k_2$$

If one quantity increases and some other quantity decreases in such a manner that the product of the two is constant, the two quantities are then said to be *inversely* proportional to each other. For example, the area of the rectangular floor that can be covered by a given amount of carpet is constant and equals the product of the length l and width W of the floor. That is,

$$l \times W = k_3$$

and

$$l = \frac{k_3}{W}$$

We say that the length of the floor is *inversely proportional* to its width for a constant area of carpeting or that the length varies *inversely* with the width.

4. Quadratic Equations

A quadratic equation of the form $ax^2 + bx + c = 0$ may be solved by using the formula

$$x = \frac{-b \pm \sqrt{b^2 - 4ac}}{2a}$$

Example. Solve the equation $2x^2 + x - 6 = 0$.

$$a = 2, \quad b = 1, \quad c = -6$$

$$x = \frac{-1 \pm \sqrt{1 + 48}}{4} = \frac{-1 \pm 7}{4} = -2 \quad \text{or} \quad +\frac{3}{2}$$

Such an equation may also be solved by *completing the square.* For example, given that $2x^2 + x - 6 = 0$:

a. By division, make the coefficient of the squared term 1; then

$$x^2 + \tfrac{1}{2}x - 3 = 0$$

b. Transpose the last term

$$x^2 + \tfrac{1}{2}x = 3$$

c. Add to each side of the equation the square of half the coefficient of the second term, that is, $(\tfrac{1}{4})^2 = \tfrac{1}{16}$. Thus the left-hand side of the equation becomes a perfect square.

$$x^2 + \tfrac{1}{2}x + \tfrac{1}{16} = 3\tfrac{1}{16} = \tfrac{49}{16}$$

d. Extract the square root of each side

$$x + \tfrac{1}{4} = \pm \tfrac{7}{4}$$

whence $x = -2$ or $+\tfrac{3}{2}$.

5. Rules for Extracting the Square Root of a Number

Problem: Extract the square root of 1,459.24.

a. Beginning at the decimal point, separate the number into periods of two figures each, counted left and right from the decimal point.

$$\overline{14}\ \overline{59.24}$$

b. Find the greatest integer (3) whose square is contained in the left-hand period.

$$\overline{14}\ \overline{59.24}(3$$

c. Subtract its square from the left-hand period and annex the next period, 59.

$$\overline{14}\ \overline{59}.\overline{24}(3$$
$$\underline{9}$$
$$\overline{5}\ \mathbf{59}$$

d. Double the root already found, and multiply by 10 to make a temporary divisor, 60.

$$\overline{14}\ \overline{59}.\overline{24}(3$$
$$\underline{9}$$
$$\overline{60|5\ 59}$$

e. Divide the remainder by this temporary divisor, and, when a quotient (8) has been decided upon, before multiplying, add the quotient to the temporary divisor, 60.

$$\overline{14}\ \overline{59}.\overline{24}(3$$
$$\underline{9}$$
$$60+8=\mathbf{68|5\ 59}$$

f. Annex the quotient (8) to the root already extracted, giving 38.

$$\overline{14}\ \overline{59}.\overline{24}(3\mathbf{8}$$
$$\underline{9}$$
$$\mathbf{68}|\overline{5\ 59}$$

g. Multiply the divisor (68) by the last figure just added to the square root (by 8 to get 544).

$$\overline{14\ 59.24}(38.\mathbf{2}$$
$$\underline{9}$$
$$68\overline{|5\ 59}$$
$$5\ 44$$
$$\mathbf{760+2=762}\,|\overline{1524}$$
$$|\overline{1524}$$

h. Subtract as in 3 and bring down the next period and continue as in 4, 5, 6, and 7. *Answer:* 38.2.

Alternatively, to approximate the square root, use the "cut-and-try" method: The square root of 1459.24 must be greater than 30 because $(30)^2 = 900$ and less than 40 because $(40)^2 = 1600$. Try 35; $(35)^2$ by multiplication is 1225. Try 38; $(38)^2 =$ 1444. Try 39; $(39)^2 = 1521$. Thus, the square root of 1459.24 lies between 38 and 39 and to two digits is 38.

6. Hints for the Solution of Algebra Problems

a. Starting with an equation with one unknown, x, get x alone on one side of the equation.

Examples:

$$\frac{4}{7} = \frac{3}{x+5}$$

Multiply both sides by 7:

$$\frac{4 \times 7}{7} = \frac{3 \times 7}{x+5}$$

Multiply both sides by $(x+5)$:

$$4(x+5) = 3 \times 7, \quad 4x+20 = 21$$
$$x = \frac{21-20}{4} = \frac{1}{4}$$

b. One may perform any algebraic operation upon an equation, such as addition, subtraction, multiplication, or division, provided that it is performed on both sides of the equation.

c. If there are two unknowns, x and y, there must be two (simultaneous) equations. Solve each of these equations for one of the unknowns, say y; then equate these two solutions.

Example:

(1) $\qquad 3x + y = 7x + 3$

(2) $\qquad 4y + 5 = 2x$

(1′) $\qquad y = 7x + 3 - 3x = 4x + 3$

(2′) $\qquad y = \dfrac{2x - 5}{4}$

From equations 1′ and 2′,

$$4x + 3 = \frac{2x - 5}{4}$$
$$16x + 12 = 2x - 5$$
$$14x = -17$$
$$x = -\tfrac{17}{14}$$

Substitute x in equation 2 to yield $y = -\tfrac{26}{14} = -\tfrac{13}{7}$.

7. Concerning Triangles

a. Area of a triangle $= \frac{1}{2}$ (Base $\times$ Altitude).

b. Two triangles are similar if their corresponding angles are equal. The corresponding sides of similar triangles are proportional.

c. **Theorem of Pythagoras:** *The sum of the squares of the sides of a right triangle is equal to the square of the hypotenuse.*

In physics many problems arise involving the solution of right triangles. If enough data are given for any triangle to permit drawing it, a solution is possible. One method, though a slow one, is to draw the triangle accurately to scale and then measure the side or angle desired.

Many problem types can be adequately illustrated by restricting the problems to a few selected right triangles. This has been done whenever possible in the present book, and you should memorize the ratio of the sides or refer frequently to the following summary.

d. Right triangles.

(1) The 30°-60° right triangle. This can be remembered by visualizing the 30°-60° right triangle as half of an equilateral triangle. In Fig. 1*a*, the ratio of the short side *bc* to the hypotenuse *ab* is 0.5. The Pythagorean theorem gives $\sqrt{3}$ cm for the base *ac*.

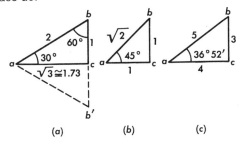

Fig. 1.

(2) The 45° right triangle. Visualize this as one-half of a square (Fig. 1*b*).

(3) The 3-4-5 right triangle. See Fig. 1*c*.

The elegant method of solving any triangle is to use trigonometric tables which systematically list the ratios of the sides of all possible right triangles. In any right-angled triangle, *ABC* (Fig. 2):

The sine of the angle θ is

$$\frac{\text{the "opposite" side, } BC}{\text{the hypotenuse, } AB} = \frac{o}{h}$$

The cosine of θ is $\dfrac{\text{the "adjacent" side, } AC}{\text{the hypotenuse, } AB} = \dfrac{a}{h}$

The tangent of θ is

$$\frac{\text{the "opposite" side, } BC}{\text{the "adjacent" side, } AC} = \frac{o}{a} = \frac{\text{sine } \theta}{\text{cosine } \theta}$$

The cotangent of θ is

$$\frac{\text{the "adjacent" side, } AC}{\text{the "opposite" side, } BC} = \frac{a}{o} = \frac{\text{cosine } \theta}{\text{sine } \theta}$$

From the ratios of the 30°, 60°, and 45° right triangles above, the following values can be deduced immediately. A table of sines, cosines, tangents, and cotangents appears on pages 785 and 786.

Angle	30°	36°52′	60°	45°
Sine	0.500	0.600	$0.866 = \frac{1}{2}\sqrt{3}$	$0.707 = \frac{1}{2}\sqrt{2}$
Cosine	$0.866 = \frac{1}{2}\sqrt{3}$	0.800	0.500	$0.707 = \frac{1}{2}\sqrt{2}$
Tangent	$0.577 = \frac{1}{3}\sqrt{3}$	0.750	$1.73 = \sqrt{3}$	1.00

e. Slope. The slope of a straight line is the tangent of the angle made by the line with the *x*-axis. Let (x_1, y_1) and (x_2, y_2) be any two points on the line; the *slope* is the ratio $(y_2 - y_1)/(x_2 - x_1)$ and equals tan θ, where θ is the angle between the line and the *x*-axis. The slope of a curve at a point is the slope of the line drawn tangent to the curve at that point.

References: Rees, Paul K., and Fred W. Sparks, *Algebra and Trigonometry,* McGraw-Hill Book Company, New York, 1962.

Vance, Elbridge P., *Modern College Algebra,* Addison Wesley Publishing Company, Reading, Mass., 1962.

8. Logarithms

a. The logarithm of a number to the base 10 is the power to which 10 must be raised to equal that number. Thus $100 = 10^2$; hence $\log_{10} 100 = 2$.

(1) To find the logarithm of a number, $N = 4567$:

    (a) Express it as a number between 1 and 10, multiplied by a power *c* of 10. $N = 4.567 \times 10^3$; hence $c = 3$.

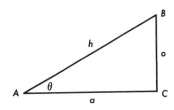

Fig. 2.

    (b) In the table on page 787 find the logarithm of N, independent of the power of 10 by which it is multiplied. Log 4.567 $\cong$ 0.6597.

    (c) Add to this the exponent c of 10.

              Log $N = 3.6597$.

  (2) To find the "antilogarithm" N of a given logarithm; log $N = 3.4567$:

    (a) Look in the table for the number corresponding to the decimal portion of the logarithm; that is, $0.4567 = 0.4564 + 0.0003 \cong$ log 2.862.

    (b) Multiply this by 10^c, in which c is the integer of the logarithm; that is, $N = 10^3 \times 2.862 = 2862$.

  (3) Use of logarithms in computations:

    (a) Multiplication: log $(M \times N) =$ log $M +$ log N.

    (b) Division: log $M/N =$ log $M -$ log N.

    (c) Raising to a power: log $M^N = N \times$ log M.

    (d) Extracting a root: log $\sqrt[N]{M} = (1/N) \times$ log M.

  b. The natural or Napierian logarithm of a number N is the logarithm of N to the base e where $e \cong 2.72$. The natural logarithm, written ln N, is the power to which e must be raised to equal N. Thus if $e^y = N$, $y =$ ln N. It can be shown that ln $N \cong 2.30$ $\log_{10} N$. Hence, you can find the (approximate) natural logarithm of a number by multiplying its logarithm to the base 10 by 2.30.

9. Three-Place Table of Sines, Cosines, Tangents, and Cotangents

Angles	Sines	Cosines	Tangents	Cotangents	
0° 00′	0.000	1.00	0.000		90° 00′
30	0.009	1.00	0.009	115	30
1° 00′	0.018	1.00	0.018	57.3	89° 00′
30	0.026	1.00	0.026	38.2	30
2° 00′	0.035	0.999	0.035	28.6	88° 00′
30	0.044	0.999	0.044	22.9	30
3° 00′	0.052	0.999	0.052	19.1	87° 00′
30	0.061	0.998	0.061	16.4	30
4° 00′	0.070	0.998	0.070	14.3	86° 00′
30	0.078	0.997	0.079	12.7	30
5° 00′	0.087	0.996	0.088	11.4	85° 00′
30	0.096	0.995	0.096	10.4	30
6° 00′	0.104	0.994	0.105	9.51	84° 00′
30	0.113	0.994	0.114	8.78	30
7° 00′	0.122	0.992	0.123	8.14	83° 00′
30	0.130	0.991	0.132	7.60	30
8° 00′	0.139	0.990	0.140	7.12	82° 00′
30	0.148	0.989	0.150	6.69	30
9° 00′	0.156	0.988	0.158	6.31	81° 00′
30	0.165	0.986	0.167	5.98	30
10° 00′	0.174	0.985	0.176	5.67	80° 00′
30	0.182	0.983	0.185	5.40	30
11° 00′	0.191	0.982	0.194	5.14	79° 00′
30	0.199	0.980	0.204	4.92	30
12° 00′	0.208	0.978	0.213	4.70	78° 00′
30	0.216	0.976	0.222	4.51	30
13° 00′	0.225	0.974	0.231	4.33	77° 00′
30	0.233	0.972	0.240	4.17	30
14° 00′	0.242	0.970	0.249	4.01	76° 00′
30	0.250	0.968	0.259	3.87	30
15° 00′	0.259	0.966	0.268	3.73	75° 00′
30	0.267	0.964	0.277	3.61	30
16° 00′	0.276	0.961	0.287	3.49	74° 00′
30	0.284	0.959	0.296	3.38	30
17° 00′	0.292	0.956	0.306	3.27	73° 00′
30	0.301	0.954	0.315	3.17	30
18° 00′	0.309	0.951	0.325	3.08	72° 00′
30	0.317	0.948	0.335	2.99	30
19° 00′	0.326	0.946	0.344	2.90	71° 00′
30	0.334	0.943	0.354	2.82	30
20° 00′	0.342	0.940	0.364	2.75	70° 00′
30	0.350	0.937	0.374	2.67	30
21° 00′	0.358	0.934	0.384	2.61	69° 00′
30	0.366	0.930	0.394	2.54	30
22° 00′	0.375	0.927	0.404	2.48	68° 00′
30	0.383	0.924	0.414	2.41	30
	Cosines	Sines	Cotangents	Tangents	Angles

9. Three-Place Table of Sines, Cosines, Tangents, and Cotangents (Continued)

Angles	Sines	Cosines	Tangents	Cotangents	
23° 00′	0.391	0.920	0.424	2.36	67° 00′
30	0.399	0.917	0.435	2.30	30
24° 00′	0.407	0.914	0.445	2.25	66° 00′
30	0.415	0.910	0.456	2.19	30
25° 00′	0.423	0.906	0.466	2.14	65° 00′
30	0.430	0.903	0.477	2.10	30
26° 00′	0.438	0.899	0.488	2.05	64° 00′
30	0.446	0.895	0.499	2.01	30
27° 00′	0.454	0.891	0.510	1.96	63° 00′
30	0.462	0.887	0.521	1.92	30
28° 00′	0.470	0.883	0.532	1.88	62° 00′
30	0.477	0.879	0.543	1.84	30
29° 00′	0.485	0.875	0.554	1.80	61° 00′
30	0.492	0.870	0.566	1.77	30
30° 00′	0.500	0.866	0.577	1.73	60° 00′
30	0.508	0.862	0.589	1.70	30
31° 00′	0.515	0.857	0.601	1.66	59° 00′
30	0.522	0.853	0.613	1.63	30
32° 00′	0.530	0.848	0.625	1.60	58° 00′
30	0.537	0.843	0.637	1.57	30
33° 00′	0.545	0.839	0.649	1.54	57° 00′
30	0.552	0.834	0.662	1.51	30
34° 00′	0.559	0.829	0.674	1.48	56° 00′
30	0.566	0.824	0.687	1.46	30
35° 00′	0.574	0.819	0.700	1.43	55° 00′
30	0.581	0.814	0.713	1.40	30
36° 00′	0.588	0.809	0.726	1.38	54° 00′
30	0.595	0.804	0.740	1.35	30
37° 00′	0.602	0.799	0.754	1.33	53° 00′
30	0.609	0.793	0.767	1.30	30
38° 00′	0.616	0.788	0.781	1.28	52° 00′
30	0.622	0.783	0.795	1.26	30
39° 00′	0.629	0.777	0.810	1.23	51° 00′
30	0.636	0.772	0.824	1.21	30
40° 00′	0.643	0.766	0.839	1.19	50° 00′
30	0.649	0.760	0.854	1.17	30
41° 00′	0.656	0.755	0.869	1.15	49° 00′
30	0.663	0.749	0.885	1.13	30
42° 00′	0.669	0.743	0.900	1.11	48° 00′
30	0.676	0.737	0.916	1.09	30
43° 00′	0.682	0.731	0.932	1.07	47° 00′
30	0.688	0.725	0.949	1.05	30
44° 00′	0.695	0.719	0.966	1.04	46° 00′
30	0.701	0.713	0.983	1.02	30
45° 00′	0.707	0.707	1.00	1.00	45° 00′
	Cosines	Sines	Cotangents	Tangents	Angles

10. Logarithms to Base Ten

Num-bers	0	1	2	3	4	5	6	7	8	9	1	2	3	4	5	6	7	8	9
											Proportional Parts								
10	0000	0043	0086	0128	0170	0212	0253	0294	0334	0374	4	8	12	17	21	25	29	33	37
11	0414	0453	0492	0531	0569	0607	0645	0682	0719	0755	4	8	11	15	19	23	26	30	34
12	0792	0828	0864	0899	0934	0969	1004	1038	1072	1106	3	7	10	14	17	21	24	28	31
13	1139	1173	1206	1239	1271	1303	1335	1367	1399	1430	3	6	10	13	16	19	23	26	29
14	1461	1492	1523	1553	1584	1614	1644	1673	1703	1732	3	6	9	12	15	18	21	24	27
15	1761	1790	1818	1847	1875	1903	1931	1959	1987	2014	3	6	8	11	14	17	20	22	25
16	2041	2068	2095	2122	2148	2175	2201	2227	2253	2279	3	5	8	11	13	16	18	21	24
17	2304	2330	2335	2380	2405	2430	2455	2480	2504	2529	2	5	7	10	12	15	17	20	22
18	2553	2577	2601	2625	2648	2672	2695	2718	2742	2765	2	5	7	9	12	14	16	19	21
19	2788	2810	2833	2856	2878	2900	2923	2945	2967	2989	2	4	7	9	11	13	16	18	20
20	3010	3032	3054	3075	3096	3118	3139	3160	3181	3201	2	4	6	8	11	13	15	17	19
21	3222	3243	3263	3284	3304	3324	3345	3365	3385	3404	2	4	6	8	10	12	14	16	18
22	3424	3444	3464	3483	3502	3522	3541	3560	3579	3598	2	4	6	8	10	12	14	15	17
23	3617	3636	3655	3674	3692	3711	3729	3747	3766	3784	2	4	6	7	9	11	13	15	17
24	3802	3820	3838	3856	3874	3892	3909	3927	3945	3962	2	4	5	7	9	11	12	14	16
25	3979	3997	4014	4031	4048	4065	4082	4099	4116	4133	2	3	5	7	9	10	12	14	15
26	4150	4166	4183	4200	4216	4232	4249	4265	4281	4298	2	3	5	7	8	10	11	13	15
27	4314	4330	4346	4362	4378	4393	4409	4425	4440	4456	2	3	5	6	8	9	11	13	14
28	4472	4487	4502	4518	4533	4548	4564	4579	4594	4609	2	3	5	6	8	9	11	12	14
29	4624	4639	4654	4669	4683	4698	4713	4728	4742	4757	1	3	4	6	7	9	10	12	13
30	4771	4786	4800	4814	4829	4843	4857	4871	4886	4900	1	3	4	6	7	9	10	11	13
31	4914	4928	4942	4955	4969	4983	4997	5011	5024	5038	1	3	4	6	7	8	10	11	12
32	5051	5065	5079	5092	5105	5119	5132	5145	5159	5172	1	3	4	5	7	8	9	11	12
33	5185	5198	5211	5224	5237	5250	5263	5276	5289	5302	1	3	4	5	6	8	9	10	12
34	5315	5328	5340	5353	5366	5378	5391	5403	5416	5428	1	3	4	5	6	8	9	10	11
35	5441	5453	5465	5478	5490	5502	5514	5527	5539	5551	1	2	4	5	6	7	9	10	11
36	5563	5575	5587	5599	5611	5623	5635	5647	5658	5670	1	2	4	5	6	7	8	10	11
37	5682	5694	5705	5717	5729	5740	5752	5763	5775	5786	1	2	3	5	6	7	8	9	10
38	5798	5809	5821	5832	5843	5855	5866	5877	5888	5899	1	2	3	5	6	7	8	9	10
39	5911	5922	5933	5944	5955	5966	5977	5988	5999	6010	1	2	3	4	5	7	8	9	10
40	6021	6031	6042	6053	6064	6075	6085	6096	6107	6117	1	2	3	4	5	6	8	9	10
41	6128	6138	6149	6160	6170	6180	6191	6201	6212	6222	1	2	3	4	5	6	7	8	9
42	6232	6243	6253	6263	6274	6284	6294	6304	6314	6325	1	2	3	4	5	6	7	8	9
43	6335	6345	6355	6365	6375	6385	6395	6405	6415	6425	1	2	3	4	5	6	7	8	9
44	6435	6444	6454	6464	6474	6484	6493	6503	6513	6522	1	2	3	4	5	6	7	8	9
45	6532	6542	6551	6561	6571	6580	6590	6599	6609	6618	1	2	3	4	5	6	7	8	9
46	6628	6637	6646	6656	6665	6675	6684	6693	6702	6712	1	2	3	4	5	6	7	7	8
47	6721	6730	6739	6749	6758	6767	6776	6785	6794	6803	1	2	3	4	5	5	6	7	8
48	6812	6821	6830	6839	6848	6857	6866	6875	6884	6893	1	2	3	4	4	5	6	7	8
49	6902	6911	6920	6928	6937	6946	6955	6964	6972	6981	1	2	3	4	4	5	6	7	8
50	6990	6998	7007	7016	7024	7033	7042	7050	7059	7067	1	2	3	3	4	5	6	7	8
51	7076	7084	7093	7101	7110	7118	7126	7135	7143	7152	1	2	3	3	4	5	6	7	8
52	7160	7168	7177	7185	7193	7202	7210	7218	7226	7235	1	2	2	3	4	5	6	7	7
53	7243	7251	7259	7267	7275	7284	7292	7300	7308	7316	1	2	2	3	4	5	6	6	7
54	7324	7332	7340	7348	7356	7364	7372	7380	7388	7396	1	2	2	3	4	5	6	6	7

10. Logarithms to Base Ten (Continued)

Num-bers	0	1	2	3	4	5	6	7	8	9	Proportional Parts								
											1	2	3	4	5	6	7	8	9
55	7404	7412	7419	7427	7435	7443	7451	7459	7466	7474	1	2	2	3	4	5	5	6	7
56	7482	7490	7497	7505	7513	7520	7528	7536	7543	7551	1	2	2	3	4	5	5	6	7
57	7559	7566	7574	7582	7589	7597	7604	7612	7619	7627	1	2	2	3	4	5	5	6	7
58	7634	7642	7649	7657	7664	7672	7679	7686	7694	7701	1	1	2	3	4	4	5	6	7
59	7709	7716	7723	7731	7738	7745	7752	7760	7767	7774	1	1	2	3	4	4	5	6	7
60	7782	7789	7796	7803	7810	7818	7825	7832	7839	7846	1	1	2	3	4	4	5	6	6
61	7853	7860	7868	7875	7882	7889	7896	7903	7910	7917	1	1	2	3	4	4	5	6	6
62	7924	7931	7938	7945	7952	7959	7966	7973	7980	7987	1	1	2	3	3	4	5	6	6
63	7993	8000	8007	8014	8021	8028	8035	8041	8048	8055	1	1	2	3	3	4	5	6	6
64	8062	8069	8075	8082	8089	8096	8102	8109	8116	8122	1	1	2	3	3	4	5	5	6
65	8129	8136	8142	8149	8156	8162	8169	8176	8182	8189	1	1	2	3	3	4	5	5	6
66	8195	8202	8209	8215	8222	8228	8235	8241	8248	8254	1	1	2	3	3	4	5	5	6
67	8261	8267	8274	8280	8287	8293	8299	8306	8312	8319	1	1	2	3	3	4	5	5	6
68	8325	8331	8338	8344	8351	8357	8363	8370	8376	8382	1	1	2	3	3	4	4	5	6
69	8388	8395	8401	8407	8414	8420	8426	8432	8439	8445	1	1	2	2	3	4	4	5	6
70	8451	8457	8463	8470	8476	8482	8488	8494	8500	8506	1	1	2	2	3	4	4	5	6
71	8513	8519	8525	8531	8537	8543	8549	8555	8561	8567	1	1	2	2	3	4	4	5	5
72	8573	8579	8585	8591	8597	8603	8609	8615	8621	8627	1	1	2	2	3	4	4	5	5
73	8633	8639	8645	8651	8657	8663	8669	8675	8681	8686	1	1	2	2	3	4	4	5	5
74	8692	8698	8704	8710	8716	8722	8727	8733	8739	8745	1	1	2	2	3	4	4	5	5
75	8751	8756	8762	8768	8774	8779	8785	8791	8797	8802	1	1	2	2	3	3	4	5	5
76	8808	8814	8820	8825	8831	8837	8842	8848	8854	8859	1	1	2	2	3	3	4	5	5
77	8865	8871	8876	8882	8887	8893	8899	8904	8910	8915	1	1	2	2	3	3	4	4	5
78	8921	8927	8932	8938	8943	8949	8954	8960	8965	8971	1	1	2	2	3	3	4	4	5
79	8976	8982	8987	8993	8998	9004	9009	9015	9020	9025	1	1	2	2	3	3	4	4	5
80	9031	9036	9042	9047	9053	9058	9063	9069	9074	9079	1	1	2	2	3	3	4	4	5
81	9085	9090	9096	9101	9106	9112	9117	9122	9128	9133	1	1	2	2	3	3	4	4	5
82	9138	9143	9149	9154	9159	9165	9170	9175	9180	9186	1	1	2	2	3	3	4	4	5
83	9191	9196	9201	9206	9212	9217	9222	9227	9232	9238	1	1	2	2	3	3	4	4	5
84	9243	9248	9253	9258	9263	9269	9274	9279	9284	9289	1	1	2	2	3	3	4	4	5
85	9294	9299	9304	9309	9315	9320	9325	9330	9335	9340	1	1	2	2	3	3	4	4	5
86	9345	9350	9355	9360	9365	9370	9375	9380	9385	9390	1	1	2	2	3	3	4	4	5
87	9395	9400	9405	9410	9415	9420	9425	9430	9435	9440	0	1	1	2	2	3	3	4	4
88	9445	9450	9455	9460	9465	9469	9474	9479	9484	9489	0	1	1	2	2	3	3	4	4
89	9494	9499	9504	9509	9513	9518	9523	9528	9533	9538	0	1	1	2	2	3	3	4	4
90	9542	9547	9552	9557	9562	9566	9571	9576	9581	9586	0	1	1	2	2	3	3	4	4
91	9590	9595	9600	9605	9609	9614	9619	9624	9628	9633	0	1	1	2	2	3	3	4	4
92	9638	9643	9647	9652	9657	9661	9666	9671	9675	9680	0	1	1	2	2	3	3	4	4
93	9685	9689	9694	9699	9703	9708	9713	9717	9722	9727	0	1	1	2	2	3	3	4	4
94	9731	9736	9741	9745	9750	9754	9759	9763	9768	9773	0	1	1	2	2	3	3	4	4
95	9777	9782	9786	9791	9795	9800	9805	9809	9814	9818	0	1	1	2	2	3	3	4	4
96	9823	9827	9832	9836	9841	9845	9850	9854	9859	9863	0	1	1	2	2	3	3	4	4
97	9868	9872	9877	9881	9886	9890	9894	9899	9903	9908	0	1	1	2	2	3	3	4	4
98	9912	9917	9921	9926	9930	9934	9939	9943	9948	9952	0	1	1	2	2	3	3	4	4
99	9956	9961	9965	9969	9974	9978	9983	9987	9991	9996	0	1	1	2	2	3	3	3	4

E. Exponential Equations

An important equation that often occurs in physics is of the form

$$y = y_0 e^{-kx}$$

where y is the dependent variable, x the independent variable, y_0 and k constants, and e the base of natural or Napierian logarithms. (See page 783. e is equal approximately to 2.72.) y is said to decrease *exponentially* as x increases, and such an equation is said to describe an exponential decrease or decay.

1. Decay.

Let us take as an example the decay of a radioactive substance. The equation describing the decay is

$$N = N_0 e^{-\lambda t}$$

where N is the number of nuclei that have *not* disintegrated in a time t, N_0 is the number of nuclei at time $t = 0$, and λ is the decay constant characteristic of the substance.

To derive this equation, we assume that the number of nuclei ΔN that disintegrate in a time Δt is proportional to two quantities: (1) the number of nuclei that exist at the beginning of the interval and (2) the length of the time interval Δt. Thus we have

$$\Delta N = -\lambda N \, \Delta t$$

where λ is a proportionality constant and the negative sign means that nuclei are disintegrating and their number N is decreasing. We then have

$$\frac{\Delta N}{N} = -\lambda \, \Delta t$$

or in the terminology of the calculus

$$\frac{dN}{N} = -\lambda \, dt$$

and

$$\int_{N_0}^{N} \frac{dN}{N} = -\lambda \int_{t=0}^{t} dt$$

Upon integrating this we have

$$\ln N - \ln N_0 = -\lambda t$$

$$\ln \frac{N}{N_0} = -\lambda t$$

$$\frac{N}{N_0} = e^{-\lambda t}$$

Figure 3 is a graph of $\ln (N/N_0)$ versus t, plotted on semilogarithmic paper on which the y scale is logarithmic. Thus at $t = 0$, $N = N_0$, $N/N_0 = 1$. At $t = 1.0$ sec in this example, we notice on the graph that $N/N_0 = 0.5$. Therefore, the *half-life* is 1.0 sec for this substance because that is the time for half of the nuclei to disintegrate. In another half-life — at

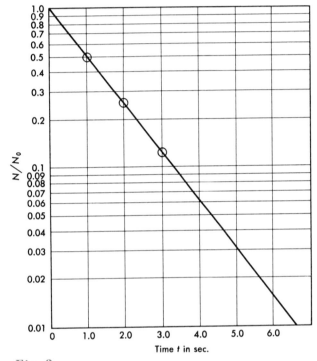

Fig. 3.

$t = 2.0$ sec — the fractional number of nuclei remaining is 0.25, and so on. The same kind of equation arises in:

a. The decrease of current in an inductive circuit (page 468):

$$i = i_0 e^{-(R/L)t}$$

where i is the current at time t after the switch is opened, i_0 the current at time $t = 0$, R the resistance, and L the self-inductance.

b. The decrease of current in charging a capacitor:

$$i = i_0 e^{-(1/RC)t}$$

where i is the current at time t after the charging has begun, i_0 the current at time $t = 0$, R the resistance, and C the capacitance.

c. The absorption of gamma rays in passing through an absorber:

$$I = I_0 e^{-\mu x}$$

where I_0 is the initial intensity of the gamma rays before they enter the absorber, I their intensity after they have traveled through a thickness x, and μ the absorption coefficient.

2. Growth.

Another common equation is that of exponential *growth* of the form

$$y = y_0(1 - e^{-kx})$$

Notice that when kx is zero, e^{-kx} is $e^0 = 1$, and

F. The Slide Rule

Use a slide rule whenever possible in solving physics problems. An inexpensive one is sufficiently accurate for most purposes and can save time and effort.

Figure 4 shows a simple device analogous to a slide rule to help you understand the principle involved. It consists of two ordinary rulers whose scales touch. Such a device is an analogue computer: distances measured along the scales correspond to numbers. For example, to compute the sum of 2 and 4, slide the top ruler along the lower one until the zero of the top ruler is in line with the number 2 on the lower one. Read the answer, 6, on the lower ruler under the number 4 on the upper one. To find the *difference* between, say, 5 and 3, one places the subtrahend (3) on the upper ruler over the diminuend (5) on the lower one and reads the answer (2) on the lower one under the zero on the upper ruler.

The slide rule is mechanically more convenient to use and serves for multiplication, division, squaring, taking square roots, and certain other operations. It has *logarithmic* scales: the distances are proportional to the logarithms of numbers to the base 10 (Fig. 5). By adding distances along the scales, the logarithms of numbers C and D are added, and the result corresponds to the *product* P of C and D, since

$$\log_{10} P = \log_{10} C + \log_{10} D.$$

Similarly, the quotient Q of C and D is obtained by

$y = 0$. Also, when kx is very large (5,362,179, for example), e^{-kx} is very small and $y = y_0$. So y increases exponentially from zero to y_0.

One example is the growth of current in a circuit containing an inductance and a resistance (page 468):

$$i = i_0(1 - e^{-(R/L)t})$$

where $t = 0$ is the time when the switch is closed.

Another example is the growth of the voltage across a capacitor that is charged by a battery through a capacitor (page 507):

$$V = V_0(1 - e^{-(1/RC)t})$$

where V_0 is the battery voltage.

subtracting the distance corresponding to $\log_{10} D$ from the corresponding to $\log_{10} C$ since

$$\log_{10} Q = \log_{10} C - \log_{10} D.$$

For example, to multiply *4.00* by *2.00*, first place the index *1* on the C scale over the multiplier *2.00* on the D scale (Fig. 5) and then slide the indicator along until its hairline is over the multiplicand *4.00* on the C scale. Read the answer — *8.00* — under the hairline on the D scale. Using scale subdivisions, which diminish in size from the left of the scale to the right, one can multiply other three-digit numbers, such as 4.18 and 2.07, following this same procedure. Note that either the left-hand or the right-hand index *1* can be used, depending on the magnitude of the product. If the product falls off the scale when one index is used, switch to the other one.

To divide *4.40* by *2.00*, place the indicator hairline over the dividend *4.40* on the D scale and place the divisor *2.00* on the C scale under the

hairline. Read the quotient — *2.20* — on the D scale under the index *1* on the C scale.

Express large numbers in scientific notation

before multiplying or dividing them on the slide rule. For example, 440,000 divided by 200 is 4.40 × 10^5/2.00 × 10^2. Using the slide rule, we find the quotient of 4.40 and 2.00 is 2.20, to which we annex the result of dividing 10^5 by 10^2, or 10^3: 2.20 × 10^3 is the answer. Always estimate the answer before making the calculation with the slide rule as a check upon the correctness of the final result. Thus

612,000 times 229 is about 6 × 10^5 times 2 × 10^2 or 12 × 10^7 or 1.2 × 10^8. The answer to three significant figures on the slide rule is 1.41 × 10^8.

In calculating (3.14 × 6.05)/(7.49 × 2.39), divide the first factor in the numerator – 3.14 – by the first factor in the denominator – 7.49 – and then multiply by the second factor in the numerator – 6.05 – and so on.

G. Dimensions

The magnitudes of physical quantities ordinarily are expressed by a *numeric* and a *unit*. For example, the width of a certain room is 3.09 meters, where 3.09 is the numeric and the meter is the unit. Exceptions are quantities that are ratios of two like quantities, such as the coefficient of sliding friction, which has no units.

sured in any other unit. Thus, if the length of the room is 6.18 meters, the ratio of the length to the width is 2.00 whether the length and width are both measured in meters, kilometers, centimeters, yards, or any other length unit.

The numerics of magnitudes expressed in the same units can be added or subtracted. Those of

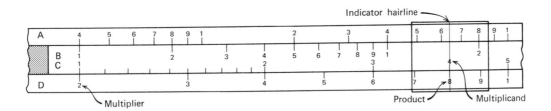

different physical quantities, or of magnitudes expressed in different units, cannot be so treated. One cannot add apples and oranges.

Secondary physical quantities are those that are functions of (determined by) primary quantities. Velocity is an example of a secondary physical quantity; it is the ratio of a length or distance to a time. Its units are length or distance units over time units. The units of acceleration are distance units over time units squared. One can multiply or divide units or raise them to powers to form the units of secondary quantities.

Any secondary quantity can be expressed as the product of fundamental quantities, each raised to the appropriate power. Such an expression is called the *dimensional formula* of the quantity. Letting M be the dimensional formula for mass, L for length, T for time, and Q for charge, we have the following examples of formulas of secondary quantities:

In mechanics, three physical quantities are *fundamental* or *primary: length, mass,* and *time.* In Chapter 1 we discussed the standard units of these fundamental quantities – the meter, the kilogram, and the second. We saw how a system of units, the mks system, was based on them. The system of units we use in electricity and magnetism includes a fourth fundamental quantity – *charge,* whose unit is the coulomb – and is called the mksq system.

A magnitude of a physical quantity can be expressed in other related units, of course. The width of the room referred to above is also 0.00309 kilometer and 309 centimeters. The numerics are always in inverse proportion to the size of the units used in such conversions. Since the kilometer equals 1000 meters, the numeric used to express a magnitude in kilometers is 1/1000 of that used to express the same magnitude in meters. Moreover, the ratio of two different lengths measured in one kind of unit is the same as their ratio when mea-

G. Dimensions (Cont.)

Quantity	Defining relationship	Dimensional formulas
Velocity	Distance/time	LT^{-1}
Acceleration	(Distance/time)/time	LT^{-2}
Force	Mass · acceleration	MLT^{-2}
Momentum	Mass · (distance/time)	MLT^{-1}
Energy	Force · distance	ML^2T^{-2}
Electric potential	Energy/charge	$ML^2T^{-2}Q^{-1}$

As an exercise, write the dimensional formulas for the following: area, pressure, torque, electric field strength, and permittivity.

Dimensional formulas are useful in checking on the correctness of equations. Each term of an equation in physics must reduce to the same dimensional formula. For example, is the equation

$$s = v_0 t + \tfrac{1}{2}at^2$$

dimensionally correct? Writing the dimensional formula for each term we have

$$L = (LT^{-1})T + (LT^{-2})(T^2)$$

or

$$L = L + L$$

and the equation checks dimensionally. Note, however, that dimensional analysis is not able to tell us what the constants should be: the factor $\tfrac{1}{2}$ in the equation must be obtained from other analysis. Use dimensional analysis to check new equations when you encounter them, either in derivations in this book or in problem solving.

H. Alternative units

1. Mechanical units.

The metric mks system uses a familiar unit of mass, the kilogram, and an unfamiliar unit of force, the newton. In the *fps system* of units, the unit of force is the familiar pound-weight. For a unit of mass the fps system uses the unfamiliar *slug*, which is defined as *the mass of a body which an unbalanced force of one pound-weight will accelerate one foot-per-second per second.*

From $F = ma$,

$$1 \text{ lb} = 1 \text{ slug} \times 1 \text{ ft/sec}^2$$

and

$$1 \text{ slug} = \frac{1 \text{ lb}}{1 \text{ ft/sec}^2}$$

Let us assume that $g = 32$ ft/sec². Then, since the weight w of a body of mass m is $w = mg$, the weight of a body whose mass is 1 slug is $w = 1$ slug $\times$ 32 ft/sec² = 32 slug-ft/sec² = 32 lb. *A body whose mass is 1 slug weighs 32 lb.*

To get the mass of a body in slugs, divide its weight in pound-weights by g in feet-per-second per second.

Example. What resultant force will accelerate a car weighing 3200 lb 4 ft/sec²?

$$m = \frac{w}{g} = \frac{3200 \text{ lb}}{32 \text{ ft/sec}^2} = 100 \text{ slugs}$$

$$F = ma = 100 \text{ slugs} \times 4 \text{ ft/sec}^2$$
$$= 400 \text{ slug-ft/sec}^2 = 400 \text{ lb}$$

2. Electrical units

a. Electrostatic Cgs System

One statcoulomb is that charge which in a vacuum exerts a force of 1 dyne on an equal charge at a distance of 1 centimeter.

One statvolt is a potential difference between points *A* and *B* such that 1 erg of work is required to transfer 1 statcoulomb from *A* to *B*. $V = W/Q$.

The electric field strength at a point is the force per unit charge at that point. (1 dyne/statcoul = 1 statv/cm.)

b. Electromagnetic Cgs System

One unit magnetic pole is a "point" pole that, in a vacuum, will exert a force of 1 dyne on an equal pole at a distance of 1 centimeter.

The magnetic field strength H at a point is the force per unit magnetic pole at that point. (1 oersted = 1 dyne/u.p.) A magnetic field strength of 1 oersted establishes a magnetic induction of 1 gauss in vacuum. (1 weber/m² = 10,000 gauss.)

One abampere is the current which, flowing in a circle of radius 1 centimeter, produces a field strength of 2π oersteds at its center. Alternatively, 1 abampere is the current which, in flowing through two long parallel wires 2 centimeters apart in a vacuum causes them to repel each other with a force of 1 dyne per centimeter of length.

One abvolt is 1 erg per abcoulomb and is a unit of potential difference.

One dyne/abcoulomb = 1 abvolt/cm, a unit of electric field strength.

c. Mksq Electrical Units and Their Equivalents in Other Systems

Quantity	Mks	Electrostatic cgs	Electro-magnetic cgs
	1 coulomb	$= 3 \times 10^9$ statcoul	$= 0.1$ abcoul
Current	1 ampere	$= 3 \times 10^9$ statamp	$= 0.1$ abamp
Potential Difference	1 volt	$= \frac{1}{300}$ statv	$= 10^8$ abv
Electric Field Strength	1 volt/cm	$= \frac{1}{300}$ statv/cm	$= 10^8$ abv/cm
Capacitance	1 farad	$= 9 \times 10^{11}$ statfarads	$= 10^{-9}$ abfarad
Inductance	1 henry	$= 1.11 \times 10^{-12}$ stathenry	$= 10^9$ abhenries
Resistance	1 ohm	$= 1.11 \times 10^{-12}$ statohm	$= 10^9$ abohms

I. The Periodic Table of the Elements

For artificially produced elements, the approximate atomic weight of the most stable isotope is given in brackets.

Group→	I	II	III	IV	V	VI	VII	VIII			O
Period 1, Series 1	1 H 1.00797										2 He 4.003
Period 2, Series 2	3 Li 6.939	4 Be 9.012	5 B 10.81	6 C 12.011	7 N 14.007	8 O 15.9994	9 F 19.00				10 Ne 20.183
Period 3, Series 3	11 Na 22.990	12 Mg 23.31	13 Al 26.98	14 Si 28.09	15 P 30.974	16 S 32.064	17 Cl 35.453				18 Ar 39.948
Period 4, Series 4	19 K 39.102	20 Ca 40.08	21 Sc 44.96	22 Ti 47.90	23 V 50.94	24 Cr 52.00	25 Mn 54.94	26 Fe 55.85	27 Co 58.93	28 Ni 58.71	
Period 4, Series 5	29 Cu 63.54	30 Zn 65.37	31 Ga 69.72	32 Ge 72.59	33 As 74.92	34 Se 78.96	35 Br 79.909				36 Kr 83.80
Period 5, Series 6	37 Rb 85.47	38 Sr 87.62	39 Y 88.905	40 Zr 91.22	41 Nb 92.91	42 Mo 95.94	43 Tc [98]	44 Ru 101.1	45 Rh 102.905	46 Pd 106.4	
Period 5, Series 7	47 Ag 107.870	48 Cd 112.40	49 In 114.82	50 Sn 118.69	51 Sb 121.75	52 Te 127.60	53 I 126.90				54 Xe 131.30
Period 6, Series 8	55 Cs 132.905	56 Ba 137.34	57–71 Lanthanide series[a]	72 Hf 178.49	73 Ta 180.95	74 W 183.85	75 Re 186.2	76 Os 190.2	77 Ir 192.2	78 Pt 195.09	
Period 6, Series 9	79 Au 196.97	80 Hg 200.59	81 Tl 204.37	82 Pb 207.19	83 Bi 208.98	84 Po [210]	85 At [210]				86 Rn [222]
Period 7, Series 10	87 Fr [223]	88 Ra [226]	89–103 Actinide series[b]	104 Rf [260]	105 Ha [260]						

[a] Lanthanide series:	57 La 138.91	58 Ce 140.12	59 Pr 140.91	60 Nd 144.24	61 Pm [147]	62 Sm 150.35	63 Eu 152.0	64 Gd 157.25	65 Tb 158.92	66 Dy 162.50	67 Ho 164.93	68 Er 167.26	69 Tm 168.93	70 Yb 173.04	71 Lu 174.97
[b] Actinide series:	89 Ac [227]	90 Th 232.04	91 Pa [231]	92 U 238.03	93 Np [237]	94 Pu [242]	95 Am [243]	96 Cm [247]	97 Bk [247]	98 Cf [251]	99 Es [254]	100 Fm [257]	101 Md [257]	102 No [255]	103 Lr [256]

J. Tables of Physical Properties and Physical Units in This Book

Subject Index

Coulomb, 345, 449
Coulomb, Charles A., 44, 344
Coulomb attraction, 643
Coulomb's law, 344 f, 365
Counter, Geiger, 735
  scintillation, 735
Counter emf, 471
Cream separator, 83
Critical angle, 546 f
Critical temperature, 263
Cross section, 755 f
Crystal lattices, 710
Crystal microphone, 504
Crystallites, 194
Crystals, 193 ff
  one-dimensional, 715 f
  tourmaline, 618
Crystallites, 194
Curie, 741
Curie, Marie and Pierre, 731
Curie temperature, 451
Curie-Joliot, Frederic and Irene, 733
Current, alternating, 470, 477 f
  Amperian, 451
  displacement, 518 f
  eddy, 473 f
  effective, 478
  electric, 339, 369 ff
  heating effect of, 401
  transient, 370
Current balance, 449
Current density, 376, 711
Curvature, radius of, 559
Curved mirrors, 559 ff
Curvilinear motion, 77
Cyclotron, 437

Daniell cell, 388
DaVinci, Leonardo, 44
Davisson, C. J., 665
Day, mean solar, 10
Death of heat, 280 f
DeBroglie, Louis, 659 f
DeBroglie wavelength, 660, 677
DeBroglie waves, 659 f
Decay, radioactive, 741
Decibel, 323
Declination, 424
Dee, 437
Degradation of energy, 280 f
Degrees of freedom, 248
Delta particle, 764
Democritus, 189
Demodulation, 503
Density, current, 376, 711

mass, 16, 245
  radial probability, 686
  weight, 15 f
Detection, 503
Detectors, 733 f
Deuterium, 337
Dew point, 260
Dewar, James, 230
Dewar flask, 230
Diamagnetic, 451
Diatomic gas, 248
Dielectric breakdown, 718
Dielectric constant, 365
Dielectrics, 363 f
Diesel engine, 276
Differential diffusion, 243
Differential expansion, 216
Diffraction, 303, 599 ff
  electron wave, 666
  Fraunhofer, 604 f
  Fresnel, 606
  single-slit, 604 f
  X-ray, 648 ff
Diffraction grating, 606
Diffraction pattern, powder, 649 f
Diffuse reflection, 540
Diffusion, differential, 243
  of gases, 242 f, 247
Diffusion pump, 239
Dilation, time, 153 ff
Diode, 494 f
Diopter, 554
Dip, 424
Dipole, electric, 364
  magnetic, 419, 450
  oscillating, 523
Dirac, P. A. M., 656, 695
Direct-current motor, 433 f
Disintegration constant, 742
Dislocations, 198 f
Dispersion, 583 ff
Displacement, angular, 130
Displacement current, 518 f
Distance of distinct vision, 572
Diverging lens, 557 f
Diverging mirror, 561 f
Docking in orbit, 177
Domains, magnetic, 451
Donor level, 720
Doppler effect, 321, 591 f
Dosimeter, 338 f, 734
Double refraction, 619
Doubtful figures, 10
Drag, 40 f, 118
Drude, P., 711

Dry cell, 389
Dyne, 15, 68
Dyne-centimeter, 92

Ear, human, 323
Earth, as a magnet, 423 f
  atmosphere of, 26
  mass of, 179
Eddy currents, 473 f
Edison, Thomas, 404 f, 408
Edison lamp, 404 f
Effective current, 478
Effective emf, 470
Effective voltage, 478
Efficiency, luminous, 529
  of heat engines, 276 f
  thermal, 277
Einstein, Albert, 152, 181 f, 613, 635,
    652, 656
Einstein-deHaas effect, 451
Einstein's equation, 161 ff
Elastic collision, 102 f
Elastic limit, 196 f
Elasticity, 196
Electric charges, 335 f
Electric current, 339, 369 ff
Electric field, 352
Electric field strength, 352 f, 711
Electric lines, 354
Electrical heating, 405 f
Electrical inertia, 467
Electrical power, 400
Electrical resonance, 485
Electrocardiogram, 394 f
Electrochemical equivalent, 378
Electrodynamics, quantum, 693
Electrolysis, 376 f
Electrolyte, 377
Electromagnetic flowmeter, 464
Electromagnetic induction, 458 f
Electromagnetic interactions, 762
Electromagnetic spectrum, 524 ff
Electromagnetic waves, 517 ff, 585 f
Electrometer, 634
Electromotive force, 390 f, 402 ff
  alternating, 470, 477
  induced, 458
Electron, 336
  charge on, 359 f, 436
  energy of, 642 ff
  mass of, 436
    potential energy of, 678
    refraction of, 677 f
    sources of, 492
    spin of, 694

Electron emission, 492 f
Electron gas, 373, 710
Electron microscope, 672 f
Electron multiplier tube, 735
Electron scattering, 754
Electron volt, 401
Electron waves, 666, 713 f
  diffraction of, 666
Electronegative substance, 389
Electronic devices, 491 f
Electronics, 491 ff
Electrophorus, 342
Electrophysiology, 393 ff
Electropositive substance, 389
Electroscope, 337 f, 343 f, 734
Electrostatics, 335 ff
Elements, 190
Elongation, 196
Emf, alternating, 470, 477
  average, 470
  counter, 471
  effective, 470
  maximum, 470
Emission, electron, 492 f
  stimulated, 728
  thermionic, 721 ff
Emittance, 229
Emulsion plate, nuclear, 739
Endothermic reaction, 754
Energy, 94 ff
  binding, 752
  conservation of, 97 ff
  degradation of, 280 f
  of electron, 642 ff
  excitation, 700 ff
  ionization, 702 f
  kinetic, 94 f
  potential, 96
  rest, 162
  of satellite, 175 f
  of sun, 281
  of vibrating body, 292
Energy bands, 717
Energy gap, 716
Energy levels, particle, 680
Energy transformations, 97
Engine, Carnot, 277 ff
  Diesel, 276
  heat, 274 ff
  internal-combustion, 274 f
  steam, 274 f
Equilibrant, 35
Equilibrium, 16, 35 f
  conditions of, 126
  neutral, 129

  stable, 129
  unstable, 129
Equivalence, principle of, 182
Erg, 92
Escape velocity, 176 f
Eta particle, 764
Ether, 152
Evaporation, 256, 259
Excitation energy, 700 ff
Exclusion principle, Pauli, 696
Exothermic reaction, 754
Expansion, adiabatic, 278
  differential, 216
  isothermal, 278
  linear, 215
  volume, 216 f
Expansivity, thermal, 215
Eye, human, 570 f
Eyepiece, 574

f-value, 568
Fahrenheit scale, 212
Falling bodies, 58 ff
Farad, 361
Faraday, 378
Faraday, Michael, 356, 404, 458
Faraday's law, 378
Fast neutron, 758
Feedback, 501, 510
Fermi, 8, 753
Fermi, Enrico, 717
Fermi level, 717
Ferrites, 454
Ferromagnetic, 451
Field, electric, 352
  magnetic, 415
  periodic, 713
Field intensity, magnetizing, 452
Field strength, electric, 352 f, 711
  gravitational, 352
Figures, significant, 10 f
Films, surface, 202
Filter circuit, 495 f
Fission, 163, 757 f
Flow, laminar, 114 ff
  tubes of, 110
  turbulent, 114
Flowmeter, electromagnetic, 464
Fluid mechanics, 109
Fluorescence, 647
Flux, 526
  luminous, 527 f
  magnetic, 423, 458
Flux density, magnetic, 423, 462
Fluxmeter, 462

Focal length, 553
Focus, principal, 552, 560
Foot-pound, 92 f
Forbidden bands, 716
Force, 14 f, 33 ff, 99 f
  centrifugal, 87
  centripetal, 82 ff
  magnetic, 420 ff, 429
Force, moment of, 125
  normal, 43
  nuclear, 762 ff
  restoring, 287
Forces, addition of, 34
  molecular, 189 ff, 201 f
Foucault, Jean, 531
Fourier analysis, 327
Frame, coordinate, 145
  of reference, 145, 148 f
Franck, James, 700
Franklin, Benjamin, 348
Fraunhofer diffraction, 604 f
Fraunhofer lines, 590
Free-body diagram, 70
Freedom, degrees of, 248
Freezing, 253 f
Frequency, 287 f
Frequency coherence, 604
Frequency modulation, 504
Fresnel diffraction, 606
Friction, 42 ff
  coefficient of, 43
  kinetic, 43
  rolling, 47
  sliding, 43, 46
  starting, 44 f
  static, 44
Fulcrum, 124
Function, wave, 671, 680 f, 686
Fundamental, 308 f, 320
Fuse, 402
Fusion, 163, 757 f
  heat of, 253

G, 173
g, 58, 179 f
Galaxies, 592
Galileo, Galilei, 27, 58 f, 169, 189 f,
  210 f, 237, 530
Galvani, Luigi, 388
Galvanometer, 431 f
  a-c, 485 f
Gamma rays, 732, 746
Gas, diatomic, 248
  electron, 373, 710
  ideal, 240

monatomic, 248
state of, 240 f
Gas gauge, 237 f
Gas law, general, 242
Gas thermometer, 212
Gasses, diffusion of, 242 f, 247
liquefaction of, 263 f
Gauge, Bourdon, 238
gas, 237 f
pressure, 238
thermocouple, 239 f
Gauss's law, 354 f
Geiger counter, 735
Generator, 469 ff
Van de Graaff, 343
Gerlach, W., 705
Germanium, pure, 719
Germer, L. H., 665 f
Gilbert, William, 335, 413, 416, 423 f
Glancing angle, 649
Glass, nonreflecting, 611
opera, 577
Glider, 67
Goudsmit, S. A., 694
Gram, 9
Gram-weight, 15
Grating, diffraction, 606
reflection, 607
transmission, 607
Grating space, 649
Gravitation, 167, 172
law of, 172
Gravitational collapse, 762
Gravitational constant, 173
Gravitational field strength, 352
Gravitational interactions, 762
Gravitational mass, 181
Gravitational potential energy, 96
Gravity, acceleration of, 58, 179 f
center of, 127 f
specific, 16
Gray, Elisha, 464
Gray, Stephen, 339
Grid, 497
Grid bias voltage, 498
Ground state, 698
Grounding, 349
Group velocity, 662
Guericke, Otto von, 25
Gyroscope, 137 f, 692 f
Gyroscopic compass, 138

Hahn, 733
Half-life, 741 f
Hall coefficient, 711 f

Hall effect, 712
Hallwachs, 634
Harmonic, 309, 320 f
Harmony, 326
Hearing, 323
Heat, 221 ff
of combustion, 223
of fusion, 253
mechanical equivalent of, 271
molecular, 248, 273 f
quantity of, 221
specific, 221 f, 247 f
of vaporization, 259 f
and work, 270 f
Heat engine, 274 ff
efficiency of, 276 f
Heat insulation, 227
Heat pump, 279
Heat transfer, 224 ff
Heating, electrical, 405 f
Heating effect of current, 401
Heisenberg, Werner, 667, 679
Helmholtz theory, 594
Hemispheres, Magdeburg, 25 f
Henry, 466
Hertz, 287, 501
Hertz, G., 700
Hertz, Heinrich, 518, 633
Hess, Victor, 732 f, 760
High-resistance direction, 724
Hofstadter, R., 754
Hole, 721, 746
Hooke, Robert, 293
Hooke's law, 288, 196
Horsepower, 94
Hue, 594
Humidity, relative, 260 f
Huygens, Christian, 293, 539, 601
Huygens' principle, 539
Hydraulic brakes, 21
Hydraulic press, 21
Hydrodynamics, 109
Hydrofoil, 118
Hydrogen atom, 682 ff
Hydrogen spectrum, 641 f
Hydroplaning, 46
Hydrostatic paradox, 20
Hyperon, 765
Hysteresis, magnetic, 454

Iceland spar, 620
Ideal crystal, 194
Ideal gas, 240, 243
Ignition system, 463 f
Illuminance, 528

Illumination, 528, 594
Image, 537 f
latent, 570
real, 555
virtual, 555
Images, resolution of, 607 f
Immersion objective, 609
Impedance, 478
Impulse, 100 f
Impurity semiconductor, 719
Incandescent lamp, 404
Incidence, angle of, 538
Inclination, angle of, 424
Index of refraction, 532 f, 543 ff,
585 f
Induced emf, 458
Inductance, mutual, 466
Induction, charging by, 340
electromagnetic, 458 f
magnetic, 422
mutual, 466
Induction coil, 462
Induction motor, 486 f
Inductive reactance, 479
Inelastic collision, 103 f
Inelastic resonance collisions, 700 f
Inertia, 65
electrical, 467
moment of, 132 f
rotational, 132
Inertial frame of reference, 149
Inertial guidance, 138
Inertial mass, 181
Insulation, 226 f
Insulator, 339, 718
Intensity, of radiation, 526
of sound waves, 322
Interactions, 762
Interference, 303, 307 f, 318 ff, 599
thin-film, 610
Interference bands, 611 f
Interference pattern, 599
Interferometer, 152, 612
Internal-combustion engine, 274 f
Internal resistance, 391
Intrinsic semiconductor, 719
Ion, 376
Ionization chamber, 734 f
Ionization energy, 702 f
Irradiance, 526 f
Isobaric process, 241
Isochoric process, 240
Isoclinic lines, 425
Isogonic lines, 425
Isothermal expansion, 278

Isothermal process, 241
Isotopes, 336, 637

Jensen, Harald, 263
Jet, liquid, 112 f
Joule, 92
Joule, James, 270
Joule-Thomson effect, 264
Junction rectifier, 723

Kappa particle, 764
Kelvin temperature scale, 213
Kepler, Johannes, 170, 237
Kepler's laws, 170 f
Kerst, 733
Kilocalorie, 221
Kilogram, 9
Kilogram-weight, 15
Kilowatt, 94, 400
Kilowatt-hour, 400
Kinematics, 52
Kinetic energy, 94 f
  relativistic, 162
  rotational, 134 f
Kinetic friction, coefficient of, 43 ff
Kinetic theory, 236
Kirchhoff's laws, 392 f
  of radiation, 229

Lag, phase, 480
Lambda particle, 764
Laminar flow, 114 ff
Lamp, Edison, 405
  incandescent, 404
Land, Edwin, 595
Langmuir, Irving, 203, 405
Laser, 604, 726 f
Latent image, 570
Latitude effect, 761
Lattices, crystal, 710
Law, Ampere's, 446
  Boyle's, 241
  Brewster's, 620
  Charles', 241
  of cosines, 35
  Coulomb's, 344 f, 365
  Faraday's, 378
  Gas, 242
  Gauss's, 354 f
  of gravitation, 172
  of heating (Joule's), 401
  Hooke's, 196 f
  Kepler's, 170 f
  Kirchhoff's, 229, 392 f
  Lenz's, 459

Newton's, 65, 68, 72, 100
  of radioactive decay, 742
  of sines, 36
  of thermodynamics, first, 272 f
    second, 280
Law, Ohm's, 371 f, 711
  parallelogram, 34
  Poiseuille's, 116
  Prevost's, 232
  Radiation, 231
  Snell's, 542 f
  Stefan-Boltzmann, 231
  Stokes', 116
  Torricelli's, 113
  Wiedemann-Franz, 376
Length, 8
Lengths, contraction of, 156 f
Lens, achromatic, 587 f
  concave, 557 f
  converging, 552
  convex, 552
  diverging, 557 f
  objective, 574 f
Lenses, 552 ff
Lens-maker's equation, 553
Lenz's law, 459
Lepton, 765
Level, acceptor, 721
  donor, 720
Level population inversion, 727
Lever, 124 ff
Lever arm, 125 f
Lift, 41, 117
Light, polarized, 617 ff
  scattered, 621 f
  speed of, 150 f, 531
Light meter, 529
Light particles, 601, 655 f
Light waves, 527, 601, 655
Light-year, 9
Lightning, 347 f
Limit, elastic, 196 f
Line, magnetic, 415
Linear expansion coefficient, 215
Linear magnification, 557
Linear momentum, 99
Lines, electric, 354
  of force, 352
  Fraunhofer, 590
Liquefaction of gases, 263 f
Liquid air, 264
Liquid drop model, 764
Liquids, 192
Load line, 498 ff
Longitudinal waves, 309 f

Looming, 548
Lorentz, H. A. 157, 711
Lorentz constant, 376
Lorentz contraction, 157
Loudness, 323
Loudspeaker, 505 f
Low-resistance direction, 724
Lubrication, 46
Lumen, 528
Luminosity of radiation, 592 ff
Luminous efficiency, 529
Luminous flux, 527 f
Lunar landing, 177 f
Lux, 528
Lyman series, 645

Mks system, 10
Magdeburg hemispheres, 25 f
Magnet, 413 f
  analyzing, 739
  permanent, 413, 455
Magnetic cavity, 440 f
Magnetic dipole, 419, 450
Magnetic domains, 451
Magnetic field, 415
  of earth, 423 f, 439 f
  sources of, 445
Magnetic flux, 423, 458
Magnetic flux density, 423, 462
Magnetic force, 420 ff, 429
Magnetic hysteresis, 454
Magnetic induction, 422
  circular loop, 446 f
  solenoid, 447 f
  straight conductor, 447
  thin coil, 447
Magnetic line, 415
Magnetic materials, 450 f
Magnetic moment, 430
  atomic, 704
Magnetic pole, 413 f
Magnetic quantum number, 685, 692
Magnetism, 413 ff
Magnetization, 418 ff
Magnetization curves, 453
Magnetizing field intensity, 452
Magnetosphere, 440
Magnification, angular, 574
  linear, 557 ff
Magnifying glass, 572 f
Manometer, closed-tube, 238
  open-tube, 237 f
Many-electron atoms, 695 f
Many-electron crystal, 716 f
Mars, motion of, 168 f

Mass, 9, 70
  atomic, 378
  of earth, 179
  gravitational, 181
  inertial, 181
  nuclear, 751 f
  point, 64
  relativistic, 160 f
  rest, 161
  of sun, 179
Mass density, 16, 245
Mass-energy, 161 f
Mass number, 336, 752
Mass spectrograph, 751
Mass unit, atomic, 9, 752
Matter, phases of, 192
Matter waves, 659 ff, 678 ff
Maxwell, Clerk, 517 ff
Maxwell-Boltzmann distribution,
    246 f, 632
McMillan, 733
Mean free path, 247
Measurement, 7
Mechanical equivalent of heat, 271
Mechanical pump, 238
Mechanics, 3
  classical, 144
  fluid, 109
  Newtonian, 144
  quantum, 646, 659, 676, 710
Meitner, Lise, 757
Melting, 253 f
Meson, 763, 765
Meson, mu, 158, 745, 763
  pi, 745, 763
Metastable state, 727
Meter, 8
Meter-candle, 528
Meters, moving-coil, 431 f, 485 f
Method of mixtures, 222 f
Metric system, 7
Michelson, Albert A., 531, 612 ff
Michelson-Morley experiment, 151 ff,
    612 ff
Microfarad, 361
Microphone, 464
  crystal, 504
Microscope, compound, 574
  electron, 672 f
  X-ray, 669 f
Millikan, Robert A., 359, 634, 760
Mirage, 548 f
Mirror, concave, 559 ff
  converging, 559 ff
  convex, 561 f

  curved, 559 ff
  diverging, 561 f
  parabolic, 562 f
  spherical, 559
Mixtures, method of, 222 f
Model, shell, 764
Models, 237
Moderator, 759
Modulation, 503 f
  frequency, 504
  plate, 504
Modulus, shear, 200 f
  volume, 201 f
  Young's, 198
Molecular forces, 189 ff, 201 f
Molecular heat, 248, 273 f
Molecules, 191
Moment, magnetic, 430
  of force, 125
  of inertia, 132 f
Momentum, 99
  angular, 134, 690
  conservation of, 101 f
  linear, 99
Monatomic gas, 248
Moon, travel to, 177 f
Morley, E. W., 152
Motion, curvilinear, 77
  orbital, 174 f
  planetary, 170 f
  rectilinear, 52 ff
  rotational, 124 ff
  simple harmonic, 288
  translational, 129 f
  wave, 299 ff
Motor, a-c, 486 ff
  compound, 472
  direct-current, 433 f
  induction, 486 f
  series, 472
  shunt, 472
  star-wheel, 422
  synchronous, 488
  universal, 435
Motor rule, 421
Moving-coil meter, 431 f, 485 f
Moving systems, 146 ff
Mu meson, 745, 763
Mu meson experiment, 158
Multimeter, 432 f
Multiplet lines, 694
Muon, 745
Music, 323 f
Mutual inductance, 466
Mutual induction, 466

N-type germanium, 720
Neddermyer, 733
Negative, photographic, 570
Negative charge, 335 ff
Negative energy states, 745 f
Negative feedback, 510 f
Negatron, 764
Neptunium, 758
Nerve conduction, 393 ff
Neutral equilibrium, 129
Neutrino, 745
Neutron, 336, 754, 764
  fast, 758
Newton, 15, 67
Newton, Sir Isaac, 4, 65, 144, 171 f,
    237, 583 f, 586, 595, 601
Newton, laws of, 65, 68, 72, 100
Newton-meter, 92
Newtonian mechanics, 144
Nicol prism, 620
Night blindness, 570
Node, 308
Nonreflecting glass, 611
Normal force, 43
North pole, magnetic, 413 f
Nuclear charge, 750 f
Nuclear emulsion plate, 739
Nuclear forces, 762 ff
Nuclear mass, 751 f
Nuclear reaction, 754 f
Nuclear reactor, 758 f
Nuclear spin, 754
Nucleon, 750, 764
Nucleus, 750 ff
  shell model of, 764
  size of, 753 f
Nuclide, 750
Null method, 387
Number, atomic, 336, 750
  mass, 336, 752
  quantum, 642, 685, 697

Objective lens, 574 f
Oersted, Hans, 416
Ohm, 372, 478 f
Ohm, Georg, 371
Ohm's law, 371 f, 711
Ohmmeter, 386
Oil drop experiment, 359 f
Omega particle, 764
One-dimensional crystal, 715 f
Open-tube manometer, 237 f
Opera glass, 577
Optical axis, 620
Optical instruments, 567 ff

Optical pumping, 727
Optical pyrometer, 211
Orbit, Bohr, 642 f
  of a satellite, 175 f
Orbital angular momentum, 693 f
Orbital motion, 174 f
Orbital quantum number, 685
Orbital velocity, 174 f
Orbits of satellites, 175 f
Oscillator, electrical, 500
Oscilloscope, cathode-ray, 507 f
Overtone, 320

P-type germanium, 721
Packets, wave, 661 ff
Pair annihilation, 746
Parabolic mirror, 562 f
Parallax, 538
Parallel, capacitors in, 362 f
  resistors in, 384
Parallel spin vector, 694
Parallelogram law, 34
Paramagnetic, 450
Particle in a box, 680, 682
Pascal, Blaise, 20
Pascal's principle, 20 f
Paschen series, 645
Pauli, Wolfgang, 679
Pauli exclusion principle, 696
Peary, Robert E., 548
Peltier effect, 408
Pendulum, ballistic, 103 f
  compensated, 215 f
  period of, 294
  physical, 294
  simple, 292 ff
Pentode, 500
Period, 287
Period, of an electrical oscillator, 500
  of a pendulum, 323
  of a vibration, 287 ff
  of reverberation, 323
Periodic field, 713
Periodic table, 703 f
Periodicity of elements, 704
Permalloy, 454
Permanent magnet, 413, 455
Permeability, 446
  relative, 452 f
Permittivity, 345
Perrin, Jean, 247
Phase, 290, 478
  change of, 253, 310
  of matter, 192
Phase lag, 480

Phase velocity, 662
Photocell, 637
Photoelectric effect, 633, 722 f
Photometer, 528 f
Photon, 151, 652 f, 765
Physical pendulum, 294
Physics, definition of, 3
Pi meson, 745, 763
Piezoelectric effect, 504
Pion, 745, 763
Pitch, 324
Planck, Max, 631 f
Planck's constant, 631, 668
Plane, principal, 554
Plane polarized waves, 617
Planetary motion, 170 ff, 175
Planetary winds, 227
Plate, 494 f
Plate modulation, 504
Plutonium, 758
Point charge, 344
  potential due to, 360 f
Point mass, 124
Poiseuille, 116
Polarization, 389
  circular, 617
  electric, 363 f, 389
  of electromagnetic waves, 523
  elliptical, 617
  of light, 617 f
  of molecules, 363 f
  plane, 617
  by reflection, 620
  of scattered light, 621 f
  of waves, 303
Polarization charge, 339
Polaroid polarizer, 619
Pole, magnetic, 413 f
Polonium, 731
Polygon method, 37 f
Positive charge, 335 ff
Positron, 745
Potential due to point charge, 360 f
Potential difference, 357, 370
Potential energy, 96
  of electrons, 678
Potential well, 713 f
Potentiometer, 392
Pound, 9
Pound-weight, 15
Powder diffraction pattern, 649 f
Power, 93 f
  apparent, 480
  electrical, 400
  of a sound, 324

  true, 480
Power factor, 480
Power steering, 22
Power supplies, 494 f
Precession, 137
Press, hydraulic, 21
Pressure, 15
  atmospheric, 25 ff
  in fluids, 16 ff
  of a gas, 245
  gauge, 238
  radiation, 652
  vapor, 256 f
  and velocity of fluids, 110 f
Prevost's law, 232
Primary coil, 482
Primary colors, 594 f
Principal axis, 552, 559
Principal focus, 552, 560
Principal plane, 554
Principal quantum number, 685
Principal series, 698
Prism, 546
  Nicol, 620
Probability, 671, 680
Probability density, 686
Progressive waves in cords, 305 f
Projectiles, 78 f
Proton, 336, 764
Ptolemy, 168, 237
Puck, air, 65
Pump, diffusion, 239
  heat, 279
  mechanical, 238
  rotary, 238
Pumping, optical, 727
Purcell, 733
Purple, visual, 570
Pyrometer, optical, 211
  radiation, 588 f

Quality, 324
Quantity of heat, 221 f
Quantization, hydrogen atom, 678 f
  space, 692
Quantum, 631 f
Quantum electrodynamics, 693
Quantum mechanics, 646, 659, 676,
  710
Quantum number, 642, 685, 697
Quantum number, magnetic, 685, 692
  orbital, 685, 691
  principal, 685
  spin, 685

Uhlenbeck, G. E., 694
Ultracentrifuge, 83
Uncertainty principle, 667 f, 676
Undercooling, 255
Units, absolute, 67
  cgs, 10
  derived, 10
  in dynamics, 67 f
  electrostatic, 345
  mks, 10
  rotational, 138
Universal gravitation, law of, 167, 172
Universal motor, 435
Universe of the Greeks, 167 f
Unstable equilibrium, 129

Van Allen, James, 440
Van de Graaff generator, 343
Van der Waals equation, 262
Vapor barrier, 227
Vapor pressure, 256 f
Vaporization, heat of, 259 f
Variable capacitor, 501
Vecksler, 733
Vector, 33
Velocity, 53, 77 f
  angular, 131
  average, 53
  escape, 176 f
  group, 662
  of a liquid jet, 112 f
  orbital, 174 f
  phase, 662
  terminal, 116
Velocity gradient, 115
Vibrating body, energy of, 292
Vibration, 287 f
  period of, 287 ff
  phase of, 290

sympathetic, 294
Virtual image, 555
Viscosity, 113 f
Viscosity, coefficient of, 115
Viscous fluid, 113 f
Vision, 593 f
Visual purple, 570
Volt, 357, 391, 478
Volta, Alessandro, 388
Voltage, alternating, 477
  effective, 478
Voltaic cell, 387 ff
Voltmeter, 358, 385 f, 431
  a-c, 486
Volume expansion, 216 f
Volume modulus, 200 f
Volume strain, 200

Water, maximum density of, 217
Water equivalent, 223
Watt, 94
Watt, James, 93 f
Wattmeter, 486
Wavelength, 299 f
  Compton, 654
  de Broglie, 660, 677
Wavelength, of X-rays, 648 f
Wave-particle duality, 655 f
Wave-particle theory, 659
Wave front, 300 ff
Wave function, 671, 680, 686 f
Wave motion, 299 ff
Wave number, 645
Wave packets, 661 ff
Waves, compressional, 309 f
  deBroglie, 659 f
  electromagnetic, 517 ff
  electron, 666, 713 f
  interference of, 307 f

light, 527, 601, 655
  longitudinal, 309 f
  matter, 659 ff, 678 ff
  polarization of, 303
  progressive, in cords, 305 f
  sound, 314
  standing, 305 f, 308
  transverse, 299, 618
Weak interactions, 762 f
Weber, 423
Weight, 14, 70 f
Weight density, 15 f
Well, potential, 713 f
Westinghouse, George, 405
Weston cell, 389
Wheatstone bridge, 386 f
Whiskers, 199 f
Wiedemann-Franz law, 376
Wien equation, 631
Wind, solar, 440
Winds, planetary, 227
Work, 92 f
Work function, 635

Xi particle, 764
X-ray microscope, 669 f
X-ray spectra, 650 f
X-ray tube, Coolidge, 647
X-rays, 646 ff
  diffraction of, 648 ff
  wavelength of, 648 f

Yard, 8
Yellow spot, 593
Young, Thomas, 601 ff
Young's modulus, 198

Zeeman effect, 706
Zone refining, 719

Some Physical Constants

c	speed of light	3.00×10^8 m/sec
e	elementary charge	1.60×10^{-19} coul
h	Planck's constant	6.63×10^{-34} joule-sec
m_e	rest mass of electron	9.11×10^{-31} kg
e/m_e	electron's charge-to-mass ratio	1.76×10^{11} coul/kg
N	Avogadro's number	6.02×10^{23} molecules/mole
F	faraday	9.65×10^4 coul
G	gravitational constant	6.67×10^{-11} newt-m^2/kg^2
μ_0	permeability of free space	1.26×10^{-6} henry/m
ϵ_0	permittivity of free space	8.85×10^{-12} farad/m
m_p	rest mass of proton	1.67252×10^{-27} kg
m_n	rest mass of neutron	1.67482×10^{-27} kg
R	gas constant	8.31 joule/mole-K°
m_{earth}	mass of earth	5.98×10^{24} kg
r_{earth}	radius of earth	6.37×10^6 m
1 amu	atomic mass unit	1.66043×10^{-27} kg